Positron Emission Tomography

Dale L Bailey, David W Townsend,
Peter E Valk and Michael N Maisey (Eds)

Positron Emission Tomography

Basic Sciences

Dale L Bailey PhD, ARCP (London), FIPEM, MACPSEM
Principal Physicist, Department of Nuclear Medicine, Royal North Shore Hospital, St Leonards, Australia;
Senior Lecturer, School of Medical Radiation Sciences, University of Sydney, Sydney, Australia;
Clinical Associate Professor, Faculty of Medicine, University of Sydney, Sydney, Australia

David W Townsend BSc, PhD, PD
Director, Cancer Imaging and Tracer Development Program, The University of Tennessee Medical Center, Knoxville, TN, USA

† Peter E Valk († Deceased) MB, BS, FRACP
Northern California PET Imaging Center, Sacramento, CA, USA

Michael N Maisey MD, BSc, FRCP, FRCR
Professor Emeritus, Department of Radiological Sciences, Guy's and St Thomas' Clinical PET Centre, Guy's and St Thomas' Hospital Trust, London, UK

British Library Cataloguing in Publication Data
Positron emission tomography : basic sciences
1. Tomography, Emission
I. Bailey, Dale L.
616'.07575

ISBN 1852337982

Library of Congress Cataloging-in-Publication Data
Positron emission tomography: basic sciences / Dale L. Bailey ... [et al.], (eds).
p. cm.
Includes bibliographical references and index.
ISBN 1-85233-798-2 (alk. paper)
1. Tomography, Emission. I. Bailey, Dale L.
RC78.7.T62 P688 2004
616.07'575–dc22 2004054968

ISBN 1-85233-798-2
Springer Science+Business Media
springeronline.com

Printed in Singapore. (EXP/KYO)

Printed on acid-free paper SPIN 10944028

Preface

In 2003 we published *Positron Emission Tomography: Basic Science and Clinical Practice*. The aim of that book was to address what we perceived of as a lack, at the time, of a comprehensive contemporary reference work on the rapidly expanding area of positron emission imaging. The scope was intentionally wide. The original proposal for a 350 page book turned into a nearly 900 page volume.

This book, *Positron Emission Tomography: Basic Sciences*, is a selected and updated version of the non-clinical chapters from the original book. In addition, a number of new chapters have been added which address the role of PET today for the scientist currently working in or entering this rapidly expanding area. The audience that this is intended for is the scientist, engineer, medical graduate or student who wants to learn more about the science of PET. Many of the chapters have been updated from the original to reflect how rapidly the technology underpinning PET is changing.

The following diagram encapsulates much of what is required in understanding the science of PET. It is taken from an introduction by Professor Terry Jones to a book of the proceedings from a PET neuroscience conference in the mid-1990s. It is the intention of this book to deal with the majority of these topics and to produce a comprehensive "science of PET" textbook which is more focussed and manageable than the original volume. We hope this book will be of use to you.

Finally, we are sad to report that the principal editor of the original work, Peter E Valk, *MB, BS, FRACP*, passed away in December 2003. Peter was a great friend and outstanding advocate for, and practitioner of, nuclear medicine and PET. He will be greatly missed by his many colleagues and friends everywhere. We are indeed fortunate that Peter left us with a truly wonderful book on PET to preserve his memory and not let us forget the debt that we owe him for the leading role he played in bringing PET into clinical patient care.

Dale L Bailey
David W Townsend
Michael N Maisey

Sydney, Knoxville, London
March 2004

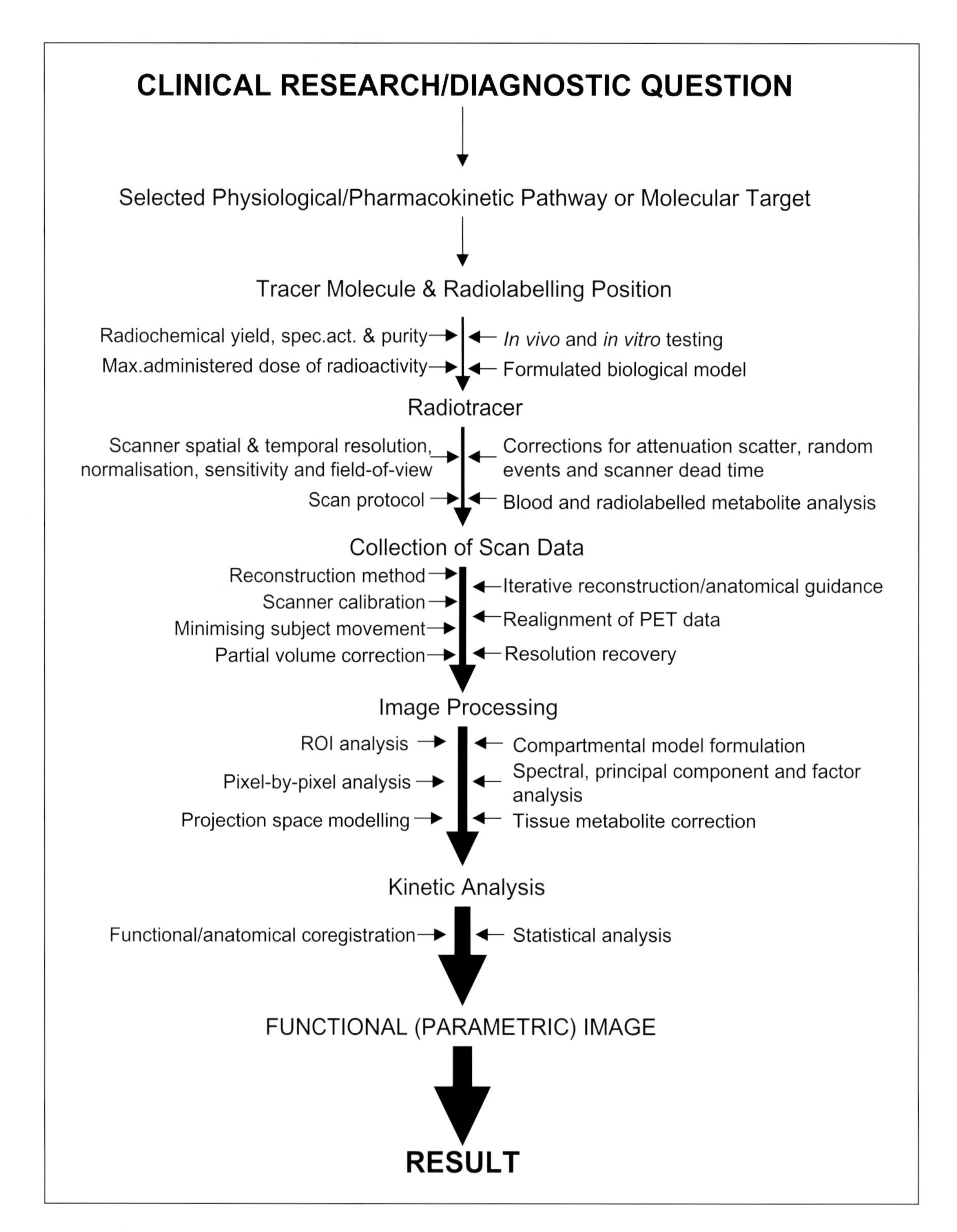

Figure 1. Jones' view of the science of PET (adapted from Myers R. Cunningham VJ, Bailey DL, Jones T (Eds): *Quantification of Brain Function with PET.* Academic Press; 1996 and used with Professor Jones' permission).

Contents

Contributors

Gunnar Antoni PhD
Uppsala Research Imaging Solutions AB
Uppsala
Sweden

Ramsey D Badawi PhD
Affiliate Physicist
Department of Radiology
UC Davis Medical Center
Sacramento, CA
USA

Dale L Bailey PhD, ARCP (London), FIPEM, MACPSEM
Department of Nuclear Medicine
Royal North Shore Hospital
St Leonards;
School of Medical Radiation Sciences
University of Sydney
Sydney;
Faculty of Medicine
University of Sydney
Sydney
Australia

Thomas Beyer PhD
University Hospital of Essen
Department of Nuclear Medicine
Essen
Germany

Richard E Carson PhD
Positron Emission Tomography Department (PET)
Warren Grant Magnuson Clinical Center (CC)
National Institutes of Health (NIH)
Bethesda, MD
USA

Gary JR Cook MBBS, MD
Department of Nuclear Medicine
Royal Marsden Hospital
Sutton
UK

Bernadette F Cronin DCR (R), DRI, FETC
Department of Nuclear Medicine
The Royal Marsden Hospital
Sutton
UK

Abhijit De MPhil PhD
Stanford University School of Medicine
Department of Radiology and Bio-X Program
The James H. Clark Center
Stanford, CA
USA

Michel Defrise PhD
Division of Nuclear Medicine
University Hospital AZ-VUB
Brussels
Belgium

William C Eckelman PhD
Intramural Program
National Institute of Biomedical Imaging and Bioengineering
Bethesda, MD
USA

Sanjiv Sam Gambhir MD, PhD
Stanford University School of Medicine
Department of Radiology and Bio-X Program
The James H. Clark Center
Stanford, CA
USA

Lucy Hallpike BSc
Division of Imaging Sciences
School of Medicine
Guy's Hospital
King's College London
London
UK

David J Hawkes BA, MSc, PhD
Computational Imaging Science Group
Radiological Science
Guy's Hospital
King's College London
London
UK

Derek LG Hill BSc, MSc, PhD
Radiological Science
Guy's Hospital
King's College London
London
UK

Joel S Karp PhD
PET Center & Physics and Instrumentation Group
Department of Radiology
University of Pennsylvania
Philadelphia, PA
USA

Paul E Kinahan PhD
Department of Radiology
University of Washington
Seattle, WA
USA

Bengt Långström PhD
Uppsala Research Imaging Solutions AB
Uppsala
Sweden

Michael N Maisey MD, BSc, FRCP, FRCR
Department of Radiological Sciences
Guy's and St Thomas' Clinical PET Centre
Guy's and St Thomas' Hospital Trust
London
UK

N Scott Mason PhD
PUH PET Facility
University of Pittsburgh Medical Center
Pittsburgh, PA
USA

Chester A Mathis PhD
PUH PET Facility
University of Pittsburgh Medical Center
Pittsburgh, PA
USA

Deborah W McCarthy PhD
Division of Radiological Sciences
Department of Radiology
Washington University School of Medicine
St Louis, MO
USA

Paul McQuade PhD
Division of Radiological Sciences
Department of Radiology
Washington University School of Medicine
St Louis, MO
USA

Steven R Meikle BAppSc, PhD
School of Medical Radiation Sciences
University of Sydney
Sydney
Australia

Christian J Michel PhD
CPS Innovations
Knoxville, TN
USA

Andrew M Scott MB, BS, FRACP
Centre for Positron Emission Tomography
Austin Hospital;
Tumour Targeting Program
Ludwig Institute for Cancer Research
Heidelberg
Germany

Paul D Shreve MD
Advanced Radiology Services
Grand Rapids, MI
USA

Suleman Surti PhD
Department of Radiology
University of Pennsylvania
Philadelphia, PA
USA

David W Townsend BSc, PhD, PD
Cancer Imaging and Tracer Development Program
The University of Tennessee Medical Center
Knoxville, TN
USA

Jocelyn EC Towson MA, MSc
Department of PET and Nuclear Medicine
Royal Prince Alfred Hospital
Sydney
Australia

Peter E Valk MB, BS, FRACP†
Northern California PET Imaging Center
Sacramento, CA
USA

Michael J Welch PhD
Division of Radiological Sciences
Department of Radiology
Washington University School of Medicine
St Louis, MO
USA

1 Positron Emission Tomography in Clinical Medicine

Michael N Maisey

Introduction

Positron emission tomography (PET) imaging is set to change the whole impact and role of Nuclear Medicine, not because it does everything better than conventional single photon imaging (planar and single photon emission computed tomography (SPECT)), but because it also has the impact and public relations of the fastest growing diagnostic speciality. PET is a powerful metabolic imaging technique utilising possibly the best radiopharmaceutical we have ever used [^{18}F]-fluorodeoxyglucose (FDG). However, in addition, it yields excellent quality images, the importance of which can be appreciated by non-nuclear medicine clinicians, and has an enormous clinical impact, as demonstrated in many well-conducted studies. Any oncologist exposed to a good PET imaging service very quickly appreciates its value. Sitting in on routine clinical PET reporting sessions, it is easy to appreciate how patient after patient is having their management changed in a very significant way as a direct result of the new information provided by the PET scan.

There is now an impressive body of data evaluating the impact of PET on patient management. These studies are showing that PET results alter management in a significant way in more than 25% of patients, with some as high as 40%[1]. Examples include changing decisions on surgical treatment for non-small cell lung cancer (both avoiding inappropriate surgery and enabling potentially curative resection), the staging and treatment of lymphoma, decisions on surgical resections for metastatic colo-rectal cancer, referral for revascularisation of high-risk coronary artery disease (CAD) patients and many others. This is a level of impact on patient care for common and life-threatening diseases not previously achieved by Nuclear Medicine. Nuclear Medicine has always improved patient care, but usually marginally, such that it has sometimes been difficult to argue that good medicine could not be practised without it. This has often resulted in limitations on the manpower and other resources being put into Nuclear Medicine, particularly in health care systems functioning at the lower end of gross national product (GNP) percentage investment, such as the National Health Service (NHS) in the United Kingdom. This is not true of PET. It is no longer possible to practice the highest standard of clinical oncology without access to PET, and it is clear that without it many patients are needlessly undergoing major surgical procedures and many are being denied potentially curative treatments. If PET and X-ray computed tomography (CT) were to be introduced simultaneously now for oncology staging, follow-up, assessment of tumour recurrence, evaluation of treatment response, *etc*, there would be no competition with PET proving vastly superior in these areas of cancer patient management.

We therefore have in clinical PET a new imaging tool as part of Nuclear Medicine which has brought the speciality to the very heart of patient management, especially for Oncology, but also in Cardiology and Neuropsychiatry. Nuclear Medicine has always been excited by the potential for new ligands for clinical application and the study of patho-physiology. Although for many reasons the potential has not been fully delivered, it may be that the future role of PET ligands will be huge, especially as we are on the brink of molecular and genetic imaging. Furthermore, for PET to be the future of Nuclear Medicine we do not need to argue on the grounds of the potential, as, with FDG, we have the most effective and powerful radiopharmaceutical of all time. Nuclear Medicine has never had a single tracer which could study brain metabolism, cardiac function, image sites of infection, and detect cancer as FDG does in thousands of scans world-wide every day.

Technical developments will also drive the widespread introduction of PET as the main developing area of Nuclear Medicine. PET scanners are becoming significantly more sensitive leading to considerably faster patient throughput, as long scanning times were one of the weaknesses of early scanners. "Fusion imaging", always a promising "new" methodology, has been kick-started by the combined PET/CT concept (see chapters 8 and 9). However, the greatest benefits of fusion imaging may eventually come from software, rather than hardware, fusion because of the flexibility of fusing multiple imaging modalities with PET (*e.g.*, magnetic resonance imaging (MRI)) as well as image fusion of sequential PET images over time, which will be of increasing importance for PET-based molecular and metabolic imaging when used for following the response to treatment. The spatial resolution of PET images is also improving, so that metabolic images with millimetre resolution are increasingly probable. The power derived from quantification will be revealed as measurement of early tumour responses becomes routine practice. Many of these benefits are because of the investment of time and money that industry is putting into PET as it is perceived as a major area of expansion.

With increased patient throughput and a greater number of PET scanners and imaging resources, there are opportunities for PET methodologies to be used for studies such as bone scans (with $[^{18}F]-F^{-}$ or FDG, or even a combination of the two), all cardiac perfusion and myocardial viability studies, and many other current SPECT-based studies (*e.g.* imaging neuroendocrine tumours using $[^{111}In]$-octreotide or $[^{131}I]$-mIBG) could be performed by PET. A lot will depend on the inventiveness and will of the cyclotron operators and radiochemists who will be responding to the clinical agenda.

Current Clinical Applications of PET

Clinical PET imaging, almost exclusively with FDG at present, is being used in three important areas of clinical diagnosis and management:

- Cancer diagnosis and management
- Cardiology and cardiac surgery
- Neurology and psychiatry.

Each of these areas will be examined in more detail.

Cancer Diagnosis and Management

Although FDG is by far the most important radiopharmaceutical at present others such as ^{11}C-labelled methionine and choline and fluorine labelled DNA proliferation markers such as fluoro-L-tyrosine (FLT) will have an increasing role in the years ahead. The applications can be classified according to the generic use for which the PET scan is applied, that is detection, staging tumour response, *etc* or by tumour types. Both are important to understand although the tumour type approach will be the method chosen for agencies responsible for agreeing reimbursements.

- *Diagnosis of malignancy*: examples will include differentiating malignant from benign pulmonary nodules, and differentiating brain scarring after treatment (surgery, chemotherapy and radiation therapy) from tumour recurrence.
- *Grading Malignancy*: as the uptake of FDG and other metabolic tracers is related to the degree of malignancy (the principle established by Warburg in the early part of the 20th century[2]) the PET scan can be used to grade tumours and therefore indirectly provide information on prognosis (the so-called "metabolic biopsy").
- *Staging disease*: staging is documenting how widespread the cancer is in the patient. The PET scan has been show to be superior to anatomical methods of staging disease and therefore planning therapy. Examples include non–small cell lung cancer, lymphoma and oesophageal tumours.
- *Residual disease*: because purely anatomical methods for deciding on the viability of residual masses after treatment has been poor, metabolic imaging is proving extremely useful *e.g.*, post-treatment mediastinal lymphoma masses and testicular abdominal masses.
- *Detection of recurrences*: good examples include the confirmation and site of recurrent colo-rectal cancer after surveillance blood testing has detected a rise in circulating tumour (CEA) markers.
- *Measuring the response to therapy*: it is often important to know how effective initial treatment has been in order to plan future therapeutic strategies. The best example is assessing response following the initial course of treatment of Hodgkin's lymphoma, when poor early response indicates that supplemen-

tary neo-adjuvant therapy may be necessary for the desired effect.

- *To identify the site of disease*: identifying the site of disease may be important to plan surgery *e.g.*, for squamous cell cancers of the head and neck, to direct biopsy when the disease is heterogeneous, in soft tissue sarcomas, and to find the site of disease when the only sign may be a raised circulating tumour marker such as in thyroid cancer or teratomas.
- *To identify the primary tumour when secondary cancers are present*: it may be critical to discover the primary cancer when a patient presents with an enlarged lymph node, as in head and neck cancers where the primary tumour may be small, or alternatively when the presentation raises suspicion of a para-neoplastic syndrome.

Cardiology and Cardiac Surgery

At present there are three major indications for PET scans using two physiological measurements in clinical practice. The two measurements are (i) to measure the myocardial perfusion using [^{13}N]-ammonia (or ^{82}Rb from an on-site generator) and (ii) to measure myocardial viability (using [^{18}F]-FDG). There is increasing interest in a third measurement, cardiac innervation by studying myocardial receptors, which may have a greater role in the future. The three applications of these measurements are:

- in the diagnosis and assessment of the functional significance of coronary artery disease (CAD) usually when the SPECT scan is not definitive. However with the increasing use of medical therapy for treating CAD the quantification of myocardial blood flow and changes will become more important in the near future.
- in the assessment of the viability of ischaemic or jeopardised myocardium. This is important because the risks and benefits of medical treatments in advanced CAD are closely related to the presence and extent of viable but hibernating myocardium *versus* non-viable infarcted/scar tissue.
- during the work-up of patients who are being considered for cardiac transplantation (although this may be regarded as a subset of viability assessment). It is of such importance it is often considered separately from assessing viability. Due to the procedural

Table 1.1. US Centers for Medicaid and Medicare Services Indications and Limitations for PET scans[3].

Indication	Date Approved	Purpose
Solitary Pulmonary Nodules (SPNs)	Jan 1, 1998	Characterisation
Lung Cancer (Non Small Cell)	Jan 1, 1998	Initial staging
Lung Cancer (Non Small Cell)	July 1, 2001	Diagnosis, staging and restaging
Esophageal Cancer	July 1, 2001	Diagnosis, staging and restaging
Colo-rectal Cancer	July 1, 1999	Determining location of tumours if rising CEA level suggests recurrence
Colo-rectal Cancer	July 1, 2001	Diagnosis, staging and restaging
Lymphoma	July 1, 1999	Staging and restaging only when used as an alternative to Gallium scan
Lymphoma	July 1, 2001	Diagnosis, staging and restaging
Melanoma	July 1, 1999	Evaluating recurrence prior to surgery as an alternative to a ^{67}Ga scan
Melanoma	July 1, 2001	Diagnosis, staging and restaging; Non-covered for evaluating regional nodes
Breast Cancer	Oct 1, 2002	As an adjunct to standard imaging modalities for staging patients with distant metastasis or restaging patients with loco-regional recurrence or metastasis; as an adjunct to standard imaging modalities for monitoring tumour response to treatment for women with locally advanced and metastatic breast cancer when a change in therapy is anticipated.
Head and Neck Cancers (excluding CNS and thyroid)	July 1, 2001	Diagnosis, staging and restaging
Thyroid Cancer	Oct 1, 2003	Restaging of recurrent or residual thyroid cancers of follicular cell origin that have been previously treated by thyroidectomy and radioiodine ablation and have a serum thyroglobulin >10ng/ml and negative ^{131}I whole body scan performed
Myocardial Viability	July 1, 2001 to Sep 30, 2002	Covered only following inconclusive SPECT
Myocardial Viability	Oct 1, 2001	Primary or initial diagnosis, or following an inconclusive SPECT prior to revascularisation. SPECT may not be used following an inconclusive PET scan.
Refractory Seizures	July 1, 2001	Covered for pre-surgical evaluation only
Perfusion of the heart using ^{82}Rb	Mar 14, 1995	Covered for non-invasive imaging of the perfusion of the heart
Perfusion of the heart using [^{13}N]-NH_3	Oct 1, 2003	Covered for non-invasive imaging of the perfusion of the heart

Table 1.2. UK Intercollegiate Committee on Positron Emission Tomography Recommended Indications for Clinical PET Studies[4]. The evidence supporting this is classified as (A) Randomised controlled clinical trials, meta-analyses, systematic reviews, (B) Robust experimental or observational studies, or (C) other evidence where the advice relies on expert opinion and has the endorsement of respected authorities.

Oncology Applications	Indicated	Not indicated routinely (but may be helpful)	Not indicated
Brain and spinal cord	•Suspected tumour recurrence when anatomical imaging is difficult or equivocal and management will be affected. Often a combination of methionine and FDG PET scans will need to be performed. (B) •Benign versus malignant lesions, where there is uncertainty on anatomical imaging and a relative contraindication to biopsy. (B) •Investigation of the extent of tumour within the brain or spinal cord. (C)	•Assess tumour response to therapy (C) •Secondary tumours in the brain. (C)	
Parotid	•Identification of metastatic disease in the neck from a diagnosed malignancy. (C)		•Differentiation of Sjögrens Syndrome from malignancy in the salivary glands. (C) •Primary tumour of the parotid to distinguish benign from malignant disease. (C)
Malignancies of the oropharynx	•Identify extent of the primary disease with or without image registration. (C) •Identify tumour recurrence in previously treated carcinoma. (C)	•Pre-operative staging of known oropharyngeal tumours. (C) •Search for primary with nodal metastases. (C)	
Larynx	•Identify tumour recurrence in previously treated carcinoma. (C)	•Staging known laryngeal tumours. (C) •Identification of metastatic disease in the neck from a diagnosed malignancy. (C)	
Thyroid	•Assessment of patients with elevated thyroglobulin and negative iodine scans for recurrent disease. (B)	•Assessment of tumour recurrence in medullary carcinoma of the thyroid. (C)	•Routine assessment of thyroglobulin positive with radioiodine uptake. (C)
Parathyroid		•Localisation of parathyroid adenomas with methionine when other investigations are negative. (C)	
Lung	•Differentiation of benign from metastatic lesions where anatomical imaging or biopsy are inconclusive or there is a relative contraindication to biopsy. (A) •Pre-operative staging of non small cell primary lung tumours. (A) •Assessment of recurrent disease in previously treated areas where anatomical imaging is unhelpful. (C)	•Assessment of response to treatment. (C)	
Oesophagus	•Staging of primary cancer. (B) •Assessment of disease recurrence in previously treated cancers. (C)	•Assessment of neo-adjuvant chemotherapy. (C)	
Stomach	•No routine indication. (C)	•Assessment of gastro-oesophageal malignancy and local metastases. (C)	
Small bowel	•No routine indication. (C)	•Proven small bowel lymphoma to assess extent of disease. (C)	

Table 1.2. Continued.

Oncology Applications	Indicated	Not indicated routinely (but may be helpful)	Not indicated
Breast cancer	•Assessment and localisation of brachial plexus lesions in breast cancer. (Radiation effects versus malignant infiltration.) (C) •Assessment of the extent of disseminated breast cancer. (C)	•Axillary node status where there is a relative contraindication to axillary dissection. (C) •Assessment of multi-focal disease within the difficult breast (dense breast or equivocal radiology). (C) •Suspected local recurrence. (C) Assessment of chemotherapy response. (C)	•Routine assessment of primary breast cancer. (C)
Liver: primary lesion			•Routine assessment of hepatoma. (C)
Liver: secondary lesion	•Equivocal diagnostic imaging (CT, MRI, ultrasound). (C) •Assessment pre and post therapy intervention. (C) •Exclude other metastatic disease prior to metastectomy. (C)		
Pancreas		•Staging a known primary. (C) •Differentiation of chronic pancreatitis from pancreatic carcinoma. (C) •Assessment of pancreatic masses to determine benign or malignant status. (C)	
Colon and rectum	•Assessment of recurrent disease. (A) •Prior to metastectomy for colo-rectal cancer. (C)	•Assessment of tumour response. (C) •Assessment of a mass that is difficult to biopsy. (C)	•Assessment of polyps. (C) •Staging a known primary. (C)
Renal and adrenal	•Assessment of possible adrenal metastases. (C)	•Paraganglionomas or metastatic phaeochromocytoma to identify sites of disease. (C)	•Assessment of renal carcinoma. (C) •Phaeochromocytoma – [^{131}I]-mIBG scanning is usually superior. (C)
Bladder	•No routine indication. (C)	•Staging a known primary in selected cases. (C) •Recurrence with equivocal imaging. (C)	
Prostate			•FDG in prostate cancer assessment. (C)
Testicle	•Assessment of recurrent disease from seminomas and teratomas. (B)	•Assessment of primary tumour staging. (C)	
Ovary	•In difficult management situations to assess local and distant spread (C)		
Uterus: cervix	•No routine indication (C)	•In difficult situations to define the extent of disease with accompanying image registration. (C)	
Uterus: body	•No routine indication. (C)		
Lymphoma	•Staging of Hodgkin's lymphoma. (B) •Staging of non-Hodgkin's lymphoma. (B) •Assessment of residual masses for active disease (B) •Identification of disease sites when there is suspicion of relapse from clinical assessment (C) Response to chemotherapy. (C)	•Assessment of bowel lymphoma. (C) •Assessment of bone marrow to guide biopsy. (C) •Assessment of remission from lymphoma. (C)	

Table 1.2. Continued.

Oncology Applications	Indicated	Not indicated routinely (but may be helpful)	Not indicated
Musculo-skeletal tumours	•Soft tissue primary mass assessment to distinguish high grade malignancy from low or benign disease. (B) •Staging of primary soft tissue malignancy to assess non-skeletal metastases. (B) •Assessment of recurrent abnormalities in operative sites. (B) •Assessment of osteogenic sarcomas for metastatic disease. (C) •Follow up to detect recurrence or metastases. (B)	•Image registration of the primary mass to identify optimum biopsy site. (C)	
Skin tumours	•Malignant melanoma with known dissemination to assess extent of disease. (B) •Malignant melanoma in whom a sentinel node biopsy was not or can not be performed in stage II. (AJCC updated classification). (C)	•Staging of skin lymphomas. (C)	•Malignant melanoma with negative sentinel node biopsy. (B)
Metastases from unknown primary	•Determining the site of an unknown primary when this influences management. (C)		•Widespread metastatic disease when the determination of the site is only of interest. (C)
Cardiac Applications	**Indicated**	**Not indicated routinely (but may be helpful)**	**Not indicated**
	•Diagnosis of hibernating myocardium in patients with poor left ventricular function prior to revascularisation procedure. (A) •Patients with a fixed SPECT deficit who might benefit from revascularisation. (B) •Prior to referral for cardiac transplantation. (B)	•Diagnosis of coronary artery disease or assessment of known coronary stenosis where other investigations (SPECT, ECG), etc) remain equivocal. (B) •Differential diagnosis of cardiomyopathy (ischaemic versus other types of dilated cardiomyopathy). (C) •Medical treatment of ischaemic heart disease in high risk hyperlipidemic patients. (C)	•Patients with confirmed coronary artery disease in whom revascularisation is not contemplated or indicated. (C) •Routine screening for coronary artery disease. (C)
Neuropsychiatry Applications	**Indicated**	**Not indicated routinely (but may be helpful)**	**Not indicated**
	•Pre-surgical evaluation of epilepsy. (B) •Suspected recurrence or failed primary treatment of primary malignant brain tumours. (Most of these patients will have had MRI and CT with equivocal results). (B) •Early diagnosis of dementia (especially younger patients and Alzheimer's disease) when MRI or CT is either normal, marginally abnormal or equivocally abnormal. (B)	•The grading of primary brain tumour. (B) •Localisation of optimal biopsy site (either primary or recurrent brain tumour). (C) •Differentiating malignancy from infection in HIV subjects where MRI is equivocal. (C)	•Diagnosis of dementia where MRI is clearly abnormal (C) •Most instances of stroke. (C) •Most psychiatric disorders other than early dementia. (C) •Pre-symptomatic or at risk Huntingdon's disease. (C) •Diagnosis of epilepsy. (C)

Table 1.2. Continued.

Miscellaneous Application	Indicated	Not indicated routinely (but may be helpful)	Not indicated
Disease assessment in HIV and other immuno-suppressed patients	•Identification of sites to biopsy in patients with pyrexia. (C) •Differentiating benign from malignant cerebral pathology. (B)	•Routine assessment of weight loss where malignancy is suspected. (C)	
Assessment of bone infection		•Assessment of bone infection associated with prostheses. (C) •Assessment of spinal infection or problematic cases of infection. (C)	
Assessment of bone metastases		•When bone scan or other imaging is equivocal. (C)	
Assessment of tumour recurrence in the pituitary		•Identifying recurrent functional pituitary tumours when anatomical imaging has not been successful. (C)	
Fever of unknown origin		•Identifying source of the fever of unknown origin. (C)	

risks of heart transplantation, costs and limitation of donors it is vital to select only those patients who, because of the lack of viable myocardium, cannot benefit from revascularisation procedures.

Neurology and Psychiatry

Applications in these medical disciplines include the management of brain tumours, the pre-surgical work-up of patients with epilepsy (complex partial seizures) resistant to medical therapies, and the identification of tumours causing para-neoplastic syndromes. Further, PET has been shown to precede all other methods for the early diagnosis and differential diagnosis of dementias. While there clearly is a role for this in management of patients it is only with the introduction of effective treatments that it will prove to be important and could become the most important clinical use of PET with time.

Currently Approved Indications

Tables 1.1 and 1.2 from the United States and the UK illustrate the current indications for clinical PET studies. While the tables use different criteria, they form a useful basis for an understanding of the present day role of PET in clinical management.

FDG-PET Cost Effectiveness Studies

In addition to being subjected to careful scrutiny, more than any other diagnostic technology, PET imaging has been required to demonstrate that it delivers cost effective diagnoses. Cost effectiveness studies in Nuclear Medicine including FDG PET studies have been reviewed by Dietlein (1999) [5] and by Gambhir (2000) [6]. These reviews also provide a detailed critique of the individual studies and in the review by Gambhir only six studies in the nuclear medicine literature were found which met all ten of their quality criteria for cost effectiveness studies and only one of these [7] was an FDG PET study. The following is not a comprehensive or detailed analysis of every cost effectiveness study in the literature but a review of FDG PET related to the more important studies in the literature including some published since the two reviews mentioned above and some that have been completed and will be published shortly. Table 1.3 shows the clinical conditions that have been analysed to date with a moderate degree of rigour which include solitary pulmonary nodules, staging non-small cell lung cancer, recurrent colo-rectal cancer, metastatic melanoma, lymphoma staging, and coronary artery disease.

Table 1.3. Reports of moderately rigorous PET cost-effectiveness studies.

Target Population	Evaluation Method (references)
Coronary artery disease	Decision Analysis Model [7], [8], [9]
Solitary Pulmonary Nodule	Decision Analysis Model [10], [11], [12]
Staging NSCLC	Decision Analysis Model [13], [14], [15]
Re-staging colo-rectal cancer	Decision Analysis Model [16]
Lymphoma staging	Retrospective costing [17], [18]
Adenosine vs Dipyridamole	Cost minimisation [19]
General oncology	Retrospective costing [20]
Neuropsychiatric	[21]

The economic modelling has been performed in different health care settings and suggests that PET is cost-effective, or even cost-saving, based on the assumptions made. Whether PET affects long term outcome remains to be fully tested in malignant conditions, but what is clear is that it can affect the short term management of patients with cancer (Table 1.4). Outcome effects may take up to 20 years to evaluate, for example, whether changes in chemotherapy or radiotherapy regimens early in the course of disease treatment will reduce second cancers. If an imaging modality is superior to another imaging modality and provides different information allowing management changes we should not wait a further 5 to 10 years to show long term outcome effects - these changes have been modelled and prospective studies are showing these models to be true. Furthermore the human costs of delay in the introduction of this modality may be large, since the management changes demonstrated suggest that unnecessary surgery can be avoided and necessary surgery expedited. There is therefore the potential to enable the appropriate treatment pathway.

Conclusion

The following examples will serve to illustrate the power of clinical PET in substantially altering patient management, thereby avoiding futile aggressive therapy and improving cost effectiveness. In Figure 1.1, the ability of PET to detect more extensive disease, as in this case, changed management by avoiding a futile thoracotomy and treating the patient appropriately with chemotherapy and palliative radiotherapy. As illustrated in Figure 1.2, although metastasis resection is clinically effective, this is only when the lesion is solitary. PET-FDG is now becoming routine before this surgery and avoiding, as in this case, many unnecessary resections. Staging of breast cancer both influences treatment and is the best guide to prognosis. Figure 1.3 very well demonstrates how the accuracy of staging is improved by the routine use of the PET scan, in this case by upstaging the disease. PET is now routinely used in certain scenarios for the initial assessment of patients with malignant melanoma. It is also valuable as in this case, Figure 1.4, as an effective means of follow-up when there is suspicion of recurrence in order that appropriate treatment can be instituted without delay. Finally, PET scanning is increasingly used because of its sensitivity for assessing early metabolic changes when early detection of tumour response, or evaluation of the success of chemotherapy, is critical. Figure 1.5 dramatically demonstrates this effect with complete resolution in a case of non-Hodgkin's lymphoma when tailoring of chemotherapy and prognosis are both a direct result of the outcome of the PET scan.

Table 1.4. Comparison of costs per life-year saved in different clinical procedures.

Procedure	Cost/Life-Year Saved ($US)
Liver transplant	$43,000–250,000
Mammography (<50 years)	$160,000
Renal dialysis	$116,000
Chemotherapy (Breast)	$46,200
Cardiac transplant	$27,200
CABG	$13,000

References and Suggested Reading

1. Gambhir S, Czernin J, Schwimmer J, Silverman DHS, Coleman RE, Phelps ME. A Tabulated Summary of the FDG PET Literature. J Nucl Med 2001;42(S1):1S-93S.
2. Warburg O. The metabolism of tumors. New York: Richard R. Smith; 1931.
3. CMS. PET Limitations and Indications - http://www.cms.gov/. Washington: Centers for Medicaid and Medicare; 2004.
4. *Positron Emission Tomography: A Strategy for Provision in the UK.* Royal College of Physicians of London, Royal College of Physicians and Surgeons of Glasgow, Royal College of Physicians of Edinburgh, Royal College of Pathologists, Royal College of Radiologists, British Nuclear Medicine Society: Intercollegiate Standing Committee on Nuclear Medicine; Jan 2003.
5. Dietlein M, Knapp WH, Lauterbach KW, Schica H. Economic Evaluation Studies in Nuclear Medicine: the Need for Standardization. Eur J Nucl Med 1999;26(6):663-680.
6. Gambhir SS. Economics of Nuclear Medicine. Introduction. Q J Nucl Med 2000;44(2):103-104.
7. Garber AM, Solomon NA. Cost-effectiveness of alternative test strategies for the diagnosis of coronary artery disease. Ann Intern Med 1999;130(9):719-728.
8. Patterson RE, Eisner RL, Horowitz SF. Comparison of cost-effectiveness and utility of exercise ECG, single photon emission computed tomography, positron emission tomography, and coronary angiography for diagnosis of coronary artery disease. Circulation 1995;92(6):1669-1670.

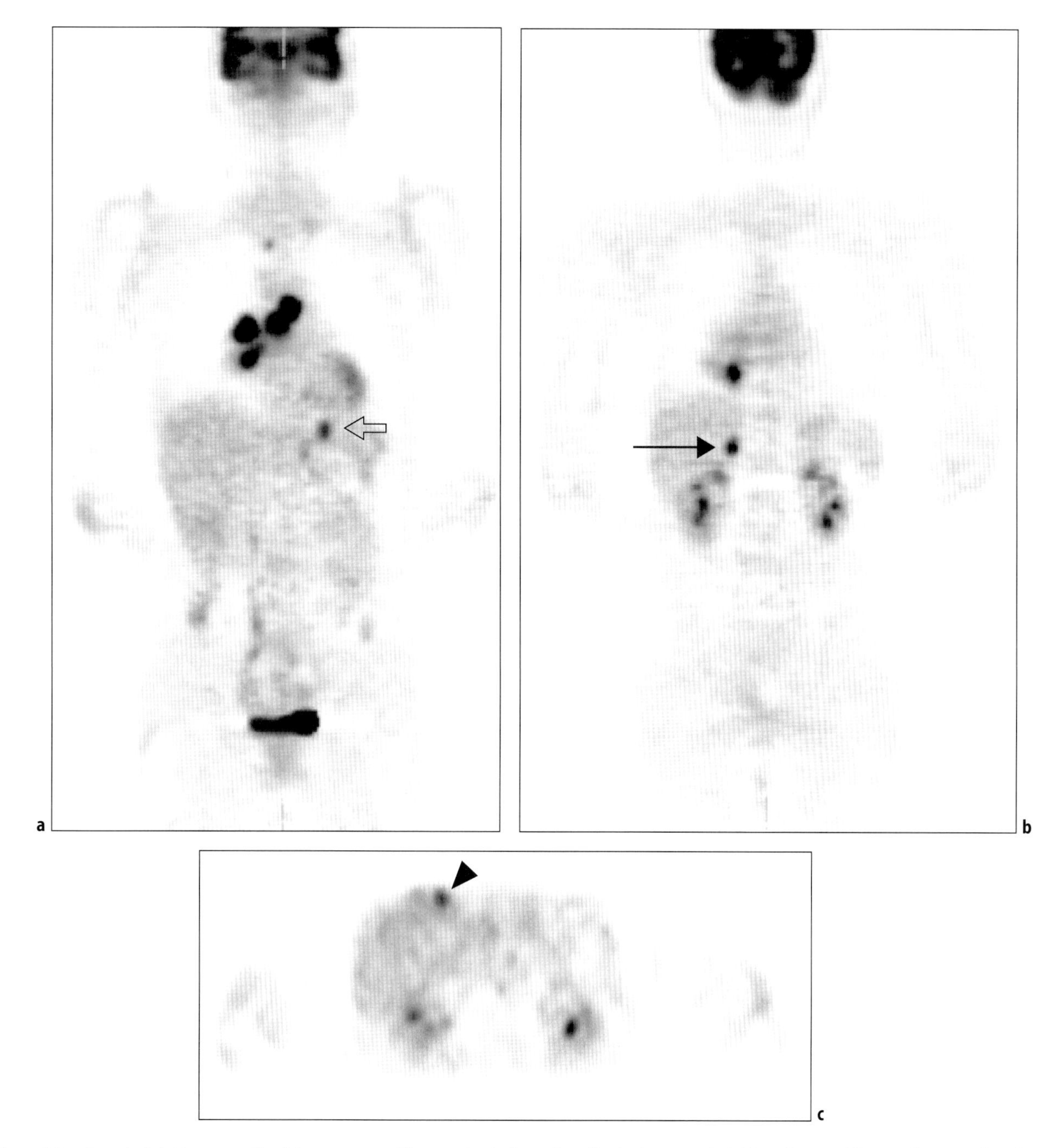

Figure 1.1. A central right non-small cell lung cancer with extensive ipsilateral mediastinal metastasis (**a**) in a 61-year-old man who was otherwise well. Staging by abdominal CT and bone scan showed 1.5 cm enlargement of the right adrenal gland and no other evidence of distant metastasis, and neoadjuvant therapy and resection were being considered. PET scan showed metastasis in the right adrenal gland(→) (**b**), left upper quadrant of the abdomen (⇨) (**a**) and the liver (▶) (**c**) (arrows) and management was changed to palliative radiation and chemotherapy. (Reproduced from Valk PE, Bailey DL, Townsend DW, Maisey MN. Positron Emission Tomography: Basic Science and Clinical Practice. Springer-Verlag London Ltd 2003, p. 527.)

9. Maddahi J, Gambhir SS. Cost-effective selection of patients for coronary angiography. J Nucl Cardiol 1997;4(2 Pt 2):S141-51.
10. Gambhir SS, Shepherd JE, Shah BD, Hart E, Hoh CK, Valk PE, et al. Analytical decision model for the cost-effective management of solitary pulmonary nodules. J Clin Oncol 1998;16(6):2113-2125.
11. Gould MK, Lillington GA. Strategy and cost in investigating solitary pulmonary nodules. Thorax 1998;53(Aug):Suppl 2:S32-S37.
12. Dietlein M, Weber K, Gandjour A, Moka D, Theissen P, Lauterbach KW, et al. Cost-effectiveness of FDG-PET for the management of solitary pulmonary nodules: a decision analysis based on cost reimbursement in Germany. Eur J Nucl Med 2000;27(10):1441-1456.
13. Gambhir SS, Hoh CK, Phelps ME, Madar I, Maddahi J. Decision tree sensitivity analysis for cost-effectiveness of FDG-PET in the staging and management of non-small-cell lung carcinoma. J Nucl Med 1996;37(9):1428-1436.
14. Dietlein M, Weber K, Gandjour A, Moka D, Theissen P, Lauterbach KW, et al. Cost-effectiveness of FDG-PET for the management of

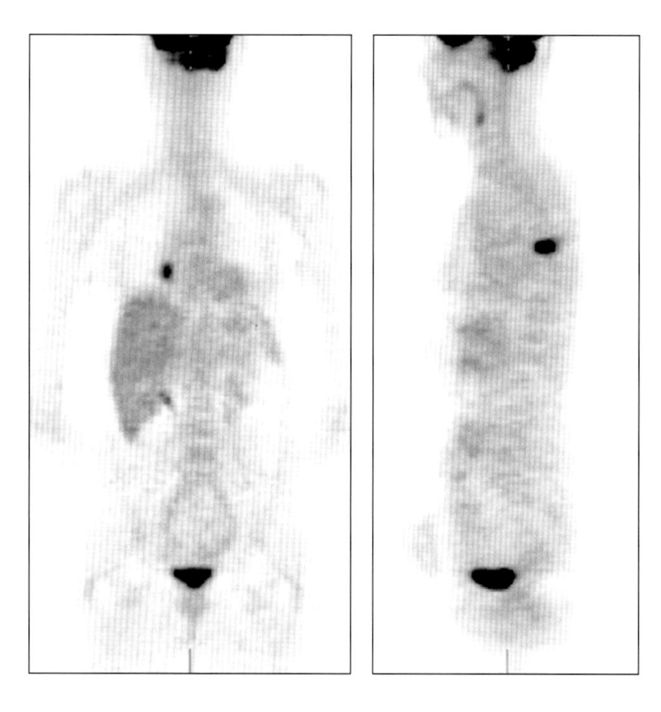

Figure 1.2. Coronal (right) and sagittal (left) FDG PET images in a 51-year-old man with a history of resection of rectal cancer three years earlier. CT demonstrated a lesion in the lower zone of the right lung and biopsy confirmed recurrent rectal cancer. CT imaging showed no other abnormality and PET study was performed for pre-operative staging. PET showed high uptake in the lung metastasis (left) and also showed metastasis in a thoracic vertebra, thereby excluding surgical resection of the lung lesion. The patient was treated by chemotherapy and irradiation. (Reproduced from Valk PE, Bailey DL, Townsend DW, Maisey MN. Positron Emission Tomography: Basic Science and Clinical Practice. Springer-Verlag London Ltd 2003, p. 565.)

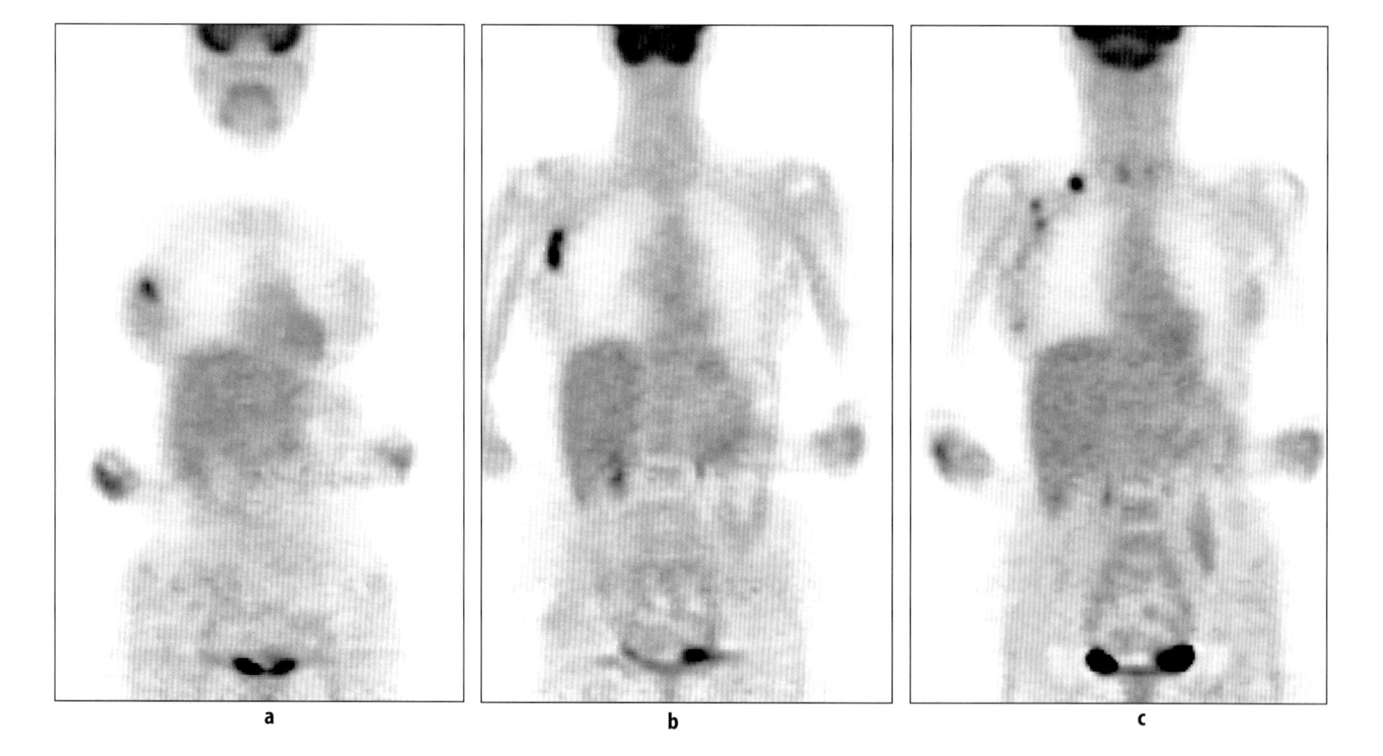

Figure 1.3. Coronal FDG PET image sections showing uptake in (**a**) right breast cancer (**b**) palpable right axillary lymph nodes (**c**) right supraclavicular and high axillary lymph nodes that were not clinically apparent. (Reproduced from Valk PE, Bailey DL, Townsend DW, Maisey MN. Positron Emission Tomography: Basic Science and Clinical Practice. Springer-Verlag London Ltd 2003, p. 599.)

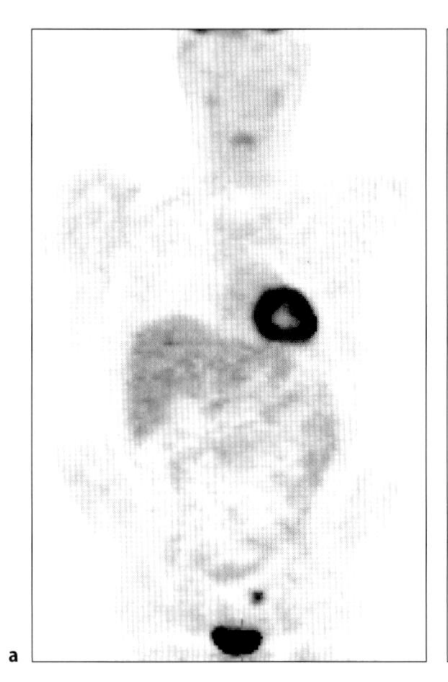

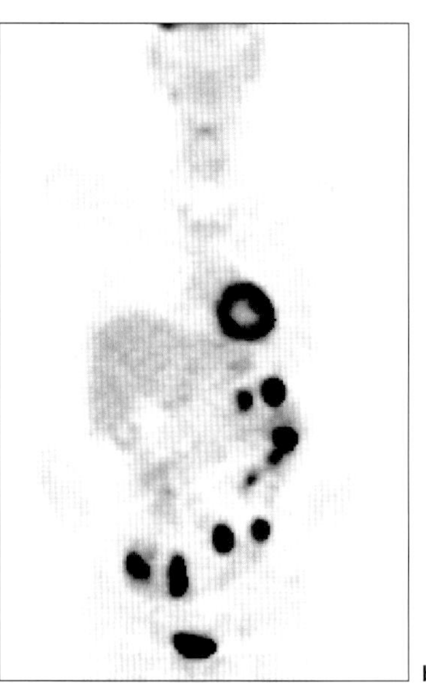

Figure 1.4. Coronal whole-body PET image section obtained in a 65-year-old man, one month after resection of a Clark's level III melanoma from the right thigh, showing a focus of increased uptake in the left pelvis (**a**). A similar focus was seen in the right pelvis. The patient was asymptomatic and CT scan of the pelvis was negative. A follow-up CT five months later also showed no pelvic abnormality. One year after the PET study, the patient presented with GI bleeding and was found to have a mass in the gastric mucosa, which proved to be recurrent melanoma on biopsy. Repeat PET scan after the biopsy showed multiple tumor masses in the abdomen and pelvis (**b**). (Reproduced from Valk PE, Bailey DL, Townsend DW, Maisey MN. Positron Emission Tomography: Basic Science and Clinical Practice. Springer-Verlag London Ltd 2003, p. 630.)

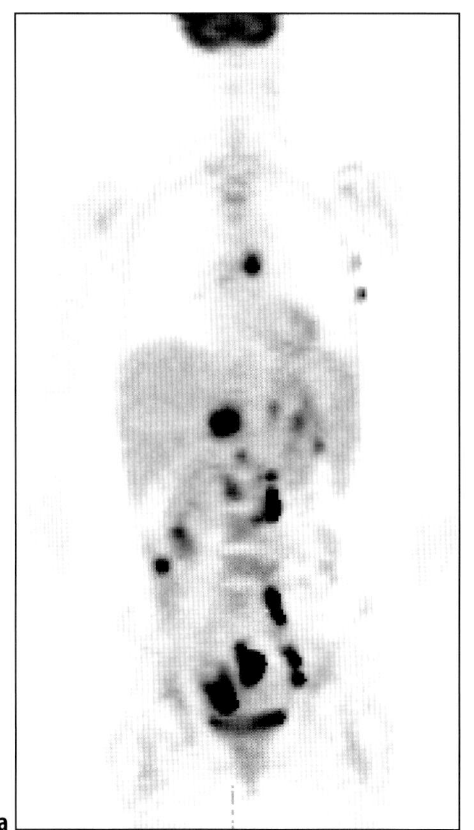

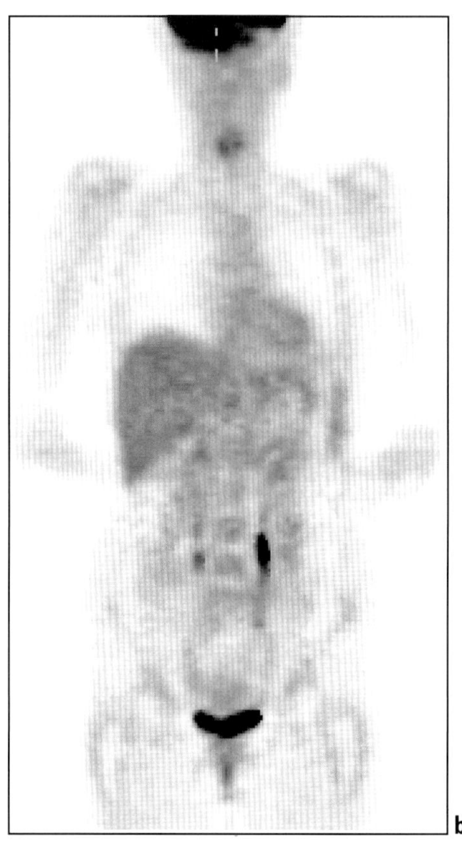

Figure 1.5. Response to treatment in a patient with non-Hodgkin's disease. The pre-treatment scan (**a**) shows extensive tumor above and below the diaphragm, whereas the post-treatment scan (**b**) shows no abnormal tracer localization, indicating complete response to therapy. (Reproduced from Valk PE, Bailey DL, Townsend DW, Maisey MN. Positron Emission Tomography: Basic Science and Clinical Practice. Springer-Verlag London Ltd 2003, p. 701.)

potentially operable non-small cell lung cancer: priority for a PET-based strategy after nodal-negative CT results. Eur J Nucl Med 2000;27(11):1598-1609.

15. Scott WJ, Shepherd J, Gambhir SS. Cost-effectiveness of FDG-PET for staging non-small cell lung cancer: a decision analysis. Ann Thorac Surg 1998;66(6):1876-1883.
16. Park KC, Schwimmer J, Shepherd JE, Phelps ME, Czernin JR, Schiepers C, et al. Decision analysis for the cost-effective management of recurrent colorectal cancer. Ann Surg 2001;233(3):310-319.
17. Hoh CK, Glaspy J, Rosen P, Dahlbom M, Lee SJ, Kunkel L, et al. Whole-body FDG-PET imaging for staging of Hodgkin's disease and lymphoma. J Nucl Med 1997;38(3):343-348.

18. Klose T, Leidl R, Buchmann I, Brambs HJ, Reske SN. Primary staging of lymphomas: cost-effectiveness of FDG-PET versus computed tomography. Eur J Nucl Med 2000;27(10):1457-1464.
19. Holmberg MJ, Mohiuddin SM, Hilleman DE, Lucas BDJ, Wadibia EC. Outcomes and costs of positron emission tomography: comparison of intravenous adenosine and intravenous dipyridamole. Clin Ther 1997;19(3):570-581.
20. Valk PE, Abella-Columna E, Haseman MK, Pounds TR, Tesar RD, Myers RW, et al. Whole-body PET imaging with [18F]fluorodeoxyglucose in management of recurrent colorectal cancer. Arch Surg 1999;134(5):503-511.
21. Small GW. Positron emission tomography scanning for the early diagnosis of dementia. West J Med 1999;171(5-6):298-294.

2 Physics and Instrumentation in PET*

Dale L Bailey, Joel S Karp and Suleman Surti

Introduction

In 1928 Paul AM Dirac postulated that a subatomic particle existed which was equivalent in mass to an electron but carried a positive charge. Carl Anderson experimentally observed these particles, which he called *positrons*, in cosmic ray research using cloud chambers in 1932. Both received Nobel Prizes in physics for their contributions. The positrons observed by Anderson were produced naturally in the upper atmosphere by the conversion of high-energy cosmic radiation into an electron–positron pair. Soon after this it was shown that when positrons interact with matter they give rise to two photons which, in general, are emitted simultaneously in almost exactly opposed directions. This sequence of events touches on many of the momentous developments in physics that occurred in the first 50 years of the twentieth century: radioactivity, Einstein's special relativity (energy–mass equivalence famously described by $E = mc^2$), quantum mechanics, de Broglie's wave–particle duality, and the laws of conservation of physical properties.

Today we produce positron-emitting radionuclides under controlled laboratory conditions in particle accelerators in the hospital setting for use in positron emission tomography (PET). In this chapter we will examine the basic physics of radioactivity and positrons and their detection as it relates to PET.

Models of the Atom

We use models, or representations, constantly in our lives. A painting, for example, is one individual's representation of a particular scene or feeling. It is clearly not the scene itself, but it is a model, or an attempt, to capture some expression of the reality as perceived by the artist. Likewise, scientists use models to describe various concepts about very-large-scale phenomena such as the universe, and very-small-scale phenomena such as the constituent components of all matter. One important feature of a model is that it usually has a restricted range over which it applies. Thus, we employ different models to account for different observations of the same entity, the classical example being the wave–particle duality of radiation: sometimes it is convenient to picture radiation as small discrete "packets" of energy that we can count individually, and at other times radiation appears to behave like a continuous entity or wave. The latter is evidenced by phenomena such as the diffraction of coherent light sources in a double-slit experiment. This could present a problem if we were to confuse the model and reality, but we emphasize again that the model is a *representation* of the underlying reality that we observe.

Amongst the ancient Greeks, Aristotle favored a continuous matter model composed of air, earth, fire, and water, where one could go on dividing matter infinitely into smaller and smaller portions. Others, though, such as Democritus, preferred a model in which matter was

* Chapter reproduced from Valk PE, Bailey DL, Townsend DW, Maisey MN. Positron Emission Tomography: Basic Science and Clinical Practice. Springer-Verlag London Ltd 2003, 41–67.

corpuscular. By the nineteenth century it was clear that chemicals combined in set proportions, thus supporting a corpuscular, or discrete, model of matter. At the turn of the twentieth century evidence was mounting that there were basic building blocks of matter called atoms (*Greek*: *indivisible*), but the question remained as to what, if anything, the atoms themselves were composed of. It was shown by JJ Thomson and, later, Ernest (Lord) Rutherford, that atoms could be broken down into smaller units in experiments using cathode ray tubes. Thomson proposed a model of the atom that was composed of a large, uniform and positively charged sphere with smaller negative charges embedded in it to form an electrostatically neutral mixture. His model of the atom is known as the "plum pudding" atom. Rutherford showed, however, that alpha particles (doubly ionized helium nuclei emitted from some unstable atoms such as radium) could pass through sheets of aluminum, and that this was at odds with the Thomson model. He proposed a model similar to that used to describe the orbit of the planets of the solar system about the sun (the "planetary" model). The Rutherford model had a central positive core – the nucleus – about which a cloud of electrons circulated. It predicted that most of the space in matter was unoccupied (thus allowing particles and electromagnetic radiation to pass through). The Rutherford model, however, presented a problem because classical physics predicted that the revolving electrons would emit energy, resulting in a spiralling of the electrons into the nucleus. In 1913, Bohr introduced the constraint that electrons could only orbit at certain discrete radii, or energy levels, and that in turn only a small, finite number of electrons could exist in each energy level. Most of what was required to understand the subatomic behavior of particles was now known. This is the Bohr (planetary) model of the atom. Later, the neutron was proposed by Chadwick (1932) as a large particle roughly equivalent to the mass of a proton, but without any charge, that also existed in the nucleus of the atom.

We shall continue to use the planetary model of the atom for much of our discussion. The model breaks down in the realm of quantum mechanics, where Newtonian physics and the laws of motion no longer apply, and as particles approach relativistic speeds (i.e., approaching the speed of light). Also, there are times when we must invoke a non-particulate model of the atom where the particles need to be viewed as waves (or, more correctly, *wave functions*). Electrons, for example, can be considered at times to be waves. This helps to explain how an electron can pass through a "forbidden" zone between energy levels and appear in the next level without apparently having passed through the forbidden area, defined as a region of space where there is zero probability of the existence of an electron. It can do so if its wave function is zero in this region. For a periodic wave with positive and negative components this occurs when the wave function takes a value of zero. Likewise, electromagnetic radiation can be viewed as particulate at times and as a wave function at other times. The planetary model of the atom is composed of nucleons (protons and neutrons in the nucleus of the atom) and circulating electrons. It is now known that these particles are not the fundamental building blocks of matter but are themselves composed of smaller particles called *quarks*. A deeper understanding of the elementary particles, and the frequently peculiar world of quantum physics, is beyond the scope of this book.

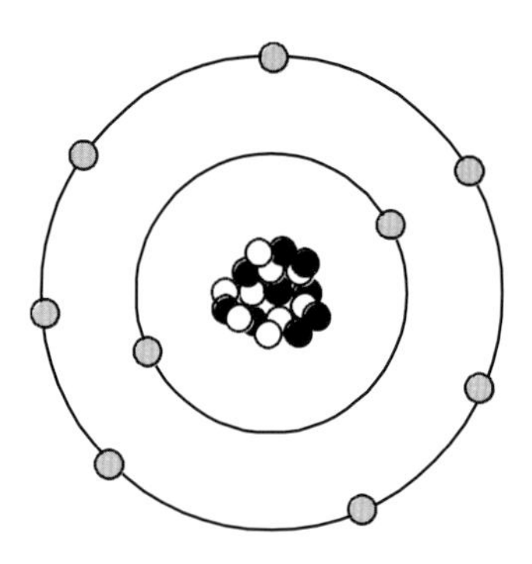

Figure 2.1. Atomic "planetary" model of radioactive fluorine-18 (^{18}F). The nucleus contains 9 protons (●) and 9 neutrons (○) and there are 9 electrons circulating in defined orbits. Stable fluorine would contain 10 neutrons.

The simple planetary model of the atom is illustrated in Fig. 2.1 for the case of radioactive fluorine-18 ($^{18}_{9}F$). Nine orbital electrons circulate in defined energy levels about a central nucleus containing nine neutrons and nine protons. Stable fluorine is $^{19}_{9}F$ i.e., the nucleus contains one more neutron than protons and this produces a stable configuration. In all non-ionized atoms the number of electrons equals the number of protons, with the difference between the atomic number (Z) and mass number (A) being accounted for by the neutrons. In practice we usually omit the atomic number when writing radionuclide species (e.g., ^{18}F) as it is implicit in the element's symbol.

Mass and Energy

In 1900 Max Planck demonstrated that the energy (*E*) of electromagnetic radiation was simply related to the frequency of the radiation (υ) by a constant (Planck's constant, h):

$$E = h\upsilon \quad (1)$$

In addition, experiments indicated that the radiation was only released in discrete "bursts". This was a startling result as it departed from the classical assumption of continuous energy to one in which electromagnetic radiation could only exist in integral multiples of the product of $h\upsilon$. The radiation was said to be *quantized*, and the discrete quanta became known as *photons*. Each photon contained an amount of energy that was an integer multiple of $h\upsilon$. The unit for energy is the joule (J), and we can calculate the energy of the radiation contained in a photon of wavelength of, for example, 450 nm as:

$$E = h\upsilon = \frac{hc}{\lambda} = \frac{6.63 \times 10^{-34}\text{J.s} \times 3 \times 10^{8}\text{m.s}^{-1}}{450 \times 10^{-9}\text{m}} \quad (2)$$
$$= 4.42 \times 10^{-19}\text{J}$$

This radiation (450 nm) corresponds to the portion of the visible spectrum towards the ultraviolet end. Each photon of light at 450 nm contains the equivalent of 4.42×10^{-19} J of energy in a discrete burst. We shall see the significance of this result later in this chapter when we discuss the emission of photons from scintillators.

The joule is the Système International d'Unites (abbreviated SI) unit of energy, however, a derived unit used frequently in discussions of the energy of electromagnetic and particulate radiation is the *electron volt* (eV). The electron volt is defined as the energy acquired when a unit charge is moved through a potential difference of one volt. Energy in joules can be converted to energy in electron volts (eV) by dividing by the conversion factor 1.6×10^{-19} J.eV^{-1}. Thus, the energy in eV for photons of 450 nm would be:

$$E = 4.42 \times 10^{-19}\text{J} \equiv \frac{4.42 \times 10^{-19}\text{J}}{1.6 \times 10^{-19}\text{J.eV}^{-1}} \quad (3)$$
$$= 2.76 \text{ eV}$$

X rays and gamma rays have energies of thousands to millions of electron volts per photon (Fig. 2.2).

Einstein's Special Theory of Relativity, published in 1905 while he was working in the patent office in Zurich, turned the physical sciences on its head. It predicted, amongst other things, that the speed of light was constant for all observers independent of their frame of reference (and therefore that time was no longer constant), and that mass and energy were equivalent. This means that we can talk about the *rest-mass equivalent energy* of a particle, which is the energy that would be liberated if all of the mass were to be converted to energy. By *rest mass* we mean that the particle

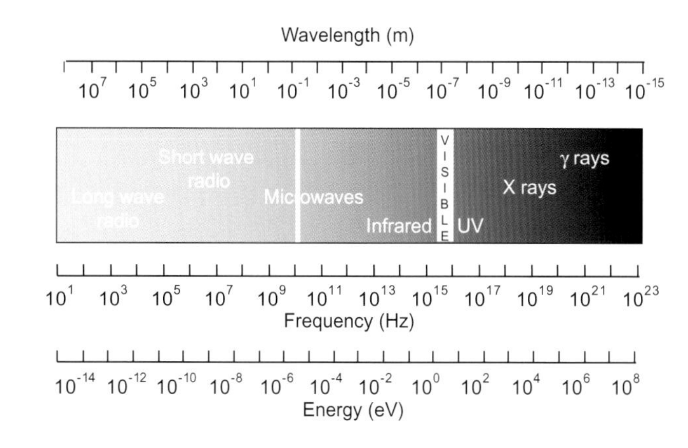

Figure 2.2. The electromagnetic spectrum showing the relationship between wavelength, frequency, and energy measured in electron volts (eV).

is considered to be at rest, i.e., it has no kinetic energy. Consider the electron, which has a rest mass of 9.11×10^{-31} kg; we can calculate the amount of energy this mass is equivalent to from:

$$\begin{aligned} E &= mc^2 \\ &= 9.11 \times 10^{-31}\text{kg} \times (3 \times 10^{8})^2 \text{ m.s}^{-1} \\ &= 8.2 \times 10^{-14}\text{ J} \quad (4) \\ &\equiv \frac{8.2 \times 10^{-14}\text{J}}{1.6 \times 10^{-19}\text{ J.eV}^{-1}} \\ &= 511 \text{ keV} \end{aligned}$$

The reader may recognize this as the energy of the photons emitted in positron–electron annihilation.

Conservation Laws

The principle of the conservation of fundamental properties comes from classical Newtonian physics. The concepts of conservation of mass and conservation of energy arose independently, but we now see that, because of the theory of relativity, they are merely two expressions of the same fundamental quantity. In the last 20–30 years the conservation laws have taken on slightly different interpretations from the classical ones: previously they were considered to be inviolate and equally applicable to all situations. Now, however, there are more conservation laws, and they have specific domains in which they apply as well as situations in which they break down. To classify these we must mention the four fundamental forces of nature. They are called the *gravitational*, *electromagnetic*, *strong*, and *weak* forces. It is believed that these forces are the only mechanisms which can act on the various

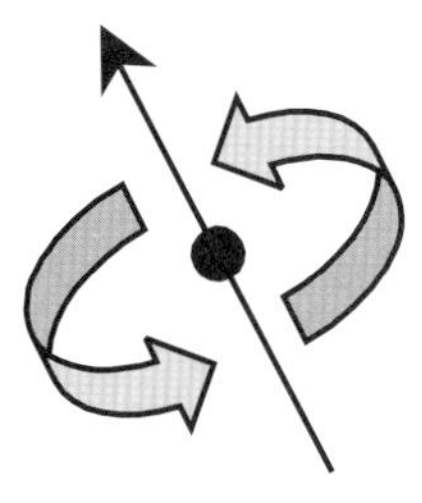

Figure 2.3. The spin quantum number for a particle can be pictured as a vector in the direction of the axis about which a particle is rotating. In this example, spin can be either "up" or "down".

properties of fundamental particles which make up all matter. These properties are electrostatic charge, energy and mass, momentum, spin and iso-spin, parity, strangeness and hypercharge (a quantity derived from strangeness and baryon numbers).

Charge is the electrostatic charge on a particle or atom and occurs in integer multiples of 1.6×10^{-19}.

Energy and *mass* conservation are well known from classical theory and are unified under special relativity.

Angular and *linear momentum* are the product of the mass (or moment of inertia) and the linear (or angular) velocity of a particle or atom.

Spin (s) and Isospin (i): Spin is the intrinsic angular momentum of a particle. It can be thought of by using the model of a ball rotating about its axis (Fig. 2.3). Associated with this rotation will be angular momentum which can take values in an arbitrary direction (labelled z) between –s to +s. The universe can be divided into two groups of particles on the basis of spin: those with spin $\frac{1}{2}$, and those with integer spin of 0, 1, or 2. The particles with spin $\frac{1}{2}$ are the mass-containing particles of the universe (fermions); the spin 0, 1, and 2 particles are the "force-carrying" particles (bosons). Some bosons, such as the pion, which serve as exchange particles for the strong nuclear force, are "virtual" particles that are very short-lived. Only spin $\frac{1}{2}$ particles are subject to the Pauli exclusion principle, which states that no two particles can have exactly the same angular momentum, spin, and other quantum mechanical physical properties. It was the concept of spin that led Dirac to suggest that the electron had an antimatter equivalent, the positron. Iso-spin is another quantum mechanical property used to describe the symmetry between different particles that behave almost identically under the influence of the strong force. In particular, the isospin relates the symmetry between a particle and its anti-particle as well as nucleons such as protons and neutrons that behave identically when subjected to the strong nuclear force. Similar to the spin, the isospin, *i*, can have half integer as well as integer values together with a special z direction which ranges in magnitude from $-i$ to $+i$. We shall see later that under certain conditions a high-energy photon (which has zero charge and isospin) can spontaneously materialize into an electron–positron pair. In this case both charge and isospin are conserved, as the electron has charge –1 and spin $+\frac{1}{2}$, and the positron has charge +1 and spin $-\frac{1}{2}$. Dirac possessed an over whelming sense of the symmetry in the universe, and this encouraged him to postulate the existence of the positron. Table 2.1 shows physical properties of some subatomic particles.

Parity is concerned with the symmetry properties of the particle. If all of the coordinates of a particle are reversed, the result may either be identical to the original particle, in which case it would be said to have *even* parity, or the mirror image of the original, in which case the parity is *odd*. Examples illustrating odd and even functions are shown in Fig. 2.4. Parity is conserved in all but weak interactions, such as beta decay.

The main interactions that we are concerned with are summarized in Table 2.2.

These are believed to be the only forces which exist in nature, and the search has been ongoing since the time of Einstein to unify these in to one all-encompassing law, often referred to as the Grand Unified Theory. To date, however, all attempts to find a grand unifying theory have been unsuccessful.

The fundamental properties and forces described here are referred to as the "Standard Model". This is the most widely accepted theory of elementary parti-

Table 2.1. Physical properties of some subatomic particles.

Particle	Symbol	Rest Mass (kg)	Charge	Spin	Isospin	Parity
Electron	e^-	9.11×10^{-31}	–1	$\frac{1}{2}$	$+\frac{1}{2}$	Even
Positron	e^+	9.11×10^{-31}	+1	$\frac{1}{2}$	$-\frac{1}{2}$	Even
Proton	p^+	1.673×10^{-27}	+1	$\frac{1}{2}$	$+\frac{1}{2}$	Even
Neutron	n^0	1.675×10^{-27}	0	$\frac{1}{2}$	$-\frac{1}{2}$	Even
Photon	Q	0	0	1	–	Odd
Neutrino	n	~0	0	$\frac{1}{2}$	$\frac{1}{2}$	Even

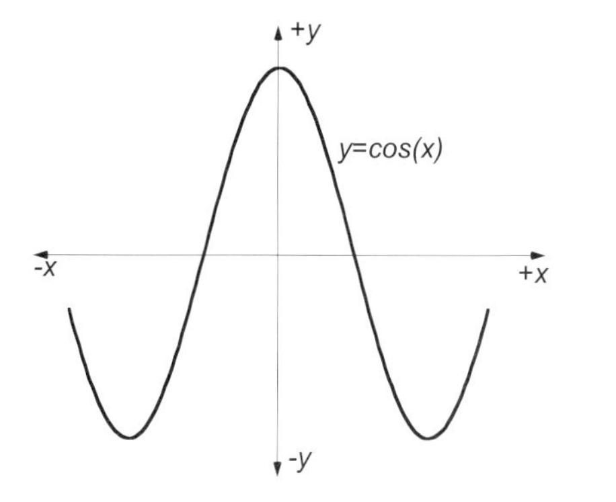

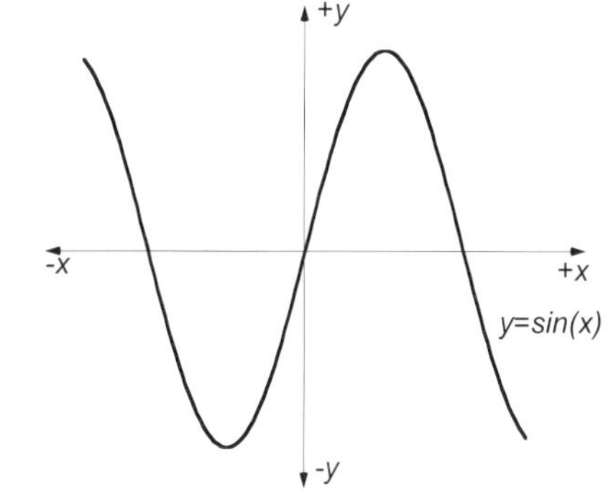

Figure 2.4. Examples of even (left) and odd (right) functions, to illustrate parity. In the even example (y = cos(x)) the positive and negative values of x have the same y-values; for the odd function (y = sin(x)) the negative x-values have opposite sign to the positive x-values.

Table 2.2. The table indicates whether the property listed is conserved under each of the fundamental interactions shown (gravity is omitted).

Property	Electromagnetic	Strong	Weak
Charge	Yes	Yes	Yes
Energy/mass	Yes	Yes	Yes
Angular momentum	Yes	Yes	Yes
Linear momentum	Yes	Yes	Yes
Iso-spin	No	Yes	No
Parity	Yes	Yes	No
Strangeness	Yes	Yes	No

cles and their interactions, which applies for all forces but gravity. The Standard Model remains a model though, and does not explain all observed phenomena, and work continues to find a grand unifying theory.

Radiation

Radiation can be classified into electromagnetic or particulate. *Ionising radiation* is radiation that has sufficient energy associated with it to remove electrons from atoms, thus causing ionisation. This is restricted to high-energy electromagnetic radiation (x and γ radiation) and charged particles (α, β^-, β^+). Examples of non-ionising electromagnetic radiation include light, radio, and microwaves. We will concern ourselves specifically with ionising radiation as this is of most interest in nuclear medicine and radiological imaging.

Electromagnetic Radiation

Electromagnetic radiation is pure energy. The amount of energy associated with each "bundle", or quantum, of energy is determined by the wavelength (λ) of the radiation. Human senses are capable of detecting some forms of electromagnetic radiation, for example, thermal radiation, or heat, ($\lambda \approx 10^{-5}$m), and visible light ($\lambda \approx 10^{-7}$m). The energy of the radiation can be absorbed to differing degrees by different materials: light can be stopped (absorbed) by paper, whereas radiation with longer wavelength (e.g., radio waves) or higher energy (γ rays) can penetrate the same paper.

We commenced our discussion at the beginning of this chapter with the comment that we are dealing with models of reality, rather than an accurate description of the reality itself; we likened this to dealing with paintings of landscapes rather than viewing the landscapes themselves. This is certainly the case when we discuss electromagnetic and particulate radiation. It had long been known that light acted like a wave, most notably because it caused interference patterns from which the wavelength of the light could be determined. Radiation was thought to emanate from its point of origin like ripples on the surface of a pond after a stone is dropped into it. This concept was not without its difficulties, most notably, the nature of the medium through which the energy was transmitted. This proposed medium was known as the "ether", and many experiments sought to produce evidence of its existence to no avail. Einstein, however, interpreted some experiments performed at the turn of the twentieth century where light shone on a photocathode could induce an electric current (known as the photoelectric effect) as showing that light acted as a particle. Einstein proposed that radiant energy was quantized into discrete packets, called *photons*. Thus, electromagnetic radiation could be viewed as having wave-like and particle-like properties. This view persists to this day and is known as the wave–particle duality. In 1924, Louis Victor, the Duc de Broglie, proposed that if wave–particle duality could apply to electromagnetic radiation, it could also apply to matter. It is now known that this is true: electrons, for example, can exhibit particle-like properties such as when they interact like small billiard balls, or wave-like properties as when they undergo diffraction. Electrons can pass from one position in space to another, separated by a "forbidden zone" in which they cannot exist, and one way to interpret this is that the electron is a wave that has zero amplitude within the forbidden zone. The electrons could not pass through these forbidden zones if viewed strictly as particles.

An important postulate proposed by Neils Bohr was that De Broglie's principle of wave–particle duality was complementary. He stated that either the wave or the particle view can be taken to explain physical phenomena, but not both at the same time.

Electromagnetic radiation has different properties depending on the wavelength, or energy, of the quanta. Only higher-energy radiation has the ability to ionize atoms, due to the energy required to remove electrons from atoms. Electromagnetic ionising radiation is restricted to x and γ rays, which are discussed in the following sections.

X rays: X rays are electromagnetic radiation produced within an atom, but outside of the nucleus. *Characteristic* X rays are produced when orbital electrons drop down to fill vacancies in the atom after an inner shell electron is displaced, usually by firing electrons at a target in a discharge tube. As the outer shell electron drops down to the vacancy it gives off energy and this is known as a characteristic X ray as the energy of the X ray is determined by the difference in the binding energies between the electron levels (Fig. 2.5).

As any orbital electron can fill the vacancy, the quanta emitted in this process can take a number of energies. The spectrum is characteristic, however, for the target metal and this forms the basis of quantitative X-ray spectroscopy for sample analysis. The spectrum of energies emerging in X-ray emission displays a continuous nature, however, and this is due to a second process for X-ray production known as Bremsstrahlung (*German*: "braking radiation").

Bremsstrahlung radiation is produced after a free electron with kinetic energy is decelerated by the influence of a heavy target nucleus. The electron and the nucleus interact via a Coulomb (electrostatic charge) interaction, the nucleus being positively charged and the electron carrying a single negative charge. The process is illustrated in Fig. 2.6. The electron loses kinetic energy after its deceleration under the influence of the target nucleus, which is given off as electromagnetic radiation. There will be a continuum of quantized energies possible in this process depending on the energy of the electron, the size of the nucleus, and other physical factors, and this gives the continuous component of the X-ray spectrum. The efficiency of Bremsstrahlung radiation production is highly dependent on the atomic number of the nucleus, with the fraction of positron energy converted to electromagnetic radiation being approximately equal to ZE/3000, where Z is the atomic number of the absorber and E is the positron energy in MeV. For this reason, low Z materials such as perspex are preferred for shielding positron emitters.

X rays generally have energies in the range of $\sim 10^3$–10^5 eV.

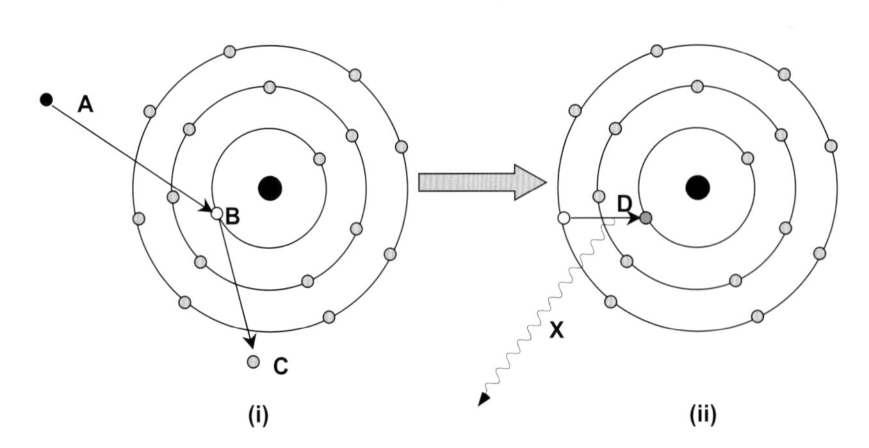

Figure 2.5. The characteristic X-ray production process is shown. In (i) an electron(A) accelerated in a vacuum tube by an electric field gradient strikes the metal target and causes ionization of the atom; in this case in the k-shell. The electron (B) is ejected from the atom (C). Subsequently (ii), a less tightly bound outer orbital electron fills the vacancy (D) and in doing so gives up some energy (X), which comes off at a characteristic energy equal to the difference in binding energy between the two energy levels. The radiation produced is an X ray.

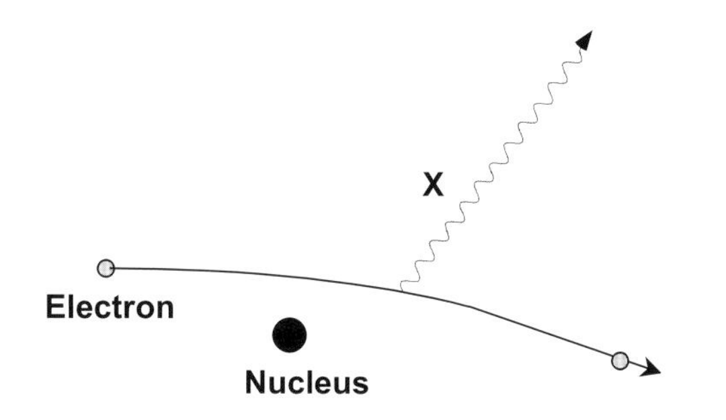

Figure 2.6. The Bremsstrahlung process is responsible for the continuous spectrum of X rays.

Gamma Radiation

Gamma rays are electromagnetic radiation emitted from the nucleus after a spontaneous nuclear decay. This is usually associated with the emission of an alpha or beta particle although there are alternative decay schemes. X and γ rays are indistinguishable after they are emitted from the atom and only differ in their site of origin. After the emission of a particle in a radioactive decay the nucleus can be left in an excited state and this excess energy is given off as a γ ray, thus conserving energy.

Gamma ray emission is characteristic, and it is determined by the difference in energy levels between the initial and final state of the energy level transitions within the nucleus.

Annihilation Radiation

As this book is primarily concerned with positrons and their applications, we include a further classification for electromagnetic radiation which is neither x nor γ. *Annihilation radiation* is the energy produced by the positron–electron annihilation process. The energy of the radiation is equivalent to the rest mass of the electron and positron, as we saw in the section on Mass and Energy, above. The mechanism of positron decay is discussed in depth in the next section.

Annihilation radiation, arising from positron–electron annihilation, is produced outside of the nucleus, and often outside of the positron-emitting atom. There are two photons produced by each positron decay and annihilation. Each photon has energy of 0.511 MeV, and the photons are given off at close to 180° opposed directions. It is this property of collinearity that we exploit in PET, allowing us to define the line-of-sight of the event without the need for physical collimation.

Particulate Radiation

Particle emission from natural radioactive decay was the first observation of radioactivity. Wilhelm Röntgen had produced X rays in 1896, and a year later Henri Becquerel showed that naturally occurring uranium produced radiation spontaneously. While the radiation was thought initially to be similar to Röntgen's x rays, Rutherford showed that some types of radiation were more penetrating than others. He called the less penetrating radiation alpha (α) rays and the more penetrating ones beta (β) rays. Soon after, it was shown that these radiations could be deflected by a magnetic field, i.e., they carried charge. It was clear that these were not electromagnetic rays and were, in fact, particles.

Radioactive Decay

The rate at which nuclei spontaneously undergo radioactive decay is characterized by the parameter called the half-life of the radionuclide. The half-life is the time it takes for half of the unstable nuclei present to decay (Fig. 2.7). It takes the form of an exponential function where the number of atoms decaying at any particular instant in time is determined by the number of unstable nuclei present and the decay constant (λ) of the nuclide. The rate of decay of unstable nuclei at any instant in time is called the *activity* of the radionuclide. The activity of the nuclide after a time t is given by

$$A_t = A_0 e^{-\lambda t} \tag{5}$$

where A_0 is the amount of activity present initially, A_t is the amount present after a time interval t, and λ is the decay constant. The decay constant is found from

$$\lambda = \frac{\log_e(2)}{t_{\frac{1}{2}}} \tag{6}$$

and the units for λ are time^{-1}. The SI unit for radioactivity is the becquerel (Bq). One becquerel (1 Bq) equals one disintegration per second.

Example: calculate the radioactivity of a 100 MBq sample of ^{18}F ($t_{\frac{1}{2}}$ = 109.5 mins) 45 minutes after calibration and from this deduce the number of atoms and mass of the radionuclide present:

$$\begin{aligned} \lambda &= \frac{0.6931}{109.5} = 6.330 \times 10^{-3}\ \text{min}^{-1} \\ A_t &= 100 \times e^{-6.330 \times 10^{-3} x\, 45} \\ &= 75.2\ \text{MBq} \end{aligned} \tag{7}$$

The total number of ^{18}F atoms present, N, can be calculated from the activity and the decay constant using:

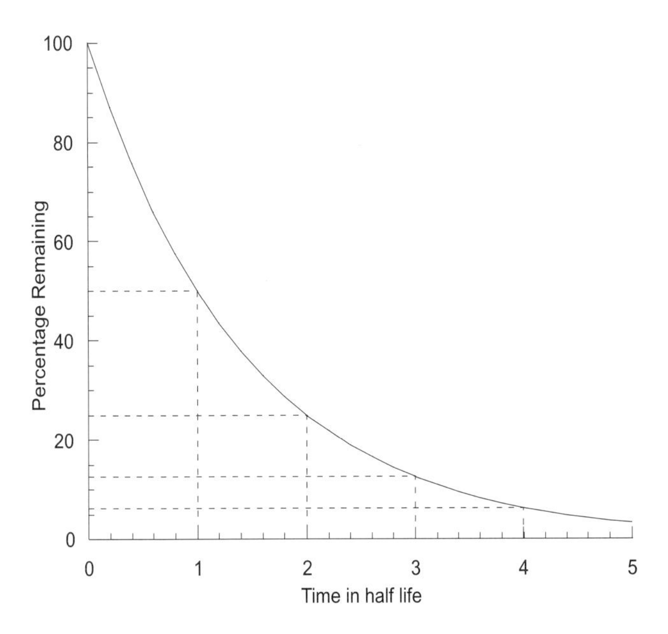

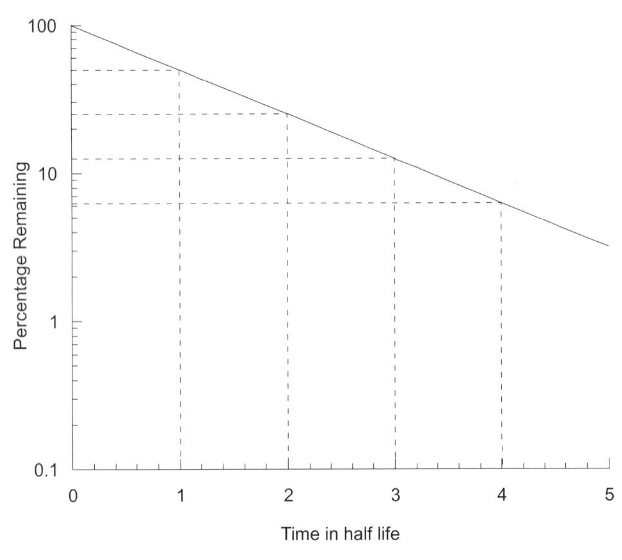

Figure 2.7. The decay of a radionuclide follows an exponential form seen in the top graph, which gives a straight line in the log-linear plot on the bottom. The dashed lines indicate the amount remaining after each half-life.

$$N = \frac{A_t}{\lambda} \tag{8}$$

In this example the total number of ^{18}F nuclei present would be:

$$N = \frac{75.2 \times 10^6}{1.055 \times 10^{-4}\ (\text{sec}^{-1})} \tag{9}$$
$$= 7.13 \times 10^{11} \text{ nuclei}$$

We can determine the mass of this number of nuclei using Avogadro's number ($N_A = 6.023 \times 10^{23}$ mole^{-1}) and the mass of a mole of ^{18}F (18 g) to be

$$m = \frac{N}{N_A} = \frac{7.13 \times 10^{11}}{6.023 \times 10^{23}} \times 18 \text{ g}$$
$$= 2.13 \times 10^{-11} g \text{ (21.3 pg)} \tag{10}$$

There are two other terms related to radioactivity that are useful. *Specific Activity* is the ratio of radioactivity to total mass of the species present. It has units of TBq/gm or TBq/mole. *Branching ratio* is the fraction of atoms that decay by the emission of a particular radiation. For example, ^{11}C is a pure positron emitter and therefore has a branching ratio of 1.00 (or 100%). ^{18}F, however, decays to ^{18}O by positron emission only 96.9% of the time, the remaining time being by electron capture (EC) which does not emit a positron. Its branching ratio is 0.969 (or 96.9%). Note that the radioactivity of a nuclide is the number of atoms decaying per second, not the number of radiation particles given off. Thus, to calculate the radioactivity from a measurement of the emitted rate of particles or photons, a correction is required to account for the non-radiative disintegrations.

Correcting for decay is often required in calculations involving radioactivity. The decay correction factor can be calculated from the point in time of an instantaneous measurement to a reference time. The decay correction factor (F) is given by:

$$F = e^{\lambda(t-t_0)} \tag{11}$$

where t is the time of the measurement and t_0 is the reference time. It is often necessary to account for decay *within* the interval of the counting period, especially with short-lived tracers as are used in positron imaging. The correction factor (F_{int}) to account for decay during a measurement is:

$$F_{int} = \frac{\lambda t}{1 - e^{-\lambda t}} \tag{12}$$

although taking the time t from the *mid-point* of the counting interval (rather than the time at the start of the measurement) to the reference time in the calculation of F introduces an error of typically less than 1% for counting intervals $<0.75 t_{\frac{1}{2}}$.

Alpha Decay

Alpha particles are helium nuclei ($^4_2He^{2+}$). They are typically emitted from high Z-number atoms and form the components of many naturally occurring radioactive decay series. Due to their large mass, alpha particles deposit large amounts of energy in a very small distance in matter. Therefore, as a radiation hazard they represent a very large problem if ingested, however, conversely, as they are relatively easy to stop, they are easily shielded. An example of alpha decay is shown in the following:

$$^{238}_{92}\mathrm{U} \rightarrow ^{234}_{90}\mathrm{Th} + ^{4}_{2}\alpha + \gamma \quad (13)$$

The half-life for this particular process is 4.5×10^9 years.

Beta Decay

Beta particles are negatively charged electrons that are emitted from the nucleus as part of a radioactive disintegration. The beta particles emitted have a continuous range of energies up to a maximum. This appeared at first to be a violation of the conservation of energy. To overcome this problem, in 1931 Wolfgang Pauli proposed that another particle was emitted which he called the neutrino (ν). He suggested that this particle had a very small mass and zero charge. It could carry away the excess momentum to account for the difference between the maximum beta energy and the spectrum of energies that the emitted beta particles displayed. In fact, we now refer to the neutrino emitted in beta-minus decay as the *antineutrino*, indicated by the '‾' over the symbol ν. β^- decay is an example of a weak interaction, and is different to most other fundamental decays as parity is not conserved.

The following shows an example of a beta decay scheme for ^{131}I:

$$^{131}_{53}I \rightarrow ^{131}_{54}Xe + ^{0}_{-1}\beta^- + \gamma + \bar{\nu} \quad (14)$$

The half-life for ^{131}I decay is 8.02 days. The most abundant β particle emitted from ^{131}I has a maximum energy of 0.606 MeV and there are many associated γ rays, the most abundant (branching ratio = 0.81) having an energy of 0.364 MeV.

Positron Decay

There are two methods of production of positrons: by pair production, and by nuclear transmutation. Pair production will be discussed in the following section. Positron emission from the nucleus is secondary to the conversion of a proton into a neutron as in:

$$^{1}_{1}p^+ \rightarrow ^{1}_{0}n + ^{0}_{1}\beta^+ + \nu \quad (15)$$

with in this case a neutrino is emitted. The positron is the antimatter conjugate of the electron emitted in β^- decay.

The general equation for positron decay from an atom is:

$$^{A}_{Z}X \rightarrow {}_{Z-1}^{A}Y + ^{0}_{1}\beta^+ + \nu + Q\,(+e^-) \quad (16)$$

where Q is energy. The atom X is proton-rich and achieves stability by converting a proton to a neutron. The positive charge is carried away with the positron. As the daughter nucleus has an atomic number one less than the parent, one of the orbital electrons must be ejected from the atom to balance charge. This is often achieved by a process known as *internal conversion*, where the nucleus supplies energy to an orbital electron to overcome the binding energy and leave it with residual kinetic energy to leave the atom. As both a positron and an electron are emitted in positron decay the daughter nucleus must be at least two electron masses lighter than the parent.

The positron will have an initial energy after emission, which, similar to the case of β^- decay, can take a continuum of values up to a maximum. After emission from the nucleus, the positron loses kinetic energy by interactions with the surrounding matter. The positron interacts with other nuclei as it is deflected from its original path by one of four types of interaction:

(i) *Inelastic collisions* with atomic electrons, which is the predominant mechanism of loss of kinetic energy,
(ii) *Elastic scattering* with atomic electrons, where the positron is deflected but energy and momentum are conserved,
(iii) *Inelastic scattering* with a nucleus, with deflection of the positron and often with the corresponding emission of Bremsstrahlung radiation,
(iv) *Elastic scattering* with a nucleus where the positron is deflected but does not radiate any energy or transfer any energy to the nucleus.

As the positron passes through matter it loses energy constantly in ionisation events with other atoms or by radiation after an inelastic scattering. Both of these situations will induce a deflection in the positron path, and thus the positron takes an extremely tortuous passage through matter. Due to this, it is difficult to estimate the range of positrons based on their energy alone, and empirical measurements are usually made to determine the mean positron range in a specific material.

The positron eventually combines with an electron when both are essentially at rest. A metastable intermediate species called positronium may be formed by the positron and electron combining. Positronium is a non-nuclear, hydrogen-like element composed of the positron and electron that revolve around their combined centre of mass. It has a mean life of around 10^{-7} seconds. As expected, positronium displays similar properties to the hydrogen atom with its spectral lines having approximately half the frequency of those of hydrogen due to the much smaller mass ratio. Positronium formation occurs with a high probability

Table 2.3. Properties of some positron-emitting nuclides of interest in PET compiled from a variety of sources.

Nuclide	E_{max} (MeV)	E_{mode} (MeV)	$t_{\frac{1}{2}}$ (mins)	Range in Water (mm) Max	Mean	Use in PET
^{11}C	0.959	0.326	20.4	4.1	1.1	Labelling of organic molecules
^{13}N	1.197	0.432	9.96	5.1	1.5	$^{13}NH_3$
^{15}O	1.738	0.696	2.03	7.3	2.5	$^{15}O_2$, $H_2{}^{15}O$, $C^{15}O$, $C^{15}O_2$
^{18}F	0.633	0.202	109.8	2.4	0.6	[^{18}F]-DG, $^{18}F^-$
^{68}Ga	1.898	0.783	68.3	8.2	2.9	[^{68}Ga]-EDTA, [^{68}Ga]-PTSM
^{82}Rb	3.40	1.385	1.25	14.1	5.9	Generator-produced perfusion tracer
^{94m}Tc	2.44	†	52	‡	‡	β^+-emitting version of ^{99m}Tc
^{124}I	2.13	†	6.0×10^3	‡	‡	Iodinated molecules

‡Not reported to date.
†Many-positron decay scheme hence no E_{mode} value given.

in gases and metals, but only in about one-third of cases in water or human tissue where direct annihilation of the electron and the positron is more favorable. Positronium can exist in either of two states, para-positronium (spin $= +\frac{1}{2}$) or orthopositronium (spin $= +\frac{3}{2}$). Approximately three-quarters of the positronium formed is orthopositronium.

Positron emission from the nucleus, with subsequent annihilation, means that the photon-producing event (the annihilation) occurs outside the radioactive nucleus. The finite distance that positrons travel after emission contributes uncertainty to the localisation of the decaying nucleus (the nucleus is the species that we wish to determine the location of in positron tomography, not where the positron eventually annihilates). The uncertainty due to positron range is a function that increases with increasing initial energy of the positron. For a high-energy positron such as ^{82}Rb (E_{max} = 3.4 MeV), the mean range in water is around 5.9 mm. Table 3.3 shows some commonly used positron emitting nuclides and associated properties.

When the positron and electron eventually combine and annihilate electromagnetic radiation is given off. The most probable form that this radiation takes is of two photons of 0.511 MeV (the rest-mass equivalent of each particle) emitted at 180° to each other, however, three photons can be emitted (<1% probability). The photons are emitted in opposed directions to conserve momentum, which is close to zero before the annihilation.

Many photon pairs are not emitted strictly at 180°, however, due to non-zero momentum when the positron and electron annihilate. This fraction has been estimated to be as high as 65% in water. This con-

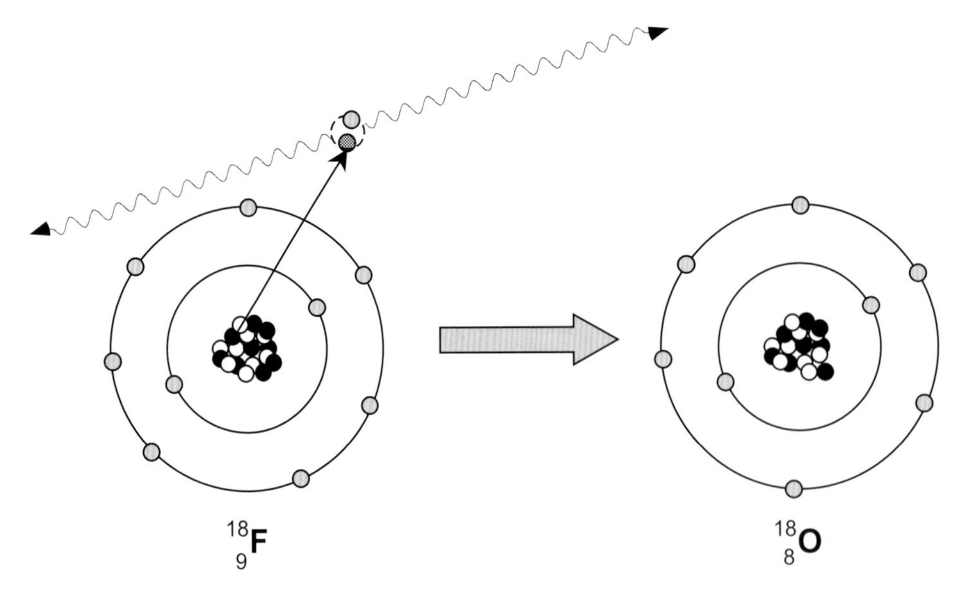

Figure 2.8. Annihilation radiation is produced subsequent to a positron being ejected from the nucleus. The positron travels a finite distance, losing energy by interaction with other electrons and nuclei as it does, until it comes to rest and combines (annihilates) with an electron to give rise to two photons, each equivalent to the rest-mass energy of the particles. The two photons are approximately anti-collinear and it is this property that is used to localize events in PET.

tributes a further uncertainty to the localisation of the nuclear decay event of 0.5° FWHM from strictly 180°, which can degrade resolution by a further 1.5 mm (dependent on the distance between the two coincidence detectors). This effect, and the finite distance travelled by the positron before annihilation, places a fundamental lower limit of the spatial resolution that can be achieved in positron emission tomography.

Interaction of Radiation with Matter

When high-energy radiation interacts with matter energy can be transferred to the material. A number of effects may follow, but a common outcome is the ionisation or excitation of the atoms in the absorbing material.

In general, the larger the mass of the particle the greater the chance of being absorbed by the material. Large particles such as alpha particles have a relatively short range in matter, whereas beta particles are more penetrating. The extremely small mass of the neutrino, and the fact that it has no charge, means that it interacts poorly with material, and is very hard to stop or detect. High-energy photons, being massless, are highly penetrating.

Interaction of Particulate Radiation with Matter

When higher energy particles such as alphas, betas, protons, or deuterons interact with atoms in an absorbing material the predominant site of interaction is with the orbital electrons of the absorber atoms. This leads to ionisation of the atom, and liberation of excited electrons by the transfer of energy in the interaction. The liberated electrons themselves may have sufficient energy to cause further ionisation of neighboring atoms and the electrons liberated from these subsequent interactions are referred to as delta rays.

Positron annihilation is an example of a particulate radiation interacting with matter. We have already examined this process in detail.

Interaction of Photons with Matter

High-energy photons interact with matter by three main mechanisms, depending on the energy of the electromagnetic radiation. These are (i) the photoelectric effect, (ii) the Compton effect, and (iii) pair production. In addition, there are other mechanisms such as coherent (Rayleigh) scattering, an interaction between a photon and a whole atom which predominates at energies less than 50 keV; triplet production and photonuclear reactions, where high energy gamma rays induce decay in the nucleus, and which require energies of greater than ~10 MeV. We will focus on the three main mechanisms which dominate in the energies of interest in imaging in nuclear medicine.

Photoelectric Effect

The photoelectric effect occupies a special place in the development of the theory of radiation. During the course of experiments which demonstrated that light acted as a wave, Hertz and his student Hallwachs showed that the effect of an electric spark being induced in a circuit due to changes in a nearby circuit could be enhanced if light was shone upon the gap between the two coil ends. They went on to show that a negatively charged sheet of zinc could eject negative charges if light was shone upon the plate. Philipp Lenard demonstrated in 1899 that the light caused the metal to emit electrons. This phenomenon was called the photoelectric effect. These experiments showed that the electric current induced by the ejected electrons was directly proportional to the intensity of the light. The interesting aspect of this phenomenon was that there appeared to be a light intensity threshold below which no current was produced. This was difficult to explain based on a continuous wave theory of light. It was these observations that led Einstein to propose the quantized theory of the electromagnetic radiation in 1905, for which he received the Nobel Prize.

The photoelectric effect is an interaction of photons with orbital electrons in an atom. This is shown in Fig. 2.9. The photon transfers all of its energy to the electron. Some of the energy is used to overcome the

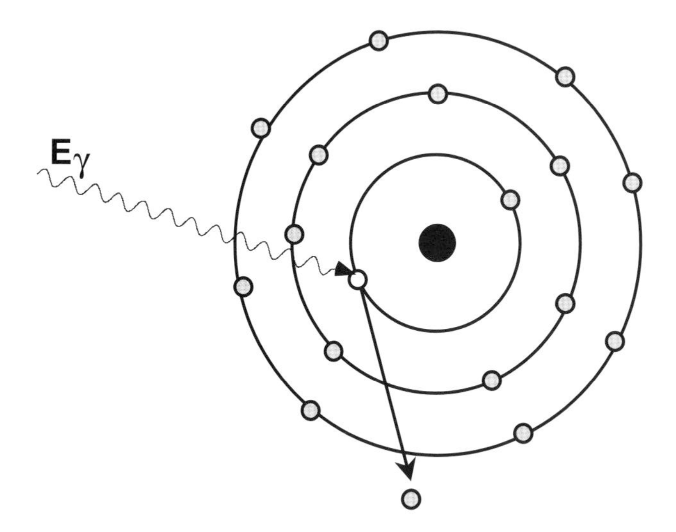

Figure 2.9. The photoelectric effect involves all of the energy from a photon being transferred to an inner shell electron, causing ionization of the atom.

binding energy of the electron, and the remaining energy is transferred to the electron in the form of kinetic energy. The photoelectric effect usually occurs with an inner shell electron. As the electron is ejected from the atom (causing ionisation of the atom) a more loosely bound outer orbital electron drops down to occupy the vacancy. In doing so it will emit radiation itself due to the differences in the binding energy for the different electron levels. This is a characteristic X ray. The ejected electron is known as a photoelectron. Alternately, instead of emitting an X ray, the atom may emit a second electron to remove the energy and this electron is known as an Auger electron. This leaves the atom doubly charged. Characteristic X rays and Auger electrons are used to identify materials using spectroscopic methods based on the properties of the emitted particles.

The photoelectric effect dominates in human tissue at energies less than approximately 100 keV. It is of particular significance for X-ray imaging, and for imaging with low-energy radionuclides. It has little impact at the energy of annihilation radiation (511 keV), but with the development of combined PET/CT systems, where the CT system is used for attenuation correction of the PET data, knowledge of the physics of interaction via the photoelectric effect is extremely important when adjusting the attenuation factors from the X-ray CT to the values appropriate for 511 keV radiation.

Compton Scattering

Compton scattering is the interaction between a photon and a loosely bound orbital electron. The electron is so loosely connected to the atom that it can be considered to be essentially free. This effect dominates in human tissue at energies above approximately 100 keV and less than ~2 MeV. The binding potential of the electron to the atom is extremely small compared with the energy of the photon, such that it can be considered to be negligible in the calculation. After the interaction, the photon undergoes a change in direction and the electron is ejected from the atom. The energy loss by the photon is divided between the small binding energy of the energy level and the kinetic energy imparted to the Compton recoil electron. The energy transferred does not depend on the properties of the material or its electron density (Fig. 2.10).

The energy of the photon after the Compton scattering can be calculated from the Compton equation:

$$E'_\gamma = \frac{E_\gamma}{1 + \frac{E_\gamma}{m_0 c^2}(1 - \cos(\theta_c))} \quad (17)$$

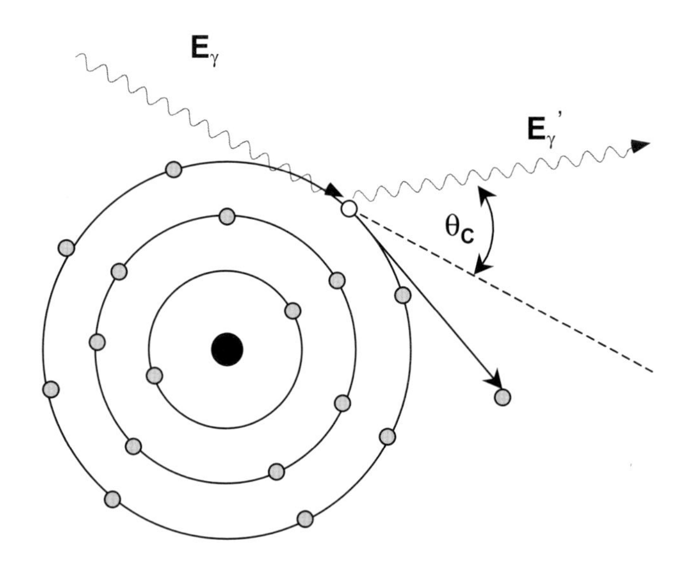

Figure 2.10. In Compton scattering, part of the energy of the incoming photon is transferred to an atomic electron. This electron is known as the recoil electron. The photon is deflected through an angle proportional to the amount of energy lost.

e.g., What is the energy of an annihilation photon after a single scatter through 60°?

$$E_\gamma = 511 \text{ keV} \quad (18)$$

$$\theta_c = 60°;\ \cos[\theta_c] = 0.5 \quad (19)$$

$$\mathrm{m_0 c^2} = 9.11 \times 10^{-31}\ \mathrm{kg} \times (3.0 \times 10^8\ \mathrm{m \cdot s^{-1}})^2 \equiv 511\ \mathrm{keV} \quad (20)$$

$$E'_\gamma = \frac{511}{1 + \frac{511}{511}(1 - 0.5)} = 341\ keV \quad (21)$$

From consideration of the Compton equation it can be seen that the maximum energy loss occurs when the scattering angle is 180° (cos (180°) = –1), i.e., the photon is *back-scattered*. A 180° back-scattered annihilation photon will have an energy of 170 keV.

Compton scattering is not equally probable at all energies or scattering angles. The probability of scattering is given by the Klein–Nishina equation [1]:

$$\frac{d\sigma}{d\Omega} = Zr_0^2 \left(\frac{1}{1 + \alpha(1 - \cos\theta_C)}\right)^2 \left(\frac{1 + \cos^2\theta}{2}\right) \left(1 + \frac{\alpha^2 (1 - \cos\theta_C)^2}{(1 + \cos^2\theta_C)(1 + \alpha\{1 - \cos\theta_C\})}\right) \quad (22)$$

where $d\sigma/d\Omega$ is the differential scattering cross-section, Z is the atomic number of the scattering material, r_0 is the classical electron radius, and $\alpha = E_\gamma/m_0c^2$. For positron annihilation radiation (where $\alpha = 1$) in tissue,

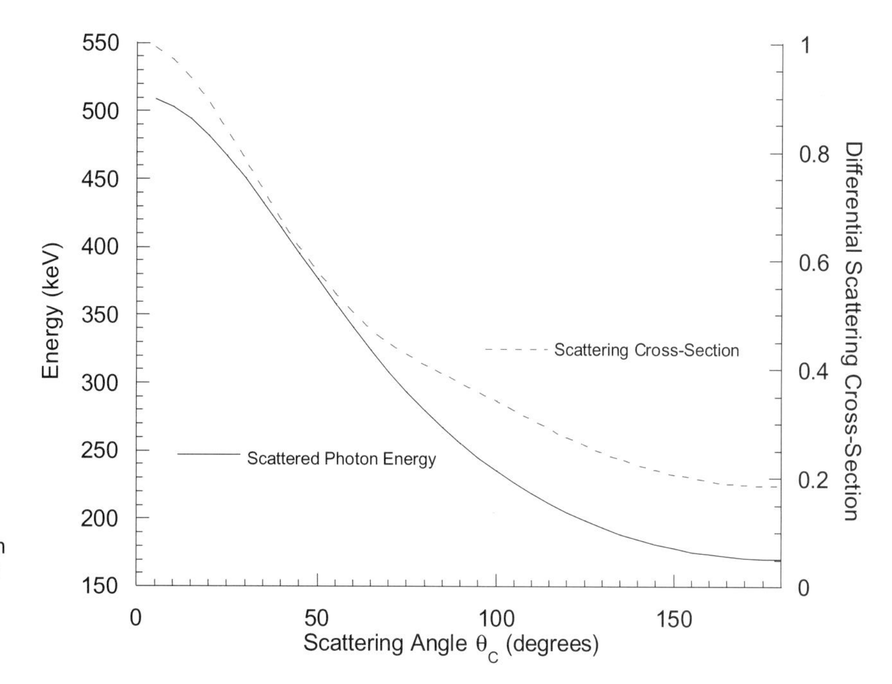

Figure 2.11. The angular probability distribution (differential scattering cross-section, broken line) and resultant energy (solid line) for Compton-scattered annihilation photons are shown.

this equation can be reduced for first-order scattered events to give the relative probability of scatter as:

$$\frac{d\sigma}{d\Omega}=\left(\frac{1}{2-\cos\theta_C}\right)^2\left(1+\frac{(1-\cos\theta_C)^2}{(2-\cos\theta_C)(1+\cos^2\theta_C)}\right) \quad (23)$$

Figure 2.11 shows the form that this function takes in the range 0–180°. A number of Monte Carlo computer simulation studies of the interaction of annihilation radiation with tissue-equivalent material in PET have shown that the vast majority (>80%) of scattered events that are detected have only undergone a single scattering interaction.

Pair production: The final main mechanism for photons to interact with matter is by pair production. When photons with energy greater than 1.022 MeV (twice the energy equivalent to the rest mass of an electron) pass in the vicinity of a nucleus it is possible that they will spontaneously convert to two electrons with opposed signs to conserve charge. This direct electron pair production in the Coulomb field of a nucleus is the dominant interaction mechanism at high energies (Fig. 2.12). Above the threshold of 1.022 MeV, the probability of pair production increases as energy increases. At 10 MeV, this probability is about 60%. Any energy left over after the production of the electron–positron pair is shared between the particles as kinetic energy, with the positron having slightly higher kinetic energy than the electron as the interaction of the particles with the nucleus causes an acceleration of the positron and a deceleration of the electron.

Pair production was first observed by Anderson using cloud chambers in the upper atmosphere, where high-energy cosmic radiation produced tracks of diverging ionisation left by the electron–positron pair.

The process of pair production demonstrates a number of conservation laws. *Energy* is conserved in the process as any residual energy from the photon left over after the electron pair is produced (given by $E_\gamma - 2m_0c^2$) is carried away by the particles as kinetic energy; *charge* is conserved as the incoming photon

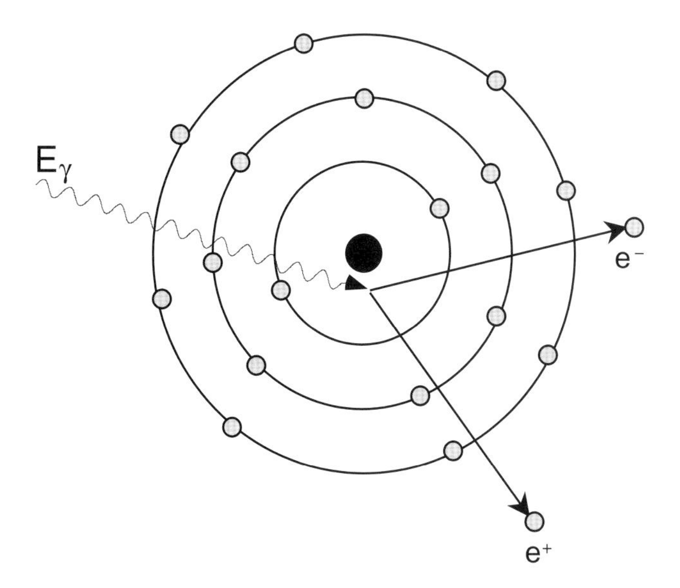

Figure 2.12. The pair production process is illustrated. As a photon passes in the vicinity of a nucleus spontaneous formation of positive and negatively charged electrons can occur. The threshold energy required for this is equal to the sum of the rest masses for the two particles (1.022 MeV).

has zero charge and the outgoing positive and negative electrons have equal and opposite charge; and *momentum* is conserved as the relatively massive nucleus absorbs momentum without appreciably changing its energy balance.

Electron–positron pair production offered the first experimental evidence of Dirac's postulated "antimatter", i.e., that for every particle in the universe there exists a "mirror image" version of it. Other particles can produce matter/antimatter pairs, such as protons, but, as the mass of the electron is much less than a proton, a photon of lower energy is required for electron–positron pair production, thus making the process more probable. The particles produced will behave like any other free electron and positron, causing ionisation of other atoms, and the positron will annihilate with an orbital electron, producing annihilation radiation as a result.

At energies above four rest-mass equivalents of the electron, pair production can take place in the vicinity of an electron. In this case it is referred to as "triplet production" as there is a third member of the interaction, the recoiling electron.

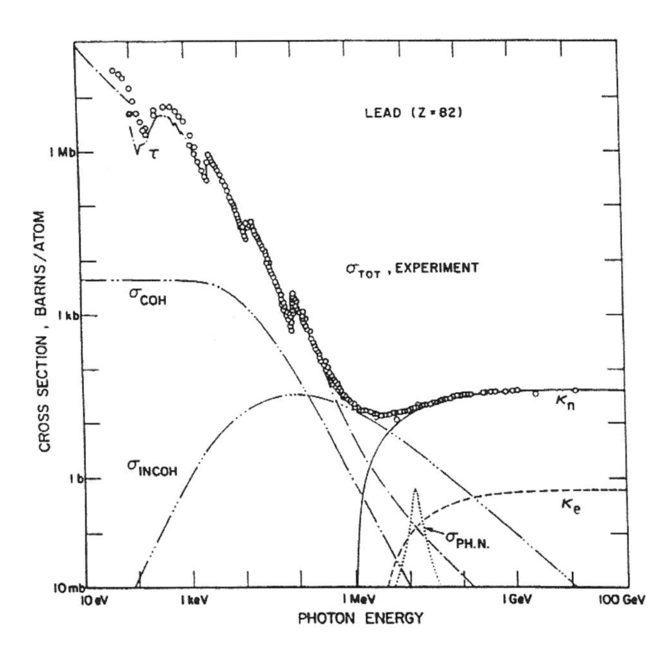

Figure 2.13. Total atomic cross-section as a function of photon energy for lead. The scattering cross-sections (σ) are given for coherent (COH), incoherent (INCOH) or Compton scattering, photonuclear absorption (PH.N.), atomic photoelectric effect (τ), nuclear field pair production (κ_n), electron field pair production (triplet) (κ_e), and the overall total cross section (TOT). (Reproduced with permission of the Institute of Physics Publishing from: Hubbell JH. Review of photon interaction cross section data in the medical and biological context. Phys Med Biol 1999;44(1):R1–22).

Attenuation and Scattering of Photons

In the previous section we have seen how radiation interacts with matter at an atomic level. In this section we will examine the bulk "macroscopic" aspects of the interaction of radiation with matter, with particular reference to positron emission and detection.

Calculations of photon interactions are given in terms of atomic cross sections (σ) with units of cm^2/atom. An alternative unit, often employed, is to quote the cross section for interaction in barns/atom (b/atom) where 1 barn = $10^{-24}cm^2$. The total atomic cross section is given by the sum of the cross sections for all of the individual processes [2], i.e.,

$$\sigma_{tot} = \sigma_{pe} + \sigma_{incoh} + \sigma_{coh} + \sigma_{pair} + \sigma_{tripl} + \sigma_{nph} \quad (24)$$

where the cross sections are for the photoelectric effect (*pe*), incoherent Compton scattering (*incoh*), coherent (Rayleigh) scattering (*coh*), pair production (*pair*), triplet production (*tripl*), and nuclear photoabsorption (*nph*). Values for attenuation coefficient are often given as *mass attenuation coefficients* (μ/ρ) with units of $cm^2.g^{-1}$. The reason for this is that this value can be converted into a linear attenuation coefficient (μ_l) for any material simply by multiplying by the density (ρ) of the material:

$$\mu_l(cm^{-1}) = \mu/\rho\ (cm^2.g^{-1})\,\rho(g.cm^{-3}) \quad (25)$$

The mass attenuation coefficient is related to the total cross section by

$$\mu/\rho\ (cm^2.g^{-1}) = \frac{\sigma_{tot}}{u(g)A} \quad (26)$$

where $u(g) = 1.661 \times 10^{-24}$g is the atomic mass unit ($1/N_A$ where N_A is Avogadro's number) defined as $1/12^{th}$ of the mass of an atom of ^{12}C, and A is the relative atomic mass of the target element [2].

An example of the total cross section as a function of energy is shown in Fig. 2.13.

Photon Attenuation

We have seen that the primary mechanism for photon interaction with matter at energies around 0.5 MeV is by a Compton interaction. The result of this form of interaction is that the primary photon changes direction (i.e., is "scattered") and loses energy. In addition, the atom where the interaction occurred is ionized.

For a well-collimated source of photons and detector, attenuation takes the form of a mono-exponential function, i.e.,

$$I_x = I_0 e^{-\mu x} \quad (27)$$

Table 2.4. Narrow-beam (scatter-free) linear attenuation coefficients for some common materials at 140 keV (the energy of ^{99m}Tc photons) and 511 keV (annihilation radiation).

Material	Density (ρ) [g.cm^{-3}]	μ (140 keV) [cm^{-1}]	μ (511 keV) [cm^{-1}]
Adipose tissue*	0.95	0.142	0.090
Water	1.0	0.150	0.095
Lung*	1.05¶	~0.04–0.06§	~0.025–0.04§
Smooth muscle	1.05	0.155	0.101
Perspex (lucite)	1.19	0.173	0.112
Cortical bone*	1.92	0.284	0.178
Pyrex glass	2.23	0.307	0.194
NaI(Tl)	3.67	2.23	0.34
Bismuth germanate (BGO)	7.13	~5.5	0.95
Lead	11.35	40.8	1.75

(Tabulated from Hubbell [3] and *ICRU Report 44 [4]).
¶This is the density of non-inflated lung.
§Measured experimentally.

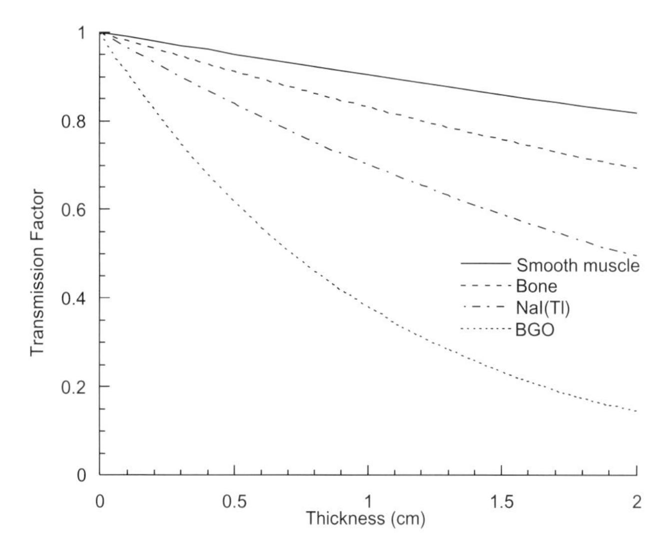

Figure 2.14. Narrow-beam transmission factors for 511 keV photons in smooth muscle, bone, NaI(Tl) and BGO as a function of the thickness of the material.

where I represents the photon beam intensity, the subscripts "0" and "x" refer respectively to the unattenuated beam intensity and the intensity measured through a thickness of material of thickness x, and m refers to the attenuation coefficient of the material (*units: cm^{-1}*). Attenuation is a function of the photon energy and the electron density (Z number) of the attenuator. The attenuation coefficient is a measure of the probability that a photon will be attenuated by a unit length of the medium. The situation of a well-collimated source and detector are referred to as *narrow-beam* conditions. The narrow-beam linear attenuation coefficients for some common materials at 140 keV and 511 keV are shown in Table 2.4 and Fig. 2.14.

However, when dealing with in vivo imaging we do not have a well-collimated source, but rather a source emitting photons in all directions. Under these uncollimated, *broad-beam* conditions, photons whose original emission direction would have taken them out of the acceptance angle of the detector may be scattered such that they are counted. The geometry of narrow and broad beam detection are illustrated in Fig. 2.15.

In the broad-beam case, an uncollimated source emitting photons in all directions contributes both unscattered and scattered events to the measurement by the detector. In this case the detector "sees" more photons than would be expected if unscattered events were excluded, and thus the transmission rate is higher than anticipated (or, conversely, attenuation appears lower). In the narrow-beam case, scattered photons are precluded from the measurement and thus the transmission measured reflects the bulk attenuating properties of the object alone.

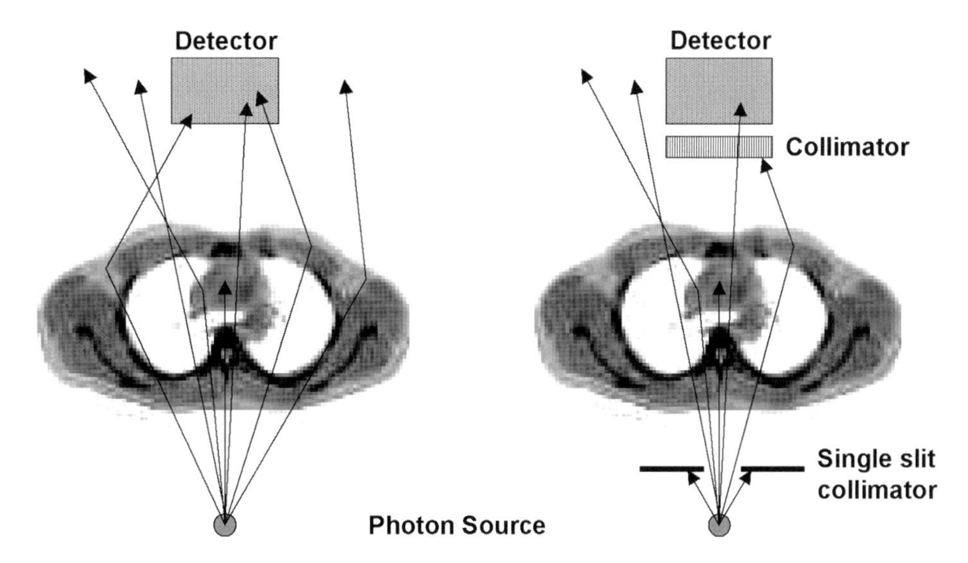

Figure 2.15. Broad-beam geometry (left) combines an uncollimated source of photons and an uncollimated detector, allowing scattered photons to be detected. The narrow-beam case (right) first constrains the photon flux to the direction towards the detector, and second, excludes scattered photons by collimation of the detector.

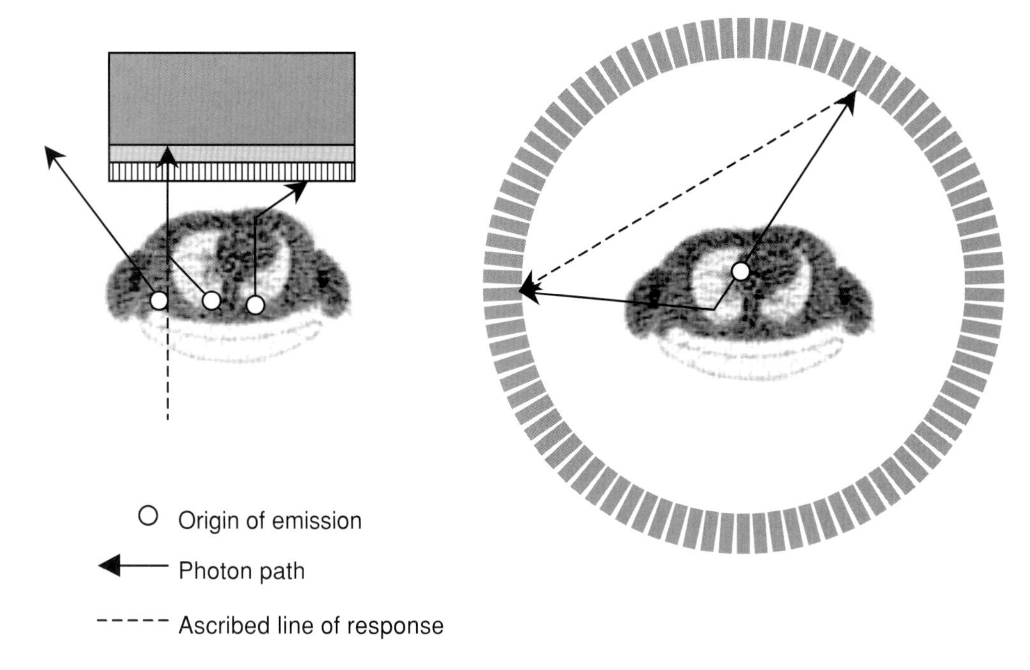

Figure 2.16. Scattered photons in SPECT and PET are shown. In SPECT, the recorded scatter is constrained within the object boundaries as there is low probability for scattering in air. In PET, as two photons are utilized, the line of response connecting the detectors may not intersect the object at all. This fact can be used to infer the underlying scatter distribution within the object by interpolation of the projections (see Ch. 6).

The geometry of scattered events is very different for PET and single photon emission computed tomography (SPECT). As PET uses coincidence detection, the line-of-sight ascribed to an event is determined by the paths taken by both annihilation photons. In this case, events can be assigned to lines of response outside of the object. This is not true in the single-photon case where, assuming negligible scattering in air, the events scattered within the object will be contained within the object boundaries. The difference in illustrated in Fig. 2.16.

Positron emission possesses an important distinction from single-photon measurements in terms of attenuation. Consider the count rate from a single photon emitting point source of radioactivity at a depth, a, in an attenuating medium of total thickness, D (see Fig. 2.17). The count rate C observed by an external detector A would be:

$$C_a = C_0 e^{-\mu a} \qquad (28)$$

where C_0 represents the unattenuatted count rate from the source, and μ is the attenuation coefficient of the medium (assumed to be a constant here). Clearly the count rate changes with the depth a. If measurements were made of the source from the 180° opposed direction the count rate observed by detector B would be:

$$C_b = C_0 e^{-\mu(D-a)} \qquad (29)$$

where the depth b is given by $(D - a)$. The count rate observed by the detectors will be equivalent when $a = b$.

Now consider the same case for a positron-emitting source, where detectors A and B are measuring coincident photons. The count rate is given by the *product* of the probability of counting both photons and will be:

$$\begin{aligned} C &= (C_0 e^{-\mu a}) \times (C_0 e^{-\mu(D-a)}) \\ &= C_0\,(e^{-\mu a}\,.e^{-\mu(D-a)}) \\ &= C_0\, e^{-\mu(a+(D-a))} \\ &= C_0\, e^{-\mu D} \end{aligned} \qquad (30)$$

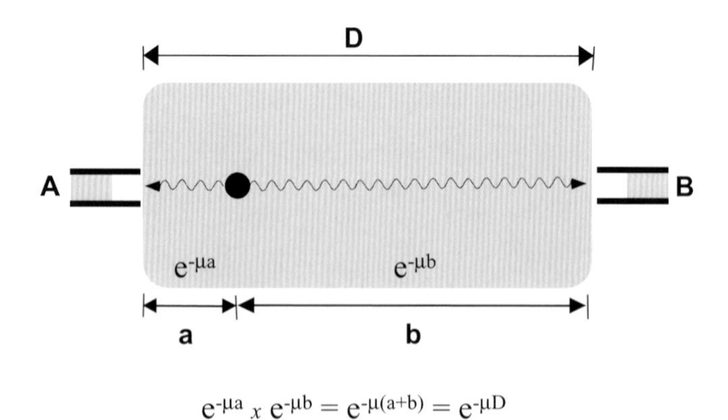

Figure 2.17. Detectors A and B record attenuated count rates arising from the source (●) located a distance a from detector A and b from detector B. For each positron annihilation, the probability of detecting both photons is the **product** of the individual photon detection probabilities. Therefore, the combined count rate observed is independent of the position of the source emitter along the line of response. The total attenuation id determined by the total thickness (D) alone.

which shows that the count rate observed in an object only depends on the total thickness of the object, D; i.e., the count rate observed is *independent* of the position of the source in the object. Therefore, to correct for attenuation of coincidence detection from annihilation radiation one measurement, the total attenuation path length ($-\mu D$), is all that is required. In single-photon measurements the depth of the source in the object, in principle, must be known as well.

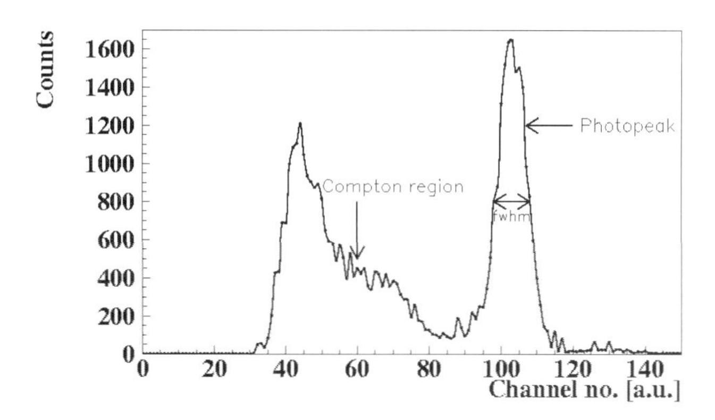

Figure 2.18. Photon energy spectrum measured by a scintillation detector.

Radiation Detection

The interactions of ionising radiation with matter form the basis upon which radiation detectors are developed. The inherent idea in these detectors is to measure the total energy lost or deposited by radiation upon passage through the detector. Typically, radiation detectors convert the deposited energy into a measurable electrical signal or charge. The integral of this signal is then proportional to the total energy deposited in the detector by the radiation. For mono-energetic incident radiation, there will be fluctuations as well as large variations in the total charge collected by the detector (see energy spectrum in Fig. 2.18). The large variations represent incomplete deposition of energy by the incident radiation. For example, in PET some of the incident 511 keV photons may undergo one or more Compton scatter, deposit a portion of their energy and then exit the detector. Multiple Compton scatter could eventually lead to deposition of almost the entire energy by the photon, thereby pushing the event into the photopeak of the energy spectrum. The continuous portion of the energy spectrum (Fig. 2.18) shows the Compton region for this measured energy spectrum with partial deposition of energy. The small fluctuations in the energy spectrum, however, arise due to several processes. The most dominant are the statistical fluctuations in the conversion process of the deposited energy into measurable charge or signal. In Fig. 2.18, the peak position marks the mean energy of the incident radiation (after complete deposition in the detector). The width of this peak (called the photopeak) shows the effect of fluctuations in the measured charge for complete deposition of energy by the mono-energetic photons. The ability of the radiation detector to accurately measure the deposited energy is of paramount importance for most of its uses. This accuracy is characterized by the width of the photopeak in the energy spectrum, and is referred to as the *energy resolution* of the detector. The energy resolution is a dimensionless number and is defined as the ratio of the full width at half maximum (FWHM) of the photopeak to its centroid position.

Radiation Detectors

Radiation detectors can generally be divided into three broad categories: proportional (gas) chambers, semiconductor detectors, and scintillation detectors.

The proportional chamber works on the principle of detecting the ionisation produced by radiation as it passes through a gas chamber. A high electric field is applied within this chamber that results in an acceleration of the ionisation electrons produced by the radiation. Subsequently, these highly energetic electrons collide with the neutral gas atoms resulting in secondary ionisations. Hence, a cascade of electrons is eventually collected at the cathode after some energy deposition by the incident radiation. Typically, inert gases such as xenon are used for detecting photons. The cathode normally consists of a single thin wire, but a fine grid of wires can be utilized to measure energy deposition as a function of position within the detector. Such position-sensitive Multi-wire Proportional Chambers (MWPC) have been used in high-energy physics for a long time, and PET scanners have been developed based upon such a detector [5, 6]. However, the disadvantage of these detectors for use in PET is the low density of the gas, leading to a reduced stopping efficiency for 511 keV photons, as well as poor energy resolution.

Another class of radiation detectors is the semiconductor or solid-state detectors. In these detectors, incident radiation causes excitation of tightly bound (valence band) electrons such that they are free to migrate within the crystal (conduction band). An applied electric field will then result in a flow of charge through the detector after the initial energy deposition by the photons. Semiconductor detectors have excellent energy resolution but because of their production process, the stopping efficiency for 511 keV photons is low.

The third category of radiation detectors, which are of most interest to us, are the scintillation detectors. These detectors consist of an inorganic crystal (scintillator) which emits visible (scintillation) light photons after the interaction of photons within the detector. A photo-detector is used to detect and measure the number of scintillation photons emitted by an interaction. The number of scintillation photons (or intensity of light) is generally proportional to the energy deposited within the crystal. Due to their high atomic numbers and therefore density, scintillation detectors provide the highest stopping efficiency for 511 keV photons. The energy resolution, though much better than the proportional chambers, is not as good as that attained with the semiconductor detectors. This is due to the inefficient process of converting deposited energy into scintillation photons, as well as the subsequent detection by the photo-detectors. However, for PET, where both high stopping efficiency as well as good energy resolution are desired, scintillation detectors are most commonly used. For a more thorough treatment of radiation detection and measurement the reader is referred to Knoll (1988) [7].

Scintillation Detectors in PET

As mentioned above, scintillation detectors are the most common and successful mode for detection of 511 keV photons in PET imaging due to their good stopping efficiency and energy resolution. These detectors consist of an appropriate choice of crystal (scintillator) coupled to a photo-detector for detection of the visible light. This process is outlined in further detail in the next two sections.

Scintillation Process and Crystals Used in PET

The electronic energy states of an isolated atom consist of discrete levels as given by the Schrödinger equation. In a crystal lattice, the outer levels are perturbed by mutual interactions between the atoms or ions, and so the levels become broadened into a series of *allowed bands*. The bands within this series are separated from each other by the *forbidden bands*. Electrons are not allowed to fill any of these forbidden bands. The last filled band is labelled the *valence band*, while the first unfilled band is called the *conduction band*. The energy gap, E_g, between these two bands is a few electron volts in magnitude (Fig. 2.19).

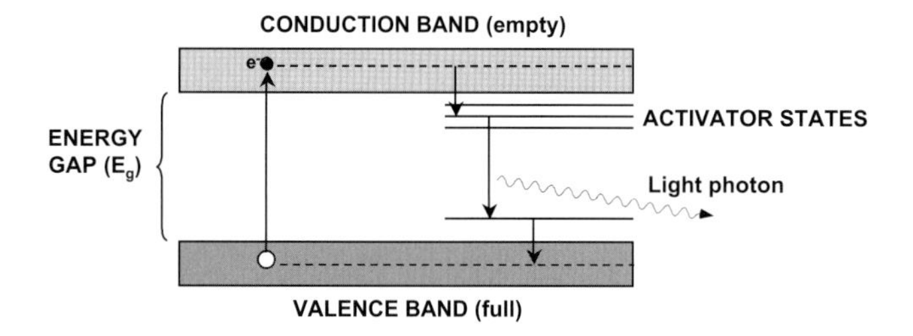

Figure 2.19. Schematic diagram of the energy levels in a scintillation crystal and the mechanism of light production after energy is absorbed. The photon energy is sufficient to move a valence band electron to the conduction band. In returning to the ground state, light photons are emitted.

Electrons in the valence band can absorb energy by the interaction of the photoelectron or the Compton scatter electron with an atom, and get excited into the conduction band. Since this is not the ground state, the electron de-excites by releasing scintillation photons and returns to its ground state. Normally, the value of E_g is such that the scintillation is in the ultraviolet range. By adding impurities to a pure crystal, such as adding thallium to pure NaI (at a concentration of ~1%), the band structure can be modified to produce energy levels in the prior forbidden region. Adding an impurity or an activator raises the ground state of the electrons present at the impurity sites to slightly above the valence band, and also produces excited states that are slightly lower than the conduction band. Keeping the amount of activator low also minimizes the self-absorption of the scintillation photons. The scintillation process now results in the emission of visible light that can be detected by an appropriate photo-detector at room temperature. Such a scintillation process is often referred to as *luminescence*. The scintillation photons produced by *luminescence* are emitted isotropically from the point of interaction. For thallium-activated sodium iodide (NaI(Tl)), the wavelength of the maximum scintillation emission is 415 nm, and the photon emission rate has an exponential distribution with a decay time of 230 ns. Sometimes the excited electron may undergo a radiation-less transition to the ground state. No scintillation photons are emitted here and the process is called *quenching*.

There are four main properties of a scintillator which are crucial for its application in a PET detector. They are: the stopping power for 511 keV photons, signal decay time, light output, and the intrinsic energy resolution. The stopping power of a scintillator is characterized by the mean distance (attenuation length = $1/\mu$) travelled by the photon before it deposits its energy within the crystal. For a PET scanner with high sensitivity, it is desirable to maximize the number of photons which interact and deposit energy in the de-

Table 2.5. Physical properties of commonly used scintillators in PET. The energy resolution and attenuation coefficients (linear (μ) and mass (μ/ρ)) are measured at 511 keV

Property	NaI(Tl)	BGO	LSO	YSO	GSO	BaF_2
Density (g/cm^3)	3.67	7.13	7.4	4.53	6.71	4.89
Effective Z	50.6	74.2	65.5	34.2	58.6	52.2
Attenuation length	2.88	1.05	1.16	2.58	1.43	2.2
Decay constant (ns)	230	300	40	70	60	0.6
Light output (photons/keV)	38	6	29	46	10	2
Relative light output	100%	15%	75%	118%	25%	5%
Wavelength λ(nm)	410	480	420	420	440	220
Intrinsic ΔE/E (%)	5.8	3.1	9.1	7.5	4.6	4.3
ΔE/E (%)	6.6	10.2	10	12.5	8.5	11.4
Index of refraction	1.85	2.15	1.82	1.8	1.91	1.56
Hygroscopic?	Yes	No	No	No	No	No
Rugged?	No	Yes	Yes	Yes	No	Yes
μ (cm^{-1})	0.3411	0.9496	0.8658	0.3875	0.6978	0.4545
μ/ρ(cm^2/gm)	0.0948	0.1332	0.117	0853	0.104	0.0929

tector. Thus, a scintillator with a short attenuation length will provide maximum efficiency in stopping the 511 keV photons. The attenuation length of a scintillator depends upon its density (ρ) and the effective atomic number (Z_{eff}). The decay constant affects the timing characteristics of the scanner. A short decay time is desirable to process each pulse individually at high counting rates, as well as to reduce the number of random coincidence events occurring within the scanner geometry (see Ch. 6). A high light-output scintillator affects a PET detector design in two ways: it helps achieve good spatial resolution with a high encoding ratio (ratio of number of resolution elements, or crystals, to number of photo-detectors) and attain good energy resolution. Good energy resolution is needed to efficiently reject events which may Compton scatter in the patient before entering the detector. The energy resolution (ΔE/E) achieved by a PET detector is dependent not only upon the scintillator light output but also the intrinsic energy resolution of the scintillator. The intrinsic energy resolution of a scintillator arises due to inhomegeneities in the crystal growth process as well as non-uniform light output for interactions within it. Table 2.5 shows the properties of scintillators that have application in PET. They are:

(i) sodium iodide doped with thallium (NaI(Tl)),
(ii) bismuth germanate $Bi_4Ge_3O_{12}$ (BGO),
(iii) lutetium oxyorthosilicate doped with cerium Lu_2SiO_5:Ce (LSO),
(iv) yttrium oxyorthosilicate doped with cerium Y_2SiO_5:Ce (YSO),
(v) gadolinium oxyorthosilicate doped with cerium Gd_2SiO_5:Ce (GSO), and
(vi) barium fluoride (BaF_2).

The energy resolution values given in this table are for single crystals. In a full PET system, variations between crystals and other factors such as light readout due to block geometry contribute to a significant worsening of the energy resolution. Typically, NaI(Tl) detectors in a PET scanner achieve a 10% energy resolution for 511 keV photons, while the BGO scanners have system energy resolution of more than 20%.

NaI(Tl) provides very high light output leading to good energy and spatial resolution with a high encoding ratio. The slow decay time leads to increased detector dead time and high random coincidences (see Energy Resolution and Scatter, below). It suffers from lower stopping power than BGO, GSO or LSO due to its lower density. BGO, on the other hand, has slightly worse timing properties than NaI(Tl) in addition to lower light output. However, the excellent stopping power of BGO gives it high sensitivity for photon detection in PET scanners. Currently, commercially produced whole-body scanners have developed along the lines of advantages and disadvantages of these two individual scintillators. The majority of scanners employ BGO and, when operating in 2D mode, use tungsten septa to limit the amount of scatter by physically restricting the axial field-of-view imaged by a detector area. This results in a reduction of the scanner sensitivity due to absorption of some photons in the septa. The low light output of BGO also requires the use of small photo-multiplier tubes to achieve good spatial resolution, thereby increasing system complexity and cost. The NaI(Tl)-based scanners [8] compromise on high count-rate performance by imaging in 3D mode in order to achieve acceptable scanner sensitivity.

LSO, a relatively new crystal, appears to have an ideal combination of the advantages of the high light output of NaI(Tl) and the high stopping power of BGO in one crystal [9]. In spite of its high light output (~75% of NaI(Tl)), the overall energy resolution of LSO is not as good as NaI(Tl). This is due to intrinsic properties of the crystal. Another disadvantage for general applications of this scintillator is that one of the naturally occurring isotopes present (^{176}Lu, 2.6% abundance), is itself radioactive. It has a half-life of 3.8×10^{10} years and decays by β^- emission and the subsequent release of γ photons with energies from 88–400 keV. The intrinsic radioactivity concentration of LSO is approximately ~280 Bq/cc; approximately 12 counts per sec per gram would be emitted that would be detected within a 126–154 keV energy window. Thus its use in low-energy applications is restricted. This background has less impact in PET measurements due to the higher energy windows set for the annihilation radiation and the use of coincidence counting.

GSO is another scintillator with useful physical properties for PET detectors. One advantage of GSO over LSO, in spite of a lower stopping power and light output, is its better energy resolution and more uniform light output. Commercial systems are now being developed with GSO detectors.

Finally, the extremely short decay time of BaF_2 (600 psec) makes it ideal for use in time-of-flight scanners (see Time-of-flight Measurement, below), which helps to partially compensate for the low sensitivity arising due to the reduced stopping power of this scintillator.

In addition to these scintillators, which have all been used in PET tomographs already, new inorganic scintillators continue to be developed. Many of the newer scintillators are based on cerium doping of lanthanide and transition metal elements. Examples include LuAP:Ce, Y_2SiO_5 (YSO), $LuBO_3$:Ce, and others based on lead (Pb), tungsten (W) and gadolinium (Gd).

Photo-detectors and Detector Designs Used in PET

Generally, the photo-detectors used in scintillation detectors for PET can be divided into two categories, the photo-multiplier tubes (PMTs) and the semiconductor-based photodiodes. Photo-multiplier tubes (Fig. 2.20) represent the oldest and most reliable technique to measure and detect low levels of scintillation light. They consist of a vacuum enclosure with a thin photocathode layer at the entrance window. An incoming scintillation photon deposits its energy at the photocathode and triggers the release of a photo-electron. Depending upon its energy, the photo-electron can escape the surface potential of the photo-cathode and in the presence of an applied electric field accelerate to a nearby dynode which is at a positive potential with respect to the photo-cathode. Upon impact with the dynode, the electron, with its increased energy, will result in the emission of multiple secondary electrons. The process of acceleration and emission is then repeated through several dynode structures lying at in-

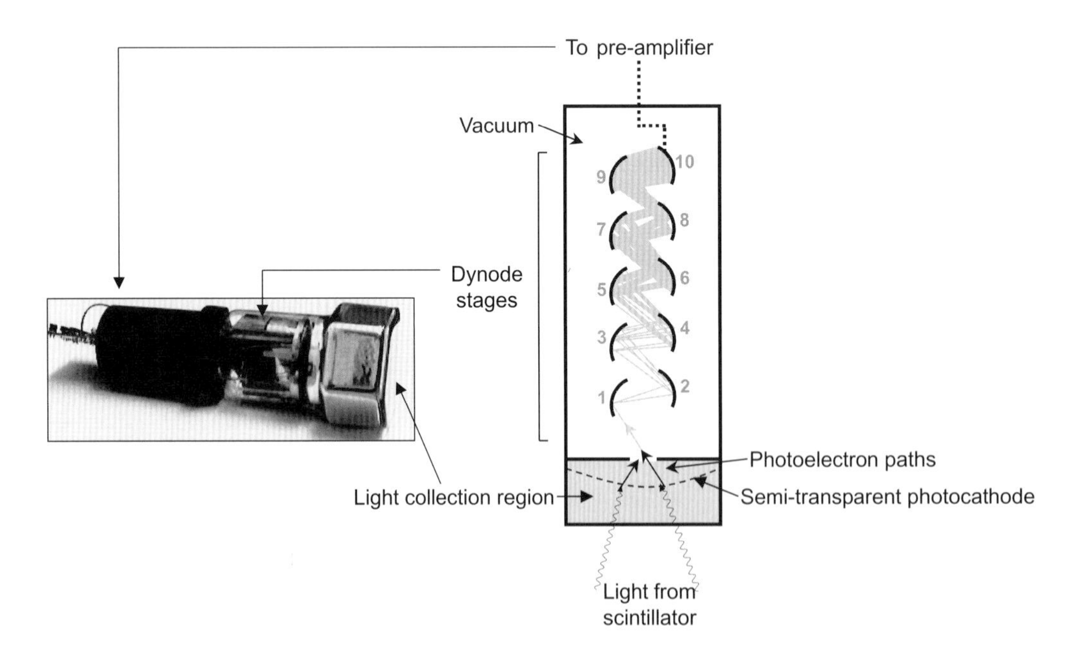

Figure 2.20. Schematic diagram of a photomultiplier tube and a photograph of a hexagonal 6 cm-diameter tube (inset). Light entering the PMT displaces a photoelectron which is electrostatically focused to the first-stage dynode. Each dynode has a positive voltage bias relative to the previous one, and so electrons are accelerated from one dynode to the next. The increase in kinetic energy acquired by this process is sufficient to displace a number of electrons at the next dynode, and so on, causing large amplification by the end-stage dynode (usually tenth or twelfth).

creasing potentials, leading to a gain of more than a million at the final dynode (anode). This high gain obtained from a photo-multiplier tube leads to a very good signal-to-noise ratio (SNR) for low light levels and is the primary reason for the success and applicability of photo-multiplier tubes for use in scintillation detectors. The only drawback of a photo-multiplier tube is the low efficiency in the emission and escape of a photo-electron from the cathode after the deposition of energy by a single scintillation photon. This property is called the *Quantum Efficiency* (QE) of the photo-multiplier tube and it is typically 25% for most of the photo-multiplier tubes. Different, complex arrangements of the dynode structure have been developed over the years in order to maximize the gain, reduce the travel time of the electrons from the cathode to the anode, as well as reduce the variation in the travel times of individual electrons. In particular, a fine grid dynode structure has been developed which restricts the spread of photoelectrons while in trajectory, thereby providing a position-sensitive energy measurement within a single photo-multiplier tube enclosure (Position Sensitive PMT or PS-PMT). More recently, a multi-channel capability has been developed which essentially reduces a single photo-multiplier tube enclosure into several very small channels. It uses a 2D array of glass capillary dynodes each of which is a few microns wide. Additionally, a multi-anode structure is used for electron collection, thereby providing a dramatically improved position-sensitive energy measurement with very little cross-talk between adjacent channels (Multi-Channel PMT, MC-PMT).

Photodiodes, on the other hand, are based upon semiconductors which, unlike the situation for detecting the photons, have high sensitivity for detecting the significantly lower energy scintillation photons. These detectors typically are in the form of PIN diodes (PIN refers to the three zones of the diode: P-type, Intrinsic, N-type). Manufacturing a PIN photodiode involves drifting an alkali metal such as lithium onto a p-type semiconductor such as doped silicon. Incident scintillation photons produce electron–hole pairs in the detector and an applied electric field then results in a flow of charge that can be measured through an external circuit. A significant disadvantage of the photodiodes is the low SNR achieved due to the presence of thermally activated charge flow and very low intrinsic signal amplification. In recent years, a new type of photodiode, called the Avalanche Photo Diode (APD), has been developed which provides an internal amplification of the signal, thereby improving the SNR. These gains are typically in the range of a few hundred and are still several orders of magnitude lower than the photo-multiplier tubes. More importantly, APD gains are sensitive to small temperature variations as well as changes in the applied bias voltage.

In general, there are three ways of arranging the scintillation crystals and coupling them to photo-detectors for signal readout in a PET detector. The first is the so-called one-to-one coupling, where a single crystal is glued to an individual photo-detector. A close-packed array of small discrete detectors can then be used as a large detector that is needed for PET imaging. The spatial resolution of such a detector is limited by the size of the discrete crystals making up the detector. In order to achieve spatial resolution better than 4 mm in one-to-one coupling, very small photo-detectors are needed. However, individual photo-multiplier tubes of this size are not currently manufactured. One solution is the use of photodiodes, or APDs instead of photo-multiplier tubes. The APDs are normally developed either as individual components or in an array, and so are ideal for use in such a detector design [10, 11]. However, as mentioned earlier, the APD gain is sensitive to variations in temperature and bias voltage that can lead to practical problems of stability in their implementation for a complete PET scanner. Another option is the coupling of individual channels of a PS-PMT or a MC-PMT to the small crystals [12]. Due to the large package size of these photo-multiplier tubes, however, clever techniques are needed to achieve a close-packed arrangement of the crystals in the scanner design. Despite the very good spatial resolution and minimal dead time achieved by the one-to-one coupling design, the inherent complexity (number of electronic channels) and cost of such PET detectors limits their use at present to research tomographs; in particular, small animal systems.

The next two detector schemes are attempts at reducing these disadvantages by increasing the encoding for the detector. Both the designs involve the use of larger photo-multiplier tubes without intrinsic position-sensing capabilities. The Anger detector, originally developed by Hal Anger in the 1950s, uses a large (*e.g.*, 1 cm thick × 30–50 cm in diameter) NaI(Tl) crystal glued to an array of photo-multiplier tubes via a light guide. This camera is normally used with a collimator to detect low-energy single photons in SPECT imaging. An application of the Anger technique to a PET detector, on the other hand, uses 2.5 cm-thick NaI(Tl) scintillators. An array of 6.5 cm-diameter photo-multiplier tubes can be used to achieve a spatial resolution of about 5 mm [8]. A weighted centroid positioning algorithm is used for estimation of the interaction position within the detector. This algorithm uses a weighted sum of the individual photo-multiplier tube signals

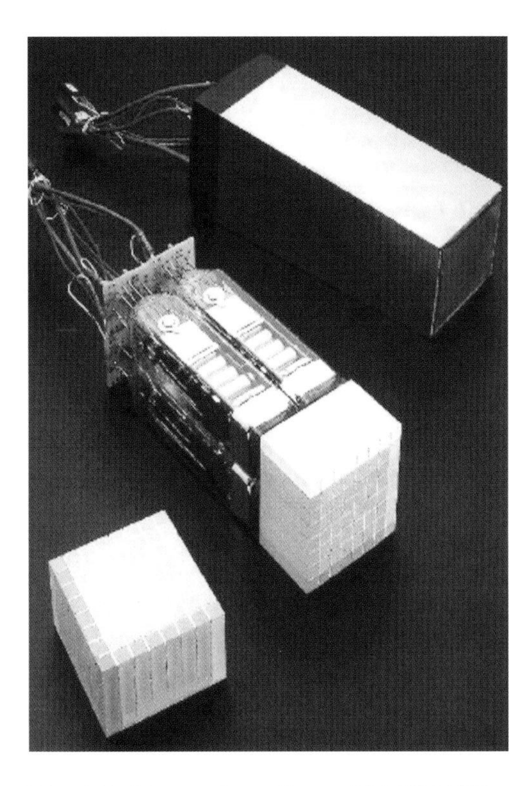

Figure 2.21. A block detector from a Siemens-CTI ECAT 951 PET scanner is shown. The sectioned (8 × 8 elements) block of BGO is in the bottom left corner, with the four square PMTs attached in the center, and the final packaged module in the top right corner. The scanner would contain 128 such modules in total, or 8192 individual detector elements. (Figure courtesy of Dr Ron Nutt, CTI PET Systems, Knoxville, TN, USA).

and normalizes it with the total signal obtained from all the photo-multiplier tubes. The weights for the photo-multiplier tube signals depend exclusively upon the photo-multiplier tube position within the array. Since these detectors involve significant light sharing between photo-multiplier tubes, a high light-output scintillator such as NaI(Tl) is needed to obtain good spatial resolution. The use of large photo-multiplier tubes produces a very high encoding ratio, leading to a simple and cost-effective design. However, a disadvantage of this detector, independent of the use of NaI(Tl) as a scintillator, is the spread of scintillation light within the crystal which leads to significant detector dead time at high count rates.

The block detector design uses Anger positioning in a restricted manner to achieve good spatial resolution and reduced dead time at the expense of a lower encoding ratio. The initial design used an 8 × 4 array of $6 \times 14 \times 30\ mm^3$ BGO crystals glued to a slotted light guide [13]. The slots in the light guide are cut to varying depths with the deepest slots cut at the detector's edge (see Fig. 2.21, left and centre).

The read-out in this block design is performed by four 25 mm-square photo-multiplier tubes. The slotted light guide allows the scintillation light to be shared to varying degrees between the four photo-multiplier tubes depending upon the position of the crystal in which the interaction takes place. The centroid calculation is performed here as well to identify the crystal of interaction. An improved design of this detector allows the identification of smaller, $4 \times 4 \times 30\ mm^3$, leading to an improved spatial resolution but with smaller 19 mm photo-multiplier tubes. Besides the advantages and disadvantages of BGO as a scintillator, the block detector design has the benefit of reduced detector dead time compared to the large-area Anger detector due to the restricted light spread. This, however, is achieved by increasing the number of detector channels (lower encoding ratio), thus leading to increased cost. A modification of the block design, called the quadrant-sharing block design [14], can distinguish smaller (half the size in either direction) crystals by straddling the 19 mm photo-multiplier tube over four block quadrants (see Fig. 2.22, right). This design, in comparison to the standard block, results in a better spatial resolution with almost double the encoding ratio, but increased detector dead time due to the use of nine photo-multiplier tubes (not four) for signal readout from an event.

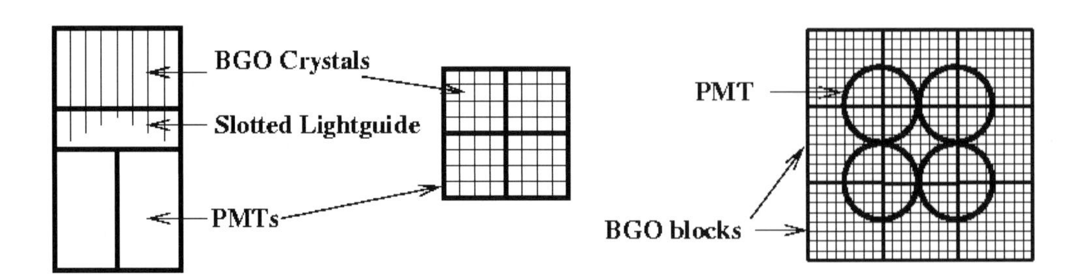

Figure 2.22. The standard block detector design from the side (left) and looking down through the crystals (middle). The quad-sharing block design as seen from the top through the crystals is shown on the right. Figures are not drawn to scale.

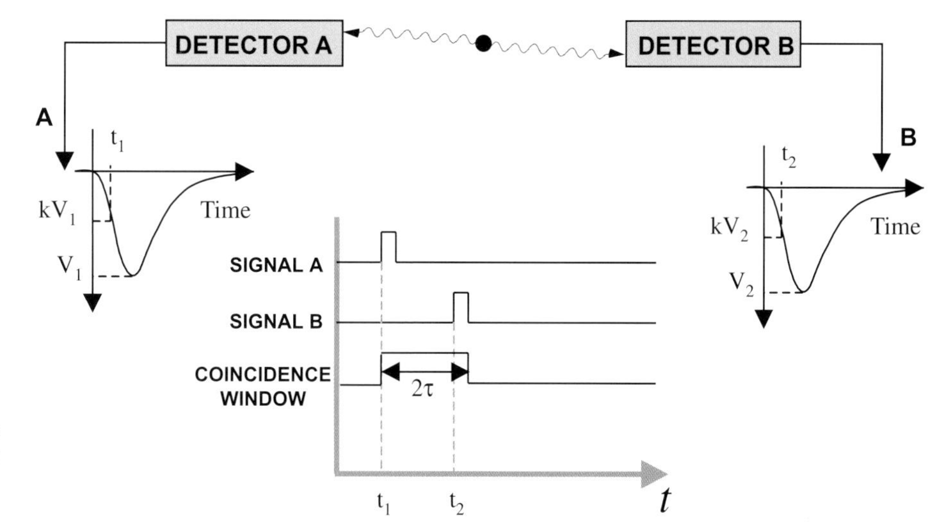

Figure 2.23. Schematic representation of detecting coincidence events in two detectors. Signal A results in a trigger pulse 1 which marks the start of the coincidence window of width Δt. Similarly, signal B results in a trigger pulse 2. A coincidence (AND) circuit then checks for coincidence between the pulse 2 and the coincidence window.

Timing Resolution and Coincidence Detection

The timing resolution of a PET detector describes the uncertainty in the timing characteristics of the scintillation detector on an event-by-event basis due to statistical fluctuations. With a fast signal (or short decay time), the timing resolution is small as well. The timing resolution of a PET detector is important because it involves the detection of two photons originating from a single coincident event. Since the timing resolution represents the variability in the signal arrival times for different events, it needs to be properly accounted for when detecting coincident events. Figure 2.23 gives a schematic representation of two detectors set up to measure coincident photons being emitted from a point equidistant from the two detectors.

The amplitude of the signal from the two detectors (V_1 and V_2 in Fig. 2.23) may be different owing to incomplete deposition of energies or varying gains of the photo-detectors in the two detectors. The coincidence circuitry, however, generates a narrow trigger pulse when the detector signals cross a certain fixed fraction of their individual amplitudes. At time t_1, signal A triggers pulse 1 which also produces a coincidence time window of a predetermined width, 2τ. Signal B, depending upon the timing resolution of the detector, will trigger at a later time, t_2. Depending upon the difference $t_2 - t_1$, the start of pulse 2 may or may not overlap with the coincidence window. For detectors with poor timing resolution, a large value for 2τ needs to be used in order to detect most of the valid coincidence events.

In a PET scanner, the two coincident photons will be emitted from anywhere within the scanner field-of-view (FOV), and so the distance travelled by each of them before interaction in the detectors will be different. For a typical whole-body scanner, this distance can be as large as the scanner diameter (about 100 cm). Using the value of speed of light ($c = 3 \times 10^8$ m/s), one can calculate an additional maximum timing difference of about 3–4 ns between the two signals (the photons travel 1 m in 3.3 ns). As a result, the coincidence timing window (2τ) of a PET detector needs to be increased even more than the requirements of the timing resolution. For an extremely fast scintillator such as BaF_2, the timing resolution is very small. However, the coincidence timing window cannot be reduced to less than 3–4 ns (in a whole-body scanner geometry) due to the difference in arrival times of two photons emitted at the edge of the scanner field of view, as this would restrict the transverse field of view.

Random Coincidences

Random coincidences are a direct consequence of having a large coincidence timing window. They arise when two unrelated photons enter the opposing detectors and are temporally close enough to be recorded within the coincidence timing window. For such events, the system produces a false coincident event. Due to the random nature of such events, they are labelled as random or accidental coincidences. Random coincidences add uncorrelated background counts to an acquired PET image and hence decrease image contrast if no corrections are applied to the acquired data. In Fig. 2.23, if signal A and signal B are unrelated, then a large coincidence timing window will result in an increased number of such events being registered as coincident events (random coincidences). The random

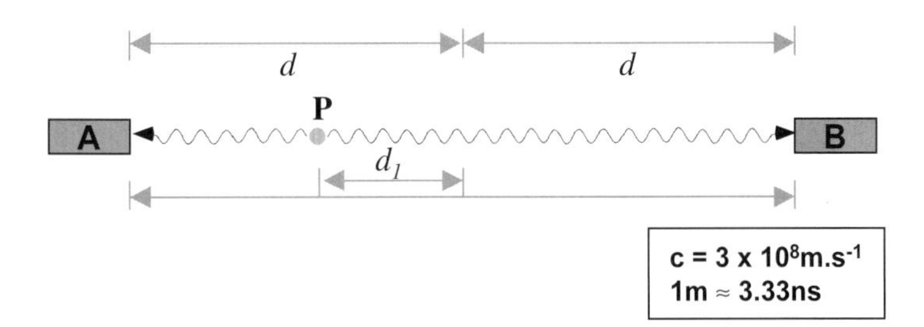

Figure 2.24. Time-of-flight measurement. P marks the annihilation point from where the two photons originate and are recorded in detectors A and B.

coincidence rate in a PET scanner is proportional to $2\tau A^2$, where A is the activity present in the scanner field of view. The true coincident rate, on the other hand, increases linearly with a given activity level in the scanner. Hence, at high activity levels, the random coincidences will overwhelm the true coincidences.

The random coincidence rate can be estimated during data collection and a correction applied to the projected data. These techniques will be outlined in further detail in the following chapters. However, it is important to point out that the random correction techniques result in a propagation of noise through the data set and so the image signal-to-noise ratio suffers. Thus, the best way to improve image contrast without reducing its signal-to-noise ratio is to minimize the collection of random coincidences. Since random coincidences are proportional to the coincidence timing window, a narrow window helps in reducing their occurrence within the detector. Hence, for PET imaging a fast scintillator with good timing resolution is desirable for reducing the number of random coincidences.

Time-of-flight Measurement

Good timing resolution of a PET detector, besides helping reduce the number of random coincidences, can also be use to estimate the annihilation point between the two detectors by looking at the difference in arrival times of the two photons. For this, an extremely fast scintillator, such as BaF_2, is needed.

In Fig. 2.24, point P marks an annihilation point which is located a distance d_1 from the point which is exactly halfway (distance d) between the two detectors. A photon moving along PA will travel a distance $d - d_1$, while the coincident photon travels a total distance $d + d_1$ along PB before entering detector B. Thus, one photon will travel an extra distance $(d + d_1) - (d - d_1) = 2d_1$ relative to the other. The coincident detectors can be used to measure the difference in arrival times (δt) of the two photons. Using the speed of light, c, for the speed of the photons, d_1 can be calculated from $2d_1 = c\delta t$. In order to obtain a good estimation of d_1, however, an accurate measurement of δt is needed, which in turn requires a fast scintillator with a timing resolution of less than 0.8 ns. Thus, the timing resolution of a PET detector introduces a blurring in the estimation of d_1. It can be shown from the above calculation that for BaF_2 with $\delta t = 0.8$ ns, a blurring of about ±6 mm is introduced in the d_1 estimation. Slow scintillators will increase this blurring significantly. Presently, only BaF_2 is feasible for use as a scintillator in time-of-flight measuring PET scanners, and such scanner designs have been successfully implemented. The advantage of estimating the location of the annihilation point is the improved signal-to-noise ratio obtained in the acquired image, arising due to a reduction in noise propagation during the image reconstruction process. However, since BaF_2 also has a very low stopping power, time-of-flight scanners have a reduced sensitivity leading to lower signal-to-noise ratios. Hence, the overall design of such scanners requires a careful trade-off between the scanner sensitivity and the time-of-flight measurement so that the overall SNR for the scanner remains high.

Energy Resolution and Scatter

The energy resolution of a radiation detector characterizes its ability to distinguish between radiation at different energies. In scintillation detectors the energy resolution is a function of the relative light output of the scintillator, as well as its intrinsic energy resolution. The intrinsic energy resolution accounts for other non-statistical effects that arise in the energy measurement process. Good energy resolution is necessary for a PET detector (especially in 3D volume imaging mode) in order to achieve good image contrast and reduce background counts in the image.

A PET scanner acquires three different kinds of coincident events: true, random, and scatter coincidences. True coincidences are emissions from single annihilation points that enter the PET detector without undergoing any significant interactions within the imaging field of view. Random coincidences, as we have already seen, arise due to the accidental detection of two unre-

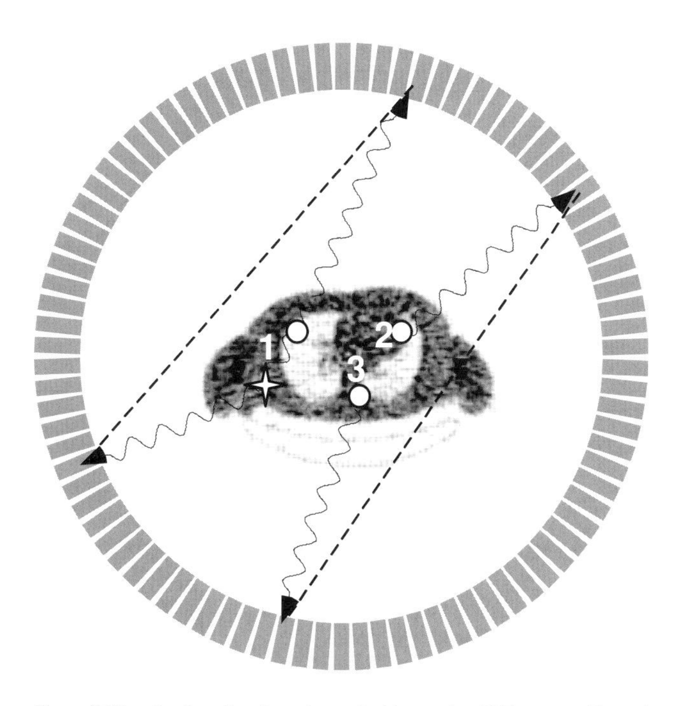

Figure 2.25. Scattered and random coincidences in a PET scanner. Event 1 shows a coincident event where one of the γ rays is scattered leading to an incorrectly assigned line-of-response (LOR, dotted) for image reconstruction (scatter coincidence). Events 2 and 3 represent two unrelated events with only one photon being detected (singles events). If they occur within the coincidence timing window, then an incorrect LOR (dotted) gets assigned (random coincidence).

lated, single events within the coincidence timing window. These coincidences add to the image background and so reduce its contrast. Finally, scatter coincidences are true coincidence events from single annihilation points, but where one or both the photons undergo Compton scatter within the imaging FOV before entering the PET detector (see Fig. 2.25). Since scattered coincidences lead to mis-positioned lines-of-response, and therefore misrepresent the true activity distribution within the FOV, the image contrast worsens.

The density of tissue in human body is approximately the same as that of water, and so the mean free path of a 511 keV photon is about 7 cm in human tissue. Since the cross-section of a human body is much greater than 7 cm, many of the photons originating inside the human body are Compton scattered before they enter the PET detectors. Since scatter involves loss of energy, in principle some of these scattered coincidences can be rejected using an energy-gating technique around the photopeak in the energy spectrum. Good energy resolution for the detector allows the application of a very narrow energy gate, and thus a more extensive and accurate rejection of scatter coincidences can be performed. However, some scattered events may be indistinguishable from true coincidences based upon the energy if they lie within the photopeak. For example, in NaI(Tl)-based detectors the good energy resolution allows the use of about 450 keV as the lower energy gate on the photopeak. Assuming only single scatter within the object, this implies that the maximum deviation from true line-of-response for scattered events within the photopeak will be about 30°. In comparison, for the BGO-based detectors, the lower energy gate is set at 300–400 keV, leading to a maximum deviation of more than 70° from the true line-of-response. Hence, additional scatter-correction techniques which estimate the distribution of scattered radiation are then employed in order to remove them from the image and improve image contrast.

Sensitivity and Depth of Interaction

The sensitivity of a PET scanner represents its ability to detect the coincident photons emitted from inside the scanner FOV. It is determined by two parameters of the scanner design; its geometry and the stopping efficiency of the detectors for 511 keV photons. Scanner geometry defines the fraction of the total solid angle covered by it over the imaging field. Small-diameter and large axial FOV typically leads to high-sensitivity scanners. The stopping efficiency of the PET detector is related to the type of detector being used. As we have seen, scintillation detectors provide the highest stopping power for PET imaging with good energy resolution. The stopping power of the scintillation detector is in turn dependent upon the density and Z_{eff} of the crystal used. Hence, a majority of commercially produced PET scanners today use BGO as the scintillator due to its high stopping power (see Table 2.4). A high-sensitivity scanner collects more coincident events in a fixed amount of time and with a fixed amount of radioactivity present in the scanner FOV. This generally translates into improved SNR for the reconstructed image due to a reduction in the effect of statistical fluctuations.

A high stopping power for the crystal is also desirable for the reduction of parallax error in the acquired images. After a photon enters a detector, it travels a short distance (determined by the mean attenuation length of the crystal) before depositing all its energy. Typically, PET detectors do not measure this point, known as the depth of interaction (DOI) within the crystal. As a result, the measured position of energy deposition is projected to the entrance surface of the detector (Fig. 2.26). For photons that enter the detector at oblique angles, this projected position can produce significant deviations from the real position, leading to a

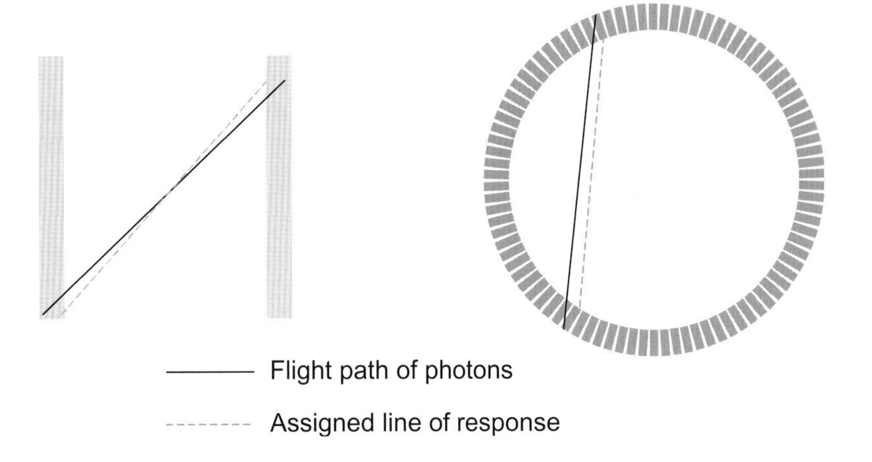

Figure 2.26. Schematic representation of parallax error introduced in the measured position due to the unknown depth-of-interaction of the photons within the detectors for a flat detector (left) and ring-based system (right).

blurring of the reconstructed image. Typically, annihilation points located at large radial distances from the scanner's central axis suffer from this parallax blurring. For a BGO whole-body scanner, measurements show that the spatial resolution worsens from 4.5 mm near the centre of the scanner to about 8.9 mm at a radial distance of 20 cm [15]. A thin crystal with high stopping power will help reduce the distance travelled by the photon in the detector and so reduce parallax effects. However, a thin crystal reduces the scanner sensitivity. Thus, to separate this inter-dependence of sensitivity and parallax error, an accurate measurement of the photon depth-of-interaction within the crystal is required.

Development of PET detectors with depth-of-interaction measurement capabilities is an ongoing research interest. Currently there are two practically feasible techniques that can be used for depth-of-interaction measurement. The first is the *phoswich* detector [16] method that involves stacking thin layers of different scintillators on top of each other, instead of using a thick layer of one crystal type. The depth-of-interaction measurement in a phoswich detector depends on the identification of interaction layer through an examination of the different signal decay times for the scintillators. As a result, the scintillators used in a phoswich detector need to have significantly different decay times in order to successfully distinguish them via pulse shape discrimination techniques. Another potential problem in its implementation is the optical coupling between the individual layers of crystals. Good optical coupling is necessary for the successful transmission of scintillation photons from the crystals into the photo-detectors, thereby achieving good spatial and energy resolution as well.

Another technique for determining the depth of interaction involves the use of photo-detectors at both the ends of a thick (or long) scintillator. This technique is based upon the physical principle according to which the relative number of scintillation photons reaching either of the end photo-detectors is a function of the photons depth of interaction in the crystal. Figure 2.27 shows a single-channel implementation of this technique. For a

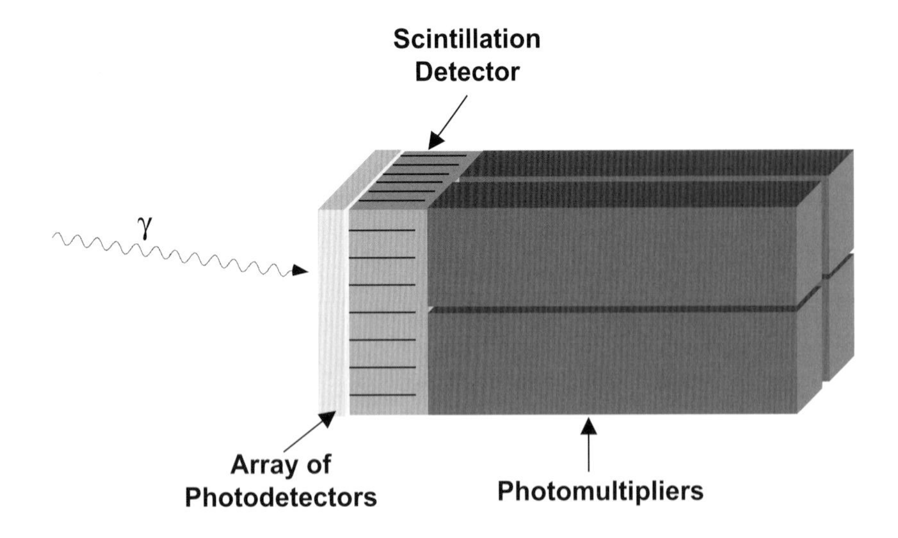

Figure 2.27. A single channel of one layer detector for DOI determination through the use of two photo-detectors at the crystal ends. In this schematic conventional light collection by PMTs are used at one end and an array of avalanche photodiodes are used on the incident face of the detector.

practical implementation in a scanner design, the use of regular photo-multiplier tubes at both ends is not feasible. As result, at least one such detector design has considered using a different type of photo-detector, such as PIN photodiodes or Avalanche photodiodes, on the crystal end that enters the scanner field of view [17].

Concluding Remarks

PET detectors and instrumentation have developed into sophisticated clinical tools, but further scope exists to develop higher-sensitivity, higher-resolution devices. There are now a number of scintillator crystals employed in commercial scanners, each with their own unique characteristics, including price. The range of scintillators may expand even further, especially if time-of-flight machines are developed. Light-collection technology may move away from photomultiplier tubes to solid-state devices (photodiodes) which will improve coupling and increase the bandwidth for data collection and processing by reducing the multiplexing of the signals.

Scanner design will continue to evolve and provide challenges in terms of photon detection, discrimination, and performance. Developments in basic physics will underpin many of these enhancements.

References

1. Klein O, Nishina Y. Uber die streuung von strahlung durch frei elektronen nach der neuen relativistischen quantendynamik von Dirac. Z Physik 1928;52:853–868.
2. Hubbell JH. Review of photon interaction cross section data in the medical and biological context. Phys Med Biol 1999;44(1):R1–22.
3. Hubbell JH. Photon cross sections, attenuation coefficients, and energy absorption coefficients from 10 keV to 100 GeV: National Bureau of Standards, US Dept of Commerce; 1969.
4. ICRU. Tissue substitutes in radiation dosimetry and measurement. Bethesda, MD, USA: International Commission on Radiation Units and Measurements; 1989. Report No. 44.
5. Townsend DW, Frey P, Jeavons A, Reich G, Tochon-Danguy HJ, Donath A, et al. High-density avalanche chamber (HIDAC) positron camera. J Nucl Med 1987;28:1554–62.
6. Jeavons AP, Chandler RA, Dettmar CAR. A 3D HIDAC-PET camera with sub-millimetre resolution for imaging small animals. IEEE Trans Nucl Sci 1999;NS-46:468–73.
7. Knoll GF. Radiation detection and measurement. 2nd ed. New York: John Wiley and Sons; 1988.
8. Karp JS, Muehllehner G, Geagan MJ, Freifelder R. Whole-body PET scanner using curve-plate NaI(Tl) detectors. J Nucl Med 1998;39:50P(abstract).
9. Melcher CL, Schweitzer JS. Cerium-doped lutetium oxyorthorthsilicate: a fast, efficient new scintillator. IEEE Trans Nucl Sci 1992;39(4):502–5.
10. Lecomte R, Cadorette J, Richard P et al. Design and engineering aspects of a high-resolution positron tomograph for small-animal imaging. IEEE Trans Nucl Sci 1996;NS-41(4):1446–52.
11. Ziegler SI, Pichler BJ, Boening G, Rafecas M, Pimpl W, Lorenz E, et al. A prototype high-resolution animal positron tomograph with avalanche photodiode arrays and LSO crystals. Eur J Nucl Med 2001;28:136–43.
12. Cherry SR, Shao Y, Silverman RW, Chatziioannou A, Meadors K, Siegel S, et al. MicroPET: A high-resolution PET scanner for Imaging small animals. IEEE Trans Nucl Sci 1997;44:1161–6.
13. Casey ME, Nutt R. A multicrystal, two-dimensional BGO detector system for positron emission tomography. IEEE Trans Nucl Sci 1986;NS-33(1):460–3.
14. Wong W, Uribe J, Hicks K, Hu G. An analog decoding BGO block detector using circular photomultipliers. IEEE Trans Nucl Sci 1995;NS-42:1095–101.
15. Adam LE, Zaers H, Ostertag H, Trojan H, Bellemann ME, Brix G. Performance evaluation of the whole-body PET scanner ECAT EXACT HR+ following the IEC standard. IEEE Trans Nucl Sci 1997;NS-44:1172–9.
16. Carrier C, Martel C, Schmitt C, Lecomte R. Design of a high-resolution positron emission tomograph using solid scintillation detectors. IEEE Trans Nucl Sci 1988;NS-35(1):685–90.
17. Huber JS, Moses WW, Derenzo SE, Ho MH, Andreaco MS, Paulus MJ, et al. Characterization of a 64-channel PET detector using photodiodes for crystal identification. IEEE Trans Nucl Sci 1997;NS-44:1197–201.

3 Data Acquisition and Performance Characterization in PET*

Dale L Bailey

Introduction

Positron emission detection systems have developed since their first use in the 1950s to the high-resolution, high-sensitivity tomographic devices that we have today. Configurations differ far more than for a gamma camera, with such variables as the choice of scintillation crystal, 2D or 3D acquisition mode capability, continuous or discrete detectors, full or partial surrounding of the patient, and a variety of transmission scanning arrangements and radioactive sources. In addition, PET instrumentation is an area that has continued to evolve rapidly, especially over the last decade, with the emphases on increasing sensitivity, improving resolution, and decreasing patient scanning times. This chapter discusses the issues that are determinants of PET system performance. Much of the discussion is based on circular tomographs with discrete detectors, however, the principles are applicable also to flat detector systems and rotating gamma camera PET systems.

Detected Events in Positron Tomography

Event detection in PET relies on electronic collimation. An event is regarded as valid if:

(i) two photons are detected within a predefined electronic time window known as the coincidence window,
(ii) the subsequent line-of-response formed between them is within a valid acceptance angle of the tomograph, and,
(iii) the energy deposited in the crystal by both photons is within the selected energy window.

Such coincident events are often referred to as prompt events (or "prompts").

However, a number of prompt events registered as having met the above criteria are, in fact, unwanted events as one or both of the photons has been scattered or the coincidence is the result of the "accidental" detection of two photons from unrelated positron annihilations (Fig. 3.1). The terminology commonly used to describe the various events in PET detection are:

(i) A *single* event is, as the name suggests, a single photon counted by a detector. A PET scanner typically converts between 1% and 10% of single events into paired coincidence events;
(ii) A *true coincidence* is an event that derives from a single positron–electron annihilation. The two annihilation photons both reach detectors on opposing sides of the tomograph without interacting significantly with the surrounding atoms and are recorded within the coincidence timing window;
(iii) *A random* (or *accidental*) *coincidence* occurs when two nuclei decay at approximately the same time. After annihilation of both positrons, four photons are emitted. Two of these photons from different annihilations are counted within the timing window and are considered to have come from the same positron, while the other two are lost. These events are initially regarded as valid, prompt events, but are spatially uncorrelated with the distribution of tracer. This is clearly a function of the number of disintegrations per second,

* Chapter reproduced from Valk PE, Bailey DL, Townsend DW, Maisey MN. Positron Emission Tomography: Basic Science and Clinical Practice. Springer-Verlag London Ltd 2003, 69–90.

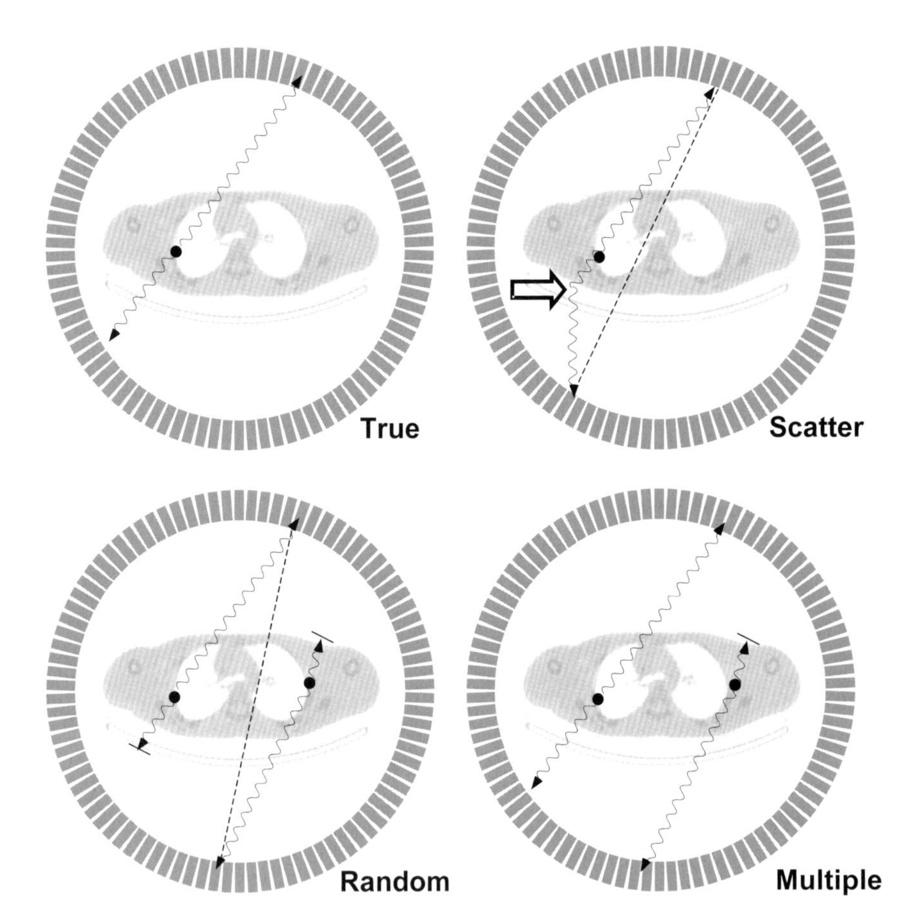

Figure 3.1. The various coincidence events that can be recorded in PET are shown diagrammatically for a full-ring PET system. The black circle indicates the site of positron annihilation. From top left clockwise the events shown are: a true coincidence, a scattered event where one or both of the photons undergo a Compton interaction (indicated by the open arrow), a multiple coincidence arising from two positron annihilations in which three events are counted, and a random or accidental coincidence arising from two positrons in which one of the photons from each positron annihilation is counted. In the case of the scattered event and the random event, the mis-assigned line of response is indicated by the dashed line.

and the random event count rate (R_{ab}) between two detectors *a* and *b* is given by:

$$R_{ab} = 2\tau N_a N_b \quad (1)$$

where *N* is the single event rate incident upon the detectors *a* and *b*, and 2τ is the coincidence window width. Usually $N_a \approx N_b$ so that the random event rate increases approximately proportionally to N^2. There are two common methods for removing random events: (i) estimating the random event rate from measurements of the single event rates using the above equation, or (ii) employing a delayed coincidence timing window. These methods are discussed in detail in Ch. 6.

(iv) *Multiple* (or triple) *events* are similar to random events, except that three events from two annihilations are detected within the coincidence timing window. Due to the ambiguity in deciding which pair of events arises from the same annihilation, the event is disregarded. Again, multiple event detection rate is a function of count rate;

(v) *Scattered events* arise when one or both of the photons from a single positron annihilation detected within the coincidence timing window have undergone a Compton interaction. Compton scattering causes a loss in energy of the photon and change in direction of the photon. Due to the relatively poor energy resolution of most PET detectors, many photons scattered within the emitting volume cannot be discriminated against on the basis of their loss in energy. The consequence of counting a scattered event is that the line-of-response assigned to the event is uncorrelated with the origin of the annihilation event. This causes inconsistencies in the projection data, and leads to decreased contrast and inaccurate quantification in the final image. This discussion refers primarily to photons scattered within the object containing the radiotracer, however, scattering also arises from radiotracer in the subject but *outside* the coincidence field of view of the detector, as well as scattering off other objects such as the gantry of the tomograph, the lead shields in place at either end of the camera to shield the detectors from the rest of the body, the floor and walls in the room, the septa, and also within the detector. The fraction of scattered events is not a function of count rate, but is constant for a particular object and radioactivity distribution.

The *prompt* count rate is given by the sum of the true plus random plus scattered event rates, as all of these events have satisfied the pulse height energy criteria for further processing. The corrections employed for random and scattered events are discussed in Ch. 6.

The sensitivity of a tomograph is determined by a combination of the radius of the detector ring, the axial length of the active volume for acquisition, the total axial length of the tomograph, the stopping power of the scintillation detector elements, packing fraction of detectors, and other operator-dependent settings (e.g., energy window). However, in general terms the overall sensitivity for true (T), scattered (S), and random (R) events are given by [1–3]:

$$T \propto \frac{Z^2}{D}$$
$$S \propto \frac{Z^3}{L \times D} \qquad (2)$$
$$R \propto \left(\frac{Z^2}{L}\right)^2$$

where Z is the axial length of the acquisition volume, D is the radius of the ring, and L is the length of the septa. For a multi-ring tomograph in 2D each plane needs to be considered individually and the overall sensitivity is given by the sum of the individual planes.

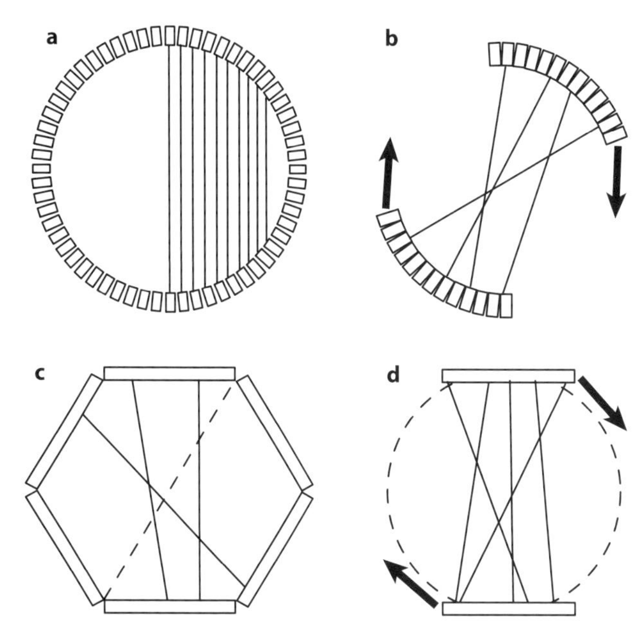

Figure 3.2. Various configurations of PET tomographs are shown in this figure. The solid lines show coincidence lines-of-response (LORs). Configurations (a) and (c) are stationary fixed systems, while (b) and (d) both need to rotate to acquire a complete data set. Configuration (a) is a full-ring circular system, (b) is a partial-ring circular system with continuous rotation, (c) consists of a number (typically 6–8) of flat detectors (LORs not measured indicated by the dashed line), and (d) is the geometry used for gamma camera PET and some other prototype systems using multi-wire proportional counters, where the detectors typically exhibit "step-and-shoot" acquisition protocols to obtain a complete data set.

Image Formation in PET

Historically, PET systems have generally developed as circular "rings". The earliest tomographs consisted of few detectors that rotated and translated to obtain a complete set of projection data, but soon full ring systems were developed. As PET uses coincidence detection, the detectors have to encompass 360° for complete sampling, unlike SPECT (single photon emission computed tomography) where 180° is sufficient. Today, PET systems use either full ring circular (or partial ring) configurations or multiple flat detector arrangements. In the case of gamma camera PET (GC-PET) systems, two or three large-area flat detectors that rotate are employed. Various configurations for PET detector systems are shown in Fig. 3.2.

Radial Sampling

The geometry and coordinate system that will be used to describe the PET systems in this section are shown in Fig. 3.3. The angle that the transaxial (x–y) plane makes with the z-axis is referred to as the polar angle, θ, and the rotated x–y plane forms an azimuthal angle, ϕ, around the object. In 2D PET, data are acquired for $\theta \approx 0°$, while in 3D PET, the polar angle can be opened

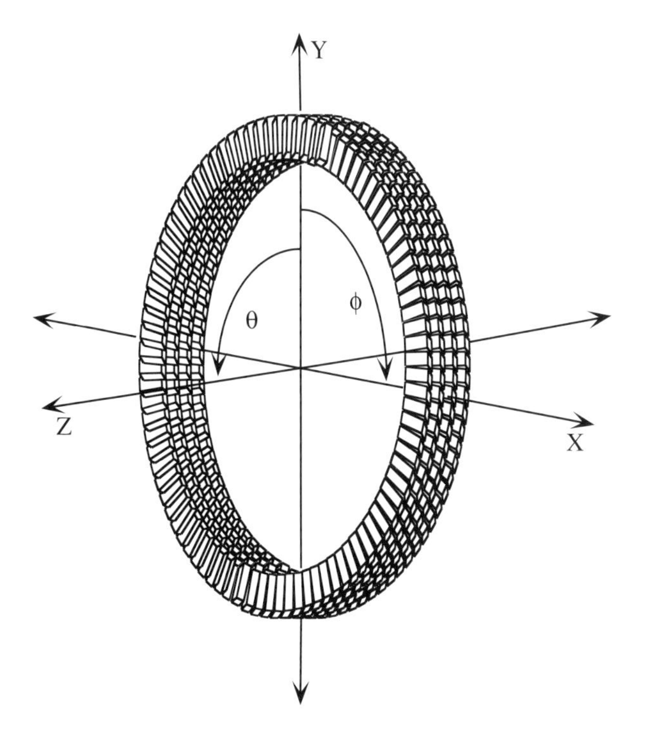

Figure 3.3. A diagram of a full-ring camera is shown with the coordinate system that describes the orientation of the camera. The azimuthal angle (ϕ) is measured around the ring, while the polar angle (θ) measures the angle between rings.

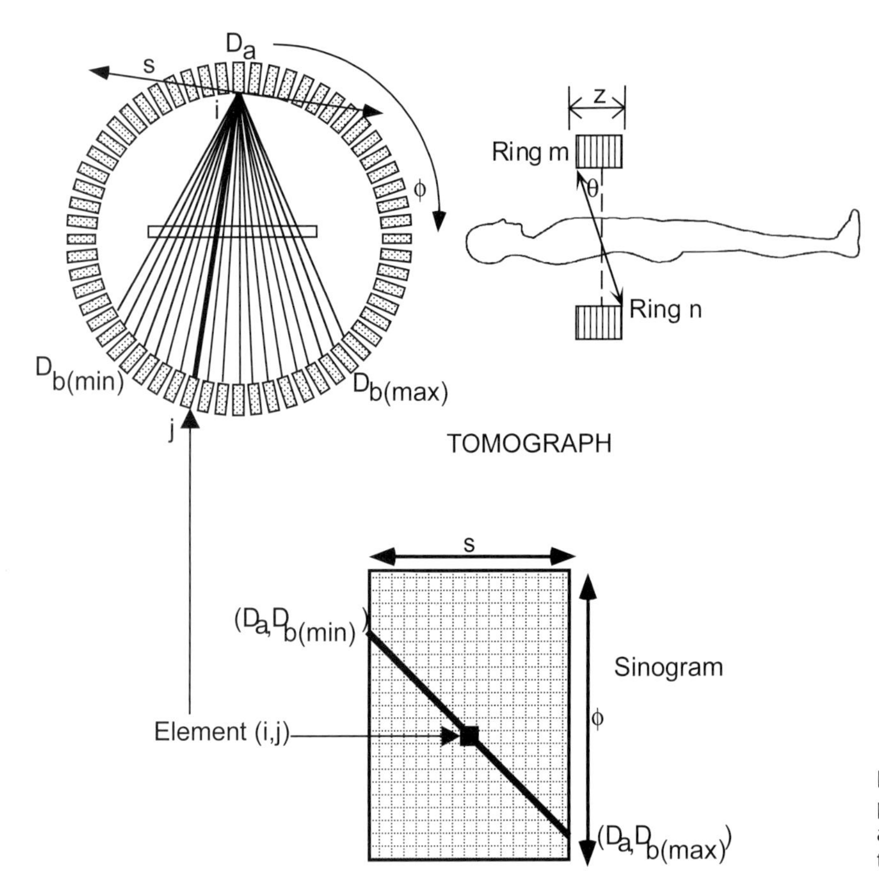

Figure 3.4. The mapping from sampling projections to sinograms is shown. The fan angle of acceptance in the ring in the top left corner maps to a diagonal line in the sinogram.

up to the desired acceptance to strike a trade-off between sensitivity gain and scatter increase.

Individual detector elements form coincidence pairs with opposing detectors (both in-plane and axially) and are mapped to the sinogram space as indicated in Fig. 3.4. Sinograms consist of approximately parallel projections; they are approximately parallel because increased sampling can be achieved by interpolation to form quasi-parallel projections between the detectors that contribute the truly parallel lines of response.

Instead of forming projections between detectors thus:

$$(\mathrm{D_a} : \mathrm{D_b}), (\mathrm{D_{a+1}} : \mathrm{D_{b+1}}), (\mathrm{D_{a+2}} : \mathrm{D_{b+2}}) \quad (3)$$

etc., in effect "double sampling" is achieved with the scheme:

$$(\mathrm{D_a} : \mathrm{D_b}), (D_{a+1} : D), (\mathrm{D_{a+1}} : \mathrm{D_{b+1}}), (D_{a+2} : D_{b+1}), (\mathrm{D_{a+2}} : \mathrm{D_{b+2}}) \quad (4)$$

etc., where the detector combinations in italics are formed between detectors with an offset of one detector between them, but assumed to be parallel to the adjacent projection formed between directly opposed detectors (Fig. 3.5).

The transaxial field of view of a PET tomograph is defined by the acceptance angle in the plane. This is de-

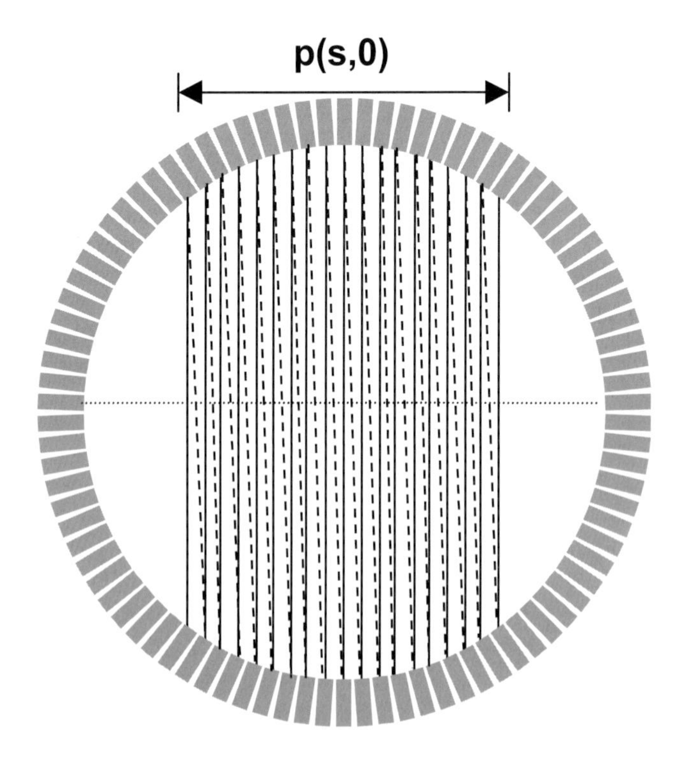

Figure 3.5. Sampling of the projections is doubled by forming coincidences between "opposite-but-one" detectors (dashed lines) as well as with the directly opposed detectors (solid lines). The azimuthal angle assumed for these interpolated lines of response is the same as for the direct lines-of-response. This effectively doubles the sampling in the projections.

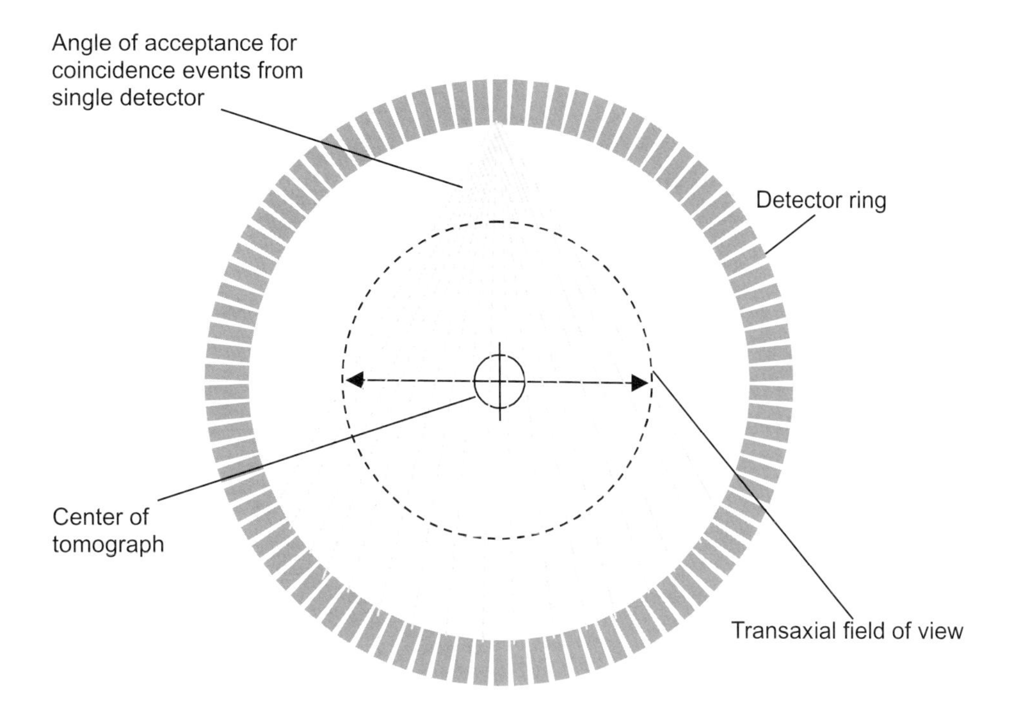

Figure 3.6. The transverse field of view of a PET tomograph is determined by the length of the chord defined by the acceptance angle of the electronics along the central axis of the system.

termined by the electronics, which permit an individual detector to be in coincidence with a finite number of detectors in the opposing side of the ring; the greater the acceptance angle the larger the number of detectors which form the "fan". The width of the fan along the diameter of the tomograph determines the width of the field of view (Fig. 3.6).

The fact that a circular ring is the geometry often used leads to a number of distortions in the sampling which require correction prior to (or as part of) the reconstruction process. The two main effects are:

(i) the distance between the opposing detectors decreases towards the edges of the sampling space (maximum distance from the central line of response). This causes an opening of the acceptance angle and effectively makes these lines of response more sensitive. However, this is offset to some extent by the decreasing surface area of the face of the detector exposed at this increasingly oblique angle, and,

(ii) the lines of response are not evenly spaced in the projection; they get closer together for the lines of response farthest from the central axis of the scanner (see Fig. 3.2(a)). This has the effect of decreasing the inter-detector spacing. Corrections for both effects are discussed in the following chapters.

Axial Sampling

The sinograms formed in PET are composed of projections p (s, ϕ, θ, z). In the 2D case all data are sampled (or assumed to be sampled) with polar angle $\theta = 0°$. In the 3D case this is extended to measuring projections at polar angles $\theta > 0°$. According to Orlov's criteria, the data acquired in 2D are sufficient for reconstructing the entire volume [4, 5]. However, in the 3D case all projections formed from angles with $\theta \neq 0°$ are redundant, as the object can be completely described by the 2D projections. The $\theta \neq 0°$ data are useful, however, as they contribute an increase in *sensitivity* and hence improve the signal-to-noise ratio of the reconstructed data. The redundancy of the oblique lines of response was exploited in the 3D reprojection algorithm [6, 7]. This is discussed further in the next chapter.

A convenient graphical representation was introduced by the Belgian scientist Christian Michel to illustrate the plane definitions used in a large multi-ring PET system, showing how the planes can be combined to optimize storage space and data-handling requirements. They have become known as "Michelogram" representations. Different modes of acquisition are shown in the Michelograms in Fig. 3.7 for a simple eight-ring tomograph.

The situation gets far more complicated for a larger number of rings, and when operating in 3D mode.

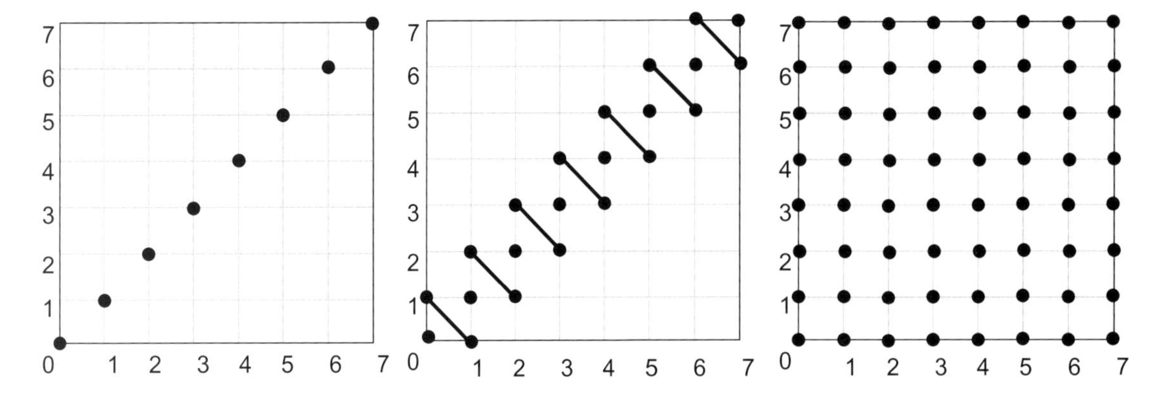

Figure 3.7. The graphical Michelogam is shown for three different acquisition modes on a simple eight-ring tomograph. Each point in the graph represents a plane of response defined between two sets of opposed detectors (a sinogram). In the graph on the left (a simple 2D acquisition with no "inter-planes"), the first plane defined is ring 0 in coincidence with the opposing detectors in the same ring, 0; ring 1 in coincidence with ring 1; etc, for all rings, resulting in a total of eight sinograms. In the middle graph, the same planes are acquired with the addition of a set of "inter-planes" formed between the rings with a ring difference of ±1 ring (ring 0 with ring 1, ring 1 with ring 0, etc). These planes are added together to form a single plane, indicated by the line joining them. This would lead to approximately twice the count rate in this plane compared with the adjacent plane which contains data from one ring only. Physically, this plane is positioned half way between detector rings 0 and 1. While the data come from adjacent rings they are assumed to be acquired with a polar angle of 0° for the purposes of reconstruction. This pattern is repeated for the rest of the rings. This results in 15 (i.e., 2N – 1) sinograms. This is a conventional 2D acquisition mode, resulting in almost twice the number of planes as the previous mode, improving axial sampling, and contributing over 2.5 times as many acquired events. In the graph on the right, a fully 3D acquisition is shown with each plane of data being stored separately (64 in total). The 3D mode would require a fully 3D reconstruction or some treatment of the data, such as a rebinning algorithm, to form 2D projections prior to reconstruction (see Ch. 4).

Examples are shown in Fig. 3.8 for the case of a 48-ring scanner in one particular 2D configuration, with planes added up to a maximum ring difference of ± 4 rings, and in a 3D acquisition configuration, where there are 48^2 (= 2,304) possible planes of response, but in this case the maximum acceptance angle between rings in limited to a ring difference of 40, with up to five axial lines of response being combined into a single plane.

The entire motivation for 3D PET is to increase sensitivity. While radionuclide emission imaging techniques in general use minute tracer amounts (usually micrograms or less), the proportion of the available signal detected is still relatively poor. A radiotracer in most cases distributes throughout the body with only a small fraction localizing in the target organ (if one exists), and collimation, attenuation, and scattering preclude many emitted photons from being detected. A conventional PET camera with interplane septa in 2D mode detects around 4,000–5,000 coincidence events per 10^6 (~0.5%) positron emissions with approximately uniform sensitivity over the axial profile, apart from the less sensitive end planes (Fig. 3.9). A gamma

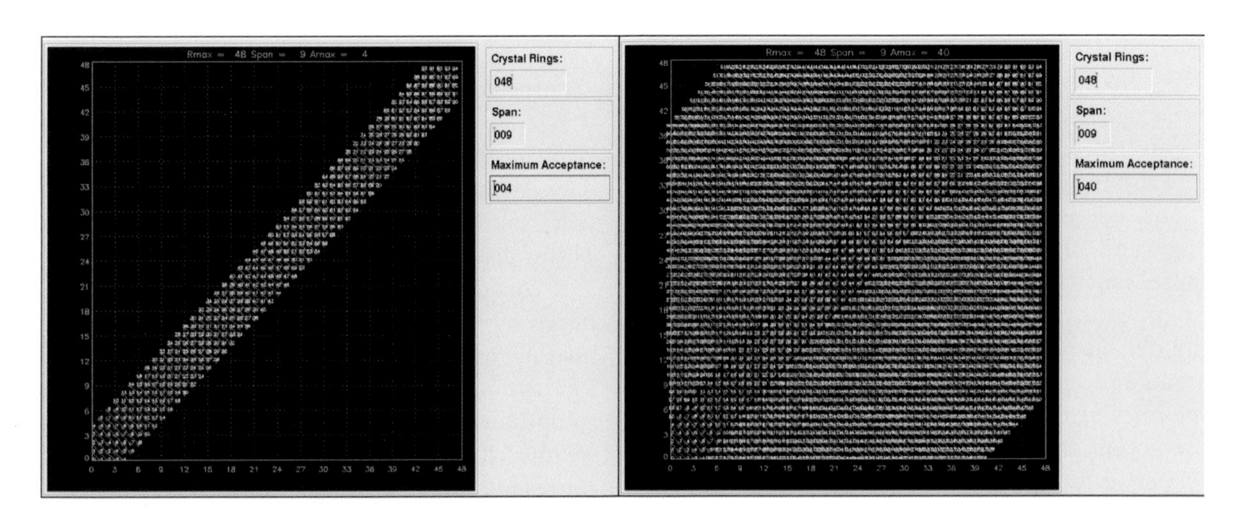

Figure 3.8. Michelograms representing the plane combinations for a 48-ring scanner are shown for the 2D case (left) and the 3D case (right). The x and y axes represent ring numbers on opposing sides of the scanner. Each point on the graph defines a unique plane of response (e.g., all lines-of-response in ring 1 in combination with ring 2).The diagonal lines joining individual dots indicate that the planes of response are combined (added together) thus losing information about each individual point's polar acquisition angle. This form of combination of data from different planes represents a "lossy" compression scheme.

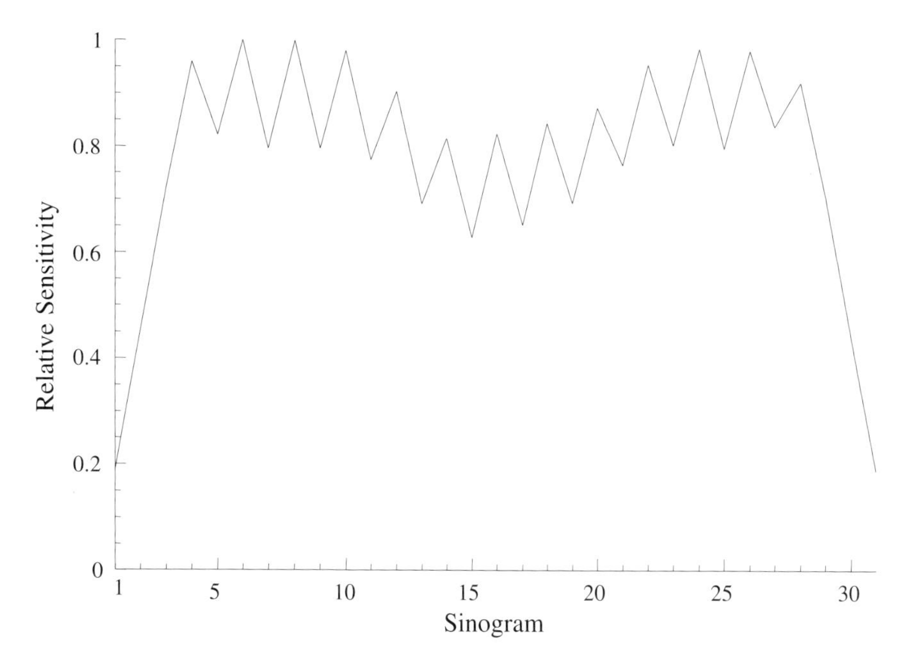

Figure 3.9. The 2D axial sensitivity profile for a line source in air on a 16-ring tomograph (CTI ECAT 951R) demonstrates both the bimodal pattern resulting from the two blocks used in this camera and the sinogram-to-sinogram variation arising from the combination of either three (odd-numbered sinograms) or four (even-numbered sinograms) axial lines-of-response in forming the sinogram. The end sinograms, which contain only one axial line-of-response, are only 20% as efficient as the sinograms formed in the center of the block detector.

camera, with its inefficient lead collimator, detects only around 200 of every 10^6 photons emitted. In spite of this modest efficiency, PET remains the most sensitive emission tomographic modality.

Constraining the allowed coincidences to a narrow plane orthogonal to the z axis of the PET camera severely restricts the overall sensitivity of the technique. Historically, the reasons for this restriction were twofold: the lack of appropriate 3D reconstruction software, and to keep the scatter fraction low. When the interplane septa are removed and all possible lines-of-response within the field-of-view are acquired in 3D, sensitivity is increased by two factors:

(i) the increased number of lines-of-response that it is now possible to acquire without the septa in place, and,
(ii) the amount by which the detector crystals are "shadowed" by the septa when they are in place [7–9]. The 3D acquisition mode leads to a non-uniform axial sensitivity profile, though, as shown in Fig. 3.10 for a 16-ring scanner and a distributed source.

In a 16-ring tomograph the sensitivity gain can be up to around thirty times greater in the center of the scanner compared to the end planes. The "average" gain over the entire axial feld-of-view is around five- to sevenfold. It is possible to separate the contributions of the two factors indicated above by scanning the same source in 2D mode both with and without the interplane septa using the usual 2D configurations of plane-defining lines-of-response. This demonstrates the effect due to septal shadowing alone, seen in Fig. 3.11. The shadowing effect of the septa is greater when the plane definition utilizes cross-planes as is usually done in a conventional 2D acquisition, as would be expected. The average sensitivity improvement due to shadowing is a factor of approximately 2.2. In the studies with a maximum ring difference (d_{max}) of zero, the component of sensitivity lost due to the thickness of the septa themselves (1 mm), and the amount of the detector that this covers is seen in isolation. The second component of the increase in sensitivity is the greater number of lines-of-response that can be accepted in 3D. When the 16 direct rings only are used ($d_{max} = 0$), this corresponds to 16 planes-of-response accepted; with the usual 31 plane definition for 2D acquisitions (ring difference $d = 0, \pm 2$ for odd-numbered planes (apart from the end detectors) and $d = \pm 1, \pm 3$ for even-numbered planes) this becomes a total of 100 planes-of-response. In a full 3D acquisition this would become 16×16 for this tomograph, i.e., 256 planes-of-response, as now each ring is in coincidence with every other ring on the opposing fan. This gives a factor of 256/100 = 2.56 increase in sensitivity due to the increased numbers of planes accepted compared with conventional 2D mode. However, there is a concomitant increase in the acceptance of scattered events axially as well.

A further effect produces a gain in coincidence count rates in 3D PET compared with 2D in addition to septal shadowing and acquiring more lines-of-response at greater polar acceptance angle. It has been shown that the 3D mode of acquisition is more efficient at converting single events into an annihilation pair which are both detected [9]. Measurements on a first-generation 2D/3D PET system have shown that the conversion rate

Figure 3.10. The axial sensitivity variation for the 3D acquisition geometry of 16-ring PET camera is shown. The center of the scanner is sampled around 32 times more than the end planes, where all possible planes of response are accepted. The plot is normalized to the first plane. Restricting the maximum acceptance angle (ring difference in this example with discrete detectors) will "flatten" the profile between defined limits (broken line), thereby achieving more uniform sampling in the central axial region of the scanner.

from single events to coincidences for a line source measured in air (i.e., no scatter) was 6.7% in 2D and 10.2% in 3D. For the same source measured in a 20 cm-diameter water-filled cylinder, the conversion rate in 2D was 2.4% and 4.8% in 3D. The ratio of these results show that, without scatter, the increase in conversion from single photons to coincidences for 3D compared to 2D is over 50% (10.2/6.7) higher, and in a scattering medium approaches 100% (4.8/2.4), although many of these events will be scattered events. The explanation is simple: more single photons can now form coincidence pairs in 3D where, in 2D, one or both would have been lost to the system by virtue of the flight angle (outside the allowed maximum ring difference) or by attenuation by the septa.

The non-uniform axial sampling in 3D, however, causes truncation of the projections, which is potentially a far greater problem for reconstruction than an axial variation in sensitivity (Fig. 3.12). This problem was solved, however, in 1989 with the development of the "reprojection" algorithm [6, 7]. This method exploits the fact that the data contains redundancy and the volume can be adequately reconstructed from the direct ring data ($d_{max} = 0$). The first step in this algo-

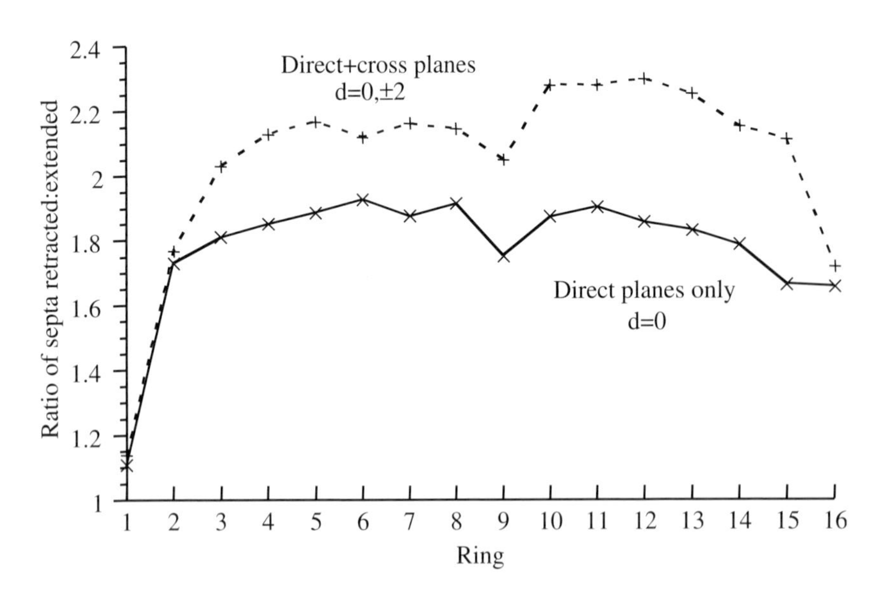

Figure 3.11. Sensitivity improvement due to removal of septa alone is demonstrated. The results are for direct planes only (solid line) and for conventional 2D planes where cross-planes contributions are included (broken line). The "dip" towards the center is at the block boundary of this two-block (axial) tomograph. This shows the effect purely due to septal shadowing.

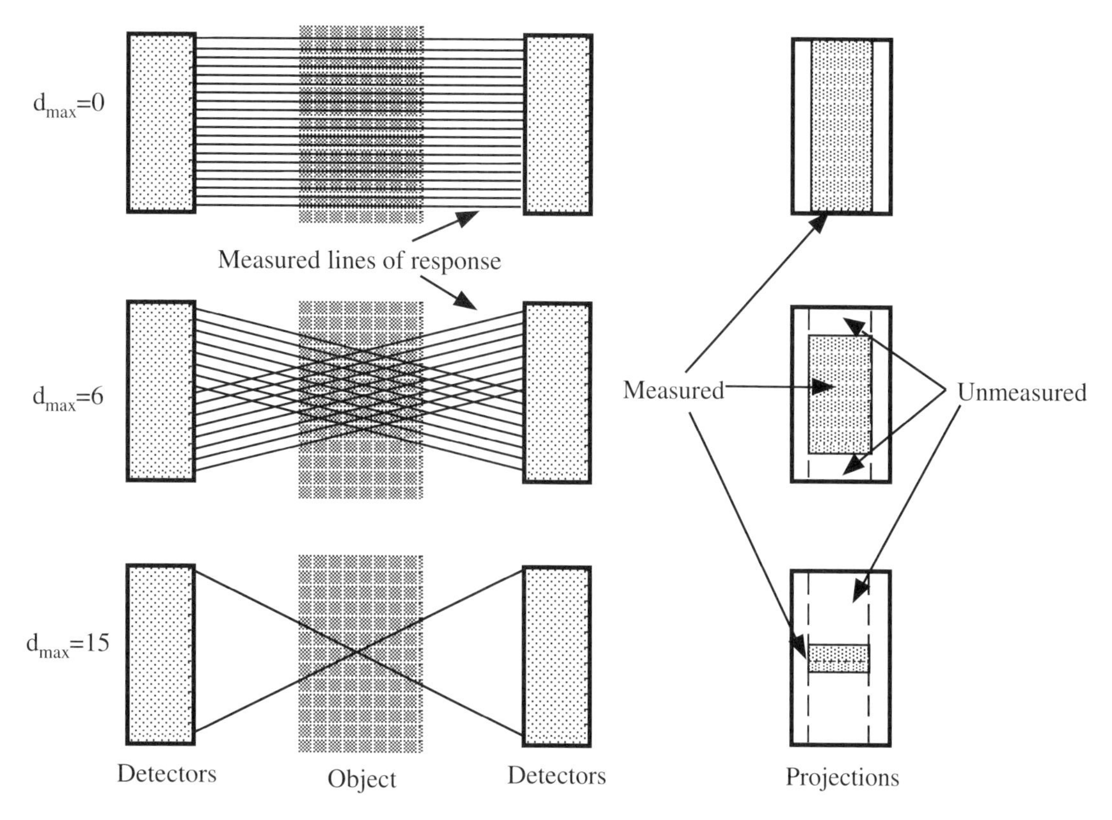

Figure 3.12. In the 3D acquisition case truncation of the projections occurs for those polar angles > 0°. At the top (d = 0) the entire field of view is sampled – this is the usual 2D case. When the ring difference is increased there is truncation of the axial field of view resulting in loss of data corresponding to the ends of the tomograph (center). In the limiting case (bottom) it results in severe truncation of the object.

rithm is to reconstruct the volume from the conventional 2D data sinograms. The unmeasured, or missing, data are then synthesized by forward projection through this volume. After this the data are complete and shift-invariant, and a fully 3D reconstruction algorithm can be used. This algorithm is discussed in depth in the next chapter.

From Projections to Reconstructed Images

Finally in this section, a brief description of how the data discussed are used to reconstruct images in positron tomography is included. The theory of reconstruction is dealt with in detail in the next chapter.

The steps involved and the different data sets required for producing accurate reconstructed images in 2D PET are shown in Fig. 3.13. All data (apart from the reconstructions) are shown as sinograms (i.e., the coordinates are (s,ϕ)). The usual data required are:

(i) the emission scan which is to be reconstructed,
(ii) a set of normalization sinograms (one per plane in 2D) to correct for differential detector efficiencies and geometric effects related to the ring detector, or a series of individual components from which such a normalization can be constructed (see Ch. 6), and,
(iii) a set of sinograms of attenuation correction factors to correct for photon attenuation (self-absorption or scattering) by the object.

The normalization factor singrams can include a global scaling component to account for the plane-to-plane variations seen in Fig. 3.9. The attenuation factor sinograms are derived from a "transmission" scan of the object and a transmission scan without the object in place (often called a "blank" or reference scan); the ratio of blank to transmission gives the attenuation correction factors. The most common method for acquiring the transmission and blank scans is with either a ring or rotating rod(s) of a long-lived positron emitter such as $^{68}Ge/^{68}Ga$, with which the object is irradiated [10]. The emission sinograms are first corrected for attenuation and normalized for different crystal efficiencies, and then reconstructed using the filtered back-projection process. During the final step, scalar corrections for dead time and decay may also be applied.

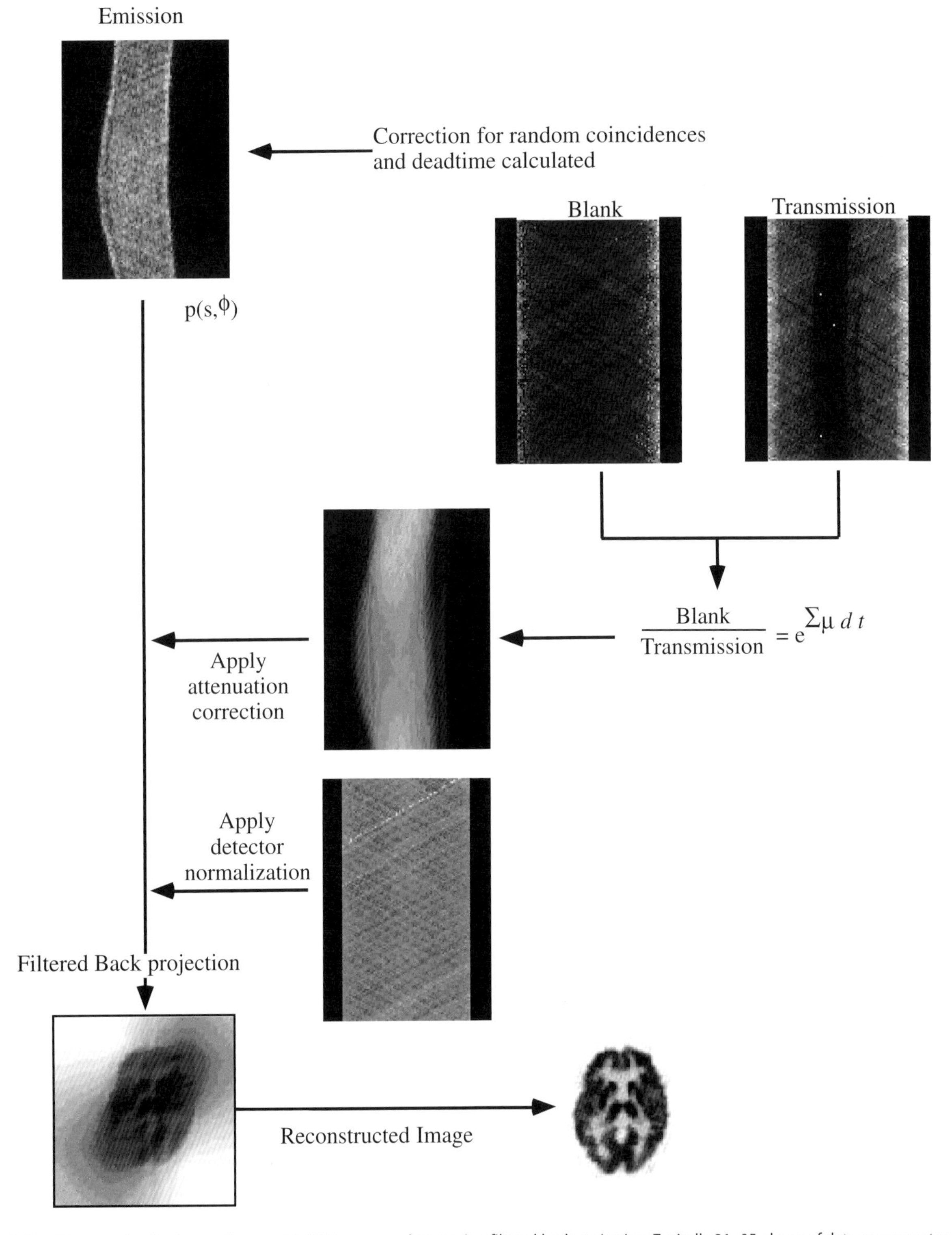

Figure 3.13. The steps involved in producing a 2D PET image are shown using filtered back-projection. Typically 31–95 planes of data are reconstructed in transverse section.

Development of Modern Tomographs

To understand the current state of commercial PET camera design, and why, for example, the development of 3D PET on BGO ring detector systems was only relatively recent, it is instructive to briefly trace the development of full ring PET systems. One of the first widely implemented commercial PET cameras was the Ortec

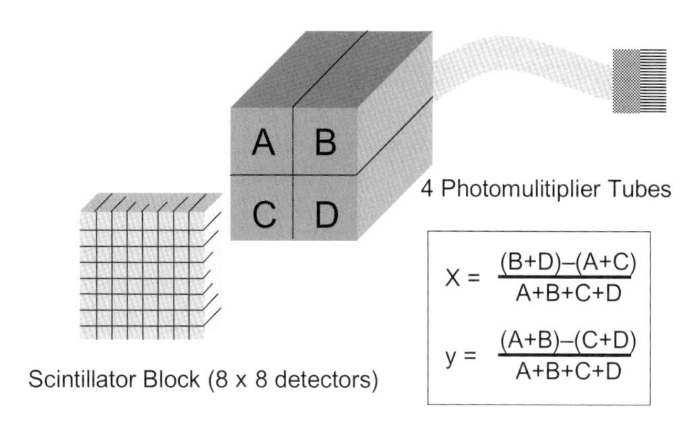

Figure 3.14. A schematic diagram of the block detector system, shown here as an 8 % 8 array of detectors, and the four PMTs which view the light produced is shown. The light shared between the PMTs is used to calculate the *x* and *y* position signals, with the equations shown.

ECAT (EG&G Ortec, Oak Ridge, Tennessee, USA) [11]. This single-slice machine used NaI(Tl) and had a hexagonal arrangement of multiple crystals with rotational and axial motion during a scan. Its axial resolution could be varied by changing the width of the slice-defining lead side shields, thereby altering the exposed detector area. This not only changed the resolution, but also the scatter and random event acceptance rates as well. In their paper of 1979, the developers of this system even demonstrated that in going from their "high-resolution" mode to "low-resolution" mode, they measured a threefold increase in scatter within the object (0.9%–2.7%), although total scatter accepted accounted for only around 15% of the overall signal [12]. In this and other early work on single-slice scanners, the relationship between increasing axial field-of-view and scatter fraction was recognized [1]. Various scintillation detectors have been used in PET since the early NaI(Tl) devices, but bismuth germanate (BGO) has been the crystal of choice for more than a decade now for non time-of-flight machines [13, 14]. BGO has the highest stopping power of any inorganic scintillator found to date.

After the adoption of BGO, the next major development in PET technology was the introduction of the "block" detector [15]. The block detector (shown schematically in Fig. 3.14) consists of a rectangular parallelepiped of scintillator, sectioned by partial saw cuts into discrete detector elements to which a number (usually four) of photomultiplier tubes are attached. An ingenious scheme of varying the depth of the cuts permits each of the four photomultiplier tubes to "see" a differential amount of the light released after a photon has interacted within the block, and from this the point where the photon deposited its energy can be localized to one of the detectors in the array. The aim of this development was to reduce crystal size (thereby improving resolution while still retaining the good-pulse-height-energy spectroscopy offered by a large scintillation detector), modularize detector design, and reduce detector cost. Small individual detectors with one-to-one coupling to photomultiplier tubes is impractical commercially due to packaging limitations and the cost of the large number of components required. The block detector opened the way for large, multi-ring PET camera development at the expense of some multiplexing of the signals. However, a stationary, full ring of small discrete detectors encompassing the subject meant that rapid temporal sequences could be recorded with high resolution, as the gantry no longer needed to rotate to acquire the full set of projections. The evolution and continuously decreasing detector and block size is shown in Fig. 3.15.

The major drawback for the block detector is count-rate performance, as the module can only process a single event from one individual detector in a particu-

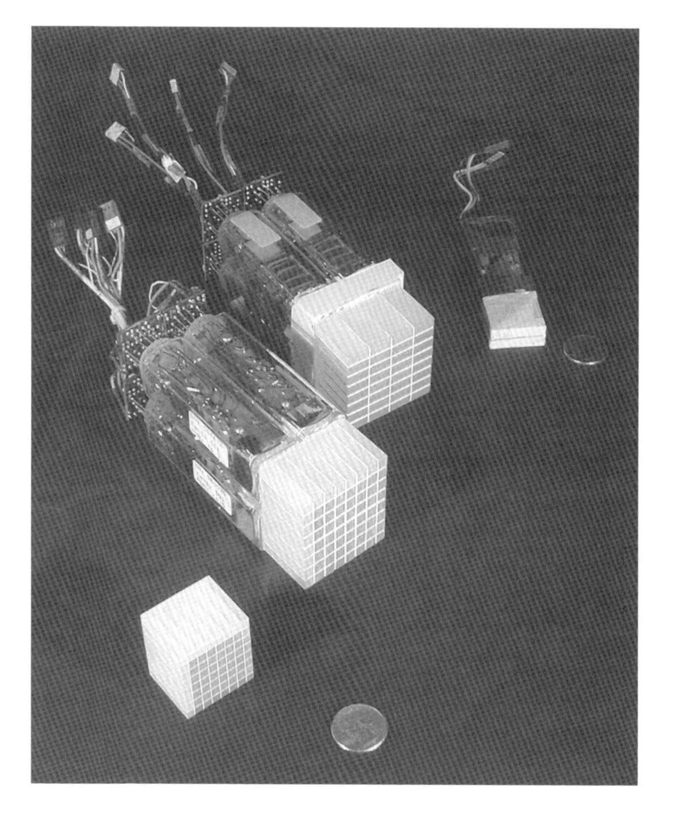

Figure 3.15. The evolution of PET detectors from CTI is shown. In the top right corner is the original ECAT 911 detector, then the first true block detector, the ECAT 93x block (8×4 detectors) with four PMTs attached, the 95x series block, which had double the number of axial saw-cuts, thus doubling the axial sampling compared with the 93x, and the high-resolution ECAT HR+ series block in the bottom left corner, where each detector element measures approximately 4 mm × 4mm × 30 mm. For scale, US25c coins are shown. (Photo courtesy of Dr Ron Nutt, CTI PET Systems).

lar block in a given time interval. Individual detectors with one-to-one coupling to the opto-electronic device would be a lot faster, however, at far greater expense and with a problem of packaging and stability of the great number of devices that would be required.

In a conventional 2D PET camera each effective "ring" in the block is separated by lead or tungsten shields known as septa. The aim was to keep the multi-ring tomograph essentially as a series of separate rings with little cross-talk between rings. This helped keep scatter and random coincidence event rates low, reduce single-photon flux from outside the field of view, and allowed conventional single-slice 2D reconstruction algorithms to be used. However, it limited the sensitivity of the camera.

Alternative systems to block-detector ring-based systems exist. Work commenced in the mid-1970s using large-area, continuous NaI(Tl) flat (or more recently curved) detectors in a hexagonal array around the subject and has resulted in commercially viable systems (GE Quest, ADAC C-PET) [16–19]. These systems have necessarily operated in 3D acquisition mode due to the lower stopping power of NaI(Tl) compared with BGO. The NaI(Tl) detectors, with their improved energy resolution, also provide better energy discrimination for improved scatter rejection based on pulse height spectroscopy. Larger detectors will always be susceptible to dead time problems, however, even when the number of photo-multiplier tubes involved in localizing the event in the crystal is restricted, and hence the optimal counting rates for these systems is lower than one with small, discrete detector elements. This affects clinical protocols by restricting the amount of radiotracer than can be injected.

PET Camera Performance

PET systems exhibit many variations in design. At the most fundamental level, different scintillators are used. The configuration of the system also varies greatly from restricted axial field of view, discrete (block-detector) systems to large, open, 3D designs. With such a range of variables, assessing performance for the purposes of comparing the capabilities of different scanners is a challenging task.

In this section, a number of the determinants of PET performance are discussed. New standards for PET performance have been published which may help to define standard tests to make the comparison of different systems more meaningful [27].

Measuring Performance of PET Systems

Spatial Resolution

Spatial resolution refers to the minimum limit of the system's spatial representation of an object due to the measurement process. It is the limiting distance in distinguishing juxtaposed point sources. Spatial resolution is usually characterized by measuring the width of the profile obtained when an object much smaller than the anticipated resolution of the system (less than half) is imaged. This blurring is referred to as the spread function. Common methods to measure this in emission tomography are to image a point source (giving a *point* spread function (PSF)), or, more usually, a line source (*line* spread function (LSF)) of radioactivity. The resolution is usually expressed as the full width at half maximum (FWHM) of the profile. A Gaussian function is often used as an approximation to this profile. The standard deviation is related to the FWHM by the following relationship:

$$\mathrm{FWHM} = \sqrt{8 \log_e 2}\,\sigma \quad (5)$$

where σ is the standard deviation of the fitted Gaussian function. There are many factors that influence the resolution in a PET reconstruction. These include:

(i) non-zero positron range after radionuclide decay,
(ii) non-collinearity of the annihilation photons due to residual momentum of the positron,
(iii) distance between the detectors,
(iv) width of the detectors,
(v) stopping power of the scintillation detector,
(vi) incident angle of the photon on the detector,
(vii) the depth of interaction of the photon in the detector,
(viii) number of angular samples, and
(ix) reconstruction parameters (matrix size, windowing of the reconstruction filter, etc.).

Resolution in PET is usually specified separately in transaxial and axial directions, as the sampling is not necessarily the same in some PET systems. In general, ring PET systems are highly oversampled transaxially, while the axial sampling is only sufficient to realize the intrinsic resolution of the detectors. The in-plane oversampling is advantageous because it partially offsets the low photon flux from the center of the emitting object due to attenuation. Transaxial resolution is often subdivided into radial (FWHM_r) and tangential (FWHM_t) components for measurements offset from the central axis of the camera, as these vary in a ring tomograph due to differential detector penetration at different locations in the x–y plane (see Fig. 3.16). Due

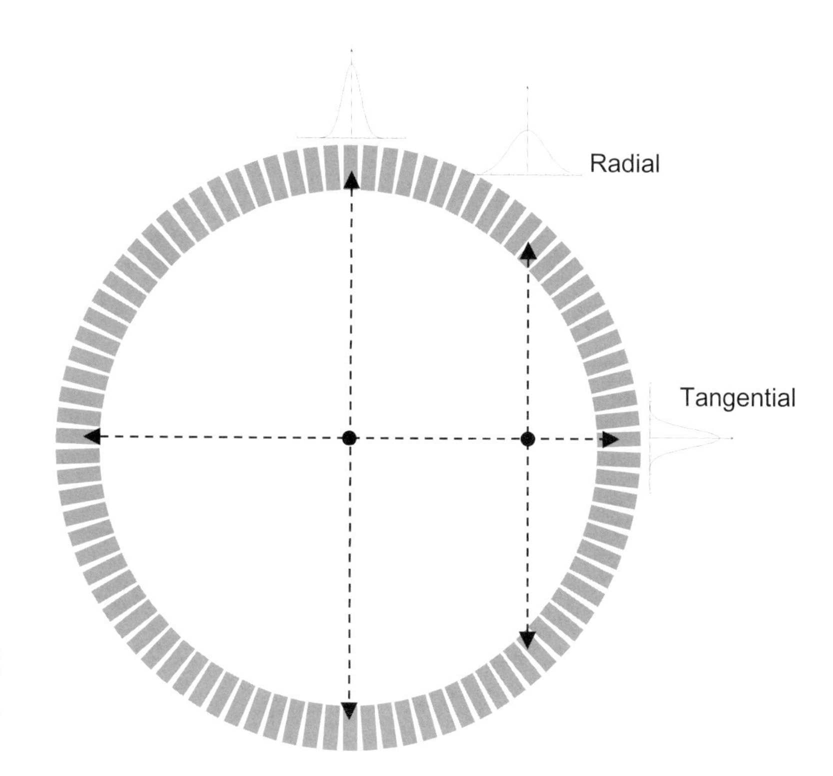

Figure 3.16. Transaxial resolution is separated into tangential and radial components. As the source of radioactivity is moved off-axis there is a greater chance that the energy absorbed in the scintillator will be spread over a number of detector elements. This uncertainty in localizing the photon interaction to one discrete detector degrades the spatial resolution in this direction.

to the limited, discrete sampling in the axial direction with block detector tomographs (one sample per plane), it is inappropriate to measure axial resolution ($FWHM_z$) on such systems from profiles of reconstructed data as there are insufficient sampling points with which it can be accurately estimated (only one point per plane). However, measurement of axial slice sensitivity of a point source as it passes in small steps through a single slice can be shown to be equivalent to 2D axial resolution, and thus can be utilized to overcome the limited axial sampling to measure the axial resolution.

Energy Resolution

Energy resolution is the precision with which the system can measure the energy of incident photons. For a source of 511 keV photons the ideal system would demonstrate a well-defined peak equivalent to 511 keV. BGO has low light yield (six light photons per keV absorbed) and this introduces statistical uncertainty in determining the exact amount of energy deposited. There are two possible ways to define the energy resolution for a PET scanner: the *single event* energy resolution, or the "*coincidence*" (i.e., both events) energy resolution.

Energy resolution is usually measured by stepping a narrow energy window, or a single lower-level discriminator, in small increments over the energy range of interest while a source is irradiating the detector(s). The count rate in each narrow window is then plotted to give the full spectrum. The data in Fig. 3.17 show the system energy resolution for single photons for a BGO tomograph for three different source geometries. An increase is seen in lower energy events in the scattering medium compared with the scatter-free air measurement.

Energy resolution is a straightforward measurement for single events, but less so for coincidence events. A method often used in coincidence measurements is to step a small window in tandem over the energy range. However, this is not the situation that is encountered in practice as it shows the spectrum when *both* events fall within the narrow energy band. It is more useful is to examine the result when the window for one coincidence of the pair is set to accept a wide range of energies (e.g., 100–850 keV) while the other coincidence channel is narrow and stepped in small increments over the energy range. This allows detection of, for example, a 511 keV event and a 300 keV event as a coincidence (as happens in practice). This is the method used in Fig. 3.18. It demonstrates energy resolution for a line source of $^{68}Ge/^{68}Ga$ in air of approximately 20% at 511 keV for a BGO scanner, similar to that obtained for the single photon counting spectrum.

Count Rate Performance

Count rate performance refers to the finite time it takes the system to process detected photons. After a photon is detected in the crystal, a series of optical

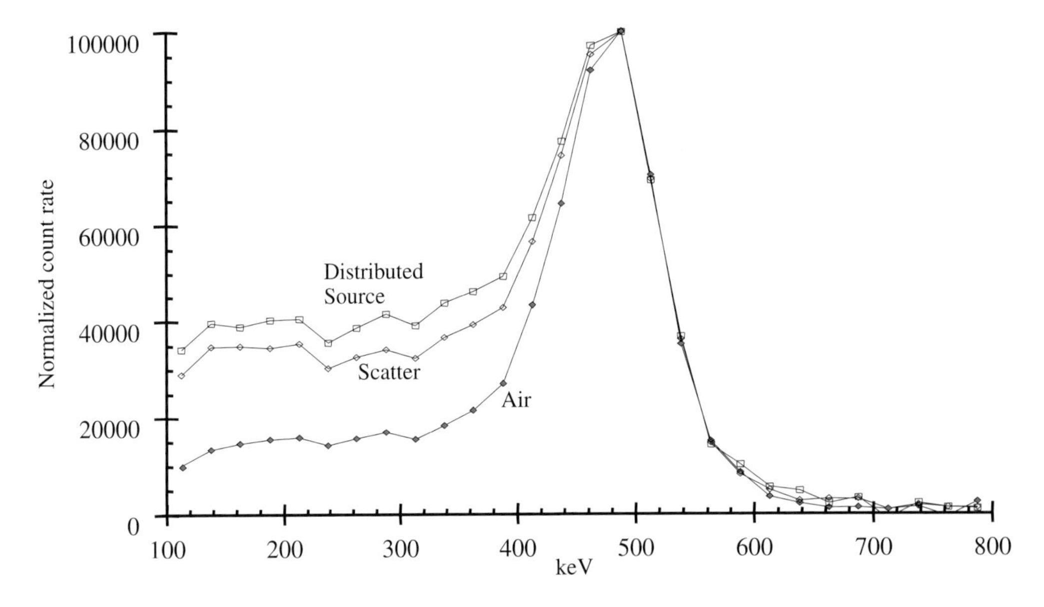

Figure 3.17. The energy spectra for single photons for a BGO PET system. The air and scatter measurements are of a ^{68}Ge line source in air and in a 20 cm-diameter water-filled cylinder respectively, while the distributed source is for a solution of ^{18}F in water in the same cylinder, to demonstrate the effect on energy spectrum of a distribution of activity. The respective energy resolutions are: air – 16.4%, line source in scatter – 19.6%, and distributed source – 21.6%.

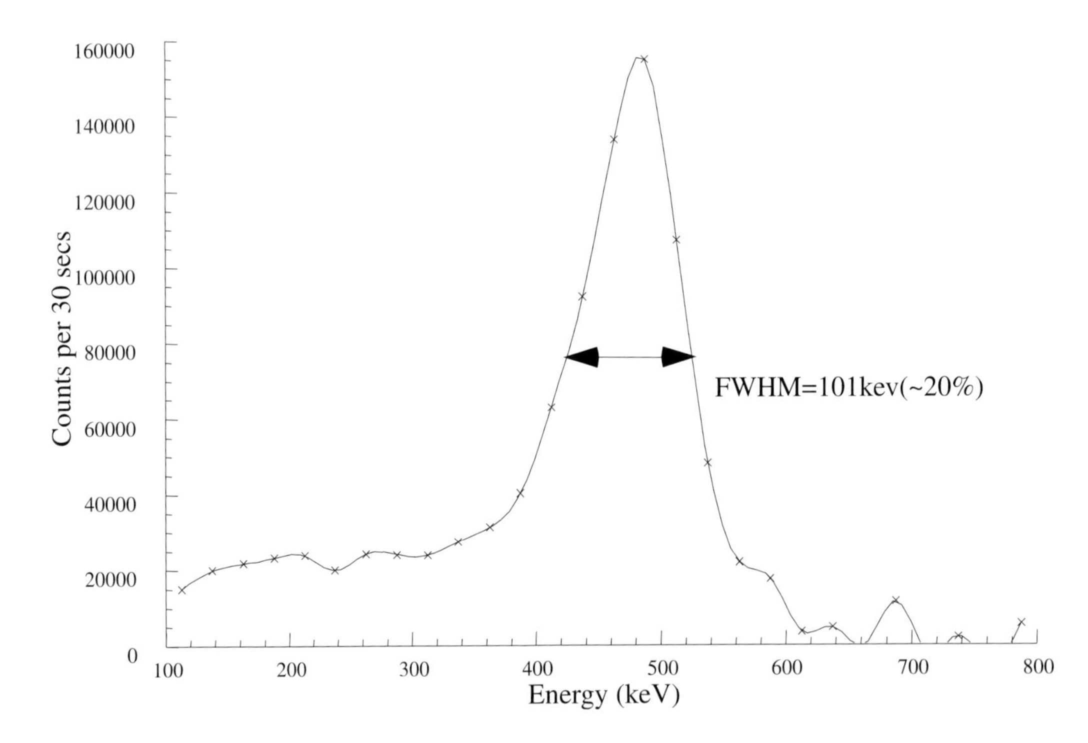

Figure 3.18. The "true" coincidence energy spectrum of a BGO full-ring scanner is shown for a ^{68}Ge line source measured in air. The spectrum is obtained by having one photon energy window set from 100–850 keV and the opposing detector window stepped in small increments of 25 keV to yield an integral coincidence spectrum. The derivative of the integral spectrum results in the above graph.

and electronic processing steps results, each of which requires a finite amount of time. As these combine in series, a slow component in the chain can introduce a significant delay. Correction for counting losses due to dead time are discussed in detail in Ch. 6. In this section we will restrict ourselves to the determination of count rate losses for PET systems for the purposes of comparing performance.

The most common method employed in PET for count rate and dead time determinations is to use a

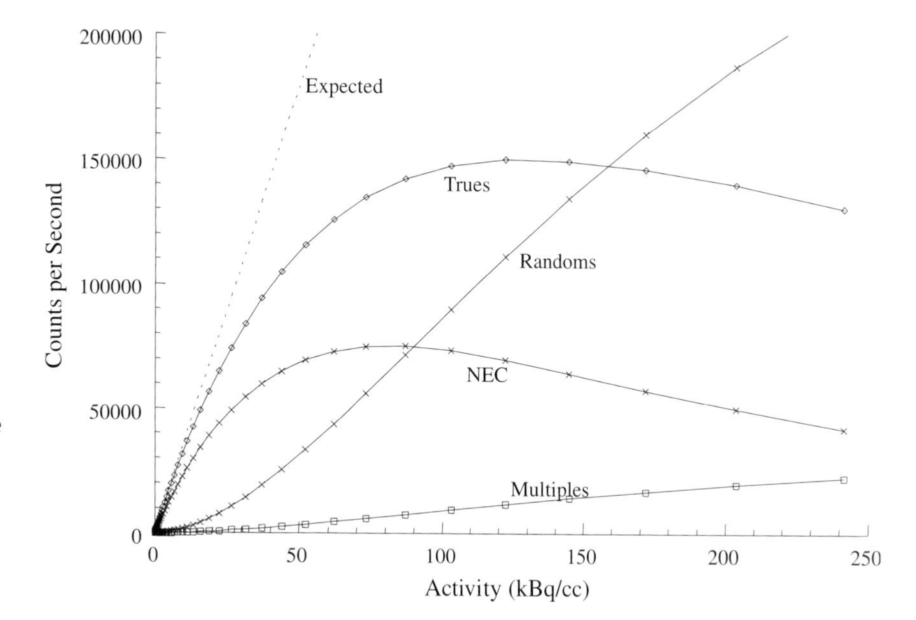

Figure 3.19. Count rate curves are shown for the measured parameters of true (unscattered plus scattered) coincidences, random coincidences, and multiple coincidences (three events within the time window), and the derived curves for expected (no counting losses) and noise equivalent count rate (NEC). The data were recorded on a CTI ECAT 953B PET camera using a 20 cm-diameter water-filled cylinder filled with ^{11}C in water.

source of a relatively short-lived tracer (e.g., ^{18}F, ^{11}C) in a multi-frame dynamic acquisition protocol and record a number of frames of data of suitably short duration over a number of half-lives of the source. Often, a cylinder containing a solution of ^{18}F in water is used. From this, count rates are determined for true, random, and multiple events. The count rates recorded at low activity, where dead time effects and random event rates should approach zero, can then be used to extrapolate an "ideal" response curve with minimal losses (observed = expected count rates). An example of the counting rates achieved for a BGO-based scanner in 2D mode is shown in Fig. 3.19.

It is possible to apply appropriate models to calculate dead time parameters. The data in Fig. 3.19 were characterized by modelling as a cascaded non-paralysable/paralysable system (Fig. 3.20) [20]. From this analysis, the non-paralysable dead-time component (τ_{np}) and the paralysable dead-time component (τ_p) were found to be approximately 3μs and 2μs respectively. Clearly, this is very different to the coincidence timing window duration (in this case $2\tau = 12$ns). The purposes of such parameter determinations might be to derive a dead-time correction factor from the observed counting rates.

The purpose of defining count rate performance is motivated by the desire to assess the impact of increasing count rates on image quality. Much of the theory behind measuring image quality derives from the seminal work of Dainty and Shaw with photographic film [21] and has been applied in a general theory of quality of medical imaging devices to measure detector quantum efficiency [22]. In PET an early suggestion for

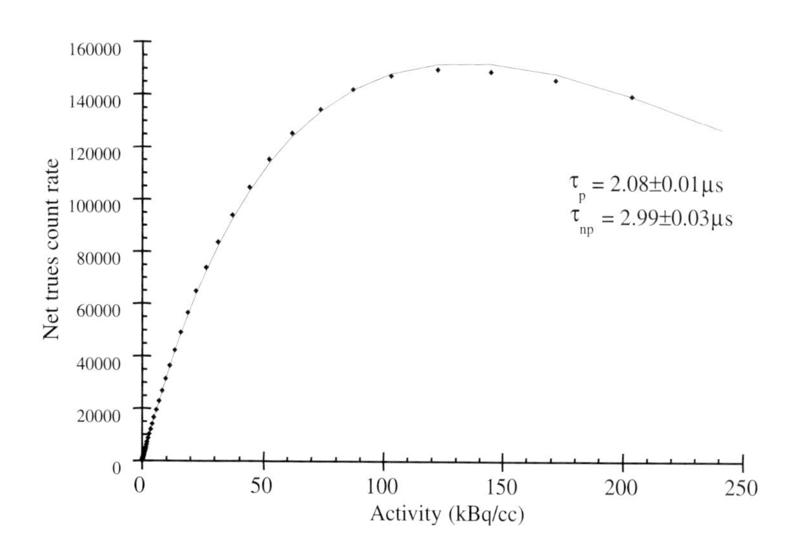

Figure 3.20. The true coincidence count rate for a 16-ring BGO scanner modelled as a combined paralysable and non-paralysable system produces the above fit to the data. From this, estimates of the dead time components can be derived.

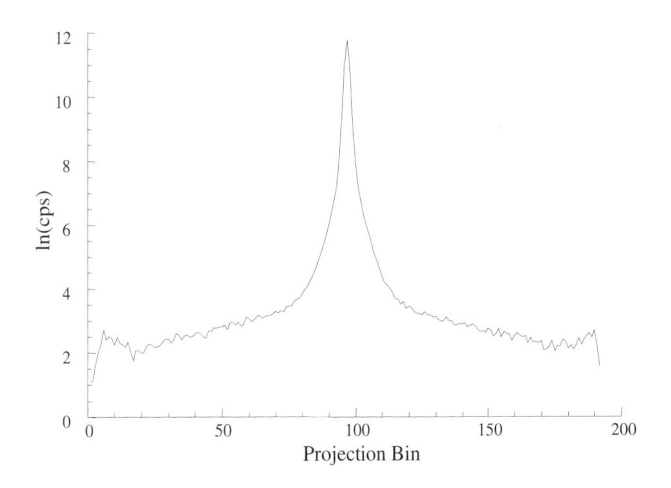

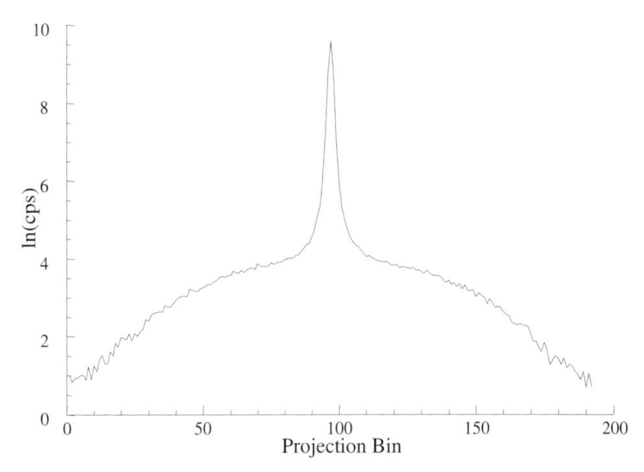

Figure 3.21. Log-linear count rate profiles from sinograms of a line source of ^{68}Ge in air (left) and centered in a 20 cm-diameter water-filled cylinder (right) demonstrate the additive scatter component outside of the central peak in the measurement in the cylinder. Interpolation of this section permits an estimate of the scatter fraction to be made. Both measurements were in 2D mode.

the use of such a figure of merit defined an '"effective" image event rate, Q to be:

$$Q = D_I\,(d_I \,/\, d_T); \qquad d_T = d_I + d_S + d_A \tag{6}$$

where d_I, d_S, and d_A are the count rates per cm from the center of a uniform cylinder containing radioactivity for the unscattered, scattered, and accidental (random) coincidences respectively, D_I is the total unscattered coincidence rate and (d_I/d_T) is the contrast. It was suggested that "*... Q may also be called an 'effective' image event rate, since the same signal-to-noise ratio would be obtained in an ideal tomograph...*." [2].

This has been further developed in recent years. Comparison of the count rate performance of different tomographs, or of the same scanner operating under different conditions (e.g., 2D and 3D acquisition mode) have been difficult to make because of the vastly different physical components of the measured data (e.g., scatter, randoms) and the strategies for dealing with these. These effects necessitate a comparison which can take account of these differences. The noise equivalent count (NEC) rate [23] provides a means for making meaningful inter-comparisons that incorporate these effects. The noise equivalent count rate is that count rate which would have resulted in the same signal-to-noise ratio in the data in the absence of scatter and random events. It is always less than the observed count rate.

The noise equivalent count rate is defined as:

$$\mathrm{NEC} = \frac{\left[T_{total}\left(\dfrac{T}{S+T}\right)\right]^2}{(T_{total} + 2fR)} \tag{7}$$

where T_{total} is the observed count rate (including scattered events), T and S are the unscattered and scattered event rates respectively, f is the "random event field fraction", the ratio of the source diameter to the tomograph's transaxial field-of-view, and R is the random coincidence event rate. This calculation assumes that the random events are being corrected by direct measurement and subtraction from the prompt event rate and that both measurements contain noise, hence the factor of 2 in the denominator (see Ch. 6). The NEC rate is shown, along with the data from which it was derived, in Fig. 3.19.

Some caution is required when comparing NECs from various systems, namely what scatter fraction was used and how it was determined, how the randoms fraction (R) was determined and how randoms subtraction was applied (delay-line method, estimation from single event rates, etc). However, the NEC does provide a parameter which can permit comparisons of count rate, and therefore an index of image quality, between systems.

Scatter Fraction

Scatter fraction is defined as that fraction of the total coincidences recorded in the photopeak window which have been scattered. The scattering may be of either, or both, of the annihilation photons, but it is predominantly scattering of one photon only. Scattering arises from a number of sources:

(i) scattering within the object containing the radionuclide,
(ii) scattering off the gantry components such as lead septa and side shields,
(iii) scattering within the detectors.

A number of methods for measuring scatter have been utilized. Perhaps the simplest method is to acquire data

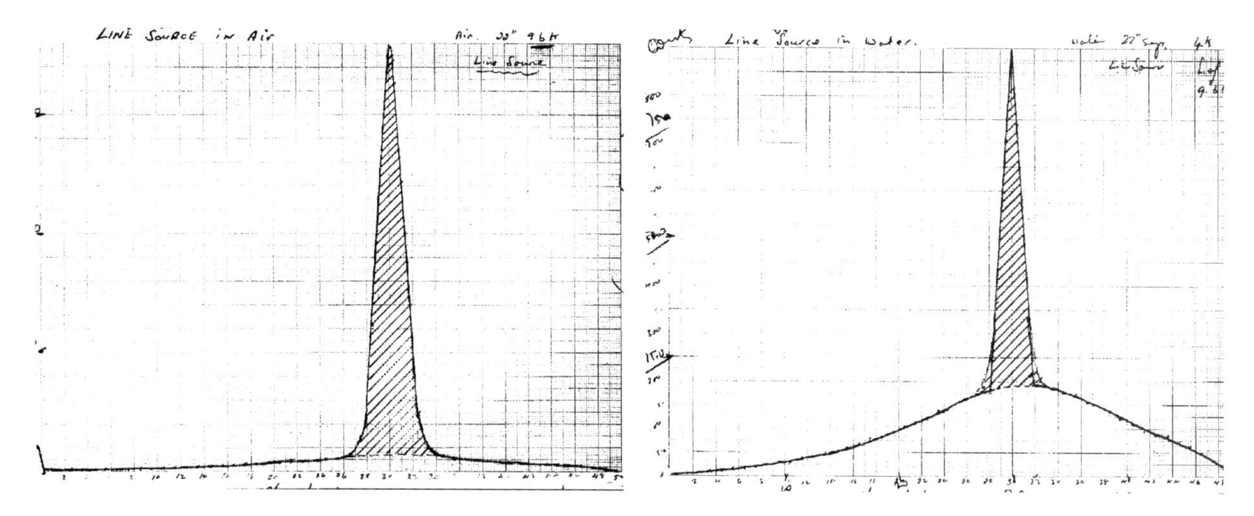

Figure 3.22. One of the earliest demonstrations of scattered radiation in coincidence PET measurements was by Jones and Burnham in 1973 on the first tomographic positron system, the PC-1, developed at the Massachusetts General Hospital in Boston. The process involved measuring a line source suspended vertically in air (left) and then immersing the source in a bucket of water (right) and repeating the measurement. The plots shown here are from the personal notebook of Terry Jones and are reproduced with his permission.

from a line source containing a suitably long-lived tracer in a scattering medium (typically a 20 cm-diameter water-filled cylinder) and produce profiles in the *s* dimension. Interpolation under the peak of the profile recorded outside the known location of the source permits an estimate of the scatter contribution, as used in the previous standard defined by the National Electrical Manufacturers' Association (NEMA) [24]. One criticism of this approach, however, is the assumption about the shape of the "wings" extending into the central section of the profile under the peak, and whether or not it be included in the scatter or non-scattered term (Fig. 3.21).

Scatter in 2D PET is usually relatively small and typically less than 15% of the total photopeak events. Thus it has been a small correction in the final image and often ignored with little impact on quantitative accuracy. The first scatter correction régimes for emission tomography were in fact developed for 2D PET [25].

The largest single difference between 2D and 3D PET after the increase in sensitivity is the greatly increased scatter that is included in the 3D measurements. Septa were originally included in PET camera designs for two reasons: (i) 3D reconstruction algorithms did not exist at the time, and (ii) to restrict random, scattered, and out of field-of-view events. One of the earliest demonstrations of scattered radiation in an open PET geometry was measured on the first positron tomograph PC-1 [26] in Boston in November 1973 shown in Fig. 3.22. Data were taken on this system which comprised two planar opposed arrays of NaI(Tl) detectors. This demonstrates clearly the increase in scatter in the profiles.

Scatter constitutes 20–50%+ of the measured signal in 3D PET. The scatter is dependent on object size, density, acceptance angle, energy discriminator settings, radiopharmaceutical distribution, and the method by which it is defined. The scatter fraction and distribution will vary for distributed versus localized sources of activity, and as such, the method for measuring and defining scatter as well as the acquisition parameters (axial acceptance angle, energy thresholds, etc) need to be quoted with the value for the measurement. In the updated NEMA testing procedures [27] a line source of ^{18}F positioned 45 mm radially from the center of a 20 cm diameter by 70 cm long water-filled cylinder is used to measure the scatter fraction. The scatter is measured on the projections by considering the events detected in the region outside of the cylinder boundary +20 mm on each side, which is interpolated to estimate the scattered events within the peak of the line source location. As mentioned in Ch. 2, scatter in PET is not strongly correlated spatially with the object boundary as it is in SPECT as the line of response from two photons is used. This is dramatically demonstrated in Fig. 3.23, which shows diagrams of the profiles of count rate obtained when a line source is moved laterally in a fixed-position water-filled cylinder. Even when the line source is centered within the object, the profile does not show any discontinuity at the boundary of the cylinder.

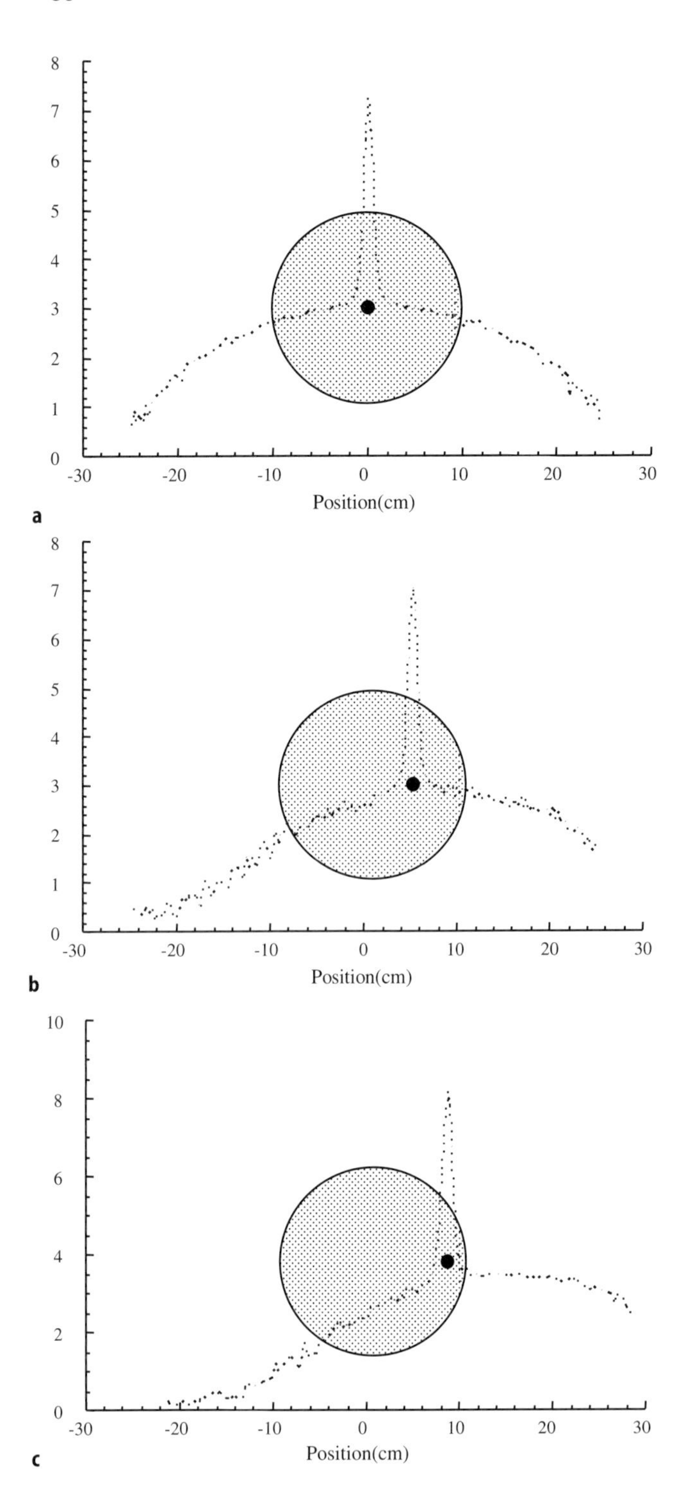

Figure 3.23. Demonstration of the spatial nature of scatter in 3D PET. The graphs show the profile from a line source in a water-filled cylinder in three different positions, with cylinder (grey circle) and line source (black dot) locations superimposed (to scale). The scatter profiles clearly demonstrate that the distribution of the scattered lines of response are only poorly correlated with the object, and can extend a large distance outside of the object. This is not true for SPECT where all scatter is constrained within the object boundaries. The reason for this is that two photons detected are ascribed a line of response joining the detectors in which they deposited their energy, and this can occur well beyond the object boundary.

Chapter 6 covers scatter correction techniques in detail.

Sensitivity of Positron Tomographs

The most commonly used mode for PET scanning at present is the 2D mode, with performance attributes as described in this chapter. Many of the corrections required (such as for dead time and crystal efficiency normalization) are well understood, making quantitative measurements accurate and precise. This has allowed PET to be used routinely as a highly sensitive tool for in vivo functional studies in spite of the 0.5% overall efficiency. However, while the sensitivity of 2D PET is unquestionably high compared with other modalities, the absolute sensitivity remains low compared with the potential signal available from the radiotracer, and consequently there remains room for improvement not only in detection efficiency, but in improving the spatial resolution of the technique as well. As Hoffman has shown, resolution improvements must be accompanied by an approximately third-power increase in sensitivity to maintain equivalent signal-to-noise ratio to realize the improvement in image quality [28]. This is intuitively seen by considering a twofold improvement in resolution: this decreases the effective resolution volume by two in each of the x, y, and z directions and therefore a 2^3 increase in sensitivity would be required to maintain equivalent signal-to-noise ratio per voxel. This is partially offset, though, by an effect known as signal amplification [29], which has guided PET detector designs for over a decade now. Signal amplification essentially means that an improvement in resolution *per se* will lead to an improvement in signal-to-noise in the reconstructed image as the higher resolution means that the reconstructed values will be "spread" over a smaller region, due purely to the higher resolution. This is turn means a higher reconstructed count within the region containing the activity, and hence better noise properties. However, increasing sensitivity still remains the main focus for improving the quality of PET data, and for these reasons the challenge in recent years has focussed on improving sensitivity.

The purpose of a sensitivity measurement on a positron tomograph is primarily to facilitate comparisons between different systems, as, in general, the higher the sensitivity the better signal-to-noise ratio in the reconstructed image (neglecting dead time effects). The sensitivity of positron tomographs has traditionally been measured using a distributed source of a relatively long-lived tracer, such as ^{18}F, in water. The value was quoted in units of counts per second per microCurie per millilitre (cts.sec^{-1}.µCi^{-1}.ml^{-1}) in non-SI units, without correction for attenuation or scattered radiation. This measurement was adequate to compare systems of similar design, e.g., 2D scanners with limited axial field of view. However, with the advent of vastly different designs emerging, and, especially, the use of 3D acquisition methods, this approach is limited for making meaningful comparisons. In 3D, scatter may constitute 20–50% or more of the recorded events and this needs to be allowed for in the sensitivity calculation. In addition, comparison of the true sensitivity compared to SPECT would be meaningless due to the differing attenuation at the different photon energies used. Thus, an absolute sensitivity measurement that is not affected by scatter and attenuation is desirable. A simple source of a suitable positron emitter could be used, however, a significant amount of surrounding medium is required for capture of the positrons within the source, which in itself causes attenuation of the annihilation photons.

A method has been developed to make absolute sensitivity measurements in PET [30], and has been adopted in the new updated NEMA testing procedures [27]. It employs the measurement of a known amount of ^{18}F (or ^{99m}Tc for SPECT) in a small source holder made from aluminum. The thickness of the aluminum wall of the source holder used is sufficient to stop all of the positrons, causing annihilation radiation to be produced, but which also causes some attenuation. The count rate for this source is found by measuring it for a defined period in the camera. Next, another tube of aluminum of known thickness is added to the holder, causing further attenuation, and this is counted again. This is done for a number of extra tubes of aluminum, all of known thickness, and an attenuation curve is produced. The extrapolated *y*-intercept from this curve gives the "sensitivity in air" for the camera. The units of this measurement are ct.sec^{-1}.MBq^{-1}. This provides an absolute measure of sensitivity. The method can also be used for PET system calibration of reconstructed counts without requiring scatter or attenuation correction [31].

In spite of the improvements in sensitivity with 3D PET, however, much of the available signal still goes undetected. Due to scatter, dead time, and random event rates, the *effective sensitivity* is far less than is measurable in an "absolute" sense. In an attempt to quantify this, a parameter combining the NEC with the absolute sensitivity measurements has been proposed [32]. At extremely low count rates where detector dead time and random events are negligible, the effective sensitivity (as it relates to the image variance) in a distributed object is simply the absolute sensitivity level with a correction for the scatter in the measurement. As the count rate increases, this effective sensitivity decreases due to the increased dead time and random events while scatter remains constant. Therefore, the effective sensitivity as a function of count rate can be expressed as the quotient of the noise equivalent rate divided by the ideal trues count rate with no scatter, dead time or random events, multiplied by the absolute sensitivity. The effective sensitivity, $C_{Eff}(a)$, is defined as:

$$C_{Eff}(a) = \frac{NEC(a)}{T_{Ideal}(a)} \times C_{Abs} \tag{8}$$

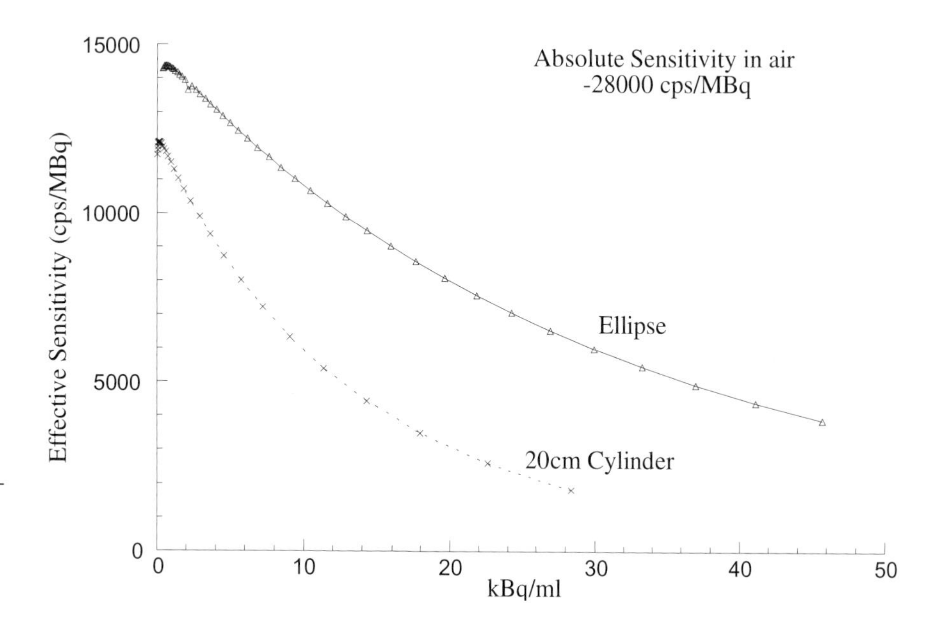

Figure 3.24. Effective sensitivity (cps/MBq) is shown as a function of activity concentration for two different elliptical phantoms (ellipse = 20.5×16.5 cm in cross-section and 20 cm axially approximating an average head size, and a 20 cm-diameter circular cylinder). The system used was a first-generation 2D/3D tomograph (ECAT 953B, CTI) operated in 3D mode. The curves demonstrate the loss of the ability to process events as activity concentration increases.

where C_{Abs} is the absolute sensitivity and $NEC(a)$ and $T_{Ideal}(a)$ are the noise equivalent and ideal (no count rate losses or random events) trues rates, respectively, which are functions of the activity concentration in the object. The effective sensitivity is a function of the activity in the object. This effective sensitivity is shown for 3D measurements using a small elliptical cylinder and a 20 cm cylinder in Fig. 3.24. The effective sensitivity demonstrates that the increase in solid angle from 3D acquisition is only one aspect of improving the sensitivity of PET, and that increasing detector performance by keeping the detectors available for signal detection for a longer proportion of the time can be thought of in a similar manner to increasing the solid angle as both improve the sensitivity of the device.

Other Performance Measures

In addition to the parameters described above (resolution, count rate, scatter, sensitivity), a number of other parameters are specified by bodies such as NEMA to assess PET scanner performance. These include accuracy of corrections for attenuation, scatter, randoms and dead time, and image quality assessments. Uniformity is another parameter that has been found to be useful to test. Energy resolution, though a major determinant of PET performance, has not been included in the latest NEMA PET tests [27]. No explicit tests for assessing transmission scan quality are specified, although a need exists with the variety of systems now available.

A difficulty in extrapolating from performance in standards test to the clinical situation is the highly unrealistic (clinically relevant) nature of the objects scanned. This has been recognized and attempts to address this have been made by employing long test objects (70 cm cylinder, NEMA) and objects which resemble the body in cross-section (EEC phantom [33]).

Impact of Radioactivity Outside the Field of View

Scanner design has traditionally included significant lead end-shields to restrict the majority of single photons emitted from outside the axial field of view of the scanner from having direct line-of-sight trajectories to the detectors. Single photons from outside the field of view will not form a true coincidence, but will increase the number of events the detector has to process leading to increased dead time and random coincidences. Some true coincidences from scattered photons may be included if the positron annihilation was just outside the axial field of view, but in general, the photons from outside the field of view will be unpaired events.

Single photons from outside the field of view were not a large problem with 2D tomographs that used interplane septa, as the septa added extra shielding for the detectors for photons from outside the field of view as well as inside. However, a number of developments over the past decade have exacerbated this situation:

(i) the move to acquire data in high-sensitivity 3D mode, thereby removing the interplane septa,
(ii) the increase in length of the axial field of view, which has the effect of increasing the acceptance angle for single photons from outside the field of view, and,
(iii) decreasing the length of the end shielding to accommodate large subjects. This has the effect of "opening up" the acceptance angle even further.

Examples illustrating this effect are shown in Fig. 3.25. It is a particular issue when using detectors such as BGO or NaI(Tl), which are not fast scintillators, and coincidence timing windows that are relatively long, of the order of 10 nsec or greater.

A number of solutions have been proposed, including "staggered" partial septa to restrict the out-of-field-of-view component without greatly decreasing the axial acceptance angle for true coincidences, shielding the subject (rather than the detectors) by placing or wrapping some form of flexible lead over the part of the body outside the field of view, and decreasing the coincidence window width. As the random coincidence rate varies linearly with window width (recall $R_{ab} = 2\tau N_a N_b$ where 2τ is the width of the coincidence timing window), a decrease by a factor of two from 12 nsec to 6 nsec would be expected to halve the random event rate. However, this would be at the expense of energy and positioning information due to the need to truncate the pulses from the detectors. One simple solution that has been widely employed in brain studies is to add a removable lead shield to the end of the tomograph on the patient side, effectively extending the end shielding [34]. Unfortunately this is only applicable for brain studies. Nevertheless, it is very effective in this application [35].

The solution would appear to be to use a fast scintillator, such as LSO, YSO, or GSO, and a shorter coincidence window. However, a time window of 4 nsec or less would require the use of time-of-flight electronics as the time window duration is now approaching the time it would take for an annihilation photon produced at the edge of the transaxial field of view (perhaps from a transmission source) to travel to the opposing

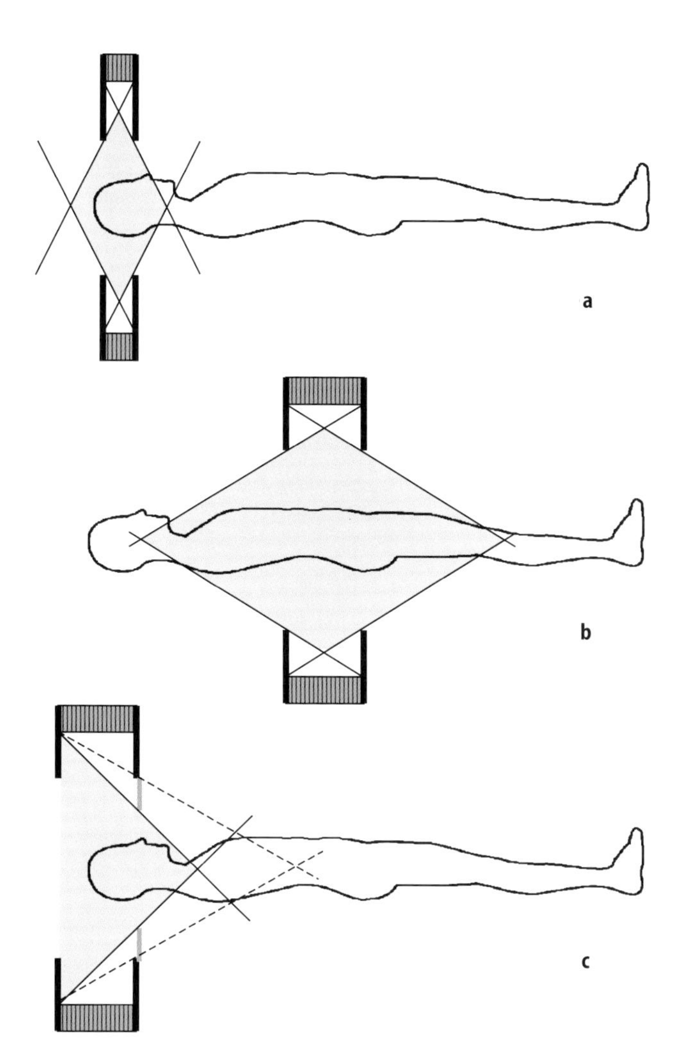

Figure 3.25. The field of view for single photons is shown schematically for three different scanner configurations in 3D. In (a), the first purpose-built 2D/3D tomograph (CTI ECAT 953B), which was designed for brain studies only, is shown. It had a 76 cm ring diameter, 10.8 cm axial field of view, and end shielding that restricted the subject aperture to 38 cm. In (b), the first purpose-built 3D-only full ring BGO tomograph (CTI EXACT3D) is shown. It had a 23.4 cm axial field of view and 82 cm ring diameter. However, as it was intended for whole-body scanning, the end shields were limited in extent to allow a large subject aperture (64 cm). This produced an enormous single-photon field of view which impacted on performance. A modification to the same tomograph, with removable lead end shields for use in brain studies (c), improved performance dramatically by restricting the single-photon field of view. The broken lines show the single-photon field of view without the shields in place.

detector, a distance close to one metre away. If non-time-of-flight electronics are employed the width of the transaxial field of view will be restricted.

There will also be an increase in true coincidences arising from outside the field of view in which one or both photons are scattered. This poses a problem for scatter-correction algorithms that use estimation methods, rather than direct measurements, to define the scatter contribution. Some algorithms combine both approaches by using the measured event rate outside the object being imaged, which must be due to scatter, to scale the estimated scatter within the object.

References

1. Derenzo SE, Zaklad H, Budinger TF. Analytical study of a high-resolution positron ring detector system for transaxial reconstruction tomography. J Nucl Med 1975;16(12):1166–73.
2. Derenzo SE. Method for optimizing side shielding in positron emission tomographs and for comparing detector materials. J Nucl Med 1980;21(10):971–7.
3. Kouris K, Spyrou NM, Jackson DF. Imaging with ionizing radiations. 1982 ed. Surrey: Surrey University Press; 1982.
4. Orlov S. Theory of three-dimensional reconstruction. 1. Conditions of a complete set of projections. Sov Phys Crystallogr 1976;20:312–4.
5. Defrise M, Townsend DW, Clack R. Three-dimensional image reconstruction from complete projections. Phys Med Biol 1989;34:573–87.
6. Kinahan PE, Rogers JG. Analytic 3-D image reconstruction using all detected events. IEEE Trans Nucl Sci 1989;NS-36:964–8.
7. Townsend DW, Spinks TJ, Jones T, Geissbühler A, Defrise M, Gilardi M-C, et al. Three-dimensional reconstruction of PET data from a multi-ring camera. IEEE Trans Nucl Sci 1989;36(1):1056–65.
8. Cherry SR, Dahlbom M, Hoffman EJ. 3D PET using a conventional multislice tomograph without septa. J Comput Assist Tomogr 1991;15:655–68.
9. Bailey DL, Jones T, Spinks TJ, Gilardi M-C, Townsend DW. Noise equivalent count measurements in a neuro-PET scanner with retractable septa. IEEE Trans Med Imag 1991;10(3):256–60.
10. Phelps ME, Hoffman EJ, Mullani NA, Ter-Pogossian MM. Application of annihilation coincidence detection to transaxial reconstruction tomography. J Nucl Med 1975;16(3):210–24.
11. Phelps ME, Hoffman EJ, Huang SC, Kuhl DE. ECAT: A new computerized tomographic imaging system for positron emitting radiopharmaceuticals. J Nucl Med 1978;19:635–47.
12. Hoffman EJ, Huang SC, Phelps ME. Quantitation in positron emission tomography: 1. Effect of object size. J Comput Assist Tomogr 1979;3(3):299–308.
13. Weber MJ, Monchamp RR. Luminescence of $Bi_4Ge_3O_{12}$: spectral and decay properties. J Appl Phys 1973;44:5495–9.
14. Cho Z, Farukhi M. Bismuth germanate as a potential scintillation detector in positron cameras. J Nucl Med 1977;18:840–4.
15. Casey ME, Nutt R. A multicrystal two-dimensional BGO detector system for positron emission tomography. IEEE Trans Nucl Sci 1986;NS-33(1):460–3.
16. Muehllehner G. Positron camera with extended counting rate capability. J Nucl Med 1975;16(7):653–7.
17. Muehllehner G, Karp JS, Mankoff DA, Beerbohm D, Ordonez CE. Design and performance of a new positron emission tomograph. IEEE Trans Nucl Sci 1988;35(1):670–4.
18. Karp JS, Muehllehner G, Geagan MJ, Freifelder R. Whole-body PET scanner using curve-plate NaI(Tl) detectors. J Nucl Med 1998;39:50P (abstract).
19. Karp JS, Muehllener G, Mankoff DA, Ordonez CE, Ollinger JM, Daube-Witherspoon ME, et al. Continuous-slice PENN-PET: a positron tomograph with volume imaging capability. J Nucl Med 1990;31:617–27.
20. Cranley K, Millar R, Bell T. Correction for deadtime losses in a gamma camera data analysis system. Eur J Nucl Med 1980;5:377–82.
21. Dainty JC, Shaw R. Image science: principles, analysis and evaluation of photographic-type imaging processes. London: Academic Press; 1974.
22. Wagner RF. Low-contrast sensitivity of radiologic, CT, nuclear medicine and ultrasound medical imaging systems. IEEE Trans Med Imag 1983;MI-2:105–21.

23. Strother SC, Casey ME, Hoffman EJ. Measuring PET scanner sensitivity: relating countrates to image signal-to-noise ratios using noise equivalent counts. IEEE Trans Nucl Sci 1990;37(2):783–8
24. NEMA. Performance measurements of positron emission tomographs. Washington: National Electrical Manufacturers Association; 1994. Report No. NU2-1994.
25. Bergström M, Eriksson L, Bohm C, Blomqvist G, Litton J-E. Correction for scattered radiation in a ring detector positron camera by integral transformation of the projections. J Comput Assist Tomogr 1983;7(1):42–50.
26. Burnham CA, Brownell GL. A multi-crystal positron camera. IEEE Trans Nucl Sci 1973;NS-19(3):201–5.
27. NEMA. Performance measurements of positron emission tomographs. Washington: National Electrical Manufacturers Association; 2001. Report No. NU2–2001.
28. Hoffman EJ, Phelps ME. Positron emission tomography: principles and quantitation. In: Phelps ME, Mazziotta JC, Schelbert HR, editors. Positron emission tomography and autoradiography. Principles and applications for the brain and heart. New York: Raven Press; 1986. pp. 237–86.
29. Phelps ME, Huang SC, Hoffman EJ, Plummer D, Carson RE. An analysis of signal amplification using small detectors in positron emission tomography. J Comput Assist Tomogr 1982;6(3):551–65.
30. Bailey DL, Jones T, Spinks TJ. A method for measuring the absolute sensitivity of positron emission tomographic scanners. Eur J Nucl Med 1991;18:374–9.
31. Bailey DL, Jones T. A method for calibrating three-dimensional positron emission tomography without scatter correction. Eur J Nucl Med 1997;24(6):660–4.
32. Bailey DL, Meikle SR, Jones T. Effective sensitivity in 3D PET: the impact of detector dead time on 3D system performance. IEEE Trans Nucl Sci 1997;NS–44:1180–5.
33. IEC. 61675-1: Radionuclide imaging devices - characteristics and test condition. Part I: Positron emission tomographs. Geneva: International Electrotechnical Commission; 1997. Report No. IEC 62C/205/FDIS.
34. Spinks TJ, Miller MP, Bailey DL, Bloomfield PM, Livieratos L, Jones T. The effect of activity outside the direct field of view in a 3D-only whole body positron tomograph. Phys Med Biol 1998;43(4):895–904.
35. Bailey DL, Miller MP, Spinks TJ, Bloomfield PM, Livieratos L, Bánáti RB, et al. Brain PET studies with a high-sensitivity fully 3D tomograph. In: Carson RE, Daube-Witherspoon ME, Herscovitch P, editors. Quantitative functional brain imaging using positron emission tomography. San Diego: Academic Press; 1998. pp. 25–31.

4 Image Reconstruction Algorithms in PET*

Michel Defrise, Paul E Kinahan and Christian J Michel

Introduction

This chapter describes the 2D and 3D image reconstruction algorithms used in PET and the most important evolutions in the last ten years: the introduction of 3D acquisition and reconstruction and the increasing role of iterative algorithms. As will be seen, iterative algorithms improve image quality by allowing more accurate modeling of the data acquisition. This model includes the detection, the photon transport in the tissues, and the statistical distribution of the acquired data, i.e. the noise properties. The popularity of iterative methods dates back to the seminal paper of Shepp and Vardi on the maximum-likelihood (ML) estimation of the tracer distribution. Practical implementation of this algorithm has long been hindered by the size of the collected data, which has increased more rapidly than the speed of computers. Thanks to the introduction of fast iterative algorithms in the nineties, such as the popular Ordered Subset Expectation Maximization (OSEM) algorithm, iterative reconstruction has become practical. Reconstruction time with iterative methods nevertheless remains an issue for very large 3D data sets, especially when multiple data sets are acquired in whole-body or dynamic studies. Speed, however, is not the only reason why filtered-backprojection (FBP) remains important: analytic algorithms are linear and thereby allow an easier control of the spatial resolution and noise correlations in the reconstruction, a control which is mandatory for quantitative data analysis.

The chapter is organized as follows. First, the organization of the data acquired in 2D mode is described, and the reconstruction problem is defined. The third section reviews the classical analytic reconstruction of 2D tomographic data and describes the FBP method, which remains a workhorse of tomography. Iterative reconstruction is presented in the following section, where the accent is set on the key concepts and on their practical implications. Owing to the wide variety of iterative methods, only the popular ML-EM and OSEM methods are described in detail, though this does not entail any claim that these algorithms are optimal. The last sections concern the reconstruction of data acquired in 3D mode. Three-dimensional FBP is described, as well as fast *rebinning* algorithms, which reduce the redundant 3D data set to *synthetic* 2D data that can be processed by analytic or iterative 2D algorithms. *Hybrid algorithms* combining rebinning with a 2D iterative algorithm are introduced, and the chapter concludes with a discussion of the practical aspects of fully 3D iterative reconstruction.

Presented here as a separate chapter, image reconstruction cannot be understood independently of the other steps of the data-processing chain, including data acquisition, data corrections (described in chapters 2, 3, 5), as well as the quantitative or qualitative analysis of the reconstructed images. The variety of algorithms for PET reconstruction arises from the fact that there is no such thing as an *optimal* reconstruction algorithm. Different algorithms may be preferred depending on factors such as the signal-to-noise ratio (number of collected coincident events in the emission and transmission scans), the static or dynamic character of the tracer distribution, the practical constraints on the processing time, and, most importantly, the specific clinical *task* for which the image is reconstructed. It is

* Figures 4.1–4.11 are reproduced from Valk PE, Bailey DL, Townsend DW, Maisey MN. Positron Emission Tomography: Basic Science and Clinical Practice. Springer-Verlag London Ltd 2003, 91–114.

important to keep this observation in mind when discussing reconstruction as an isolated topic.

2D Data Organization

Line of Responses

A PET scanner counts coincident events between pairs of detectors. The straight line connecting the centers of two detectors is called a *line of response* (LOR). Unscattered photon pairs recorded for a specific LOR arise from annihilation events located within a thin volume centered around the LOR. This volume typically has the shape of an elongated parallelipiped and is referred to as a *tube of response*.

To each pair of detectors d_a, d_b is associated an LOR $\mathcal{L}_{d_a,d_b}$ and a sensitivity function $\psi_{d_a,d_b}(\vec{r} = (x, y, z))$ such that the number of coincident events detected is a Poisson variable with a mean value

$$< p_{d_a,d_b} > = \tau \int_{FOV} d\vec{r} f(\vec{r}) \psi_{d_a,d_b}(\vec{r}) \quad (1)$$

where τ is the acquisition time and $f(\vec{r})$ denotes the tracer concentration. We assume that the tracer concentration is stationary and that $f(\vec{r}) = 0$ when $\sqrt{(x^2 + y^2)} > R_F$, where R_F denotes the radius of the field-of-view (FOV). The reconstruction problem consists of recovering $f(\vec{r})$ from the acquired data p_{d_a,d_b}, $\{d_a, d_b\} = 1 \cdots, N_{LOR}$, where N_{LOR}, the number of detector pairs in coincidence, can exceed 10^9 with modern scanners.

The model defined by Eq. (1) is *linear* and hence implies that nonlinear effects due to random coincidences and dead time be pre-corrected. In the absence of photon scattering in the tissues, the sensitivity function vanishes outside the tube of response centered on the LOR. In such a case, the accuracy of the spatial localization of the annihilation events is determined by the size of the tube of response, which in turn depends on the geometrical size of the detectors and on other factors such as the photon scattering in the detectors, or the variable depth of interaction of the gamma rays within the crystal (*parallax error, figure 2.26*).

We have so far considered a scanner comprising multiple small detectors. Scanners based on large-area, position-sensitive detectors such as Anger cameras can be described similarly if viewed as consisting of a large number of very small virtual detectors.

Analytic reconstruction algorithms assume that the data have been pre-corrected for various effects such as randoms, scatter and attenuation. In addition, these algorithms model each tube of response as a mathematical line joining the center of the front face of the two crystals[1]. This means that the sensitivity function $\psi_{d_a,d_b}(\vec{r})$ is zero except when $\vec{r} \in \mathcal{L}_{d_a,d_b}$. With this approximation, the data are modeled as *line integrals* of the tracer distribution:

$$\langle p_{d_a,d_b} \rangle = \int_{\mathcal{L}_{d_a,d_b}} d\vec{r} f(\vec{r}) \quad (2)$$

Sinogram Data and Sampling

The natural parameterization of PET data uses the indices (d_a, d_b) of the two detectors in coincidence, as in Eq. (1). However, there are several reasons to modify this parameterization:

- The natural parameterization is often poorly adapted to analytic algorithms. This is why raw data are usually interpolated into an alternative *sinogram* parameterization described below.
- The number of recorded coincidences N_{events} in a given scan may be too small to take full advantage of the nominal spatial resolution of the scanner. In such a case, undersampling by grouping neighboring LORs reduces the data storage requirements and the reconstruction time without significantly affecting the reconstructed spatial resolution, which is primarily limited by the low count density.

Another approach to reduce data storage and processing time when $N_{LOR} \gg N_{events}$ consists of recording the coordinates (d_a, d_b) of each coincident event in a sequential data stream called a *list-mode* data set. Additional information such as the time or the energy of each detected photon can also be stored. In contrast to undersampling, list-mode acquisition does not compromise the accuracy of the spatial localization of each event. But the fact remains that the number of measured coincidences may be too low to exploit the full resolution of the scanner.

Let us define the standard parameterization of 2D PET data into *sinograms*. Consider a transaxial section $z = z_0$ measured using a ring of detectors. Figure 4.1 defines the variables s and ϕ used to parameterize a straight line (an LOR) with respect to a Cartesian coordinate system (x, y) in the plane. The *radial* variable s is

[1] When the depth of interaction is accounted for, LORs are defined by connecting photon interaction points projected on the long axis of the crystals [1].

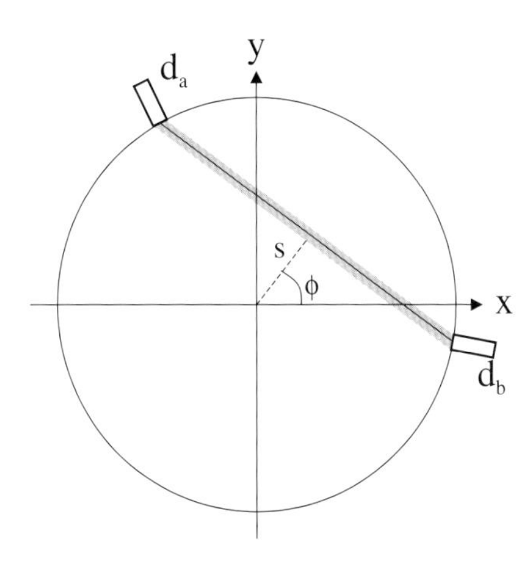

Figure 4.1. Schematic representation of a ring scanner. A tube of response between two detectors d_a and d_b is represented in grey with the corresponding LOR, which connects the center of the front face of the two detectors. The sinogram variables s and ϕ define the location and orientation of the LOR.

the signed distance between the LOR and the center of the coordinate system (usually the center of the detector ring). The *angular variable* ϕ specifies the orientation of the LOR. Line integrals of the tracer distribution are then defined as

$$p(s, \phi, z_0) = \int_{-\infty}^{\infty} dt\, f(x = s\cos\phi - t\sin\phi, \quad y = s\sin\phi + t\cos\phi,\, z = z_0) \tag{3}$$

where t, the integration variable, is the coordinate along the line. In the presentation of the 2D reconstruction problem below, we will omit the z arguments in the functions p and f.

The next section describes how a function $f(x, y)$ can be reconstructed from its line integrals measured for $|s| < R_F$ and $0 \le \phi < \pi$. The mathematical operator mapping a function $f(x, y)$ onto its line integrals $p(s, \phi)$ is called the *x-ray transform*[2], and this operator will be denoted X, so that $p(s, \phi) = (Xf)(s, \phi)$. The function $p(s, \phi)$ is referred to as a *sinogram*, and the variables (s, ϕ) are called sinogram variables. This name was coined in 1975 by the Swedish scientist Paul Edholm because the set of LORs containing a fixed point (x_0, y_0) are located along a sinusoid $s = x_0 \cos\phi + y_0 \sin\phi$ in the (s, ϕ) plane, as can be seen from Eq. (3). For a fixed angle $\phi = \phi_0$, the set of parallel line integrals $p(s, \phi_0)$ is a *1D parallel projection* of f.

At the line integral approximation, and after data pre-correction, the PET data provide estimates of the x-ray transform for all LORs connecting two detectors, i.e., $p_{d_a,d_b} \simeq p(s, \phi)$, where the parameters (s, ϕ) correspond to the radial position and angle of $_{d_a d_b}$. Thus, the geometrical arrangement of discrete detectors in a scanner determines a set of samples (s, ϕ) in sinogram space. The most common arrangement is a ring scanner: an even number N_d of detectors uniformly spaced along a circle of radius $R_d > R_F$[3]. Each detector, in coincidence with an arc of detectors on the opposite side of the ring, defines a *fan* of LORs (figure 3.6), and the corresponding sampling of the sinogram is:

$$\begin{aligned} s_{j,k} &= R_d \cos((2k - j)\pi / N_d) & k &= 0,\ldots,N_d - 1 \\ \phi_j &= j\pi / N_d & j &= 0,\ldots,N_d - 1 \end{aligned} \tag{4}$$

where the pair of indices j, k corresponds to the coincidences between the two detectors with indices $d_a = j - k$ and $d_b = k$. Due to the curvature of the ring, each parallel projection j is sampled non-uniformly in the radial variable, with a sampling distance $\Delta s \simeq 2\pi R_d/N_d$ near the center of the FOV (i.e. for $s \simeq 0$). The radial samples of two adjacent parallel projections j and $j+1$ are shifted by approximately $\Delta s/2$, as can be seen by shifting only one end of a LOR (Fig. 4.2).

For practical and historical reasons, it is customary in PET to reorganize the data on a rectangular sampling grid

$$\begin{aligned} s_k &= k\Delta s & k &= -N_s,\ldots,N_s \\ \phi_j &= j\Delta\phi & j &= 0,\ldots N_\phi - 1 \end{aligned} \tag{5}$$

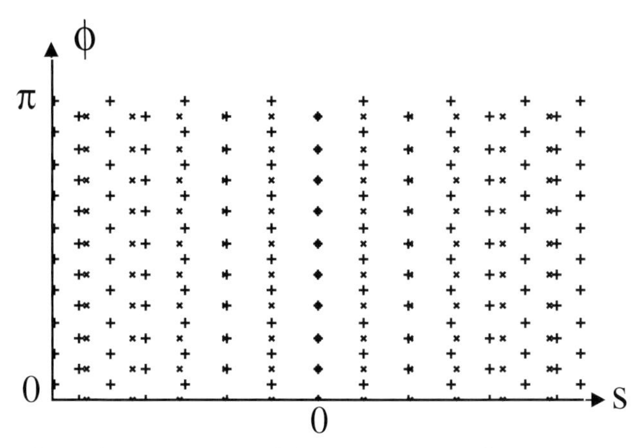

Figure 4.2. Representation of the sinogram sampling for a ring scanner with 20 detectors. The interleaved pattern provided by the LORs connecting detector pairs is shown by +'s. Note the decrease of the radial sampling distance at large values of s, which is exaggerated here because the plot extends to 90% of the ring radius. PET acquisition systems reorganize these data into the rectangular sampling pattern (see equation (5)) shown by ×'s.

---

[2] In 2D, the x-ray transform coincides with the Radon transform, see [2].

[3] If the depth of interaction is not measured, an effective value of R_d is used that accounts for the mean penetration of the 511 keV gamma rays into the crystal.

with $\Delta\phi = 2\pi/N_d$, $N_\phi = N_d/2$, and a uniform radial sampling interval $\Delta s = R_d\pi/N_d$ equal to half the spacing between adjacent detectors in the ring. The *parallel-beam sampling* defined by Eq. (5) will be used in the rest of the chapter. In this scheme the line defined by a sample (j, k) no longer coincides with a measured LOR connecting two detectors. The reorganization into parallel-beam data therefore requires an interpolation (usually linear interpolation) to redistribute the counts on the rectangular sampling grid (Eq. (5)). This interpolation entails a loss of resolution, which is usually negligible owing to the relatively low SNR in PET[4]. In addition, the geometry of some scanners is not circular, but hexagonal or octagonal. Resampling is then needed anyway if standard analytic algorithms are to be used.

When the average number of detected coincidences per sinogram sample is small, undersampling is often applied to reduce the storage and computing requirements. Angular undersampling (increasing $\Delta\phi$) is called *transaxial mashing* in the PET jargon. The mashing factor defined by $m = \Delta\phi N_d/(2\pi)$ is usually an integer so that undersampling simply amounts to summing groups of m consecutive rows (j's) in the sinogram. Angular undersampling results in a loss of resolution, which is smallest at the center of the FOV and maximum at its edge. Therefore, the maximum allowed mashing factor depends not only on the SNR but also on the radius R_F of the reconstructed FOV: for a fixed SNR, we can allow more mashing for a brain scan than for a whole-body study. Radial undersampling (increasing Δs) tends to generate more severe artifacts, and is rarely used. A rule of thumb to match the radial and angular sampling is the relation $\Delta\phi \simeq \Delta s/R_F$, which is derived using Shannon's sampling theory [2].

Multi-slice 2D Data

So far we have discussed data sampling for a single ring scanner located in the plane $z = z_0$. Multi-ring scanners are stacks of N_R rings of detectors spaced axially by Δz and indexed as $r = 0, \cdots, N_R - 1$ [3]. The coincidences between two detectors belonging to the same ring r are organized in a *direct* sinogram $p(s, \phi, z = r\Delta z)$ as described in the previous section. This is the sinogram of the function $f(x, y, z = r\Delta z)$ (Fig. 4.3). Multi-ring scanners also collect coincidences between detectors located in a few adjacent rings, i.e. between one detector in some ring r and another detector in one of the rings $r + d$, with $d = -d_{2D,max}, \cdots, d_{2D,max}$. The LORs connecting such detector pairs are not transaxial, but the maximum *ring difference* $d_{2D,max}$ is chosen to be small enough (typically 5) that the angle between these oblique LORs and the transaxial planes ($\delta\theta \simeq d_{2D,max}\,\Delta z/(2R_d)$) can be neglected[5].

Consider first the LORs between detectors in adjacent rings r and $r + 1$. These data are assembled in a 2D sinogram $p(s, \phi, z = (r + 1/2)\Delta z)$ and used to reconstruct a transaxial slice that is approximated as lying midway, axially, between the two detector rings. Each sample in this *cross-plane* sinogram is the average of two LORs: on the one hand the LOR connecting a detector d_a in ring r to a detector d_b in ring $r + 1$, and on the other hand the LOR connecting detectors d_a in ring $r + 1$ and d_b in ring r. Indeed, these two LORs coincide if we neglect the small angles $\pm\delta\theta$ they form with the transaxial plane. One effect of the introduction of the cross-plane sinograms is to increase the sampling rate in the axial direction so that instead of reconstructing N_r image planes of thickness Δz, we end up with $2N_r - 1$ image planes separated by $\Delta z/2$.

More generally, the LORs between rings $r - j$ and $r + j$, with $j = 0, 1, 2, .. \leq d_{2D,max}/2$ are added to form the direct sinogram of slice $z = r\Delta z$, and the LORs between rings $r - j + 1$ and $r + j$, with $j = 0, 1, 2, .. \leq (d_{2D,max} + 1)/2$ are added to form the cross-plane sinogram of slice $z = (r + 1/2)\Delta z$. There are an odd number of ring pairs contributing to the direct plane sinograms and an even number of ring pairs contributing to the cross-plane sinograms (Fig. 4.3). For small values of $d_{2D,max}$,

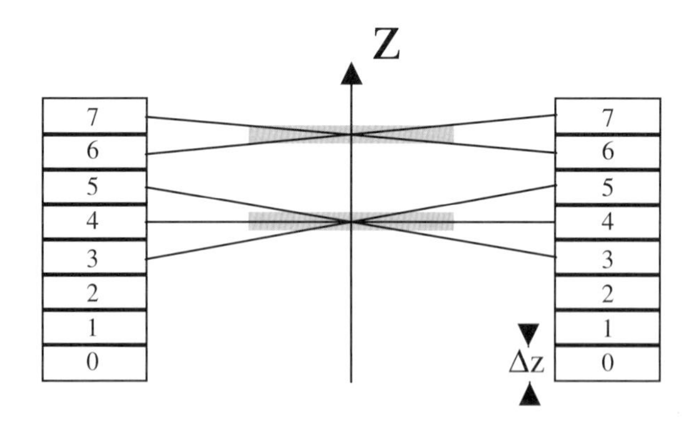

Figure 4.3. Longitudinal view of a multi-ring scanner with $N_r = 8$ rings, operated in 2D mode, illustrating the formation of sinograms for two transaxial slices (in grey), with $d_{2D,max} = 2$. The sinogram for the cross slice at $z = 13\triangle z/2$ (top) is obtained by averaging the coincidences between two rings pairs $(r_a, r_b) = (6, 7)$ and $(7, 6)$. The sinogram for the direct slice at $z = 8\triangle z/2$ (bottom) is obtained by averaging the coincidences between three rings pairs $(r_a, r_b) = (3, 5)$, $(5, 3)$ and $(4, 4)$.

[4] Parallel-beam resampling is used by some CT scanners despite more severe requirements in terms of spatial resolution.
[5] For $d = \pm 1$ this approximation is of the same order as when resampling the sinogram to parallel beam.

there is thus a significant difference in the number of LORs contributing to the different types of sinograms, and therefore also a difference in the corresponding SNRs (see also figure 3.9). As $d_{2D,max}$ increases, the SNR of both types of sinograms increases and the differences diminish; however, there is a degradation in the image resolution as we will see later in the single-slice rebinning algorithm. In practice, the value of $d_{2D,max}$ is chosen to balance these trade-offs, with typical values ranging from 3 to 11.

Analytic 2D Reconstruction

Properties of the X-ray Transform

In this section, we solve the inverse 2D x-ray transform. A closed-form solution of the integral equation, Eq. (3) is first derived assuming a continuous sampling of the sinogram variables over $(s, \phi) \in [-R_F, R_F] \times [0, \pi]$. An approximation to this exact solution will then be written in terms of the discrete data samples (defined by Eq. (5)), leading to the standard *filtered-backprojection* algorithm (FBP). We refer for this section to the comprehensive books by Natterer [2, 4], Kak and Slaney [5], Barrett and Swindell [6], and Barrett and Myers [7].

First, two properties of Eq. (3) should be stressed:

- The problem is *invariant for translations* in the sense that the x-ray transform of a translated image $f_t(x, y) = f(x - t_x, y - t_y)$ is $(Xf_t)(s, \phi) = (Xf)(s - t_x \cos\phi - t_y \sin\phi, \phi)$. Translating the image simply shifts each sinogram row.
- The problem is *invariant for rotations* in the sense that the x-ray transform of a rotated image $f_\theta(x, y) = f(x\cos\theta - y\sin\theta, x\sin\theta + y\cos\theta)$ is $(Xf_\theta)(s, \phi) = (Xf)(s, \phi + \theta)$.

These two invariances, and also the algorithms described in the next sections, are valid only when the scanner measures *all* line integrals crossing the support of the image (the disc of radius R_F), so that the sinogram is sampled over the complete range $(s, \phi) \in [-R_F, R_F] \times [0, \pi]$. When this condition is not satisfied, the problem is called an *incomplete data problem* (among many references, see [2] Ch. VI, [4, 8, 9]). This happens in particular with hexagonal or octagonal scanners such as the Siemens/CPS HHRT, where the gaps between adjacent flat panel detectors cause unmeasured diagonal bands in the sinogram [10]. Before applying the FBP algorithm presented below, the incompletely measured sinograms must first be completed by estimating the missing LOR data. When the gaps in the sinogram are not too wide, simple interpolation can be used, but more sophisticated techniques have been proposed [11, 12]. An alternative is to apply iterative reconstruction techniques, which are less sensitive to the specific geometry. We note, however, that the use of iterative methods does not provide a solution for the missing data problem. Rather it simplifies the introduction of prior knowledge which can partially compensate for the missing data.

The Cornerstone of Tomographic Reconstruction: The Central Section Theorem

Tomographic reconstruction relies on Fourier analysis. Recall that the Fourier transform of a function $f(x, y)$ is defined by

$$\begin{aligned}(\mathcal{F}f)(\nu_x,\nu_y) &= F(\nu_x,\nu_y)\\ &= \int_{\mathbb{R}^2} dx\,dy\, f(x,y)\exp(-2\pi i(x\nu_x + y\nu_y))\end{aligned} \quad (6)$$

and is inverted by changing the sign of the argument of the complex exponential

$$\begin{aligned}(\mathcal{F}^{-1}F)(x,y) &= f(x,y)\\ &= \int_{\mathbb{R}^2} d\nu_x\, d\nu_y\, F(\nu_x,\nu_y)\exp(2\pi i(x\nu_x + y\nu_y))\end{aligned} \quad (7)$$

We use ν_x and ν_y to denote the frequencies associated to x and y respectively, and denote the Fourier transform of a function, e.g., f, by the corresponding upper case character, e.g., F. These definitions are extended in the obvious way to N dimensions.

A key property of the Fourier transform is the *convolution theorem*, which states that the Fourier transform of the convolution of two functions f and h,

$$(f * h)(x,y) = \int_{\mathbb{R}^2} dx'dy' f(x',y')h(x-x',y-y') \quad (8)$$

is the product of their Fourier transforms:

$$(\mathcal{F}(f * h))(\nu_x,\nu_y) = (\mathcal{F}f)(\nu_x,\nu_y)\cdot(\mathcal{F}h)(\nu_x,\nu_y) \quad (9)$$

In signal- or image-processing terms, convolving f with h amounts to *filtering* f with a shift-invariant (i.e. invariant for translations) point spread function h. The convolution theorem simplifies convolution by reducing it to a product in frequency space. In general, the Fourier transform is useful for all problems that are invariant for translation, and therefore also for tomographic reconstruction as will now be shown.

The *central section* theorem, also called the projection slice theorem, states that the 1D Fourier transform of the x-ray transform Xf with respect to the radial variable s is related to the 2D Fourier transform of the image f by

$$P(\nu,\phi) = \mathrm{F}(\nu_x = \nu\cos\phi, \nu_y = \nu\sin\phi) \tag{10}$$

where

$$P(\nu,\phi) = (\mathcal{F}p)(\nu,\phi) = \int_{\mathbb{R}} ds\, p(s,\phi)\exp(-2\pi i s\nu) \tag{11}$$

and ν is the frequency associated to the radial variable s.

This theorem is easily proven by replacing the x-ray transform $p(s, \phi) = (Xf)(s, \phi)$ in the right hand side of Eq. (11) by its definition (Eq. (3)) as a line integral of f. Thus the 1D Fourier transform of a parallel projection of an image f at an angle ϕ determines the 2D Fourier transform of that image along the radial line in frequency plane (ν_x, ν_y) that forms an angle ϕ with the ν_x axis. The implication for reconstruction is the following: if we measure all projections $\phi \in [0, \pi]$, the radial line sweeps over the whole frequency plane and thereby allows the recovery of $F(\nu_x, \nu_y)$ for all frequencies $(\nu_x, \nu_y) \in \mathbb{R}^2$. The image f can then be reconstructed by inverse 2D Fourier transform (Eq. (7)).

The discrete implementation of the inversion formula combining Eqs. (11), (10) and (7) is referred to as the *direct Fourier reconstruction*. This algorithm is numerically efficient because the discretized 2D Fourier transform (Eq. (7)) can be calculated with the FFT algorithm. The 2D FFT requires as input the values of F on a square grid $(\nu_x = k\Delta, \nu_y = l\Delta)$, $(k, l) \in Z^2$, which does not coincide with the polar grid of samples provided by the data (see the right hand side of Eq. 10). Direct Fourier reconstruction therefore involves a 2D interpolation to map the polar grid onto the square grid. This interpolation is often based on gridding techniques similar to those used for magnetic resonance imaging [13, 14, 15].

The Filtered Backprojection Algorithm

The FBP algorithm is the standard algorithm of tomography. It is equivalent to the direct Fourier reconstruction in the limit of continuous sampling, but its discrete implementation differs.

The FBP inversion explicitly combines Eqs. (11), (10) and (7). Straight-forward manipulations involving changing from Cartesian (ν_x, ν_y) to polar (ν, ϕ) coordinates lead to a two-step inversion formula (Fig. 4.4):

$$f(x,y) = (X^* p^F)(x,y) = \int_0^{\pi} d\phi\, p^F(s = x\cos\phi + y\sin\phi, \phi) \tag{12}$$

where the *filtered projections* are

$$p^F(s,\phi) = \int_{-R_F}^{R_F} ds' p(s',\phi) h(s-s') \tag{13}$$

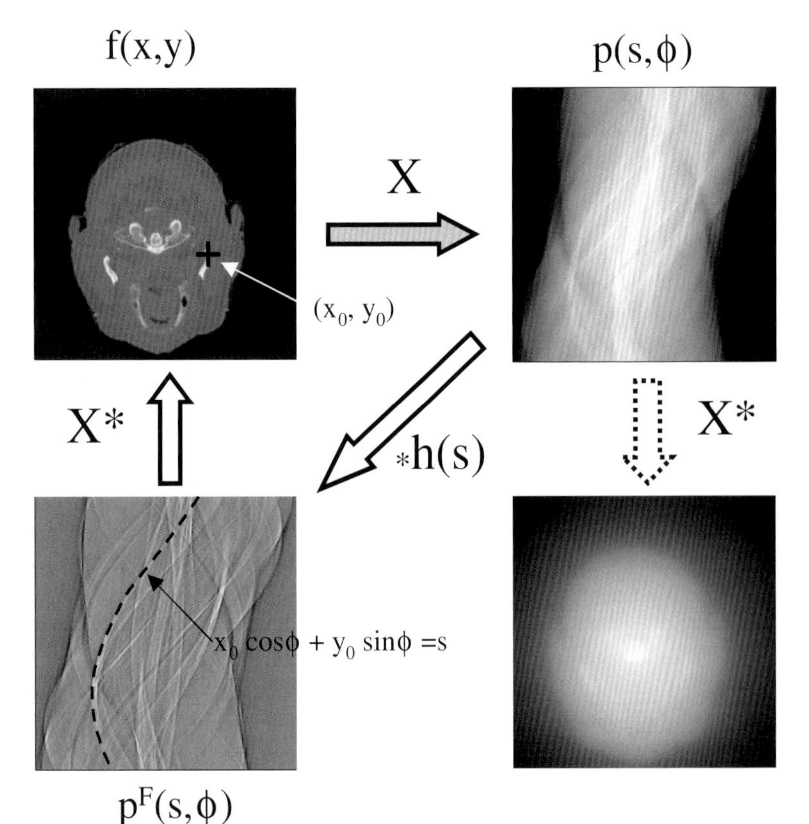

Figure 4.4. Illustration of 2D filtered backprojection. The top row shows a brain section and its sinogram $p = Xf$. The backprojection X^*p of the sinogram (bottom right) is the 2D convolution of f with the point spread function $1/\sqrt{(x^2 + y^2)}$ and illustrates the blurring effect of line integration. The filtered sinogram p^F obtained by 1D convolution with the ramp filter kernel has enhanced high frequencies, and when backprojected, yields the original image f, up to noise and discretization errors.

and the *ramp filter* kernel is defined as

$$h(s) = \int_{-\infty}^{\infty} d\nu \, |\nu| \, \exp(2\pi i s \nu) \qquad (14)$$

Three remarks are in order.

(i) The operator X^* mapping p^F onto f in Eq. (12) is called the *backprojection* and is the dual of the x-ray transform. Geometrically, $(X^* p^F)(x, y)$ is the sum of the filtered data p^F for all lines that contain the point (x, y).
(ii) The convolution (Eq. (13)) can be expressed using the convolution theorem as $P^F(\nu, \phi) = |\nu| \, P(\nu, \phi)$.
(iii) The integral (Eq. (14)) defines the kernel h as the inverse 1D Fourier transform of the ramp filter function $|\nu|$. This integral does not converge in the usual sense, and h is only defined as a generalized function (see chapter 2 in [7]).

Discrete Implementation of the FBP

The discrete implementation of Eqs. (12) and (13) using the measured samples of $p(s, \phi)$ described in the section on sinogram data and sampling, above (Eq. (5)), involves four approximations:

(i) The approximation of the kernel $h(s)$ by an *apodized* kernel

$$h_w(s) = \int_{-\infty}^{\infty} d\nu \, |\nu| \, w(\nu) \exp(2\pi i s \nu) \qquad (15)$$

where $w(\nu)$ is a low-pass filter which suppresses the high spatial frequencies, and will be discussed later in the section on the ill-posedness of the inverse X-ray transform.
(ii) The approximation of the convolution integral by a discrete quadrature. Usually standard trapezoidal quadrature is used:

$$p^F(k\Delta s, \phi_j) \simeq \Delta s \sum_{k'=-N_s}^{N_s} p(k'\Delta s, \phi_j) h_w((k-k')\Delta s) \qquad (16)$$
$$k = -N_s, \ldots, N_s$$

The calculation of this discrete convolution can be accelerated using the discrete Fourier transform (FFT) (see [16] section 13.1). In this case, some care is needed when defining the discrete filter: to avoid bias, this filter must be calculated as the FFT of the sampled convolution kernel $h_w(k\Delta s)$, $k = 0, \pm 1, \pm 2, \ldots$, and not by simply sampling the continuous filter function $|\nu| w(\nu)$.
(iii) The approximation of the backprojection by a discrete quadrature

$$f(x, y) \simeq \Delta\phi \sum_{j=0}^{N_\phi - 1} p^F(s = x\cos\phi_j + y\sin\phi_j, \phi_j) \qquad (17)$$

for a set of image points (x, y) (usually a square pixel grid)[6].
(iv) The estimation of p^F $(s = x\cos\phi_j + y\sin\phi_j, \phi_j)$ in Eq. (17) from the available samples p^F $(k\Delta s, \phi_j)$. This is usually done using linear interpolation:

$$p^F(s, \phi_j) \simeq \left(k + 1 - \frac{s}{\Delta s}\right) p^F(k\Delta s, \phi_j) + \left(\frac{s}{\Delta s} - k\right) p^F((k+1)\Delta s, \phi_j) \qquad (18)$$

where k is the integer index such that $k\Delta s \le s < (k + 1)\Delta s$. Instead of linear interpolation some implementations apply a faster nearest-neighbor interpolation to filtered projections which have first been linearly interpolated on a finer grid (typically sampled at a rate $\Delta s/4$).

Remarkably, most FBP implementations only use simple tools of numerical analysis, such as linear interpolation and trapezoidal quadrature, despite many attempts to demonstrate the benefits of more sophisticated techniques.

The Ill-posedness of the Inverse X-ray Transform

Like many problems in applied physics, the inversion of the x-ray transform is an *ill-posed problem*: the solution f defined by Eqs. (11), (10) and (7) does not depend continuously on the data $p(s, \phi)$. Concretely, this means that an arbitrarily small perturbation of p due to measurement noise can cause an arbitrarily large error on the reconstructed image f. We refer to Bertero and Boccacci [20] and Barrett and Myers [7] for an introduction to the concept of ill-posedness and its implication in tomography. Intuitively, ill-posedness can be understood by noting that the ramp filter $|\nu|$ amplifies the high frequencies during the filtering step $P(\nu, \phi) \rightarrow P^F(\nu, \phi) = |\nu| P(\nu, \phi)$. The power spectrum of a typical image decreases rapidly with increasing frequencies, whereas the noise power spectrum decreases in general slowly[7]. Consider a hypothetical perturbation of the data $p(s, \phi) \rightarrow p(s, \phi) + \cos(2\pi\nu_0 s)/\sqrt{\nu_0}$ for some $\nu_0 > 0$. This perturbation becomes arbitrarily small when ν_0 tends to ∞, but the corresponding

[6] Alternative and faster implementations of the backprojection have been proposed [17, 18, 19].
[7] In the so-called *white noise* limit, the noise power spectrum is constant.

perturbation of the filtered projection is easily seen to be $p^F(s, \phi) \to p^F(s, \phi) + \sqrt{\nu_0}\cos(2\pi\nu_0 s)$ and is arbitrarily large for large ν_0. This artificial example illustrates the fact that the ill-posedness of the inverse x-ray transform (and of most inverse problems) arises from high-frequency perturbations.

This discussion suggests that the reconstruction can be stabilized by filtering out the high frequencies. This is achieved by introducing a low-pass apodizing window $w(\nu)$ as in Eq. (15). A window frequently used in tomography is the Hamming window

$$\begin{aligned} w_{ham}(\nu) &= (1+\cos(\pi\nu/\nu_c))/2 \quad & |\nu| < \nu_c \\ &= 0 & |\nu| \geq \nu_c \end{aligned} \tag{19}$$

where ν_c is some cut-off frequency. The rectangular window

$$\begin{aligned} w_{rec}(\nu) &= 1 \quad & |\nu| < \nu_c \\ &= 0 & |\nu| \geq \nu_c \end{aligned} \tag{20}$$

results in a better spatial resolution, but introduces ringing artifacts near sharp boundaries. Figure 4.5 illustrates the apodized window and the convolution kernel $h_w(s)$. The choice of the cut-off frequency must take two factors into account:

- Given the radial sampling distance Δs in the sinogram, Shannon's sampling theory states that the maximum frequency that can be recovered without aliasing is $1/2\Delta s$. The cut-off frequency is therefore constrained by $\nu_c \leq 1/2\Delta s$.
- As we have seen, stabilization requires suppressing high frequencies. Therefore, lower values of ν_c are selected when the signal-to-noise ratio (i.e. the number of detected coincidences) is low.

The stability of the discrete FBP can be analyzed assuming a Poisson distribution for the measurement noise. Consider the reconstruction of a disc of radius R containing a uniform tracer distribution, from 2D PET data comprising N_{events} coincident events. Neglecting attenuation, scatter and random, the relative variance of the reconstructed image at the center of the disc can be shown [21] to be

$$\text{variance } f(x=0, y=0) \simeq \frac{\pi^3}{6}\frac{(R/\Delta s)^3}{N_{events}} \tag{21}$$

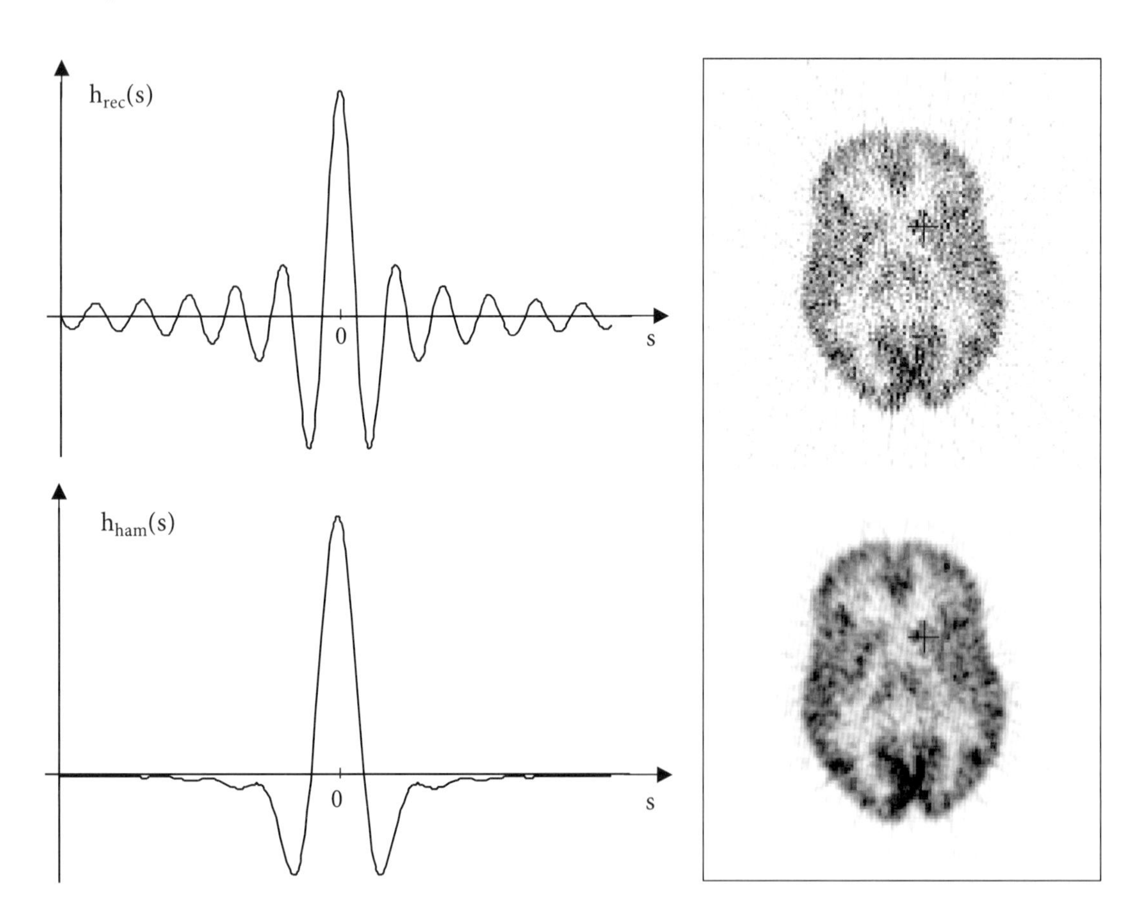

Figure 4.5. The convolution kernels corresponding to the rectangular window in equation (20) (top), and to the Hamming window (19) (bottom) are shown with arbitrary vertical scales. The smaller width of the central lobe of $h_{rec}(s)$ results in higher spatial resolution in the reconstruction, while the larger side lobes, compared to $h_{ham}(s)$ indicate a higher sensitivity to noise. A transaxial slice of an FDG brain scan reconstructed using FBP with these two windows is shown on the right.

where Δs is the radial sampling and a rectangular window with $\nu_c = 1/2\Delta s$ has been used. This result means that the number of detected events, hence the scanner sensitivity, should be multiplied by a factor of 8 when the spatial resolution Δs is halved. This is to be compared with the factor 4 increase that suffices in the absence of tomographic reconstruction, e.g. if perfect time-of-flight information is available, or if f is obtained from planar scintigraphy as in single photon imaging. The supplementary factor of 2 reflects the ill-posed character of the inverse x-ray transform. For a multi-slice 2D reconstruction, an additional factor of 2 must be included if the axial resolution is also halved, leading to a 16 fold increase of the number of counts when the isotropic resolution is halved. When an improvement in detector resolution is not matched by an increase in sensitivity, a cut-off frequency ν_c smaller than the Nyquist frequency $1/2\Delta s$ must be used to limit noise. In such a case, the improvement in detector resolution is not fully translated in the reconstructed image resolution. The improvement nevertheless remains beneficial because the modulation transfer function is enlarged at the lower frequencies $|\nu| \le \nu_c$, allowing better recovery coefficients for small structures.

Iterative Reconstruction

This section introduces the major concepts of the iterative reconstruction algorithms, which play an increasingly important role in clinical PET. These algorithms rely on a discrete representation of both the data and the reconstructed image, in contrast with the analytic algorithms, which are derived assuming a continuous data sampling and introduce the discrete character of the data a posteriori. We begin this section with a general discussion of the ingredients of an iterative algorithm: the data model, the image model, the objective function, and the optimization algorithm. We refer to [22, 23] for more details. The various possible choices for each of these ingredients explains the wide variety of iterative algorithms in the literature. One specific algorithm will be described in detail in the section on ML-EM and OSEM (below).

One of the strengths of iterative algorithms is that they are largely independent of the acquisition geometry. Therefore, the concepts presented below apply equally to 2D and to 3D PET data.

The General Ingredients of Iterative Reconstruction Algorithms: Data Model

The data are represented using Eq. (1). To simplify notations, a single index j is used to denote the detector pair (d_a, d_b), and the mean number of events detected for one LOR is then rewritten as

$$\langle \vec{p} \rangle = \{\langle p_j \rangle = \tau \int_{FOV} d\vec{r} f(\vec{r}) \Psi_j(\vec{r}), j = 1,\ldots, N_{LOR}\} \quad (22)$$

Any linear physical effect can be modeled in the sensitivity function Ψ_j : attenuation and scatter (assuming a known density map), gaps in the detectors, non-uniform resolution of the detectors, etc. The accuracy of the physical model ultimately determines the accuracy of the reconstruction. Nevertheless, approximate models are often used to limit the computational burden, and these approximations are justified for low-count studies where image quality is primarily limited by noise. Many approaches can be found, ranging from a simple line integral model (as for FBP) up to a highly accurate model required for high SNR studies with small-animal scanners. A clever exploitation of the symmetries of the scanner and the use of lookup tables, as described in Qi et al. [24], allows the computational costs of such a complex modeling to be kept to a reasonable level.

Eq. (22) represents the *mean* value of the data. The *statistical distribution* of each LOR data p_j around its mean value $< p_j >$ must also be modeled. An inaccurate statistical model results not only in a sub-optimal variance, but also in a bias. Usually, the "raw data" p_j are counted numbers of detected photon pairs and are distributed as independent Poisson variables. The *likelihood function* then has the form

$$\Pr\{\vec{p} \mid f\} = \prod_{j=1}^{N_{LOR}} \exp(-\langle p_j \rangle) \langle p_j \rangle^{p_j} / p_j! \quad (23)$$

Due to the various forms of data pre-processing, the actual distribution of the data presented to the algorithm often deviates from the Poisson model. If the number of counts per bin is high enough, the distribution is approximately Gaussian

$$\Pr\{\vec{p} \mid f\} = \prod_{j=1}^{N_{LOR}} \frac{1}{\sqrt{2\pi}\,\sigma_j} \exp\left(-\frac{(p_j - \langle p_j \rangle)^2}{2\sigma_j^2}\right) \quad (24)$$

and the variance σ_j^2 of each LOR can be estimated knowing the data pre-processing steps. A more general Gaussian model with a non-diagonal covariance matrix may be needed if the pre-processing introduces correlations between LORs.

Another variant has been proposed to model data pre-corrected for random coincidences[8]. When this is done by subtracting delayed coincidences, the subtracted data (the "prompts" minus the "delayed") are no longer Poisson variables. An approximate model, the *shifted Poisson model*, for the distribution of such pre-corrected data has been proposed in [25, 26].

We conclude this section on data modeling with a few words on the reconstruction of transmission data acquired with monoenergetic photons of energy E. Typically, E = 511 keV if a positron source such as $^{68}Ge/^{68}Ga$ is used or E = 662 keV for a ^{137}Cs single photon source. These data are also distributed as Poisson variables but with mean values

$$\langle \vec{p} \rangle = \{\langle p_j \rangle = p_j^0 \exp(-\int_{FOV} d\vec{r}\mu(\vec{r},E)\Psi_j^{tr}(\vec{r})) \quad j = 1,\ldots,N_{LOR}\} \tag{25}$$

instead of Eq. (22). Here, p_j^0 is the mean number of coincident events in the reference (blank) scan, Ψ_j^{tr} is the sensitivity function for the transmission data, and $\mu(\vec{r}, E)$ is the attenuation coefficient to be reconstructed. The difference between this model and Eq. (22) shows that specific iterative algorithms are needed for transmission data [27, 28, 29, 30]. An alternative is to apply algorithms developed for emission data to the logarithm $\log(p_j^0 / p_j)$, but this approach introduces significant biases because the logarithm of the data is not a Poisson variable any more. Examples of these biases are given in [31].

The Image Model: Basis Functions and Prior Distribution

Iterative algorithms model the image as a linear combination of basis functions

$$f(x,y) \simeq \sum_{i=1}^{P} f_i b_i(x,y) \tag{26}$$

Most algorithms use contiguous and non-overlapping pixel basis functions, which partition the field of view:

$$\begin{aligned} b_i(x,y) &= 1 \quad |x-x_i| < \Delta x/2 \text{ and } \quad |y-y_i| < \Delta x/2 \\ &= 0 \quad |x-x_i| \geq \Delta x/2 \text{ or } \quad |y-y_i| \geq \Delta x/2 \end{aligned} \tag{27}$$

with $i = (i_x, i_y)$ and the center of the ith pixel is $(x_i = i_x\Delta x, y_i = i_y\Delta x)$. The pixel size is $\Delta x = \Delta s/Z$, where Z is the *zoom factor*.

The pixel basis function is not band-limited: its Fourier transform $(\mathcal{F}b_i)(\nu_x,\nu_y)$ decreases slowly at large frequencies due to the discontinuity at the boundary of the pixel. This property is at odds with the fact that the frequencies larger than $\nu_c = 1/2\Delta s$ cannot be recovered from sampled data (see The Ill-posedness of the Inverse X-ray Transform, above). An alternative proposed by Lewitt [32] consists of using smooth basis functions which are essentially band-limited. Significant improvements in image quality have been demonstrated using truncated Kaiser–Bessel functions, dubbed *blobs* [33]. These radially symmetrical basis functions have a compact support, but they do overlap, which increases the processing time unless the spacing and size of the basis functions are carefully chosen. At the time of writing, most iterative algorithms are still based on discontinuous basis functions, but at least one clinical scanner implements blobs.

In principle, the choice of the basis functions determines the image model and reduces image reconstruction to the estimation of a vector $\{f_i, i = 1, \cdots, P\}$, usually with the constraint $f_i \geq 0$. The constraint implicit in this discrete representation[9] helps to stabilize the reconstruction, but may be insufficient. In such a case, a small perturbation of the data vector $\vec{p}$ still causes an unacceptably large perturbation of the reconstruction $\vec{f}$. The set of admissible images must then be further restricted. Several techniques can be used for this purpose, we focus here on the popular *Bayesian scheme* (see, for example, [34, 35, 7]).

In the Bayesian scheme, regularization is achieved by considering the image as a random vector with a prescribed probability distribution $\Pr(\vec{f})$. This distribution is called the *prior distribution* (or simply "the prior"). Typically, the prior enforces *smoothness* by assigning a low probability to images having large differences $|f_{i_x,i_y} - f_{i_x \pm 1, i_y \pm 1}|$ between neighboring pixels. One says that large differences between neighboring pixels are *penalized* by the prior. In practice, priors are defined empirically because the clinically relevant prior information is usually too complex to be expressed mathematically. We will see in the section on the cost function (below) how the prior is incorporated in the reconstruction.

Priors based on a Gaussian distribution with a uniform (i.e. shift-invariant) covariance are in essence equivalent to the linear smoothness constraint introduced in the FBP algorithm by low pass windows $w(\nu)$ discussed earlier. More sophisticated priors can

[8] In contrast with the randoms, the contribution of scattered coincidences is linearly related to true coincidences and hence can in principle be included in the model $\Psi_j(\vec{r})$.However, the scatter background is more often subtracted from the data prior to reconstruction for the sake of numerical efficiency. See chapter 5.

[9] Mathematically we constrain $f(x, y)$ to belong to the P-th dimensional space of functions spanned by the b_i.

improve image quality, especially sharpness, in specific situations and for specific tasks, but may introduce subtle nonlinear biases and noise correlations.

An attractive class of prior distributions exploits a registered anatomic, MR or CT, image of the patient [36, 37, 38, 39, 40]. This image defines likely boundaries between regions in which uniform tracer concentration is expected. These boundaries can be incorporated in a prior that enforces smoothness only between pixels belonging to the same anatomical region. Despite promising results, that approach still needs further validation and comparison with the alternative approach in which the MR or CT prior information is exploited visually using, for example, image fusion techniques.

Let us finally stress that the 2D or 3D nature of the image model is independent of the fact that the data are acquired in 2D or 3D mode. Indeed, true 3D image models based, for example, on 3D blobs and on 3D smoothness constraints are useful even when the data are collected independently for each slice (or rebinned, see section on 3D analytic reconstruction by rebinning (below)) [41]. For dynamic or gated PET studies, mixed basis functions depending on both the time and the spatial coordinates can be defined to model the expected behavior of the tracer kinetics [42, 43].

The System Matrix

We can now summarize the assumptions in the two previous sections. Putting the image model (Eq. (26)) into the data model (Eq. (22)) reduces the problem to a *set of linear equations*:

$$\langle p_j \rangle = \sum_{i=1}^{P} a_{j,i} f_i \qquad j = 1,\ldots,N_{LOR} \tag{28}$$

where the elements of the *system matrix* are

$$a_{j,i} = \tau \int_{FOV} d\vec{r}\, b_i(\vec{r}) \Psi_j(\vec{r}) \qquad j = 1,\ldots,N_{LOR}; i = 1,\ldots,P \tag{29}$$

A line integral model including only attenuation correction and normalization generates a sparse system matrix a with elements simple enough to be calculated on the fly. More accurate models that include scatter lead to densely populated matrices, which are complex to calculate. A practical algorithm then requires a compromise between accuracy, required storage, and speed. A useful approach is to factor a as a product of matrices, each of which models a specific aspect of the data acquisition [24].

A direct inversion of the linear system (Eq. (28)) with the $< p_j >$ replaced by the measured data p_j is impractical for two reasons:

- The discrete system is *ill-conditioned*: the condition number of a is large[10]. Consequently, the solution[11] of Eq. (28) is unstable for small perturbations $p_j - < p_j >$ of the data. Ill-conditioning is the discrete equivalent of the ill-posedness of the inverse x-ray transform discussed in the section on the ill-posedness of the inverse X-ray transform (above).
- Numerically, the inversion of matrix a is hindered by its very large size (typically $P = 10^6$ unknowns and $N_{\mathrm{LOR}} \simeq 10^6$ up to $N_{\mathrm{LOR}} \simeq 10^9$ in 3D PET).

The first problem is solved by incorporating prior knowledge in a cost function. The second, numerical problem, is solved by optimizing the cost function by successive approximations.

The Cost Function

The key ingredient of an iterative algorithm is a cost function $Q(\vec{f} = (f_1, \cdots, f_p), \vec{p})$, which depends on the unknown image coefficients and on the measured data. $Q(\vec{f}, \vec{p})$ is also called the objective function. The reconstructed image estimate f^* is defined as one that maximizes Q:

$$\vec{f}^* = \arg\max_{\vec{f}} Q(\vec{f}, \vec{p}) \tag{30}$$

with usually the constraint $f_j \geq 0$. The role of the cost function is to enforce (i) a good fit with the data, i.e., Eq. (28) should be approximately satisfied, (ii) the prior conditions on the image model.

In the Bayesian framework, the cost function is the *posterior probability distribution*

$$\Pr\{\vec{f} \mid \vec{p}\} = \frac{\Pr\{\vec{p} \mid \vec{f}\}\Pr\{\vec{f}\}}{\Pr\{\vec{p}\}} \tag{31}$$

The first factor in the numerator of the right hand side is the data likelihood (given, for example, by the Poisson model (Eq. (23)), and the second factor is the prior probability discussed above in the section on the image model. The denominator is independent of $\vec{f}$ and can be dropped. Maximizing the posterior proba-

[10] The condition number of a matrix is the ratio between its largest and smallest singular values.
[11] Or the generalized Moore–Penrose solution if a is singular or $\vec{p} \notin$ range a, see [20].

bility is equivalent to maximizing its logarithm, and the cost function becomes

$$Q(\vec{f} \mid \vec{p}) = \log \Pr\{\vec{p} \mid \vec{f}\} + \log \Pr\{\vec{f}\} \tag{32}$$

The first term penalizes images which do not well fit the data, whereas the second one stabilizes the inversion by penalizing images which are deemed a priori "unlikely". An image maximizing $Q(\vec{f}, \vec{p})$ is called a *maximum a posteriori* (MAP) estimator. When the log-likelihood is Gaussian (Eq. (24)), the first term in Eq. (32) is a quadratic function and the algorithm maximizing $Q(\vec{f}, \vec{p})$ is called a *penalized weighted least-square* method [44,45].

Ideally, the maximum of Q should be unique. Uniqueness is guaranteed when the cost function is convex, i.e., when the Hessian matrix

$$H_{i,j} = \frac{\partial^2 Q(\vec{f},\vec{p})}{\partial f_i \partial f_j} \qquad i,j = 1,\ldots,P \tag{33}$$

is negative definite for all feasible $\vec{f}$. Non-convex cost functions may still have an unique global maximum, but they can also have local maxima, which complicate the optimization.

Optimization Algorithms

The cost function (assuming it has an unique global maximum) defines the looked-for estimate $\vec{f}^*$ of the tracer distribution. To actually calculate $\vec{f}^*$, an optimization algorithm is needed. Such an algorithm is a prescription to produce a sequence of image estimates $\vec{f}^n$, $n = 0, 1, 2, \cdots$, which should converge asymptotically to the solution:

$$\lim_{n \to \infty} \vec{f}^n = \vec{f}^* \tag{34}$$

Asymptotic convergence is not the only requirement: the optimization algorithm should be stable, efficient numerically, and ensure fast convergence independently of the choice of the starting image $\vec{f}^0$. A further property is that of monotonic convergence, which guarantees that $Q(\vec{f}^{n+1}, \vec{p}) \geq Q(\vec{f}^n, \vec{p})$ at each iteration. Though not strictly needed, monotonic convergence is useful in practice and is often the key property used to prove asymptotic convergence.

In principle, the choice of the optimization algorithm should not influence the solution, which is defined by Eq. (30). In practice, however, the image that will be used is produced by a necessarily finite number of iterations and thereby does depend on the algorithm.

When the cost function is differentiable and a non-negative solution is required, the solution $\vec{f}^*$ must satisfy the *Karush–Kuhn–Tucker conditions*:

$$\begin{aligned}(\nabla Q(\vec{f}^*, \vec{p}))_j &= 0 \qquad f_j^* > 0, j = 1,\ldots,P \\ &\leq 0 \qquad f_j^* = 0\end{aligned} \tag{35}$$

where the gradient of the cost function is the vector with components

$$\nabla Q(\vec{f}^*, \vec{p}))_j = \frac{\partial Q(\vec{f},\vec{p})}{\partial f_j}\Big|_{\vec{f}=\vec{f}^*} \tag{36}$$

When positivity is not enforced, the Karush–Kuhn–Tucker condition reduces to the first line of Eq. (35). If in addition the cost function is quadratic (e.g., with a Gaussian log-likelihood), optimization reduces to a set of P linear equations in P unknowns. With a Gaussian likelihood without prior, these equations are the so-called *normal equations* corresponding to Eq. 28 [7,20].

There is a considerable literature on optimization, and even within the field of tomography a wide variety of methods have been proposed. A detailed overview (see [7, 16, 22, 46]) is beyond the scope of this chapter, but it may be useful to briefly list a few basic tools that can be used to develop iterative methods. The major difficulty is that the system of equations (Eq. (35)) is large, strongly coupled, and often non-linear. Many algorithms are based on the replacement at each iteration of the original optimization problem (Eq. (30)) by an alternative problem which is easier to solve because

- it has a much smaller dimensionality, and/or
- the modified cost function is quadratic in its unknowns, or even better separable in the sense that its gradient is a sum of functions each depending on a single unknown parameter f_j.

Standard examples include:

(i) *Gradient-based methods*. The prototype is the steepest-ascent method, which reduces the problem to a one-dimensional optimization along the direction defined by the gradient. The nth iteration is defined by:

$$\begin{aligned}\vec{f}^{n+1} &= \vec{f}^n + \alpha_n \nabla Q(\vec{f}^n, \vec{p}) \\ \alpha_n &= \arg\max_{\alpha} Q(\vec{f}^n + \alpha \nabla Q(\vec{f}^n, \vec{p}), \vec{p})\end{aligned} \tag{37}$$

The step length α_n maximizes the cost function along the gradient direction, taking into account possible constraints such as positivity.

(ii) *Methods using subsets of the image vector.* Only a subset of the components of the unknown image vector (i.e., a subset of voxels) is allowed to vary at each iteration, while the value of the other components is kept constant. A different subset of voxels is allowed to vary at each iteration. In the *coordinate ascent algorithm*, a single voxel is varied at each iteration, according to:

$$\begin{aligned} f_j^{n+1} &= f_j^n & j \neq J(n) \\ &= \arg\max_{f_j} Q((f_1^n,\ldots,f_{j-1}^n, f_j, f_{j+1}^n,\ldots,f_P^n),\vec{p}) & j = J(n) \end{aligned} \quad (38)$$

where $J(n)$ defines the order in which voxels are accessed in successive iterations, e.g., $J(n) = n \bmod P$.

(iii) *Methods based on surrogate cost functions.* The original cost function $Q(\vec{f}, \vec{p})$ is replaced at each step by a modified objective function $\tilde{Q}(\vec{f}, \vec{f}^n, \vec{p})$ that satisfies the following conditions [47]:

- $\tilde{Q}(\vec{f}, \vec{f}^n, \vec{p})$ can easily be maximized with respect to $\vec{f}$, e.g., it is quadratic or separable,
- $\tilde{Q}(\vec{f},\vec{f},\vec{p}) = Q(\vec{f},\vec{p})$
- $\tilde{Q}(\vec{f},\vec{f}^n,\vec{p}) \leq Q(\vec{f},\vec{p})$

The two last conditions ensure that the next image estimate

$$\vec{f}^{n+1} = \arg\max_{\vec{f}} \tilde{Q}(\vec{f},\vec{f}^n,\vec{p}) \quad (39)$$

monotonically increases the value of the cost function: $Q(\vec{f}^{n+1}, \vec{p}) \geq Q(\vec{f}^n, \vec{p})$. The ML-EM algorithm (next section), the least-square ISRA algorithm [48, 49], and Bayesian variants [50] can be derived using surrogate functions.

(iv) *Block-iterative methods* use at each iteration only a subset of the data. They are called *row-action* methods when a single datum is used at each iteration as in the ART algorithm. The OSEM method (see next section) and its variants are also block-iterative methods. While allowing significant acceleration of the optimization, these methods do not guarantee a monotonic increase of the cost function. In addition, the iterated image estimates tend asymptotically to cycle between S slightly different solutions, where S is the number of subsets. Appropriate under-relaxation can be used to alleviate the problem.

ML-EM and OSEM

The most widely used iterative algorithms in PET are the ML-EM (maximum-likelihood expectation maximization) algorithm and its accelerated version OSEM (Ordered Subset EM). The ML-EM method was introduced by Dempster et al in 1977 [51] and first applied to PET by Shepp and Vardi [52] and Lange and Carsson [27]. The algorithm is akin to the Richardson–Lucy algorithm developed for image restoration in astronomy (see, for example, [20]). The OSEM variation of the ML-EM algorithm, proposed in 1994 by Hudson and Larkin was the first iterative algorithm sufficiently fast for clinical applications.

The cost function in the ML-EM and OSEM algorithms is the Poisson likelihood (Eq. (23)). Putting Eq. (28) into Eq. (23), taking the logarithm, and dropping the terms that do not depend on the unknowns f_i, we get

$$Q(\vec{f},\vec{p}) = \sum_{j=1}^{N_{LOR}} \{-\sum_{i=1}^{P} a_{j,i} f_i + p_j \log(\sum_{i=1}^{P} a_{j,i} f_i)\} \quad (40)$$

If the matrix a is non-singular, this cost function is convex and defines a unique image.

The EM iteration is a mapping of the current image estimate $\vec{f}^n$ onto the next estimate $\vec{f}^{n+1}$:

$$\vec{f}_i^{n+1} = f_i^n \frac{1}{\sum_{j'=1}^{N_{LOR}} a_{j',i}} \sum_{j=1}^{N_{LOR}} a_{j,i} \frac{p_j}{\sum_{i'=1}^{P} a_{j,i'} f_{i'}^n} \qquad i = 1,\ldots,P \quad (41)$$

Usually, the first estimate is a uniform distribution $f_i^1 = 1$, $i = 1, \ldots, P$. The sum over i' in the denominator of the second factor in the right hand side is a forward projection and corresponds to Eq. (28): therefore the denominator is the average value $<p_j^n>$ that would be measured if $\vec{f}^n$ was the true image. The sum over j in the numerator is a multiplication with the transposed system matrix and represents the backprojection of the ratio between the measured and estimated data. Finally, the denominator in the first factor is equal to the sensitivity of the scanner for pixel i.

The ML-EM iteration has several remarkable properties:

- The cost function increases monotonically at each iteration, $Q(\vec{f}^{n+1},\vec{p}) \geq Q(\vec{f}^n,\vec{p})$,
- The iterates $\vec{f}^n$ converge for $n \to \infty$ to an image $\vec{f}^*$ that maximizes the loglikelihood,
- All image estimates are non-negative if the first one is,
- The algorithm can easily be implemented with list-mode data [53, 54, 55, 56] because the only LORs that contribute to the backprojection sum over j in Eq. 41 are those for which at least one event has been detected ($p_j \geq 1$).

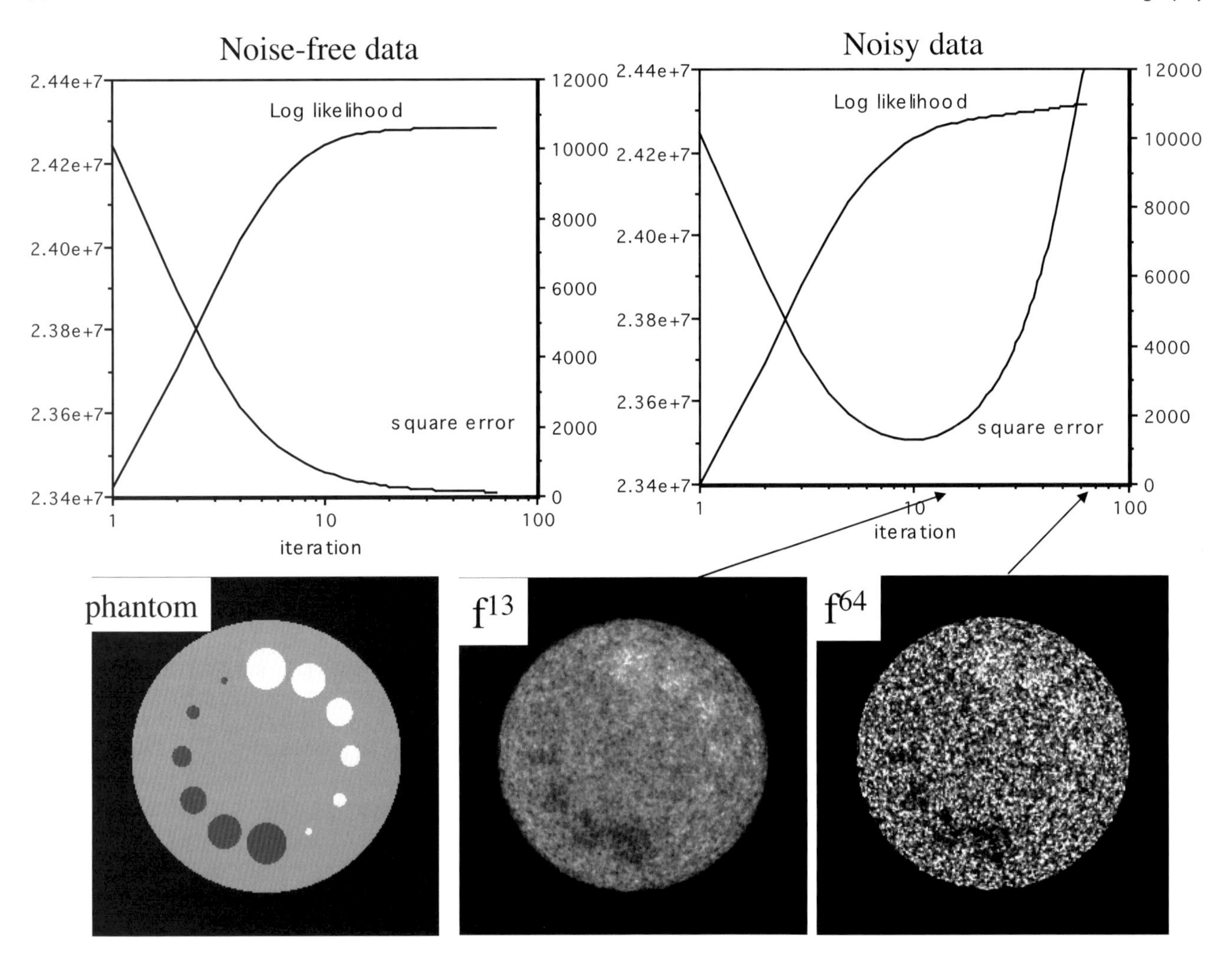

Figure 4.6. 2D reconstruction of a mathematical phantom with the ML-EM algorithm ($N_r = 128, N_\phi = 256, N_x = N_y = 256$). The Poisson log-likelihood (left scale) and the square reconstruction error with regard to the reference image (right scale) are plotted versus the iteration number. The left plot is for ideal noise-free data. For the right plot, pseudo-random Poisson noise has been added for a total of 400,000 coincidences. The cost function increases monotonically in contrast with the error, which reaches a minimum around 10 iterations. The 13th and 64th image estimates obtained from noisy data are shown.

What about stability? The ML-EM cost function does not include any prior. The algorithm converges therefore to the image that "best" fits the data ("best" in the sense defined by the Poisson likelihood). But fitting too closely the *noisy* data of an ill-conditioned problem induces instabilities (Fig. 4.6). In practice, this instability corrupts the image estimates $\vec{f}^n$ by high-frequency "checkerboard-like" artifacts when the number of iterations exceeds some threshold [57, 58]. Various methods can remedy this problem:

- Introduce a Bayesian prior term into the cost function (see [59] and references therein),
- apply a post-reconstruction filter, typically a 3D Gaussian filter with a FWHM related to the spatial resolution that is deemed achievable given the SNR,
- filter the data before applying the ML-EM algorithm,
- stop the algorithm after n_{max} steps, and use $\vec{f}^{n_{max}}$ as solution estimate. Methods to automatically estimate an appropriate number of iterations have been proposed [60, 61] though all clinical implementations determine n_{max} empirically.

The Ordered Subset Expectation Maximization algorithm [62] is based on a simple modification of Eq. (41), which has a significant impact on clinical PET imaging by making iterative reconstruction practical. The LOR data are partitioned in S disjoint subsets $J_1, \cdots, J_S \subset [1, \cdots, N_{LOR}]$. For 2D sinogram data (see Eq. (5)), one usually assigns the 1D parallel projections $m, m + S, m + 2S, \cdots, \leq N_\phi$ to the subset J_{m+1}.

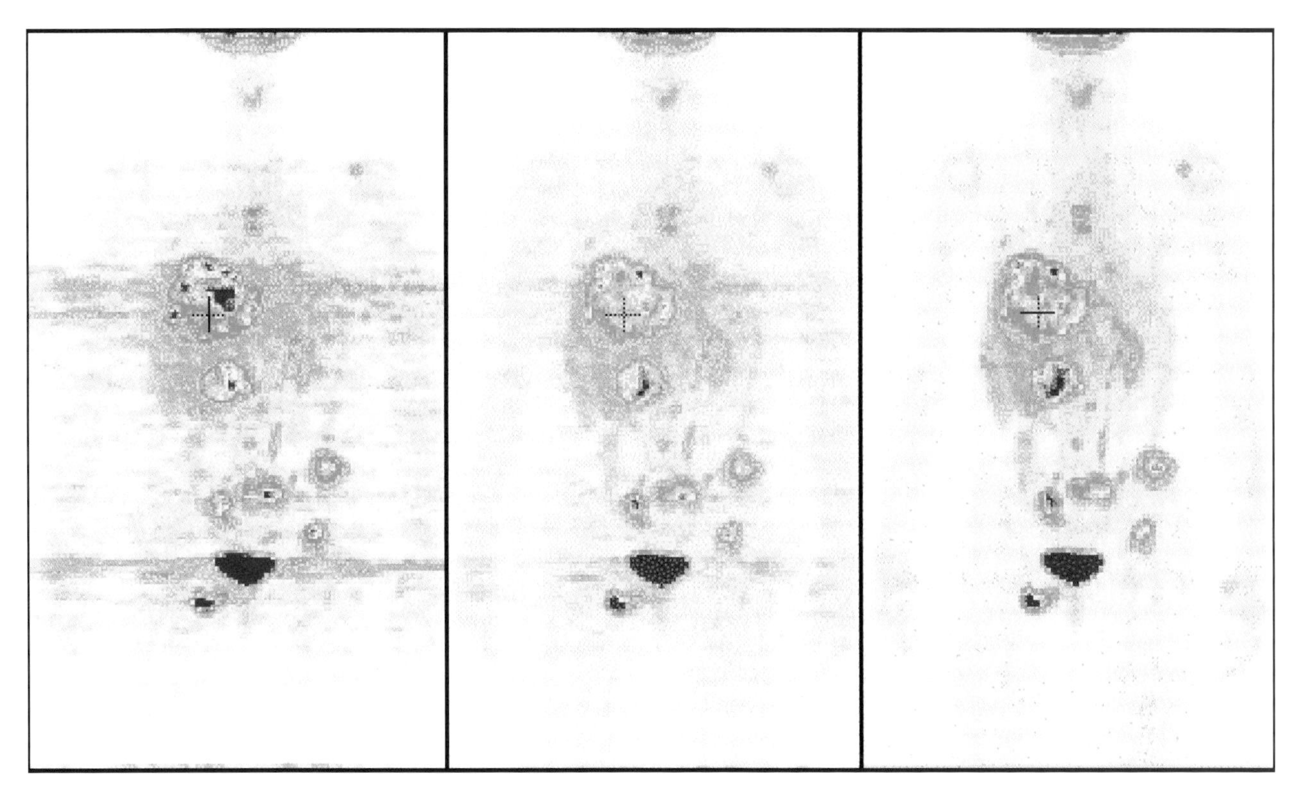

Figure 4.7. Comparison between the FBP and the OSEM reconstruction of a 2D FDG whole-body study, showing a frontal section. The algorithms used for the reconstruction of the transmission scan and of the emission scan are FBP-FBP (left), OSEM-FBP (center), OSEM-OSEM (right).

The ML-EM iteration (Eq. (41)) is then applied incorporating the data from one subset only. Each subset is processed in a well-defined order, usually in a periodic pattern where subset $J_{n \bmod S}$ is used at iteration n[12]:

$$f_i^{n+1} = f_i^n \frac{1}{\sum_{j' \in J_{n \bmod S}} a_{j',i}} \sum_{j \in J_{n \bmod S}} a_{j,i} \frac{p_j}{\sum_{i'=1}^{P} a_{j,i'} f_{i'}^n}$$
$$i = 1, \ldots, P \tag{42}$$

Empirically, the convergence is accelerated by a factor $\simeq S$ with respect to ML-EM. But the asymptotic convergence to the maximum-likelihood estimator is no longer guaranteed. In fact, OSEM tends to cycle between S slightly different image estimates. To minimize the adverse effects of this behavior it is recommended to keep the number of 1D parallel projections in each subset equal to at least 4. In addition, several authors suggest progressively decreasing the number of subsets during iteration. Finally, we only mention here the *row action maximum likelihood* (RAMLA) algorithm [63] and the *rescaled block-iterative ML-EM* algorithm [64]. These two algorithms for maximum-likelihood estimation with a Poisson distribution are closely related to OSEM, but guarantee asymptotic convergence under certain conditions.

Compared to FBP reconstructions, some qualitative characteristics of images reconstructed from Poisson data using ML-EM or OSEM are:

- Reduced streak artifacts
- A better SNR in regions of low tracer uptake, resulting in particular in a better visibility of the contours of the body
- Some non-isotropy and non-uniformity of the spatial resolution, especially when the range of values of the attenuation correction factor is large, as e.g. in the chest
- A slower convergence for regions of low tracer uptake than for regions of high tracer uptake.

Figure 4.7 illustrates some of these properties.

Finally, some comments are in order about data corrections prior to reconstruction with ML-EM or OSEM. Physical effects such as detector efficiency variations, attenuation, scattered and random coincidences, etc., must be accounted for to obtain quantitatively correct

[12] In the OSEM jargon, such an iteration is called a *sub-iteration*, and an *iteration* denotes a set of S consecutive sub-iterations, corresponding to one pass through the whole data set.

images. With analytical algorithms such as FBP the data are corrected before reconstruction to comply with the line integral model. With the ML-EM algorithm, on the contrary, pre-correction must be avoided because it would destroy the Poisson character of the data and thereby could bias the reconstruction. This means that the ML-EM algorithm should be applied to the raw data, and that all physical effects should be included in the system matrix as described in the section on the system matrix (above). A full modeling, however, can be impractical when the system matrix is too large to be pre-computed. A faster, approximate, procedure consists of including only the most significant effect – attenuation – in the system matrix. Corrections for scattered and random coincidences can be of the order of 50%, but are often less. Attenuation correction, however, involves multiplication by factors ranging from 5 to more than 100 ! The attenuation correction is multiplicative and can easily be incorporated in the ML-EM iteration as shown by Hebert and Leahy [65],

$$f_i^{n+1} = f_i^n \frac{1}{\sum_{j'=1}^{N_{LOR}} a_{j',i} / \alpha_{j'}} \sum_{j=1}^{N_{LOR}} a_{j,i} \frac{p_j}{\sum_{i'=1}^{P} a_{j,i'} f_{i'}^n} \quad i = 1,\ldots,P \tag{43}$$

where the p_j are the data corrected for all effects except attenuation, α_j is the pre-computed attenuation correction factor[13] for LOR j, and the system matrix a does not include the effect of attenuation. This attenuation-weighting (AW) of the ML-EM algorithm is easily extended to the attenuation-weighted OSEM algorithm (AW-OSEM). The AW-OSEM approach has been shown to perform almost as well as algorithms that model all physical effects, with only modest increases in computation time over OSEM applied to pre-corrected sinogram data [66].

The previous approach can also be applied to other multiplicative corrections such as the normalization for detector efficiency variations. For more complex, e.g., non-linear, relations between the raw data p_j and the corrected data p_j^c, an approximate statistical modeling can be achieved by applying the ML-EM algorithm to *scaled* data $p_j^s = \beta_j p_j^c$, where $\beta_j = \langle p_j^c \rangle / \mathrm{var}(p_j^c)$ is a low-variance (smoothed) estimate of the ratio between the mean and the variance of the corrected data. With this choice of β_j the scaled data satisfy the same relation $\langle p_j^s \rangle \simeq \mathrm{var}(p_j^s)$ as data obeying Poisson statistics, and it is therefore reasonable to reconstruct them using the ML-EM algorithm [30]. This yields the following iteration:

$$f_i^{n+1} = f_i^n \frac{1}{\sum_{j'=1}^{N_{LOR}} \beta_{j'} a_{j',i}} \sum_{j=1}^{N_{LOR}} a_{j,i} \frac{p_j^s}{\sum_{i'=1}^{P} a_{j,i'} f_{i'}^n} \quad i = 1,\ldots,P \tag{44}$$

In the case of the attenuation correction, $p_j^c = \alpha_j p_j$, and one easily checks that $\beta_j = 1/\alpha_j$, *and* $p_j^s = p_j$, so that Eqs. (44) and (43) coincide.

When the data are acquired in true mode as the difference between the prompt and delayed coincidences, the shifted Poisson model (described earlier) leads to the following modified ML-EM algorithm [26],

$$f_i^{n+1} = f_i^n \frac{1}{\sum_{j'=1}^{N_{LOR}} a_{j',i} / \alpha_{j'}} \sum_{j=1}^{N_{LOR}} a_{j,i} \frac{p_j + 2\bar{r}_j + \bar{s}_j}{\sum_{i'=1}^{P} a_{j,i'} f_{i'}^n + \alpha_j (2\bar{r}_j + \bar{s}_j)} \quad i = 1,\ldots,P \tag{45}$$

where the p_j are the data corrected for random and scatter (but not attenuation), α_j and $a_{j,i}$ are as in equation (43), and $\bar{r}_j$ and $\bar{s}_j$ are low-variance estimates of the random and scatter background in LOR j. The mean random $\bar{r}_j$ is generally estimated using variance reduction techniques or from the single photon data (see sections Randoms Variance Reduction and Estimation from Single Rates in next chapter). The mean scatter $\bar{s}_j$ is estimated using a model based scatter model (see section Simulation-based Scatter Correction in next chapter).

Variance and resolution with non-linear reconstruction algorithms

Predicting and controling the statistical properties and the resolution of reconstructed PET images is of paramount importance for quantitative applications of PET and for task oriented performance studies using numerical observers. For clinical PET, a good awareness of these properties helps minimizing the probability of erroneous image interpretations.

Denote the "true" image by $\vec{f}$, the measured data vector by $\vec{p}$, and the mean data by $\langle \vec{p} \rangle = A\vec{f}$, where A is the system matrix (see Eq. (28)). Consider any

[13] The ratio between the blank and transmission scans.

specific algorithm denoted by $\mathcal{T}$ (for example 100 ML-EM iterations with a uniform initial image estimate). The reconstruction is then $\vec{f}^* = \mathcal{T}(\vec{p})$, and the reconstruction error is

$$\vec{f}^* - \vec{f} = (\mathcal{T}(\vec{p}) - < \mathcal{T}(\vec{p}) >) + (< \mathcal{T}(\vec{p}) > - \vec{f}) \quad (46)$$

where $< \mathcal{T}(\vec{p}) >$ denotes the mean value of the reconstructed image, which could be estimated by averaging a large number of images reconstructed by applying the algorithm $\mathcal{T}$ to statistically independent realizations of the random data vector $\vec{p}$. The first term in the RHS of Eq. (46) is the *statistical error* due to the fluctuations of data $\vec{p}$ around its mean value $< \vec{p} >$. The statistical error is characterized by the covariance matrix

$$V_{j,j'} = < (\mathcal{T}(\vec{p})_j - < \mathcal{T}(\vec{p})_j >) (\mathcal{T}(\vec{p})_{j'} - < \mathcal{T}(\vec{p})_{j'} >) > \quad j,j' = 1, \dots, P \quad (47)$$

the diagonal elements of which give the variance of each reconstructed pixel value. The second term in the RHS of Eq. (46) is the *systematic error* or *bias*: even the mean value of the reconstructed image is not exact because of sampling, apodization, finite number of iterations, etc.

For a linear reconstruction algorithm the image covariance can easily be determined once we know the statistical (e.g. Poisson) properties of the data. In addition, with a linear algorithm, the systematic error is fully characterized by the *point response* defined as the reconstruction of the mean data of a point source located in a voxel $j_0 \in [1, \dots, P]$. While the point response depends in general on the position of the voxel j_0 relative to the scanner, it does not depend on the strength of the point source, or on whether that source is sitting or not over some background. This allows an unambiguous definition of the resolution, using parameters such as the FWHM of the point response. For the FBP algorithm, in particular, the statistical error and the bias are determined by the apodized ramp filter, and the trade-off between these two errors is well understood (see the section Ill-posedness of the Inverse X-ray Transform).

For non-linear algorithms such as ML-EM, the derivation of analytical expressions for the covariance matrix is complex. More importantly, the point response becomes object dependent. To understand this important point, consider any data set $\vec{p}$, measured e.g. as a "normal" whole-body tracer distribution. Consider also some additional point source $\Delta\vec{f}$ located in voxel j_0: $\Delta f_j = \delta_{j,j_0}$, and denote the corresponding mean contribution to the data by $\Delta\vec{p} = A\Delta\vec{f}$. Then the non-linearity of the algorithm $\mathcal{T}$ means that, in general,

$$\mathcal{T}(\vec{p} + \Delta\vec{p}) \neq \mathcal{T}(\vec{p}) + \mathcal{T}(\Delta\vec{p}) \quad (48)$$

A concrete consequence of this non-linearity can be observed when the ML-EM algorithm is used with the small number of iterations typical of clinical practice: the reconstruction of a unit "point source" sitting on top of a uniform background broadens when the strength of the background is increased. Similarly, the anisotropy of the attenuation correction factors for an elongated object such as the chest at the level of the shoulders is translated by ML-EM into an anisotropy of the point response: if the point source is located in an ellipsoidal attenuating medium with long axis along the x-axis, the point response takes an ellipsoidal shape with long axis along the y-axis. One should therefore interpret with care results on the "reconstructed resolution of ML-EM" obtained for isolated point or line sources. Similar observations hold for MAP or other non-linear algorithms.

A local infinitesimal point response function, depending both on the data $\vec{p}$ and on the position at voxel j_0, can be defined as the image

$$\vec{\delta}_{\vec{p},j_0} = \lim_{\varepsilon \to 0} \frac{1}{\varepsilon}(\mathcal{T}(\vec{p} + \varepsilon\Delta\vec{p}) - \mathcal{T}(\vec{p})) = \frac{\partial \mathcal{T}(A\vec{f})}{\partial f_{j_0}} \in \mathbb{R}^P \quad (49)$$

Approximate expressions and efficient numerical techniques have been developed [67] to calculate this point response, as well as methods to design a penalty term $\log Pr\{\vec{f}\}$ in Eq. (32)[14] that guarantee homogeneous resolution [68]. An alternative approach to improve the homogeneity of the resolution consists in pursuing the ML-EM iteration beyond the point where the image is deemed acceptable, and in post-filtering this image with an appropriate filter [69].

The image covariance (Eq. (47)) can be estimated numerically by reconstructing a large number of data sets simulated with statistically independent pseudorandom noise realizations. An alternative for maximum-likelihood algorithms is to calculate the Fisher information matrix, the inverse of which is related by the Cramer-Rao theorem to the covariance of the ML estimator (see e.g. [7]). An approximate expression of the covariance, for the more relevant case where ML-EM iteration is stopped well before convergence, was derived in [57], and validated numerically in [58].

[14] This penalty depends on the data and can no longer be interpreted as a real Bayesian prior. The algorithm is then better referred to as a penalized likelihood method.

The major conclusion is that the variance of the ML-EM reconstruction is roughly proportional to the image itself, i.e.

$$V_{j,j} \simeq C < f_j >^2 \qquad j = 1, \dots, P \tag{50}$$

for some constant C depending on the object and on the number of iterations. Thus, the ML-EM reconstructions have lower variance in regions of low tracer uptake, thereby allowing good detectability in these regions. This is in contrast with FBP reconstructions, in which the noise arising from the high uptake regions spreads more uniformly over the whole FOV, resulting in particular in the well-known streak artefacts.

3D Data Organization

Two-dimensional Parallel Projections

We have seen in the previous section that 2D data acquired with a ring scanner can be stored in a sinogram $p(s, \phi)$. If the data are modeled as line integrals, as for analytic algorithms, the sinogram is a set of 1D parallel projections of $f(x, y)$ for a set of orientations $\phi \in [0, \pi]$. Similarly, the LORs measured by a volume PET scanner can be grouped into sets of lines parallel to a direction specified by a unit vector $\vec{n} = (n_x, n_y, n_z) = (-\cos\theta \sin\phi, \cos\theta \cos\phi, \sin\theta) \in S^2$ where S^2 denotes the unit sphere. The angle θ is the angle between the LOR and the transaxial plane, so that the data acquired in a 2D acquisition therefore correspond to $\theta = 0$. The set of line integrals parallel to $\vec{n}$ is a *2D parallel projection* of the tracer distribution:

$$p(\vec{s},\vec{n}) = \int_{\mathbb{R}} dt\ f(\vec{s} + t\vec{n}) \tag{51}$$

where the position of the line is specified by the vector $\vec{s} \in \vec{n}^{\perp}$, which belongs to the *projection plane* $\vec{n}^{\perp}$ orthogonal to $\vec{n}$.

Consider a cylindrical scanner with N_r rings of radius R_d, extending axially over $0 \le z \le L$, where $L = N_r \Delta z$. Assuming continuous sampling, this scanner measures all LORs such that the line defined by $(\vec{s}, \vec{n})$ has two intersections with the lateral surface of the cylinder (these intersections are the positions of the two detectors in coincidence). The set of measured orientations is

$$\Omega(\theta_{\max}) = \{\vec{n} = (\phi,\theta) \mid \phi \in [0,\pi), \theta \in [-\theta_{\max}, +\theta_{\max}]\} \tag{52}$$

with $\tan\theta_{\max} = L / 2\sqrt{R_d^2 - R_F^2}$, where R_F is the radius of the transaxial FOV. However, for each $\theta \neq 0$, not all LORs parallel to $\vec{n}$ and crossing the FOV of the scanner are measured. That is, the parallel projection $p(\vec{s}, \vec{n})$ is measured only for some subset of LORs $\vec{s} \in M(\vec{n}) \subset \vec{n}^{\perp}$. One says that this projection is *truncated*.

Two important properties of the 3D data can already be stressed:

(i) 3D data are *redundant* since four variables are required to parameterize $p(\vec{s}, \vec{n})$ (two for the orientation $\vec{n}$ and two for the vector $\vec{s}$) whereas the image only depends on three variables (x, y, z).

(ii) 3D data are not invariant for translation as in the 2D case because the cylindrical detector has a finite length and the measured projections are truncated.

The vector $\vec{s}$ can be defined by its components (s, u) on two orthonormal basis vectors in $\vec{n}^{\perp}$.

$$\vec{s} = s(\cos\phi, \sin\phi, 0) + u(\sin\theta\sin\phi, -\sin\theta\cos\phi, \cos\theta) \tag{53}$$

The variable s coincides with the 2D radial sinogram variable of Eq. (3). We will thus write $p(\vec{s}, \vec{n}) = p(s, u, \phi, \theta)$. The subset $p(s, u, \phi, 0)$ is the 2D sinogram of the slice $z = u$.

The LORs measured by a PET scanner do not uniformly sample the variables (s, u, ϕ, θ), and therefore interpolation is needed to reorganize the raw data into parallel projections. This holds both for multi-ring scanners and for scanners based on flat panel detectors.

Oblique Sinograms

Some analytic algorithms use an alternative parameterization of the parallel projections, where the vari-

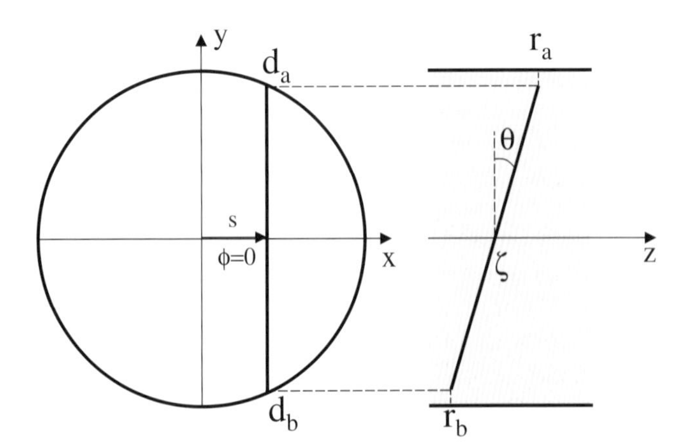

Figure 4.8. A transverse and a longitudinal view of a multi-ring scanner. An LOR connecting a detector d_a in ring r_a to a detector d_b in ring r_b is shown, with the four variables (s, ϕ, ζ, θ) used for the oblique sinogram parameterization. The particular LOR represented has $\phi = 0$.

able u in Eq. (53) is replaced by the axial coordinate $\zeta = u/\cos\theta$, the average of the axial coordinates of the two detectors in coincidence. One defines weighted parallel projections

$$\begin{aligned} p_s(s,\phi,\zeta,\theta) &= p(s,\zeta\cos\theta,\phi,\theta)\cos\theta \\ &= \int_{\mathbb{R}} dt' f(s\cos\phi - t'\sin\phi, s\sin\phi + \\ &\qquad t'\cos\phi, \zeta + t'\tan\theta) \end{aligned} \tag{54}$$

The domain of the variables is $|s| \le R_F$, $\phi \in [0, \pi)$, $|\theta| \le \arctan\left(L/2\sqrt{R_d^2 - s^2}\right)$, and $\zeta \in \left[|\tan\theta|\sqrt{R_d^2 - s^2}, L - |\tan\theta|\sqrt{R_d^2 - s^2}\right]$ (Fig. 4.8). For each pair ζ, θ the function $p_s(.,.,\zeta,\theta)$ is called an *oblique sinogram* by analogy with Eq. (3). The similarity with the 2D format makes this oblique sinogram format suited to the analytic rebinning algorithms, which reduce the 3D data to 2D data.

Consider now the discrete sampling of the oblique sinograms. The measured LORs connecting detector d_a in ring r_a to detector d_b in ring r_b corresponds to parameters (s, ϕ, ζ, θ) in Eq. (48), where s and ϕ are determined as in the 2D case (Eq. (4)), and the axial variables are determined by

$$\begin{aligned} \tan\theta &= (r_b - r_a)\Delta z / \left(2\sqrt{R_d^2 - s^2}\right) \\ \zeta &= (r_a + r_b)\Delta z / 2 \end{aligned} \tag{55}$$

If the radius of the FOV is small, θ in Eq. (55) is approximately independent of s. With this approximation, the coincidences between two rings r_a and r_b can be used to build an oblique sinogram $p_s(.,.,\zeta,\theta)$ with $\zeta = (r_a + r_b)\Delta z/2$ and $\tan\theta = (r_b - r_a)\Delta z/(2R_d)$.

To save storage and computation, some volume scanners use axial angular undersampling by averaging sets of sinograms with adjacent values of θ. The degree of undersampling is characterized by an odd integer parameter S, called the *span*. The resulting sampling is non-interleaved:

$$\begin{aligned} \tan\theta &= i_\theta S\Delta z/(2R_d) & i_\theta &= -i_{\max},\ldots,+i_{\max} \\ \zeta &= i_z\Delta z/2 & z_{\min}(i_\theta) &\le i_z \le 2N_r - 2 - z_{\min}(i_\theta) \end{aligned} \tag{56}$$

where $z_{min}(i_\theta) = \max(0, |i_\theta|S - S/2)$. Each sample (i_θ, i_z) is obtained by averaging data from all pairs of rings such that

$$\begin{aligned} i_\theta S - S/2 \le r_b - r_a \le i_\theta S + S/2 \\ i_z = r_b + r_a \end{aligned} \tag{57}$$

The sampling scheme is often illustrated on a 2D diagram, the "Michelogram", in which each grid point represents one ring pair and each sampled oblique sinogram (i_θ, i_z) is represented by a line segment connecting the contributing pairs r_a, r_b (Fig. 4.9). Just as for the azimuthal undersampling ("mashing", see end of sinogram data and sampling section, above), a good

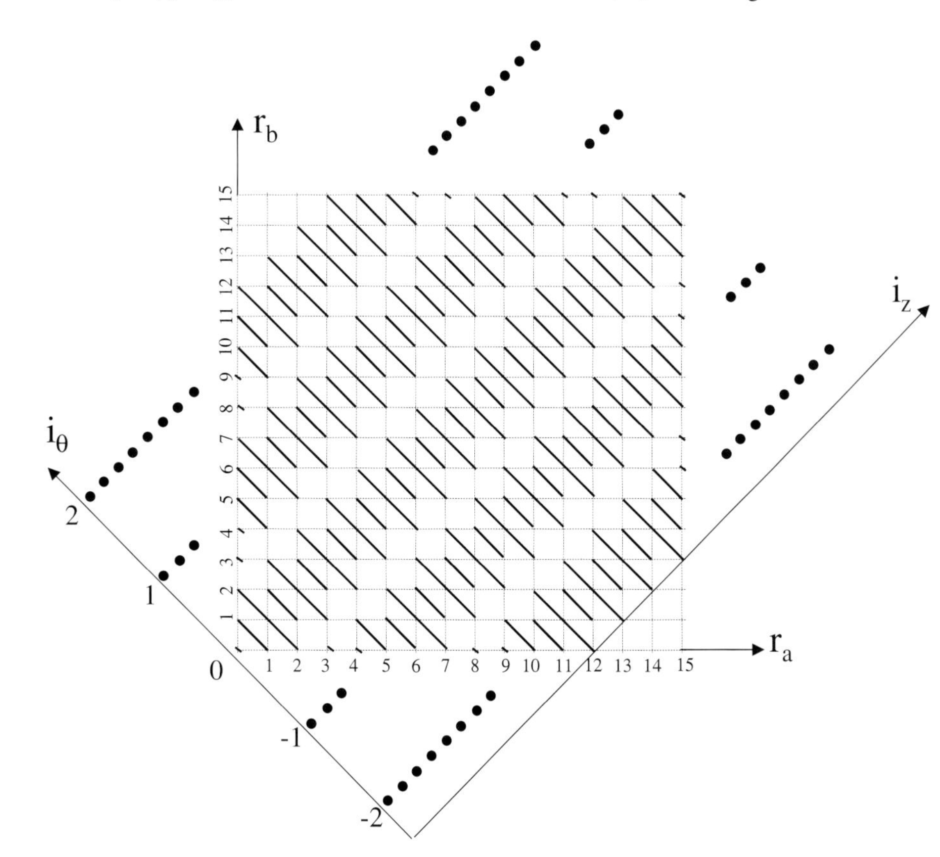

Figure 4.9. A Michelogram for a 16-ring scanner, illustrating the axial sampling with a span $S = 5$ and a maximum ring difference defined by $i_{max} = 2$. Each grid point corresponds to a ring pair, and each diagonal line segment links the ring pairs (2 or 3 except at the edge of the FOV) that are averaged to form one oblique sinogram. The samples located outside the square (dots) are the unmeasured oblique sinograms needed to obtain a shift-invariant response. In the 3DRP algorithm, these missing sinograms are estimated by forward-projecting an initial 2D reconstruction of the direct segment $i_\theta = 0$.

choice of the span S depends both on the SNR and on the radius of the FOV. Values between 3 (for high-statistics brain studies) and 9 (for low-count, whole-body studies) are standard.

When the radius of the FOV is large, more accurate interpolation is needed to reorganize the raw data into parallel projections according to Eq. (55), but the sampling pattern (Eq. (56)) can be kept.

3D Analytic Reconstruction by Filtered-backprojection

The Central Section Theorem

The central section theorem (Eq. 10) can be generalized to 3D, and states that

$$P(\vec{v},\vec{n}) = F(\vec{v}) \qquad \vec{v} \in \vec{n}^{\perp} \tag{58}$$

where

$$P(\vec{v},\vec{n}) = \int_{n^{\perp}} d\vec{s}\, p(\vec{s},\vec{n}) \exp(-2\pi i \vec{s}\cdot\vec{v}) \tag{59}$$

is the 2D Fourier transform of a parallel projection and F is the 3D Fourier transform of the image. Note that as the integral in Eq. (59) is over the whole projection plane $\vec{n}$, the central section theorem is only valid for non-truncated parallel projections.

Geometrically, this theorem means that a projection of direction $\vec{n}$ allows the recovery of the Fourier transform of the image on the central plane orthogonal to $\vec{n}$ in 3D frequency space. A corollary is that the image can be reconstructed in a stable way from a set of non-truncated projections $\vec{n} \in \Omega \subset S^2$ if and only if the set Ω has an intersection with any equatorial circle on the unit sphere S^2. This condition is due to Orlov [70]. The equatorial band $\Omega(\theta_{\max})$ in Eq. (52) satisfies Orlov's condition for any $\theta_{\max} > 0$.

The *direct 3D Fourier reconstruction* algorithm is a direct implementation of Eq. (58) [71]. This technique involves a complex interpolation in frequency space, and has not so far been used in practice. However, Matej [15] recently demonstrated a significant gain of reconstruction time compared to the standard FBP.

3D Filtered Backprojection

Following the same lines as for the 2D FBP inversion, Eq. (58) leads to a two-step inversion formula for a set of non-truncated 2D projections with orientations $\vec{n} \in \Omega$, where Ω is a subset of the unit sphere that satisfies Orlov's condition. The reconstructed image is a 3D backprojection

$$f(\vec{r}) = \int_{\Omega} d\vec{n}\, p^{F}(\vec{s} = \vec{r} - (\vec{r}\cdot\vec{n})\vec{n}, \vec{n}) \tag{60}$$

which, as in 2D, is the sum of the *filtered projections* p^F for all lines containing the point $\vec{r}$. The filtered projections are given by

$$p^{F}(\vec{s},\vec{n}) = \int_{\vec{n}^{\perp}} d\vec{s}'\, p(\vec{s}',\vec{n}) h_C(\vec{s} - \vec{s}',\vec{n}) \tag{61}$$

In this equation, the 2D convolution kernel $h_C(\vec{s})$ is the 2D inverse Fourier transform of the filter function due to Colsher [72]:

$$H_C(\vec{v},\vec{n}) = \{\int_{\Omega} d\vec{n}'\,\delta(\vec{v}\cdot\vec{n}')\}^{-1} = \frac{|\vec{v}|}{L_{\Omega}(\vec{v})} \qquad \vec{v} \in \vec{n}^{\perp} \tag{62}$$

where δ is the Dirac delta function, and $L_\Omega(\vec{v})$ is the arc length of the intersection between Ω and the great circle normal to $\vec{v}$ (Fig. 4.10). Orlov's condition ensures that $L_\Omega(\vec{v}) > 0$. An expression of this filter in terms of the variables v_s, v_u, ϕ, θ can be found in [72]. Like the ramp filter, Colsher's filter is proportional to the modulus of the frequency. In contrast to the 2D case, however, the filter depends on the angular part of the

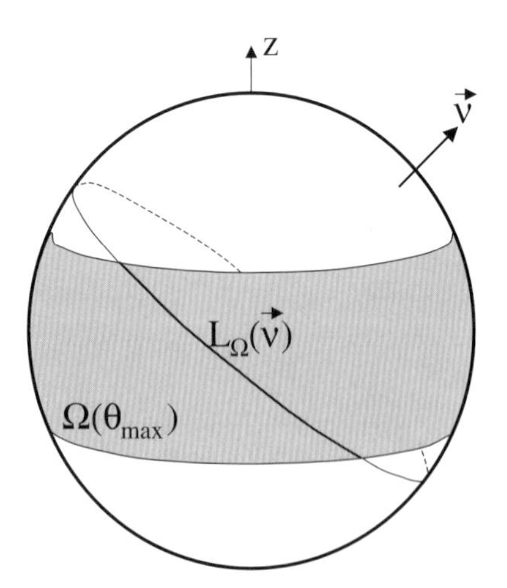

Figure 4.10. Each vector $\vec{n}$ on the unit sphere S^2 is the direction of one 2D parallel projection $p(\vec{s},\vec{n})$. The set of directions $\Omega(\theta_{max})$ measured by a cylindrical scanner (equation (52)) is shown as a grey subset. The Fourier transform $F(\vec{V})$ can be recovered from any projection along the measured (thick line) segment of the great circle orthogonal to $\vec{V}$. The reciprocal l/L_Ω of the length of this segment is the angular part of the reconstruction filter.

frequency. Another specificity of 3D reconstruction, due to the redundancy of the 3D data, is that the reconstruction filter is not unique [73]. Colsher's filter, however, yields the reconstructed image with the *minimal variance* under fairly general assumption on the data statistics [74].

The discretization of the 3D FBP algorithm is based as in 2D on replacing integrals by trapezoidal quadratures and on linear interpolation in $\vec{s}$ for the 3D backprojection. The 3D backprojection is the most time-consuming step in the algorithm and various techniques have been proposed to accelerate this procedure (see [17] and references therein). The 2D convolution is implemented in frequency space as:

$$p^F(\vec{s},\vec{n}) = \int_{\vec{n}^\perp} d\vec{v}\; h_C(\vec{v},\vec{n})\; w(\vec{v})\; P(\vec{v},\vec{n})\; \exp(2\pi i\vec{s}\cdot\vec{v}) \quad (63)$$

where $P(\vec{v}, \vec{n})$ is the 2D Fourier transform of the non-truncated projection and $w(\vec{v})$ is an apodizing window, which plays the same stabilizing role as in 2D (see the remark below Eq. (16) and reference [75] for details on the discrete implementation using the 2D FFT).

The Reprojection Algorithm

The 3D FBP algorithm is valid only for non-truncated parallel projections. In almost all PET studies, the tracer distribution extends axially over the whole FOV of the scanner, and the only non-truncated parallel projections are those with $\theta = 0$. For sampled data, the equality $\theta = 0$ is replaced by $\theta < \theta_0$ for some small maximum oblicity angle θ_0, which corresponds typically to the maximum ring difference $d_{2D,max}$ incorporated in a 2D acquisition.

The standard analytic reconstruction algorithm for volume PET scanners is the *3D reprojection algorithm* (3DRP) [76], which consists of four steps:

(i) Reconstruct a first image estimate $f_{2D}(\vec{r})$ by applying the 2D FBP algorithm to the non-truncated data subset $\theta < \theta_0$.
(ii) Forward project $f_{2D}(\vec{r})$ to estimate the unmeasured parts $p(\vec{s} \notin M(\vec{n}), \vec{n})$ of a set of 2D parallel projections $\vec{n} \in \Omega(\theta_{max})$,
(iii) Merge the measured and estimated data to form non-truncated projections,
(iv) Reconstruct these merged data with the 3D FBP algorithm described in the previous section.

In general, a value of θ_{max} smaller than the scanner maximum axial acceptance angle is used to limit the amount of missing data, which must be estimated and backprojected. With a 24-ring scanner, using $d_{max} = 19$ instead of the maximum value $N_r - 1 = 23$ still incorporates 95% of the data.

Images reconstructed with the 3DRP algorithm share many features with 2D FBP reconstructions, including linearity (the reconstructed FWHM in a given point is the same for a cold and for a hot spot) and the prevalence of streak artifacts in low-count studies. One difference with 2D reconstructions is the axial dependence of the spatial resolution, due to the increasing contribution of the estimated data near the edges of the axial FOV (see Fig. 4.9). This property of 3DRP reflects the non-uniform sensitivity of the volume PET scanner. Clearly, *any* analytic or iterative algorithm has to somehow reflect this property in the reconstruction. With the rebinning algorithms described below, the lower sensitivity in the edge slices is translated in an increased variance rather than in a degraded spatial resolution.

3D Analytic Reconstruction by Rebinning

The high sensitivity of a PET scanner operated in 3D mode is directly related to the large number of sampled LORs, which is much larger than the number of reconstructed pixels: $N_{LOR} >> P$ (by a factor proportional to N_r). We have already mentioned in the previous section that this data redundancy results in the non-uniqueness of the reconstruction filter. From the practical point of view, redundancy increases the data storage requirements and the computational load for reconstruction and data correction.

This observation has motivated the development of *rebinning* algorithms. A rebinning algorithm is an algorithm that estimates the ordinary sinogram (Eq. (3)) of each sampled transaxial section $z \in [0, L]$, i.e.

$$p_{reb}(s,\phi,z) = \int_{-\infty}^{\infty} dt\; f(x = s\cos\phi - t\sin\phi, y = s\sin\phi + t\cos\phi, z) \quad (64)$$

from the measured oblique sinograms $p_s(s, \phi, \zeta, \theta)$ defined by Eq. (54). Each rebinned sinogram is then reconstructed separately using a 2D reconstruction algorithm. This procedure is illustrated in Fig. 4.11.

Rebinning would be trivial for noise-free data because one easily checks by comparing Eqs. (54) and (3) that

$$p_{reb}(s,\phi,z) = p_s(s,\phi,\zeta = z,\theta = 0) \quad (65)$$

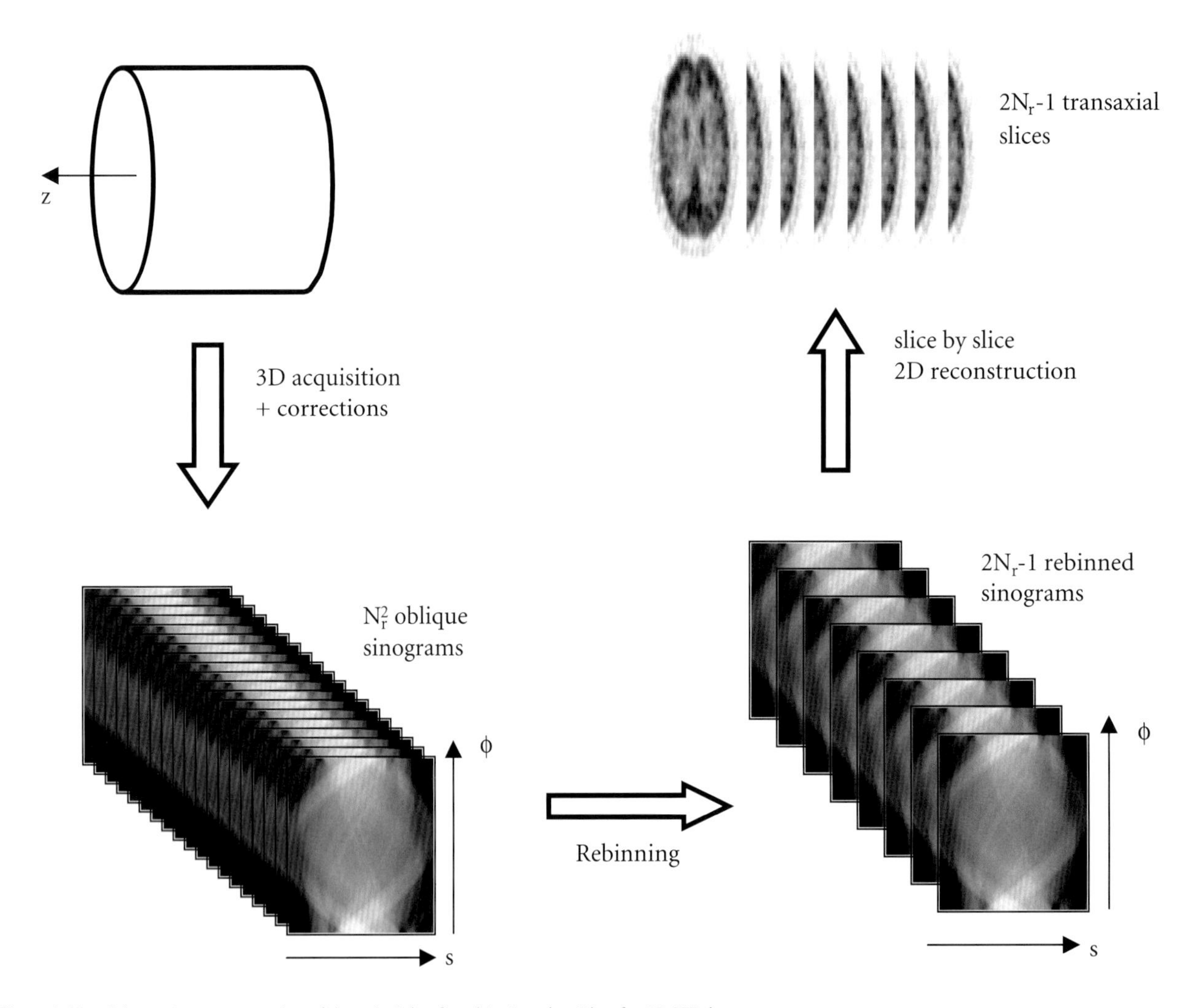

Figure 4.11. Schematic representation of the principle of a rebinning algorithm for 3D PET data.

In the presence of noise, however, an efficient rebinning method should optimize the SNR by exploiting the whole set of oblique sinograms to estimate p_{reb}.

Several approximate [77, 78, 79, 80, 81] and exact [82, 83] rebinning methods have been published. We only summarize the two algorithms that have been most used in practice.

The Single-slice Rebinning Algorithm (SSRB)

This approximate algorithm [77] is based on the assumption that each measured oblique LOR only traverses a single transaxial section within the support of the tracer distribution. Referring to the third argument of f in Eq. (54), this assumption amounts to neglecting the product $R_F \tan\theta$, where R_F, the radius of the FOV, is the maximum value of the variable t'. Using this approximation, Eq. (65) can be extended to

$$p_{reb}(s,\phi,z) \simeq p_s(s,\phi,\zeta = z,\theta = 0) \tag{66}$$

and by averaging all available estimates, SSRB defines the rebinned sinograms by

$$p_{ssrb}(\mathrm{s},\phi,\mathrm{z}) = \frac{1}{2\theta_{\max}(\mathrm{s},\mathrm{z})} \int_{-\theta_{\max}(s,z)}^{\theta_{\max}(s,z)} d\theta\, p_s(s,\phi,\zeta = z,\theta) \tag{67}$$

where $\theta_{\max}(s,z) = \arctan\left(\min\left[z, L-z\right] / \sqrt{R_d^2 - s^2}\right)$ is the maximum axial aperture for an LOR at a distance s from the axis in slice z. The algorithm is exact for tracer distributions which are linear in z, of the type $f(x, y, z) = a(x, y) + zb(x, y)$. For realistic distributions, the accuracy of the approximation will decrease with increasing R_F and $\theta_{\max}$. Axial blurring and transaxial distortions increasing with the distance from the axis of the scanner are the main symptoms of the SSRB approximation.

The discrete implementation of the SSRB algorithm is simply the extension of the technique described in the multi-slice 2D data section (above) to build 2D data with a multi-ring scanner operated in 2D mode, with d_{2Dmax} replaced by a larger value d_{max}. The choice of d_{max} entails a compromise between the systematic errors (which increase with d_{max}) and the reconstructed image variance (which increases with decreasing d_{max}).

The Fourier Rebinning Algorithm (FORE)

The approximate Fourier rebinning algorithm [81] is more accurate than the SSRB algorithm and extends the range of 3D PET studies that can be processed using rebinning algorithms. The main characteristics of FORE is that it proceeds via the 2D Fourier transform of each oblique sinogram, defined as

$$P_s(\nu,k,\zeta,\theta) = \int_0^{2\pi} d\phi \exp(-ik\phi) \int_{\mathbb{R}} ds \exp(-2\pi i s\nu) \times p_s(s,\phi,\zeta,\theta) \quad k \in Z, \nu \in \mathbb{R} \tag{68}$$

where k is the azimuthal Fourier index. Rebinning is based on the following relation between the Fourier transforms of oblique and direct sinograms:

$$P_s(\nu,k,z,0) \simeq P_s(\nu,k,\zeta = z + k\tan\theta/(2\pi\nu),\theta) \tag{69}$$

For each θ such that the oblique sinogram ζ, θ is measured (see Eq. (54)), the RHS yields an independent estimate of the direct data $\theta = 0$. FORE then averages all these estimates to optimize the SNR. The accuracy of the approximation (Eq. (69)) breaks down at low frequencies ν. Therefore, for all frequencies below some small threshold, the Fourier transform of the rebinned data is estimated using the SSRB approximation.

The main steps of the FORE algorithm are:

(i) Initialize a stack of Fourier transformed sinograms $P_{fore}(\nu, k, z)$,
(ii) For each oblique sinogram ζ, θ
  a. Calculate the 2D Fourier transform $P_s(\nu, k, \zeta, \theta)$,
  b. For each frequency component (ν, k), increment $P_{fore}(\nu, k, \zeta - k \tan \theta/(2\pi\nu))$ by $P_s(\nu, k, \zeta, \theta)$,
(iii) Normalize $P_{fore}(\nu, k, z)$ for the varying number of contributions it has received,
(iv) Take the 2D inverse Fourier transform to get the rebinned data $p_{fore}(s, \phi, z)$.

Like all analytic algorithms, FORE assumes that the data p_s (s, ϕ, ζ, θ) are line integrals of the tracer distribution and that each oblique sinogram is sampled over the whole range $(s, \phi) \in [-R_F, R_F] \times [0, \pi]$. Therefore, the raw data must be corrected for all effects including detector efficiency variations, attenuation, and scattered and random coincidences, before applying FORE. Also, when the data are incomplete due to gaps in the detector assembly, the sinograms must be filled as discussed in the section on properties of the inverse 2D radon transform (above). Refer to [81] for a detailed description and for the derivation of FORE.

In practice, FORE is sufficiently accurate when the axial aperture θ_{max} is smaller than about 20°, though the limit depends on the radius of the FOV and on the type of image. Beyond 20°, artifacts similar to those observed with SSRB (at lower apertures) appear [84]: degraded image quality at increasing distance from the axis. Two variations of FORE, the FOREJ and FOREX rebinning algorithms [82, 83], are exact in the limit of continuous sampling, and have been shown to overcome this loss of axial resolution when reconstructing high statistics data acquired with a large aperture scanner [85]. However, the current implementation of the FOREJ algorithm [82] is more sensitive to noise than FORE since the correction term involves a second derivative of the data with respect to the axial coordinate ζ, and the application to low statistics data remains questionable.

Hybrid Reconstruction Algorithms for 3D PET

The future evolution of image reconstruction in PET will most probably lead to the generalized utilization of iterative algorithms, both for 2D and for 3D data. As shown in the next section, it is straightforward to extend iterative methods, such as OSEM, to fully 3D scanning. These algorithms have the potential to model accurately the data acquisition, the measurement noise, and also the prior information on the tracer distribution. In contrast, analytic algorithms are bound to the line integral representation of the data. Even though some physical effects can be incorporated in pre- or post-processing steps, an accurate modeling of the Poisson statistics of the data is difficult with analytic methods. To date, however, the computational burden of fully 3D iterative algorithms remains a major issue for some applications involving multiple acquisitions, or for research scanners such as the HRRT which sample a very large number of LORs. The current practice of undersampling these data (see above) to accelerate reconstruction is contradictory with the aim of accurate modeling claimed by iterative methods.

This limitation has led to the application of *hybrid algorithms* for 3D PET data [41, 66, 91]. These algorithms first rebin the 3D data into a multi-slice set of ordinary sinogram data, using e.g. the SSRB method, or, more often, FORE. Each rebinned sinogram is then reconstructed using some 2D iterative algorithm. This hybrid approach provides a significant time gain with respect to fully 3D iterative reconstruction.

The two components of hybrid algorithms, rebinning and iterative methods, have been discussed in previous sections. In this section, we briefly discuss the interplay between these two elements, the main difficulty being to model the rebinned data that are presented to the 2D iterative algorithm. We focus on the application of FORE followed by a 2D OSEM reconstruction but the same problems would arise with other combinations, such as SSRB followed by an iterative minimization of a 2D penalized weighted least-square (PWLS) cost function [86].

One of the major benefits of iterative reconstruction arises from a correct modeling of the data statistics, which allows to weight each LOR according to its variance. This is the reason why improved image quality is obtained by reconstructing the raw, uncorrected data with a system matrix incorporating the effects of attenuation, normalisation and scatter, rather than by reconstructing pre-corrected data with a system matrix modeling only the detector's geometric response. Ideally, therefore, we would like to develop a hybrid algorithm in which un-corrected rebinned data are reconstructed by means of a 2D iterative algorithm including the effects of attenuation, etc. This approach is impossible because the FORE Eq. (69) must be applied to fully pre-corrected data as discussed at the end of the previous section. The rebinned data must then be reconstructed with a 2D iterative algorithm which does *not* model the pre-corrected physical effects.

One solution to improve the statistical model is to *de-correct* the data for the physical effects after the rebinning. This de-correction restores Poisson-like statistics to the rebinned data, and the physical effects can then be reintroduced in the system matrix. If we hypothesize that the most important effect is that of attenuation, we can decorrect for attenuation only and then reconstruct the de-corrected rebinned data with AW-OSEM (see Eq. (43)). This approach is referred to as the FORE+OSEM(AW) algorithm. Note that this algorithm is still approximate: even in the absence of attenuation and scatter, the rebinned sinograms are not independent Poisson variables because of the complex linear combination of the 3D data during FORE rebinning. Strictly speaking, it is inappropriate to reconstruct the rebinned data using the OSEM algorithm derived for independent Poisson data, and it is preferable to use a weighted least-square method [87] or the NEC scaling technique [30] (Eq. (44)). In each case, one needs to estimate the variance of the rebinned data [88] and also, ideally, the covariance [89].

Finally, modeling the shift-variant detector response (e.g. due to crystal penetration) has not yet been attempted with hybrid methods. One approach would be to apply sinogram restoration prior to rebinning.

A related problem occurs with scanners such as the Siemens/CPS HRRT [10], which has gaps between adjacent flat panel detector heads. Since Fourier rebinning requires complete sinogram data, these gaps must be filled before rebinning. Gap filling techniques may range in complexity from linear interpolation to forward projection of an image reconstructed from the 2D segment by using a system matrix which accounts for the missing data [12]. In general, however, a 3D iterative reconstruction is preferable to an hybrid one because the gap filling procedure followed by the rebinning is sensitive to noise propagated from regions with high attenuation.

Despite these difficulties, fast hybrid algorithms such as FORE+OSEM(AW) have been applied to whole-body FDG scans, and shown to provide for these studies an image quality comparable to fully 3D iterative reconstruction (see [90, 92] and the example in Fig. 4.13 below).

Fully 3D Iterative Reconstruction

Axial and transaxial undersampling techniques were developed to reduce the data to a manageable size while hybrid algorithms were developed to achieve fast reconstruction for clinical PET scanners with limited computer resources. With sufficient CPU power and disk capacity these early approaches are not needed. The application of fully 3D iterative reconstruction methods then allows to overcome the limitations of the hybrid algorithms discussed in the previous section.

We have seen that iterative reconstruction methods are conceptually independent of the 2-D or 3-D nature of the data. Several implementations have been described for 3-D data, based on the Space Alternating Generalized EM [93], on ML-EM and OSEM [94], on Bayesian estimation [26, 95], and on the row-action maximum likelihood [90]. All algorithms of the ML-EM type described above, for instance, can be readily generalized to 3D PET by replacing the system matrix $a_{j,i}$ describing the acquisition geometry (equation (29)) by its 3D equivalent, which takes into account the axial coordinates of the LORs. For block-iterative methods such as OSEM (see Eq. (42)), the set of LORs parameterized by the two transaxial sinogram indices s_k, ϕ_j in Eq. (5) and by the two axial coordinates i_θ, i_z in Eq. (56) must be divided into subsets. Most implementations simply subdivide the azimuthal index ϕ_j, exactly as in the 2D case. Each subset then contains all axial samples i_θ, i_z.

The benefit expected from fully 3D iterative reconstruction is easily demonstrated for scanners with large polar aperture, particularly in the presence of gaps. Fig. 4.12, for example, shows a high resolution phantom measured with the HHRT brain scanner. The bottom image was reconstructed with FORE+OSEM (AW), while the top one was reconstructed with OSEM3D(ANW), where "ANW" indicates that both the normalization and attenuation corrections are incorporated in the system matrix. The horizontal streak artifacts in the coronal section of the FORE+OSEM(AW) image are attributed to the gap filling step prior to FORE. Blurring can also be observed on the 3 brightest rods at the edge of the cylinder. When the polar angle is smaller, as with many clinical scanners, the bias introduced by FORE is small, and the benefit of 3D reconstruction is harder to visualize, especially at

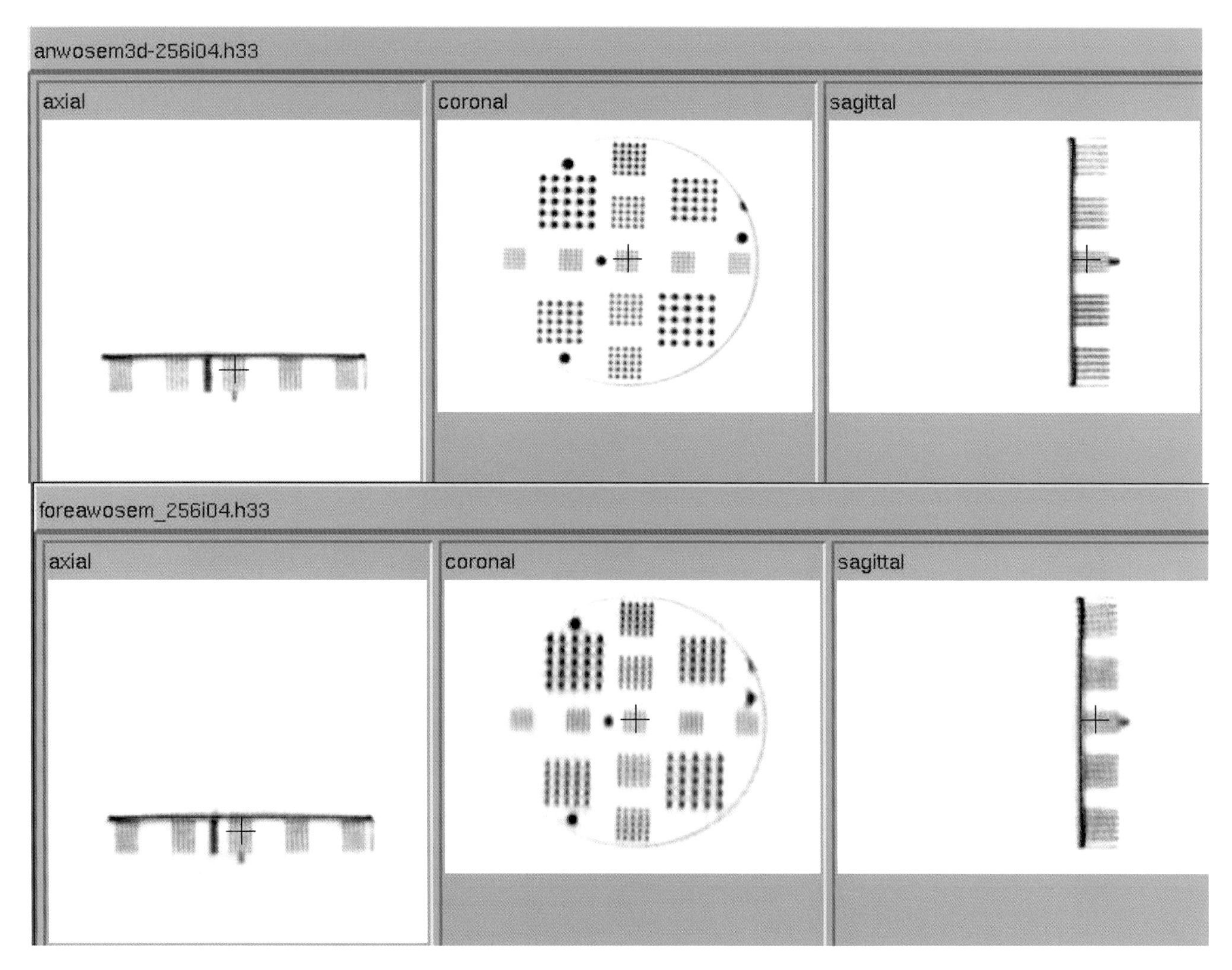

Figure 4.12. High resolution phantom data acquired on the HRRT: comparison of a fully 3D iterative reconstruction using OSEM3D(ANW) (top) and of a hybrid reconstruction with FORE+OSEM(AW) (bottom). Both images were reconstructed with 4 iterations and 16 subsets. The phantom is oriented vertically in the FOV of the scanner and the vertical axis on the coronal section is parallel to the axis of the scanner (courtesy K. Wienhard, Köln).

low count statistics and when regularization is achieved by post-reconstruction smoothing. This is illustrated by Fig. 4.13, which shows whole-body patient data processed with FORE+OSEM(AW) and OSEM3D(ANW). Note the similarity between the two reconstructions, even in regions with high attenuation (shoulder and neck for this patient with arms up).

In contrast with the algorithms illustrated above, the fully-3D image reconstruction developed by Leahy et al. [24, 26, 95, 96] is based on an extensive system model. The algorithm incorporates a shifted Poisson model that includes the statistics of true, scattered and random coincidences, as well as positron range, annihilation photon acolinearity, attenuation, sinogram sampling, detector dead-time and efficiency, block detector effects, and the spatially varying detector resolution due to parallax (depth of interaction) and Compton scatter in the scintillators (Chapter 2). Although the size of the system matrix is reduced using a factorized model and by taking advantage of symmetries, the computation time is necessarily longer than with simplified system models. This lead us to considerations of the potential for parallel-processing of image reconstructions on processor arrays.

Parallel Implementation of Iterative Reconstruction

The need for parallel implementation of the ML-EM algorithm was already recognized in the mid-eighties. Pioneering work proposed the use of a cluster of commodity PCs [97] or dedicated hardware [98]. But as soon as commercial parallel systems became available, dedicated algorithms were developed on high-end computers such as transputers [99, 100], hypercubes [101], meshes [102], rings [103], fine-grain message-

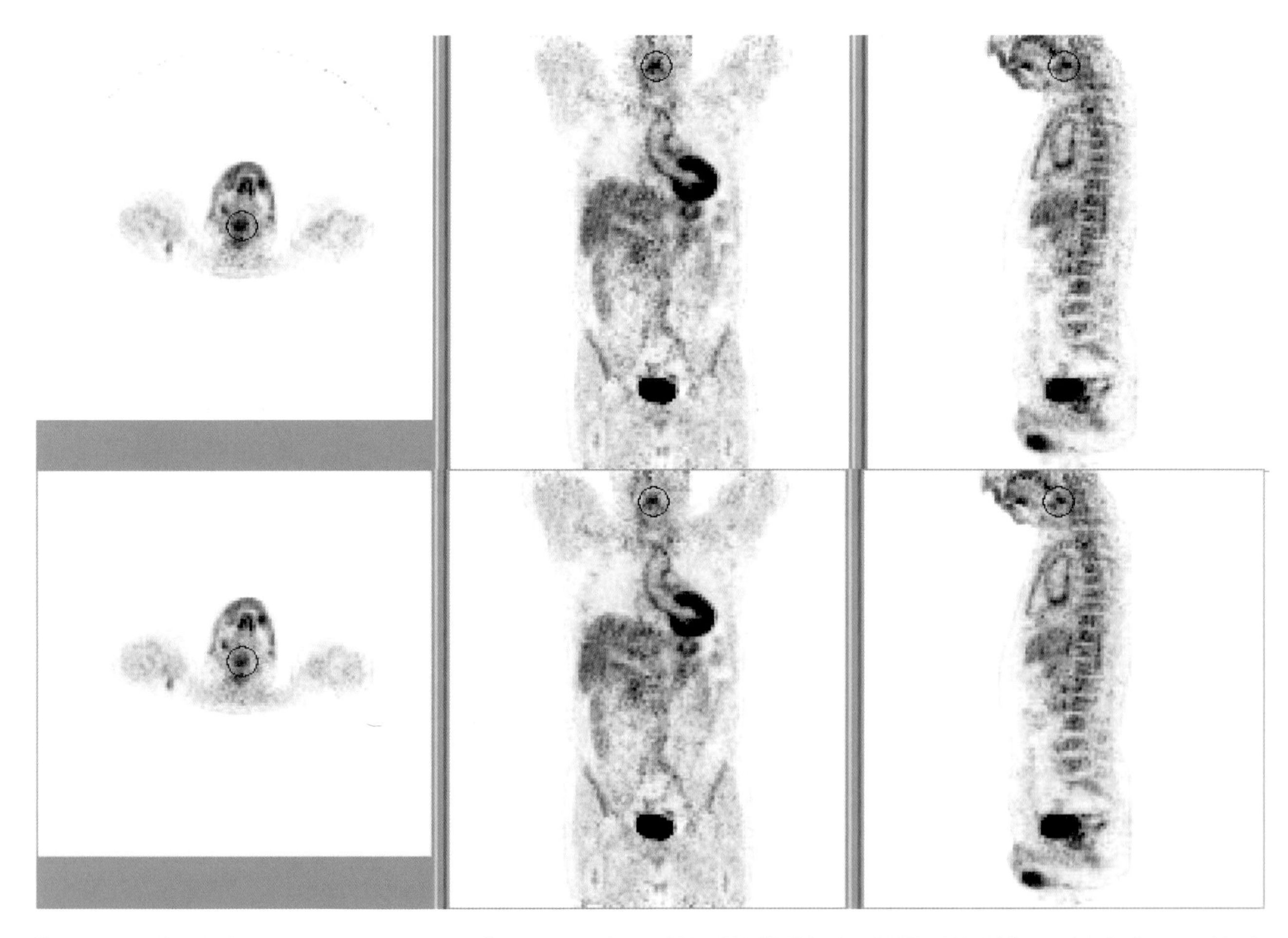

Figure 4.13. Whole-body FDG scan on an HR+ tomograph, reconstructed using FORE+OSEM(AW) (top) and OSEM3D(ANW) (bottom), in both cases with 4 iterations and 16 subsets. A 3D gaussian filter with FWHM 4 mm was applied after reconstruction. The orthogonal views are passing through the cursor (small circle in the neck area).

passing machines [104], linear arrays of DSPs [105] to cite a few examples. Recent efforts concentrated on using clusters of multi-processor PCs, sometimes called component off the shelf (COS), and combine both shared and distributed memory approaches. This choice is dictated by the cost/performance ratio of the hardware, by its flexibility and by the possibility to upgrade the system with faster and cheaper hardware in this very competitive market. One key problem in distributed computing is to optimize the balance between computation and communication amongst the nodes. The ultimate goal is to keep individual processors busy all the time by interleaving I/O and computation. A good measure of the performance of a parallel algorithm is how well the speed-up factor scales linearly with the number of nodes. In their work, Shattuck et al [106] describe a parallel implementation of the MAP-PCG reconstruction [26] using a master-slave model with 9 dual PC nodes. The work of Vollmar [107] describes a parallel extension of the OSEM3D reconstruction [92] and is also using a masterslave model with 7 quad PC nodes. By calculating the system matrix on the fly and neglecting the physics of the detection system these authors could handle very large reconstruction problems on the HRRT. The HRRT scanner acquires generally data in span 3(9) with a maximum ring difference of 67, which generates 3D data of 983 MB (326MB). The work of Jones et al [108] is another parallel extension of the OSEM3D reconstruction [92]. It uses a single program multiple-data (SPMD) rather than a master-slave model. These authors have shown that image space decomposition (ISD) and projection space decomposition (PSD) were roughly equivalent since the communication burden was large at forward projection when using ISD but was also large at backprojection when using PSD. However, by developing an efficient I/O subsystem and reorganizing the data, these authors finally favored the PSD model [109]. The performance of this parallel implementation of OSEM3D was shown to scale relatively well up to 16 nodes (32 processors). A commercial implementation of this computing cluster uses 8 nodes of dual Pentium 4 Xeon at 3.0 Ghz, and performs one iteration of OSEM3D in about 20 min for a 3D sinogram set of 983 MB and an image size of 256×256×207 (27 MB). Finally, the PARAPET initiative, currently known as the STIR project [110], has developed a generic, multi-platform and multi-scanner, implementation of OSEM3D using an object-oriented library [111, 112]. The parallel implementation uses a master-slave model and a PSD scheme. On a 12-node Parsytec CC system it provides a factor 7 speed-up compared to serial mode.

Acknowledgment

This work has been supported in part by NIH Grant CA-74135 and by the grant G.0174.03 of the FWO (Belgium).

References

1. Virador PRG, Moses WW, Huesman RH. Reconstruction in PET cameras with irregular sampling and depth of interaction capability. IEEE Trans Nucl Sc 1998;NS-45:1225–30.
2. Natterer F. The mathematics of computerized tomography. New York: Wiley, 1986.
3. Townsend DW, Isoardi RA, Bendriem B. Volume imaging tomographs. In: Bendriem B, Townsend DW (eds) The theory and practice of 3D PET, Dordrecht: Kluwer Academic, pp 111–32, 1998.
4. Natterer F,Wuebbeling F. Mathematical methods in image reconstruction. SIAM Monographs on Mathematical Modeling and Computation, 2001.
5. Kak AC, Slaney M. Principles of computerized tomographic imaging. New York: IEEE Press, 1988.
6. Barrett HH, Swindell W. Radiological Imaging. New York: Academic Press, 1981.
7. Barrett HH, Myers KJ, Foundations of Image Science. New York: Wiley, 2004.
8. Ramm AG, Katsevich AI. The radon transform and local tomography. New York: CRC Press, 1996.
9. Quinto ET. Exterior and limited-angle tomography in non-destructive evaluation. Inverse Problems 1998;14:339–53.
10. Wienhard K et al. The ECAT HRRT: Performance and first clinical application of a new high-resolution research tomograph. In: Records of the 2000 IEEE Medical Imaging Symposium, Lyon, France.
11. Karp JS, Muehllehner G, Lewitt RM. Constrained Fourier space method for compensation of missing data in emission-computed tomography. IEEE Trans Med Imag 1988;MI-7:21–5.
12. de Jong, H, Boellaard R, Knoess C, Lenox M, Michel C, Casey M, Lammertsma A. Correction methods for missing data in sinograms of the HRRT PET scanner. IEEE Trans Nucl Sc 2003;NS-50:1452–1456.
13. O'Sullivan JD. A fast sinc function gridding algorithm for Fourier inversion in computer tomography. IEEE Trans Med Imag 1985;MI-4:200–7.
14. Schomberg P, Timmer J. The gridding method for image reconstruction by Fourier transforms. IEEE Trans Med Imag 1995;MI-14:596–607.
15. Matej S, Lewitt RM. 3D FRP: Direct Fourier reconstruction with Fourier reprojection for fully 3D PET. IEEE Trans Nucl Sc 2001;NS-48:1378–85.
16. Press WH, Teukolsky SA, Vetterling WT, Flannery BP. Numerical recipes in C, Cambridge University Press 1992.
17. Egger ML, Joseph C, Morel VC. Incremental beamwise backprojection using geometrical symmetries for 3D PET reconstruction in a cylindrical scanner geometry. Phys Med Biol 1998; 43:3009–24.
18. Brady ML. A fast discrete approximation algorithm for the Radon transform. SIAM J Comp 1998;27:107–19.
19. Brandt A, Mann J, Brodski M, Galun M. A fast and accurate multilevel inversion of the Radon transform. SIAM J Appl Math 1999;60:437–62.
20. Bertero M, Boccacci P. Introduction to inverse problems in imaging. Bristol: IOP Publishing, 1998.
21. Huesman RH. The effects of a finite number of projection angles and finite lateral sampling of projections on the propagation of statistical errors in transverse section reconstruction. Phys Med Biol 1977;22:511–21.

22. J Fessler, 2001 NSS/MIC statistical image reconstruction short course notes, www.eecs.umich.edu/ fessler/papers/talk.html
23. Leahy R, Qi J. Statistical approaches in quantitative positron emission tomography. Statistics and Computing 2000;10:147–65.
24. Qi J, Leahy RM, Cherry SR, Chatziioannou, Farquhar TH. High-resolution 3D Bayesian image reconstruction using the micro-PET small animal scanner. Phys Med Biol 1998;43:1001–13.
25. Yavuz M, Fessler J. New statistical models for random precorrected PET scans. In: Duncan J, Gindi G (eds), Information Processing in Medical Imaging, 15th International Conference, Berlin: Springer, pp 190–203, 1998.
26. Qi J, Leahy R, Hsu C, Farquhar T, Cherry S. Fully 3D Bayesian image reconstruction for ECAT EXACT HR+. IEEE Trans Nucl Sci 1998;NS-45:1096–1103.
27. Lange K, Carson R. EM reconstruction algorithms for emission and transmission tomography. J Comp Ass Tomo 1984;8:306–16.
28. Mumcuoglu EU, Leahy R, Cherry SR, Zhou Z. Fast gradient-based methods for Bayesian reconstruction of transmission and emission PET images. IEEE Trans Med Im 1994;MI-13:687–701.
29. Erdogan H, Fessler JA. Accelerated monotonic algorithms for transmission tomography. In Proc IEEE Intl Conf on Image Processing 1998;2:6804.
30. Nuyts J, Michel C, Dupont P. Maximum-likelihood expectation maximization reconstruction of sinograms with arbitrary noise distribution using NEC transformations. IEEE Trans Med Imag 2001;MI-20:365–75.
31. Bai C, Kinahan P, Brasse D, Comtat C, Townsend D. Postinjection single photon transmission tomography with ordered-subset algorithms for whole-body PET. IEEE Trans Nucl Sci 2002;NS-49:74–81.
32. Lewitt RM. Alternatives to voxels for image representation in iterative reconstruction algorithms. Phys Med Biol 1992;37:705–16.
33. Matej S, Herman GT, Narayanan TK, Furie SS, Lewitt RM, Kinahan PE. Evaluation of task-oriented performance of several fully 3D PET reconstruction algorithms. Phys Med Biol 1994; 39:355–67.
34. Geman S, Geman D. Stochastic relaxation, Gibbs distributions, and the Bayesian restoration of images. IEEE Trans Patt Anal Mach Intel 1984;PAMI-6:721–41.
35. Mumcuoglu EU et al. Bayesian reconstruction of PET images, methodology and performance analysis. Phys Med Biol 1996; 41:1777–807.
36. Leahy RM, Yan X. Incorporation of anatomical MR data for improved functional imaging with PET. In: Information Processing in Medical Imaging: 12th International Conference, Berlin: Springer-Verlag, pp 105–20, 1991.
37. Gindi G, Lee M, Rangarajan A, Zubal G. Bayesian reconstruction of functional images using anatomical information as priors. IEEE Trans Med Imag 1993;12:670–80.
38. Bowsher JE, Johnson VE, Turkington TG, Jaszczak RJ, Floyd Jr CE, Coleman RE. Bayesian reconstruction and use of anatomical a priori information for emission tomography. IEEE Trans Med Imag 1996;MI-15:673–86.
39. Comtat C, Kinahan P, Fessler F, Beyer T, Townsend D, Defrise M, Michel C. Clinically feasible reconstruction of whole-body PET/CT data using blurred anatomical labels. Phys Med Biol 2002;47:1–20.
40. Baete K, Nuyts J, Van Paesschen W, Suetens P, Dupont P. Anatomical based FDG-PET reconstruction for the detection of hypo-metabolic regions in epilepsy. To appear in IEEE Trans Med Imag 2004.
41. Obi T, Matej S, Lewitt RM, Herman GT. 2.5D simultaneous multislice reconstruction by iterative algorithms from Fourier-rebinned PET data. IEEE Trans Med Imag 2000;MI-19:474–84.
42. Nichols TE, Qi J, Leahy RM. Continuous time dynamic pet imaging using list mode data. In: Colchester A et al. (eds), Information Processing in Medical Imaging, 12th International Conference, Berlin: Springer, pp 98–111, 1999.
43. Asma E, Nichols TE, Qi J, Leahy RM. 4D PET image reconstruction from list-mode data. Records of the 2000 IEEE Medical Imaging Symposium, Lyon (France), paper 15–57.
44. Fessler J. Penalized weighted least-square image reconstruction for PET. IEEE Trans Med Imag 1994;MI-13:290–300.
45. Lalush D, Tsui B. A fast and stable maximum a posteriori conjugate gradient reconstruction algorithm. Med Phys 1995; 22:1273–1284.
46. Leahy R, Byrne C. Recent developments in iterative image reconstruction for PET and SPECT. IEEE Trans Med Imag 2000;19:257–60.
47. Lange K, Hunter D, Yang I. Optimization Transfer algorithms using surrogate objective functions. J Comp Graph Stat 2000; 9:1–59.
48. Daube-Witherspoon ME, Muehllehner G. An iterative image space reconstruction algorithm for volume ECT. IEEE Trans Med Imag 1986;MI-5:61–6.
49. De Pierro A. On the relation between the ISRA and the EM algorithm for positron emission tomography. IEEE Trans Med Imag 1993;MI-12:328–33.
50. Levitan E, Herman GT. A maximum a posteriori expectation maximization algorithm for image reconstruction in emission tomography. IEEE Trans Med Imag 1987;MI-6:185–92.
51. Dempster A, Laird N, Rubin D. Maximum likelihood from incomplete data via the EM algorithm. Journal of the Royal Statistical Society 1977;39:1–38.
52. Shepp L A, Vardi Y. Maximum likelihood reconstruction for emission tomography. IEEE Trans Med Imag 1982;MI-1:113–22.
53. Barrett HH, White T, Parra L. List-mode likelihood. J Opt Soc Am 1997;14:2914–23.
54. Reader A, Erlandsson K, Flower M, Ott R. Fast, accurate, iterative reconstruction for low-statistics positron volume imaging. Phys Med Biol 1998;43:835–46.
55. Levkovitz R, Falikman D, Zibulevsky M, Ben-Tal A, Nemirovski A. The design and implementation of COSEM, an iterative algorithm for fully 3D list-mode data. IEEE Trans Med Imag 2001;MI-20:633–42.
56. Huesman RM, Klein G, Moses W, Qi J, Reutter B, Virador P. List-mode maximum-likelihood reconstruction applied to positron emission mammography (PEM) with irregular sampling. IEEE Trans Med Imag 2000;MI-19:532–7.
57. Barrett HH, Wilson DW, Tsui BMW. Noise properties of the EM algorithm: I. Theory. Phys Med Biol 1994;39:833–46.
58. Wilson DW, Tsui BMW, Barrett HH. Noise properties of the EM algorithm: II. Monte Carlo simulations. Phys Med Biol 1994; 39:847–71.
59. De Pierro AR, Yamagishi MEB. Fast EM-like methods for maximum "a posteriori" estimates in emission tomography. IEEE Trans Med Imag 2001;MI-20:280–8.
60. Veklerov E, Llacer J. Stopping rule for the MLE algorithm based on statistical hypothesis testing. IEEE Trans Med Imag 1987;MI-6:313–19.
61. Selivanov V, Lapointe D, Bentourkia M, Lecomte R. Cross-validation stopping rule for ML-EM reconstruction of dynamic PET series: effect on image quality and quantitative accuracy. IEEE Trans Nucl Sc 2001;NS-48:883–9.
62. Hudson H, Larkin R. Accelerated image reconstruction using ordered subsets of projection data. IEEE Trans Med Imag 1994;MI-13:601–9.
63. Browne J, De Pierro AR. A row action alternative to the EM algorithm for maximizing likelihoods in emission tomography. IEEE Trans Med Imag 1996;MI-15:687–99.
64. Byrne CL. Convergent block-iterative algorithms for image reconstruction from inconsistent data. IEEE Trans Imag Proc 1997;IP-6:1296–304.
65. Hebert T, Leahy RM. Fast methods for including attenuation in the EM algorithm. IEEE Trans Nucl Sc 1990;NS-37:754–8.
66. Comtat C, Kinahan PE, Defrise M, Michel C, Townsend DW. Fast reconstruction of 3D PET with accurate statistical modeling. IEEE Trans Nucl Sc 1998;NS-45:1083–9.
67. Stayman J W, Fessler J A. Fast methods for approximating the resolution and covariance for SPECT, in IEEE Nuclear Science and Medical Imaging Conference; 2002, Norfolk, VA, paper M3-146.
68. Stayman JW, Fessler JA. Regularization for uniform spatial resolution properties in penalized-likelihood image reconstruction. IEEE Trans Med Imag 2000;MI-19:601–615.
69. Nuyts J, Fessler JA. A penalized-likelihood image reconstruction method for emission tomography, compared to post-smoothed

maximum-likelihood with matched spatial resolution. IEEE Trans Med Imag 2003;MI-22:1042–1052.
70. Orlov SS. Theory of three-dimensional reconstruction. 1. Conditions of a complete set of projections. Sov Phys Crystallography 1976;20:429–33.
71. Stearns CW, Chesler DA, Brownell GL. Accelerated image reconstruction for a cylindrical positron tomograph using Fourier domain methods. IEEE Trans Nucl Sci 1990;NS-37:773–7.
72. Colsher JG. Fully three-dimensional PET. Phys Med Biol 1980; 25:103–115.
73. Defrise M, Townsend DW, Clack R. Image reconstruction from truncated two-dimensional projections. Inverse Problems 1995;11:287–313.
74. Defrise M, Townsend DW, Deconinck F. Statistical noise in three-dimensional positron tomography. Phys Med Biol 1990; 35:131–138.
75. Stearns CW, Crawford CR, Hu H. Oversampled filters for quantitative volumetric PET reconstruction. Phys Med Biol 1994;39:381–8.
76. Kinahan PE, Rogers JG. Analytic three-dimensional image reconstruction using all detected events. IEEE Trans Nucl Sci 1990;NS-36:964–8.
77. Daube-Witherspoon ME, Muehllehner G. Treatment of axial data in three-dimensional PET. J Nucl Med 1987;28:1717–24.
78. Erlandsson K, Esser PD, Strand S-E, van Heertum RL. 3D reconstruction for a multi-ring PET scanner by single-slice rebinning and axial deconvolution. Phys Med Biol 1994;39:619–29.
79. Tanaka E, Amo Y. A Fourier rebinning algorithm incorporating spectral transfer efficiency for 3D PET. Phys Med Biol 1998; 43:739–46.
80. Lewitt RM, Muehllehner G, Karp JS. Three-dimensional reconstruction for PET by multi-slice rebinning and axial image filtering. Phys Med Biol 1994;39:321–40.
81. Defrise M, Kinahan PE, Townsend DW, Michel C, Sibomana M, Newport DF. Exact and approximate rebinning algorithms for 3D PET data. IEEE Trans Med Imag 1997;MI-16:145–58.
82. Defrise M, Liu X. A fast rebinning algorithm for 3D PET using John's equation, Inverse Problems 1999;15:1047–65.
83. Liu X, Defrise M, Michel C, Sibomana M, Comtat C, Kinahan PE, Townsend DW. Exact rebinning methods for three-dimensional positron tomography. IEEE Trans Med Imag 1999;MI-18:657–64.
84. Matej S, Karp JS, Lewitt RM, Becher AJ. Performance of the Fourier rebinning algorithm for 3D PET with large acceptance angles. Phys Med Biol 1998;43:787–97.
85. Michel C, Hamill J, Panin V, Conti M, Jones J, Kehren F, Casey M, Bendriem B, Byars L, Defrise M. FORE(J)+OSEM2D versus OSEM3D reconstruction for Large Aperture Rotating LSO, In: IEEE Nuclear Science and Medical Imaging Conference; 2002, Norfolk, VA, paper M7-93.
86. Kinahan PE, Fessler JA, Karp JS. Statistical image reconstruction in PET with compensation for missing data. IEEE Trans Nucl Sc 1997;NS-44:1552–7.
87. Stearns CW, Fessler JA. 3D PET reconstruction with FORE and WLS-OS-EM. In: IEEE Nuclear Science and Medical Imaging Conference; 2002, Norfolk, VA, paper M5-5.
88. Janeiro L, Comtat C, Lartizien C, Kinahan P, Defrise M, Michel C, Trébossen R, Almeida P. NEC-scaling applied to FORE+OSEM. In: IEEE Nuclear Science and Medical Imaging Conference; 2002, Norfolk, VA, paper M3-71.
89. Alessio A, Sauer K, Bouman C. PET statistical reconstruction with modeling of axial effects of FORE. In: IEEE Nuclear Science and Medical Imaging Conference; 2003, Portland, OR, paper M11-194.
90. Daube-Witherspoon ME, Matej S, Karp JS, Lewitt RM. Application of the Row Action Maximum Likelihood Algorithm with spherical basis functions to clinical PET imaging. IEEE Trans Nucl Sc 2001;NS-48:24–30.
91. Kinahan PE, Michel C, Defrise M, Townsend DW, Sibomana M, Lonneux M, Newport DF, Luketich JD. Fast iterative image reconstruction of 3D PET data. In: IEEE Nuclear Science and Medical Imaging Conference; 1996, Anaheim, CA, 1918–22.
92. Liu X, Comtat C, Michel C, Kinahan PE, Defrise M, Townsend DW. Comparison of 3D reconstruction with 3D OSEM and with FORE+OSEM for PET. IEEE Trans Med Imag 2001;MI-20:804–14.
93. Ollinger J. Maximum likelihood reconstruction in fully 3D PET via the SAGE algorithm. In Proc. 1996 IEEE Nucl. Sci. Symp. Medical Imaging Conf., Anaheim, CA, 1594–1598.
94. Johnson C, Seidel S, Carson R, Gandler W, Sofer A, Green M, Daube-Witherspoon M. Evaluation of 3-D reconstruction algorithms for a small animal PET camera. IEEE Trans Nucl Sci 1997;NS-44:1303–1308.
95. Qi J, Leahy RM. Resolution and noise properties of MAP reconstruction for fully 3D PET. IEEE Trans Med Imag MI-19:493–506.
96. Bai B, Li Q, Holdsworth C, Asma E, Tai Y, Chatziioannou A, Leahy R. Modelbased normalization for iterative 3D PET image reconstruction. Phys Med Biol 2002;47:2773–2784.
97. Llacer J and Meng J. Matrix-based image reconstruction methods for tomography. IEEE Trans Nucl Sci 1985; NS-32:855–864.
98. Jones W, Byars L, Casey M. Design of a super fast three-dimensional projection system for positron emission tomography. IEEE Trans Nucl Sci 1990;NS-37:800–804.
99. Barresi S, Bollini D, Del Guerra A. Use of a transputer system for fast 3-D image reconstruction in 3-D PET. IEEE Trans Nucl Sci 1990;NS-37:812–816.
100. Atkins S, Murray D, Harrop R. Use of transputers in a 3-D positron emission tomograph. IEEE Trans Med Imag 1991; MI-10:276–283.
101. Chen C.-M., Lee S.-Y., Cho Z. Parallelization of the EM algorithm for 3-D PET image reconstruction. IEEE Trans Med Imag 1991;MI-10:513–522.
102. Rajan K, Patnaik L, Ramakrishna J. High-speed computation of the EM algorithm for PET image reconstruction. IEEE Trans Nucl Sci 1994;NS-41:1721–1728.
103. Johnson C, Yan Y, Carson R, Martino R, Daube-Witherspoon M. A system for the 3-D reconstruction of retracted-septa PET data using the EM algorithm. IEEE Trans Nucl Sci 1995; NS-42:1223–1227.
104. Cruz-Rivera J, DiBella E, Wills D, Gaylord T, Glytsis E. Parallelized formulation of the maximum likelihood expectation maximization algorithm for fine-grain message-passing architectures. IEEE Trans Med Imag 1995;MI-14:758–762.
105. Rajan K, Patnaik L, Ramakrishna J. Linear array implementation of the EM reconstruction algorithm for PET image reconstruction. IEEE Trans Nucl Sci 1995;NS-42:1439–1444.
106. Shattuck D, Rapela J, Asma E, Chatzioannou A, Qi J, Leahy R. Internet2-based 3-D PET image reconstruction using a PC cluster. Phys Med Biol 2002;47:2785–2795.
107. Vollmar S, Michel C, Treffert J, Newport D, Casey M, Knoss C, Wienhard K, Liu X, Defrise M, Heiss W-D. HeinzelCluster: accelerated reconstruction for FORE and OSEM3D. Phys Med Biol 2002;47:2651–2658.
108. Jones J, Jones W, Kehren F, Newport D, Reed J, Lenox M, Baker K, Byars L, Michel C, Casey M. SPMD cluster-based parallel 3D OSEM. IEEE Trans Nucl Sci 2003;NS-50:1498–1502.
109. Jones J, Jones W, Kehren F, Burbar Z, Reed J, Lenox M, Baker K, Byars L, Michel C, Casey M. Clinical Time OSEM3D: Infrastructure Issues. In: IEEE Nuclear Science and Medical Imaging Conference; 2003, Portland, OR, paper M10-244.
110. The current home page of the PARAPET/STIR project is http://stir.sourceforge.net/homepage.shtm
111. Bettinardi V, Pagani E, Gilardi M-C, Alenius S, Thielemans K, Teras M, Fazio F. Implementation and evaluation of a 3D One Step Late reconstruction algorithm for 3D Positron Emission Tomography studies using Median Root Prior. Eur J Nucl Med 2002;29:7–18.
112. Jacobson M, Levkovitz R, Ben Tal A, Thielemans K, Spinks T, Belluzzo D, Pagani E, Bettinardi V, Gilardi MC, Zverovich A, Mitra G. Enhanced 3D PET OSEM reconstruction using inter-update Metz filtering. Phys Med Biol 2000;45:2417–2439.

5 Quantitative Techniques in PET*

Steven R Meikle and Ramsey D Badawi

Introduction

PET has long been regarded as a quantitative imaging tool. That is, the voxel values of reconstructed images can be calibrated in absolute units of radioactivity concentration with reasonable accuracy and precision. The ability to accurately and precisely map the radiotracer concentration in the body is important for two reasons. First, it ensures that the PET images can be interpreted correctly since they can be assumed to be free of physical artefacts and to provide a true reflection of the underlying physiology. Second, it enables the use of tracer kinetic methodology to model the time-varying distribution of a labelled compound in the body and quantify physiological parameters of interest.

The reputation of PET as a quantitative imaging tool is largely based on the fact that an exact correction for attenuation of the signal due to absorption of photons in the body is theoretically achievable. However, accurate attenuation correction is not so easy to achieve in practice and there are many other factors, apart from photon attenuation, that potentially impact on the accuracy and precision of PET measurements. These include count-rate losses due to dead time limitations of system components, variations in detector efficiency, acceptance of unwanted scattered and random coincidences and dilution of the signal from small structures (partial volume effect). The ability to accurately measure or model these effects and correct for them, while minimizing the impact on signal-to-noise ratio, largely determines the accuracy and precision of PET images.

This chapter discusses the various sources of measurement error in PET. Methodological approaches to correct for these sources of error are described, and their relative merits and impact on the quantitative accuracy of PET images are evaluated. The sequence of the following sections corresponds approximately to the order in which the various corrections are typically applied. It should be noted, however, that the particular sequence of corrections varies from scanner to scanner and depends on the choice of algorithms.

Randoms Correction

Origin of Random Coincidences

Random coincidences, also known as "accidental" or "chance" coincidences, arise because of the finite width of the electronic time window used to detect true coincidences. This finite width allows the possibility that two uncorrelated single detection events occurring sufficiently close together in time can be mistakenly identified as a true coincidence event, arising from one annihilation. This is shown schematically in Fig. 5.1.

The rate at which random coincidences occur between a detector pair is related to the rate of single events on each detector and to the width of the time window. The exact relationship is dependent upon the implementation of the counting electronics. Figure 5.2 shows an implementation whereby each timing signal opens a gate of duration τ; if gates on two channels are open at the same time, a coincidence is recorded. If there is a timing signal on channel i at time T, there will be a coincidence on the relevant line-of-response L_{ij} if there is a timing signal on channel j at any time between $T - \tau$ and $T + \tau$. Therefore, the total time during which a coincidence may be recorded with the event on channel i (a

* Figures 1–3, 5, 6, 12–16 and 19–21 are reproduced from Valk PE, Bailey DL, Townsend DW, Maisey MN. Positron Emission Tomography: Basic Science and Clinical Practice. Springer-Verlag London Ltd 2003, 91–114.

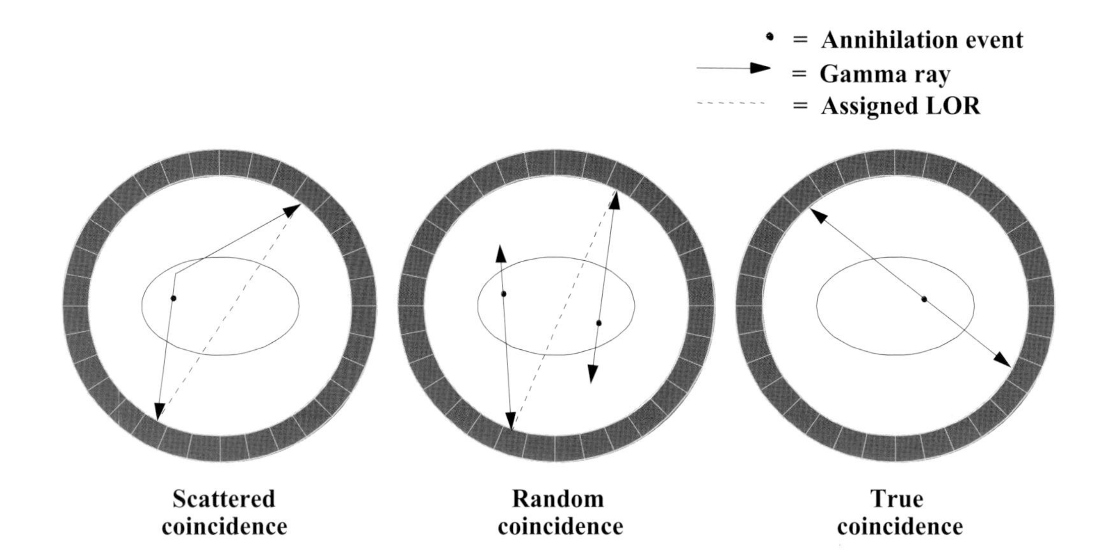

Figure 5.1. Types of coincidence event recorded by a PET scanner.

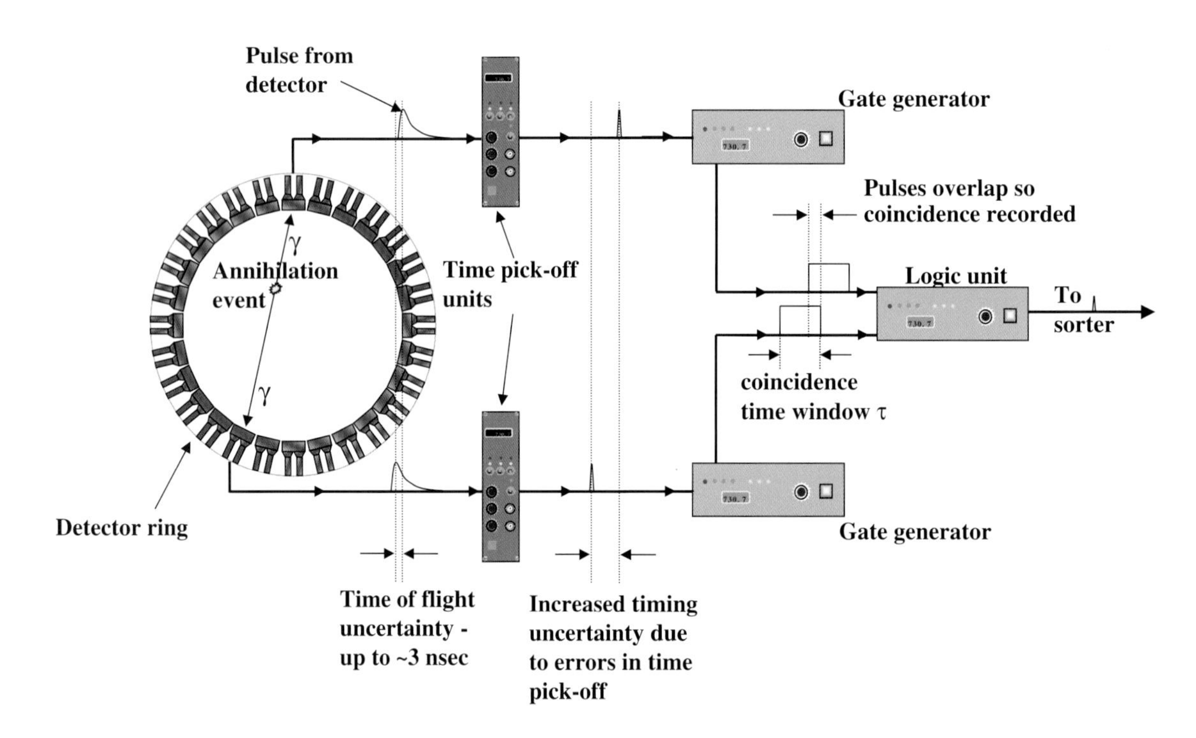

Figure 5.2. Figure 5.2 Example coincidence circuitry. Each detector generates a pulse when a photon deposits energy in it; this pulse passes to a time pick-off unit. Timing signals from the pick-off unit are passed to a gate generator which generates a gate of width τ. The logic unit generates a signal if there is a voltage on both inputs simultaneously. This signal then passes to the sorting circuitry.

parameter known as the resolving time of the circuit, or the coincidence time window) is 2τ. So, if the rate of single events in channel i is r_i counts per second, then in one second the total time during which coincidences can be accepted on L_{ij} will be $2\tau r_i$. If we can assume that the single events occurring on channel j are uncorrelated with those on channel i (i.e., there are no true coincidences), then C_{ij}, the number of random coincidences on L_{ij} per second, will be given by

$$C_{ij} = 2\tau\, r_i r_j \tag{1}$$

where r_j is the rate of single events on channel j. While it is obviously not generally true that there is no correlation between the single events on channel i and the

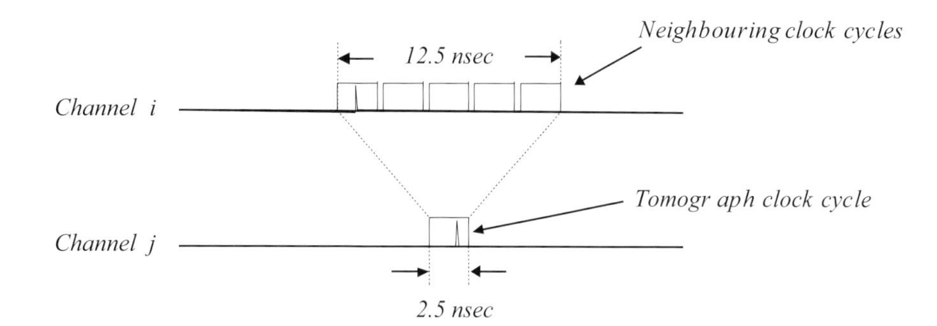

Figure 5.3. Detecting coincidences using the tomograph clock.

single events on channel j, the number of single events acquired during a PET acquisition is typically 1 to 2 orders of magnitude greater than the number of coincidences. In such an environment, equation (1) provides a good estimate of the random coincidence rate.

The timing of commercial tomographs is usually governed by a system clock. A timing signal on channel i is thus assigned to a particular clock cycle. If there is a timing signal on channel j within a certain range of, say, n neighbouring clock cycles, a coincidence is recorded on L_{ij} (Fig. 5.3). Therefore the randoms rate on L_{ij} would be given by

$$C_{ij} = nt_c r_i r_j \tag{2}$$

where t_c is the duration of a single clock cycle. A typical BGO tomograph might have a 2.5 nanosecond clock cycle, and $n = 5$ clock cycles. Thus, the total coincidence time window nt_c (equivalent to 2τ for an analog system) would equal 12.5 nanoseconds.

Equations (1) and (2) indicate that the overall randoms rate for an acquisition will change at a rate proportional to the square of the overall singles rate. Provided dead time is small, this means that for a given source distribution the randoms will change roughly in proportion to the square of the activity concentration.

Random coincidences can form a significant fraction of all recorded coincidences in PET imaging, particularly if large amounts of activity are used or if scans are performed in 3D mode. The number of randoms detected may be reduced by shortening the coincidence window. However, the window must be large enough to prevent loss of true coincidences due to the difference in arrival times (which may be up to 2 ns for an annihilation pair originating 30 cm from the centre of the tomograph) or statistical variations in the triggering of the event timing circuitry. Thus, selection of the coincidence window is a trade-off between minimising acceptance of randoms and loss of sensitivity to true coincidences. The coincidence window is typically set to 3 to 4 times the full width half maximum (FWHM) timing resolution of the tomograph.

The use of fast scintillators such as LSO or GSO reduces timing uncertainty (compared to that obtainable with slower scintillators such as BGO or NaI), but the window width cannot be less than 3 nsec to 4 nsec without accounting for time-of-flight effects. Randoms may also be reduced by shielding the detectors from activity that lies outside the tomograph field of view–this reduces the singles rates without adversely affecting sensitivity to true coincidences [1, 2].

Randoms tend to be fairly uniformly distributed across the field of view. This contrasts with true coincidences, which follow activity concentration and are reduced in regions of high attenuation. Thus, the fraction of random coincidences in regions of high attenuation can become very large and, if uncorrected, substantial quantitative errors can arise.

Corrections for Random Coincidences

Tail Fitting

Because the distribution of random coincidences in sinogram or projection space tends to be a slowly changing function, it may be possible to estimate the distribution within the object by fitting a function such as a paraboloid or Gaussian to the tails falling outside the object. This method requires that the object subtend only a fraction of the field of view, so that the tails are of reasonable length and contain a reasonable number of counts--otherwise small changes in the tails will result in large changes in the randoms estimate. In some systems this method has been used to correct for both scatter and randoms simultaneously [3].

Estimation from Singles Rates

The total number of randoms on a particular line of response L_{ij} can, in principle, be determined directly from the singles rates r_i and r_j using equation (1) or (2). Consider an acquisition of duration T. The random coincidences R_{ij} in the data element corresponding to the line of response L_{ij} may be found by integrating equation (1) or (2) over time:

$$R_{ij} = \int_0^T C_{ij}(t)dt = 2\tau \int_0^T r_i(t) r_j(t)dt \tag{3}$$

If $r_i(t)$ and $r_j(t)$ change in the same way over time, we can factor out this variation to obtain

$$R_{ij} = 2\tau s_i s_j \int_0^T f(t)dt = k s_i s_j \tag{4}$$

where k is a constant and s_i and s_j are the single event rates at, say, the start of the acquisition. For an emission scan, $f(t)$ is simply the square of the appropriate exponential decay expression, provided that tracer redistribution can be ignored. R_{ij} can then be determined from the single events accumulated on channels i and j over the duration of the acquisition. It should be noted that the randoms total is proportional to the integral of the product of the singles rates, and not simply the product of the integrated singles rates. Failure to account for this leads to an error of about 4% when the scan duration T is equal to the isotope half-life $T_{\frac{1}{2}}$, and about 15% when $T = 2T_{\frac{1}{2}}$.

For coincidence-based transmission scans, where positron-emitting sources are rotating in the field of view, $f(t)$ becomes a complicated function dependent on position as well as time, and equation (4) is no longer valid. However, in principle the total number of randoms could still be obtained by sampling the singles rates with sufficiently high frequency.

The singles rates used for calculating randoms should ideally be obtained from data that have already been qualified by the lower energy level discriminator – they are not the same as the singles rates that determine the detector dead-time. Correction schemes have been implemented which use detector singles rates prior to LLD qualification [4], but the differences between the energy spectrum of events giving rise to randoms and that giving rise to trues and scatter must be carefully taken into account. These differences are dependent upon the object being imaged and upon the count-rate, since pulse pile up can skew the spectra [5].

In its simplest form this method does not account for the electronics dead-time arising from the coincidence processing circuitry (to which the randoms in the coincidence data are subject).

Delayed Coincidence Channel Estimation

The most accurate (and currently the most commonly implemented) method for estimating random coincidences is the delayed channel method. In this scheme, a duplicate data stream containing the timing signals from one channel is delayed for several times the duration of the coincidence window before being sent to the coincidence processing circuitry. This delay removes the correlation between pairs of events arising from actual annihilations, so that any coincidences detected are random. The resulting coincidences are then subtracted from the coincidences in the prompt channel to yield the number of true (and scattered) coincidences. The coincidences in the delayed channel encounter exactly the same dead-time environment as the coincidences in the prompt channel, and the accuracy of the randoms estimate is not affected by the time-dependence of the activity distribution.

While accurate, this method has two principal drawbacks. Firstly, the increased time taken to process the delayed coincidences contributes to the overall system dead time. Secondly, and more importantly, the estimates of the randoms on each line-of-response are individually subject to Poisson counting statistics. The noise in these estimates propagates directly back into the data, resulting in an effective doubling of the statistical noise due to randoms. This compares poorly to the estimation from singles method, since the singles rates are typically two orders of magnitude greater than the randoms rates, so that the fractional noise in the resulting randoms estimate is effectively negligible. To reduce noise, the delayed channel can be implemented with a wider coincidence time window. However, this will further increase the contribution of delayed channel coincidences to system dead time.

Randoms Variance Reduction

Where randoms form a significant fraction of the acquired events, as is frequently the case in 3D imaging, it becomes desirable to obtain randoms estimates that are accurate but contain less noise than those obtained using the delayed channel method. Most delayed channel implementations allow the acquisition of separate datasets from the prompt and delayed coincidence channels – this allows the possibility of post-processing the randoms estimate to reduce noise, prior to subtraction from the prompt coincidence channel data.

The simplest form of variance reduction is to smooth the delayed data. The success of this approach will depend somewhat on the architecture of the scanner. In full-ring block detector systems, there are significant differences between the efficiency of adjacent detectors. This information is lost during smoothing, and if unaccounted for, high-frequency circular artefacts can appear in the reconstructed images [6]. However, in rotating systems, lines of response may be sampled by many detectors (particularly in the centre of the field of view), so that efficiency differences become less important. Caution must still be exercised, because rotational sampling effects can result in varying sensitivity to randoms across the field of view [7]. One solution is to smooth only over lines of response which share a common radius.

More accurate methods of variance reduction can be envisaged. A randoms sinogram consists of noisy estimates of the R_{ij}, the randoms in the prompt data. A typical data acquisition may consist of a few million such estimates (one for each LOR), but there may only be a few thousand of the singles values s_i (one for each detector element). There is therefore substantial redundancy in the data, which may be used to reduce the effects of statistical noise. Let us consider two opposing groups of N detectors, A and B. Detector i is a member of group A and j is a member of B (Fig. 5.4).

If the singles flux varies in the same way for all detectors for the duration of the acquisition, so that equation (4) is valid, then R_{iB}, the sum of the randoms on all the lines of response joining detector i and group B may be written

$$R_{iB} = ks_i \sum_{j=1}^{N} s_j \tag{5}$$

similarly, R_{jA}, the sum of the randoms on all the lines of response joining detector j and group A may be written

$$R_{jA} = ks_j \sum_{i=1}^{N} s_i \tag{6}$$

Now R_{AB}, the sum of all the randoms over all possible lines of response between groups A and B is simply the sum of R_{jA} over all possible j:

$$R_{AB} = \sum_{j=1}^{N}\left[ks_j \sum_{i=1}^{N} s_i\right] = k\left[\sum_{i=1}^{N} s_i\right]\left[\sum_{j=1}^{N} s_j\right] \tag{7}$$

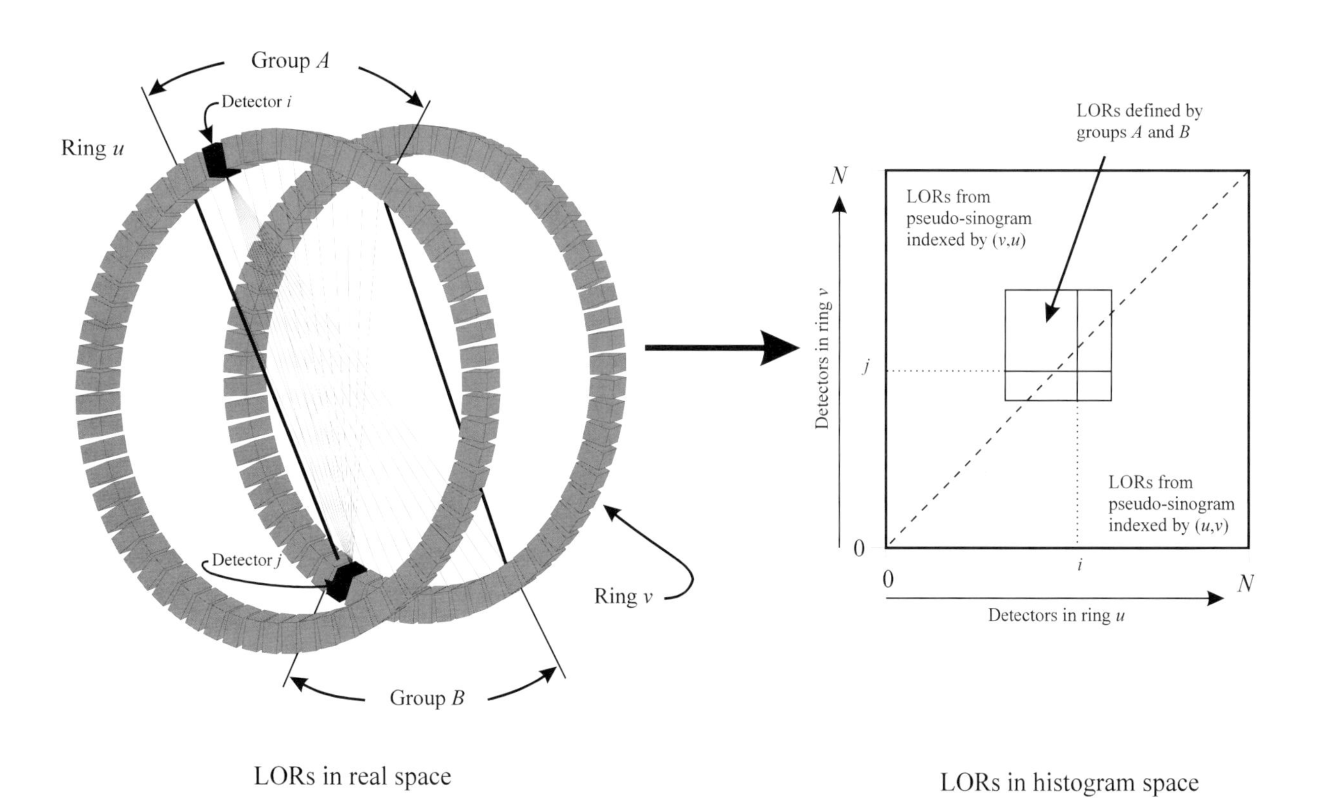

Figure 5.4. Accurate randoms variance reduction. To obtain a variance-reduced estimate of the number of random coincidences in the LOR joining detectors i and j, the product of the mean values of the LORs in each of the two LOR fans shown is calculated, and divided by the mean value of all possible LORs between detectors in groups A and B. For ease of implementation, the LOR data from the relevant sinograms can be re-binned into histograms as shown. (From [9], with kind permission from Kluwer Academic publishers.)

If we multiply equations (5) and (6) and divide by equation (7), we get

$$\frac{R_{jA} \times R_{iB}}{R_{AB}} = ks_i s_j = R_{ij} \tag{8}$$

All of the terms on the left hand side of equation (8) can be obtained from the data, and we have obtained another estimate of R_{ij}. However, if N is large enough the variance of this estimate is less than that of the original estimate, since the line-of-response sums R_{jA}, R_{iB} and R_{AB} are all larger than R_{ij} by factors of approximately N, N and N^2 respectively (assuming that there are roughly the same number of randoms on each line of response). This method was devised by Casey and Hoffman [8], who also showed that the ratio Q of the variance of the noise-reduced estimate and the original estimate of R_{ij} is given by

$$Q = \frac{2N+1}{N^2} \tag{9}$$

so that there is an improvement in the noise provided N, the number of detectors per group, is three or more.

Several related algorithms have been developed and applied to the problem of randoms variance reduction, but the one described here has been shown to be the most accurate [9]. The only significant drawback of this method (compared to direct subtraction of the delayed channel data) is that acquiring a separate randoms sinogram doubles the size of the dataset. This can be a particular problem for fast dynamic scanning in 3D mode, where sorter memory and data transfer time can be a limiting factor.

Normalisation

Lines of response in a PET dataset have differing sensitivity for a variety of reasons including variations in detector efficiency, solid angle subtended and summation of neighbouring data elements. Information on these variations is required for the reconstruction of quantitative and artefact-free images – indeed, most algorithms require that these variations be removed prior to reconstruction. The process of correcting for these effects is known as normalisation, and the individual correction factors for each LOR are referred to as normalisation coefficients.

Causes of Sensitivity Variations

Summing of Adjacent Data Elements

It is common practice to sum adjacent data elements in order to simplify reconstruction or to reduce the size of the dataset. This is usually performed axially, but may also be performed radially (a process known as "mashing"). Summation of data elements axially cannot be performed uniformly across the entire field of view and image planes at the ends of the field of view have substantially reduced sensitivity compared to those in the centre. This effect is fairly simple to account for, since the degree of summing is always known. However, it can complicate the process of correcting for other effects if the summing is performed prior to normalisation [10, 11].

Rotational Sampling

In a rotating system, LORs at the edge of the field of view are sampled just once per half-rotation, while those near the centre are sampled many times (see Fig. 5.5). As a result, sensitivity falls as radius increases.

Detector Efficiency Variations

In a block detector system, detector elements vary in efficiency because of the position of the element in the block, physical variations in the crystal and light guides and variations in the gains of the photomultiplier tubes. These variations result in substantial high-frequency non-uniformities in the raw data. In particular there is a systematic variation in detector efficiency with the crystal position within the block (the "block profile") which results in significant variations in the sensitivity of the tomograph in the axial direction. Radially the effect is not so great, because any one pixel in the image is viewed by many detectors and there is a tendency for these effects to cancel out during reconstruction. Nevertheless, failure to correct for them leads to radial streaking in the image, and the systematic block profile effects can reinforce during reconstruction, resulting in circular "saw-tooth" artefacts. Detector efficiency, and in particular the block profile, can be affected by count rate. One result of pulse pileup within a block detector is the shifting of detected events towards the centre of the block [12]. This is not really a normalisation effect in the conventional sense, but since it results in a systematic change in the apparent efficiency of the lines of response with position in the block it manifests itself in a very similar way. The

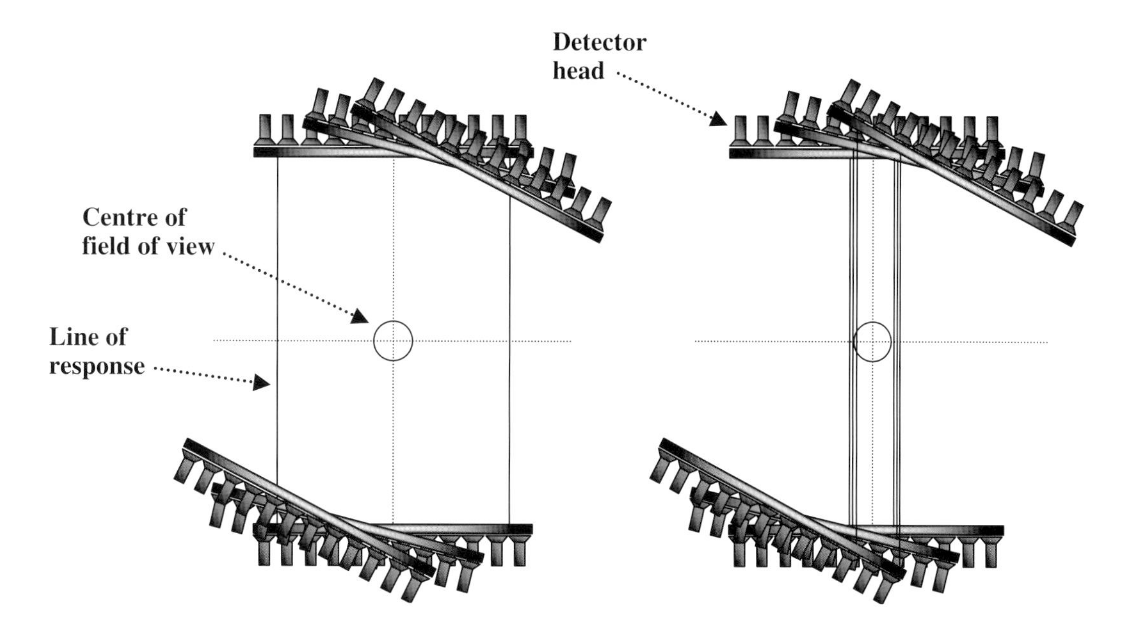

Figure 5.5. Rotational sampling. (Left) Lines of response at the edge of the transaxial field of view are sampled once per detector half-rotation. (Right) lines of response close to the centre of the field of view are sampled many times, as more detector elements are brought to bear.

effect can be reduced by measuring normalisation coefficients at a similar count-rate to that used during data acquisition, or by creating a rate-dependent look-up table of normalization coefficients [13].

If this is not possible, any resulting image artefacts may be reduced by extracting systematic effects from the raw data after normalisation but prior to reconstruction.

Geometric and Solid Angle Effects

Figure 5.6 shows that in a system with segmented detectors, such as a block-detector based system, lines of response close to the edge of the field of view are narrower and more closely spaced than those at the centre. This geometric effect is also apparent axially and can be significant for large area tomographs operating in 3D mode. The narrowing of the LORs results in a tighter acceptance angle and in reduced sensitivity, although in the transaxial plane this effect is partially compensated by the fact that the separation between opposing detectors is less towards the edge of the field of view, so that the acceptance angle is changed in the opposite direction. The narrowing of LORs also results in reduced sampling distance. However, this effect is easily describable analytically and can be corrected for at reconstruction time – a process known as "arc correction". Arc correction may not be an issue for systems that employ continuous detectors, as it is usually possible to bin the data directly into LORs of uniform width.

An effect that is relevant for systems employing either continuous or discrete detectors, and that is not so easy to describe analytically, is related to the angle of incidence of the line of response at the detector face. A photon entering a crystal at an angle will usually have more material in its path than one entering normally, thus having an increased probability of interaction. In the case of a ring scanner, this results in measurable changes in sensitivity as the radial position of the line of response is increased and is known as the

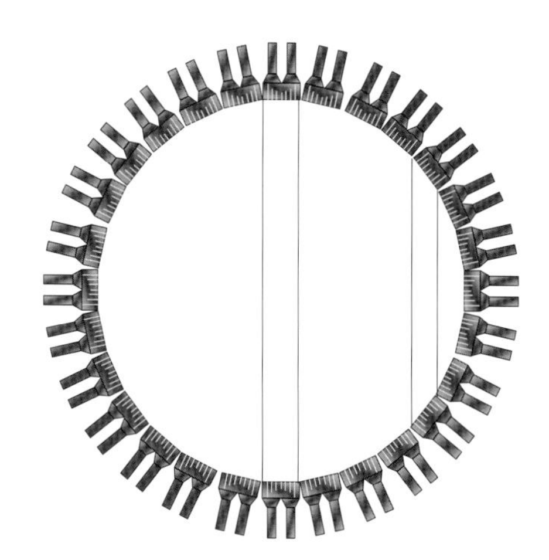

Figure 5.6. Lines of response narrow as the radial distance increases.

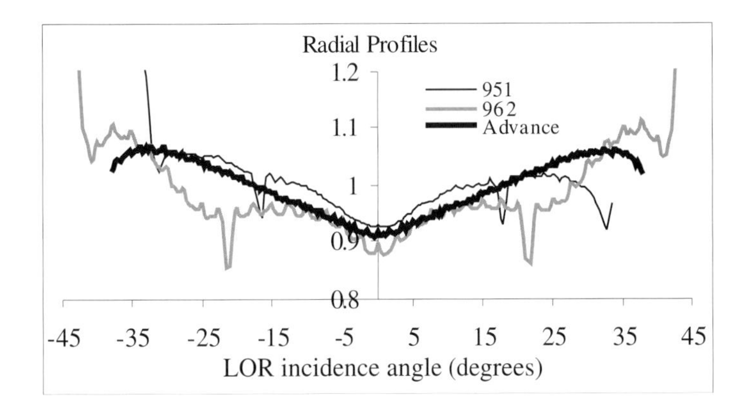

Figure 5.7. Figure 5.7. Mean radial geometric profiles for three block-detector tomographs – the Siemens/CTI ECAT 951, the Siemens/CTI ECAT 962 and the GE Advance, measured using the rotating transmission sources. The 951 data shows asymmetry due to the fact that the centre of rotation of the transmission sources is not coincident with the centre of the detector ring. (From [11], with permission.)

radial, or transaxial, geometric effect (Fig. 5.7). However, a photon entering a detector close to its edge and at an angle may have significantly less material along its path and may therefore be more likely to escape. For block detector systems this results in a pattern of sensitivity change which varies both with radial position and with the position of the line of response with respect to the block (Fig. 5.8). This has become known as the "crystal interference" effect [10]. Again, similar effects can be found in the axial direction [11].

It should be noted that the photon incidence angle is most strongly correlated with the line of response for true coincidences – these geometric effects would be expected to be much weaker or non-existent for random and scattered coincidences [14].

Time Window Alignment

For coincidence detection to work efficiently, timing signals from each detector must be accurately synchronised. Asynchronicity between detector pairs results in an offset and effective shortening of the time window for true and scattered (but not random) coincidences. This, in turn, results in variations in the sensitivity to true and scattered coincidences. For block detector systems, the greatest source of such variations occurs at the block level. Figure 5.9 shows the variations in efficiency resulting from time alignment effects in a block tomograph plotted as a sinogram. Each diamond corresponds to a different block combination.

Structural Alignment

In a ring tomograph, the accuracy with which the detectors are aligned in the gantry can affect line of response efficiency. Such variations will manifest in different ways depending on the exact design of the tomograph, the detectors and any casing in which the detectors are contained. Frequently, block detectors are mounted in modules or cartridges, each containing

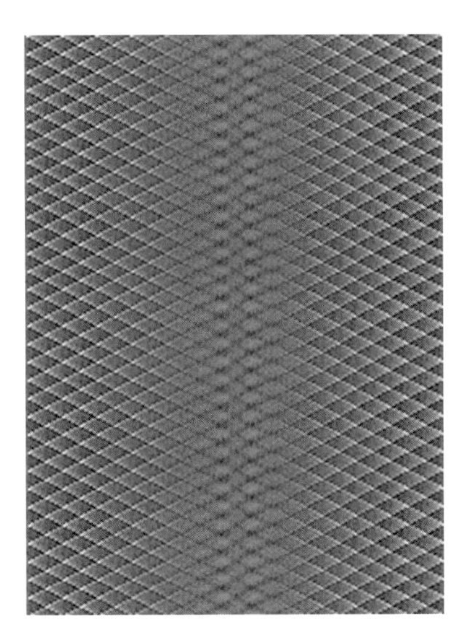

Figure 5.8. Crystal interference factors for the Siemens/CTI ECAT 951. (From [15], with permission.)

Figure 5.9. Time-window alignment factors for the Siemens/CTI ECAT 951. The factors range in value from 0.872 to 1.120. (From [15], with permission.)

several units. Misalignments of these modules can have noticeable affects on LOR sensitivity [11, 15]. Some full-ring systems have a "wobble" feature designed to improve spatial sampling – this feature allows the detectors to describe a small orbit about the mean detector position. As a result, it is possible that the transmission sources can rotate about a point which is not actually the centre of the detector ring, and if they are used to perform normalisation measurements, erroneous asymmetries can be introduced into the normalisation coefficients [11].

Septa

Septa can affect LOR sensitivity in a variety of ways. They have a significant shadowing effect on the detectors, which can reduce sensitivity by 40% or more [16]. For block detector systems, they also preferentially shadow the edges of the detectors, which may change their relative performance. On systems which can operate either with or without septa, it is therefore preferable to have a separate normalisation measurement for each case.

Direct Normalisation

The simplest approach to normalisation is to illuminate all possible LORs with a planar or rotating line positron source (usually ^{68}Ge). Once an analytical correction for non-uniform radial illumination has been applied, the normalisation coefficients are assumed to be proportional to the inverse of the counts in each LOR. This process is known as "direct normalisation". Problems with this approach include:

1. To obtain adequate statistical quality in the normalisation dataset, scan times are long, typically several hours.
2. The sources used must have a very uniform activity concentration or the resultant normalisation coefficients will be biased.
3. The amount of scatter and its distribution in the normalisation acquisition may be substantially different from that encountered in normal imaging, particularly if the tomograph is operating in 3D mode. This can result in bias and possibly artefacts.

To reduce normalisation scan times, variance reduction techniques similar to those devised for randoms correction can be applied. However, in order to implement these, the normalisation coefficients must be factored into a series of components, each reflecting a particular source of sensitivity variation. A drawback of this approach is that the accuracy of the normalisation is dependent on the accuracy of the model used to describe the tomograph. However, it has the advantage that a more intelligent treatment of the different properties of scattered and true coincidences is possible, which can be very helpful in 3D imaging.

A Component-based Model for Normalization

Consider a tomograph where detectors are indexed using the coordinate system shown in Fig. 5.4. A general expression for the activity contained in a particular LOR joining a detector i in ring u and detector j in ring v can be written as follows:

$$A_{uivj} \propto \left(P_{uivj} - S_{uivj} - R_{uivj}\right) \cdot AC_{uivj} \cdot DT_{uivj} \cdot \eta_{uivj}^{true} \qquad (10)$$

where A_{uivj} is the activity within the LOR, P_{uivj}, S_{uivj} and R_{uivj} are the prompt, scattered and random count rates respectively, AC_{uivj} is the attenuation correction factor for the LOR, DT_{uivj} is the dead time correction factor for the LOR and η_{uivj}^{true} is the normalization coefficient for true coincidences. We will assume that R_{uivj}, AC_{uivj} and DT_{uivj} can be measured accurately for each LOR. However, S_{uivj} cannot be measured directly and must be calculated. Most algorithms for calculating scatter result in a smoothly varying function that does not include normalization effects. Where scatter is only a small proportion of the signal (e.g., 2D imaging) this is probably unimportant. In 3D imaging, where scatter can make up a significant fraction of detected events, we can modify equation (10) as follows:

$$A_{uivj} \propto \left(P_{uivj} - \frac{S_{uivj}^{calculated}}{\eta_{uivj}^{scatter}} - R_{uivj}\right) \cdot AC_{uivj} \cdot DT_{uivj} \cdot \eta_{uivj}^{true} \quad (11)$$

where the $\eta_{uivj}^{scatter}$ are the normalization coefficients for scattered coincidences.

As a first approximation, we could say that $\eta^{scatter} \approx \eta^{trues}$. However, this leads to bias because some of the more important normalization effects for true coincidences arise because photons resulting in coincidences on a particular LOR have a tightly constrained angle of incidence at the detector face, a condition which is clearly not met for scatter. Allowing $\eta^{scatter}$ to take different values to η^{trues} was first proposed by Ollinger [14]. This is still an approximation, because the distribution of incidence angles and photon energies for scattered photons will be dependent on the source and attenuation distribution, so that, in general, there will be no unique value for $\eta_{uivj}^{scatter}$. However, at the present time, errors in this formulation are likely to be small compared to errors in the scatter estimate itself.

The task of normalization is to obtain values for the $\eta_{uivj}^{scatter}$ and the η_{uivj}^{true}. It is clear from the discussion above that there is no generally applicable model which will yield the $\eta_{uivj}^{scatter}$ and the η_{uivj}^{true} for all tomograph designs. We can, however, write down an example expression for a block detector system:

$$\eta_{uivj}^{true} = \varepsilon_{ui}\varepsilon_{vj}b_{ui}^{tr}b_{vj}^{tr}b_{u}^{ax}b_{v}^{ax}t_{uivj}g_{uivj}^{tr}g_{uv}^{ax}m_{uivj} \qquad (12)$$

where

- the ε are the intrinsic detector efficiency factors, describing the random variations in detector efficiency due to effects such as crystal non-uniformity and variations in PMT gain
- the b^{tr} are the transaxial block profile factors, describing the systematic transaxial variation in detector efficiency with position in the block detector. These are frequently incorporated into the ε; however, it can be useful to consider them separately if count-rate dependent effects are to be included in the normalization process
- the b^{ax} are the axial block profile factors – they are the relative efficiencies of each axial ring of detectors. Again, these are rate dependent to a degree – however, the primary reason for separating them from the ε is to simplify the process of measurement (see section on axial block profile factors below)
- the t are the time-window alignment factors
- the g^{tr} are the transaxial geometric factors, describing the relationship between LOR efficiency, photon incidence angle and detector position within the block. In this formulation they include the crystal interference effect.
- the g^{ax} are the axial geometric factors. There is one factor for each ring combination. As with the axial block profile factors, they are separated from their transaxial counterparts simply for ease of measurement.
- the m are the structural misalignment factors. These are similar to the geometric factors in that they will usually vary with photon incidence angle.

The analytically derivable components are missing from this model since they do not need to be measured.

The normalization coefficients for scatter may be written as follows:

$$\eta_{uivj}^{scatter} = \varepsilon_{ui}\varepsilon_{vj}b_{ui}^{tr}b_{vj}^{tr}b_{u}^{ax}b_{v}^{ax}t_{uivj} \qquad (13)$$

The geometric components have been removed and the efficiency components retained. This model makes the assumption that scattered photons have a random distribution of incidence angles for any particular LOR [14], and that the efficiency factors are the same for trues and scatter. Thus, any dependence of the distribution of incidence angles for scattered photons on the source and attenuation distribution is ignored, as are any changes in detection efficiency with photon energy.

Measurement of the Components

Although several components must be accounted for in component-based normalisation, they can be measured from just two separate scans using a relatively simple protocol. A typical protocol involves scanning a rotating rod source with nothing in the field of view and a uniform cylindrical source. Both scans are performed with low activity concentrations to minimise dead time effects and the scan times are quite long, typically several hours, to ensure adequate counting statistics. The rod scan is used to calculate the geometric effects while the uniform cylinder scan is used to calculate the crystal efficiencies. The details of how the various factors are extracted from each of these scans are given in the following sections (Fig. 5.10).

Axial Block Profile Factors, b_u^{ax} and Axial Geometric Factors, g_{uv}^{ax}

The axial block profile factors may be calculated from an acquisition of a central uniform right cylinder source. If scatter is not significant, the calculation is straightforward – the total counts C_u in each of the direct plane (i.e., ring difference = 0) sinograms are computed, and the b_u^{ax} are then given by

$$b_u^{ax} = \sqrt{\frac{\overline{C_u}}{C_u}} \qquad (14)$$

where $\overline{C_u}$ is the mean value of the total counts in each sinogram. In 3D imaging, the amount of scatter can be large, and more importantly, the distribution can vary in the axial direction. The data should therefore be scatter corrected prior to the calculation of the b_u^{ax}. A simple algorithm such as fitting a Gaussian to the scatter tails is usually sufficient for this purpose, but care must be taken to ensure that high-frequency variations in detector efficiencies do not bias the results [11].

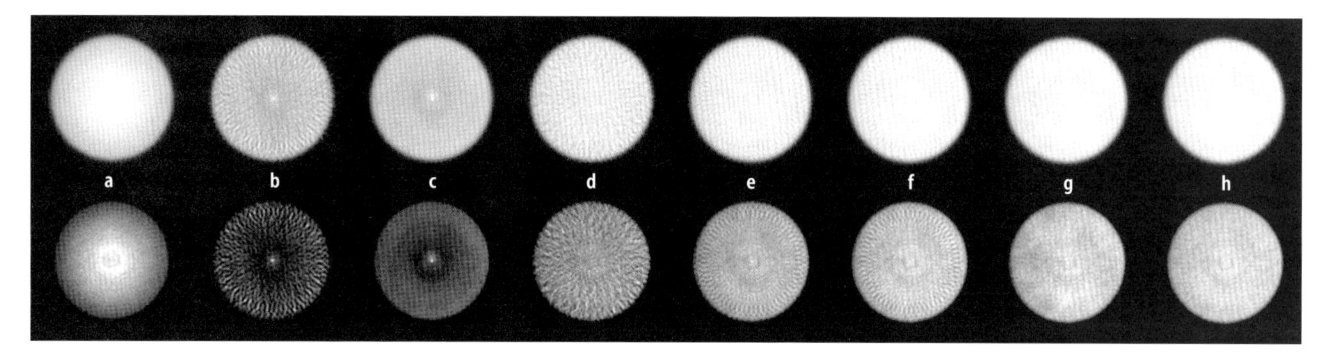

Figure 5.10. Effects of normalisation on image uniformity. Images (summed over all axial planes) from a low-variance 20 cm cylinder acquisition, performed in 3D mode on a Siemens/CTI ECAT 951. (From [15], with permission.)

(Upper row) linear grey scale covering entire dynamic range.
(Lower row) linear grey scale, zero-point set to 70% of image maximum.

(a) no scatter correction;
(b) no normalisation;
(c)no correction for the radial profile;
(d) no crystal efficiency correction;
(e) no transaxial block profile correction;
(f) no crystal interference correction;
(g) no time alignment correction;
(h) fully normalised and scatter corrected.

The axial geometric factors g_{uv}^{ax} are also computed from cylinder data, after they have been corrected for scatter and for the axial block profile. If C_{uv} is the sum of the counts in the sinogram indexed by ring u and ring v, the corresponding axial geometric factor g_{uv}^{ax} is obtained simply by dividing C_{uv} by the mean value of all the C_{uv} and inverting the result.

If the axial acceptance angle is large, it may be necessary to correct for the variation in source attenuation between sinograms corresponding to large and small ring differences prior to calculating the g_{uv}^{ax}.

An unfortunate consequence of calculating the values of the axial components in this way is that errors in the scatter correction give rise to bias in the normalisation coefficients [15].

In some implementations, the axial block profile and geometric factors are not calculated directly. Instead, the cylinder data are reconstructed and correction factors are computed by comparing the counts in each image plane with the mean for all planes. This works well in 2D imaging, where the data used to reconstruct any one image plane is effectively independent of those used to reconstruct any other. In 3D imaging this is not the case, and the use of post-reconstruction correction factors entangles effects due to normalisation, reconstruction and source distribution.

Intrinsic Detector Efficiencies, ε_{ui}, and Transaxial Block Profile, b_u^{tr}

The intrinsic detector efficiencies are again usually computed from an acquisition of a central uniform right cylinder source, although planar or rotating line sources can also be used. Variance reduction may be effected using the *fan-sum* algorithm, which is essentially a simplified version of that used in randoms variance reduction. In the fan-sum algorithm, the fans of LORs emanating from each detector and defining a group A of opposing detectors are summed (see Fig. 5.11). It is assumed that the activity distribution intersected by each fan is the same, and that the effect of all normalisation components apart from detector efficiency is also the same for each fan. The total counts in each fan C_{ui} then obeys the following relation:

$$C_{ui} \propto \sum_{v \in A} \sum_{j \in A} \varepsilon_{ui}\varepsilon_{vj} \quad \text{or} \quad C_{ui} \propto \varepsilon_{ui} \sum_{v \in A} \sum_{j \in A} \varepsilon_{vj} \tag{15}$$

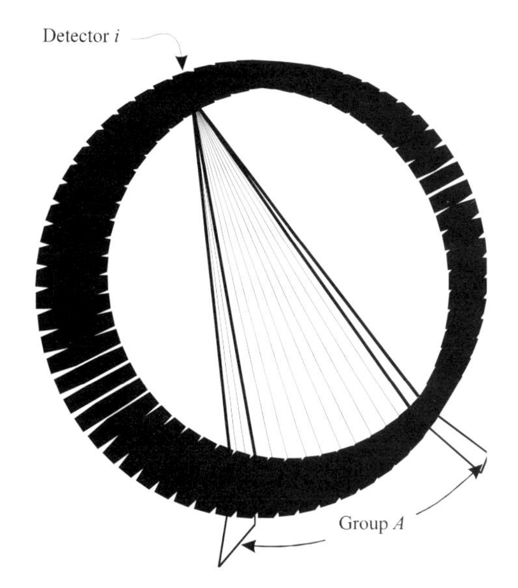

Figure 5.11. Lines of response in the fan-sum algorithm.

If A contains a sufficiently large number of detectors, it can be assumed that the expression $\sum_{v \in A} \sum_{j \in A} \varepsilon_{vj}$ is also a constant (the fan-sum approximation, attributable to 17). The ε_{ui} are then given by the following expression:

$$\varepsilon_{ui} \approx \frac{\overline{C_{ui}}}{C_{ui}} \tag{16}$$

where $\overline{C_{ui}}$ is the mean value of all the fan-sums for detector ring u. Note that the efficiencies are not determined using the mean value of the C_{ui} computed over all detector rings as the numerator in equation (16). This avoids potential bias arising from the fact that the mean angle of incidence of the LORs at the detector face varies from the axial centre to the front or back of the tomograph.

If the C_{ui} are calculated by summing only over LORs lying within detector ring u, the method is known as the 2D fan-sum algorithm. This method is quite widely implemented because of its simplicity, and because it can be used for both 2D and 3D normalization. However, in the 3D case it is both less accurate and less precise than utilizing all possible LORs [18]. The accuracy of the fan-sum approximation also depends crucially on utilizing an accurately centered source distribution [19–21]. Other algorithms for calculating the ε_{ui} also exist (see, for example, [17, 18, 20–22]).

The ε_{ui} calculated in this way incorporate the transaxial block profile factors b_u^{tr}. If required, they can be extracted from the ε_{ui} very simply – they are just the mean values of the detector efficiencies calculated for each position in the block detector:

$$b_{ui}^{tr} = \left(\sum_{n=0}^{\frac{N}{D}-1} \varepsilon_{u,nD+i \bmod D} \right) \Bigg/ \frac{N}{D} \tag{17}$$

where N is the number of detectors around the ring and D is the number of detectors across a detector block. In practice, the b_u^{tr} are obtained by averaging data across such a large and evenly sampled proportion of the field of view that they are effectively independent of the source distribution. As a result, changes in the transaxial block profile factors due to count-rate effects can be computed directly from the emission data, a process known as *self-normalisation* [24].

The Transaxial Geometric Factors g_{uivj}^{tr}

Rotating transmission sources, planar sources and scanning line sources have all been used to generate data for calculating the transaxial geometric factors [e.g., 16, 22, 25, 26]. Once the data have been collected, an analytic correction is applied to compensate for non-uniform illumination of the LORs by the source. The data are then corrected for variations in detector efficiency and block profile. For systems where "crystal interference" is not expected to be a problem, the transaxial geometric factors can be obtained by averaging the data in each sinogram over all LORs sharing a common radius. Thus, one "radial profile" describing the transaxial geometric effect is obtained for each sinogram. Otherwise, the data are averaged over LORs which share a common radius and a common position within block detectors, resulting in D radial profiles per sinogram, where D is the number of detectors across a block. Each radial profile is then divided by its mean and inverted to yield the transaxial geometric factors.

The Time-Window Alignment Factors t_{uivj}

As with the transaxial geometric factors, time-window alignment factors can be derived from the data acquired using rotating transmission sources, planar sources or scanning line sources. Non-uniform illumination is compensated for, and the data are then corrected for intrinsic detector efficiency, block profile and all geometric effects. Data elements with common block detector combinations are summed to produce an array with one element for each block combination. This array is then divided by the mean of all its elements and inverted to yield the t_{uivj}.

The Structural Misalignment Factors, m_{uivj}

The effects of structural misalignment are not easy to predict. They can often be determined by examining data used for calculating the transaxial geometric factors after normalisation for all other known components. On the GE Advance (GE Medical Systems, Milwaukee,WI), this process reveals high-frequency non-uniformities which are consistent in every sinogram, regardless of ring difference. These non-uniformities are correlated with rotational misalignments of the block modules, which extend for the entire length of the tomograph. However, examination of data from the Siemens/CTI ECAT 962 tomograph (CTI Inc., Knoxville, TN), which also has block modules that extend for the entire length of the tomograph, does not reveal these consistent non-uniformities. On the Advance, the consistency of the non-uniformities can be exploited in a simple manner to yield the required correction factors. Data from a rotating line source is corrected for all known normalisation effects and then summed over all ring differences, yielding a matrix with the same dimensions as a single sinogram. Each element in the matrix is then divided by the mean over all the matrix elements and inverted to yield the m_{uivj}.

Frequency of Measurement

The geometric factors do not normally change with time and need only be measured once. Depending on their nature, the misalignment factors may either be fixed, or may need to be re-measured as components are replaced. The time window alignment factors should be re-measured whenever detector components are replaced. The detector efficiency and block profile components can change with time, as photomultiplier tube gains drift, and should be re-measured routinely (usually monthly or quarterly, but possibly more often in a less stable environment such as that found in a mobile PET system). The rate-dependent component of the transaxial block profile can, if necessary, be determined for each individual scan using self-normalisation.

Dead Time Correction

Definition of Dead Time

PET scanners may be regarded as a series of subsystems, each of which requires a minimum amount of time to elapse between successive events for them to be registered as separate. Since radioactive decay is a random process, there is always a finite probability that successive events will occur within this minimum time, and at high count-rates, the fraction of events falling in this category can become very significant. The principle effect of this phenomenon is to reduce the number of coincidence events counted by the PET scanner, and since the effect becomes stronger as the photon flux increases, the net result is that the linear response of the system is compromised at high count-rates. The parameter that characterises the counting behaviour of the system at high event rates is known as the "dead time". The fractional dead time of a system at a given count-rate is defined as the ratio of the measured count-rate and the count-rate that would have been obtained if the system behaved in a linear manner.

Sources of Dead Time

The degree to which a system suffers from dead time and the sources of dead time within a system are highly dependent on its design and architecture. We now describe three sources of dead time typically found in clinical PET scanners. A more detailed discussion of this topic can be found in [12] and [27].

Within a well-designed scintillation detector subsystem, the primary factor affecting the minimum time between separable events is the integration time, that is, the time spent integrating charge from the photomultiplier tubes arising from a scintillation flash in the detector crystal. If a photon deposits energy in the detector crystal while charge is still being integrated from the previous event (a phenomenon known as "pulse pileup"), there are two possible outcomes. Either the total collected charge is sufficiently great that the upper energy level discriminator threshold is exceeded, in which case both events will be rejected, or the two events are treated as one, with incorrect position and energy (Fig. 5.12). In addition to the integration time, the detector electronics will usually have a "reset" time, during which the sub-system is unable to accept further events. The effects of pulse pileup in block detectors have been investigated by Germano and Hoffman [28], and in large-area PET detectors by Wear et al [29]. To reduce the limiting effect of integration time, several groups have implemented schemes for fast digitisation of the detector output signal. This signal can then be post-processed to separate overlapped pulses [e.g., 30].

Within the coincidence detection circuitry, there is the possibility that more than two events might occur during the coincidence time window. This is known as a "multiple" coincidence, and since it is impossible to ascertain which is the correct coincidence pair, all events comprising the multiple coincidence are rejected.

Processing a coincidence event also takes time, during which no further coincidences may be accepted. Although the number of coincidences is usually small compared to the number of single events, dead time arising from coincidence processing can be significant because of the architecture of the coincidence electronics. There are too many detector pairs in a PET scanner for each to have its own coincidence circuit. To overcome this problem, the data channels are multiplexed into a much smaller number of shared circuits. These shared circuits have commensurately higher data rates, and as a result become important contributors to overall system dead time.

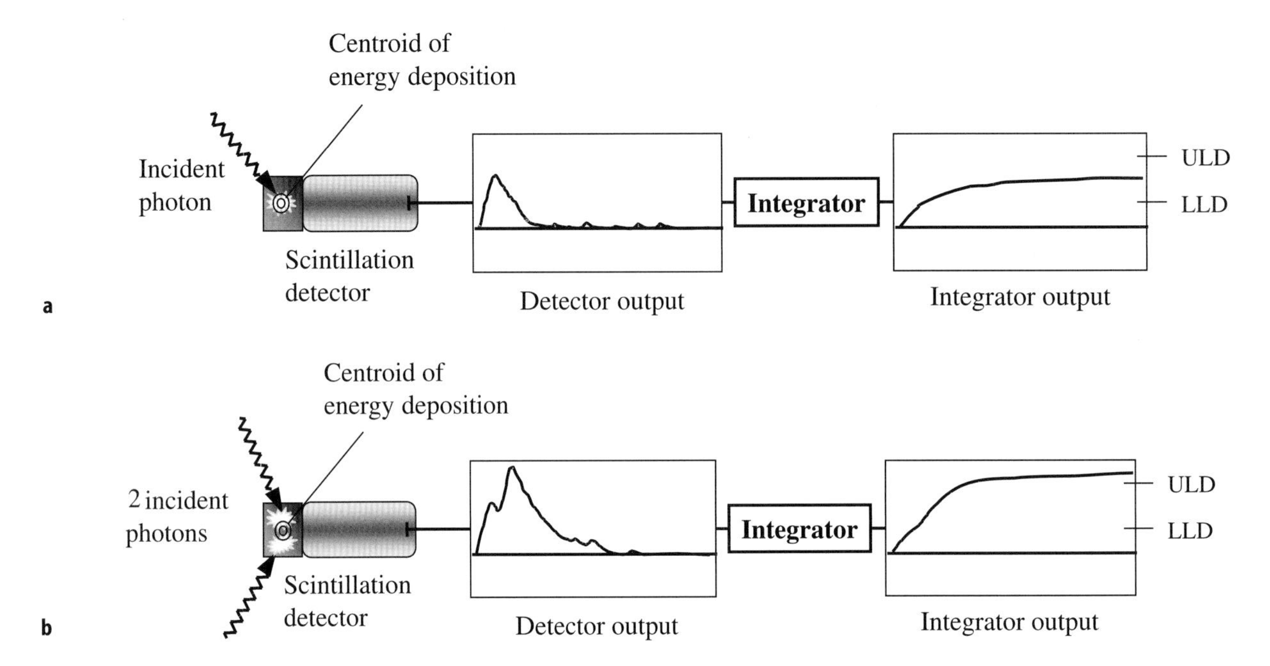

Figure 5.12. Effects of pulse pileup. (a) Single incident photon interacting with scintillation detector. (b) Two photons incident within integration time, resulting in pulse pileup. If the total deposited energy is greater than the upper energy level discriminator (ULD), pileup event is rejected – otherwise it is assigned incorrect energy and position. In either case at least one event is lost.

Measurement of Dead Time

To measure the dead time behaviour for a PET scanner as a function of count-rate, a "decaying source" experiment is performed. A uniform source containing a known quantity of a short-lived positron emitter such as ^{18}F or ^{11}C is placed in the field of view of the PET scanner. Repeated measurements of the singles, prompt and random coincidence rates are then made as the activity in the field of view decays. The incident count-rate for a given level of activity in the field of view is obtained by linear extrapolation from the count-rate response measured when most of the activity has decayed away and dead time effects are small. The ratio between the incident and measured count-rate then gives the fractional count-rate loss.

Figure 5.13a shows simulated count-rate curves for a current generation BGO PET scanner based on a validated count-rate model [31], including the extrapolated ideal trues rate. For a scanner operating in 3D mode, the count losses reach 20% at approximately 10 kBq/ml. Note that the total observed count-rate (trues + randoms) plateaus at 2.6×106 counts/sec which corresponds to the bandwidth limit of the coincidence electronics on this scanner. Figure 5.13b illustrates the effect of shortening the integration time on dead time and the resulting count loss curve. These curves assume no loss of sensitivity to true coincidences as a result of shortening the integration time as would be expected if a faster scintillator, such as LSO, was used instead of BGO.

Approaches to Dead Time Correction

The simplest method for dead time correction involves constructing a look-up table of dead time correction factors derived from decaying source measurements. However, this approach does not account for spatial variations in source distribution that may alter the relative count-rate load in the different sub-systems within the scanner. In practice more accurate dead time correction schemes are constructed in which, where possible, the "live time" (= acquisition time × [1– fractional dead time]) is measured for each sub-system. For those sub-systems where it is impractical to measure the live time, an analytic model incorporating knowledge of the system architecture is constructed and fitted to data obtained from decaying source experiments. The decay correction scheme then consists of applying a series of measured and modelled correction factors to the acquired data.

The live time in a sub-system may be measured in a variety of ways. One possibility is to implement a second circuit parallel to the measurement circuit for which the live time estimate is to be made. Regular

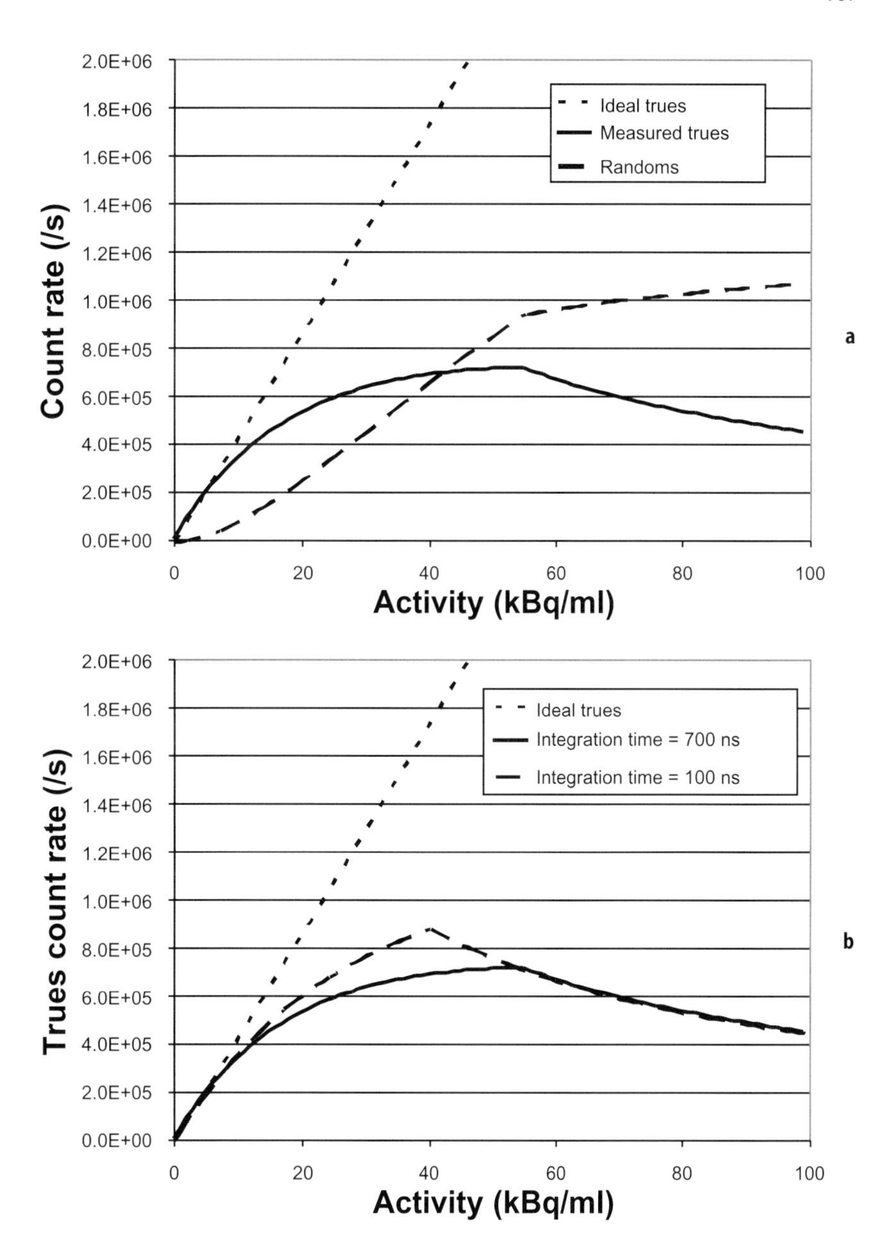

Figure 5.13. Count-rate curves as a function of radioactivity concentration in a phantom. The curves demonstrate the effect of (a) dead time on count-rate linearity and (b) shortening the signal integration time. Note the divergence of the measured true coincidence rate from the ideal trues rate at high radioactivity concentrations and that dead time is reduced when the integration time is shortened. Also note the discontinuity in the count-rate curves at approximately 55 kBq/ml which is caused by the bandwidth limit of the coincidence processing electronics.

pulses are sent down the second circuit to a counter. However, if a signal is being processed on the measurement circuit, a gate is closed on the second circuit for the duration of the processing, preventing pulses from reaching the counter. The number of pulses observed by the counter is then approximately proportional to the live time of the measurement circuit. Other schemes for measuring dead time are described in [27].

Dead time models usually treat system dead time as being separable into two components, described as "paralysable" and "non-paralysable" [e.g., 31, 32, 33]. The paralysable component describes the situation where the system is unable to process events for a fixed amount of time τ after each event, regardless of whether or not the system is dead. For example, if an event arrives while the system is dead due to a preceding event, the system remains dead for a further τ seconds from the time of arrival of the second event. Thus, at a sufficiently high count-rate, the system becomes saturated, and the recorded count-rate will actually decrease as the incident count-rate increases. The dead time behaviour of the detector sub-system has a substantial paralysing component, because every time a photon interacts with the crystal, more light is deposited, which must decay away before the detector can process the next event. If the time-of-arrival distribution of the events obeys Poisson statistics, the relationship between the measured event rate m, the actual event rate n, and the dead time resulting from a single event τ is given by

$$m = ne^{-n\tau} \tag{18}$$

In the non-paralysing case, the system is again rendered "dead" for a time τ after each event, but while the system is dead, further events have no effect. The dead time behaviour of the coincidence processing subsystem is essentially non-paralysing, because events arriving while a coincidence is being processed are simply ignored. For such systems, the measured count-rate tends asymptotically to a limiting value of τ^{-1} as the actual count-rate increases, and the relationship between m, n and τ is given by

$$m = \frac{n}{1 - n\tau} \tag{19}$$

A more detailed treatment of this topic may be found in [34]. The two components can be present in the system in series and this has been shown to be the case for PET systems (see Chapter 3).

The corrections discussed so far address factors mainly related to detector sub-system performance, including timing resolution, detector uniformity and count-rate performance. In the following sections, we discuss factors whose magnitude is also affected by aspects of tomograph performance but which arise from measurement errors, including those due to photon interactions in the body.

Scatter Correction

Characteristics of Scattered Radiation

When a positron annihilates in the body, there is a reasonable chance that one or both of the annihilation photons will scatter in the body or in the detector itself. At the energy of annihilation photons (0.511 MeV), the most likely type of interaction is Compton scattering in which the photon transfers some of its energy to loosely bound electrons and deviates from its initial path [35]. Since the coincidence LOR formed after one or both photons undergo Compton scattering is no longer colinear with the site of annihilation (Fig. 5.1), such events degrade the PET measurement. Furthermore, the Compton equation that relates the photon energy before (E) and after scattering (E_{sc}) to the scattering angle (Ω) tells us that an annihilation photon may scatter through as much as 45 degrees and lose only 115 keV of its energy to the recoil electron:

$$E_{sc} = \frac{E}{1 + \frac{E}{m_0 c^2}\left(1 - \cos\Omega\right)} \tag{20}$$

where m_0c^2 is the resting energy of the electron before scattering. Because of the poor energy resolution of PET scanners, particularly BGO scanners, a coincidence event involving the scattered photon in this example would most likely be accepted within the energy window, which is typically set between 350 and 650 keV. Thus, scattered coincidences are not easily discriminated from unscattered coincidences based on their energy and may significantly degrade both image quality (due to loss of contrast) and quantitative accuracy.

The proportion of accepted coincidences which have undergone Compton scattering is referred to as the scatter fraction and its magnitude depends on several factors, including the size and density of the scattering medium, the geometry of the PET scanner and the width of the energy acceptance window (which is mainly determined by the energy resolution of the detectors). The scatter fraction typically ranges from about 15% in a ring tomograph with slice-defining septa (2D mode, or septa extended) to 40% or more for the same tomograph operated without slice-defining septa (3D mode, or septa retracted). Indeed, a major function of slice-defining septa is to minimise scatter by preventing photons which scatter out of the plane defined by a ring of detectors from being detected in an adjacent detector ring and forming an oblique LOR.

Although the underlying physics describing Compton scattering of annihilation photons is reasonably complex, there are several characteristics of the resultant LORs which can be exploited to estimate their distribution and potentially correct the measured data. For example:

- LORs recorded outside the object boundary can only be explained by scatter in the object (assuming that randoms have been subtracted) since LORs arising from unscattered trues must be collinear with the point of annihilation,
- The scatter distribution is very broad (i.e., it contains mainly low spatial frequencies) and relatively featureless,
- The portion of the coincidence energy spectrum below the photopeak has a large (but not exclusive) contribution from scattered events, and
- Scattered coincidences that fall within the photopeak window are mainly due to photons that have only scattered once (Fig. 5.14).

These various characteristics have given rise to a wide variety of approaches for estimating and correcting scattered coincidences in PET data. They can be broadly divided into four categories: empirical approaches, methods based on two or more energy windows, convolution (or, equivalently, deconvolution) methods and methods which model the scatter distrib-

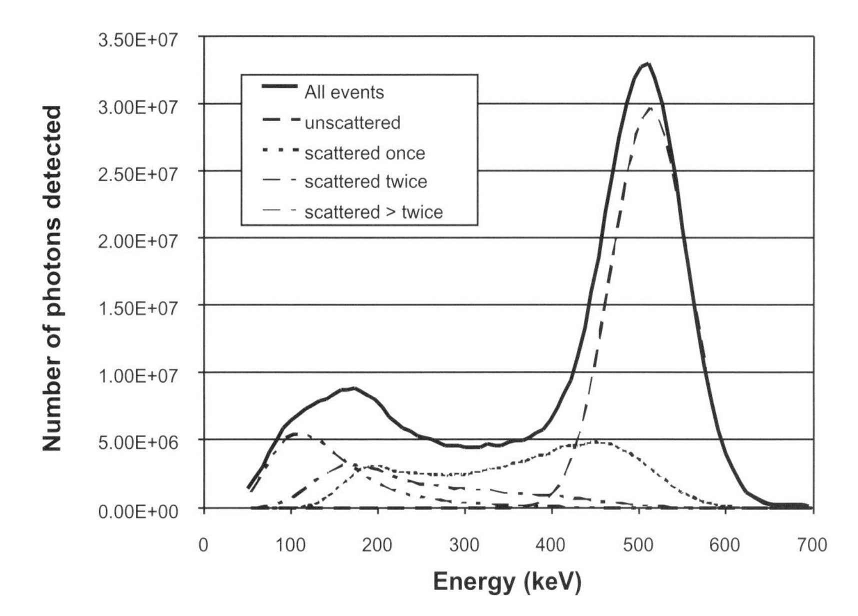

Figure 5.14. Energy spectra indicating the spectral distribution of scattered 511 keV photons according to the number of times each photon scatters in the source object. The spectra were obtained by Monte Carlo simulation of a 20 cm diameter cylinder uniformly filled with positron emitter in water. The simulated PET scanner has a ring diameter of 86 cm, an axial field of view of 16 cm and energy resolution of 20% FWHM.

ution during forward projection based on knowledge of tissue densities (or equivalently attenuation coefficients) in the body and an initial estimate of the scatter-free image.

Empirical Scatter Corrections

Fitting the Scatter Tails

Perhaps the simplest approach to scatter correction is to fit an analytical function to the scatter tails outside the object in projection space. For example, a second order polynomial [3] and a 1D Gaussian [36] have both been used to fit the scatter tails. This approach is based on the observations that coincidences recorded outside the object boundary are entirely due to scatter (assuming that randoms have previously been subtracted) and the scatter distribution contains mainly low spatial frequencies.

The method is effective for neurological PET studies because it guarantees that the scatter recorded outside the object is reduced to approximately zero and it inherently corrects for scatter arising from activity outside the axial field of view, something some of the more complex methods are unable to do. It also has the advantages of being simple to implement and computationally very efficient. The main drawback of this approach is that the scatter distribution is not always well approximated by a smooth analytical function, particularly in the thorax where tissue density is heterogenous, which may lead to over- or under-subtraction. A further problem in the thorax is that the body occupies a large portion of the field of view leaving relatively small scatter tails to fit. This reduces the accuracy of the fit and may lead to over- or under-subtraction of scatter in the centre of the body.

A Direct Measurement Technique

Another approach takes advantage of differences between the scatter distribution with septa extended and the scatter distribution with septa retracted [37]. This method is only applicable to PET scanners with retractable septa and it was intended primarily as a means of characterising scatter in 3D PET by direct measurement. However, it can also be used as an effective method of scatter correction. The first step is to make a measurement of the same object with septa extended and with septa retracted (in 3D mode). After scaling the septa extended projections to account for differences in detection efficiency due to septa shadowing, they are subtracted from the projections corresponding to polar angle $\theta = 0\infty$ in the 3D dataset to yield a measurement of the scatter contribution to the direct plane data. The scatter contribution to oblique planes is then estimated by interpolation of the direct plane scatter corresponding to the detector rings from which the oblique plane was formed. The assumption in this last step is that the scatter distribution does not vary markedly with changes in polar angle up to the maximum polar angle allowed. This assumption may break down for scanners with a large axial field of view and a large acceptance angle for oblique sinograms.

This method has the advantages that it makes few assumptions, it is relatively simple to implement and it inherently corrects for scatter arising from activity outside the field of view. It also enables direct measurement of the scatter distribution in complex objects which may be used to obtain a better understanding of scattering in 3D PET or to validate other methods of scatter correction. The main drawbacks are that it requires an additional measurement which may be impractical (for example, in dynamic studies where the scatter distribution may change throughout the study) and that it is only applicable to PET scanners with retractable septa.

A related scheme has been designed for scanners with coarse septa – in this method, coincidence data are acquired for lines of response that intersect the septa and thus cannot contain trues. These so-called "shady" lines of response are assumed to provide an estimate of the scatter that can then be subtracted from the "sunny" lines of response where the septa are not intersected [38]. This method has the advantage that the scatter estimate is acquired contemporaneously with the data containing the true signal, but again relies on the assumption that the scatter distribution does not change significantly with polar angle.

Multiple Energy Window Techniques

Multiple energy window techniques make use of the observations that (1) a greater proportion of Compton scattered events are recorded in the region of the single photon energy spectrum below the photopeak compared with those recorded near the photopeak and (2) there exists a critical energy above which only unscattered photons are recorded [39] (Fig. 5.14). Thus, data recorded in energy windows set below or above the photopeak window, or both, can be used to derive an estimate of the scatter contribution within the photopeak window. Such techniques have been extensively employed and investigated for single photon emission computed tomography (SPECT) (40). Interest in multiple energy window approaches for PET was stimulated by two advances: the development of energy lookup tables and threshold setting for individual crystals in the block detector leading to improved energy resolution, and improvements in the electronics for NaI(Tl) PET systems that took advantage of the intrinsically high energy resolution of NaI(Tl).

Dual Energy Window Methods

There have been two distinct approaches to the use of dual energy windows for scatter estimation. The dual energy window (DEW) method uses an energy window set below the photopeak and abutting it [41] (Fig. 5.15a) while the estimation of trues method (ETM) uses an energy window whose lower level discriminator is set above 511 keV and which overlaps the upper portion of the photopeak window [42] (Fig. 5.15b). These methods both make use of measurements in the auxiliary energy window to estimate the scatter contribution to the photopeak.

In the DEW method, the unscattered events in the photopeak energy window, C_{unsc}^{pw}, are defined in terms of the total coincidence events recorded in the photopeak and lower energy windows, C^{pw} and C^{lw} respectively, and the ratios R_{sc} and R_{unsc} as follows:

$$C_{unsc}^{pw} = \frac{C^{pw} R_{sc} - C^{lw}}{R_{sc} - R_{unsc}} \tag{21}$$

where R_{sc} is the ratio of scattered events ($C_{sc}^{lw} / C_{sc}^{pw}$) and R_{unsc} is the ratio of unscattered events ($C_{unsc}^{lw} / C_{unsc}^{pw}$). These ratios were determined experimentally using line and point sources and by Monte Carlo simulation. It was observed that R_{unsc} was almost constant across the transaxial field of view, whereas R_{sc} was not and both R_{unsc} and R_{sc} exhibited nonuniformity in the axial direction, which could be explained by the block structure of the detector rings used in their tomograph [41]. Despite the nonuniformity of these ratios, they were essentially independent of object size and shape in the limited range of phantoms studied. Using this method, the activity concentration values in a multicompartment phantom were recovered to within 10% of their correct levels. However, some studies suggest that this method may be prone to bias when applied to more complex source distributions or objects with nonuniform density [43, 44].

The ETM assumes that coincidences recorded above a certain energy threshold include only unscattered events. This is a reasonable approximation in the case of a PET scanner with energy resolution of approximately 20% when the lower energy discriminator is placed above 511 keV. In the original implementation of ETM, an auxiliary window was set which accepts coincidences between 550 and 650 keV. This upper window overlaps with the main photopeak window which typically accepts coincidences between 350 and 650 keV. Data recorded in the upper window are scaled to match the total true coincidences recorded in the photopeak window. Subtracting the scaled upper window data from the photopeak data yields an estimate of the scatter contribution to the photopeak. This estimate is smoothed and subtracted from the measurement made in the photopeak energy window.

The main advantage of the dual energy window methods is that they take into account scatter arising

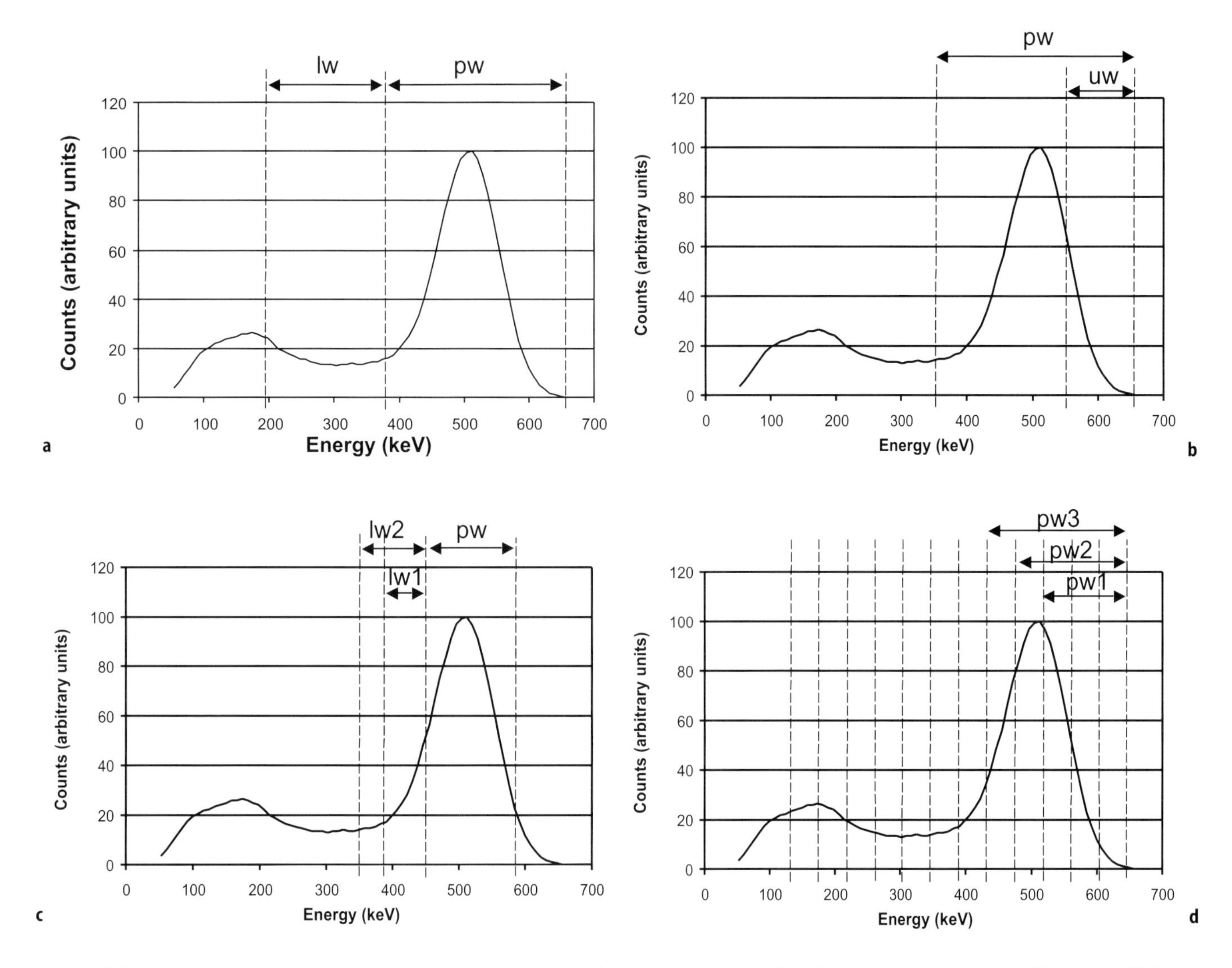

Figure 5.15. Energy spectra showing the window settings typically used in four energy window based methods: (a) dual energy window (DEW) method [41], (b) estimation of trues method (ETM) [42]), (c) triple energy window (TEW) method [45] and (d) multispectral method (indicating energy ranges of first 3 windows) [46].

from activity beyond the axial field of view. These methods should also be well suited to tomographs with better energy resolution than is typically achieved with BGO, such as those based on NaI(Tl), LSO and GSO detectors. This was demonstrated in the case of NaI(Tl) volume PET scanners where an adaptation of the ETM method performed well for a range of source distributions [44]. The main drawback is that scatter estimates are based on Poisson measurements which are noisy, particularly when derived from the early frames of a dynamic study and other count limited acquisitions. For this reason, scatter estimates derived from the auxiliary energy window are normally smoothed considerably before subtraction from the photopeak data. This may not be a major drawback since the scatter distribution typically contains mainly low spatial frequency components.

Multiple Energy Window Methods

The triple energy window (TEW) method is a straightforward extension of DEW which introduces a modification factor that accounts for source size and distribution dependencies in R_{sc} [45].Under the assumption that $R_{sc} >> R_{unsc}$, the TEW method can be written

$$C_{unsc}^{uw} = C^{uw} - M\left(\frac{C^{lw}}{R_{sc}}\right) \tag{22}$$

$$\text{where } M = \left(\frac{R_{obj}}{R_{calib}}\right)^{b}$$

R_{obj} and R_{calib} are the ratios of counts in the two lower energy windows for the object being imaged and a calibration phantom respectively. The parameter b is a relaxation factor that controls the amount of feedback of the modification term into the correction and when $b = 0$, the TEW technique becomes the DEW technique. The energy window used in the C^{lw}/R_{sc} term may be either of the two lower energy windows (Fig. 5.15c). In the implementation of Shao et al., the lower energy windows spanned the energies 385–450 keV and 350–450 keV, the calibration phantom was a 20 cm diameter uniform cylinder and the relaxation factor was 0.5 [45]. As in the DEW technique, the ratios R_{obj} and R_{calib} and the modification factor M are calculated for each sinogram element. The TEW method has many of the same advantages and drawbacks of the dual energy methods. However, it improves on the DEW method in particular by reducing the sensitivty of the scatter correction to variations in source distribution and size.

In the methods discussed so far, a relatively narrow energy window is set over the photopeak and events recorded below the lower energy threshold are assumed to be unwanted events, mainly due to scatter in the object being imaged. However, when small discrete detectors are used in high resolution tomographs, such as those designed for animal imaging studies, a large proportion of events recorded in the low energy range are due to scattering in the detectors and these are potentially useful events.

Bentourkia et al demonstrated that with careful characterisation and correction of scatter in multiple energy windows, it is possible to extend the useful energy range for acceptance of coincidences without degrading the image [46]. Specifically, they showed by Monte Carlo simulation and measurements on a PET simulator that up to 80% of events recorded above a threshold of 129 keV are either trues or detector scatters and, therefore, potentially useful for image formation. The approach developed was to correct for object scatter using position-dependent convolution subtraction while detector scatter is handled by nonstationary restoration. First, they carefully characterised the slope and amplitude of scatter components as a function of energy and position by measuring coincidence data in 16 × 16 energy window pairs. They summed the coincidence data into energy windows which had a common upper energy threshold of 645 keV and a variable lower energy threshold spanning the range 129 keV to 516 keV (Fig. 5.15d). Count profiles derived from these energy windows were fitted with multi-exponential functions. During imaging, coincidence data are recorded in the same energy windows and the scatter subtraction-restoration is effected using:

$$P_{unsc} = \left\{ P_{obs} \otimes \left(\partial - F_0 \right) \right\} \otimes R_d \tag{23}$$

where P_{obs} are the measured projection data, F_0 is the nonstationary object scatter kernel and R_d is the nonstationary restoration kernel that corrects for detector scatter. R_d is defined as:

$$R_d = \Im^{-1} \left\{ \frac{f_g + f_d}{f_g + \Im\left(f_d\right)} \right\} \tag{24}$$

where f_g and f_d are the fractions of geometric and detector scatter components and $\Im$ and $\Im^{-1}$ are the forward and inverse Fourier transforms respectively.

The multiple energy window approach is not straightforward to implement as it requires specialised hardware and extensive measurements to characterise the scatter components. However, the technique makes better use of coincidence data measured over a wider energy range than in conventional imaging, resulting in an effective increase in sensitivity of approximately 60%. The method is particularly well suited to high resolution tomographs with small discrete detectors.

Convolution and Deconvolution Approaches

Whereas the energy based methods derive information about the scatter distribution from auxiliary measurements, convolution based methods model it with an integral transformation of the projections recorded in the photopeak window. Initially, the method was developed for a ring type PET scanner operated in 2D mode [47] and the projected scatter distribution in a given slice took the following analytical form:

$$P_{sc}\left(s\right) = \int_{-\infty}^{\infty} P_{unsc}\left(t\right) h\left(s-t,t\right) dt \tag{25}$$

where p_{unsc} is the one dimensional projection of the true activity distribution and $h(s,x)$ is the scatter contribution to radial position s (along the projection) due to a source positioned at x. If h is spatially invariant, equation (26) is a straightforward convolution integral. In the initial implementation, however, the scatter response was assumed to be position dependent and described by the following function:

$$h\left(s,x\right) = A\left(x\right) e^{-b\left(x\right)\left|s\right|} + C\left(x\right) \tag{26}$$

The scatter response was measured using a line source positioned at regular intervals across the scanner's field of view and the parameters A, b and C corresponding to each position were determined by least-squares fitting.

The model described by equation (26) does not provide a means for affecting scatter correction since

p_{unsc} cannot be measured. However, we can substitute the measured projection data, p_{obs}, for p_{unsc} in equation (26) and the model still holds to a reasonable degree of accuracy. Thus, the method consists of convolving the measured projection data, line by line (i.e., view angle by view angle), with an experimentally determined scatter function and then subtracting the resulting scatter estimate from the measured projections.

The method is sufficiently accurate for 2D slice-oriented scanners but does not take into account scattering between adjacent planes which is a significant component in 3D PET. The method was extended to take into account cross-plane scattering for both large area PET scanners [48] and multi-ring scanners operated with the septa retracted [49]. This was done by defining a two dimensional scatter response function and performing a two dimensional convolution operating on the projections:

$$P^{n}_{unsc}(s,z) = P_{obs}(s,z) - k\left\{P^{n-1}_{unsc}(s,z) \otimes h(s,z)\right\} \qquad (27)$$

Here, the projection data and the scatter response function are defined in terms of the radial (s) and axial (z) position variables and $\otimes$ denotes the two dimensional convolution operator. The scatter correction described by equation (28) is written as an iterative improvement method which was suggested by the developers of both implementations of the 2D method [49, 49] and is equally applicable to the 1D case. This overcomes the problem that the scatter model is defined in terms of an unobservable quantity, p_{unsc}, by substituting it with the previous estimate of the scatter-free data. As in the 1D method, the parameters that define the scatter response function are derived from point or line source measurements made under carefully controlled scattering conditions.

This approach has been demonstrated to perform reliably in neurological studies where the scattering medium is relatively homogeneous, providing results comparable to energy window based methods [50]. Indeed, it has an advantage in dynamic studies since the scatter estimate is essentially noise-free and, therefore, does not contribute additional noise to the scatter-corrected projections. However, the assumptions break down in both the SPECT and PET cases when more complex objects are studied, such as the thorax [37, 51, 52]. Also, the method does not take account of scatter arising from activity outside the scanners field of view. The possibility of incorporating information derived from transmission measurements to determine object/position dependent scatter fractions has been suggested (49), similar to an approach that was successfully applied in SPECT (53, 54). However, this has not been fully explored in the PET case. Alternative and potentially more accurate approaches that take into account information derived from both emission and transmission data are described in the following sections.

Simulation-based Scatter Correction

Since the physics of photon interactions in matter is well understood, it is possible to model these processes and estimate the scatter contribution to projections given an accurate map of attenuation coefficients in the scattering medium and an initial estimate of the scatter-free radioactivity distribution. The scatter can be estimated analytically or numerically (for example, using Monte Carlo techniques).

Analytical Simulation

Consider first the analytical approach. If we make the assumption that only one of the annihilation photons forming a coincidence accepted within the photopeak undergoes a Compton interaction, the processes involved in forming such coincidences can be readily modelled. This assumption has been shown to be reasonable as 75 to 80% of scattered coincidences arise from single scattered events in a ring tomograph with a 10 cm axial field of view [39, 55]. With reference to Fig. 5.16, the scatter

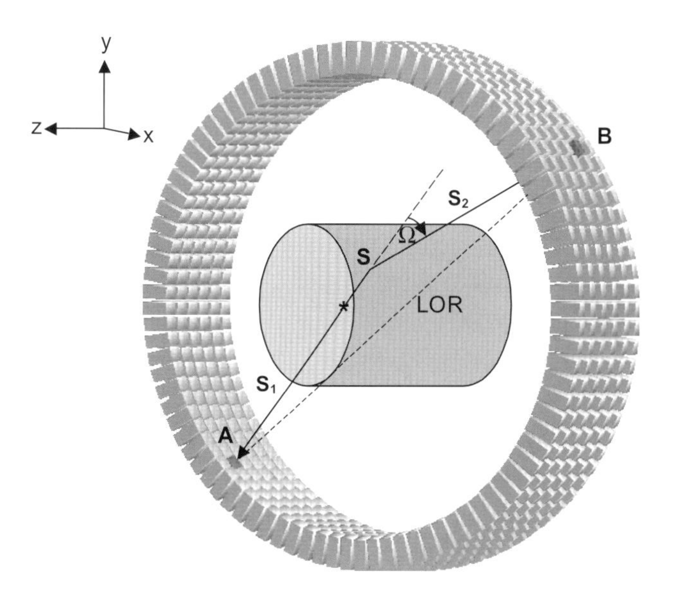

Figure 5.16. Geometry of the single scattering model used in simulation based scatter correction.

contribution to the LOR formed by detectors A and B can be calculated by considering:

1. Attenuation of the annihilation pair along the unscattered path s_1 which can be determined by ray tracing through the reconstructed attenuation volume,
2. The integrated emitter intensity along the path s_1 which can be obtained from an initial estimate of the scatter-free radioactivity distribution,
3. The probability of scattering at point S through angle Ω towards detector B which can be determined by integrating the Klein-Nishina formula,
4. Attenuation of the scattered photon along the path s_2 towards detector B, and
5. The efficiencies of detectors A and B as a function of incidence angle and photon energy. These should take into account the solid angle subtended by the detectors, their stopping power, energy resolution and discriminator settings.

Using the formulation of Watson et al, the single scatter coincidence rate at LOR AB due to one of the annihilation photons having scattered at S is calculated by integrating over the volume V_s [56, 57]:

$$P_{scat}^{AB} = \int_{V_s} dV_s \left(\frac{\sigma_{AS}\sigma_{BS}}{4\pi R_{AS}^2 R_{BS}^2} \right) \frac{\mu d\sigma_c}{\sigma_c d\Omega} \left(I^A + I^B \right) \tag{28}$$

where

$$I^A = \varepsilon_{AS}\varepsilon'_{BS} e^{-\left(\int_S^A \mu ds + \int_S^B \mu' ds \right)} \int_S^A \lambda ds,$$

$$I^B = \varepsilon'_{AS}\varepsilon_{BS} e^{-\left(\int_S^A \mu' ds + \int_S^B \mu ds \right)} \int_S^B \lambda ds,$$

λ is the emitter intensity,
μ is the attenuation coefficient,
σ_{AS} and σ_{BS} are the geometric cross sections for detectors A and B as seen from S,
R_{AS} and R_{BS} are their respective distances from the scatter point,
ε_{AS} and ε_{BS} are their respective efficiencies for photons arriving from the point S,
σ_c is the Compton scattering cross-section calculated from the Klein-Nishina formula, and
Ω is the scattering solid angle.

The primed variables correspond to the Compton scattered photon and are evaluated at the scattered photon's energy, whereas unprimed variables are evaluated at 511 keV. Note that the annihilation leading to the scatter event depicted in Fig. 5.16 could just as easily have occurred along path SB as along path AS. This explains the final term in equation (29) which calculates the attenuated unscattered emitter intensities for both possibilities (i.e., I^A and I^B).

Ollinger developed a similar formulation that describes the single scatter contribution to each LOR [58]. However, he extended the technique by extrapolating from the distribution of single scattered coincidences to a distribution that includes both single scatters and multiple scatters. This was done by convolving the single scatter distribution with a one dimensional Gaussian kernel. Ollinger also took into account scatter arising from activity outside the axial field of view in his implementation by extrapolating the initial scatter-free activity estimate and including detector side shielding in his forward projection model.

The methods of Ollinger and Watson et al yield estimates of the scatter distribution that are reasonably accurate under most circumstances (Fig. 5.17), although there is some evidence that problems can arise in clinical studies of obese patients [59]. However, they are also computationally demanding if the volume is integrated over every possible scattering point and the scatter contribution calculated for every LOR. A more efficient approach, which has been demonstrated to be practical for clinical PET, is to sample the object volume on a regular grid of sparsely spaced scattering points and to calculate the scatter for only a subset of all LORs [56, 57]. The full scatter distribution is then interpolated from the calculated LORs. This approach results in little bias due to the broad scatter distribution in 3D PET.

The main advantage of the model-based methods of scatter correction is that they make use of well understood physical principles to produce accurate scatter estimates. Their main drawbacks are the complexity of implementation, their computational demand and the assumptions required to model scatter arising from activity outside the axial field of view.

Monte Carlo Simulation

Monte Carlo methods are frequently used to evaluate scatter correction techniques since this approach allows separation of the simulated scattered and unscattered contributions to the projections which is not possible using phantom experiments. Furthermore, many Monte Carlo codes are able to simulate the scatter distribution for any specified emission and attenuation distribution and several different PET scanner geometries.

As well as providing a powerful method of evaluating the accuracy of scatter correction techniques,

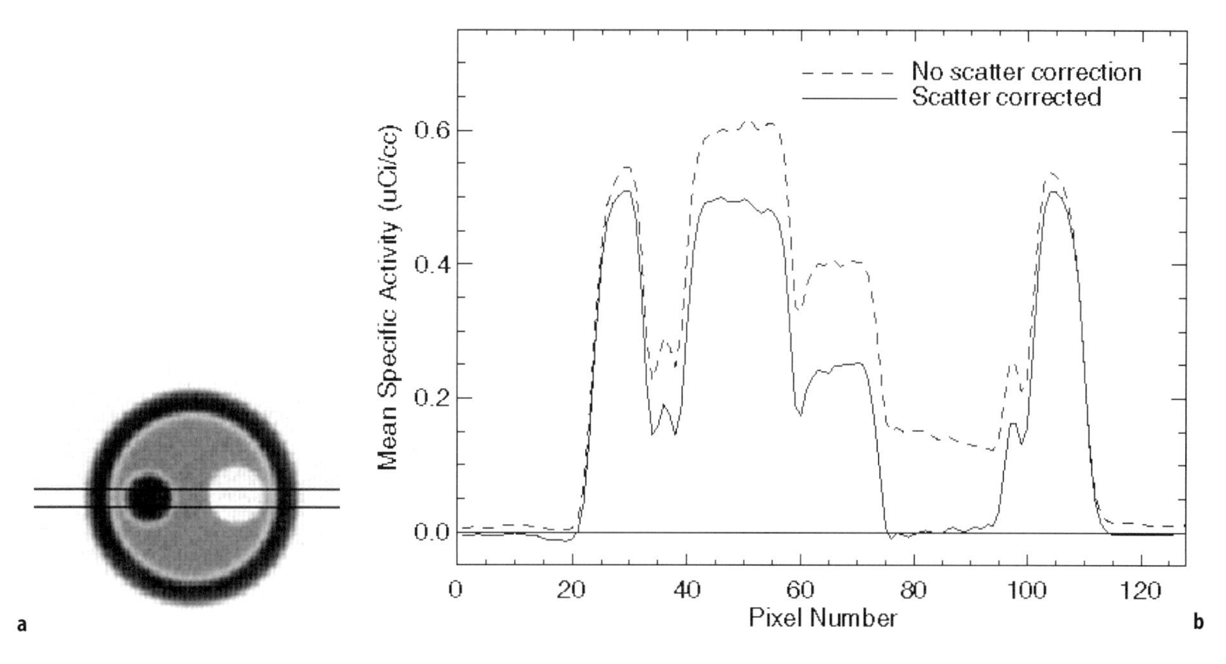

Figure 5.17. (a) Reconstructed transaxial slice of the Utah phantom after simulation based scatter correction. (b) Count profiles through the image in (a) at the level indicated. (From [57], with kind permission from Kluwer Academic Publishers).

Monte Carlo simulation can itself be used to perform scatter correction [60, 61]. Like analytical simulation, the estimation of scatter by Monte Carlo techniques is based on well understood physical principles that govern photon interactions in matter. Where this approach differs is that, rather than calculating the scatter contribution to a given LOR, photon pairs are generated at their point of origin (defined by the initial estimate of the activity distribution) with random orientation and "tracked" as they traverse through the scattering medium which may be defined by the attenuation map. Tracked photons have a random chance of interaction in each voxel they traverse, with the type and likelihood of interaction being determined by the same equations as those used in analytical scatter simulation.

Most standard codes take too long to compute the scatter distribution with sufficient counting statistics to be practical for routine scatter correction, even when executed on the most powerful computers currently available. This is because a large proportion of tracked photons may never contribute to the projections either because they undergo photoelectric absorption in the object or because they emerge from the object with an energy or a trajectory that does not permit detection. However, recent work demonstrated that dramatic improvements can be made in the computational efficiency of Monte Carlo simulation by making reasonable approximations and using implementation techniques that don't compromise the accuracy of scatter estimation [61, 62]. With such improvements, the Monte Carlo method is potentially a very accurate and practical approach to scatter correction in PET.

Implementation of Simulation Based Scatter Correction

There are five main steps which are common to the simulation based scatter correction methods, including analytical and Monte Carlo approaches. These are:

1. **Reconstruct the attenuation volume:** This is normally done using conventional 2D reconstruction of the blank and transmission data. However, any method that produces an accurate map of linear attenuation coefficients (μ, in units cm^{-1}) in the body can be used, including appropriately registered and scaled CT data if available.
2. **Reconstruct an initial estimate of the emission volume:** Different approaches have been adopted for this step. Watson et al use 3D reconstruction of the measured projections which include scatter [57], while Ollinger determines the initial emission estimate iteratively from direct plane data only [58]. He showed that the process converges rapidly and requires only a small number of iterations. In the Monte Carlo implementation due to Holdsworth et al [61], the initial estimate is obtained by performing the analytical scatter simulation technique (57),

as implemented on the EXACT HR+ scanner (CTI PET Systems, Knoxville, TN).

3. **Estimate the scatter contribution to projections:** This is the main step which involves estimating the scatter contributing to direct and oblique emission sinograms as described above for the various simulation techniques. Each of the implementations described includes some means of estimating the scatter contribution due to activity outside the field of view.
4. **Scale the scatter estimate:** Here the scatter distribution is scaled globally to ensure a good fit between the estimated scatter and the measured projections in regions not sampled by the object (i.e., regions where only scatter is present). Alternatively, if the detection system is modelled accurately including detector efficiencies and energy response, it may be possible to compute a scatter distribution which is intrinsically scaled relative to the measured projections as in a more recent implementation by Watson [56]. In the Monte Carlo approach, the total coincidences in each projection can be simulated as well as the scatter coincidences. Thus, the scaling step simply involves determining the scale factor that yields the same total coincidences in both the estimated and measured projections. This global scale factor is less prone to noise in low count studies than the factor calculated using the scatter tails.
5. **Correct 3D emission projections for scatter:** The final step is to subtract the estimated scatter from direct and oblique sinograms. In some cases, the scatter estimate is smoothed before subtraction without loss of accuracy since the scatter distribution contains only low spatial frequency components.

Attenuation Correction

Definition of the Problem

A coincidence event requires the simultaneous detection of both photons arising from the annihilation of a positron. If either photon is absorbed within the body or scattered out of the field of view, a coincidence will not occur. The probability of detection, therefore, depends on the combined path of both photons. We saw in Chapter 2 that, since the total path length is the same for all sources lying on the line that joins two detectors, the probability of attenuation is the same for all such sources, independent of source position.

This is true even if the source is positioned outside the body. In this case, the probability terms are e^0 and $e^{-\mu D}$ for the near and far detectors respectively (where D is the total thickness of the body), and the number of detected coincidences is:

$$C = C_0 e^0 e^{-\mu D} = C_0 e^{-\mu D} \tag{29}$$

which is the same as that obtained for an internal source. Therefore, the problem of correcting for photon attenuation in the body is that of determining the probability of attenuation for all sources lying along a particular line of response.

Measured Attenuation Correction

The probability of attenuation for each line of response can be determined by comparing the count rate from an external (transmission) source with the unattenuated count rate from the same source when the patient is not in the tomograph, referred to as a blank scan. Transmission measurements are routinely performed in PET to correct for attenuation of the annihilation photons within the body. These measurements can be performed using several different source and detector configurations and these are discussed in the following sections.

Attenuation Correction Using Coincidence Transmission Data

The most common approach has been to use a long-lived positron emitter, such as ^{68}Ge-^{68}Ga (^{68}Ga is the positron emitter and 68Ge is its parent isotope with a half-life of 271 days), and measure the annihilation photons in coincidence as they pass through the body from an external source. A transmission scan typically takes 2–10 minutes to acquire and may be performed before or after the PET tracer is administered. However, it is not uncommon to perform transmission scans after tracer administration, nor is it uncommon to acquire transmission scans of much shorter duration for certain types of clinical studies.

In early PET scanner designs, the most common transmission source arrangement was a ring or multiple rings containing positron emitter. The rings surrounded the patient and were retracted behind lead shielding at the back of the scanner when not in use. When extended into the field of view, coincidences are recorded by detecting annihilation photons arising from anywhere on the ring(s), with one photon being detected on the near side and the other being detected on the far side after traversing the patient's body. Despite the relatively high energy of annihilation photons (511 keV), the likelihood of an annihilation photon passing through the body unattenuated can be very low. For example, for lines of response which pass through the long axis of the shoulders, typically only 1 photon in 50 may reach the far detector. As a result, transmission (and emission) count-rates for these lines of response are very low. Therefore, transmission measurements with this source-detector geometry are a major source of noise in reconstructed PET images [63–66].

In later generation PET scanners of the ring detector design, a more common transmission source geometry is the rotating rod source (Fig. 5.18a) [67]. This approach has the potential to provide improved signal to noise ratio (SNR) compared with ring sources due to reduced random and scatter coincidence fractions [68–70]. With this source geometry, the contribution of random and scattered coincidences to the transmission measurement can be further reduced using sinogram windowing [68, 71]. This is a technique in which the acquisition sinogram is electronically masked to distinguish those lines of response that are approximately collinear with the rod source at a given moment during its orbit from those that are not. The events recorded by lines of response that are not collinear with the rod source are rejected, as these mostly comprise scattered and random coincidences.

To maintain reasonable counting statistics in the transmission measurement, radioactivity in the rods must be more concentrated than in a ring source. This produces very high single event rates in the detectors nearest to the source, resulting in large dead time losses [12]. Therefore, the radioactivity is normally distributed among more than one rod source – typically two or three are used. Sinogram windowing (or "rod windowing") can be applied to one or more rotating rods, provided the exact location of each rod source is known at all times.

Sinogram windowing also has the potential to enable the transmission study to be performed after tracer administration [72–75]. The transmission scan may be performed either before the emission study (during the tracer uptake period) or after the emission scan. In either case, the time the patient spends on the scanning bed is reduced compared with the typical pre-injection transmission protocol, and patient throughput may be increased.

Furthermore, the feasibility of performing simultaneous emission and transmission scans using sinogram windowing has been demonstrated. For example, Thompson et al [76, 77] used two point sources of ^{68}Ge, one for each detector ring, encapsulated in lead subcollimators. The lead collimators were designed to shape the photon beams into a fan within the imaging plane, corresponding to the edge of the transaxial field of view. Sinogram windowing was applied to separate those lines of response which are collinear with the transmission source (primarily transmission events) from those that are far from collinear (primarily emission events). A rejection band on either side of the transmission window was also applied, in which neither emission nor transmission coincidences were recorded. This combination of physical and electronic collimation is similar to the approach used with the scanning line source in SPECT [78]. An alternative approach uses conventional sinogram windowing of rotating rod sources, but with a substantially reduced amount of radioactivity in the rod sources [79, 80]. The reduced source activity is to minimise their impact on dead time and the randoms contribution to the emission data. It was demonstrated using this approach, that the noise equivalent count-rate (NEC) of the emission data acquired at the same time as transmission data is only 10 to 15% lower than that of an equivalent emission source imaged using a separate emission-transmission scan protocol [79].

The use of simultaneous emission and transmission scanning has been applied to whole body PET imaging and, when combined with segmentation of attenuation images and iterative reconstruction, yields high quality diagnostic scans in a practical time frame [79]. However, the impact of transmission sources on the emission data is not negligible and results in bias which may not be acceptable for quantitative analyses, particularly when estimating tracer uptake in small tumours [81]. A more common approach in this clinical setting is to interleave short duration (2–4 minutes) emission and transmission scans as the patient couch is translated through the PET scanner gantry in discrete increments. Both approaches result in transmission data with poor SNR and some form of post reconstruction image processing, such as segmentation, is required to produce acceptable data for attenuation correction. Such post-processing methods are discussed below.

Attenuation Correction Using Singles Transmission Data

In tomographs with a dual 2D to 3D imaging capability, the slice-defining septa can be extended during transmission scanning and the acquisition performed in 2D mode using one of the coincidence detection ap-

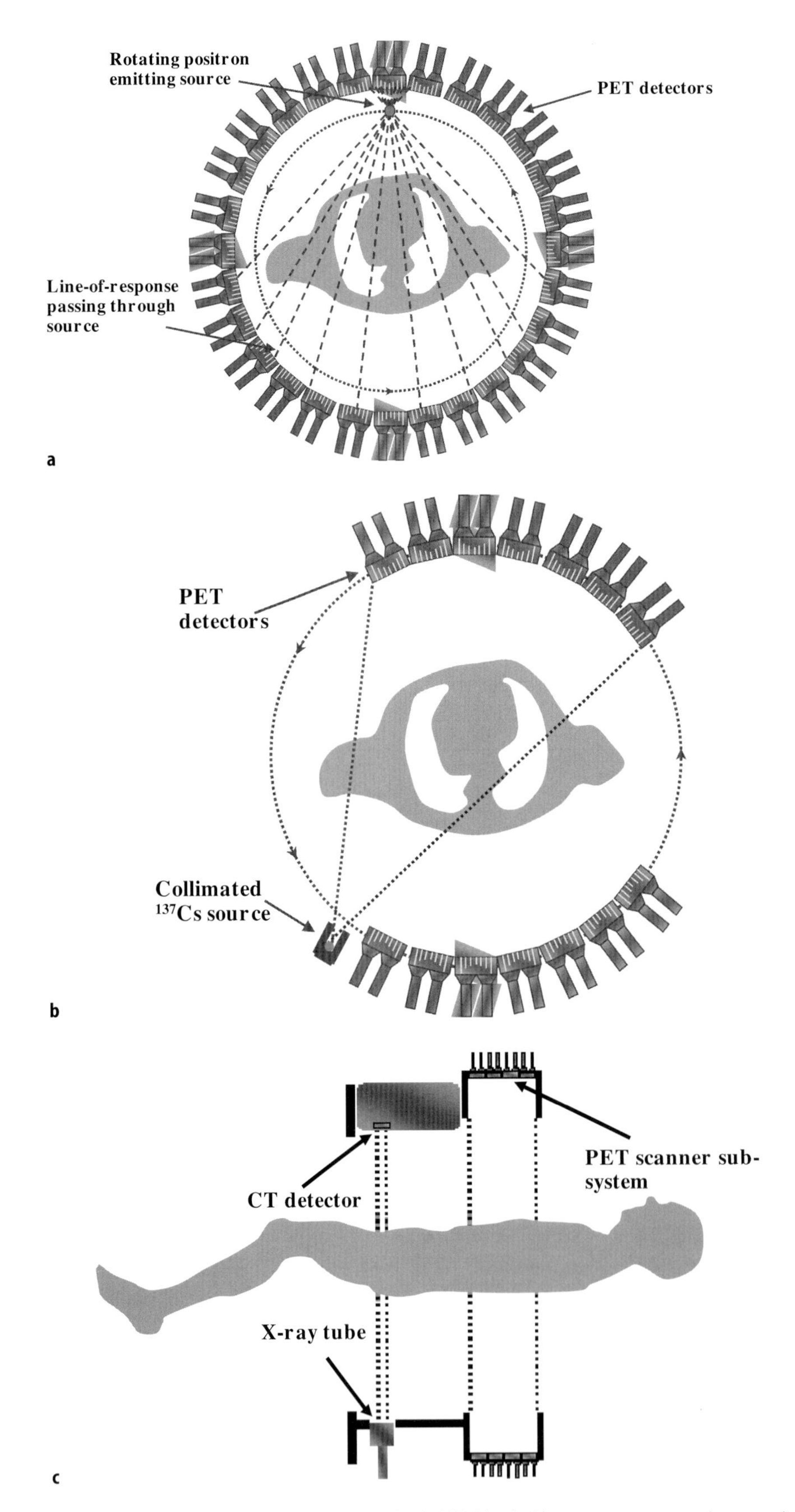

Figure 5.18. Methods of acquiring transmission data for attenuation correction in PET: (a) coincidence measurements using a rotating positron-emitting rod source, (b) singles measurements using a rotating single photon emitting point source, and (c) X-ray CT measurements performed on a dual modality PET/CT scanner.

proaches described above. However, in a scanner with 3D capability only, coincidence transmission scanning is impractical. This is because in 3D the near detector is exposed to an extremely high single photon flux arising from the transmission source. The source activity can not be reduced sufficiently without compromising the quality of the transmission data acquired on the far detector. The exception to this is when a fast detector with dedicated electronics is used as the near side transmission detector and rotates with the source [82, 83]. This promising approach is currently under development.

An alternative approach is to employ a point transmission source and shield the near detector from it, as suggested by Derenzo et al (Fig. 5.18b) [84]. Data are acquired with the far detector in "singles" rather than coincidence mode [85–87]. The shielded point source can be rotated around the patient and translated along the axial length of the scanner (or rotated along a spiral path) and, provided that the source location is known at all times, LORs can be formed between the source position and the position of the single photon event on the far detector. As with coincidence transmission studies, a separate blank scan is performed and the event rates compared with those recorded during the transmission scan along common LORs to determine the attenuation factors.

Use of single photon imaging results in better counting statistics in the transmission data than does coincidence imaging for two reasons. First, single photon counting is inherently more efficient than coincidence counting since the count-rate is dependent only on the efficiency of a single detector, whereas the count-rate in coincidence counting is dependent on the combined efficiencies of two detectors which are multiplied, resulting in an overall decrease in efficiency. Second, since the near detector is shielded from the point source, the activity can be substantially increased without severely impacting on dead time.

However, there are also drawbacks with the use of single photon transmission scanning. In coincidence transmission scanning, rod windowing substantially reduces the scatter contribution to the measurement and the resulting attenuation coefficients are very close to those expected for a narrow beam geometry. In the singles case, windowing cannot be employed since any detected photon could have originated from the source. Therefore, a significant scatter component may be included in the measured data and the attenuation coefficients are considered broad beam. In the case of NaI(Tl) volume PET scanners, this problem is offset by the relatively good energy resolution of NaI(Tl) compared with BGO PET scanners, which allows for better scatter rejection by energy discrimination. However, a further problem is that the transmission source normally emits photons at a different energy than the energy of annihilation photons. In the case of ^{137}Cs, the photon energy is 662 keV.

The problems of scaling the attenuation coefficients to those corresponding to 511 keV photons and a narrow beam geometry are normally addressed by segmenting the reconstructed attenuation image into a small number of tissues and assigning coefficients that are assumed to be known a priori (Fig. 5.19). Uncertainties associated with such assumptions are discussed below.

Attenuation Correction Using CT Data

With the advent of dual modality scanners capable of acquiring PET and CT data during the same imaging session, there has been considerable effort put into developing methods to make use of CT data for PET attenuation correction (Fig. 5.18c). The potential advantages of this approach arise because the statistical quality and spatial resolution of CT data is far superior to conventional transmission data used in PET, and because a whole body CT can be acquired in less than 1 minute using current generation multi-slice spiral scanners (compared to approximately 20 minutes for conventional transmission scanning), resulting in a significant reduction in scan time [88].

However, the fact that CT scanning is so much faster than PET scanning is also a potential pitfall, because in CT a snapshot of respiratory motion is obtained, rather than a time-averaged image. Without due care, this can lead to substantial artefacts in the reconstructed images [89]. A further problem arises because the transaxial field of view for CT scanners may be insufficient to accommodate the arms of the patient (if they are held by the sides), resulting in missing data. Artefacts are also caused by misregistration between the CT and PET data when the patient moves between scans – for example, positioning the arms above the head is not well tolerated by many patients and discomfort increases the likelihood of movement.

More minor problems include the fact that μ values do not scale linearly from the low energy of X-rays (approximately 60 keV) to the relatively high energy of annihilation photons (511 keV) – an issue which may be further complicated when contrast agents are used as an adjunct to the CT study. Finally, CT images are normally calibrated in Hounsfield units and must first be converted to μ values. The last two challenges can be addressed by segmenting the CT images into a discrete set of tissue types (see Chapter 8 and [90]). Once the

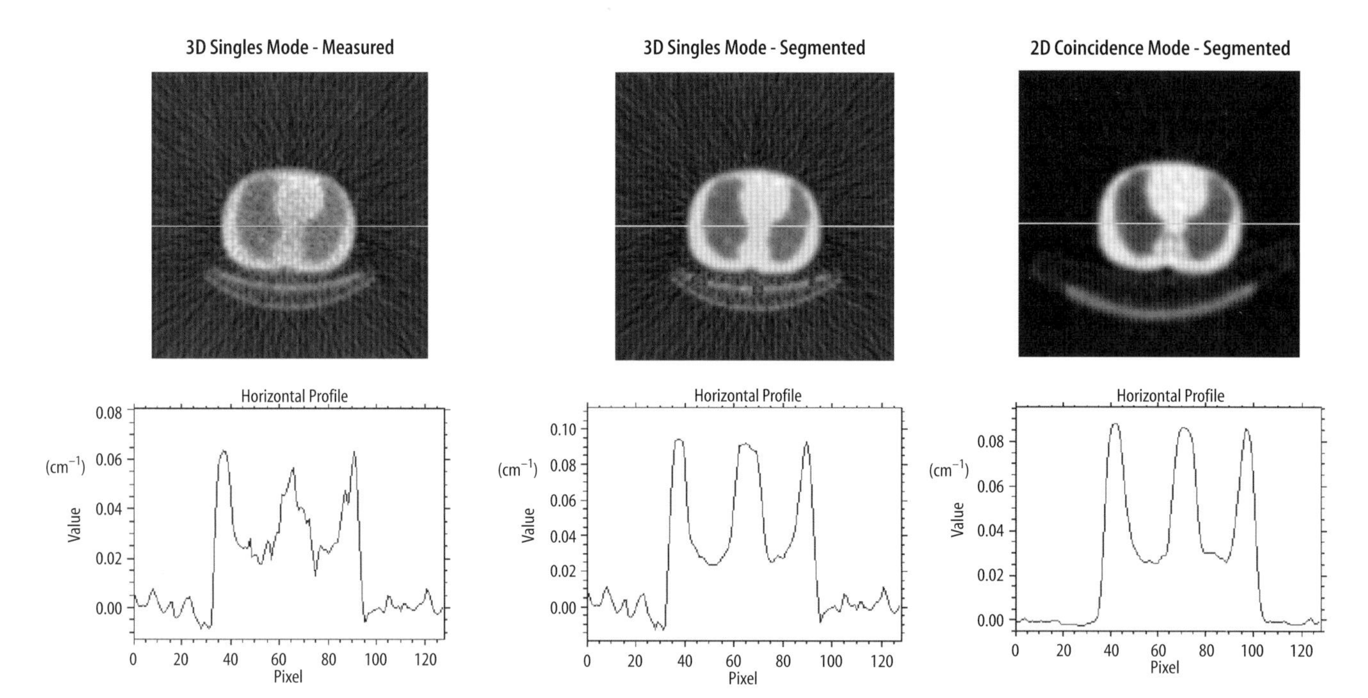

Figure 5.19. Transmission images of an anthropomorphic thorax phantom obtained using: (left) a singles transmission source in 3D mode; (middle) a singles transmission source with segmentation applied to the reconstruction; and (right) a coincidence transmission source in 2D mode with segmentation applied. The profiles of μ values below each image indicate the reduction in apparent attenuation observed with the singles source due to the greater photon energy and increased contribution of scatter compared with the coincidence measurement. This problem is overcome by segmentation. (These data were kindly provided by Dr Lefteris Livieratos, MRC Cyclotron Unit, Hammersmith Hospital, London.)

images are classified in this way, each discrete tissue type can be treated separately by applying an empirical scale factor that scales the corresponding voxels to a μ value appropriate for 511 keV imaging. The segmentation algorithm should ideally be sufficiently general to be able to identify tissue types in the presence of contrast agents and metallic prosthetic implants. However, most current implementations ignore these effects. After segmenting and scaling the CT data, the images are normally smoothed to a resolution that matches the PET data before forward projecting and calculating the attenuation correction factors.

Calculated Attenuation Correction

Since a transmission scan adds significantly to the time it takes to perform a PET study, alternative methods of attenuation correction have been investigated. One method, which assumes a regular geometric body outline and constant tissue density, is commonly referred to as calculated (as opposed to measured) attenuation correction. This may be valid for brain studies, particularly if the skull is taken into account as well as the soft tissue of the brain. The attenuation factors are calculated for each coincidence line, based on the constant attenuation along a chord through the object. The chord is typically calculated from an ellipse selected by the operator to fit the body outline. An improvement on this method is to define the body outline automatically using edge detection methods operating on the emission sinogram [91, 92].

While calculated attenuation correction produces accurate noise-free attenuation factors when imaging regularly shaped phantoms of uniform density, it is prone to bias when applied to most regions of the body, even the brain where it is often considered applicable. In particular, this approach leads to systematic underestimation of attenuation by up to 20% in the parietal and occipital lobes due to thickening of the adjacent skull bones, and overestimation by up to 12% in the gyrus recti due to the close proximity of the frontal sinuses [75].

Segmented Attenuation Correction

Conventionally, 2D smoothing is applied to transmission sinograms before dividing into the blank scan to determine attenuation correction factors (ACF) [65, 93]. However, this method of processing has the undesirable effect of causing a mismatch between the resolution of the emission and transmission data and is not completely effective in controlling noise propagation [66]. An alternative approach is to reconstruct attenuation coefficient images derived from the transmission and blank data and then segment the images into a small number of tissue types with a priori known attenuation coefficients. For example, Huang et al [94] described a method where the operator manually defines the body and lung outlines on the attenuation images. After assigning attenuation coefficients to these regions, noiseless attenuation correction factors are then calculated by forward projection. This method was shown to provide equivalent results to measured attenuation correction using a transmission scan acquired for approximately one quarter of the time.

This approach has been extended by several investigators by automating the determination of lung and body outlines using various morphological operators and heuristics. For example, Xu et al used a simple thresholding method to segment attenuation images into three discrete regions: air, lung and soft tissue [95]. Unsupervised image segmentation is more practical than the manual approach, particularly for large data sets such as those encountered in whole body PET.

The main problem with segmented attenuation correction is that there is a large degree of variability in tissue densities from patient to patient, particularly in the lungs. Assigning the same population average value to the lung regions of each patient may lead to significant bias. An alternative approach is to calculate the histogram of μ values for each patient study and assign values based on an assumed probability distribution for the lung and soft tissue components of the histogram [66]. More recently, Bettinardi et al described an adaptive segmentation method based on a fuzzy clustering algorithm [96]. This method is also based on the histogram of μ values but it automatically determines both the number of tissue classes that can be supported by the data (based on the variance in the images) and their centroids. The method is sufficiently general that it can be applied to any region of the body and is able to distinguish bony structures from soft tissue given adequate counting statistics (Fig. 5.20).

As segmentation is a non-linear, non-stationary process, the various sources of error discussed are difficult to predict. Therefore, not only is the accuracy of the method slightly inferior to measured attenuation correction, but it is also less predictable. This may be a serious drawback in studies where reliable estimates of quantitative values, including physiological variables, are required. However, given the improvement in SNR that segmentation provides compared with conventional transmission processing, these disadvantages may not be important in many clinical applications.

Partial Volume Correction

The Partial Volume Effect

In quantitative PET, the reconstructed image should map the radiotracer concentration with uniform accuracy and precision throughout the field of view. However, due to the partial volume effect, the bias in reconstructed pixel values may vary depending on the size of the structure being sampled and its radioactivity concentration relative to surrounding structures. The partial volume effect may be described as follows. When the object or structure being imaged only partially occupies the sensitive volume of the PET scanner, its signal amplitude becomes diluted with signals from surrounding structures. The sensitive volume has dimensions approximately equal to twice the FWHM resolution of the reconstructed image. For example, if a tomograph has isotropic reconstructed resolution of 6 mm FWHM, then a structure which has any dimension less than 12 mm will have its signal diluted and the degree of underestimation of radioactivity concentration will depend not only on its size but also on the relative concentration in surrounding structures.

There are several possible approaches to correcting or minimising the partial volume effect. These include methods that attempt to recover resolution losses before or during image reconstruction and methods that use side information from anatomical imaging modalities such as CT and MRI.

Resolution Recovery

One can attempt to improve the resolution of the reconstructed images either by applying resolution recovery techniques to the data before reconstruction, through the use of inverse filtering for example, or by extending the imaging model during image reconstruction to include resolution effects. The latter is done within an iterative reconstruction framework using a Bayesian approach where the additional information is treated as a prior [97, 98]. This approach is discussed in detail in Chapter 4 but the general outline is as follows. Projection data are estimated based on a model of the imaging system, which may include resolution effects, and an initial estimate of the radiophar-

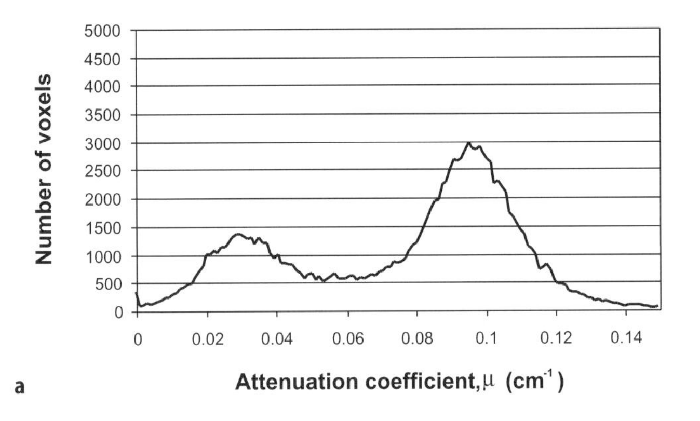

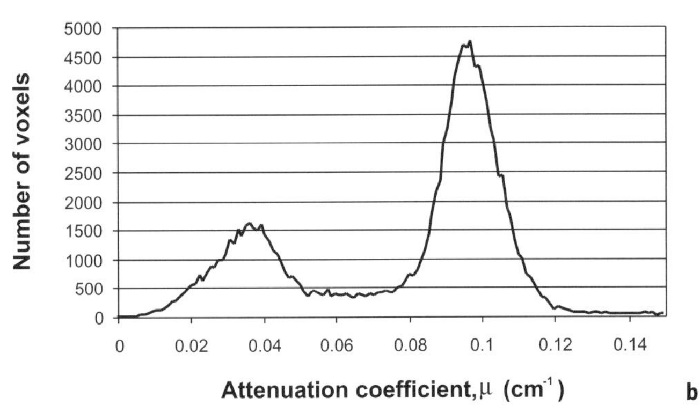

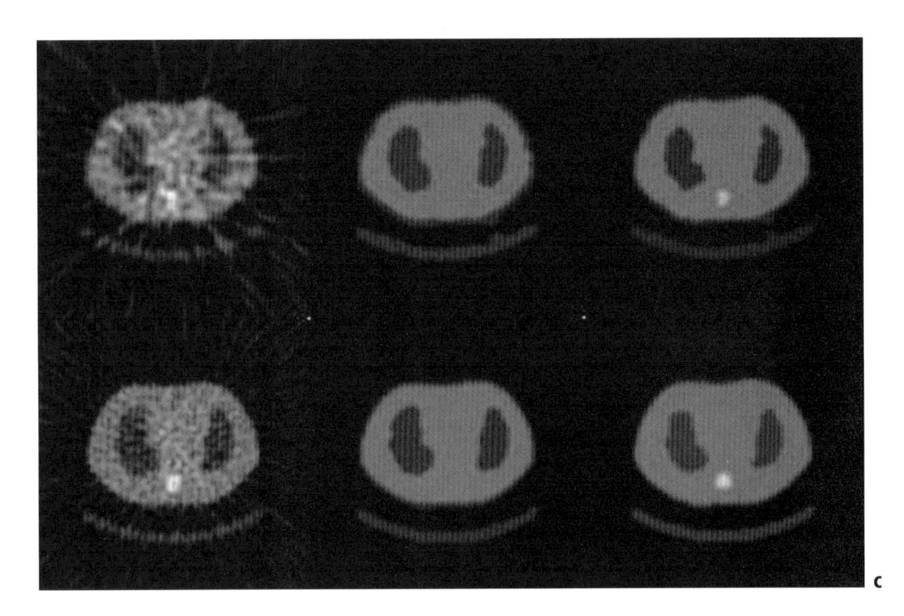

Figure 5.20. Histograms of attenuation coefficients obtained from a 3-minute transmission scan (a) and a 16-minute transmission scan (b) of an anthropomorphic thorax phantom (Data Spectrum, Chapel Hill, NC). The original (first column) and segmented images of the same phantom are shown in (c). The 3-minute scan is on the top row and the 16-minute scan is on the bottom row. The images in the middle column were segmented using a histogram-based technique [66], while the images in the 3rd column were segmented using an adaptive classification technique [96]. Note that the adaptive technique correctly classifies the more highly attenuating spine insert.

maceutical distribution in the body. The estimated projections are compared with the measured projections and the errors are back projected and used to improve the image estimate. The process is repeated iteratively until a very close match between the estimated and measured projections is achieved. Clearly, if the model of the imaging system is a good one, the image estimate after convergence will closely resemble the underlying radiopharmaceutical distribution in the body and if the model includes resolution effects, the impact of the partial volume effect should also be minimised.

The method has the potential to account for a number of factors affecting the spatially varying resolution of PET images, including geometric effects and physical effects such as positron range and non-collinearity of the annihilation photons. Therefore, it is potentially more accurate than simple inverse filtering but the accuracy depends on the quality of information incorporated into the imaging model.

Use of Anatomical Imaging Data

An example of how anatomical information can be used to effect partial volume correction is illustrated in Fig. 5.21 using the method described by Müller-Gartner et al [99]. This method, like most others based on post reconstruction image analysis, only applies to brain imaging. The PET image volume is first spatially coregistered to the corresponding MR image of the same subject. Then, the MR image is segmented into grey and white matter regions of the brain. Separate images representing grey and white matter regions are convolved with a smoothing kernel which is derived from the point spread function (PSF) of the PET scanner. The smoothed white matter image is normalised to the counts in a white matter region of interest on the PET image and then subtracted from the PET image to remove spill over of white matter signal into grey matter regions. The final step is to divide the

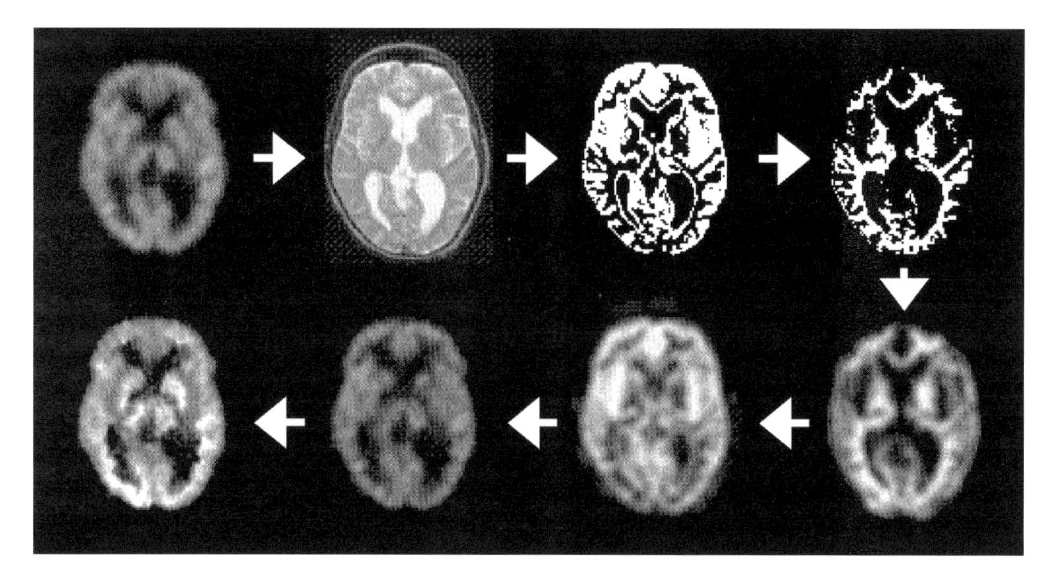

Figure 5.21. An FDG PET study of a patient with AIDS dementia illustrating the application of partial volume correction. The images in order of processing steps applied (clockwise from top left) are: (a) original uncorrected PET image, (b) coregistered MR image of the same patient, (c) segmented grey matter image, (d) segmented white matter image, (e) blurred white matter image, (f) blurred grey matter image, (g) PET grey matter image obtained by subtracting e from a, and (h) corrected PET image obtained by dividing g by f.

resulting image by the smoothed grey matter image. The effect of this final step is to preferentially enhance the signal in regions where the smoothing step resulted in greatest dilution of the signal (i.e., small and/or thin structures).

Other methods which are also based on high resolution anatomical imaging data and a model of the PET scanner PSF have been described. For example, the geometric transfer matrix (GTM) method assumes that the PET image can be divided into a discrete number of tissue domains or ROIs, each of which has uniform radiotracer concentration [100]. The ROIs are defined on an anatomical atlas or a MR image coregistered with the PET image and the mean value within the ROI is modelled as the weighted sum of the true activity values in surrounding voxels belonging to the same domain. The weights are independent of the radiotracer concentration and can be determined from knowledge of the position dependent PSF. The correction is applied by inverting the matrix of weights to recover the voxel values of the true radioactivity concentration.

An alternative approach to that of modelling the PET scanner PSF is to incorporate information from anatomical imaging into the reconstruction model [101, 102]. As in the case of resolution recovery, the incorporation of side information from anatomical imaging data is normally done in a Bayesian context where the additional information is treated as a prior. This method requires that the anatomical and functional images are spatially coregistered [103] and normally involves segmenting the MRI or CT image into a small number of tissue classes. The segmented image is then used to constrain the reconstruction such that smoothing is allowed within boundaries defined on the segmented image but not across boundaries [101, 104, 105]. This has the effect of controlling noise in the image while at the same time preserving high resolution information at functional/anatomical boundaries. While this approach does not directly address the partial volume effect, it has the potential to minimise its impact on signal recovery from small structures and, hence, quantitative accuracy. Problems with this approach may potentially arise if there is mismatch between the functional and anatomic characteristics of the tissue in question.

PET Scanner Calibration

Once the corrections for the various sources of bias described in this chapter have been applied to PET data, images can be reconstructed which are free of artefacts and which reflect the regional concentration of radiotracer in the body. In most clinical applications of PET this is sufficient as the images are interpreted visually without reference to the absolute voxel values. Indeed, this degree of "quantification" is sufficient in some types of kinetic research study also. For example, methods which take a reference tissue as the input function for the kinetic model do not require the PET images to be calibrated in absolute units of tracer concentration. However, for most kinetic study protocols PET image values must be related to the tracer concentration in blood samples withdrawn during the study. These blood samples are normally counted in a well counter. Thus, it is essential in these studies to have an accurate calibration between the PET scanner and the well counter. This is usually achieved by scanning a phantom with uniform radioactivity concentration and then counting an aliquot taken from the phantom in the well counter. The phantom images are reconstructed using the same corrections as are applied in research studies and the voxel values directly compared with the counted aliquot to determine a calibration factor.

Note that the above procedure still does not necessarily provide a reading in absolute units of radioactivity concentration for the PET image voxels unless the radioactivity concentration in the aliquot is accurately known. Once again, this is not usually necessary for quantitative tracer kinetic studies. It is usually sufficient to have an accurate calibration factor that relates PET image values to well counter measurements. If absolute units of radioactivity concentration are required, a procedure is available which results in a measurement of the detection efficiency of the tomograph in air [106]. When PET images are corrected for all the effects described above, the image values (in units of counts/sec/voxel) can be divided by the tomograph efficiency (in units of counts/sec/kBq) and then divided by the voxel volume to yield images calibrated in units of kBq/ml.

References

1. Sossi V, Pointon B, Cohen P, Johnson RR, Ruth TJ. Effect of shielding the radioactivity outside the field of view on image quality in a dual head coincidence. IEEE Trans Nucl Sci 2000;47(4):1561–1566.
2. Spinks TJ, Miller MP, Bailey DL, Bloomfiled PM, Livieratos L, Jones T. The effect of activity outside the direct field of view in a 3D-only whole-body positron tomograph. Phys Med Biol 1998;43:895–904.
3. Karp JS, Muehllehner G, Mankoff DA, Ordonez CE, Ollinger JM, Daube-Witherspoon ME, et al. Continuous-slice PENN-PET: a positron tomograph with volume imaging capability. J Nucl Med 1990;31(617–627).
4. Michel C, Sibomana M, Bol A, Bernard X, Lonneux M, Defrise M, et al. Preserving Poisson characteristics of PET data with weighted OSEM reconstruction. In: IEEE Nuclear Science Symposium and Medical Imaging Conference; 1998; Toronto: IEEE; 1998. p. 1323–1329.
5. Smith RJ, Karp JS. A practical method for randoms subtraction in volume imaging PET from detector singles countrate measurements. IEEE Trans Nucl Sci 1996;43(3):1981–1987.
6. Hoffman EJ, Huang S-C, Phelps ME, Kuhl DE. Quantitation in positron emission computed tomography: 4. Effect of accidental coincidences. J Comput Assist Tomogr 1981;5(3):391–400.
7. Swan WL, Vannoy SD, Harrison RL, Miyaoka RS, Lewellen TK. Randoms simulation for dual head coincidence imaging of cylindrically symmetric source distributions" IEEE Trans Nucl. Sci. 46(4) II: 1156 –1164. IEEE Trans Nucl Sci 1999;46(4):1156–1164.
8. Casey ME, Hoffman EJ. Quantitation in positron emission tomography: 7. A technique to reduce noise in accidental coincidence measurements and coincidence efficiency calibration. J Comput Assist Tomogr 1986;10(5):845–850.
9. Badawi RD, Miller MP, Bailey DL, Marsden PK. Randoms variance reduction in 3D PET. Phys Med Biol 1999;44(4):941–954.
10. Casey ME, Gadagkar H, Newport D. A component based method for normalisation in volume PET. In: Proceedings of the 3rd International Meeting on Fully Three-Dimensional Image Reconstruction in Radiology and Nuclear Medicine; 1995; Aix-les-Bains, France; 1995. p. 67–71.
11. Badawi RD, Ferreira NC, Kohlmyer SG, Dahlbom M, Marsden PK, Lewellen TK. A comparison of normalization effects on three whole-body cylindrical 3D PET systems. Phys Med Biol 2000;45(11):3253–3266.
12. Germano G, Hoffman EJ. Investigation of count-rate and dead-time characteristics of a high resolution PET system. J Comput Assist Tomogr 1988;12(5):836–846.
13. Bai B, Li Q, H Holdsworth CH, Asma E, Tai YC, Chatziioannou A, et al. Model-based normalization for iterative 3D PET image reconstruction. Phys Med Biol 2002;47(15):2773–2784.
14. Ollinger JM. Detector efficiency and Compton scatter in fully 3D PET. IEEE Trans Nucl Sci 1995;42(4):1168–1173.
15. Badawi RD, Marsden PK. Developments in component-based normalization for 3D PET. Phys Med Biol 1999;44(2):571–594.
16. Cherry SR, Dahlbom M, Hoffman EJ. 3D PET using a Conventional Multislice Tomograph without Septa. J Comput Assist Tomogr 1991;15(4):655–668.
17. Hoffman EJ, Guerrero TM, Germano G, Digby WM, Dahlbom M. PET system calibration and corrections for quantitative and spatially accurate imags. IEEE Trans Nucl Sci 1989;36(1): 1108–1112.
18. Badawi RD, Lodge MA, Marsden PK. Algorithms for calculating detector efficiency normalization coefficients for true coincidences in 3D PET. Phys Med Biol 1998;43(1):189–205.
19. Chesler DA, Stearns CW. Calibration of detector sensitivity in positron cameras. IEEE Trans Nucl Sci 1990;37(2):768–772.
20. Chatziioannou A-XF. Measurements and calculations towards quantitative whole body PET imaging. Los Angeles: UCLA School of Medicine; 1996.
21. Ferreira NC, Trebossen R, Comtat C, Gregoire M-C, Bendriem B. Iterative crystal efficiency calculation in fully 3-D PET. IEEE Trans Med Imag 2000;19(5):485–492.
22. Defrise M, Townsend DW, Bailey DL, Geissbuhler A, Michel C, Jones T. A normalisation technique for 3D PET data. Phys Med Biol 1991;36:939–952.
23. Patiwan Y, Kohlmyer S, Lewellen T, O'Sullivan F. PET system calibration and attenuation correction. IEEE Trans Nucl Sci 1997;44(3):1249–1253.
24. Badawi RD, Marsden PK. Self-normalization of emission data in 3D PET. IEEE Trans Nucl Sci 1999;46(3):709–712.
25. Bailey DL, Townsend DW, Kinahan PE, Grootoonk S, Jones T. An investigation of factors affecting detector and geometric correction in normalisation of 3-D PET data. IEEE Trans Nucl Sci 1996;43(6):3300–3307.
26. Oakes TR, Sossi V, Ruth TJ. Normalization for 3D PET with a low-scatter planar source and measured geometric factors. Phys Med Biol 1998;43:961–972.
27. Casey ME. An analysis of counting losses in positron emission tomography [PhD]: University of Tennessee; 1992.
28. Germano G, Hoffman EJ. A study of data loss and mispositioning due to pileup in 2-D detectors in PET. IEEE Trans Nucl Sci 1990;37(2):671–675.

29. Wear JA, Karp JS, Freifelder R, Mankoff DA, Muehllehner G. A model of the high count-rate performance of NaI(Tl)-based PET detectors. IEEE Trans Nucl Sci 1998;45(3):1231–1237.
30. Wong W-H, Li H, Uribe J, Baghaei H, Wang Y, Yokoyama S. Feasibility of a high-speed gamma-camera design using the high-yield-pileup-event-recovery method. J Nucl Med 2001;42(4): 624–632.
31. Moisan C, Rogers JG, Douglas JL. A count-rate model for PET and its application to an LSO HR plus scanner. IEEE Trans Nucl Sci 1997;44:1219–1224.
32. Daube-Witherspoon ME, Carson RE. Unified deadtime correction model for PET. IEEE Trans Med Imag 1991;10(3):267.
33. Eriksson L, Wienhard K, Dahlbom M. A simple data loss model for positron camera systems. IEEE Transactions on Nuclear Science. IEEE Trans Nucl Sci 1994;41(4):1566–1570.
34. Knoll GF. Radiation Detection and Measurement. 3rd ed. New York: Wiley and Sons; 2000.
35. Compton AH. A quantum theory of the scattering of X-rays by light elements. Phys Rev 1923;21:483–502.
36. Cherry SR, Huang SC. Effects of scatter on model parameter estimates in 3D PET studies of the human brain. IEEE Trans Nucl Sci 1995;NS-42:1174–1179.
37. Cherry SR, Meikle SR, Hoffman EJ. Correction and characterization of scattered events in three-dimensional PET using scanners with retractable septa. J Nucl Med 1993;34:671–678.
38. Hasegawa T, Tanaka E, Yamashita T, Watanabe M, Yamaya T, Murayama H. A Monte Carlo simulation study on coarse septa for scatter correction in 3-D PET. IEEE Trans Nucl Sci 2002;49(5):2133–2138.
39. Thompson CJ. The problem of scatter correction in positron volume imaging. IEEE Trans Med Imaging 1993;MI-12:124–132.
40. Buvat I, Benali H, Todd-Pokropek A, Di Paola R. Scatter correction in scintigraphy: the state of the art. Eur J Nucl Med 1994;21(7):675–694.
41. Grootoonk S, Spinks TJ, Jones T, Michel C, Bol A. Correction for scatter using a dual energy window technique with a tomograph operated without septa. In: IEEE Nuclear Science Symposium and Medical Imaging Conference; 1991; Santa Fe: IEEE; 1991. p. 1569–1573.
42. Bendriem B, Trebossen R, Froulin V, Syrota A. A PET scatter correction using simultaneous acquisitions with low and high lower energy thresholds. In: Klaisner L, editor. IEEE Nuclear Science Symposium and Medical Imaging Conference; 1993; San Francisco: IEEE; 1993. p. 1779–1783.
43. Harrison RL, Haynor DR, Lewellen TK. Dual energy window scatter corrections for positron emission tomography. In: Conference Record of the 1991 IEEE Nuclear Science Symposium and Medical Imaging Conference; 1991; Santa Fe, NM: IEEE; 1991. p. 1700–1704.
44. Adam L-E, Karp JS, Freifelder R. Scatter correction using a dual energy window technique for 3D PET with NaI(Tl) detectors. In: IEEE Nuclear Science Symposium and Medical Imaging Conference; 1998; Toronto: IEEE; 1998. p. 2011–2018.
45. Shao L, Freifelder R, Karp JS. Triple energy window scatter correction technique in PET. IEEE Trans Med Imag 1994;13(4): 641–648.
46. Bentourkia M, Msaki P, Cadorette J, Lecomte R. Energy dependence of scatter components in multispectral PET imaging. IEEE Trans Nucl Sci 1995;NS-14(1):138–145.
47. Bergstrom M, Eriksson L, Bohm C, Blomqvist G, Litton J. Correction for scattered radiation in a ring detector positron camera by integral transformation of the projections. J Comput Assist Tomogr 1983;7:42–50.
48. Shao L, Karp JS. Cross-plane scattering correction – point source deconvolution in PET. IEEE Trans Med Imag 1991;10:234–239.
49. Bailey DL, Meikle SR. A convolution-subtraction scatter correction method for 3D PET. Phys Med Biol 1994;39:411–424.
50. Sossi V. Evaluation of the ICS and DEW scatter correction methods for low statistical content scans in 3D PET. In: Del Guerra A, editor. IEEE Nuclear Science Symposium and Medical Imaging Conference; 1996; Anaheim: IEEE; 1996. p. 1537–1541.
51. Yanch JC, Flower MA, Webb S. Improved quantification of radionuclide uptake using deconvolution and windowed subtraction techniques for scatter compensation in single photon emission computed tomography. Med Phys 1990;17(6): 1011–1022.
52. Meikle SR, Hutton BF, Bailey DL, Fulton RR, Schindhelm K. SPECT scatter correction in non-homogeneous media. In: Colchester ACF, Hawkes DJ, editors. Information Processing in Medical Imaging, 12th International Conference. Berlin: Springer-Verlag; 1991. p. 34–44.
53. Bailey DL, Hutton BF, Meikle SR, Fulton RR, Jackson CB. An attenuation dependent scatter correction technique for SPECT. Phys Med Biol 1989;34:152.
54. Meikle SR, Hutton BF, Bailey DL. A transmission dependent method for scatter correction in SPECT. J Nucl Med 1994;35(2): 360–367.
55. Barney JS, Rogers JG, Harrop R, Hoverath H. Object shape dependent scatter simulations for PET. IEEE Trans Nucl Sci 1991;NS-38:719–725.
56. Watson CC. New, faster, image-based scatter correction for 3D PET. In: IEEE Nuclear Science Symposium and Medical Imaging Conference; 1998; Toronto: IEEE; 1998.
57. Watson CC, Newport D, Casey ME. A single scatter simulation technique for scatter correction in 3D PET. In: Grangeat P, Amans J-L, editors. Three-Dimensional Image Reconstruction in Radiology and Nuclear Medicine. Dordrecht: Kluwer Academic; 1996. p. 255–268.
58. Ollinger JM. Model-based scatter correction for fully 3D PET. Phys Med Biol 1996;41:153–176.
59. Holdsworth CH, Badawi RD, Santos PA, Van den Abbeele AD, El Fakhri G. Evaluation of a Monte Carlo Scatter Correction in Clinical 3D PET. In: Metzler SD, editor. Conference Record of the IEEE Nuclear Science Symposium and Medical Imaging Conference; 2003; Portland: IEEE; 2003. (in press).
60. Levin CS, Dahlbom M, Hoffman EJ. A Monte Carlo correction for the effect of Compton scattering in 3D PET brain imaging. IEEE Trans Nucl Sci 1995;42:1181–1185.
61. Holdsworth CH, Levin CS, Farquhar TH, Dahlbom M, Hoffman EJ. Investigation of accelerated Monte Carlo techniques for PET simulation and 3D PET scatter correction. IEEE Trans Nucl Sci 2001;48:74–81.
62. Holdsworth CH, Levin CS, Janecek M, Dahlbom M, Hoffman EJ. Performance analysis of an improved 3D PET Monte Carlo simulation and scatter correction. IEEE Trans Nucl Sci 2002;49(1):83–89.
63. Gullberg GT, Huesman RH. Emission and transmission noise propagation in positron emission computed tomography: Lawrence Berkely Laboratory; 1979. Report No.: LBL-9783.
64. Huang SC, Hoffman EJ, Phelps ME, Kuhl DE. Quantitation in positron emission tomography: 2. Effect of inaccurate attenuation correction. J Comput Assist Tomogr 1979;3(6):804–814.
65. Dahlbom M, Hoffman EJ. Problems in signal-to-noise ratio for attenuation correction in high resolution PET. IEEE Trans Nucl Sci 1987;34(1):288–293.
66. Meikle SR, Dahlbom M, Cherry SR. Attenuation correction using count-limited transmission data in positron emission tomography. J Nucl Med 1993;34(1):143–150.
67. Derenzo SE, Budinger TF, Huesman RH, Cahoon JL, Vuletich T. Imaging properties of a positron tomograph with 280 BGO crystals. IEEE Trans Nucl Sci 1981;28(1):81–89.
68. Carroll LR, Kretz P, Orcutt G. The orbiting rod source: Improving performance in PET transmission correction scans. In: Esser PD, editor. Emission Computed Tomography – Current Trends. New York: Society of Nuclear Medicine; 1983. p. 235–247.
69. Thompson CJ, Dagher A, Lunney DN, Strother SC, Evans AC. A technique to reject scattered radiation in PET transmission scans. In: Nalcioglu O, Cho ZH, Budinger TF, editors. International Workshop on Physics and Engineering of Computerized Multidimensional Imaging and Processing: Proc. SPIE; 1986. p. 244–253.
70. Ostertag H, Kubler WK, Doll J, Lorenz WJ. Measured attenuation correction methods. Eur J Nucl Med 1989;15(11):722–726.
71. Huesman RH, Derenzo SE, Cahoon JL, Geyer AB, Moses WW, Uber DC, et al. Orbiting transmission source for positron emission tomography. IEEE Trans Nucl Sci 1988;NS-35:735–739.
72. Carson RE, Daube-Witherspoon ME, Green MV. A method for postinjection PET transmission measurements with a rotating source. J Nucl Med 1988;29:1558–1567.

73. Daube-Witherspoon ME, Carson RE, Green MV. Post-injection transmission attenuation measurements for PET. IEEE Trans Nucl Sci 1988;35(1):757–761.
74. Ranger NT, Thompson CJ, Evans AC. The application of a masked orbiting transmission source for attenuation correction in PET. J Nucl Med 1989;30:1056–1068.
75. Hooper PK, Meikle SR, Eberl S, Fulham MJ. Validation of post injection transmission measurements for attenuation correction in neurologic FDG PET studies. J Nucl Med 1996;37:128–136.
76. Thompson CJ, Ranger N, Evans AC, Gjedde A. Validation of simultaneous PET emission and transmission scans. J Nucl Med 1991;32:154–160.
77. Thompson CJ, Ranger NT, Evans AC. Simultaneous transmission and emission scans in positron emission tomography. IEEE Trans Nucl Sci 1989;36(1):1011–1016.
78. Tan P, Bailey DL, Meikle SR, Eberl S, Fulton RR, Hutton BF. A scanning line source for simultaneous emission and transmission measurements in SPECT. J Nucl Med 1993;34(10):1752–1760.
79. Meikle SR, Bailey DL, Hooper PK, Eberl S, Hutton BF, Jones WF, et al. Simultaneous emission and transmission measurements for attenuation correction in whole body PET. J Nucl Med 1995;36:1680–1688.
80. Meikle SR, Eberl S, Hooper PK, Fulham MJ. Simultaneous emission and transmission (SET) scanning in neurological PET studies. J Comput Assist Tomogr 1997;21(3):487–497.
81. Lodge MA, Badawi RD, Marsden PK. A clinical evaluation of the quantitative accuracy of simultaneous emission/transmission scanning in whole-body positron emission tomography. Eur J Nucl Med 1998;25(4):417–423.
82. Jones WF, Moyers JC, Casey ME, Watson CC, Nutt R. Fast-channel LSO detectors and fiber-optic encoding for excellent dual photon transmission measurements in PET. IEEE Trans Nucl Sci 1999;46(4):979–984.
83. Watson CC, Eriksson L, Casey ME, Jones WF, Moyers JC, Miller S, et al. Design and performance of collimated coincidence point sources for simultaneous transmission measurements in 3-D PET. IEEE Trans Nucl Sci 2001;48(3):673–679.
84. Derenzo SE, Zaklad H, Budinger TF. Analytical study of a high-resolution positron ring detector system for transaxial reconstruction tomography. J Nucl Med 1975;16(12):1166–1173.
85. deKemp RA, Nahmias C. Attenuation correction in PET using single photon transmission measurement. Med Phys 1994;21(6): 771–778.
86. Karp JS, Muehllehner G, Qu H, Yan XH. Singles transmission in volume-imaging PET with a 137Cs source. Phys Med Biol 1995;40(5):929–944.
87. Yu SK, Nahmias C. Single-photon transmission measurements in positron tomography using 137Cs. Phys Med Biol 1995;40(7): 1255–1266.
88. Beyer T, Townsend DW, Brun T, Kinihan PE, Charron M, Roddy R, et al. A combined PET/CT scanner for clinical oncology. J Nucl Med 2000;41:1369–1379.
89. Beyer T. Design, construction and validation of a combined PET/CT tomograph for clinical oncology [PhD thesis]: University of Surrey (UK), University of Pittsburgh (USA); 1999.
90. Kinahan PE, Townsend DW, Beyer T, Sashin D. Attenuation correction for a combined 3D PET/CT scanner. Med Phys 1998; 25:2046–2053.
91. Bergstrom M, Litton J, Eriksson L, Bohm C, Blomqvist G. Determination of object contour from projections for attenuation correction in cranial positron emission tomography. J Comput Assist Tomogr 1982;6(2):365–372.
92. Siegel S, Dahlbom M. Implementation and evaluation of a calculated attenuation correction for PET. IEEE Trans Nucl Sci 1992;NS-39:1117–1121.
93. Palmer MR, Rogers JG, Bergstrom M, Beddoes MP, Pate BD. Transmission profile filtering for positron emission tomography. IEEE Trans Nucl Sci 1986;33(1):478–481.
94. Huang SC, Carson R, Phelps M, Hoffman E, Schelbert H, Kuhl D. A boundary method for attenuation correction in positron emission tomography. IEEE Trans Nucl Sci 1981;22:627–637.
95. Xu EZ, Mullani NA, Gould KL, Anderson WL. A segmented attenuation correction for PET. J Nucl Med 1991;32:161–165.
96. Bettinardi V, Pagani E, Gilardi MC, Landoni C, Riddell C, Rizzo G, et al. An automatic classification technique for attenuation correction in positron emission tomography. Eur J Nucl Med 1999;26(5):447–458.
97. Green PJ. Bayesian reconstruction from emission tomography data using a modified EM algorithm. IEEE Trans Med Imag 1990;9(1):84–93.
98. Qi J, Leahy RM, Hsu C, Farquar TH, Cherry SR. Fully 3D bayesian image reconstruction for the ECAT EXACT HR+. IEEE Trans Nucl Sci 1998;45(3):1096–1103.
99. Muller-Gartner HW, Links JM, Prince JL, Bryan RN, McVeigh E, Leal JP, et al. Measurement of radiotracer concentration in brain gray matter using positron emission tomography: MRI-based correction for partial volume effects. J Cereb Blood Flow Metab 1992;12(4):571–583.
100. Rousset OG, Ma Y, Evans AC. Correction for partial volume effects in PET: principle and validation. J Nucl Med 1998;39:904–911.
101. Ouyang X, Wong WH, Johnson VE, Hu X, Chen C-T. Incorporation of correlated structural images in PET image reconstruction. IEEE Trans Med Imag 1994;MI-14(4):627–640.
102. Baete K, Nuyts J, Van Paesschen W, Suetens P, Dupont P. Anatomical based FDG-PET reconstruction for the detection of hypometabolic regions in epilepsy. In: Metzler SD, editor. 2002 IEEE Nuclear Science Symposium Conference Record; 2003; Norfolk, VA: IEEE; 2003. p. 1481–1485.
103. Ardekani BA, Braun M, Hutton BF, Kanno I, Iida H. A fully automatic multimodality image registration algorithm. J Comput Assist Tomogr 1995;19(4):615–623.
104. Ardekani BA, Braun M, Hutton BF, Kanno I, Iida H. Minimum cross-entropy reconstruction of PET images using prior anatomical information. Phys Med Biol 1996;41(11):2497–2517.
105. Som S, Hutton BF, Braun M. Properties of minimum cross-entropy reconstruction of emission tomography with anatomically based prior. IEEE Trans Nucl Sci 1998;45(6):3014–3021.
106. Bailey DL, Jones T, Spinks TJ. A method for measuring the absolute sensitivity of positron emission tomographic scanners. Eur J Nucl Med 1991;18:374–379.

6 Tracer Kinetic Modeling in PET*

Richard E Carson

Introduction

The use of radiopharmaceuticals and the imaging of their biodistribution and kinetics with modern instrumentation are key components to successful developments in PET. Clever design and synthesis of sensitive and specific radiopharmaceuticals is the necessary first step. Each tracer must be targeted to measure a physiological parameter of interest such as blood flow, metabolism, receptor content, etc., in one or more organs or regions. State-of-the-art PET instrumentation produces high-quality 3-dimensional images after injection of tracer into a patient, normal volunteer, or research animal. With an appropriate reconstruction algorithm and with proper corrections for the physical effects such as attenuation and scatter, quantitatively accurate measurements of regional radioactivity concentration can be obtained. These images of tracer distribution can be usefully applied to answer clinical and scientific questions.

With the additional use of tracer kinetic modeling techniques, however, there is the potential for a substantial improvement in the kind and quality of information that can be extracted from these biological data. The purpose of a mathematical model is to define the relationship between the measurable data and the physiological parameters that affect the uptake and metabolism of the tracer.

In this chapter, the concepts of mathematical modeling as applied to PET are presented. Many of these concepts can be applied to radioactivity measurements from small animals made by tissue sampling or quantitative autoradiography. The primary focus in this chapter will be on methods applicable to data that can be acquired with PET imaging technology. The advantages and disadvantages of various modeling approaches are presented. Then, classes of models are introduced, followed by a detailed description of compartment modeling and of the process of model development and application. Finally, the factors to be considered in choosing and using various model-based methods are presented.

Overview of Modeling

PET imaging produces quantitative radioactivity measurements throughout a target structure or organ. A single static image may be collected at a single specific time post-injection or the full time-course of radioactivity can be measured. Data from multiple studies under different biological conditions may also be obtained. If the appropriate tracer is selected and suitable imaging conditions are used, the activity values measured in a region of interest (ROI) in the image should be most heavily influenced by the physiological characteristic of interest, be it blood flow, receptor concentration, etc. A model attempts to describe in an exact fashion this relationship between the measurements and the parameters of interest. In other words, an appropriate tracer kinetic model can account for all the biological factors that contribute to the tissue radioactivity signal.

The concentration of radioactivity in a given tissue region at a particular time post-injection primarily depends upon two factors. First, and of most interest, is the local tissue physiology, for example, the blood flow or metabolism in that region. Second is the input function, i.e., the time-course of tracer radioactivity concentration in the blood or plasma, which defines the

* Chapter reproduced from Valk PE, Bailey DL, Townsend DW, Maisey MN. Positron Emission Tomography: Basic Science and Clinical Practice. Springer-Verlag London Ltd 2003, 147–179.

availability of tracer to the target organ. A model is a mathematical description (i.e., one or more equations) of the relationship between tissue concentration and these controlling factors. A full model can predict the time-course of radioactivity concentration in a tissue region from knowledge of the local physiological variables and the input function. A simpler model might predict only certain aspects of the tissue concentration curve, such as the initial slope, the area under the curve, or the relative activity concentration between the target organ and a reference region.

The development of a model is not a simple task. The studies that are necessary to develop and validate a model can be quite complex. There are no absolute rules defining the essential components of a model. A successful model-based method must account for the limitations imposed by instrumentation, statistics, and patient logistics. To determine the ultimate form of a useful model, many factors must be considered and compromises must be made. The complexity of a "100%-accurate" model will usually make it impractical to use or may produce statistically unreliable results. A simpler, "less accurate" model tends to be more useful.

A model can predict the tissue radioactivity measurements given knowledge of the underlying physiology. At first, this does not appear to be useful, since it requires knowledge of exactly the information that we seek to determine. However, the model can be made useful by inverting its equations. In this way, measurements of tissue and blood concentration can be used to estimate regional physiological parameters on a regional or even pixel-by-pixel basis. There are many ways to invert the model equations and solve for these parameters. Such techniques are called model-based methods. They may be very complex, requiring multiple scans and blood samples and using iterative parameter-estimation techniques. Alternatively, a model-based method may be a simple clinically oriented procedure. With the knowledge of the behavior of the tracer provided by the model, straight-forward study conditions (tracer administration scheme, scanning and blood data collection, and data processing) can be defined to measure one or more physiological parameters.

This chapter provides an overview of the wide assortment of ways to develop a useful model and to use the models to obtain absolute or relative values of physiological parameters.

The Modeling Process

Once a radioactive tracer has been selected for evaluation, there are a number of steps involved in developing a useful model and a model-based method. Figure 6.1 gives an overview of this process. Based on prior information of the expected in vivo behavior of the tracer, a "complete" model can be specified. Such a model is usually overly complex and will have many more parameters than can be determined from PET data due to the presence of statistical noise. Based on initial modeling studies, a simpler model whose parameters can be determined (identified) can be developed. Then, validation studies can be performed to refine the model and verify that its assumptions are correct and that the estimates of physiological parameters are accurate. Finally, based on the understanding of the tracer provided by these modeling studies, a simpler protocol can be defined and applied for routine patient use. This method may involve limited or no blood measurements and simpler data analysis procedures. Under many conditions, such a protocol may produce physiological estimates of comparable precision and accuracy as those determined from the more complex modeling studies.

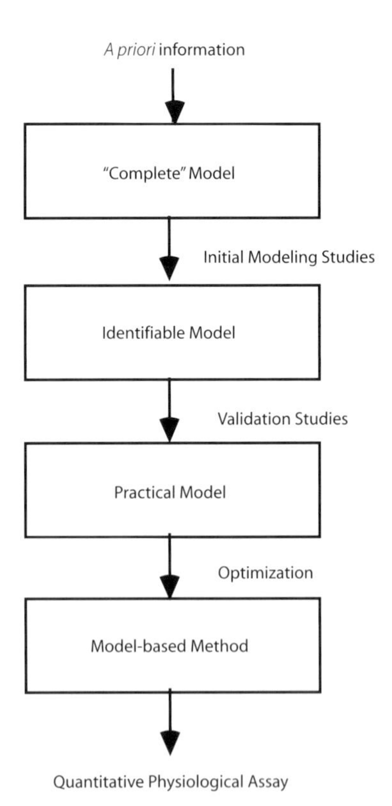

Figure 6.1. Steps in developing a model. A priori information concerning the expected biochemical behavior of the tracer is used to specify a complete model. Initial modeling studies will define an identifiable model, i.e., a model with parameters that can be determined from the measurable data. Validation studies are used to refine the model, verify its assumptions, and test the accuracy of its estimates. After optimization procedures and error analysis and accounting for patient logistical considerations, a model-based method can be developed that is both practical and produces reliable, accurate physiological measurements.

Many factors will affect the ultimate form of a useful model. In addition to the biological characteristics of the tracer, the characteristics of the instrumentation are important. It is essential to understand the accuracy of the reconstruction algorithm and its corrections, as well as the noise level in the measurements, which depends on the injected dose, camera sensitivity, reconstruction parameters, scan time, and ROI size. It may be of little use to develop a sophisticated model if there are significant inaccuracies in the radioactivity measurements due to improper corrections for attenuation or scatter. The noise level in the data also affects the number of parameters that may be estimated. It also is the primary determinant of the precision (variability) in the estimated parameters.

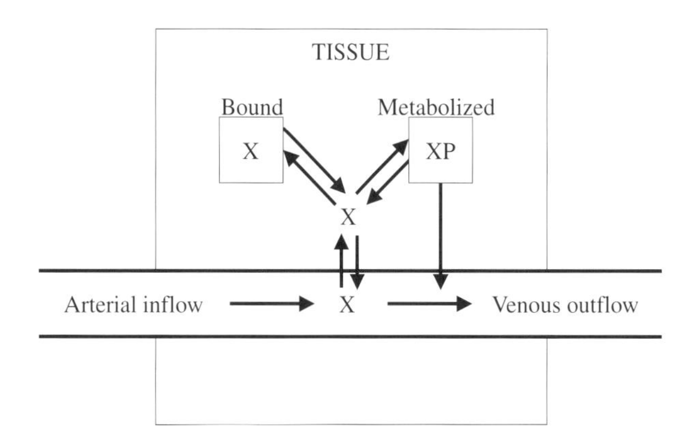

Figure 6.2. Overview of processes associated with delivery, uptake, binding, and clearance of a radioactive tracer X. Arterial inflow delivers X to the region of interest and venous outflow carries it away. The tracer may cross the capillary membrane and enter the tissue. From the tissue, it may be bound irreversibly or reversibly to intra- or extracellular sites, or may be metabolized (XP) into one or more chemical forms. The original labeled tracer or the metabolites may exit the tissue to the blood.

Tracers and Models

In this chapter, the labeled compounds will be referred to as tracer, radiotracer, or radiopharmaceutical. The term *tracer* implies that the injected compound, including both labeled and unlabelled molecules, is present in the tissue at negligible mass concentrations, so that little or no change in the saturation of relevant enzymes or receptors occurs. For this discussion, we assume that tracer levels are appropriate, except where explicitly noted.

Figure 6.2 provides an overview of the various paths that a tracer X may follow after delivery by intravenous injection. Arterial inflow delivers X to the region of interest and venous outflow carries it away. The tracer may cross the capillary membrane and enter the tissue. From the tissue, it may be bound irreversibly or reversibly to intra- or extracellular sites, or may be metabolized into one or more chemical forms. The original labeled tracer or the metabolites may exit the tissue to the blood.

Characteristics of Radiotracers

Before discussing models, it is important to consider the basic characteristics of radioactive tracers. A tracer is designed to provide information about a particular physiological function of interest, such as blood flow, blood volume, a metabolic process, a transport step, a binding process, etc. However, since any given tracer will likely have many biochemical fates following injection, great care and judgment are required to choose an appropriate compound. Ideally, the only factor controlling the uptake and distribution of the tracer will be the physiological process under study. Realistically, other factors will always affect a tracer's distribution and kinetics. For example, for a receptor-binding radiotracer, regional radioactivity concentration data are affected by regional blood flow, plasma protein binding, capillary permeability, nonspecific tissue binding, receptor association and dissociation rates, free receptor concentration, tracer clearance from blood (controlled by whole-body uptake), tracer metabolism (throughout the body), and regional uptake of any radioactive metabolites. For a well-designed tracer, the net effect of these extraneous factors is minor.

A tracer may either be a direct radiolabeled version of a naturally occurring compound, an analog of a natural compound, or a unique compound, perhaps a radiolabeled drug. An analog is a compound whose chemical properties are slightly different from the natural compound to which it is related. For example, [^{11}C]glucose is identical to glucose except for the replacement of a ^{12}C atom with ^{11}C. Analogs of glucose are deoxyglucose [1] and fluorodeoxyglucose (FDG) [2-4], which are chemically different from glucose. Often, because the naturally occurring compound has a very complex biochemical fate, a model describing the tissue radioactivity data of a directly labeled compound may need to be quite complex. A carefully designed analog can dramatically simplify the modeling and improve the sensitivity of the model to the parameter of interest. Deoxyglucose and FDG are good examples. Deoxyglucose and glucose enter cells by the same transport enzyme and are both phosphorylated by the enzyme hexokinase. However, deoxyglucose is not a substrate for the next enzyme in the glycolytic pathway, so deoxyglucose-6-phosphate accumulates in tissue. In this way, the tissue signal directly reflects the rate of metabolism, since there is little clearance of metabolized tracer. One important disadvantage of using an analog is that the measured kinetic parameters are those of the analog itself, not of the natural compound of interest. To correct for this, the relationship between the native compound and the radioactive analog must be determined. For deoxyglucose and FDG, this relationship is summarized by the lumped constant [1, 5]. To make the analog approach widely applicable, it is necessary to test if this constant changes over a wide range of pathological conditions [5-9].

Ideally, the parameter of interest is the primary determinant of the uptake and retention of a tracer, i.e., the tissue uptake after an appropriate period is directly (i.e., linearly) proportional to this parameter. This is the case for radioactive microspheres [10]. Many other compounds are substantially trapped in tissue shortly after uptake and are called chemical microspheres [11, 12]. For this class of compounds, a single scan at an appropriate time post-injection can give sufficient information about the parameter of interest. For other tracers, which both enter and exit tissue, scanning at multiple time points post-injection may be necessary to extract useful physiological information.

It is obvious that another important attribute of a tracer is that there be sufficient uptake in the organ of interest, i.e., the radioactivity concentration must provide sufficient counting statistics in a scan of reasonable length after injection of an allowable dose. Thus, the size of the structure of interest and the characteristics of the imaging equipment can also affect the choice of an appropriate tracer.

Types of Models

There are a wide variety of approaches to extract meaningful physiological data from PET tissue radioactivity measurements. All modeling approaches share some basic assumptions, in particular the principle of conservation of mass. A number of sources provide a comprehensive presentation of modeling alternatives [13–18]. Some approaches are termed stochastic or non-compartmental, and require minimal assumptions concerning the underlying physiology of the tracer's uptake and metabolism [19]. These methods permit the measurement of certain physiological parameters, such as mean transit time and volume of distribution, without an explicit description of all of the specific pools or compartments that a tracer molecule may enter.

Alternatively, there are distributed models that try to achieve a precise description of the fate of the radiotracer. These models not only specify the possible physical locations and biochemical forms of the tracer, but also include the concentration gradients that exist within different physiological domains. In particular, distributed models for capillary–tissue exchange of tracer have been extensively developed [20–26]. Since this is the first step in the uptake of any tracer into tissue, a precise model for delivery of tracer at the capillary is important. Distributed models are also used to account for processes, such as diffusion, where concentration gradients are present [27].

A class of models whose complexity lies between stochastic and distributed is the compartmental models. These models define some of the details of the underlying physiology, but do not include concentration gradients present in distributed models. The development and application of these models is the principal focus of this chapter. The most common application of compartmental modeling is the mathematical description of the distribution of a tracer throughout the body [28, 29]. Here, different body organs or groups of organs are assigned to individual compartments, and the model defines the kinetics into and out of each compartment. This type of model is useful when the primary measurable data is the time-concentration curve of the tracer in blood and urine. If there are many measurements with good accuracy, fairly complex models with many compartments and parameters can be used.

In PET, compartmental modeling is applied in a different manner. Here, scanners provide one or more measurements of radioactivity levels in a specific organ, region, or even pixel. If the tracer enters and leaves the organ via the blood, the tracer kinetics in other body regions need not be considered to evaluate the physiological traits of the organ of interest. In this way, each region or pixel can be analyzed independently. Generally, there must be some knowledge of the time-course of blood radio activity. Since each region can be evaluated separately, the models can be relatively simple, and can therefore be usefully applied to determine regional physiological parameters from PET data.

Compartmental Modeling

Compartmental modeling is the most commonly used method for describing the uptake and clearance of radioactive tracers in tissue [28, 30, 31]. These models specify that all molecules of tracer delivered to the system (i.e., injected) will at any given time exist in one of many compartments. Each compartment defines one possible state of the tracer, specifically its physical location (for example, intravascular space, extracellular space, intracellular space, synapse) and its chemical state (i.e., its current metabolic form or its binding state to different tissue elements, such as plasma proteins, receptors, etc.). Often, a single compartment represents a number of these states lumped together. Compartments are typically numbered for mathematical notation.

The compartmental model also describes the possible transformations that can occur to the tracer, allowing it to "move" between compartments. For example, a molecule of tracer in the vascular space may enter the extracellular space, or a molecule of receptor-binding tracer that is free in the synapse may become bound to its receptor. The model defines the fraction or proportion of tracer molecules that will "move" to a different compartment within a specified time. This fractional rate of change of the tracer concentration in one compartment is called a rate constant, usually expressed as "k", and has units of inverse time, e.g., min^{-1}. The inverse minute unit reflects the fraction per minute, i.e., the proportion of tracer molecules in a given compartment that will "move" to another compartment in one minute. To distinguish the various rate constants in a given model, subscripts are used to define the source and destination compartment numbers. In much of the compartmental modeling literature, k_{12}, for example, reflects the rate of tracer movement to compartment 1 from 2. This nomenclature is especially convenient for large models and is motivated by the nature of matrix algebra notation. In PET applications, the number of compartments is small (1–3), as is the number of rate constants (1–6), so it is typical to use a notation with one subscript (e.g., k_3) where the source and destination compartments associated with each constant are explicitly defined.

The physiological interpretation of the source and destination compartments defines the meaning of the rate constants for movement of tracer between them. For example, the rate constant describing tracer movement from a receptor-bound compartment to the unbound compartment will reflect the receptor dissociation rate. For a freely diffusible inert tracer, the rate constant of transfer from arterial blood to the tissue compartment will define local blood flow. By determining these rate constants (or some algebraic combination of them), quantitative estimates or indices of local physiological parameters can be obtained. The underlying goal of all modeling methods is the estimation of one or more of these rate constants from tissue radioactivity measurements.

Examples of Compartmental Models

Figure 6.3 shows examples of compartmental model configurations. In many depictions of models, a rectangular box is drawn for each compartment, with arrows labeled with the rate constants placed between the boxes. In most whole-body compartmental models, the blood is usually counted as a compartment. Measurements from blood are often the primary set of data used to estimate the model rate constants. In the PET applications described here, we are most interested in the model constants associated with the tissue regions that are being imaged. Typically, measurements will be made from the blood to define the "input function" to the first tissue compartment (see Input Functions and Convolution, below). In this presentation, we will treat these blood input measurements as known values, not as concentration values to be pre-

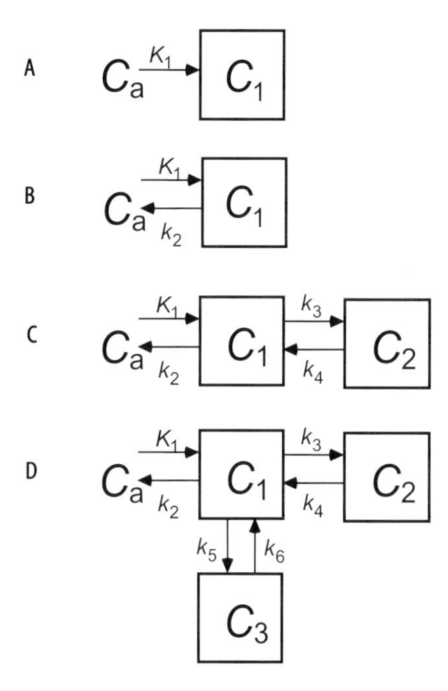

Figure 6.3. Examples of compartmental models. C_a is the concentration of tracer in arterial blood, C_1, C_2, and C_3 are the tracer concentrations in compartments 1–3, and K_1, k_2, etc., are the rate constants that define the rate of tracer movement between compartments. **A** the simplest compartmental model having one tissue compartment with irreversible uptake of tracer, e.g., microspheres. **B** a model with one tissue compartment appropriate for a tracer that exhibits reversible tissue uptake, e.g., a diffusible blood flow tracer. **C** a model with two tissue compartments, e.g., FDG. **D** a three tissue-compartment model for a receptor-binding ligand where the three compartments represent 1) free tracer, 2) tracer specifically bound to receptor, and 3) tracer nonspecifically bound to other tissue elements

dicted by the model. Thus, blood will *not* be counted as a compartment.

Figure 6.3A shows the simplest model having one tissue compartment with irreversible uptake of tracer. This irreversible uptake is shown by the presence of a rate constant K_1 for tracer moving from the blood to compartment 1, but with no rate constant for exit of tracer back to blood. Such a model is appropriate for radioactive microspheres [10] or for a tracer that is irreversibly trapped in tissue. This model is often used as an approximation when tissue trapping is nearly irreversible [11]. Figure 6.3B shows a one-tissue-compartment model, appropriate for a tracer that exhibits reversible tissue uptake. This is a common model for inert tracers used to measure local blood flow [13]. Here, the rate at which the tracer exits the tissue compartment and returns to the blood is denoted k_2. Figure 6.3C shows a model with two tissue compartments. This model may be appropriate for a tracer that enters tissue from blood, and then is either metabolized to a form that is trapped in the tissue (at a rate defined by k_3) or returns to blood (at a rate defined by k_2), such as deoxyglucose [1]. Compartment 1 represents the unmetabolized tracer and compartment 2 the metabolized tracer. Figure 6.3D shows a three-tissue-compartment model for a receptor-binding ligand where the three compartments represent free tracer, tracer specifically bound to receptor, and tracer nonspecifically bound to other tissue elements [32].

Compartmental Modeling Assumptions

The successful application of simple compartmental models to a complex biological system requires that many assumptions be true. These assumptions are typically not completely valid, so that successful use of these models depends upon whether errors in these assumptions produce acceptable errors in model measurements (see Error Analysis, below). Compartmental models, by their nature, assume that each compartment is well mixed, i.e., there are no concentration gradients within a single compartment. Therefore, all tracer molecules in a given compartment have equal probability of exchange into other compartments. This well-mixed assumption has the great advantage of producing relatively simple mathematical relationships. However, it limits the ability of compartmental models to provide an accurate description of some biological structures. For example, a compartmental model cannot include the change of activity concentration in a capillary from arterial to venous ends, or the heterogeneous distribution of receptors in a patch of tissue. Often, in PET applications, the "well-mixed" assumption is also violated by the nature of the imaging process. Due to low resolution, even single-pixel data from reconstructed images represent a mixture of underlying tissues. When larger ROIs are used to improve the statistical precision of the measurements, heterogeneity in the measurements increases.

A primary assumption of most compartmental models is that the underlying physiological processes are in steady state. Mathematically, this means that the rate constants of the system do not change with time during a study, and causes the mathematics of the model to be linear differential equations (see Model Implementation). If these rate constants reflect local blood flow or the rate of a metabolic or binding process, then the rate at which these processes occur should remain constant during a study. Since the rates of many biological processes are regulated by substrate and product concentrations, maintaining processes in steady state usually requires constant concentrations of these regulating molecules. In practice, this requirement is never precisely met. However, these assumptions are adequately met so long as any changes in the underlying rates of flow, metabolism, receptor binding, etc., are slow with respect to the time scale of the data being analyzed. Note that the concentrations of the injected radiopharmaceuticals may change dramatically during a study; however, this does not violate the steady-state assumption so long as the radioactive species exists at a negligible (tracer) concentration with respect to the non-radioactive natural biological substrates (see Biochemical Reactions and Receptor–Ligand Binding). For studies using injections of radiopharmaceuticals with low specific activity, saturation of receptors or enzymes can be significant, and nonlinear modeling techniques are required.

To generate the equations of a model, the magnitude of tracer movement from compartment A to compartment B per unit of time must be defined. This is called the flux (J_{AB}). If tracer concentration is expressed in units of kBq per mL, then flux has units of kBq per mL per min (or another appropriate time unit). The assumptions of well-mixed compartments and physiological processes in steady state lead to the mathematical relationship that the flux J_{AB} is a linear multiple of the amount, or concentration, of tracer in the source compartment A (C_A), i.e.,

$$J_{AB} = k\,C_A \quad (1)$$

where k is a rate constant with units of inverse minute and which is independent of the concentration in any compartment. This simple equation is the basis of the differential equations that describe compartmental models (see Model Implementation).

Interpretation of Model Rate Constants

The physiological interpretations of the rate constants (such as k in Eq. 1) depend upon the definition of the source and destination compartments. A single compartment of a model may often lump a number of physiological entities together, for example, tracer in extracellular and intracellular spaces or tracer that is free in tissue and nonspecifically bound. This section discusses the physiological meaning of model rate constants.

Blood Flow and Extraction

The first step in most in vivo models is the delivery of tracer to the target region from the blood. The flux of tracer into the first tissue compartment from the blood is governed by the local blood flow and the rate of extraction of the tracer from the capillary into the tissue. Conventional fluid flow describes the volume of liquid passing a given point per unit of time and has units of mL per min. A more useful physiological measure is perfusion flow, the volume of blood passing in and out of a given volume (or weight) of tissue per unit of time, which has units of mL per min per mL of tissue or mL per min per gram of tissue. In the physiological literature, the term blood flow usually means perfusion flow.

Determining blood flow and extraction information from model parameters begins with the Fick Principle (see, for example, Lassen and Perl [15]). The net flux (J) of tracer into or out of a tissue element equals the difference between the influx (J_{in}) and outflux (J_{out}), i.e.,

$$J = J_{in} - J_{out} = F\,C_a - F\,C_v \tag{2}$$

where the influx is the product of the blood flow (F) and the arterial concentration (C_a), and the outflux is the product of the blood flow and the venous concentration (C_v). The unidirectional (or first-pass) extraction fraction E is the fraction of tracer that exits the blood and enters the tissue on *one* capillary pass, or

$$E = \frac{C_a - C_v}{C_a} \tag{3}$$

A tracer with low extraction has a small arterial–venous difference on first pass. Equation 2 can then be rewritten as

$$J = \left(F \cdot E\right) C_a = k\,C_a \tag{4}$$

Equation 4 describes the unidirectional delivery of tracer from blood to tissue. The rate constant k defining this uptake process is the product of blood flow and unidirectional extraction fraction. The interpretation of the extraction fraction was further developed by Kety [13], Renkin [33], and Crone [34] by considering the capillary as a cylinder to produce the following relationship:

$$E = 1 - e^{-\frac{PS}{F}} \tag{5}$$

where P is the permeability of the tracer across the capillary surface (cm per min), S is the capillary surface area per gram of tissue (cm^2 per gram), and F is the blood flow (mL per min per gram). For highly permeable tracers, the product PS is much greater than the flow F, so the exponential term in Eq. 5 is small, and the extraction fraction is nearly 1.0. In this case, the rate constant for delivery is approximately equal to flow. Such tracers are therefore useful to measure regional blood flow and not useful to measure permeability, i.e., they are flow-limited. For tracers with permeability much lower than flow, the relationship in Eq. 5 can be approximated as

$$E \simeq \frac{PS}{F} \tag{6}$$

and the rate constant k ($F{\cdot}E$) becomes PS. Such tracers are useful to measure permeability and not useful to measure flow. Most tracers lie between these two extremes, so that the rate constant for delivery from arterial blood to tissue is affected by both blood flow and permeability. These relationships are directly applicable to tracers that enter and leave tissue by passive diffusion. For tracers transported into and out of tissue by facilitated or active transport, the PS product is mathematically equivalent to the transport rate, which depends upon the concentration and reaction rate of the transport enzymes (See Biochemical Reactions).

The interpretation of a delivery rate constant k as the product of flow and extraction fraction may depend upon whether the blood activity concentration C_a is measured in whole blood or in plasma. If there is very rapid equilibration between plasma and red blood cells, then the whole blood and plasma concentrations will be identical. However, if equilibrium is slow with respect to tracer uptake rates into tissue, or if there is trapping or metabolism of the tracer in red blood cells, than the plasma concentration should be used. In the extreme of no uptake of tracer into red cells, then the delivery rate constant k is the product of extraction fraction and plasma flow, where plasma flow is related to whole blood flow based on the hematocrit. If binding of tracer to plasma proteins is significant, similar changes in interpretation of the rate constants may also be required.

Diffusible Tracers and Volume of Distribution

One of the simplest classes of tracers is those that enter tissue from blood and then later return to blood. The net flux of tracer into a tissue compartment can be expressed as follows:

$$J = K_1 C_a - k_2 C \tag{7}$$

K_1 is the rate of entry of tracer from blood to tissue and is equal to the product of extraction fraction and blood flow, and C_a is the concentration of tracer in arterial blood. The rate constant k_2 describes the rate of return of tracer from tissue to blood, where C is the concentration of tracer in tissue. The physiological interpretation of k_2 can best be defined by introducing the concept of the *volume of distribution*. Suppose the concentration of tracer in the blood remained constant. Ultimately, the concentration of the diffusible tracer in the tissue compartment would also become constant and equilibrium would be achieved. The ratio of the tissue concentration to the blood concentration at equilibrium is called the volume of distribution (or alternatively the partition coefficient). It is termed a volume because it can be thought of as the volume of blood that contains the same quantity of radioactivity as 1 mL (or 1 gram) of tissue. Once the blood and tissue tracer concentrations have reached constant levels, i.e., equilibrium, the net flux J into the tissue compartment is 0, so the volume of distribution V_D can be expressed as

$$V_D = \frac{C}{C_a} = \frac{K_1}{k_2} \tag{8}$$

where the last equality is derived by setting the flux J in Eq. 7 to 0. Therefore, the physiological definition for the rate constant k_2 is the ratio of K_1 to V_D. Thus, k_2 has information concerning flow, tracer extraction, and partition coefficient.

Biochemical Reactions

Often, two compartments of a model represent the substrate and product of a chemical reaction. In that case, the rate constant describing the "exchange" between these compartments is indicative of the reaction rate. For enzyme-catalyzed reactions [35], the flux from substrate to product compartments is the reaction velocity v:

$$v = \frac{V_m C}{K_m + C} \tag{9}$$

V_m is the maximal rate of the reaction, C is the concentration of substrate, and K_m is the concentration of substrate that produces half-maximum velocity. This is the classic Michaelis–Menten relationship. It shows that the velocity is not a linear function of the substrate concentration, as in Eq. 1. However, when using tracer concentrations of a radioactive species and if the concentrations of the native substrates are in steady state (see Compartmental Modeling Assumptions), the linear form of Eq. 1 still holds. In the presence of a native substrate with concentration C, and the radioactive analog with concentration C^*, the reaction rate for the generation of radioactive product v^* is as follows:

$$v^* = \frac{V_m^* C^*}{K_m^* \left(1 + \frac{C}{K_m} + \frac{C^*}{K_m^*}\right)} \tag{10}$$

V_m^* and K_m^* are the maximal velocity and half-maximal substrate concentration for the radioactive analog. If the radioactive species has high specific activity (the concentration ratio of labeled to unlabelled compound in the injectate) so that its total concentration (labeled and unlabelled) is small compared to the native substrate, i.e., $C^*/K_m^* \ll C/K_m$, then Eq. (6.10) reduces to

$$v^* = \left(\frac{V_m^*}{K_m^* \left(1 + \frac{C}{K_m}\right)}\right) C^* = kC^* \tag{11}$$

The term in large brackets in Eq. 11 is composed of terms that are assumed to be constant throughout a tracer experiment. Therefore, when using radiopharmaceuticals at tracer concentrations, enzyme-catalyzed reactions can be described with a linear relationship as the product of a rate constant k and the radioactive substrate concentration C^*. The rate constant k includes information about the transport enzyme and the concentration of unlabelled substrate.

Receptor–ligand Binding

For radiotracers that bind to receptors in the tissue (see, for example, Eckelman [36]), the rate of binding, i.e., the rate of passage of tracer from the free compartment to the bound compartment, can also be described by the linear form of Eq. 1 under tracer concentration assumptions. For many receptor systems, the binding rate is proportional to the product of the concentrations of free ligand and free receptor. This classical bi-molecular association can be described mathematically as

$$v = k_{on}\left(B_{\max} - B\right) F \tag{12}$$

where k_{on} is the bi-molecular association rate ($nM^{-1}min^{-1}$), B_{max} is the total concentration of receptors (nM), B is the concentration of receptors currently bound (either by the injected ligand or by endogenous molecules), and F is the concentration of free ligand. When a radioactive species is added and competes with endogenous compound for receptor binding, the radiopharmaceutical binding velocity is

$$v^* = k_{on}^*\left(B_{max} - B - B^*\right)F^* \tag{13}$$

where k_{on}^* is the association rate of the radiopharmaceutical and B^* is the mass concentration of the bound radiopharmaceutical. If the radioactive compound has high specific activity, then $B^* \ll B$, and Eq. 13 becomes

$$v^* = k_{on}^* B'_{max} F^* = kF^* \tag{14}$$

where B'_{max} is the free receptor concentration ($B_{max} - B$). Thus, using a high specific activity receptor-binding ligand, measurement of the reaction rate constant k provides information about the product of k_{on}^* and B'_{max}, but cannot separate these parameters. Since B'_{max} is sensitive to change in total receptor and occupancy by endogenous or exogenous drugs, receptor-binding ligands can be extremely useful to measure receptor occupancy or dynamic changes in neurotransmitter levels [37]. Note that the description of receptor-binding radioligands is mathematically identical to that for enzyme-catalyzed reactions, although the conventional nomenclature is different.

Model Implementation

This section presents an overview of the mathematics associated with compartmental modeling. This includes the mathematical formulation of these models into differential equations, the solution equations to a few simple models, and a summary of parameter estimation techniques used to determine model rate constants from measured data. Here, we concentrate on applications where we have made measurements in an organ or region of interest which we wish to use to ascertain estimates of the underlying physiological rates of this region.

Mathematics of Compartmental Models

This section describes the process of converting a compartmental model into its mathematical form and determining its solution. For a more complete discussion of these topics, consult basic texts on differential equations [38] as well as a number of specialized texts on mathematical modeling of biological systems [16, 28].

First, we start with a particular model configuration like those in Fig. 6.3. The compartments are numbered 1, 2,…, and the radioactivity concentration in each compartment is designated C_1, C_2, …. Radioactivity measurements in tissue are typically of a form such as counts per mL or kBq per gram. The volume or weight unit in the denominator reflects the full tissue volume. However, the tracer may exist only in portions of the tissue; for example, just the extracellular space. In this case, the concentration of the tracer within its distribution space will be higher than its apparent concentration per gram of tissue. When these concentration values are used to define reaction rates, instead of the true local concentration, the interpretation of the relevant rate constant should include a correction for the fraction of total tissue volume in which the tracer distributes.

Differential Equations

The net flux into each compartment can be defined as the sum of all the inflows minus the sum of all the outflows. Each of these components is symbolized by an arrow into or out of the compartment, and the magnitude of each flux is the product of the rate constant and the concentration in the source compartment. The net flux into a compartment has units of concentration (C) per unit time and is equal to the rate of change (d/dt) of the compartment concentration, or dC/dt. Consider the simple one-tissue-compartment model in Fig. 6.3B. The differential equation describing the rate of change of the tissue concentration C_1 is

$$\frac{dC_1}{dt} = K_1 C_a(t) - k_2 C_1(t) \tag{15}$$

Here, $C_a(t)$ is the time course of tracer in the arterial blood, also called the input function. K_1 is the rate constant for entry of tracer from blood to tissue, and k_2 is the rate constant for return of tracer to blood. The capitalization of the rate constant K_1 is not a typographical error. K_1 is capitalized because it has different units than other rate constants. The blood radiotracer measurements are typically made per mL of blood or plasma. In non-imaging studies in animals, tissue concentration measurements are made per gram of tissue. Thus, C_1 had units of kBq per gram, and C_a had units of kBq per mL. Therefore, K_1 must have units of mL blood per min per gram tissue (usually written as mL/min/g).

The other rate constants have units of inverse minute. PET scanners actually acquire tissue radioactivity measurements per mL of tissue. Thus, to present results in comparable units to earlier work, corrections for the density of tissue must be applied to convert kBq per mL tissue to kBq per gram tissue.

Before solving Eq. 15 for a general input function $C_a(t)$, first consider the case of an ideal bolus input, i.e., the tracer passes through the tissue capillaries in one brief instant at time $t = 0$, and there is no recirculation. If C_a is the magnitude of this bolus, the model solution for the time-concentration curve for compartment 1 is as follows:

$$C_1(t) = C_a K_1 \exp(-k_2 t) \tag{16}$$

Thus, at time zero, the tissue activity jumps from 0 to a level $K_1 C_a$ and then drops towards zero exponentially with a rate k_2 per min or a half-life of $0.693/k_2$ min.

Now consider the two tissue-compartment model in Fig. 6.3C. For this model there will be two differential equations, one per compartment:

$$\frac{dC_1}{dt} = K_1 C_a(t) - k_2 C_1(t) - k_3 C_1(t) + k_4 C_2(t) \tag{17}$$

$$\frac{dC_2}{dt} = k_3 C_1(t) - k_4 C_2(t) \tag{18}$$

Note that there is a term on the right side of Eqs. 17, and 18 for each of the connections between compartments in Fig. 6.3C. An outflux term in Eq. 17 [e.g., $-k_3C_1(t)$] has a corresponding influx term in Eq. 18 [$+k_3C_1(t)$]. The solution to these coupled differential equations, again for the case of an ideal bolus input, is as follows:

$$C_1(t) = C_a\left[A_{11}\exp(-\alpha_1 t) + A_{12}\exp(-\alpha_2 t)\right] \tag{19}$$

$$C_2(t) = C_a A_{22}\left[\exp(-\alpha_1 t) - \exp(-\alpha_2 t)\right] \tag{20}$$

A_{11}, A_{12}, A_{22}, α_1, and α_2 are algebraic functions of the model rate constants K_1, k_2, k_3 and k_4 [4]. Here, the time course of each compartment is the sum of two exponentials. One special case of interest is when the tracer is irreversibly bound in tissue so that the rate of return of tracer from compartment 2 to compartment 1, k_4, is zero. In this case, the solution becomes

$$C_1(t) = C_a K_1 \exp\left[-(k_2 + k_3)t\right] \tag{21}$$

$$C_2(t) = C_a \frac{K_1 k_3}{k_2 + k_3}\left(1 - \exp\left[-(k_2 + k_3)t\right]\right) \tag{22}$$

Note that in most cases, the measured tissue activity will be the total in both compartments, so that the model prediction will be the sum $C_1(t) + C_2(t)$.

These solutions for tissue concentration are linearly proportional to the magnitude of the input, C_a. Doubling the magnitude of the input (injecting more) will double the resultant tissue concentration. The equations are non-linear with respect to many of the model rate constants (those that appear in the exponents) but is linear in K_1.

Input Functions and Convolution

In the previous section, mathematical solutions were presented for simple models under the condition of an ideal bolus, i.e., the tracer appears for one capillary transit with no recirculation. In reality, the input to the tissue is the continuous blood time–activity curve. The equations above are linear with respect to the input function C_a. This permits a direct extension of these bolus equations to be applied to solve the case of a continuous input function. Fig. 6.4 illustrates this concept. Figure 6.4a and 6.4c show ideal bolus input functions of different magnitudes at different times. Figure 6.4b and 6.4d show the corresponding tissue responses for the model with one tissue compartment (Fig. 6.3b). Suppose, as in Fig. 6.4e, the combination of the two inputs is given, i.e., there is a bolus input of magnitude A at time $t = T_1$, and a second bolus of magnitude B at $t = T_2$. The resulting tissue activity curve is:

$$C_1(t) = K_1 A \exp\left[-k_2(t - T_1)\right] \quad \text{for } T_1 \le t < T_2 \tag{23}$$

$$C_1(t) = K_1 A \exp\left[-k_2(t - T_1)\right] + K_1 B \exp\left[-k_2(t - T_2)\right] \quad \text{for } t \ge T_2 \tag{24}$$

In other words, the tissue response is a sum of the individual responses to each bolus input. The responses are scaled in magnitude and shifted in time to match each bolus input.

Suppose now there is a series of bolus administrations at times T_i, $i = 1,\ldots$, each of magnitude $C_a(T_i)$ as depicted by the square waves in Fig. 6.4g. The total tissue response (Fig. 6.4h) can be written as the summation:

$$C_1(t) = \sum_i C_a(T_i) K_1 \exp\left[-k_2(t - T_i)\right] \tag{25}$$

where the exponentials are defined to have zero value for negative arguments (i.e., $t<T_i$). If we now consider the continuous input function $C_a(t)$ (bold line in Fig. 6.4g) as an infinite summation of individual boluses,

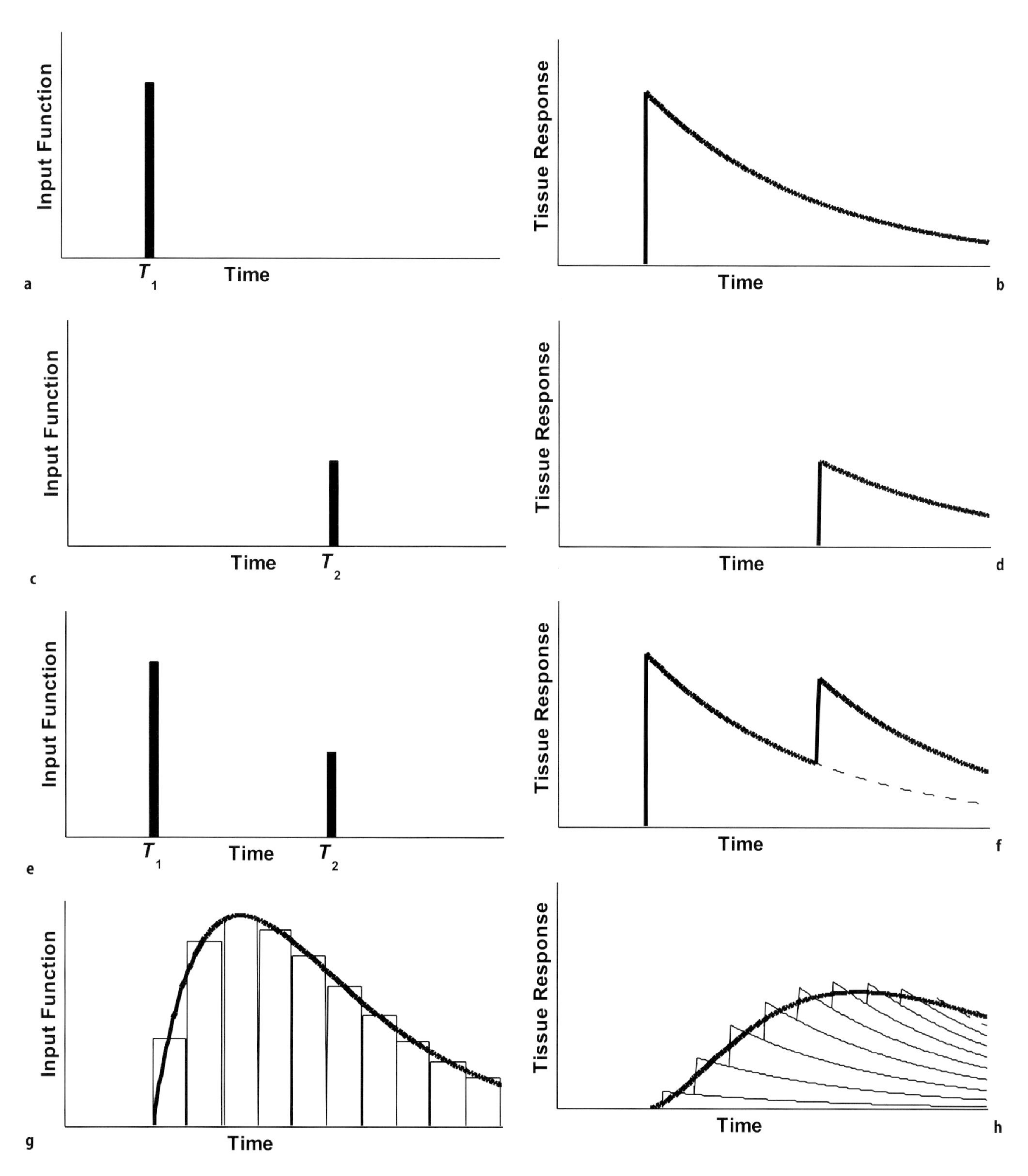

Figure 6.4. Convolution. a, c, e, and g show the input functions that produce the tissue responses in b, d, f, and h, respectively. **a** an ideal bolus input at time T_1. **b** the corresponding tissue response to the one-compartment model of Fig. 6.3b, i.e., exponential clearance following Eq. (16). **c** a single bolus of half the magnitude in a at time T_2. **d** the tissue response to c, which is altered in time and magnitude in a corresponding manner from that in b. **e** combination of inputs in a and c. **f** the tissue response to e, which is the sum of the tissue responses from each bolus administered separately (b+d). This demonstrates the linearity of the model equations. **g** a continuous input function (bold line) which can be interpreted as a series of bolus injections of varying magnitudes (fine lines). **h** the tissue activity response to g (bold line) which can be interpreted as the sum of the responses to each bolus (fine lines).

the summation in Eq. 25 becomes an integral, and the tissue response (bold line in Fig. 6.4h) is:

$$C_1(t) = \int_0^t C_a(s) K_1 \exp[-k_2(t-s)] ds \qquad (26)$$

Here, s is the integration variable. This is called a *convolution* integral, and is often written as

$$C_1(t) = C_a(t) \otimes K_1 \exp(-k_2 t) \qquad (27)$$

with the symbol ⊗ denoting convolution. This presentation corresponds to the one-compartment model of Fig. 6.3B and extends the bolus solution (Eq. 16) to the case of a general input function (Eq. 27). However, convolution applies to any compartmental model whose solution has a linear relationship to its input function. Let $h_i(t)$ be the *impulse response function* for compartment i, i.e., the time course of tissue response from a bolus input of magnitude 1 [$K_1\exp(-k_2t)$ for the one-compartment model]. Then the tissue activity resulting from the general input function $C_a(t)$ is written as

$$C_i(t) = C_a(t) \otimes h_i(t) \qquad (28)$$

Thus, for linear compartmental models, the tissue time–activity curve is the convolution of the input function with the impulse response function. For compartmental models, the latter is a sum of exponentials, typically one exponential per compartment. A number of approaches have been used to implement and solve Eq. 28 on a computer if the arterial input function is determined from serial samples. One approach is to fit the measured input function data to a suitable model [39] and then solve the convolution integral by standard mathematical methods. Alternatively, a continuous input function can be approximated by linear interpolation between the sample data values, and then Eq. 28 can be solved by analytical integration over each time period between blood samples.

Figure 6.5 shows the effects that variations in the input function can produce on the resulting tissue response. Figure 6.5a shows three input functions. The solid line is a measured arterial input function. The other two input curves were calculated based on the measured data so that the area under all curves is a constant. The tissue concentration curves produced in response to each input function are shown as the corresponding curves in Fig. 6.5b. In all cases, the tissue response is calculated from the one-compartment model, Eq. 27, using the same parameters (K_1 = 0.1 mL/min/mL and k_2 = 0.1 min^{-1}). The difference in shape between the input functions produces comparable differences in the tissue concentration curves. These differences in shape do *not* reflect differences in the local physiological parameters of the tissue, since the rate constants were the same in all cases. Thus, the main point of Fig. 6.5 is that a time–activity curve in a tissue region cannot be interpreted without knowledge of the input function.

The linear compartmental models discussed to this point have the tremendous advantage of providing exact mathematical solutions, predicting the tissue response in the form of Eq. 28. In some cases, the flux between compartments cannot be described mathematically as the product of a rate constant times the concentration of tracer in the source compartment ($J = kC$). For example, in modeling receptor-binding ligands, the linear flux assumption holds if the radiopharmaceutical is administered at tracer levels and does not produce detectable saturation of the receptor sites (Eq. 14). If such a ligand is administered in low specific activity so that it produces a change in receptor occupancy during the data collection period, the

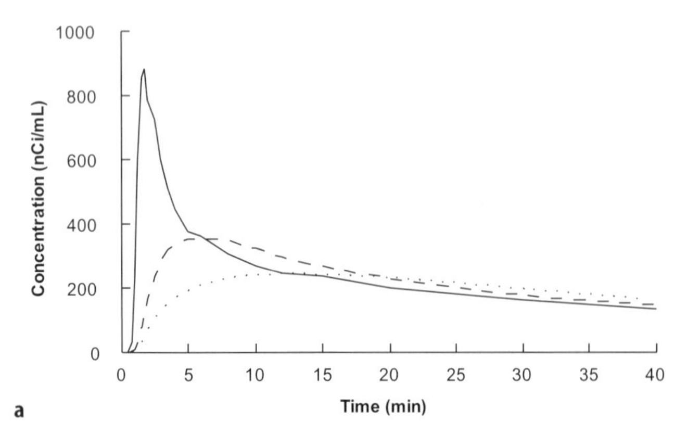

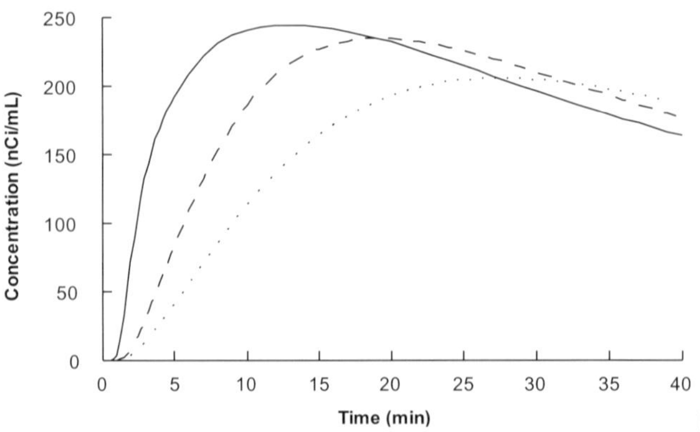

Figure 6.5. Effect of the input functions on tissue time–activity curves. **a** three input functions. The solid line is a measured arterial input function. The dashed and dotted lines represent other input functions derived from the first curve, so that the area under these curves is the same. **b** the tissue response curves from the three input functions in a calculated with a model with one tissue compartment and the fixed rate constants (see text for details).

differential equations governing the model are no longer linear. This can be seen in Eq. 13 where the flux from the free to the bound compartments v^* cannot be described as a constant multiplied by F^*, since the bound concentration B^* also appears in the relationship. To solve the model in this and most other nonlinear cases requires techniques of numerical integration of differential equations [40, 41]. The basic idea to numerically estimate the activity in each compartment is to take small steps in time and use the differential equations to determine how much each compartment's concentration should change over each time step. The most commonly used method for numerical integration is called Runge–Kutta, which provides increased accuracy with longer time steps by averaging multiple estimates of the derivative dC/dt.

In implementing models, it is important that the model formulation matches the nature of PET scan data. The models presented above predict the tissue concentration at an instant in time. Image values represent the average tissue activity collected over each scan interval. The instantaneous model value can be used to determine the integrated scan value. For example, for the one-compartment model (Eq. 15 and Eq. 27), the integrated scan value from time T_1 to T_2 is as follows:

$$\int_{T_1}^{T_2} C_1(t)dt = \frac{K_1 \int_{T_1}^{T_2} C_a(t)dt - \left(C_1(T_2) - C_1(T_1)\right)}{k_2} \tag{29}$$

Another practical issue is radioactive decay. This can be handled either by explicit decay correction of both the tissue and blood data or by incorporating decay into the model formulation. The latter approach can be accomplished by adding an additional rate constant corresponding to the decay rate (0.693/half life) to each compartment. This method is slightly more accurate than explicit decay correction for short-lived tracers, since decay correction does not account for biological change in tracer concentration within one scan interval.

Parameter Estimation

The previous sections presented the mathematical techniques necessary to solve the model equations. Thus, with knowledge of the input function $C_a(t)$, the model configuration, and its rate constants, the tissue concentration curve can be predicted mathematically. This section provides an overview of the inverse problem, i.e., given measurements of the tissue activity and the input function and a proposed model configuration, one can produce estimates of the underlying rate constants. Many references are available on the topic of parameter estimation [42–44].

There are many ways to accomplish the estimation of model parameters. The choices available and the success of any given method depend upon the form of the model and the sampling and statistical quality of the measured data. If only a single measurement of tissue radioactivity is made, obviously only a single parameter can be determined. Collection of multiple time points permits the estimation of some or all of the parameters of a model. Since measured data always have some associated noise, the estimates of model parameters from such data will also be noisy. It is often the goal of a statistical estimation method to minimize the variability of the resulting parameter estimates. Note also that the values of the parameters will affect the statistical quality of the results. For example, blood flow estimates produced by a particular method may be reliable for high-flow regions but unreliable for low-flow regions.

When many tissue measurements are collected after radionuclide administration, the most commonly used method of parameter estimation is called *least-squares estimation*. Qualitatively, the goal of this technique is to find values for the model rate constants that, when inserted into the model equations, produce the "best" fit to the tissue measurements. Quantitatively, the goal is to minimize an optimization function, specifically the sum of the squared differences between the measured tissue concentration data and the model prediction, i.e.,

$$\sum_{i=1}^{N} \left(C_i - C(T_i)\right)^2 \tag{30}$$

where there are N tissue measurements, C_i, i=1,…,N, at times T_i, and $C(T_i)$ is the model prediction of tissue activity at each of these times. This particular form is used because of the nature of the noise in the measured data. Parameter estimates produced by minimizing the sum of squared differences have minimum variability if the noise in each scan measurement is statistically independent, additive, Gaussian, and of equal magnitude. Additive and independent statistical noise is usually a good assumption for PET image data, however, often the variance of the measurements will not be constant across different scans in one multiple-scan acquisition, particularly for short-lived isotopes such as ^{15}O or ^{11}C. In this case, the least-squares function can be modified to accommodate variable noise levels as follows:

$$\sum_{i=1}^{N} w_i \left(C_i - C(T_i)\right)^2 \tag{31}$$

where w_i is a weight assigned to data point i. This method is called *weighted least-squares estimation*, and the optimal weight for each sample is the inverse of the variance of the data [43]. For simple count data, the variance of the data can be estimated from the count data itself based on its Poisson distribution [45]. For reconstructed data, many algorithms have been proposed to calculate or approximate the noise in pixel or region-of-interest data [46–52].

It is important to recognize that there are many non-random or deterministic error sources in the modeling process that cause inconsistencies between the model and the measured data (see section on random and deterministic errors). When fitting data to a model, the parameter estimation procedure is naïve in that it believes that the specified model is absolutely correct. The algorithm will do its best to minimize the optimization function. Therefore, if there are deterministic errors in the model or the input function, the estimation algorithm can produce unsuitable results. It may be appropriate in some situations to adjust the weights of some data points (e.g., early time points where errors in the model due to intravascular activity are most significant) to reduce the sensitivity of the model to the presence of deterministic errors.

Once an optimization function (Eq. 30 and Eq. 31) has been defined, there are many algorithms available to determine the values of the model parameters that minimize it [41, 43]. Unfortunately, in most cases with compartmental models, there are no direct solutions for the parameters. This is true because, although the models themselves are linear (i.e., all fluxes between compartments are linear multiples of the concentration in the source compartment), the solutions to these models are functions that are non-linear in at least one of the model parameters. For example, Eq. 27, the solution to the one-compartment model (Fig. 6.3B) is linear with respect to the parameter K_1 but is non-linear with respect to the parameter k_2. To solve for the parameters, iterative algorithms are required. First, an initial guess is made for the parameter values. Then the algorithm repeatedly modifies the parameters, at each step reducing the value of the optimization function. Convergence is reached when changes to the parameters from one iteration to the next become exceedingly small. Great care is required in the use of iterative algorithms, because incorrect solutions can be obtained, particularly if the initial guess is not appropriate.

Figure 6.6 provides an example of the process of parameter estimation applied to time-activity data collected after a bolus injection of fluorodeoxyglucose (FDG) [53]. Figure 6.6a shows a plot of region-of-interest values (occipital cortex) taken from reconstructed PET images. The solid line through the data points is the best fit obtained by minimizing the weighted sum of squared differences between the data and the two-compartment model (Figure 6.3c). Figure 6.6b shows a plot of the weighted residuals versus time. The residual is the difference between each data point and the model prediction. When weighted least squares is used, the residuals are scaled by the square root of each weight, w_i, so that the sum-of-squares optimization function equals the sum of squared residuals. Ideally the residuals would be random, have zero mean, and uniform variance. If a good estimate of the noise level in the data is known, the weighted residuals should have a standard deviation of approximately 1. Thus, when plotting the residuals versus time or versus concentration, the residuals would appear as a uniform band centered on zero. The residuals in Fig. 6.6b reasonably satisfy these expectations.

Many parameter estimation algorithms provide estimates of the uncertainties of the parameter estimates

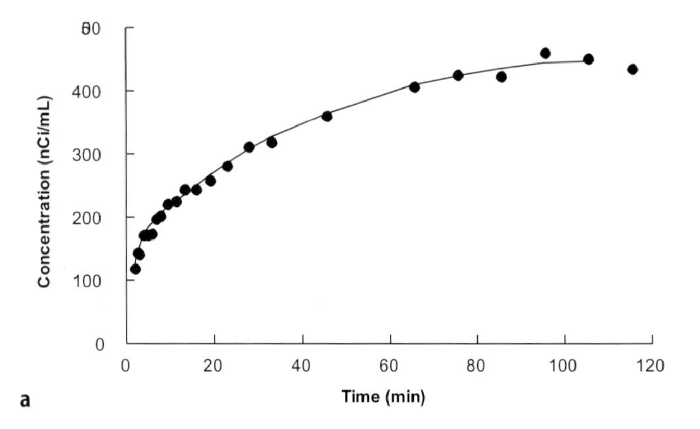

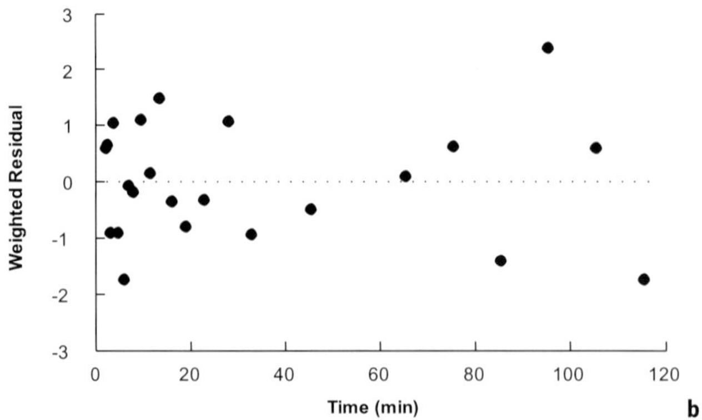

Figure 6.6. Examples of parameter estimation results from PET data after bolus injection of FDG. **a**: tissue time–activity curve from region of interest in occipital cortex. Symbols are measured data points. Solid line is fitted function for the two-compartment model (Fig. 6.3C) based on weighted least-squares parameter estimation. **b** Plot of weighted residuals (difference between data and fitted value scaled by regression weight) versus time.

called standard errors. These values can be used as estimates of the minimum random uncertainty in the estimate. The algorithm determines these standard errors based on the structure of the model, the parameter estimates, and the magnitude of the residual sum of squares. However, this measure is an underestimate of the true uncertainty of the parameters, since there are usually many sources of "real-world" errors that are not explicitly included.

It is often useful to calculate functions of the rate constants which provide different physiological information. For example, in the one-compartment model (Fig. 6.3B), a parameter estimation problem may be posed to estimate the rate constants K_1 and k_2. From these parameters, the distribution volume ($V = K_1/k_2$) can be calculated. To determine the uncertainty in the distribution volume estimate, information about the individual standard errors in K_1 and k_2 is required along with the correlation between them. The parameter estimates will be correlated since both values are determined simultaneously from the same noisy data. The coefficient of variation (CV, the ratio of the standard error to the parameter value) of the distribution volume can be calculated by propagation of errors calculations:

$$CV^2(V) = CV^2(k_1) + CV^2(k_2) - 2\rho_{12}CV(K_1)CV(k_2) \quad (32)$$

where ρ_{12} is the estimated correlation coefficient between the parameter estimates K_1 and k_2.

Least squares is the best optimization criterion for estimating parameters when a large number of assumptions are met. If any of these assumptions are not true, better estimates may be obtained by other methods (see section on error analysis). A better estimate is one that may be more accurate (less biased) or more precise (less variable). In addition, iterative least-squares algorithms may be very computationally intensive, particularly if it must be carried out individually for every pixel in an imaging volume. Often, iterative least-squares procedures are used only for a small number of regions of interest. However, it is often more useful if the data analysis procedure produces functional images where each pixel represents a physiological parameter of interest. To do this, rapid computation schemes are required. Rapid implementations of iterative least-squares procedures have been developed for the simplest non-linear models with just one non-linear parameter, e.g., the one-compartment model with solution in Eq. 27. These techniques have been applied to the measurement of cerebral blood flow [54, 55] and total volume of distribution of receptors [56–58]. In addition, a number of methods have been derived that allow direct non-iterative calculation of the parameter estimates by reformulating the problem in terms of integrals of the tissue and blood data [59–66]. These methods do not minimize the sum-of-squares optimization function, but in many cases have been shown to have comparable statistical quality to the least-squares techniques and often have less sensitivity to deterministic errors in the model. Another interesting approach for parameter estimation from non-linear models is called spectral analysis and uses the methods of linear programming with the knowledge that all the exponential clearance terms (α_i in Eq. 19 and Eq. 20, for example) are positive [67].

As shown above, the measured tissue activity is the convolution of the input function with the underlying impulse-response function (Eq. 28). This impulse-response function has a much simpler mathematical form (usually a sum of exponentials) and is therefore more easily analyzed. Some investigators have used the approach of deconvolution, whereby an estimate of the impulse-response function is determined from measurements of the tissue response and the input function [68]. However, because the process of deconvolution greatly amplifies noise in the tissue measurements and is often mathematically unstable, great care is required in the application of these techniques.

Development of Mathematical Models

The primary factor affecting the form of a model is the nature of the tracer itself. Usually, a priori information can be used to predict all of the relevant metabolic paths of the tracer in tissue, i.e., a complete model. However, technical and statistical limitations of the available data will prevent the use of such a comprehensive model, which includes all steps in the physiological uptake, metabolism, and clearance of a tracer.

Figure 6.1 shows the process of development and application of a model in PET [69, 70]. This section presents the steps starting with a complete model, then generating an identifiable model, and ultimately a practical model. An identifiable model is one which can be applied to regional kinetic data and used to extract estimates of model parameters. Such a model is a simplified version of a comprehensive description of the interactions of a radiotracer in tissue. However, this model may not be workable if its parameter estimates are too variable or inaccurate. A useful model may be derived by further simplification of the identifiable model. The useful model provides reproducible and accurate estimates of model parameters. Validation studies are necessary to demonstrate these characteristics.

Model Identifiability

The first step in defining a model is to determine identifiability, meaning that the parameters of a model can be uniquely determined from measurable data. There is an extensive literature on this topic [16, 19, 71–77], including studies with particular attention to PET applications [78–80]. In some cases, the structure of the model itself does not permit the unique definition of parameter values, even with noise-free data. One example of this is the case of high specific-activity studies with receptor-binding ligands. Here, the association rate k_{on}^* and the free receptor concentration B'_{max} appear as a product in the model differential equations (Eq. 14) and therefore can not be separated [32, 81].

A more significant problem in many applications is that of numerical identifiability. Here, parameter estimation can be successfully performed with low-noise data, but, with realistic noise levels the uncertainties in the resulting parameter estimates are large. This issue can be complicated by the fact that the values of the model parameters themselves may affect the ability to distinguish kinetic compartments. This is a common problem in brain neuroreceptor studies where the same model cannot be applied to brain regions with varying concentrations of receptor [82]. In addition, small deterministic errors in the model or in tissue radioactivity quantification can produce large changes in the parameter estimates. Thus, while an identifiable model is essential, it is not necessarily a useful model.

A common approach resulting from this form of model instability is to determine those parameters which are common to a set of models and are estimated with good precision, no matter what the model form. For example, in a complete receptor model, the free receptor concentration B'_{max} appears in the rate constant describing the movement of tracer from a free to a bound compartment. Ideally, therefore, receptor information can be obtained from this rate constant, but, in fact, functions that include this rate constant are also sensitive to B_{max}. One example of a useful lumped parameter is the total volume of distribution V described above for diffusible tracers (Eq. 8). For receptor-binding radiotracers, V represents the ratio at equilibrium between total tracer in tissue to that in plasma. Instead of trying to use individual parameter estimates, the total volume of distribution, which can be derived from the model rate constants, has been found to be a particularly useful and reliable measure for receptor quantification [56, 83, 84]. V is an algebraic function of the rate constants and has smaller uncertainty due to the positive correlation between estimated model parameters (Eq. 32). For a model with one tissue compartment, V can be calculated by setting the derivative in the differential equation (Eq. 15) to 0, resulting in $V = K_1/k_2$ (Eq. 8). For a model with two tissue compartments, setting the derivatives in Eq. 17 and Eq. 18 to zero yields, $V = K_1/k_2(1 + k_3/k_4)$. In addition, if V becomes the primary parameter of interest, simpler methods to directly estimate this parameter can be developed (see Model-based Methods).

The process of model selection proceeds as follows: Tissue measurements after injection of the radiopharmaceutical are collected. Then, a number of possible model configurations are proposed. Usually, the number of compartments covers a range from very complex to very simple, and there is a range of different numbers of parameters to be estimated in these models. Parameter estimation procedures are performed with the measured data using each model. The goodness-of-fit of each model to the data is assessed from the residual sum of squares (Eq. 31) using statistical tests such as the F-test, the Akaike information criterion [85], or the Schwarz criterion [86] to determine which model is most appropriate. In general, the use of a more complex model with additional parameters will produce a better fit to the data and a smaller residual sum of squares. However, this will be the case even if the additional parameters added by the more complex model are only providing a better fit to the noise in the data and have no relationship to the underlying true tissue model. The statistical tests used for model comparison determine whether the residual sum of squares has been reduced using the more complex model by an amount that is significantly greater than what is expected by random chance.

Another very useful approach for model comparison is the examination of the pattern of residuals as in Fig. 6.6b [43]. If the residuals from a fit of one model configuration do not appear as randomly distributed around zero, then a more complex model may be appropriate. However, given all the error sources (see section on random and deterministic errors), no model will ever be perfect. It will therefore often be the case that an overly complex model will still provide a statistically significant improvement in the fit compared to a simpler model. The modeler must have a good understanding of the degree of accuracy in the data in order to avoid an unduly complicated model.

To simplify models, pairs of compartments can be combined together. Two compartments can be collapsed into one by assuming that the rate constants "connecting" them are large enough so that the two compartments remain in continuous equilibrium. When this occurs, the physiological interpretation of the remaining rate constants in the reduced model must be changed. In this way, a set of "nested" models can be defined. Figure 6.3 shows some examples of nested models. Here, a simple model with a few rate constants can be considered to be a special case of a more complex model with more parameters. For example, the model in Fig. 6.3B is a simplified version of Fig. 6.3C which is itself a simplified version of Fig. 6.3D.

It is good practice to test a set of nested models to determine which one best characterizes a set of measured data [56]. An example of this process is shown for the opiate receptor antagonist [^{18}F]cyclofoxy [84, 87–89]. Figure 6.7a shows a typical time–activity curve measured in the thalamus with PET after bolus injection. The symbols are measured data points. The solid line is the best-weighted least squares fit using a model with two tissue compartments (Fig. 6.3c). The dashed line is the best fit using a model with one tissue compartment (Fig. 6.3B). Both models also included an additional parameter to account for radioactivity present in the tissue vascular space, so five and three parameters were estimated, respectively. The plots of weighted residuals versus time from these fits are shown in Fig. 6.7b (one compartment) and Fig. 6.7c (two compartment). The one-compartment results show a deterministic pattern of residuals. The residual points are not randomly distributed about zero, but instead show runs of sequential values that are all positive or all negative. The residual pattern is more random when using the two-compartment model. In this case, the more complex model was found to have produced a statistically significant reduction in the residual sum of squares. However, this improvement was not large and was not found uniformly for all patients or for all brain regions.

The absolute magnitude of the residual noise can also be useful in determining if a particular model configuration is appropriate. If the model is exactly correct and the magnitude of data noise is known,

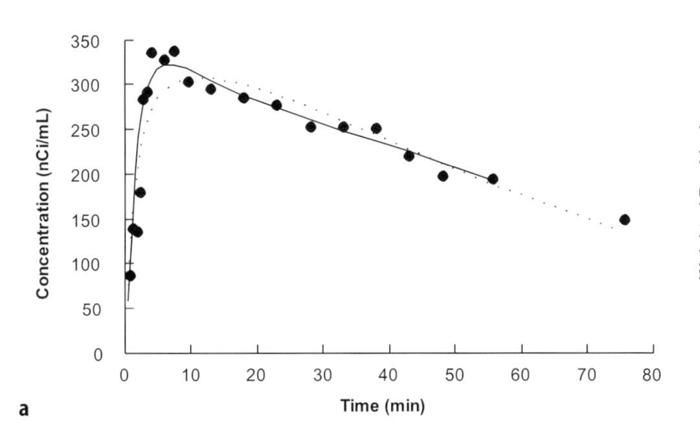

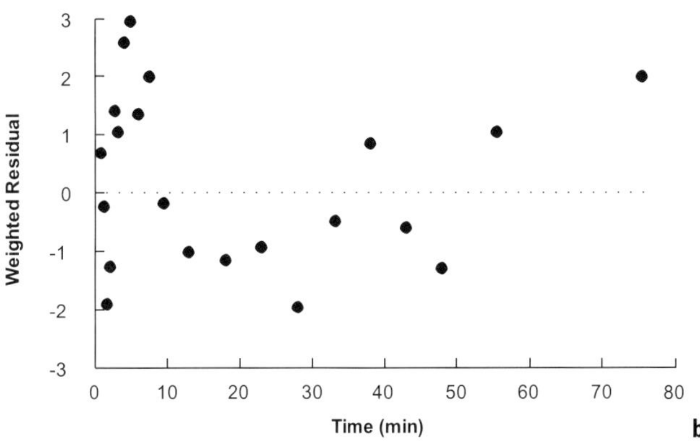

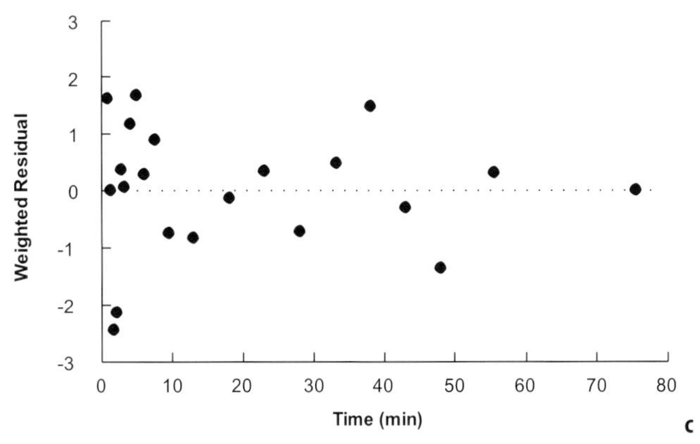

Figure 6.7. Comparison of model fits. **a** PET tissue concentration data from the thalamus acquired after bolus injection of the opiate antagonist [^{18}F]cyclofoxy. Symbols are measured data points. Solid line is weighted least-squares fit using a model with two compartments and five parameters (K_1, k_2, k_3, k_4, and blood volume fraction). Dashed line is best fit using a model with one compartment and three parameters (K_1, k_2, and blood volume fraction). **b** plot of weighted residuals versus time for the three-parameter model. **c** plot of weighted residuals versus time for the five-parameter model. The residuals for the three-parameter model show a non-random pattern that is reduced with the five-parameter model.

the weighted sum of squares (Eq. 31) will be approximately equal to $N - n_p$, where N is the number of data points and n_p is the number of model parameters being estimated. If the actual sum of squares from data fits are close to this value, the modeler gains additional confidence that the chosen model configuration is appropriate.

Model Constraints

A typical situation in PET modeling problems is that a simple model with few parameters is often not adequate to describe the tissue concentration curve. However, a more complex model that does adequately describe the data frequently produces parameter estimates that have large uncertainties (standard errors). Specifically, a simple one-compartment, two-parameter model is often insufficient, whereas a two-compartment, four-parameter model is "better" by various statistically significant measures. A number of authors have dealt with this conflict by applying constraints. These entail specifying exact values for certain parameters or defining relationships between the parameters that must be met. In either case, the effect is to reduce the number of parameters that must be determined from the model. If the constraints are accurate (or reasonably so), then the sensitivity of the model data to the remaining parameters is increased and the uncertainty in their estimation is reduced. Often the constraint equations use a priori values for physiological constants based on the presumed interpretation of the model parameters in terms of Michaelis–Menten parameters [90–92]. Alternatively, some parameters may be constrained based on measurements made in other regions [93]. For example, a common approach for receptor-binding tracers is first to analyze a reference region known to have little or no specific binding to determine parameters associated with the magnitude of nonspecific binding. Then, regions with specific binding are analyzed with nonspecific-binding rate constants constrained to equal those estimated in the reference region [81, 94]. Alternatively, additional studies can be performed to aid the estimation process by constraining parameters to be common to the analysis of both studies. For receptor-binding tracers, a study with an inactive enantiomer can be used to determine parameters of nonspecific binding [95, 96]. In addition, paired studies with high and low specific activity injections and/or displacement can be performed and analyzed simultaneously with some parameters shared in the models for the two studies [97, 98].

Validation of Physiological Measures

In the process of developing and selecting a suitable model formulation and methodology, it is important to perform validation studies which prove that the parameter estimates produced by a model are correct. These studies determine the precision and accuracy of model estimates, verify the legitimacy of the model assumptions, and help choose between various approaches. Such an evaluation invariably must be done in animals because of constraints on experimental design, scan duration, and radiation dosimetry in humans. Practical limits on animal studies include limitations on total blood sampling for input function measurements and the effects of anesthesia.

Although much of the work of model development and validation is performed using small or large animals, it is important to realize that there are considerable differences between PET image data and autoradiographic or tissue-sampling measurements, as well as the species differences among rodents, large animals, and humans, which may limit the applicability of the information obtained in the animal experiments. For example, measurement of tissue concentration data at multiple time points in rodents requires multiple animals. PET studies allow acquisition of multiple time points in a single study, avoiding inter-individual variability. However, the spatial resolution and statistical reliability of scan data are substantially worse than measurements from tissue samples in rats. Therefore, kinetic parameters that can be reliably determined from rat data may not be numerically identifiable from human scan data. Therefore, many validation studies should be repeated wherever practical with human subjects.

The simplest test of a model is reproducibility, i.e., the variability of the model parameters under identical conditions, either on the same day or different days [99]. Repeating studies on the same day will generally produce smaller differences in scan data results, since there will be less variation in subject positioning and scanner calibration. Measurements of population variability of model estimates provide information concerning the most useful model configurations. Clearly, model parameters with large coefficients of variation will not generally be useful. Also, models should provide physiologically reasonable values. Although in vivo measurements can certainly produce different results than in vitro tests, it is up to the investigator to demonstrate the accuracy of a model that produces parameter estimates inconsistent with previous results in the literature. Additionally, one can compare parameters from one tracer to another when there is reason to believe they should be similar, e.g., comparing the K_1 values of two tracers with high extraction, in which case both K_1 values should approximate flow [58].

The next steps in validation of a model are intervention studies. Here, one or more of the physiological parameters that affect tracer uptake are altered, and the model is tested to verify that the parameters change in the proper direction and by an appropriate magnitude in response to a variety of biological stimuli. For example, brain blood flow can be altered by changing arterial pCO_2, or free receptor concentration can be reduced by administration of a cold ligand. It is also useful to test whether the parameters of interest do *not* change in response to a perturbation in a different factor, e.g., does an estimate of receptor number remain unchanged when blood flow is increased [100]? Alternatively, changing the form of the input function should ideally have no effect on the model parameters [84, 101–103]. Model assumptions, e.g., parameters whose values have been constrained, should be tested. At a minimum, computer simulations of the effects of errors in various assumptions upon model results can be performed (see section on error analysis). The limitation of these simulations is that they are only as good as the models on which they are based. Therefore, experimental validation of model assumptions should be performed where possible.

Finally, the absolute accuracy of model parameters can be tested by direct comparison with a "gold standard." To test the accuracy of regional measurements, such a validation study can only be carried out with animals. While this validation step is very appealing, it is often very difficult to achieve. There is often no gold standard available for the measurement of interest. Even if such a standard is available, the comparison will require careful matching of scan data with tissue sample data. If the regions being compared are small, the effects of inaccurate registration and scanner resolution can make evaluation of the model's accuracy difficult at best. Even without a gold standard, other validations of the model can be performed. For example, model predictions of concentrations in separate compartments can be compared to biochemical measurements of tissue samples [104]. Also, microdialysis provides a method to assess extracellular tracer concentration directly for comparison with model predictions [105].

Model-based Methods

To this point, the design, development, and validation of a tracer kinetic model have been presented. Ideally, the modeling effort generates a complete, validated model that describes the relationship between tissue measurements and the underlying physiological parameters. With this knowledge, we can design a method of data acquisition and processing suitable for human studies. This section concerns this final step in the modeling process shown in Fig. 6.1: the adaptation of such a useful model to produce a practical patient protocol [69, 106]. It is often the case that the original modeling studies are complex and may not be suitable for human subjects, particularly certain patient populations. For example, arterial blood sampling may not be feasible, or a long data acquisition period may not be practical, or the statistical quality of data in humans may limit the number of parameters that can be reliably estimated. From the understanding of the characteristics of the tracer and with knowledge of the limitations imposed by instrumentation and logistical considerations, a model-based method can be developed that can achieve a useful level of physiological accuracy and reliability.

Many questions must be considered in converting a model into a model-based method. To what extent are the extra complexities of a full modeling study necessary or useful? What are the best trade-offs to maintain an adequate signal-to-noise ratio in the data? Can an appropriate input function be measured less invasively than from arterial samples, e.g., from direct scan measurements, from venous samples, or from a reference region? What is a practical data collection period that is compatible with the time availability on the scanner, the statistical requirements of the collected images, and the characteristics of the patients? Which parameters are of prime importance? Can parameter estimates be calculated on a pixel-by-pixel basis to generate functional images or must time-consuming iterative nonlinear methods be applied to region-of-interest data? What reasonable assumptions can be incorporated into the model to reduce the number of parameters to a workable set that can be determined with reasonable precision? Is the method overly sensitive to measurement errors or to inaccuracies in model assumptions, particularly in patient groups? If the method is simplified too much, could differences between patients and controls be exaggerated or hidden because physiological factors properly included in the original model are now ignored?

This section presents some approaches that have been used to produce model-based methods. These methods are generally simpler than the full parameter estimation studies, use additional assumptions, and typically allow production of functional images of the physiological estimates. In addition, sources of error in model approaches are discussed along with error analysis methodology. Finally, the trade-offs between using model-based techniques and simple empirical methods are examined.

Graphical Analysis

One increasingly common method applied to tracer kinetic data is that of graphical analysis [90, 107–112]. The basic concept of this method is that after appropriate mathematical transformation, the measured data can be converted into a straight-line plot whose slope and/or intercept has physiological meaning. This approach has advantages, since it is simple to verify visually the linearity of the data and it is simple to determine the slope and intercept by non-iterative linear regression methods. It is also generally easy to determine these values on a pixel-by-pixel basis, thus producing a functional image of the parameter [113]. For many models, the simplified equations of graphical analysis will not apply for all times post-injection, e.g., at early times when the blood activity is changing rapidly and some tissue compartments have not yet reached equilibrium with the blood. Therefore, care must be taken in selecting the time period for determination of the slope and intercept. However, it is also true that avoiding the time periods where the kinetics are rapid also makes the method less sensitive to errors introduced by oversimplifications in the model, particularly those dealing with tracer exchange between capillary, extracellular space, and intracellular space.

The most widely used graphical analysis technique is the Patlak plot [107–109]. This approach is appropriate when there is an irreversible or nearly irreversible trapping step in the model. Conceptually, the transformations of the Patlak plot convert a bolus injection experiment to a constant infusion. A simple example of this model is the two-compartment model (Fig. 6.3C), in which the rate constant for return of tracer from compartment 2 to compartment 1, k_4, is zero or is small, i.e., irreversible trapping. In this case, the model solution (from Eqs. 21, 22 and 28) for the total tissue tracer concentration $C(t)$ for an arbitrary input function $C_a(t)$ is

$$C(t) = C_a(t) \otimes \left(\frac{K_1 k_2}{k_2 + k_3} \exp[-(k_2 + k_3)t] + \frac{K_1 k_3}{k_2 + k_3} \right) \quad (33)$$

If the arterial input function were held constant (C_a), the solution to Eq. 33 would be

$$C(t) = C_a \left(\frac{K_1 k_2}{(k_2 + k_3)^2} (1 - \exp[-(k_2 + k_3)t]) + \frac{K_1 k_3}{k_2 + k_3} t \right) \quad (34)$$

After an appropriate time, t^*, after which the exponential term in Eq. 34 becomes sufficiently small, the ratio of tissue to blood activity becomes

$$\frac{C(t)}{C_a} = \frac{K_1 k_2}{(k_2 + k_3)^2} + Kt \quad (35)$$

which is a linear equation. The slope of this equation, K, is

$$K = \frac{K_1 k_3}{k_2 + k_3} \quad (36)$$

The term K is the net uptake rate of tracer into the irreversibly bound compartment 2. It is the product of two terms: K_1, the rate of entry into the tissue from the blood, and $k_3/(k_2 + k_3)$, the fraction of the tracer in the tissue that reaches the irreversible compartment (Fig. 6.3C).

For the case when the input function is not a constant, the Patlak transformation is as follows:

$$\frac{C(t)}{C_a(t)} = V_0 + K \left(\frac{\int_0^t C_a(s)ds}{C_a(t)} \right) \quad (37)$$

The term in brackets in Eq. 37 is often called stretched time or normalized time, since it has units of time and it distorts time based on the shape of the input function. If the ratio of tissue to blood activity, which is called the apparent volume of distribution, is plotted versus stretched time, under the appropriate conditions a linear plot is obtained with slope K and intercept V_0 (the initial volume of distribution). Note that in the case of a constant arterial input, stretched time becomes exactly equal to true time.

In applying this graphical method, it is important to verify that the Patlak plot is in fact linear over the range of time used, an assumption that can often be evaluated in animal studies, where longer experiments can be performed [114]. For purposes of fitting data to estimate K, instead of fitting Eq. 37, it is equivalent to use multiple linear regression to fit the measured tissue data directly:

$$C(t) = V_0 C_a(t) + K \int_0^t C_a(s) ds \quad (38)$$

This approach is better if the later values of the input function are noisy (e.g., due to metabolite correction), and more easily allows regression weights based directly on image noise estimates to be added to the estimation process.

Figure 6.8 provides an example of the use of a Patlak plot as applied to brain PET data after the injection of FDG [53]. In this study, subjects were studied on two occasions, approximately one week apart. For one scan, the subjects underwent a hyperinsulinemic euglycemic clamp, whereby high levels of insulin were infused, and simultaneously blood glucose levels were maintained at a constant level, thus maintaining the steady-state assumption of the tracer kinetic model. On the second occasion, a sham clamp was performed, i.e., a control study. The high insulin levels in the clamp study caused a dramatic change in the plasma input function, i.e., the rate of FDG clearance from plasma was much higher. Figure 6.8a shows the tissue curves for an average of gray matter regions in one individual. There is clearly a dramatic difference in the two curves. The Patlak transformation of Eq. 37 was applied to these data and is shown in Fig. 6.8b with a plot of the apparent volume of distribution versus stretched time. The two plots nearly overlay each other, demonstrating that most of the difference between the two tissue time–activity curves of Fig. 6.8a can be accounted for by the differences in the input function, not by differences in the tissue kinetic parameters. Note that the hyperinsulinemic study covers a longer period in stretched time than the control study.

A second graphical approach is that developed for measurement of parameters for reversible neuroreceptor ligands, i.e., those that approach equilibrium during the time period of the experiment. As described above, the total volume of distribution V is the most commonly estimated parameter for these types of tracers. The Logan graphical relationship [111] allows the estimation of V from the slope of a plot produced by a transformation of the data, like the Patlak plot described above. The Logan relationship can be derived exactly from the one tissue compartment model, Eq. 15, and integrating:

$$C(t) = K_1 \int_0^t C_a(s) ds - k_2 \int_0^t C(s) ds \quad (39)$$

Dividing Eq. 39 by k_2 and $C(t)$, and rearranging yields:

$$\frac{\int_0^t C(s) ds}{C(t)} = V \frac{\int_0^t C_a(s) ds}{C(t)} - \frac{1}{k_2} \quad (40)$$

where the slope of this relationship V is the volume of distribution for the one tissue compartment model (K_1/k_2). In cases where the data are not consistent with a one-compartment model, the graph becomes linear after an appropriate time, and the linear regression is performed for those later data. In that case, the slope is the estimate of the total volume of distribution. Figure 6.9 shows an example of Logan graphical analysis [111] as applied to PET time–activity data for the 5-HT_{1A} antagonist [^{18}F]FCWAY [115] as measured in the rhesus monkey. The three curves show regions with different receptor levels, with the highest slope (frontal cortex) corresponding to a region with high specific binding

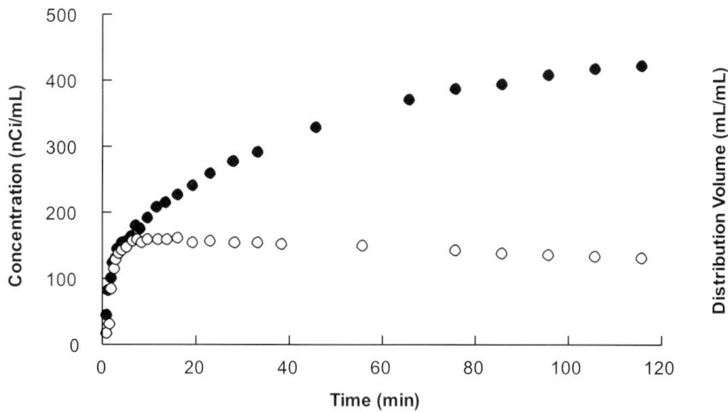

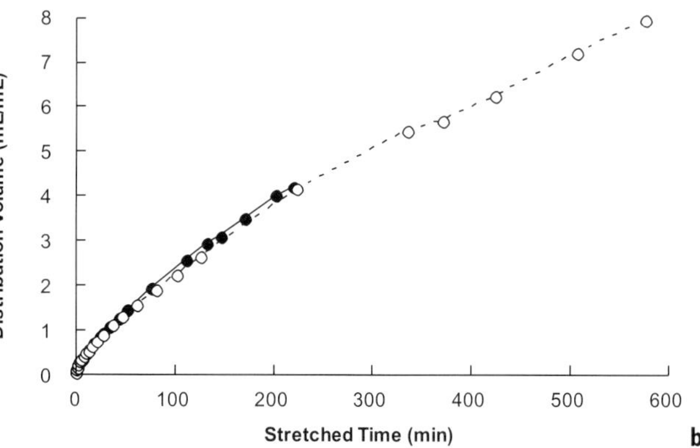

Figure 6.8. Example of graphical analysis (Patlak plot) from FDG PET data. The study involved a control scan on one day and a hyperinsulinemic euglycemic clamp on another day. **a** Average tissue time concentration curves in cortical gray matter. Filled and open symbols are scan data values from the control and clamp studies, respectively. **b** Patlak plots from the control (filled symbols and solid line) and clamp (open symbols and dashed line) studies computed from the data in A and B using Eq. 37. Despite the large differences in tissue data between the two studies, the tissue kinetics in both cases, as shown by graphical analysis, are very similar, i.e., there is at most a small effect of insulin on gray matter metabolism of FDG.

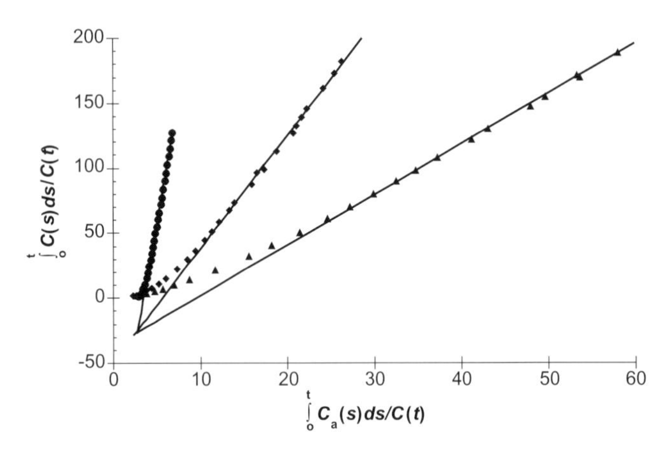

Figure 6.9. Logan graphical analysis of regional PET data acquired from the 5-HT_{1A} antagonist [^{18}F]FCWAY in rhesus monkey in frontal cortex (•), thalamus (♦), and cerebellum (▲). Data are transformed as specified in Eq. 40. Following a certain time, the graphs become straight lines with slopes equal to the volumes of distribution for each region. Regions with greater specific binding have higher slopes.

and a high value of V. Note that the time to achieve linearity of these plots differs between regions due to different receptor levels/kinetic parameters.

Reference Region Methods

The emphasis in this chapter has been the determination of kinetic rate constants using the relationship between tissue data measured with the PET scanner and the input function, usually derived from arterial blood samples. For studies in the chest with tracers that do not metabolize, the input function can be measured from the imaging data in the left ventricle, atrium, or the aorta [116–119]. Other approaches have been used where smaller blood vessels can be imaged but corrections for partial volume effect are required [120, 121]. However, in a number of other cases, approaches have been developed to avoid the measurement of the arterial input function and still deduce kinetic parameter information by comparison of the time–activity curve in the region of interest to that in a reference region. The most significant application of this approach has been in receptor modeling where the comparison of regions with and without receptors provides a natural application [122–124], which can often be extended to pixel-by-pixel analysis [125]. The general idea of these approaches is to use the mathematics of the model to infer the shape of the arterial input function based on the time-course measured in the reference region. This permits a mathematical relationship to be developed for the region-of-interest concentration in terms of the reference region data and the kinetic parameters of both regions. Usually, the number of available parameters is reduced, e.g., in this situation the uptake constant K_1 for either the region of interest or the reference region cannot be determined, but the ratio between them can be estimated.

In addition, there are reference region methods adapted for graphical analysis, either for irreversible [109] or reversible [126, 127] tracer uptake. As with all the graphical methods, only a subset of the kinetic parameters can be determined, and with the use of reference regions, the estimated parameters are typically ratios of the original parameters between their values in the region of interest and that in the reference region. However, it is often the case that the most sensitive biological parameter is a normalized model value. Normalization tends to eliminate certain methodological errors which add common variance to both the region-of-interest and the reference region results [106]. Therefore, these reference-region graphical methods tend to directly estimate the parameter ratios of interest.

Single-scan Techniques

A common approach to produce simplified model-based methods is the use of single-scan techniques. Here, based on a good understanding of the relationship between tissue radioactivity and the underlying physiological parameters, tissue radioactivity information is acquired during one scan interval. This single measurement permits the estimation of a single unknown physiological parameter. Since most models have multiple rate constants, some corrections must be applied to account for these other unknowns. Careful design of a single-scan technique ensures that variation in these nuisance parameters produces only minor errors in the parameter of interest.

For the measurement of cerebral blood flow with [^{15}O]water or comparable diffusible tracers, two approaches have been taken to produce single-scan methods. Some of the earliest studies used continuous inhalation of [^{15}O]CO_2 [128], which is rapidly converted to [^{15}O]water in the lungs. By achieving constant radioactivity levels, the derivative in the differential equation of uptake of the tracer (Eq. 15 with additional terms for radioactive decay) can be set to zero, and K_1 can be determined from an algebraic formula in terms of tissue and blood radioactivity. A different approach uses a bolus injection followed by a single short scan [129, 130]. This autoradiographic method uses the explicit solution of the model (Eq. 27) to determine K_1 from the integrated tissue radioactivity and a measured input function. Both of these methods treat the estimated K_1 values as equal to blood flow, assuming a large permeability-surface (*PS*) area product for the tracer (Eq. 5). Both methods also require the use of an assumed value for the tracer distribution volume V in order to specify k_2 as K_1/V. As only one tissue measurement is made, only one unknown parameter can be determined. The short scan of the autoradiographic method was designed in part to minimize the sensitivity of this method to errors in the assumed value of the distribution volume.

Another example of single-scan, model-based techniques is the autoradiographic method for measurement of glucose metabolism, which was developed in rats with [^{14}C]deoxyglucose [1] and extended to PET using [^{18}F]2-fluoro-2-deoxy-D-glucose [2–4]. These methods take advantage of the fact that most of the radioactivity in the tissue by 45 min post-injection has been phosphorylated, so that the total tissue radioactivity can be used to estimate the net flux into tissue of deoxyglucose, K. This is the same rate constant as determined from the slope of the Patlak plot. Effectively, these methods estimate the slope of a Patlak plot by using the measured tissue value at one data point and by using population values of the model rate constants to estimate the y-intercept of the straight line. A number of other formulations of this approach have been developed [131–133], each with different sensitivities to errors in the assumed rate constants. Finally, since FDG is an analog of glucose, the metabolic rate of glucose is estimated from the measured net flux of FDG using the measured plasma glucose level and an assumed scaling factor, the lumped constant [1, 5–9].

Equilibrium Methods

Another single-scan technique has been developed for quantification of receptors by using infusion to produce true equilibrium [84, 134, 135]. By administering tracer as a combination of bolus plus continuous infusion (B/I), constant radioactivity levels can be reached in blood and in all regions of interest. The total tissue volume of distribution can be determined from the ratio of tissue activity to metabolite-corrected plasma activity. This value will include free, nonspecifically bound, and specifically bound tracer. Estimates of the nonspecific component, e.g., from a region with low receptor binding, from measurements with an inactive enantiomer, or from data acquired after displacement with excess cold ligand, can be subtracted to estimate the binding potential, B_{max}/K_D [136] (K_D is the dissociation equilibrium constant). Multiple infusions at different specific activities can be used to determine B_{max} [137, 138].

This infusion approach can be extended to provide receptor-binding data in two states: at baseline and post-stimulus (e.g., drug-induced neurotransmitter changes), with a single administration of tracer. Without infusion, such data are conventionally acquired with paired studies, each with a bolus injection. In the first study, control levels of binding are measured, for example, by determining V by compartment modeling [56] or graphical analysis [111]. Then, following the pharmacological intervention, a second measurement of binding is made with a second injection of tracer. This approach has been used successfully with the D_2 ligand [^{11}C]raclopride [95, 139] as well as with a number of other tracers. For example, Dewey et al. have demonstrated the effects of changes in synaptic dopamine by direct effects on the dopamine system itself [140] and by indirect pharmacological interventions [141, 142]. In humans, this paired-study approach has been used to measure drug occupancy [143–146].

The alternative study design is to administer the tracer as a combined bolus plus continuous infusion (B/I) to measure short-term changes in free receptor concentration [101, 147, 148]. First, the B/I administration of tracer is performed to achieve constant radioactivity levels in blood and all brain regions. Once equilibrium is achieved, control binding levels can be determined. For example, the volume of distribution V can be measured directly from the tissue-to-plasma concentration ratio. Then, a stimulus is administered while the infusion of radiotracer continues, and the change in specific binding of the tracer can be monitored. An example of B/I data is shown in Fig. 6.10 assessing the effects of amphetamine-induced dopamine release with [^{11}C]raclopride. By comparing the pre- and post-amphetamine levels of specific binding determined directly from the tissue concentration values (Basal Ganglia/Cerebellum −1), the change in specific

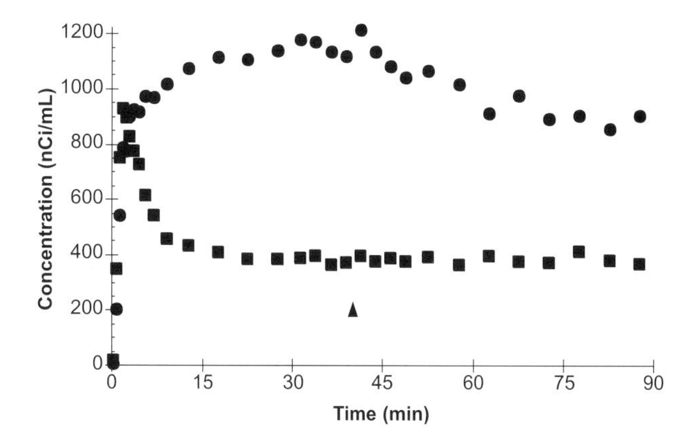

Figure 6.10. ROI data from basal ganglia (•) and cerebellum (■) following combined bolus plus infusion administration of the D_2 dopamine ligand [^{11}C]raclopride. At 40 min (arrow), 0.4 mg/kg of amphetamine was administered intravenously, producing displacement of raclopride due to competition with increased synaptic dopamine.

binding from amphetamine can be measured. This B/I study design permits the measurement of pre- and post-intervention binding levels from a single administration of tracer. It is particularly well adapted to tracers with longer half-lives.

Random and Deterministic Errors

In making the choices necessary to implement a tracer method, it is important to be aware of the many sources of error that affect the precision and accuracy of these physiological measurements [106]. A good understanding of what effects are more or less significant to a given tracer and to the biological question of interest is essential in designing a sensitive, reliable technique that is not overly complex.

One aspect to consider is random errors, i.e., the effects of random statistical noise in the data on model parameters. The duration of data acquisition, the amount of smoothing of the images, the use of pixel or region-of-interest data, the number of parameters in the model, the mathematical structure of the model, and the actual parameter values affect the statistical accuracy of the parameter estimates. Many investigators have assessed the sensitivity of PET data to model parameters and methods to optimize the statistical quality of the model estimates [149–152]. In addition, noise in measured data can directly introduce bias in parameter estimates when non-linear methods are used [66, 153].

A primary source of deterministic error is the measurement of regional radioactivity from the PET scanner. Although the quantitative accuracy of PET continues to improve, there are still many sources of inaccuracies. For example, the accuracy of the scatter correction is limited, particularly for whole-body imaging and for 3D acquisition. A key effect corrupting PET imaging data is finite resolution, i.e., the partial volume effect [154]. The magnitude of bias in concentration measurements depends on the size of the underlying structure, the distribution of radiotracer within and around the structure, the resolution of the scanner, the reconstruction algorithm, and the strategy for extracting regional concentration values. Definition of the regions of interest using registered anatomical images (MR or CT) is important, as long as registration errors are minimized.

The partial volume effect produces heterogeneity, i.e., the tissue response measured from even a single pixel will represent a weighted average of the tissue in the surrounding region, and is thus a combination of different kinetic responses. This can have minimal to large effects on model results depending on the magnitude of heterogeneity and how the parameter of interest affects the tissue concentration measurements. This effect has been studied in great detail for a number of methods [155–161]. Since finite resolution is unavoidable in real imaging data, ideally application of modeling techniques will not introduce artifactual changes in the data. In other words, suppose a heterogeneous region was composed of two tissue types. Ideally, the final kinetic estimates from that area would be the weighted averages of the appropriate values for each tissue type, weighted by the fraction of the region occupied by each tissue type. If the parameter is estimated in a linear fashion from the data, this will be the case. For non-linear methods, heterogeneity will introduce a bias. An important approach to deal with the partial volume effect is to correct the PET data for this effect [162–166]. Recently, investigators have begun to assess the effects of these corrections on kinetic modeling [167] with tendencies toward major increases in the parameter values and the noise level of the estimates.

Another source of error in model applications is the presence of intravascular radioactivity in the tissue measurements [168–173]. Some fraction of the measured counts originates from radioactivity in the blood within the tissue. Since the radioactivity time-course in blood differs from that in tissue, errors in model measurements will occur unless this effect is properly handled. In some cases, the fraction of tissue volume occupied by blood can be measured in a separate tracer experiment. Alternatively, this vascular fraction is added as a parameter to account for this effect. Obviously, these errors are most important in regions with large blood volumes or in regions near the heart chambers or large blood vessels. Typically, errors due to vascular radioactivity are more significant when data collected immediately after injection are included in the analysis. However, these early data are often most sensitive to the parameter of interest, such as in the case of blood flow tracers where the rate constant for movement of tracer from blood into tissue (K_1) is of prime importance. Various strategies involving selection of time intervals for analysis or optimal region-of-interest placement have been proposed to handle these effects [171, 174].

A key to successful quantitative methods is the accurate measurement of the input function. Typically, the blood time–activity curve is measured in a peripheral blood vessel (usually radial artery) unless the heart chambers can be imaged directly [116–118]. When individual blood samples are drawn by hand, they must be taken at a sufficiently rapid rate to characterize the curve accurately. Careful attention is required for accurate sample timing, centrifugation, pipetting or weighing, radioactivity counting, counting corrections (background, decay, dead time, etc.), and data handling. Some investigators have developed devices for automatic withdrawal and measurement of whole-blood radioactivity [175]. These devices provide consistent data, but they may have increased statistical noise depending upon their counting geometry. Timing and dispersion differences between the brain and the peripheral artery require correction, particularly for studies of short duration with sharp bolus inputs [176–179]. A number of studies have been undertaken to assess the effects of statistical noise in the input function on estimated parameters and to develop appropriate estimation methodology [180–183]. If there are radioactive metabolites of the tracer in blood, it is important to determine the fraction of blood radioactivity that corresponds to the original tracer as well as the extent to which these metabolites pass into tissue. Since metabolite determinations are often complex, particularly for short-lived tracers, metabolite measurements are made at only a small number of samples. Appropriate interpolation or modeling schemes are necessary to generate a continuous estimate of the metabolite fraction throughout the study [101, 184]. Alternatively, other modeling approaches can be used to infer the metabolite correction [185].

Error Analysis

Error analysis is a useful tool in the development of an appropriate model-based method. Performance of a thorough error analysis is a critical step in the assessment of the utility of a given method. Papers dedicated solely to error analysis are common in the literature [155, 156, 171, 186–196]. These analyses usually proceed as follows. Choose a particular source of error. Select values for the model parameters and use the model equations to simulate tissue data including this error effect, usually covering a range of effect magnitudes. Then, analyze these simulated measurements with one or more methods, compare the derived parameter estimates to their original values, and determine the magnitude of error that is produced.

Figure 6.11 provides an example of the results of an error analysis. Cerebral blood flow (CBF) measurements with the tracer [^{15}O]water are altered in the presence of errors in correction for the time delay between the measured arterial input function and the actual input to the brain. Using an actual measured input function, tissue time-activity data were simulated

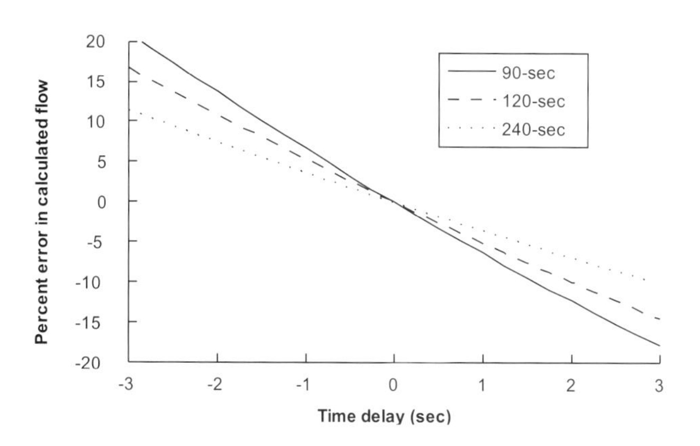

Figure 6.11. Example of error analysis – the effect of errors in time delay corrections between the brain and peripheral artery on measurement of cerebral blood flow with [^{15}O]water. A positive time delay means that the tissue data has been shifted forward in time with respect to the arterial input function. The three curves show the percent error in estimated flow, based on data collection periods of 90 sec, 120 sec, and 240 sec. See text for additional details.

over a 4-min period using the model of Eq. 27, with a flow value of 0.5 mL/min/g and a distribution volume of 0.8 mL/g. CBF (K_1) was then calculated by direct estimation of the two model parameters for total time intervals of 90, 120, and 240 sec. In each case, the tissue data were shifted with respect to the arterial input function by −3 to +3 sec (a positive shift means that the tissue data have been shifted later in time with respect to the blood data). The figure shows the percent error as a function of time delay. Positive time shifts produce underestimation of blood flow. This error is larger for shorter total acquisition times. This analysis suggests that the effect of time shift errors can be reduced by using longer data-acquisition periods. Even then, errors as large as 10% occur with time shifts of 3 sec, so care should be taken to measure or estimate time delays between tissue and blood data [178, 197].

A careful analysis of all the relevant error sources can be used to optimize methodology or to choose one approach over another. For example, various studies have been performed to choose optimal total scanning times and scan schedules [170, 198–201]. Unfortunately, it is difficult to determine the total error of a method based on the independent error analyses of a number of measurements or assumptions. First, error analyses are only as good as their ability to simulate biological reality, i.e., recognizing and analyzing all potential error sources and making appropriate choices for the magnitude of each error term. Even then, many error sources are not independent, i.e., errors in one term affect other terms. Thus, actual errors may be larger or smaller than those predicted from independent error analyses. Therefore, it is best if the ultimate choice of a method can be made by analyzing many studies with a variety of techniques and choosing the approach that has the best reproducibility, the minimum population variability, or the maximum statistical power to extract a particular physiological signal.

Selection of Model-based Methods

This chapter has presented an overview of modeling methods, from the most complex dynamic data acquisition with iterative parameter estimation to simplified methods including Patlak and Logan plots or single-scan techniques. Choosing the best approach is not simple, and other options are available when selecting a tracer method. In some studies, investigators normalize the physiological measurements. Instead of using the absolute values provided by a method, the results are scaled in some manner by a reference value, such as the average value in the entire organ or in a particular reference structure. This procedure may significantly reduce intersubject variation introduced by instrumentation, reconstruction, errors in the measurement of the input function, as well as variability due to global flow, metabolism, etc. In some cases where the model equations are linear (or nearly so) with respect to the parameter of interest, investigators can avoid the measurement of the input function and use normalized tissue concentration measurements as equivalent to a normalized model-based method [129, 202, 203]. Interpretation of results from normalized methods must be performed with care, however, since changes in ratios may be caused by changes in the numerator, denominator, or both.

Another example of choosing a normalized measure is the use of binding potential [136], B_{max}/K_D, for receptor-binding agents. This measure is usually derived from the total volumes of distribution V in regions with and without specific receptor binding. In some cases, the difference of the V values is used and in other cases a ratio is used. These different formulations have different characteristics in terms of biological interpretation as well as within-subject and between-subject variability. For example, the ratio formulation is more common because it can be estimated without measurement of the plasma input function. However, in that case, the results depend upon the assumption that the level of nonspecific tracer binding is unchanged between regions and between subject groups.

An alternative to using a model-based method is to use a simple empirical approach. Such approaches make no explicit attempt to estimate the physiological parameter(s) of interest. Instead, an index based on tissue measurements is used and presumed to reflect the underlying physiology. Empirical indices include absolute radioactivity values, radioactivity values corrected for dose and/or subject weight, and ratios of radioactivity values between target and reference regions (normalized values).

How can an investigator determine the best approach when using a tracer? Many trade-offs must be considered in designing a study, and there are no simple answers [106]. As an example, consider the use of a receptor-binding radiotracer for measurements in the brain with PET. Suppose the tracer binds reversibly, i.e., its dissociation rate from the receptor is sufficiently fast to approach equilibrium during the study period. Possible model-based quantification approaches include the following: 1) complete modeling study with iterative parameter estimation; 2) use of a simplified model with estimation of the volume of distribution [56]; and 3) use of a linearization formula to derive the volume of distribution from the later portion of the data [111]. Empirical alternatives to model-based methods include the following: 1) ratio of tissue region of interest to (metabolite-corrected) blood (apparent volume of distribution); or 2) ratio of tissue region of interest to reference region with few receptors during the apparent equilibrium phase.

Although the empirical approaches are the simplest, they can provide misleading results. For tracers that can reversibly bind with receptors, indices derived from ratios of tissue concentration to reference regions or to plasma levels can be significantly distorted due to lack of true equilibrium [84]. This effect is demonstrated in Fig. 6.12 with a "bolus plus infusion" protocol using the opiate antagonist [^{18}F]cyclofoxy (see section on single-scan techniques). Radioactivity in the tissue regions (Fig. 6.12a) reached steady levels by ~20 min. At 70 min post-injection (arrow), the infusion was discontinued, and plasma and tissue concentrations dropped. Figure 6.12b shows the apparent volume of distribution plotted against time. Discontinuing the infusion caused a dramatic increase in the values for the receptor-rich thalamus with smaller increases in frontal cortex and cerebellum. The magnitude of this effect depends upon the relative magnitudes of the rate of tracer clearance from plasma and the receptor dissociation rate. The change in the apparent distribution volume value (Fig. 6.12b) is due solely to the change in clearance of radiotracer from plasma and demonstrates that this ratio measure can be significantly affected by the plasma clearance rate.

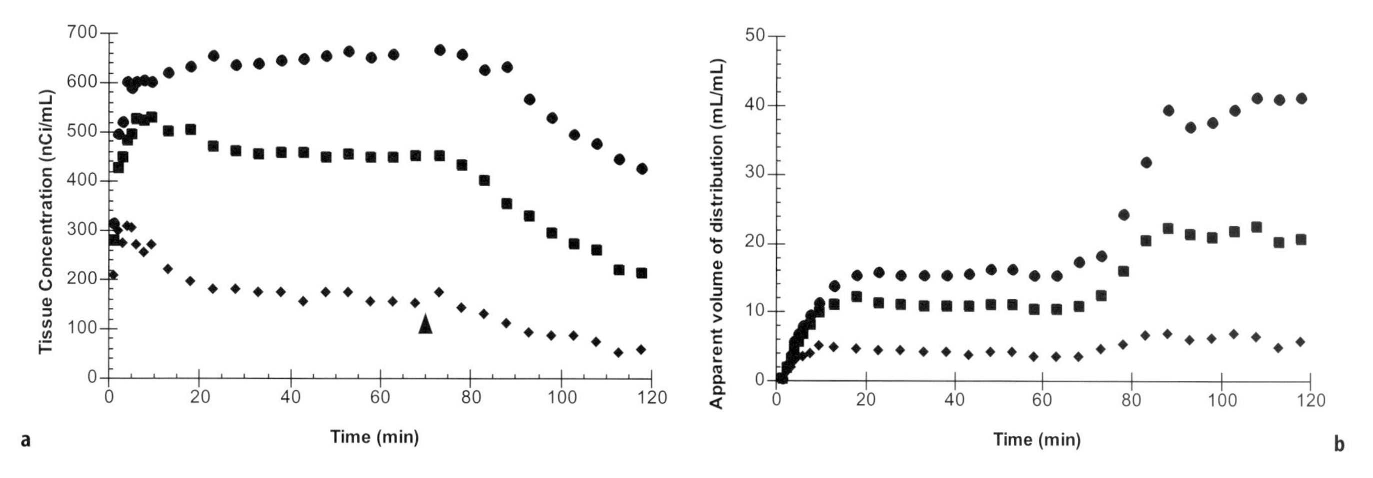

Figure 6.12. Effect of plasma clearance on tissue concentration and apparent volume of distribution (ratio of tissue to metabolite-corrected plasma). **a**: Tissue time–activity data for thalamus (●), frontal cortex (■), and cerebellum (♦). [^{18}F]cyclofoxy was administered according to a bolus/infusion protocol, but the infusion was discontinued at 70 min (arrow). **b**: apparent volume of distribution for regions in "a". There is a dramatic increase in apparent distribution volume due to increased plasma clearance beginning at 70 min. See text for additional details.

Another choice to be made with this type of study is whether tracer administered by continuous infusion is a good idea [84, 134]. With infusions, data analysis is greatly simplified, since the volume of distribution can be obtained directly from the ratio of tissue radioactivity to metabolite-corrected blood. Scans need only be collected during the equilibrium period providing more patient comfort. Fewer measurements in blood are required. Also, the technique is model-independent and only relies on equilibrium conditions. However, if true equilibrium is not obtained, errors that could have been eliminated by a more complex modeling procedure will occur. Due to normal variation in plasma clearance rates of the tracer, deviations from equilibrium will add variability to the results, although biases here will be smaller than those following bolus injections [204]. The time interval corresponding to true equilibrium must be carefully assessed and ideally verified in each subject. There are also increased logistical requirements due to a long infusion of radioactivity compared to a simple bolus injection. It is also not at all clear whether bolus or infusion approaches provide better statistical quality in the final physiological measurements.

Does the use of model-based methods improve the signal-to-noise characteristics of data? In other words, can small biological signals be detected more easily by using modeling methodology? Use of appropriate quantification methodology can reduce intersubject variability by accounting for factors affecting the raw concentration measurements that are unrelated to the physiological measure of interest. If inter-subject variability is decreased, the power of the study to detect group differences is typically increased. However, if this extraneous variability is small, then use of a model-based method may produce little improvement in the signal-to-noise ratio. In fact, since there are a large number of potential sources of error in applying modeling techniques, errors in these corrections or in the implementation of these procedures can actually increase variability over simpler, empirical methods. The net effect of applying a model on measurement variability thus depends upon the magnitude of physiological variation in the patient groups that can be removed by the model versus the accuracy of the model and the reliability of the additional measurements that it requires.

Model-based methods have one important advantage over empirical approaches. With model-based results, it is easier to justify the conclusion that any significant findings are in fact due to real differences in the biological function of interest and not due to extraneous physiological factors. When empirical methods detect significant differences, these other physiological factors may contribute substantially to the measured differences. Thus, interpretation of the results is less straightforward. This is particularly true when there are known differences in physiology between subject groups in a study. For example, if plasma tracer clearance differs between patients and control subjects, substantial errors may be made if tissue radioactivity values are directly interpreted as reflecting the relevant physiological process. On these grounds, model-based methods, which usually require a more complicated study procedure, are superior to empirical approaches. It is important, however, to remember that model-based methods rely on many assumptions, which can produce misleading results when applied inappropriately.

Summary

This chapter discussed the use of mathematical models to extract physiological information from PET studies with radioactive tracers. Modeling methods offer a number of advantages. Application of a model can explain to what extent the tissue radioactivity measurements reflect the physiological function of interest. It can produce quantitative estimates of one or more physiological parameters. Use of a model can explain the cause of different levels of uptake between subjects. It may improve the signal-to-noise characteristics of the data by removing additional variation caused by extraneous physiological factors. The application of modeling methodology also has disadvantages. Usually, modeling procedures are more complex, often requiring longer scanning sessions, blood sampling, metabolite analyses, and complex data processing. Violations in the assumptions made by models can produce misleading results.

Validation studies can demonstrate that a model-based method accurately measures the parameter(s) of interest and is not influenced by other factors. The understanding provided by a model allows the development of study procedures that maximize sensitivity to key parameters and minimize the effects of violations in model assumptions. Ideally, the understanding provided by the model will allow the design of a simple straightforward study procedure. In that way, the radiopharmaceutical can be applied to the appropriate patient groups without a complex procedure while still generating an accurate regional physiological assay. The final configuration of a model-based method may be as simple as an empirical technique but as accurate as a more complex study procedure.

It is essential to have a good understanding of the relationship between the tissue measurements and the underlying physiology, i.e., a model. A useful model will provide a mathematical description that is sufficient to predict the tracer's physiology and biochemistry within the limitations of available instrumentation and the logistics of a practical patient procedure. In addition, the assumptions and limitations of the technique must be clearly delineated. Without a model, it is difficult to assess how physiological differences between study populations affect an empirical method. Ideally, use of a model will significantly improve the physiological significance of the resulting data and may also improve the sensitivity of the tracer to the underlying physiological processes under study.

References

1. Sokoloff L, Reivich M, Kennedy C, Des Rosiers MH, Patlak CS, Pettigrew KD, et al. The [^{14}C]deoxyglucose method for the measurement of local cerebral glucose utilization; theory, procedure, and normal values in the conscious and anesthetized albino rat. J Neurochem 1977;28:897–916.
2. Reivich M, Kuhl D, Wolf A, Greenberg J, Phelps M, Ido T, et al. The [^{18}F]fluorodeoxyglucose method for the measurement of local cerebral glucose utilization in man. Circ Res 1979;44:127–37.
3. Phelps ME, Huang SC, Hoffman EJ, Selin C, Sokoloff L, Kuhl DE. Tomographic measurement of local cerebral glucose metabolic rate in humans with (F-18) 2-fluoro-2-deoxy-D-glucose: Validation of method. Ann Neurol 1979;6:371–88.
4. Huang SC, Phelps ME, Hoffman EJ, Sideris K, Selin CJ, Kuhl DE. Non-invasive determination of local cerebral metabolic rate of glucose in man. Am J Physiol 1980;238:E69–82.
5. Reivich M, Alavi A, Wolf A, Fowler J, Russell J, Arnett C, et al. Glucose metabolic rate kinetic model parameter determination in humans: the lumped constants and rate constants for [^{18}F]fluorodeoxyglucose and [^{11}C]deoxyglucose. J Cereb Blood Flow Metab 1985;5:179–92.
6. Gjedde A, Wienhard K, Heiss WD, Kloster G, Diemer NH, Herholz K, et al. Comparative regional analysis of 2-fluorodeoxyglucose and methylglucose uptake in brain of four stroke patients. With special reference to the regional estimation of the lumped constant. J Cereb Blood Flow Metab 1985;5(2):163–78.
7. Spence A, Graham M, Muzi M, Abbott G, Krohn K, Kapoor R, et al. Deoxyglucose lumped constant estimated in a transplanted rat astrocytic glioma by the hexose utilization index. J Cereb Blood Flow Metab 1990;10:190–8.
8. Dienel GA, Cruz NF, Mori K, Holden JE, Sokoloff L. Direct measurement of the lambda of the lumped constant of the deoxyglucose method in rat brain: determination of lambda and lumped constant from tissue glucose concentration or equilibrium brain/plasma distribution ratio for methylglucose. J Cereb Blood Flow Metab 1991;11(1):25–34.
9. Holden JE, Mori K, Dienel GA, Cruz NF, Nelson T, Sokoloff L. Modeling the dependence of hexose distribution volumes in brain on plasma glucose concentration: implications for estimation of the local 2-deoxyglucose lumped constant. J Cereb Blood Flow Metab 1991;11(2):171–82.
10. Heymann M, Payne B, Hoffman J, Rudolph A. Blood flow measurements with radionuclide-labeled particles. Prog Cardiovasc Dis 1977;20:55–79.
11. Phelps ME, Huang S-C, Hoffman EJ, Selin C, Kuhl DE. Cerebral extraction of N-13 ammonia: Its dependence on cerebral blood flow and capillary permeability – surface area product. Stroke 1981;12:607–19.
12. Neirinckx R, Canning L, Piper I, Nowotnik D, Pickett R, Holmes R, et al. Tc-99m d,1-HM-PAO: A new radiopharmaceutical for SPECT imaging of regional cerebral blood perfusion. J Nucl Med 1987;28:191–202.
13. Kety SS. The theory and applications of the exchange of inert gas at the lungs and tissues. Pharmacol Rev 1951;3:1–41.
14. Zierler KL. Circulation times and the theory of indicator-dilution methods for determining blood flow and volume. In: Handbook of physiology. Baltimore: Waverly Press; 1962. p. 585–615.
15. Lassen NA, Perl W. Tracer kinetic methods in medical physiology. New York: Raven Press; 1979.
16. Carson ER, Cobelli C, Finkelstein L. The mathematical modeling of metabolic and endocrine systems. New York: Wiley; 1983.
17. Lambrecht R, Rescigno A, editors. Tracer kinetics and physiological modeling. Berlin: Springer-Verlag; 1983.
18. Peters A. A unified approach to quantification by kinetic analysis in Nuclear Medicine. J Nucl Med 1993;34:706–13.
19. DiStefano JJ. Non-compartmental vs. compartmental analysis: Some basis for choice. Am J Physiol 1982;243:R1–6.
20. Johnson J, Wilson T. A model for capillary exchange. Am J Physiol 1966;210:1299–303.

21. Bassingthwaighte JB. A concurrent flow model for extraction during transcapillary passage. Circ Res 1974;35:483–503.
22. Bassingthwaighte JB, Holloway GA. Estimation of blood flow with radioactive tracers. Semin Nucl Med 1976;6:141–61.
23. Goresky CA, Ziegler WH, Bach GG. Capillary exchange modeling: Brain-limited and flow-limited distribution. Circ Res 1970;27:739–64.
24. Rose CP, Goresky CA. Constraints on the uptake of labeled palmitate by the heart. Circ Res 1977;41:534–45.
25. Larson KB, Markham J, Raichle ME. Comparison of distributed and compartmental models for analysis of cerebral blood flow measurements. J Cereb Blood Flow Metab 1985;5(Suppl 1):S649–50.
26. Larson KB, Markham J, Raichle ME. Tracer-kinetic models for measuring cerebral blood flow using externally detected radiotracers. J Cereb Blood Flow Metab 1987;7: 443–63.
27. van Osdol W, Sung C, Dedrick R, Weinstein J. A distributed pharmacokinetic model of two-step imaging and treatment protocols using streptavidin-conjugated monoclonal antibodies and radiolabeled biotin. J Nucl Med 1993;34:1552–64.
28. Jacquez JA. Compartmental analysis in biology and medicine. Amsterdam, Holland: Elsevier/North; 1972.
29. Wagner JG. Fundamentals of clinical pharmacokinetics. Hamilton, Ill.: Drug Intelligence Publications; 1975.
30. Anderson D. Compartmental modeling and tracer kinetics. Berlin: Springer-Verlag; 1983.
31. Robertson J, editor. Compartmental distribution of radiotracers. Boca Raton, FL: CRC Press; 1983.
32. Huang SC, Barrio JR, Phelps ME. Neuroreceptor assay with positron emission tomography. J Cereb Blood Flow Metab 1986;6:515–21.
33. Renkin EM. Transport of potassium-42 from blood to tissue in isolated mammalian skeletal muscles. Am J Physiol 1959;197:1205–10.
34. Crone C. Permeability of capillaries in various organs as determined by use of the indicator diffusion method. Acta Physiol Scand 1964;58:292–305.
35. Lehninger A. Biochemistry. New York: Worth Publishers; 1975.
36. Eckelman W, editor. Receptor-binding radiotracers. Boca Raton, FL: CRC Press; 1982.
37. Laruelle M. Imaging synaptic neurotransmission with in vivo binding competition techniques: A critical review. J Cereb Blood Flow Metab 2000;20(3):423–51.
38. Braun M. Differential equations and their applications. New York: Springer-Verlag; 1975.
39. Feng D, Huang S, Wang X. Models for computer simulation studies of input functions for tracer kinetic modeling with positron emission tomography. Int J Biomed Comput 1993;32:95–110.
40. Gear C. Numerical initial value problems in ordinary differential equations. Englewood Cliffs, NJ: Prentice-Hall; 1971.
41. Press W, Flannery B, Teukolsy S, Vetterling W. Numerical Recipes: The art of scientific computing. Cambridge: Cambridge University Press; 1986.
42. Bard Y. Nonlinear parameter estimation: Academic Press, New York; 1974.
43. Beck JV, Arnold KJ. Parameter estimation in engineering and science. New York: John Wiley & Sons; 1977.
44. Carson RE. Parameter estimation in positron emission tomography. In: Phelps ME, Mazziotta JC, Schelbert HR, editors. Positron emission tomography and autoradiography. New York: Raven Press; 1986. p. 347–90.
45. Sorenson JA, Phelps ME. Physics in nuclear medicine. 2nd ed. Orlando: Grune & Stratton; 1987.
46. Budinger TF, Derenzo SE, Greenberg WL, Gullberg GT, Huesman RH. Quantitative potentials of dynamic emission computed tomography. J Nucl Med 1978;19:309–15.
47. Alpert NM, Chesler DA, Correia JA, Ackerman RH, Chang JY, Finklestein S, et al. Estimation of the local statistical noise in emission computed tomography. IEEE Trans Med Imag 1982;1:142–6.
48. Huesman RH. A new fast algorithm for the evaluation of regions of interest and statistical uncertainty in computed tomography. Phys Med Biol 1984;29(5):543–52.
49. Alpert NM, Barker WC, Gelman A, Weise S, Senda M, Correia JA. The precision of positron emission tomography: theory and measurement. J Cereb Blood Flow Metab 1991; 11(2):A26–30.
50. Haynor DR, Harrison RL, Lewellen TK. The use of importance sampling techniques to improve the efficiency of photon tracking in emission tomography simulations. Med Phys 1991; 18(5):990–1001.
51. Carson RE, Yan Y, Daube-Witherspoon ME, Freedman N, Bacharach SL, Herscovitch P. An approximation formula for the variance of PET region-of-interest values. IEEE Trans Med Imag 1993;12:240–50.
52. Pajevic S, Daube-Witherspoon ME, Bacharach SL, Carson RE. Noise characteristics of 3-D and 2-D PET images. IEEE Trans Med Imaging 1998;17(1):9–23.
53. Eastman R, Carson R, Gordon M, Berg G, Lillioja S, Larson S, et al. Brain glucose metabolism in non-insulin-dependent diabetes mellitus: A study in Pima Indians using positron emission tomography during hyperinsulinemia with euglycemic glucose clamp. J Clin Endocrinol Metab 1990;71:1602–10.
54. Holden JE, Gatley SJ, Hichwa RD, Ip WR, Shaughnessy WJ, Nickles RJ, et al. Cerebral blood flow using PET measurements of fluoromethane kinetics. J Nucl Med 1981;22:1084–8.
55. Koeppe RA, Holden JE, Ip WR. Performance comparison of parameter estimation techniques for the quantitation of local cerebral blood flow by dynamic positron computed tomography. J Cereb Blood Flow Metab 1985;5:224–34.
56. Koeppe RA, Holthoff VA, Frey KA, Kilbourn MR, Kuhl DE. Compartmental analysis of [^{11}C]Flumazenil kinetic for the estimation of ligand transport rate and receptor distribution using positron emission tomography. J Cereb Blood Flow Metab 1991;11:735–44.
57. Frey KA, Holthoff VA, Koeppe RA, Jewett DM, Kilbourn MR, Kuhl DE. Parametric in vivo imaging of benzodiazepine receptor distribution in human brain. Ann Neurol 1991;30(5):663–72.
58. Carson RE, Kiesewetter DO, Jagoda E, Der MG, Herscovitch P, Eckelman WC. Muscarinic cholinergic receptor measurements with [18F]FP-TZTP: control and competition studies. J Cereb Blood Flow Metab 1998;18(10):1130–42.
59. Huang S, Carson R, Phelps M. Measurement of local blood flow and distribution volume with short-lived isotopes: A general input technique. J Cereb Blood Flow Metab 1982;2:99–108.
60. Huang S-C, Carson RE, Hoffman EJ, Carson J, MacDonald N, Barrio JR, et al. Quantitative measurement of local cerebral blood flow in humans by positron computed tomography and ^{15}O-water. J Cereb Blood Flow Metab 1983;3:141–53.
61. Alpert NM, Eriksson L, Chang JY, Bergstrom M, Litton JE, Correia JA, et al. Strategy for the measurement of regional cerebral blood flow using short-lived tracers and emission tomography. J Cereb Blood Flow Metab 1984;4:28–34.
62. Blomqvist G. On the construction of functional maps in positron emission tomography. J Cereb Blood Flow Metab 1984;4:629–32.
63. Carson RE, Huang SC, Green MV. Weighted integration method for local cerebral blood flow measurements with positron emission tomography. J Cereb Blood Flow Metab 1986;6(2):245–58.
64. Blomqvist G, Pauli S, Farde L, Eriksson L, Persson A, Halldin C. Maps of receptor binding parameters in human brain – a kinetic analysis of PET measurements. Eur J Nucl Med 1990;16:257–65.
65. Yokoi T, Kanno I, Iida H, Miura S, Uemura K. A new approach of weighted integration technique based on accumulated images using dynamic PET and $H_2{}^{15}O$. J Cereb Blood Flow Metab 1991;11:492–501.
66. Carson RE. PET parameter estimation using linear integration methods: Bias and variability considerations. In: Uemura K, Lassen NA, Jones Y, Kanno I, editors. Quantification of brain function. Tracer kinetics and image analysis in brain PET. Amsterdam: Elsevier Science Publishers; 1993. p. 499–507.
67. Cunningham VJ, Jones T. Spectral analysis of dynamic PET studies. J Cereb Blood Flow Metab 1993;13(1):15–23.
68. Howman-Giles R, Moase A, Gaskin K, Uren R. Hepatobiliary scintigraphy in a pediatric population: Determination of hepatic extraction fraction by deconvolution analysis. J Nucl Med 1993;34:214–21.

69. Huang SC, Phelps ME. Principles of tracer kinetic modeling in positron emission tomography and autoradiography. In: Phelps M, Mazziotta J, Schelbert H, editors. Positron emission tomography and autoradiography: principles and applications for the brain and heart. New York: Raven Press; 1986. p. 287–346.
70. Carson RE. The development and application of mathematical models in nuclear medicine [editorial]. J Nucl Med 1991;32(12):2206–8.
71. Berman M, Schoenfeld R. Invariants in experimental data on linear kinetics and the formulation of models. J Appl Physiol 1956;27:1361–70.
72. Berman M. The formulation of testing models. Ann N Y Acad Sci 1963;108:192–4.
73. Carson ER, Jones EA. Use of kinetic analysis and mathematical modeling in the study of metabolic pathway in vivo. N Engl J Med 1979;300:1016–27.
74. Carson ER, Cobelli C, Finkelstein L. Modeling and identification of metabolic systems. Am J Physiol 1981;240:R120–9.
75. Cobelli C, Ruggerin A. Evaluation of alternative model structures of metabolic systems: Two case studies on model identification and validation. Med Biol Eng Comput 1982;20:444–50.
76. DiStefano J, Landaw E. Multiexponential, multicompartmental, and non-compartmental modeling. I. Methodological limi tations and physiological interpretations. Am J Physiol 1984;246:R651–64.
77. Landaw EW, DiStefano JJ. Multiexponential, multicompartmental and noncompartmental modeling. II. Data analysis and statistical considerations. Am J Physiol 1984;246:R665–77.
78. Vera D, Krohn K, Scheibe P, Stadalnik R. Identifiability analysis of an in vivo receptor-binding radiopharmacokinetic system. IEEE Trans Biomed Eng 1985;32:312–22.
79. Delforge J, Syrota A, Mazoyer BM. Experimental design optimisation: theory and application to estimation of receptor model parameters using dynamic positron emission tomography. Phys Med Biol 1989;34(4):419–35.
80. Delforge J, Syrota A, Mazoyer BM. Identifiability analysis and parameter identification of an in vivo ligand-receptor model from PET data. IEEE Trans Biomed Eng 1990;37(7):653–61.
81. Wong DR, Gjedde A, Wagner HM. Quantification of neuroreceptors in the living human brain. I. Irreversible binding of ligands. J Cereb Blood Flow Metab 1986;6:137–46.
82. Watabe H, Channing MA, Der MG, Adams HR, Jagoda E, Herscovitch P, et al. Kinetic analysis of the 5-HT2A ligand [C-11]MDL 100,907. J Cereb Blood Flow Metab 2000;20(6):899–909.
83. Salmon E, Brooks DJ, Leenders KL, Turton DR, Hume SP, Cremer JE, et al. A two-compartment description and kinetic procedure for measuring regional cerebral [^{11}C]nomifensine uptake using positron emission tomography. J Cereb Blood Flow Metab 1990;10:307–16.
84. Carson RE, Channing MA, Blasberg RG, Dunn BB, Cohen RM, Rice KC, et al. Comparison of bolus and infusion methods for receptor quantitation: Application to [^{18}F]-cyclofoxy and positron emission tomography. J Cereb Blood Flow Metab 1993;13:24–42.
85. Akaike H. An information criterion (AIC). Math Sci 1976;14:5–9.
86. Schwarz G. Estimating the dimension of a model. Ann Stat 1978;6:461–4.
87. Theodore WH, Carson RE, Andreason P, Zametkin A, Blasberg R, Leiderman DB, et al. PET imaging of opiate receptor binding in human epilepsy using [^{18}F]cyclofoxy. Epilepsy Res 1992;13:129–39.
88. Cohen RM, Andreason PJ, Doudet DJ, Carson RE, Sunderland T. Opiate receptor avidity and cerebral blood flow in Alzheimer's disease. J Neurol Sci 1997;148(2):171–80.
89. Kling MA, Carson RE, Borg L, Zametkin A, Matochik JA, Schluger J, et al. Opioid receptor imaging with positron emission tomography and [(18)F]cyclofoxy in long-term, methadone-treated former heroin addicts. J Pharmacol Exp Ther 2000;295(3):1070–6.
90. Gjedde A, Reith J, Dyve S, Leger G, Guttman M, Diksic M, et al. Dopa decarboxylase activity of the living human brain. Proc Natl Acad Sci USA 1991;88(7):2721–5.
91. Kuwabara H, Evans AC, Gjedde A. Michaelis-Menten constraints improved cerebral glucose metabolism and regional lumped constant measurements with [18F]fluorodeoxyglucose. J Cereb Blood Flow Metab 1990;10(2):180–9.
92. Kuwabara H, Cumming P, Reith J, Leger G, Diksic M, Evans AC, et al. Human striatal L-dopa decarboxylase activity estimated in vivo using 6-[18F]fluoro-dopa and positron emission tomography: error analysis and application to normal subjects. J Cereb Blood Flow Metab 1993;13(1):43–56.
93. Shoghi-Jadid K, Huang SC, Stout DB, Yee RE, Yeh EL, Farahani KF, et al. Striatal kinetic modeling of FDOPA with a cerebellar-derived constraint on the distribution volume of 3OMFD: A PET investigation using non-human primates. J Cereb Blood Flow Metab 2000;20(7):1134–48.
94. Frost JJ, Douglass DH, Mayberg HS, Dannals RF, Links JM, Wilson AA, et al. Multicompartmental analysis of [^{11}C]-carfentanil binding to opiate receptors in humans measured by positron emission tomography. J Cereb Blood Flow Metab 1989;9:398–409.
95. Farde L, Eriksson L, Blomquist G, Halldin C. Kinetic analysis of central [^{11}C]raclopride binding to D_2-dopamine receptors studied by PET: A comparison to the equilibrium analysis. J Cereb Blood Flow Metab 1989;9:696–708.
96. Carson RE, Blasberg RG, Channing MA, Yolles PS, Dunn BB, Newman AH, et al. A kinetic study of the active and inactive enantiomers of ^{18}F-cyclofoxy with PET. J Cereb Blood Flow Metab 1989;9:S16.
97. Delforge J, Syrota A, Bottlaender M, Varastet M, Loc'h C, Bendriem B, et al. Modeling analysis of [11-C]flumazenil kinetics studied by PET: Application to a critical study of the equilibrium approaches. J Cereb Blood Flow Metab 1993;13:454–68.
98. Price J, Mayberg H, Dannals R, Wilson A, Ravert H, Sadzot B, et al. Measurement of benzodiazepine receptor number and affinity in humans using tracer kinetic modeling, positron emission tomography, and [11C]flumazenil. J Cereb Blood Flow Metab 1993;13:656–67.
99. Parsey RV, Slifstein M, Hwang DR, Abi-Dargham A, Simpson N, Mawlawi O, et al. Validation and reproducibility of measurement of 5-HT1A receptor parameters with [carbonyl-C-11]WAY-100635 in humans: Comparison of arterial and reference tissue input functions. J Cereb Blood Flow Metab 2000;20(7):1111–33.
100. Holthoff VA, Koeppe RA, Frey KA, Paradise AH, Kuhl DE. Differentiation of radioligand delivery and binding in the brain: validation of a two-compartment model for [11C]flumazenil. J Cereb Blood Flow Metab 1991;11(5):745–52.
101. Carson RE, Breier A, de Bartolomeis A, Saunders RC, Su TP, Schmall B, et al. Quantification of amphetamine-induced changes in [^{11}C]raclopride binding with continuous infusion. J Cereb Blood Flow Metab 1997;17(4):437–47.
102. Koeppe RA, Frey KA, Kume A, Albin R, Kilbourn MR, Kuhl DE. Equilibrium versus compartmental analysis for assessment of the vesicular monoamine transporter using (+)-alpha-[c- 11] dihydrotetrabenazine (dtbz) and positron emission tomography. J Cereb Blood Flow Metab 1997;17(9):919–31.
103. Ito H, Hietala J, Blomqvist G, Halldin C, Farde L. Comparison of the transient equilibrium and continuous infusion method for quantitative PET analysis of [C- 11]raclopride binding. J Cereb Blood Flow Metab 1998;18(9):941–50.
104. Nelson T, Lucignani G, Goochee J, Crane AM, Sokoloff L. Invalidity of criticisms of the deoxyglucose method based on alleged glucose-6-phosphatase activity in brain. J Neurochem 1986;46:905–19.
105. Benveniste H. Brain microdialysis. J Neurochem 1989;52:1667–79.
106. Carson RE. Precision and accuracy considerations of physiological quantitation in PET. J Cereb Blood Flow Metab 1991;11:A45–50.
107. Gjedde A. High- and low- affinity transport of D-glucose from blood to brain. J Neurochem 1981;36:1463–71.
108. Patlak CS, Blasberg RG, Fenstermacher JD. Graphical evaluation of blood-to-brain transfer constants from multiple-time uptake data. J Cereb Blood Flow Metab 1983;3:1–7.

109. Patlak CS, Blasberg RG. Graphical evaluation of blood-to-brain transfer constants from multiple-time uptake data. Generalizations. J Cereb Blood Flow Metab 1985;5:584–90.
110. Martin W, Palmer M, Patlak C, Calne D. Nigrostriatal function in humans studied with positron emission tomography. Ann Neurol 1989;20:535–42.
111. Logan J, Fowler JS, Volkow ND, Wolf AP, Dewey SL, Schyler DJ, et al. Graphical analysis of reversible radioligand binding from time-activity measurements applied to [N-^{11}C-methyl]-(-)-Cocaine: PET studies in human subjects. J Cereb Blood Flow Metab 1990;10:740–7.
112. Yokoi T, Iida H, Itoh H, Kanno I. A new graphic plot analysis for cerebral blood flow and partition coefficient with Iodine-123-iodoamphetamine and dynamic SPECT validation studies using oxygen-15-water and PET. J Nucl Med 1993;34:498–505.
113. Choi Y, Hawkins RA, Huang SC, Gambhir SS, Brunken RC, Phelps ME, et al. Parametric images of myocardial metabolic rate of glucose generated from dynamic cardiac PET and 2-[18F]fluoro-2-deoxy-d-glucose studies. J Nucl Med 1991;32(4):733–8.
114. [Shoaf SE, Carson RE, Hommer D, Williams WA, Higley JD, Schmall B, et al. The suitability of [C-11]-alpha-methyl-L-tryptophan as a tracer for serotonin synthesis: Studies with dual administration of [C-11] and [C-14] labeled tracer. J Cereb Blood Flow Metab 2000;20(2):244–52.
115. Carson RE, Lan LX, Watabe H, Der MG, Adams HR, Jagoda E, et al. PET evaluation of [F-18]FCWAY, an analog of the 5-HT1A receptor antagonist, WAY-100635. Nucl Med Biol 2000;27(5):493–7.
116. Weinberg I, Huang S, Hoffman E, Araujo L, Nienaber C, Grover-McKay M, et al. Validation of PET-acquired input functions for cardiac studies. J Nucl Med 1988;29:241–7.
117. Iida H, Rhodes CG, de Silva R, Araujo LI, Bloomfield PM, Lammertsma AA, et al. Use of the left ventricular time-activity curve as a noninvasive input function in dynamic oxygen-15-water positron emission tomography. J Nucl Med 1992;33(9):1669–77.
118. Germano G, Chen BC, Huang SC, Gambhir SS, Hoffman EJ, Phelps ME. Use of the abdominal aorta for arterial input function determination in hepatic and renal PET studies. J Nucl Med 1992;33(4):613–20.
119. Wu HM, Hoh CK, Choi Y, Schelbert HR, Hawkins RA, Phelps ME, et al. Factor-analysis for extraction of blood time-activity curves in dynamic FDG-PET studies. J Nucl Med 1995;36(9):1714–22.
120. Green LA, Gambhir SS, Srinivasan A, Banerjee PK, Hoh CK, Cherry SR, et al. Noninvasive methods for quantitating blood time-activity curves from mouse PET images obtained with fluorine-18-fluorodeoxyglucose. J Nucl Med 1998;39(4): 729–34.
121. Chen K, Bandy D, Reiman E, Huang SC, Lawson M, Feng D, et al. Noninvasive quantification of the cerebral metabolic rate for glucose using positron emission tomography, F-18-fluoro-2-deoxyglucose, the Patlak method, and an image-derived input function. J Cereb Blood Flow Metab 1998;18(7):716–23.
122. Hume SP, Myers R, Bloomfield PM, Opacka JJ, Cremer JE, Ahier RG, et al. Quantitation of carbon-11-labeled raclopride in rat striatum using positron emission tomography. Synapse 1992;12(1):47–54.
123. Lammertsma AA, Hume SP. Simplified reference tissue model for PET receptor studies. Neuroimage 1996;4:153–8.
124. Watabe H, Hatazawa J, Ishiwata K, Ido T, Itoh M, Iwata R, et al. Linearized method – a new approach for kinetic analysis of central dopamine D-2 receptor-specific binding. IEEE Trans Med Imag 1995;14(4):688–96.
125. Gunn RN, Lammertsma AA, Hume SP, Cunningham VJ. Parametric imaging of ligand-receptor binding in PET using a simplified reference region model. Neuroimage 1997;6(4):279–87.
126. Ichise M, Ballinger JR, Golan H, Vines D, Luong A, Tsao S, et al. Noninvasive quantification of dopamine D2 receptors with Iodine-123-IBF SPECT. J Nucl Med 1996;37:513–20.
127. Logan J, Fowler JS, Volkow ND, Wang GJ, Ding YS, Alexoff DL. Distribution volume ratios without blood sampling from graphical analysis of PET data. J Cereb Blood Flow Metab 1996;16(5):834–40.
128. Frackowiak RSJ, Lenzi G-L, Jones T, Heather JD. Quantitative measurement of regional cerebral blood flow and oxygen metabolism in man using ^{15}O and positron emission tomography: Theory, procedure and normal values. J Comput Assist Tomogr 1980;4:727–36.
129. Herscovitch P, Markham J, Raichle ME. Brain blood flow measured with intravenous $H_2{}^{15}O$. I. Theory and error analysis. J Nucl Med 1983;24:782–9.
130. Raichle ME, Martin WRW, Herscovitch P, Mintun MA, Markham J. Brain blood flow measured with intravenous $H_2{}^{15}O$. II. Implementation and validation. J Nucl Med 1983;24:790–8.
131. Brooks RA. Alternative formula for glucose utilization using labeled deoxyglucose. J Nucl Med 1982;23:538–9.
132. Hutchins GD, Holden JE, Koeppe RA, Halama JR, Gatley SJ, Nickles RJ. Alternative approach to single-scan estimation of cerebral glucose metabolic rate using glucose analogs, with particular application to ischemia. J Cereb Blood Flow Metab 1984;4:35–40.
133. Wilson PD, Huang SC, Hawkins RA. Single-scan Bayes estimation of cerebral glucose metabolic rate: comparison with non-Bayes single-scan methods using FDG PET scans in stroke. J Cereb Blood Flow Metab 1988;8(3):418-25.
134. Frey KA, Ehrenkaufer RLE, Beaucage S, Agranoff BW. Quantitative in vivo receptor binding. I. Theory and application to the muscarinic cholinergic receptor. J Neurosci 1985;5:421–8.
135. Laruelle M, Abi-Dargham A, Rattner Z, Al-Tikriti M, Zea-Ponce Y, Zoghbi S, et al. Single photon emission tomography measurement of benzodiazepine receptor number and affinity in primate brain: a constant infusion paradigm with [^{123}I]iomazenil. Eur J Pharmacol 1993;230:119--23.
136. Mintun MA, Raichle ME, Kilbourn MR, Wooton GF, Welch MJ. A quantitative model for the in vivo assessment of drug binding sites with positron emission tomography. Ann Neurol 1984;15:217–27.
137. Kawai R, Carson RE, Dunn B, Newman AH, Rice KC, Blasberg RG. Regional brain measurement of B_{max} and K_D with the opiate antagonist cyclofoxy: Equilibrium studies in the conscious rat. J Cereb Blood Flow Metab 1991;11(4):529–44.
138. Carson RE, Doudet DJ, Channing MA, Dunn BB, Der MG, Newman AH, et al. Equilibrium measurement of B_{max} and K_D of the opiate antagonist ^{18}F-cyclofoxy with PET: Pixel-by-pixel analysis. J Cereb Blood Flow Metab 1991;11:S618.
139. Farde L, Hall H, Ehrin E, Sedvall G. Quantitative analysis of D2 dopamine receptor binding in the living human brain by PET. Science 1986;231:258–61.
140. Dewey SL, Smith GS, Logan J, Brodie JD, Fowler JS, Wolf AP. Striatal binding of the PET ligand ^{11}C-raclopride is altered by drugs that modify synaptic dopamine levels. Synapse 1993;13:350–6.
141. Dewey SL, Smith GS, Logan J, Brodie JD, Yu DW, Ferrieri RA, et al. GABAergic inhibition of endogenous dopamine release measured in vivo with 11C-raclopride and positron emission tomography. J Neurosci 1992;12(10):3773–80.
142. Dewey SL, Smith GS, Logan J, Alexoff D, Ding YS, King P, et al. Serotonergic modulation of striatal dopamine measured with positron emission tomography (PET) and in vivo microdialysis. J Neurosci 1995;15:821–9.
143. Nordstrom AL, Farde L, Halldin C. Time course of D2-dopamine receptor occupancy examined by PET after single oral doses of haloperidol. Psychopharmacology 1992;106:433–8.
144. Nyberg S, Farde L, Eriksson L, Halldin C, Eriksson B. 5-HT_2 and D_2 dopamine receptor occupancy in the living human brain. Psychopharmacology 1993;110(3):265–72.
145. Farde L, Nordström A-L, Wiesel FA, Pauli S, Halldin C, Sedvall G. Positron emission tomography analysis of central D1 and D2 dopamine receptor occupancy in patients being treated with classic neuroleptic and clozapine. Arch Gen Psych 1992;49:538–44.
146. Fischman AJ, Bonab AA, Babich JW, Alpert NM, Rauch SL, Elmaleh DR, et al. Positron emission tomographic analysis of central 5- hydroxytryptamine(2) receptor occupancy in healthy volunteers treated with the novel antipsychotic agent, ziprasidone. J Pharmacol Exp Ther 1996;279(2):939–47.

147. Breier A, Su T-P, Saunders R, Carson RE, Kolachana BS, de Bartolomeis A, et al. Schizophrenia is associated with elevated amphetamine-induced synaptic dopamine concentrations: Evidence from a novel positron emission tomography method. Proc Natl Acad Sci USA 1997;94(6):2569–74.
148. Laruelle M, Abi-Dargham A, vanDyck CH, Rosenblatt W, Zea-Ponce Y, Zoghbi SS, et al. SPECT imaging of striatal dopamine release after amphetamine challenge. J Nucl Med 1995;36(7):1182–90.
149. Morris ED, Fisher RE, Alpert NM, Rauch SL, Fischman AJ. In vivo imaging of neuromodulation using positron emission tomography - optimal ligand characteristics and task length for detection of activation. Human Brain Mapping 1995;3(1):35–55.
150. Muzic RF, Nelson AD, Saidel GM, Miraldi F. Optimal experiment design for PET quantification of receptor concentration. IEEE Trans Med Imag 1996;15(1):2-12.
151. Huang SC, Zhou Y. Spatially coordinated regression for image-wise model fitting to dynamic PET data for generating parametric images. IEEE Trans Nucl Sci 1998;45(3):1194–9.
152. Watabe H, Endres CJ, Breier A, Schmall B, Eckelman WC, Carson RE. Measurement of dopamine release with continuous infusion of [11C]raclopride: optimization and signal-to-noise considerations. J Nucl Med 2000;41(3):522–30.
153. Slifstein M, Laruelle M. Effects of statistical noise on graphic analysis of PET neuroreceptor studies. J Nucl Med 2000;41(12):2083–8.
154. Hoffman EJ, Huang S-C, Phelps ME. Quantitation in positron emission computed tomography: 1. Effect of object size. J Comput Assist Tomogr 1979;3:299–308.
155. Herscovitch P, Raichle ME. Effect of tissue heterogeneity on the measurement of cerebral blood flow with the equilibrium $C^{15}O_2$ inhalation technique. J Cereb Blood Flow Metab 1983;3:407–15.
156. Herscovitch P, Raichle ME. Effect of tissue heterogeneity on the measurement of regional cerebral oxygen extraction and metabolic rate with positron emission tomography. J Cereb Blood Flow Metab 1985;5 (Suppl 1):S671–2.
157. Herholz K, Patlak CS. The influence of tissue heterogeneity on results of fitting nonlinear model equations to regional tracer uptake curves: with an application to compartmental models used in positron emission tomography. J Cereb Blood Flow Metab 1987;7:214–29.
158. Huang SC, Mahoney DK, Phelps ME. Quantitation in positron emission tomography: 8. Effects of nonlinear parameter estimation on functional images. J Comput Assist Tomogr 1987;11(2):314–25.
159. Schmidt K, Mies G, Sokoloff L. Model of kinetic behavior of deoxyglucose in heterogeneous tissues in brain: a reinterpretation of the significance of parameters fitted to homogeneous tissue models. J Cereb Blood Flow Metab 1991;11:10–24.
160. Schmidt K, Lucignani G, Moresco R, Rizzo G, Gilardi M, Messa C, et al. Errors introduced by tissue heterogeneity in estimation of local cerebral glucose utilization with current kinetic models of the [^{18}F]fluorodeoxyglucose method. J Cereb Blood Flow Metab 1992;12:823–34.
161. Blomqvist G, Lammertsma AA, Mazoyer B, Wienhard K. Effect of tissue heterogeneity on quantification in positron emission tomography. Eur J Nucl Med 1995;22(7):652–63.
162. Muller-Gartner HW, Links JM, Prince JL, Bryan RN, McVeigh E, Leal JP, et al. Measurement of radiotracer concentration in brain gray matter using positron emission tomography: MRI-based correction for partial volume effects. J Cereb Blood Flow Metab 1992;12(4):571–83.
163. Meltzer CC, Zubieta JK, Links JM, Brakeman P, Stumpf MJ, Frost JJ. MR-based correction of brain PET measurements for heterogeneous gray matter radioactivity distribution. J Cereb Blood Flow Metab 1996;16(4):650–8.
164. Meltzer CC, Kinahan PE, Greer PJ, Nichols TE, Comtat C, Cantwell MN, et al. Comparative evaluation of MR-based partial-volume correction schemes for PET. J Nucl Med 1999;40(12):2053–65.
165. Labbe C, Koepp M, Ashburner J, Spinks T, Richardson M, Duncan J, et al. Absolute PET quantification with correction for partial volume effects within cerebral structures. In: Carson RE, Daube-Witherspoon ME, Herscovitch P, editors. Quantitative functional brain imaging with positron emission tomography. San Diego, CA: Academic Press; 1998. p. 67–76.
166. Rousset OG, Ma Y, Evans AC. Correction for partial volume effects in PET: principle and validation. J Nucl Med 1998;39(5):904–11.
167. Rousset OG, Deep P, Kuwabara H, Evans AC, Gjedde AH, Cumming P. Effect of partial volume correction on estimates of the influx and cerebral metabolism of 6-[(18)F]fluoro-L-dopa studied with PET in normal control and Parkinson's disease subjects. Synapse 2000;37(2):81–9.
168. Lammertsma AA, Jones T. Correction for the presence of intravascular oxygen-15 in the steady state technique for measuring regional oxygen extraction ratio in the brain: 1. Description of the method. J Cereb Blood Flow Metab 1983;13:416–24.
169. Evans AC, Diksic M, Yamamoto YL, Kato A, Dagher, Redies C, et al. Effect of vascular activity in the determination of rate constants for the uptake of ^{18}F-labeled 2-fluoro-2-deoxy-D-glucose: error analysis and normal values in older subjects. J Cereb Blood Flow Metab 1986;6:724–38.
170. Hawkins RA, Phelps ME, Huang SC. Effects of temporal sampling, glucose metabolic rates, and disruptions of the blood–brain barrier on the FDG model with and without a vascular compartment: studies in human brain tumors with PET. J Cereb Blood Flow Metab 1986;6(2):170–83.
171. Koeppe RA, Hutchins GD, Rothley JM, Hichwa RD. Examination of assumptions for local cerebral blood flow studies in PET. J Nucl Med 1987;28(11):1695–703.
172. Iida H, Kanno I, Takahashi A, Miura S, Murakami M, Takahashi K, et al. Measurement of absolute myocardial blood flow with $H_2{}^{15}O$ and dynamic positron emission tomography. Strategy for quantification in relation to the partial-volume effect. Circulation 1988;78(1):104–15.
173. Herrero P, Markham J, Shelton ME, Weinheimer CJ, Bergmann SR. Noninvasive quantification of regional myocardial perfusion with rubidium-82 and positron emission tomography. Exploration of a mathematical model. Circulation 1990;82(4):1377–86.
174. Hutchins GD, Caraher JM, Raylman RR. A region of interest strategy for minimizing resolution distortions in quantitative myocardial PET studies. J Nucl Med 1992;33(6):1243–50.
175. Eriksson L, Kanno I. Blood sampling devices and measurements. Med Prog Technol 1991;17(3-4):249–57.
176. Dhawan V, Conti J, Mernyk M, Jarden JO, Rottenberg DA. Accuracy of PET RCBF measurements: Effect of time shift between blood and brain radioactivity curves. Phys Med Biol 1986;31:507–14.
177. Dhawan V, Jarden JO, Strother S, Rottenberg DA. Effect of blood curve smearing on the accuracy of parameter estimates for [82-Rb] PET studies of blood–brain barrier permeability. Phys Med Biol 1988;33(1):61–74.
178. Iida H, Kanno I, Miura S, Murakami M, Takahashi K, Uemura K. Evaluation of regional differences of tracer appearance time in cerebral tissues using [^{15}O]water and dynamic positron emission tomography. J Cereb Blood Flow Metab 1988;8:285–8.
179. Meyer E. Simultaneous correction for tracer arrival delay and dispersion in CBF measurements by the $H_2{}^{15}O$ autoradiographic method and dynamic PET. J Nucl Med 1989;30:1069–78.
180. Huesman RH, Mazoyer BM. Kinetic data analysis with a noisy input function. Phys Med Biol 1987;32(12):1569–79.
181. Chen KW, Huang SC, Yu DC. The effects of measurement errors in plasma radioactivity curve on parameter estimation in positron emission tomography. Phys Med Biol 1991;36(9):1183–200.
182. Markham J, Schuster DP. Effects of non-ideal input functions on PET measurements of pulmonary blood flow. J Appl Physiol 1992;72(6):2495–500.
183. Feng D, Wang X. A computer simulation study on the effects of input function measurement noise in tracer kinetic modeling with positron emission tomography. Comput Biol Med 1993;23:57–68.
184. Huang SC, Barrio JR, Yu DC, Chen B, Grafton S, Melega WP, et al. Modelling approach for separating blood time–activity curves in

positron emission tomographic studies. Phys Med Biol 1991;36(6):749–61.
185. Burger C, Buck A. Tracer kinetic modeling of receptor data with mathematical metabolite correction. Eur J Nucl Med 1996;23(5):539–45.
186. Huang S-C, Phelps ME, Hoffman EJ, Kuhl DE. A theoretical study of quantitative flow measurements with constant infusion of short-lived isotopes. Phys Med Biol 1979;24:1151–61.
187. Huang S-C, Phelps ME, Hoffman EJ, Kuhl DE. Error sensitivity of fluorodeoxyglucose method for measurement of cerebral metabolic rate of glucose. J Cereb Blood Flow Metab 1981;1: 391–401.
188. Lammertsma AA, Jones T, Frackowiak RSJ, Lenzi G-L. A theoretical study of the steady-state model for measuring regional cerebral blood flow and oxygen utilization using oxygen-15. J Comput Assist Tomogr 1981;5:544–50.
189. Lammertsma AA, Heather JD, Jones T, Frackowiak RSJ, Lenzi G-L. A statistical study of the steady state technique for measuring regional cerebral blood flow and oxygen utilization using ^{15}O. J Comput Assist Tomogr 1982;6:566–73.
190. Brownell G, Kearfott K, Kairentoi A, Elmaleh D, Alpert N, Correia J, et al. Quantitation of regional cerebral glucose metabolism. J Comput Assist Tomogr 1983;7:919.
191. Wienhard K, Pawlik G, Herholz K, Wagner R, Heiss W-D. Estimation of local cerebral glucose utilization by positron emission tomography of $[^{18}F]$2-fluoro-2-deoxy-D-glucose: A critical appraisal of optimization procedures. J Cereb Blood Flow Metab 1985;5:115–25.
192. Huang S-C, Feng D, Phelps ME. Model dependency and estimation reliability in measurement of cerebral oxygen utilization rate with oxygen-15 and dynamic positron emission tomography. J Cereb Blood Flow Metab 1986;6:105–19.
193. Iida H, Kanno I, Miura S, Murakami M, Takahashi K, Uemura K. Error analysis of a quantitative cerebral blood flow measurement using $H_2{}^{15}O$ autoradiography and positron emission tomography with respect to the dispersion of the input function. J Cereb Blood Flow Metab 1986;6:536–45.
194. Jagust WJ, Budinger TF, Huesman RH, Friedland RP, Mazoyer BM, Knittel BL. Methodologic factors affecting PET measurements of cerebral glucose metabolism. J Nucl Med 1986;27:1358–61.
195. Senda M, Buxton RB, Alpert NM, Correia JA, Mackay BC, Weise SB, et al. The ^{15}O steady-state method: correction for variation in arterial concentration. J Cereb Blood Flow Metab 1988;8:681–90.
196. Millet P, Delforge J, Pappata S, Syrota A, Cinotti L. Error analysis on parameter estimates in the ligand-receptor model – application to parameter imaging using PET data. Phys Med Biol 1996;41(12):2739–56.
197. Lammertsma AA, Cunningham VJ, Deiber MP, Heather JD, Bloomfield PM, Nutt J, et al. Combination of dynamic and integral methods for generating reproducible functional CBF images. J Cereb Blood Flow Metab 1990;10:675–86.
198. Mazoyer BM, Huesman RH, Budinger TF, Knittel BL. Dynamic PET data analysis. J Comput Assist Tomogr 1986;10(4):645–53.
199. Jovkar S, Evans AC, Diksic M, Nakai H, Yamamoto YL. Minimisation of parameter estimation errors in dynamic PET: choice of scanning schedules. Phys Med Biol 1989;34(7):895–908.
200. Kanno I, Iida H, Miura S, Murakami M. Optimal scan time of oxygen-15-labeled water injection method for measurement of cerebral blood flow. J Nucl Med 1991;32:1931–4.
201. Hoshon K, Feng DG, Hawkins RA, Meikle S, Fulham MJ, Li XJ. Optimized sampling and parameter estimation for quantification in whole-body PET. IEEE Trans Biomed Eng 1996;43(10):1021–8.
202. Fox PT, Mintun MA, Raichle ME, Herscovitch P. A noninvasive approach to quantitative functional brain mapping with $H_2{}^{15}O$ and positron emission tomography. J Cereb Blood Flow Metab 1984;4:329–33.
203. Mazziotta JC, Huang SC, Phelps ME, Carson RE, MacDonald NS, Mahoney K. A noninvasive positron computed tomography technique using oxygen-15-labeled water for the evaluation of neurobehavioral task batteries. J Cereb Blood Flow Metab 1985;5(1):70–8.
204. Carson RE. PET physiological measurements using constant infusion. Nucl Med Biol 2000;27(7):657–860.

7 Coregistration of Structural and Functional Images*

David J Hawkes, Derek LG Hill, Lucy Hallpike and Dale L Bailey

Introduction – Why Register Images?

Images are spatial distributions of information. Accurately relating information from several images requires image registration. Alignment of a PET image with a high-resolution image such as a Magnetic Resonance (MR) image has successfully allowed anatomical or structural context to be inferred from the coarser-resolution PET image. PET to MRI registration was one of the earliest successful examples of image registration to find widespread application. Since then, image registration has become a major area of research in medical imaging, spawning a wide range of applications and a large number of papers in the medical and scientific literature. Recent reviews are provided in Maintz et al. [1] and Hill et al. [2]. Much of this chapter is a summary of information in the latter plus a recent textbook on image registration [3]. Further algorithmic and implementation details are contained in these two sources.

This chapter addresses the software approach to image registration. The first section classifies registration applications and outlines the process of registration. It then discusses some concepts of correspondence inherent in image registration and summarises frequently used transformations. Methods for aligning images based on landmarks or geometric features and recent advances using the statistics of image intensities directly to align images – the so-called "voxel similarity" measures – are described. Some details are given on image preparation, optimization, image sampling, common pitfalls and validation.

A Classification of Registration Applications

Image registration applications divide into:

(a) *Intra-subject registration* – those that require registration of images taken of the same individual, and
(b) *Inter-subject registration* – those that relate information between subjects.

Examples of the former can be classified as follows:

(i) *Multi-modality registration*, where several medical images are taken of the same part of the human anatomy with different imaging technologies or "modalities" to reveal complementary information. Figures 7.1 and 7.2 show examples from the head and pelvis respectively. Image registration is particularly useful when the PET image contains very little anatomical information, as is the case in [^{11}C]-methionine or [^{18}F]-L-DOPA scans.
(ii) *Correcting PET emission data*, where aligned anatomical images (usually MRI) or attenuation maps derived from CT are used to improve the accuracy of regional uptake, correct for photon attenuation or improve reconstruction accuracy. Figure 7.3 provides an example of an aligned MR image improving spatial resolution of a PET image of the brain.
(iii) *Serial image registration* or *intra-modality registration*, where several images are taken over time in order to monitor subtle changes. Figure 7.4 shows aligned whole-body serial [^{18}F]-FDG images.
(iv) *Registration of images to physical space*, where interventional, surgical, or therapeutic technologies rely on images to guide treatment to specific

* Chapter reproduced from Valk PE, Bailey DL, Townsend DW, Maisey MN. Positron Emission Tomography: Basic Science and Clinical Practice. Springer-Verlag London Ltd 2003, 181–197.

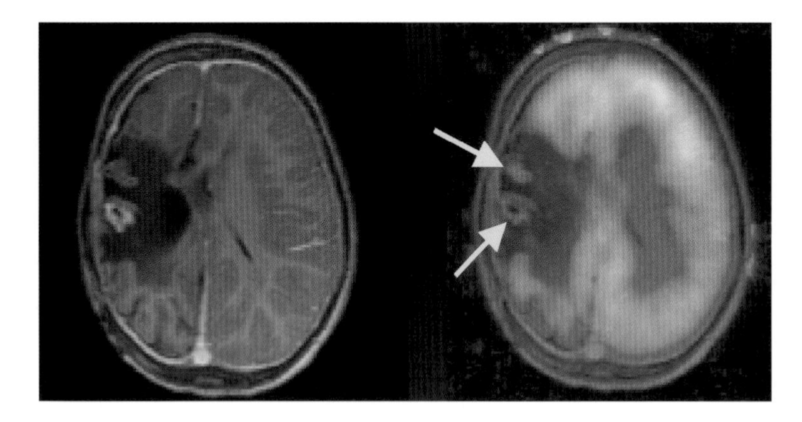

Figure 7.1. A slice from an MR volume (left) with the corresponding slice (right) of the PET [^{18}F]-FDG volume aligned in 3D and overlaid on the MR using a "hot-body" intensity scale [3]. This image shows that the suspicious bright region (lower arrow) is unlikely to be a recurrence of the astrocytoma that has been surgically removed followed by radiotherapy. The small bright region (upper arrow) anterior to this corresponds to normal cortex.

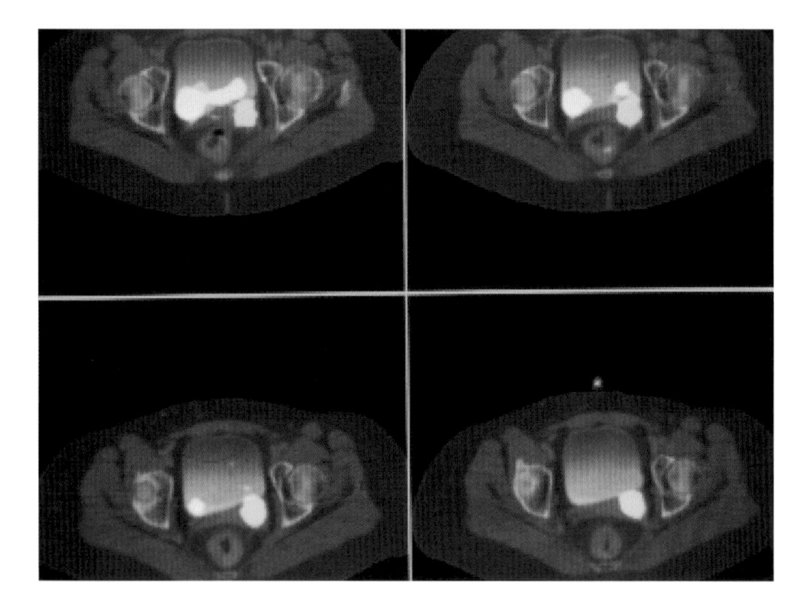

Figure 7.2. Four consecutive slices through the pelvis from a CT volume, with aligned [^{18}F]-FDG PET images overlaid, showing concentration of FDG both in the bladder and in a region of dense tissue near the cervix [3]. This indicated that the dense mass was recurrent tumor rather than fibrotic changes associated with previous radiotherapy. This was confirmed at surgery.

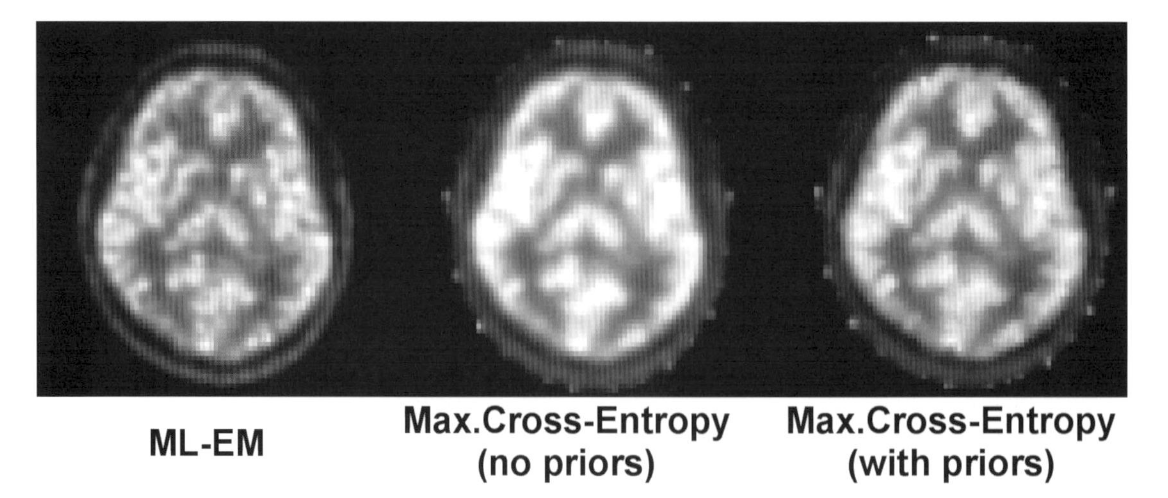

Figure 7.3. One slice of a PET [^{18}F]-FDG scan of the brain is shown, reconstructed (left) with the ML–EM algorithm, (middle) using maximum cross entropy and (right) using anatomical priors (gray matter, white matter, skull, CSF, and subarachnoid space) from aligned MR images [3] (Images courtesy of Dr Babek Ardekani, University of Technology, Sydney).

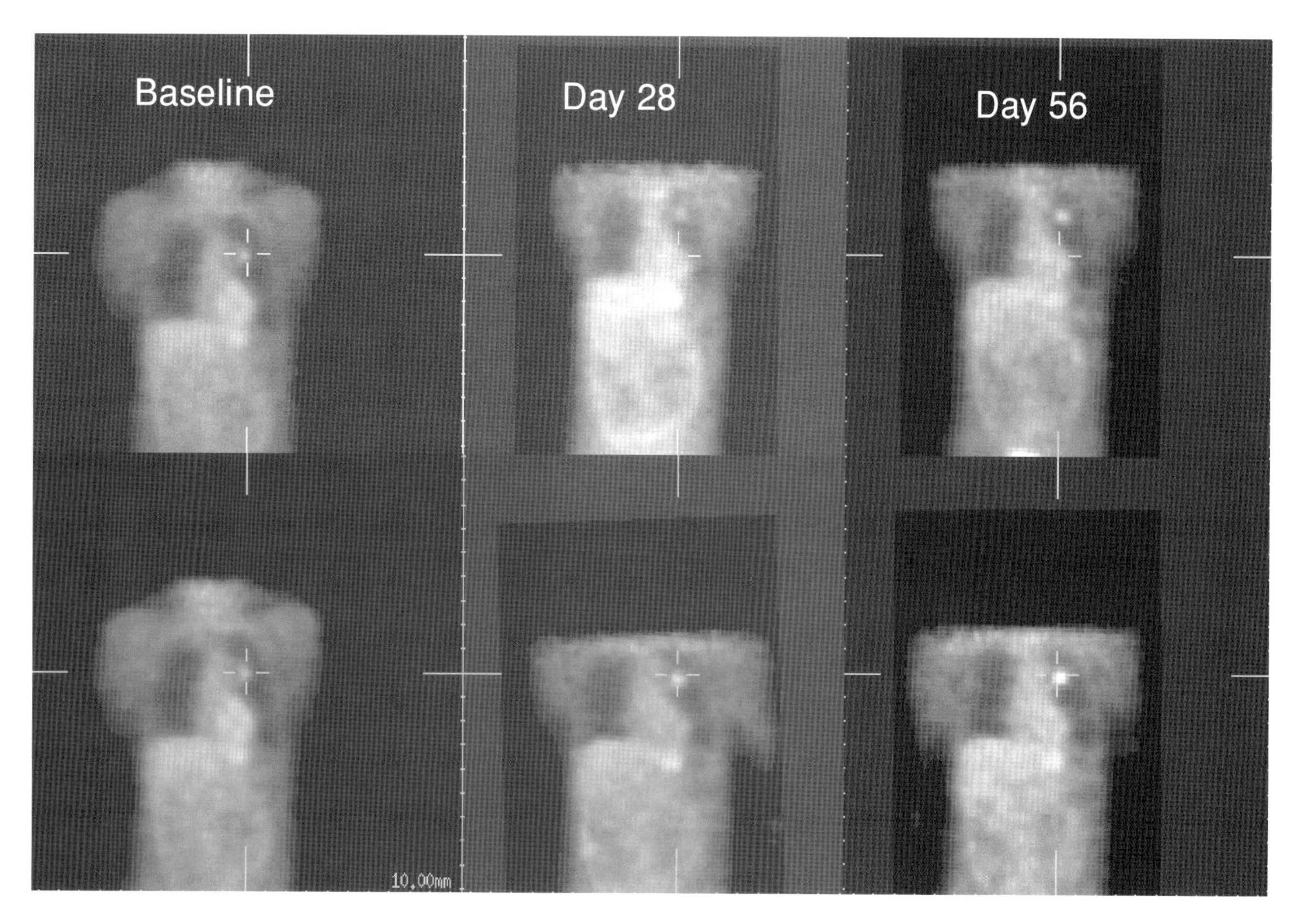

Figure 7.4. An example of registration of [^{18}F]-FDG scans obtained with a gamma camera PET system [3]. The top row shows the original reconstructed data before alignment and the bottom row shows the studies on days 28 and 56 realigned to the baseline scan (left) by maximizing mutual information [50]. The increase in activity of the lung lesion is clearly seen. Registration was successful despite different fields of view and bladder activities.

targets. The image-derived information must be aligned with physical space during treatment, again a process of registration.

Inter-subject registration has two main application areas:

(a) *Cohort studies*, in which images from a group or cohort are aligned to improve sensitivity,
(b) *Alignment of images to an atlas* to help delineate anatomy of interest.

Both of these require alignment of images from different individuals, a process of image registration. An atlas itself may be created by aligning images from a large number of individuals, for example, the Montreal Brain Atlas derived from the MR images of more than 300 normal young individuals [4].

An image registration application comprises a number of decisions and processes, summarized below:

(i) Choice of transformation – rigid-body with or without scaling, affine, non-rigid; if non-rigid, type and number of parameters,
(ii) Choice of the measure of alignment (or misalignment) and the method of optimizing this measure,
(iii) A decision on which will be the target image and which the source image to be transformed to the target,
(iv) A pre-processing step to delineate corresponding structures or transform intensities, if necessary,
(v) Computation of the transformation by optimization of the measure of alignment,
(vi) Transformation of the source image, or discrete points in the source image, to the coordinate system of the target image,
(vii) Viewing and manipulation of the results.

This sequence of processes assumes that all images to be registered are in digital form and are accessible to the computing system performing the registration.

The Concept of Spatial Correspondence

Image registration establishes spatial correspondence. Almost all of the images that are used in image registration are digital and three dimensional (3D), consisting of an array of volume elements or voxels. To each voxel is assigned a number, the image intensity at that position. The registration transformation is a mapping, T, that transforms a position x_A in image A to a point x_B in image B, or

$$T: x_A \mapsto x_B \Leftrightarrow T(x_A) = x_B \tag{1}$$

where the two images A and B are mappings of points in the patient, within their field of view or domain Ω, to intensity values:

$$A: x_A \in \Omega_A \mapsto A(x_A) \tag{2}$$

$$B: x_B \in \Omega_B \mapsto B(x_B) \tag{3}$$

Images A and B represent one object X – the patient. Image A maps position $x \notin X$ to x_A and image B maps x to x_B. The registration process derives T so that both $A(x_A)$ and $B^T(x_A)$ represent the same location in the object (within some error depending on the accuracy of T). Strictly, T defines a spatial mapping while we need a mapping that maps an accurate estimate of intensity, not just position, by including interpolation [2]. The registration process recovers T over the domain of overlap between the two images, $\Omega^T{}_{A,B}$, which depends on the domains of the original images A and B as well as the spatial transformation T.

$$\Omega^T_{A,B} = \left\{ x_A \in \Omega_A \middle| T^{-1}(x_A) \in \Omega_B \right\} \tag{4}$$

By spatial correspondence between two images, we mean that a voxel in one image corresponds to the same physical location in the patient as the corresponding voxel in an image that has been registered to it. Although this may seem an obvious definition, problems can easily arise which may lead to errors in interpretation. For example, a voxel in a PET image will usually be much larger than one in MRI and the effective spatial resolution may be up to an order of magnitude coarser in PET than in MRI. One PET voxel may therefore contain information that is spread over many hundreds of MRI voxels. This is often called the partial volume effect and registered images must be interpreted with care. Spatial correspondence may not exist when:

(i) Tissue is gained or lost between source and target image acquisition, for example, due to tumor growth or surgical removal.
(ii) Organ contents change (for example, bladder or bowel filling and emptying).
(iii) One (or both) of the images are corrupted, for example, by motion artifacts.
(iv) Structure present in one individual is absent in another (for example, detailed sulcal and gyral patterns in the cerebral cortex).

In inter-subject registration, correspondence may be defined structurally (i.e., geometrically), functionally, or histologically. Current non-rigid registration algorithms implicitly establish correspondence that does not necessarily conform to any of these definitions.

Transformations

If we can assume that the structure imaged does not change shape or size between the different images, we use a rigid body transformation. This has six degrees of freedom – namely rotations about each of the three Cartesian axes and translations along them – and the transformation is described fully by these six values. A rigid body transformation is usually assumed to be sufficient when registering head images from the same individual. It will also be sufficient when registering images of individual bones, and may even be sufficient when registering images of soft tissue structures that are securely attached to bony structure in, for example, the mandible, neck, or pelvis. The generalization of this transformation to include shears is called the affine transformation and has twelve degrees of freedom. An affine transformation transforms parallel lines to parallel lines.

If soft tissue motion is repetitive and reproducible, for example, cardiac or breathing motion, then gating techniques may be used to ensure that images of the same part of the breathing or cardiac cycle are registered. In this case, the rigid body transformation may still suffice. If soft tissue deformation is not constrained then many more parameters or degrees of freedom are required to describe the transformation. One well-known method uses approximating B-splines on a grid of control points and may require ~1000 degrees of freedom to describe tissue deformation [5].

Transformations describing the mapping of images between individuals in cohort studies or atlas registration may also require many degrees of freedom. A popular way of doing this spatial renormalization is to use Talairach space, a piecewise rigid body transformation with scaling in which brains are aligned to an axis, defined by the intersection of the interhemispheric fissure with the anterior and posterior commissure, and scaled to fit within a bounding box [6]. Often, corresponding MR images are collected in cohort PET studies. In this case, MR images can be aligned to a common reference or atlas, perhaps itself generated from an alignment of a large number of MR images [4]. A number of non-rigid registration algorithms have been proposed for this task [5, 7–11]. Alignment of the PET images is then achieved by concatenation of the non-rigid MRI to atlas with the rigid-body, patient-specific PET to MRI transformation. Each non-rigid algorithm produces slightly different transformations and validation remains a research task.

Methods for Aligning Images

Image registration algorithms can be categorized by how much user interaction is required. They can also be divided into those that use predefined features such as points or surfaces (feature based) and those that operate directly on voxel intensities (voxel-similarity based).

Interactive

The simplest conceptual method for aligning two images is to move the source image interactively with respect to the target image. Combined with a means of displaying the two images so as to allow the quality of the resulting alignment to be judged, this can provide an effective alignment method. Pietrzyk et al. [12] have described a method for the alignment of PET and MR images of the head by interactively aligning contours derived from PET [^{18}F]-FDG image with slices of MR images. While reasonably quick with the appropriate user interface and fast reslicing of PET contours, the method is prone to user error and has largely been superseded by more automated methods.

Interactive alignment remains very useful when aligning images that show gross abnormality, when attempting alignment in applications for which a specific algorithm is not designed, or when registering images where registration has failed for whatever reason. The skilled observer will bring into play expert knowledge of anatomy and radiological appearances that are not captured by even the most sophisticated computer algorithm. Care must, of course, be taken to verify that the accuracy of registration is sufficient for the application, as the purpose of registration is to extract useful and perhaps surprising information from aligned images.

Even when a fully automated algorithm is used, a certain amount of user interaction is desirable. Careful visual inspection of the registration results is always strongly recommended as no such algorithm will produce an accurate result one hundred percent of the time, and, similarly, methods for detecting failure are never one hundred percent reliable.

Corresponding Point – Point Landmarks or Fiducials

Point-based registration is one of the earliest successful examples of image registration for clinical purposes. Markers that will be visible in both images are attached to the patient before imaging. Although the use of internal landmarks such as surgically implanted markers has been proposed, these are rarely used and markers are usually attached to the skin surface or, when high accuracy is required, are attached to posts screwed directly into bone. One of the earliest examples is the stereotactic frame, which is screwed into the skull prior to all imaging and provides a reference between imaging and subsequent guidance for biopsy and tightly targeted external beam radiotherapy. Imaging markers attached to the base ring are used for registration. The stereotactic frame is large and cumbersome and extremely uncomfortable for the conscious patient. Although relocatable frames based on an acrylic bite block have been devised [13,14], stereotactic frames are usually confined to cases where patients, under general anesthetic, proceed directly from imaging to intervention. This is rarely a practical proposition for PET applications.

More practical are bone screws with imaging caps containing fluid that is visible in each of the imaging modalities. One such example is the design in [15], in which the fluid-filled cap contains a mixture of fluoride $^{18}F^{-}$, iodinated contrast material for CT, and dilute Gd-DTPA for MRI. The accuracy of alignment using this method is sufficient to provide a "gold standard" against which other registration methods can be tested [16]. The imaging marker posts can be left in place for several days with only minimal risk of infection and few complications have been reported. The process is, however, uncomfortable for the patient and is usually only justified if the patient is to proceed to surgery or other high-precision therapy.

An alternative system is the locking acrylic dental stent (LADS) [17], in which imaging markers are fastened to a patient-specific dental stent attached to upper teeth. This device has been used in PET, MRI, and CT scanners. The LADS has a reported accuracy approaching that of the bone-implanted markers, but its use is currently limited to applications involving image-guided interventions due to the expense and inconvenience in manufacture of the stent. Obviously, it is also only suitable for the dentate patient, although accurate stents have been made with as few as four healthy teeth.

For minimal invasiveness, imaging markers can be attached to the patient's skin. These markers must be in position throughout all acquisitions of the images to be registered. With the best organization this will take several hours and often more than a day will elapse between images. Skin is mobile and can move over the skull surface by 5–10 mm as the head is positioned in the different head holders of various scanners. Markers have been attached to clamps that fix to the nasion and both external auditory meatus and these have a slightly higher reported accuracy. A variety of shapes of marker have been proposed so that a 3D point can be accurately defined independent of slice orientation and voxel dimension anisotropy. These include cross and V-shaped markers [18] but the most widely used are spherical markers of much larger dimension than the voxel sizes of MRI or CT. Their coordinates are found by determining the center of gravity of the image intensities in the vicinity of the marker. Marker design has developed significantly over the last 10 years, driven by the requirements of image-guided surgery, but in imaging of the head accuracy is still limited to between 3 and 5 mm for skin markers due to skin movement between acquisitions [19]. While a minimum of three non-collinear markers are sufficient, as many markers as possible should be used to reduce the effect of random marker location errors. Unfortunately, the presence of markers has the added complication that small movements of the head during scanning and reconstruction errors can lead to significant streak artifacts, to the detriment of image quality.

All these applications can be termed *prospective* registration methods. This means that the decision to register the images had to be taken before either of the images was acquired. This frequently is not possible, which means that images have to be retaken with markers, with the concomitant increase in cost and radiation dose to the patient.

Point-based *retrospective* registration is possible using anatomical landmarks within the images themselves. Provided that sufficient landmarks are acquired in each image and that there is no bias in determining point correspondence, reasonable accuracy can be achieved [20]. However, the process is time consuming and requires a skilled operator. Although it was used at several sites for many years, it has now largely fallen into disuse due to the significant advances in more automated methods of retrospective image registration described below.

The point-based registration algorithm, however, remains an important and widely used algorithm, in particular for image-guided interventions and when exploring novel image registration applications. The algorithm is derived from the solution to the orthogonal Procrustes problem. This name derives from Greek mythology, in which the robber Procrustes would offer travelers hospitality in his roadside house, promising a bed that would fit every visitor perfectly. However, this was achieved by ensuring instead that the travelers fitted the bed, either by stretching them if they were too small or by amputation if they were too big, both of which transformations were invariably fatal. The hero Theseus put a stop to this bizarre practice by subjecting Procrustes to his own treatment. In statistics, the name Procrustes became an implied criticism of the practice of forcing one dataset to fit another, but in shape analysis this practice has now achieved widespread use. The Procrustes problem is an optimal fitting problem of least squares type. Given two configurations of the same number of non-coplanar points the algorithm derives the transformation, which minimizes the sum of the squared distances between corresponding points. The algorithm relies on prior establishment of point correspondence. We can consider 3D rigid body [21,22], rigid body plus scaling [23], and affine transformations [24].

The solution for 3D rigid body transformations can be computed directly; no iterative or optimization scheme is required. The transformation comprises a translation vector and a rotational matrix. It is straightforward to compute the translation vector from the vector joining the centers of gravity of the two point distributions. The rotational matrix is calculated using singular-value decomposition (SVD).

Given two configurations of N non-coplanar points $P = \{p_i\}$ and $Q = \{q_i\}$, we seek the transformation that minimizes $G(T) = \|T(P) - Q\|^2$. P, Q are the N-by-D matrices whose rows are the coordinates of the points p_i and q_i respectively, while $T(P)$ is the matrix of transformed points. $\|\ldots\|$ is a matrix norm, the simplest being the Frobenius $(\sum_i \|(T(p_i) - q_i)^2\|)^{1/2}$. When T is a rigid body transformation we replace the values of P and Q by the difference of each value from the mean:

$$p_i \mapsto p_i - \overline{p} \quad (5)$$

$$q_i \mapsto q_i - \overline{q} \quad (6)$$

Writing $P = [p_1, \ldots, p_N]^t$ as a matrix of row vectors and similarly for Q, and letting $K = \sum_i K_i$ where $K_i := p_i q_i^t$, we have

$$K = UDV^t \Rightarrow R = V\Delta U^t \qquad \Delta := diag(1, 1\det(VU^t)) \quad (7)$$

where $K = UDV^t$ is the SVD of K.

Finally, the translation t is given by $t = \bar{q} - R\bar{p}$.

Corresponding point landmarks can also be used to define non-rigid transformations. The most widely used is the thin-plate spline approach of Bookstein [25] in which the transformation derived is that of a thin, perfectly elastic plate. As with the rigid body solution above, the transformation can be calculated quickly and directly. Non-rigid transformations are rarely used in intra-subject PET image registration as there is usually insufficient information content in PET images to derive such transformations accurately.

Surfaces, Lines and Points

Boundaries or surfaces between organs and between the skin surface and the surrounding air can provide strong features for image registration. Automated image segmentation algorithms – or, failing that, manual or user-defined segmentation – can be used to define visible boundaries in the images to be registered. If these boundaries correspond to the same physical surface in the images to be registered then they can be used to derive the registration transformation. A number of algorithms have been based on registration between surfaces and lines or between points demarcated on surfaces. These algorithms are usually only used to determine rigid body transformations. Many surfaces in images have a high degree of symmetry (for example, the outline of the cranial vault has an almost circular outline in sagittal section) and non-rigid transformation solutions are likely to be highly susceptible to noise and hence error prone.

The Head-and-Hat Algorithm

The earliest multi-modality surface-based registration algorithm was the "head-and-hat" algorithm proposed by Pelizzari et al. [26]. This algorithm was used to align MRI, CT, and PET images of the head. The high-resolution CT or MR image was represented as a stack of disks, referred to as the "head". The second surface was represented as a list of unconnected 3D points, the "hat". The registration transformation was then determined by iteratively transforming the "hat" points until the closest fit of hat on head was found. The measure of closeness of fit was the sum of squared distances between a point on the hat and the nearest point on the head, in the direction of the centroid of the head. The original algorithm used Powell optimization [27], which involves a sequence of one-dimensional optimizations along each of the six degrees of freedom of the rigid-body transformation. The algorithm stopped when it failed to find a solution in any of the degrees of freedom that improved the measure of fit by more than a predefined tolerance. The corresponding surfaces most commonly used were the skin surface from MRI and PET transmission images or the brain from MRI and PET emission images. The algorithm proved to be reasonably robust but was prone to error with convoluted surfaces and was particularly susceptible to errors in cranio-caudal rotation due to the natural symmetry of the cranium mentioned above. It requires a reasonably good first guess or "starting estimate" of the correct registration transformation. In most cases, the known patient orientations in the scanners will suffice. The method has also been applied to cardiac MRI and PET images of the heart [28].

Distance Transform Based Surface Registration

A modification of the head-and-hat algorithm precomputes a distance transform of the source image surface. A distance transform is applied to a binary image in which voxels inside an object have the value 1 and voxels outside an object have the value 0. The distance transform labels each voxel in the image with its distance from the surface. Computation of the transform proceeds by taking a starting estimate of the transformation and looking up the distance from the surface in the distance transform image for each surface point in the target image. The cost of this transformation is computed as the sum of squares of these distances. A process of optimization is used to find the transformation that minimizes this cost. The chamfer filter defined by Borgefors [29] is widely used and efficient and has been successfully used in image registration applications [30–32]. More recently exact distance transforms have been used in place of the chamfer transform [33]. Surface-based registration is prone to finding local minima during the optimization process and presenting these as the solution rather than the true global optimum. The physical analogy is that small features on the surface "lock together" at the incorrect registration. To partially alleviate the risk of this occurring, multiscale representations have been used [30].

The Iterative Closest Point (ICP) Algorithm

The ICP algorithm was proposed by Besl and McKay [34] for the registration of 3D shapes. Although the authors did not have medical images in mind, the algorithm has been very successful in medical applications and has largely superseded the other surface-based algorithms due to its robustness to starting estimate and surface shape. The algorithm is designed to work with seven different representations of surface data: sets of points, sets of line segments, implicit curves, parametric curves, sets of triangles, implicit surfaces, and parametric surfaces. For medical imaging the most useful representations are sets of points and sets of triangles. The algorithm has two stages and iterates. The first stage involves identifying the closest point in the target image surface for each point in the source image surface. In the second stage a rigid body registration (such as the Procrustes method described above) is computed between these two point sets. This is used as the transformation for re-computing the closest points and the process continues until convergence, which is defined to occur when the change in the transformation after each iteration drops below a predetermined tolerance. The algorithm is still prone to errors caused by local minima and the original authors proposed multiple starts to estimate the global optimum. Again, multi-resolution techniques could be used to avoid becoming trapped in local minima.

Crest Lines and Other Geometric Features

An alternative to using pre-segmented surfaces is to use distinctive surface features defined by their local geometry. Using the tools of differential geometry it is possible to define two principal curvatures of a surface in 3D space. A crest line is an indication of a ridge in the surface and is defined as the loci of points where the value of the largest curvature is locally maximal in its principal direction [35]. Registration between two images proceeds by applying the ICP algorithm to these crest lines. Gueziec et al. [36] have proposed using hash tables of geometric invariants for each curve together with the Hough transform and a modified ICP algorithm. These methods, while useful for intramodality registration of high-resolution images, are less applicable to lower-resolution PET images.

Image Intensity or Voxel Similarity-based Registration

A measure of image alignment is computed directly from the voxel intensities and an optimization process used to search for the transformation that maximizes this measure. Although the number of computations required is high, modern computing power means that reasonably high-resolution image volumes can be registered sufficiently quickly to be useful. The successful methods can be fully automatic and recent validation studies have shown that for the particular case of registration of images of the head, voxel similarity-based registration methods can outperform feature-based methods in terms of accuracy and robustness [16].

As the methods operate directly on voxel intensities, different methods are required for the alignment images from the same modality and that of images from different modalities. In intramodality image registration we would expect the difference in the images at alignment to be dominated by image noise and little structure should be present. This suggests a number of possible measures of alignment that are outlined in more detail below. As it is only meaningful to compute a measure where there is overlap of data and as the volume of overlap will change with each trial alignment, care must be taken to normalize the registration measures.

Minimizing Intensity Difference

This is one of the simplest voxel similarity measures involving subtracting the two images and computing the mean sum of squares of this difference (*SSD*) image in the region of overlap. For N voxels in the overlap domain $\Omega^T_{A,B}$ this is given by

$$SSD = \frac{1}{N} \sum_{x_A \in \Omega^T_{A,B}} \left| A(x_A) - B^T(x_A) \right|^2 \qquad (8)$$

It can be shown that this measure is optimal when two measures differ only by Gaussian noise [37]. Although we are usually interested in finding differences between the images, these are often so small that this measure remains the most effective. Image noise may not have a Gaussian distribution but this is unlikely to have a significant effect on performance. The measure is frequently used although it is sensitive to a small number of voxels having very different intensities – as might occur, for example, in contrast-enhanced serial MR imaging or during a dynamic sequence of PET images. Using the sum of absolute differences (*SAD*) rather than the sum of squared differences can reduce the effect of outliers.

$$SAD = \frac{1}{N} \sum_{x_A \in \Omega_{A,B}^T} \left| A(x_A) - B^T(x_A) \right| \quad (9)$$

Correlation of Voxel Intensities

A slight relaxation of the assumption that registered images differ only by noise is that image intensities are strongly correlated. The correlation coefficient has been widely used in intramodality registration, for example [38], and is given by:

$$CC = \frac{\sum_{x_A \in \Omega_{A,B}^T} \left(A(x_A) - \bar{A}\right)\left(B^T(x_A) - \bar{B}\right)}{\left\{ \sum_{x_A \in \Omega_{A,B}^T} \left(A(x_A) - \bar{A}\right)^2 \sum_{x_A \in \Omega_{A,B}^T} \left(B^T(x_A) - \bar{B}\right)^2 \right\}^{\frac{1}{2}}} \quad (10)$$

where $\bar{A}$ is the mean voxel value in image $A|_{\Omega^T_{A,B}}$ and $\bar{B}$ is the mean of $B^T|_{\Omega^T_{A,B}}$.

Ratio of Image Uniformity (RIU)

The RIU algorithm was originally introduced by Woods et al. [39] for the registration of serial PET studies but has now been applied to serial MR images as well. It is available in the AIR registration package from UCLA. The RIU algorithm finds the transformation that minimizes the standard deviation of the ratio of image intensities. This ratio is computed on a voxel-by-voxel basis from the target image and the transformed source image that results from the current estimate of the registration transformation. The RIU measure is most easily thought of in terms of an intermediate ratio image R comprising N voxels within the overlap domain $\Omega^T_{A,B}$.

$$R(x_A) = \frac{A|_{\Omega^T_{A,B}}(x_A)}{B^T|_{\Omega^T_{A,B}}(x_A)} \quad \bar{R} = \frac{1}{N} \sum_{x_A \in \Omega_{A,B}^T} R(x_A) \quad (11)$$

$$RIU = \frac{\sqrt{\frac{1}{N} \sum_{x_A \in \Omega_{A,B}^T} \left(R(x_A) - \bar{R}\right)^2}}{\bar{R}} \quad (12)$$

Multi-modality Registration by Intensity Re-mapping

With multi-modality registration there is, in general, no simple relationship between intensities in the two images to be registered. One solution is to transform or re-map the intensities in the image from one modality so that the two images look similar to each other. This has been used with some success by re-mapping high intensities in CT, corresponding to bone, to low intensities [40]. The resulting image has the approximate appearance of an MR image and registration proceeds by maximizing cross correlation. An alternative is to compute differentials of image intensity that should correlate between the two images. Van den Elsen et al. [41] and Maintz et al. [42] compute the *edgeness* of an image from differential operators applied directly to the image intensities. If the two images have boundaries at corresponding locations then cross correlation of the edgeness measure should be maximal at registration.

Multi-modality Registration by Partitioned Intensity Uniformity

This was the first purpose-designed, widely used, multimodality registration algorithm to use a voxel similarity measure. It was proposed by Woods et al. [43] for MRI-PET registration soon after they proposed the RIU algorithm. We refer to this algorithm as partitioned intensity uniformity (PIU). The algorithm is a remarkably simple modification to the original RIU algorithm, involving the change of only a line or two of source code, but with transformed functionality. The implicit assumption here is that all voxels with a particular MR image intensity represent the same tissue type and are therefore likely to have similar PET image intensities. The algorithm partitions the MRI voxels into 256 separate bins by intensity and seeks to maximize the uniformity of the PET voxels within each bin. The uniformity within each bin is measured from the standard deviation of the corresponding PET voxel intensities. The alignment measure, *PIU,* is a weighted sum of the normalized standard deviations. For registration of images A and B (MRI and PET respectively) we can write

$$PIU = \sum_a \frac{n_a}{N} \frac{\sigma_B(a)}{\mu_B(a)} \quad (13)$$

where

$$n_a = \sum_{\Omega_a^T} 1 \quad (14)$$

$$\mu_B(a) = \frac{1}{n_a} \sum_{\Omega_a^T} B^T(x_A) \quad (15)$$

$$\sigma_B^2(a) = \sqrt{\frac{1}{n_a} \sum_{\Omega_a^T} \left(B^T(x_A) - \mu_B(a)\right)^2} \quad (16)$$

and the domain Ω_a^T is the iso-intensity set in image A with intensity value a within $\Omega_{A,B}^T$ defined as:

$$\Omega_a^T = \left\{ x_A \in \Omega_{A,B}^T \middle| A(x_A) = a \right\} \qquad (17)$$

Although the statistical basis of the algorithm is somewhat tenuous, the algorithm is widely used for registration of MRI and PET images of the head and performed very well in the Vanderbilt registration assessment study [16]. For the most reliable results the bright region of the scalp is usually removed from the MR image, a procedure known as scalp editing. The scalp is a bright region in MRI but often corresponds to lower image intensities in PET, unlike the grey and white matter in the brain.

Local Correlation for Multi-modality Registration

Recently, Netsch et al. [44] have proposed a new measure, local correlation. This measure assumes that for a local region of the image there will be a strong correlation between image intensities at registration. The correlation coefficient is computed in a region local to each voxel in the destination image and the normalized sum of these local correlations is calculated for each trial transformation. The transformation that yields the maximum normalized sum of local correlations should correspond to registration.

This algorithm is reported to have comparable accuracy to mutual information (see below) for registration of CT and MR images of the head, and is well suited for numerical optimization. Local correlation can be recast as a least squares criterion, which allows the use of dedicated methods. Such an algorithm should be able to run extremely quickly, especially as it has been shown to be effective when the similarity is calculated for only a small fraction of the image voxels. Application to PET images has yet to be reported.

Information Theoretic Measures for Multi-modality Registration

Plotting the joint histogram of the two images provides a useful insight into how voxel similarity measures might be used for multi-modality registration [45]. Figure 7.5 shows plots of the joint histogram computed for identical MR images and for an MRI and a PET image of the same subject. The joint histograms are plotted at registration and at two levels of mis-registration. A distinctive pattern emerges at registration of each pair and this pattern diffuses as mis-registration increases. This suggests certain statistical measures of mis-registration. Interestingly, the MRI-PET joint histogram also explains why the PIU measure works and why it works better when the scalp is edited out of the image. The scalp corresponds to the horizontal line in these plots, i.e., a low PET intensity and a wide range of MRI intensities with partial voluming up to very bright values. Scalp editing removes this line and the resulting plot shows a narrow distribution of PET intensities for each MRI intensity.

It can also be useful to think of image registration as trying to maximize the amount of shared information in two images. Qualitatively, the combined image of, say, two identical images of the head will contain just two eyes at registration but four eyes at mis-registration.

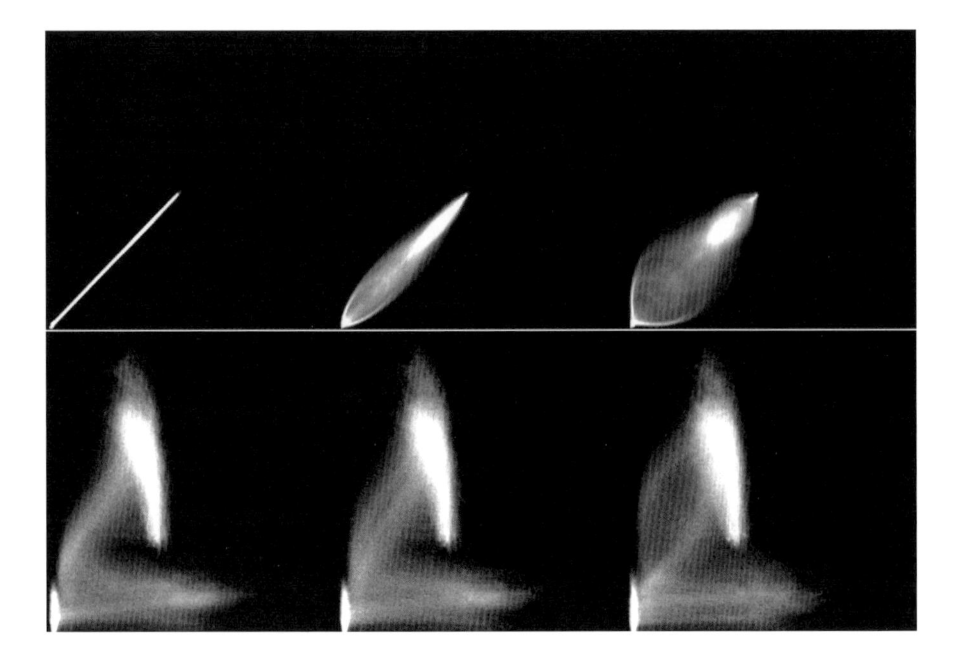

Figure 7.5. Example 2D image intensity histograms from Hill et al. [45] for identical MR images of the head (top row) and MR and PET images (bottom row, PET intensity plotted on y-axis and MR intensity on x-axis) of the same individual at registration and at alignment (left), mis-registration of 2 mm (middle) and mis-registration of 5 mm (right). Although the histograms for different image pairs have very different appearances, both sets show a dispersion of the histogram with increasing mis-registration.

This suggests the use of a measure of joint information as a measure of mis-registration. The signal processing literature contains such a measure of information, the Shannon–Weiner entropy *H(A)* and its analog for dual signals, the joint entropy *H(A,B)* [46]. These can be defined over the region of overlap between two images as follows:

$$H(A) = -\sum_{a} p_A^T(a)\log p_A^T(a) \quad \forall A(x_A) = a \,|\, x_A \in \Omega_{A,B}^T \tag{18}$$

$$H(B) = -\sum_{b}^{a} p_B^T(b)\log p_B^T(b) \quad \forall B^T(x_A) = b \,|\, x_A \in \Omega_{A,B}^T \tag{19}$$

and

$$H(A,B) = -\sum_{a}\sum_{b} p_{AB}^T(a,b)\log p_{AB}^T(a,b) \tag{20}$$

where $p_A^T(a)$ is the marginal probability of intensity a occurring at position x_A in image A, $p_B^T(b)$ the marginal probability of intensity b occurring at position x_A in image B, and $p_{AB}^T(a,b)$ the joint probability of both intensity a occurring in image A and intensity b occurring in image B. These probabilities are estimated from the histograms of images A and B (marginal probabilities) and their joint histogram (joint probability) in the overlap domain with transformation T.

The joint entropy measures the information contained in the combined image. This suggests a registration metric. Registration should occur when the joint entropy is a minimum. This measure had been proposed independently by Collignon [47] and Studholme [48]. Unfortunately, the volume of overlap between the two images to be registered also changes as they are transformed relative to one another and because of this joint entropy was found not to provide a reliable measure of alignment.

The solution, spotted independently by Collignon et al. [47] and Wells et al. [49], was to use mutual information (MI) as the registration metric instead. MI is the difference between the joint entropy and the sum of the individual (marginal) entropies of the two images. According to information theory, this difference is zero when there is no statistical relationship between the two images. If there is a statistical relationship, the mutual information will be greater than zero, and the stronger the statistical relationship the larger its value will be. This suggests that it could be used as a measure of alignment, since registration should maximize the statistical dependence of one image on the other. This measure largely overcomes the problem of volume of overlap changing as the transformation is changed. Mutual information is expressed as

$$I(A,B) = H(A) + H(B) - H(A,B) = \sum_{a}\sum_{b} p_{AB}^T(a,b)\log\frac{p_{AB}^T(a,b)}{p_A^T(a)p_B^T(b)} \tag{21}$$

This measure has proved to be remarkably robust in a very wide range of applications. The Vanderbilt study [16] showed that it performed with comparable accuracy to the PIU algorithm for PET-MRI registration and does not require scalp editing. Studholme et al. [50] showed that the measure was the most robust and accurate of a range of different measures tested.

However, mutual information sometimes fails to find the correct solution when the volume of overlap varies significantly with alignment or when there is a large volume of background (i.e., air) in the field of view. Studholme et al. [51] showed that a simple reformulation of the measure as the ratio of the joint entropy and the sum of the marginal entropies in the overlapping region provided a more robust measure of alignment,

$$\tilde{I}(A,B) = \frac{H(A)+H(B)}{H(A,B)} \tag{22}$$

This measure and mutual information have now been widely adopted and registration software based on these measures has been extensively validated for rigid body registration of images of MRI and PET images of the head [16, 50,51].

While qualitative arguments, such as those outlined above, have been used to justify mutual information or its normalized version, there is as yet no firm theoretical basis for these measures and further research is required to provide this theoretical underpinning. This research may lead to measures that outperform the information theoretic measures outlined above.

Optimisation, Precision, Capture Ranges and Robustness

All registration algorithms except the Procrustes method of point registration rely on a process of optimization to compute the transformation that best aligns the two images. The algorithms require an initial guess, or "starting estimate", of the correct transformation. They then compute image similarity for voxel intensity-based methods or distances for feature-based methods, and use this to compute a new (and hopefully better) estimate of the transformation. This process repeats until it converges to a solution, usually defined to be when the similarity or distance change after each iteration falls below a certain preset tolerance. There is a large range of different types of optimization methods available and an excellent review can be found in [27].

A major limitation of many optimization methods is that they find "local optima"; that is, they find a solution that is far from the correct one but fail to improve it further because all nearby solutions have a similarity or distance worse than that of the current transformation. To avoid finding these local optima, multi-resolution methods are often used [50]. A coarse representation of the images (or surfaces) is used to find an initial estimate of the registration, and resolution is progressively increased to finer levels until the full resolution of the data is used. These methods have the effect of blurring out local minima during optimization and have the added advantage that they allow a significant speed up in computation. However, multi-resolution methods do not always solve the problem of local optima and in these cases multi-start techniques are used. These involve giving multiple starting estimates to the registration algorithm and then selecting the solution that has the best optimum.

It may be the case that the global optimum is not the correct registration solution. For example, some algorithms find a global optimum when the images to be registered are completely separated. However, a good algorithm will have a range of registration solutions within which the correct registration is an optimum of similarity or distance. This region is known as the "capture range" and the size of it determines the "robustness" of the algorithm. This is the measure of how close the starting estimate needs to be for the optimization method to find the correct solution. In practice, image acquisition protocols allow a reasonable starting estimate to be defined, i.e., we know whether the patient was imaged supine, prone, or lateral and a standard radiographic setup should give a reasonable idea of where the patient is with respect to the scanner's field of view. In order to be clinically useful a registration algorithm should be able to find the correct solution if the starting estimate is within about 30 mm and 30 degrees.

There may be multiple optima that are all very close to the global optimum. In this case the algorithm may find any number of these solutions if multiple registrations are carried out. If a registration algorithm is started from multiple starting points the distribution of these local optima yields the "precision" of the algorithm. If all starts yield the same answer then the precision is limited by the final step size of the search.

Image Transformation and Display

Having determined what the transformation is between the target and source image, we now need to transform the information in the source image to the target image. This may be done with a simple user interface that allows a point located in the target image to be displayed in the appropriate slice of the source image. This involves a single matrix multiplication of the target image point by the derived transformation. Usually, however, we need to transform the source image into the space of the target image. This is done on a voxel-by-voxel basis. For each voxel in the target image within the volume of overlap of the two images, the voxel coordinate is multiplied by the transformation matrix and the resulting voxel coordinate used to find the appropriate intensity in the source image. This computed coordinate is very unlikely to be an integer voxel coordinate so the appropriate position needs to be computed. The most straightforward method is "nearest neighbor" interpolation, where the intensity of the nearest voxel to the computed coordinate is found. A slightly more accurate approach is to use trilinear interpolation. The weighted sum of the eight nearest voxels is computed, where the weighting is inversely proportional to the distance of the required coordinate from each voxel. For the highest accuracy, "sinc" interpolation is used. However, for applications involving PET, trilinear interpolation or even nearest neighbor should suffice.

MRI and PET images have very different spatial resolutions so a decision has to be made whether to choose the MRI or PET image as the target image. The advantage of choosing the latter is that spatial resolution is conserved in the final registered images, although the transformed PET image will have the same voxel dimensions as the MR image and a concomitant increase in storage requirements. This can be avoided by only resampling the PET image "on-the-fly" when it is needed for display. Reslicing using nearest neighbor or trilinear interpolation can be achieved in near real time with modern PCs.

Having transformed the images to a common coordinate system, the images may be displayed side by side, with a linked cursor indicating spatial correspondence, or may be overlaid or fused using a range of widely available display methods. The most common is the use of different colors to represent the different modalities as shown on Figs. 7.1 and 7.2. In a "moving curtains" display, an interactively controlled vertical or horizontal line divides the display of one image from the other.

In serial acquisitions for assessing subtle changes, difference images are often displayed. In neuro-activation studies the transformed images might be analyzed for temporal correlation of intensities with stimuli and appropriate statistics of significance displayed on a voxel-by-voxel basis.

Registered PET images can be used in image-guided surgery. Figure 7.6 shows a view through the operating microscope of a tumor with the outline of the PET [^{18}F]-FDG scan superimposed using the virtual reality display of the MAGI system (Microscope Assisted Guided Interventions) developed at Guy's Hospital [52]. This image was registered to the corresponding CT scan prior to surgery and the CT scan was in turn registered to physical space in the operating room. Registration was achieved with a modified LADS system [53].

Image Acquisition Pitfalls and Correction of Scanner Errors for Image Registration

Image Acquisition

Registration algorithms perform better with images of approximately isotropic spatial resolution and cubic voxels. Conventional multislice MR images are not acquired with overlapping slices and the resulting missing data can cause some data corruption when resampled. As a result, true 3D MR image acquisition is preferred, in which Fourier encoding is applied along all three ordinates. MR images should be acquired so as to maximize contrast between anatomical structures of interest. While radiation dose to the patient may prohibit CT image acquisition with isotropic voxels, contiguous slices should be acquired with conventional CT scanners. Modern spiral scanners and, in particular, new multislice devices offer more scope for improving axial resolution while keeping x-ray dose fixed. This will produce noisier, lower-contrast images, but it has been shown that voxel similarity-based registration is remarkably resilient to image noise [54]. Lower-resolution data can be reformatted later to produce higher-contrast diagnostic scans with larger slice thickness.

When using dual-headed gamma cameras for PET acquisition, care must be taken to ensure accurate measurement and correction of center-of-rotation errors. Full-ring PET systems have a varying distance between adjacent parallel projections, decreasing towards the edge of the field of view (see Ch. 3). This would produce spatial distortions but is usually corrected prior to reconstruction. Full-ring PET scanners have a spatial resolution that is radially dependent, being worse at the periphery. However, this is unlikely to have a significant effect on image registration accuracy. Also, non-uniform detector response can give rise to ring artifacts in both designs of PET scanner. Careful calibration and quality assurance are required to ensure their removal.

Great care must be taken to minimize patient movement during image acquisitions. Patient movement produces different characteristic artifacts in each imaging modality. These artifacts may take the form of streaks in PET or CT images produced by filtered back projection reconstruction, or of wraparound or ghosting effects in MRI. Motion between slice acquisitions in conventional CT may not produce visible artifacts in-plane but can produce significant axial distortions or discontinuities. Patient movement may produce other geometric distortions in images that are more difficult to identify and correct. Registration can be used to reduce motion arti-

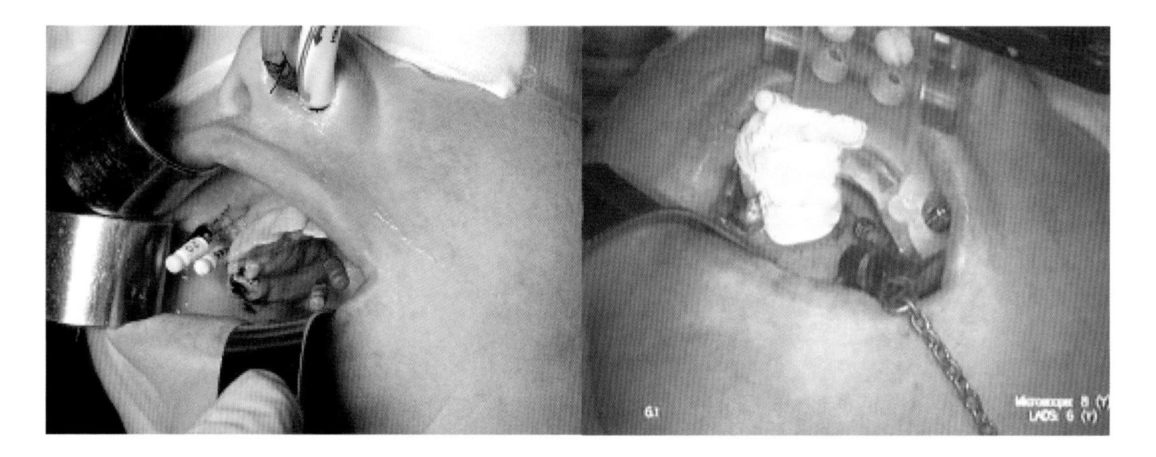

Figure 7.6. A picture of a patient's jaw immediately prior to excision of a squamous cell carcinoma (left) and with an aligned PET [^{18}F]-FDG superimposed [53] using the augmented reality displays of the MAGI system [52] at the beginning of the operation (right).

facts, for example, by patient tracking using an external locating device during acquisition [55]. Usually, good radiographic practice and patient immobilization suffices, but patient movement will affect the accuracy and robustness of image registration algorithms because poor images produce poor registration results.

Geometric Scaling and Distortion

The integrity and accuracy of image registration is dependent on the spatial accuracy of the original images. Scanners that supply images for image registration must be calibrated so that their voxel dimensions are known. Many scanners have a specified spatial accuracy no better than 1% and errors as high as 5% can occur. A 5% error on a 250 mm field of view corresponds to an error of 12.5 mm over the field of view, which will be unacceptable for most registration applications. Older CT scanners can have errors in axial dimensions due to inaccuracies in monitoring bed movement. Also, CT gantry tilt can be erroneously reported in image headers. A five-degree error will result in a shear of the data resulting in more than 20 mm error over a 250 mm field of view.

Scanner distortions can be categorized by those that distort the spatial integrity of the images and those that produce intensity shading effects. Geometric distortion is more likely in MR imaging. Distortion arises from gradient field non-linearity and from magnetic field inhomogeneity. The former results from imperfections in gradient coil design but is static and can be measured and corrected. In practice, most modern MRI scanners will incorporate such a correction. Magnetic field inhomogeneity arises from imperfections in the high B_0 field, eddy currents induced by the switching gradients and spatial variations in the magnetic susceptibility within the imaged volume. Metal objects within or close to the field of view can result in severe distortions making image registration impossible. Smaller errors can be reduced by double acquisitions with inverted gradients [56] or phase unwarping methods [57]. The former reportedly reduces errors by 30–40% while the latter reduced errors from 3.7 mm to 1.1 mm. Ramsey and Oliver [58] reported, in a recent study on modern MRI scanners, that linear distortions ranged between 0 and 2 mm. Modern MRI scanners are much less susceptible to geometric distortion than older machines.

Intensity distortion can arise from RF inhomogeneity in MRI, in particular with surface coils. Beam hardening effects in CT can produce intensity shading and photon scatter and incorrect attenuation correction can produce shading effects in PET. Highly attenuating objects will produce significant shading artifacts in PET and CT. Objects of very different magnetic susceptibility to tissue can produce shading effects in MRI, and if made of metal will usually preclude acquisition of an MR image for registration. For this reason, patients with prostheses or surgical implants near the region to be imaged may not be suitable for image registration studies.

Mutual information-based registration has been shown to be reasonably immune to a gradual drop of intensity across the field of view. Studholme et al. [59] presented an adaptation of mutual information-based registration that improves robustness of MRI-PET registration of the pelvis based on intensity partitioning of the MR data, in cases where there is severe shading across the MRI field of view. Methods that rely on automated or semi-automated segmentation of surfaces may produce biased results in the presence of shading. Shading across an MR image can easily misplace a boundary by 1 or 2 mm over the field of view.

Ideally, image geometry should be measured during image acquisition, for example, by using markers in the patient-immobilization device. Great care must be taken that these markers do not themselves induce distortion or intensity artifacts. They are also used to measure geometric distortion on the periphery of the field of view, where it is most severe, running the risk of over-correction. Of more practical importance is careful quality assurance (QA) of scanners used for image registration. QA protocols should be similar to those implemented when images are used for high-precision radiotherapy or image-guided surgery. One method is to scan a simple geometric phantom regularly to check for distortions and scaling. Hill et al. [60] have proposed using an image registration method based on mutual information to measure scanner scaling and skew errors. A digital voxel model of the phantom is created and this is registered to the image volume using a full 12-degree of freedom affine transformation. This has been shown to be extremely robust and accurate and has been applied to PET, CT, and MRI scanner calibration. Scaling errors may be deduced in a similar manner by registration of a patient's CT scan and MRI scan and using a 9 degree of freedom registration (6 degrees of freedom rigid body plus scaling along the three ordinates).

Validation and Quantifying Registration Accuracy

Registration algorithms, especially automated ones, involve a large number of computations that may be opaque to the user. Registration accuracy is the metric used to determine the quality of a specific registration. Any registration method must be sufficiently accurate for the application envisaged and sufficiently robust to be clinically useful. No algorithm is 100% robust so any clinical protocol must have the means to detect failure with close to 100% reliability.

Registration accuracy is determined quantitatively by the target registration error (TRE). If we can define a point in the target image and the same point in the transformed source image, the TRE is the distance in millimeters between these two points. TRE can be defined at anatomically relevant points or can be computed throughout the volume of interest and the mean, range, or standard deviation of TRE values can be computed. TRE can, in certain circumstances, be estimated analytically for point-based registration. Fitzpatrick et al. [61] derived a formula that predicts TRE at any position within a registered volume given the error with which a point landmark can be located; the so-called fiducial localization error (FLE). FLE can be estimated by repeatedly locating the points in space or by analyzing repeat scans of the markers. The squared expectation value of *TRE* at position x (coordinates $(x_1, \ldots, x_k)$ is given by:

$$\langle TRE(x)^2 \rangle \cong \langle FLE \rangle^2 \left(\frac{1}{N} + \frac{1}{k} \sum_{i}^{k} \sum_{j \neq i}^{k} \frac{x_i^2}{\Lambda_i^2 + \Lambda_j^2} \right) \tag{23}$$

where k is the number of spatial dimensions (usually three in medical applications) and Λ_i the singular values of the configuration matrix of the markers

This formula is only applicable if FLE is isotropic and the same for each registration point.

In all other cases, TRE can only estimated if there is a more accurate "gold standard" registration available. This is very hard to achieve in practice. West et al. [16] undertook a careful study using patients scanned with bone-implanted markers in place in the skull prior to image-guided neurosurgery. This study enabled various PET, MRI, and CT registration algorithms to be compared and provides conclusive evidence that currently available voxel-based registration methods generally outperform surface-based methods. Normalized mutual information achieved mean TRE values of between 0.6 mm and 2.5 mm for CT-MRI registration and between 1.4 mm and 3.9 mm for MRI-PET registration, with one failure with a mean TRE between 6.2 and 6.5 mm.

Accurate validation for applications outside the head remains a research task. Consistency measures have been proposed whereby three (or more) sets of images are registered and the transformations between all permutations of pairs of images are compared. This allows independent computation of the TRE of, say, image *A* to image *B* by computing the transformation of image *A* to Image *C* and Image *C* to Image *B*. If the errors in the computed transformations are unbiased and uncorrelated then an estimate of one application of the algorithm should be $1/\sqrt{3}$ times the error for the whole circuit *A* to *B* to *C*. Unfortunately, registration algorithms are highly likely to produce both correlated results and bias so the method is not particularly reliable. It can, however, give some indication of registration accuracy.

Assessment of a particular registration result could be estimated by examining the optimal similarity or distance. This is dangerous as a very small residual distance may give falsely optimistic estimates of registration accuracy. As an extreme example, two surfaces may align perfectly if spherical yet the rotations about the center of the sphere are completely undetermined. Measures delivered by voxel similarity-based registration are rarely useful and vary significantly from patient to patient.

Assessment of a particular registration result is best left to visual inspection. If image quality is good, observers can be trained to detect very small mis-registration errors, for example, the study of MRI-CT registration undertaken by Fitzpatrick et al. [62]. Holden et al. [63] have shown that mis-registration errors of as little as 0.2 mm can be detected when examining subtracted, registered serial MR images. It is strongly recommended that visual assessment is part of any clinical protocol involving image registration. Observers must be trained to estimate visually any residual mis-registration, preferably at predefined anatomical locations. An acceptable tolerance should be defined and any registration error outside this tolerance should be treated as a registration failure.

Conclusions

Medical image registration technology has developed at a rapid pace in the last five to ten years and robust methods are now widely available for applications in the head. Algorithms based on voxel similarity have been shown to be sufficiently accurate and robust for most clinical applications in neuro-imaging. Software is now widely available in the academic community for research purposes, for example, from *www.image-registration.com* under the free software foundation license running with the visualization toolkit (vtk) [*www.kitware.com*]. These algorithms are likely to be integrated in commercial medical imaging software in the next few years.

Registration technology will allow PET images to be used to guide sophisticated image-guided interventions and will be a key enabling technology in delivery and monitoring of new molecularly based therapeutics. Non-rigid algorithms will become more robust for applications in which soft tissues deform between imaging procedures. These algorithms will become more firmly linked to computer models of the biomechanics of tissue deformation.

Image registration methodologies are allowing fundamental discoveries to be made in cohort studies of brain function and are helping scientists to untangle spatio-temporal distributions of cognitive processes. Provided a definition of correspondence is agreed, non-rigid registration algorithms provide an accurate way to combine information across study groups, the analysis of function across populations and the analysis of differences between study groups.

As we will see in the next chapter, Integrated PET/CT devices and, in the future PET/MRI devices, will make image alignment more straight-forward. These technologies may open up opportunities for time-synchronized analysis of data as well as spatial registration.

References

1. Maintz JBA, Viergever MA. A survey of medical image registration. Med Image Anal 1998;2:1–36.
2. Hill DLG, Batchelor PG, Holden M, Hawkes DJ. Medical image registration. Phys Med Biol 2001;46:R1–45.
3. Hajnal J, Hill DLG, Hawkes DJ. Medical image registration. CRC Press LLC (The Biomedical Engineering Series), 2001.
4. Evans AC, Kamber M, Collins D. An MRI-based stereotactic atlas of neuroanatomy. In: Shorvon S, Fish D, Andermann F, Bydder G, Stefan H eds. Magnetic resonance scanning and epilepsy, Boston: Plenum Press (NATO ASI Series, vol 264), 1994.
5. Rueckert D, Sonoda LI, Hayes C, Hill DLG, Leach MO, Hawkes DJ. Non-rigid registration using free-form deformations: application to breast MR images. IEEE Trans Med Imaging 1999;18:712–21.
6. Talairach J, Tournoux P. Co-planar stereotactic atlas of the human brain: 3-dimensional proportional system: an approach to cerebral imaging. Stuttgart: Georg Thieme Verlag, 1988.
7. Thirion J-P. Image matching as a diffusion process: an analogy with Maxwell's demons. Medical Image Analysis 1998;2:243–60.
8. Collins DL, Peters TM, Evans AC. An automated 3D non-linear image deformation procedure for determination of gross morphometric variability in human brain. Visualization in Biomedical Computing 1994, (SPIE) 2359:180–90.
9. Gee JC, Reivich M, Bajcsy R. Elastically deforming 3D atlas to match anatomical brain images. J Comput Assist Tomogr 1993;17:225–36.
10. Collins DL, Homes CJ, Peters TM, Evans AC. Automatic 3D model-based neuroanatomical segmentation. Human Brain Mapping 1995;3:190–208.
11. Christensen G, Joshi S, Miller M. Volumetric transformation of brain anatomy. IEEE Trans Med Imaging 1996;16:864–77.
12. Pietrzyk U, Herholz K, Fink G, Jacobs A, Mielke R, Slansky I et al. An interactive technique for three-dimensional image rgistration: validation for PET, SPECT, MRI and CT brain images. J Nucl Med 1994;35:2011–18.
13. Gill SS, Thomas DGT. A relocatable frame. Journal of Neurosurgery and Psychiatry 1989;52:1460–1.
14. Hauser R, Westermann B, Probst R. Non-invasive tracking of patients' head movement during computer assisted intranasal microscopic surgery. Laryngoscope 1997;211:491–9.
15. Maurer CR Jr, Fitzpatrick MJ, Wang MY, Galloway RL, Maciunas RJ, Allen GS. Registration of head volume images using implantable fiducial markers. IEEE Trans Med Imaging 1997;16:447–61.
16. West J, Fitzpatrick JM, Wang MY, Dawant BM, Maurer CR, Kessler RM et al. Comparison and evaluation of retrospective intermodality image registration techniques. J Comput Assist Tomogr 1997;21:554–66.
17. Fenlon MR, Jusczyzk AS, Edwards PJ, King AP. Locking acrylic resin dental stent for image-guided surgery. J Prosthetic Dentistry 2000;83:482–5.
18. Van den Elsen PA,Viergever MA. Marker-guided registration of electomagnetic dipole data with tomographic images. In: Colchester ACF, Hawkes DJ eds. Information processing in medical imaging. Heidelburg: Springer Verlag, 1991;142–53.
19. Wadley JP, Doward NL, Breeuwer M, Gerritsen FA, Kitchen ND, Thomas DGT. Neuronavigation in 210 cases: further development of applications and full integration into contemporary neuosurgical practice. In: Lemke H, Vannier M, Inamura K, Farman A eds. Proceedings of computer-assisted radiology and surgery. Amsterdam: Elsevier, 1998;635–40.
20. Evans AC, Marret S, Collins L, Peters TM. Anatomical–functional correlative analysis of the human brain using three dimensional imaging systems. SPIE 1989;1092:264–74.
21. Schoenemann PH. A generalized solution of the orthogonal Procrustes problem. Psychometrika 1966;31:1–10.
22. Arun KS, Huang TS, Blostein SD. Least squares fitting of two 3D point sets. IEEE Transactions on Pattern Analysis and Machine Intelligence 1987;9:698–700.
23. Dryden I, Mardia K. Statistical shape analysis. New York, 1998.
24. Golub GH, van Loan CF. Matrix computations, 3rd edn. Baltimore: Johns Hopkins University Press, 1996.
25. Bookstein FL. Thin-plate splines and the atlas problem for biomedical images. In: Colchester ACF, Hawkes DJ eds. Information processing in medical imaging: Proc 12th International Conference. Berlin: Springer (LNCS vol. 511), 1991.
26. Pelizzari CA, Chen GTY, Spelbring DR, Weichselbraum RR, Chen C. Accurate three dimensional registration of CT, PET and/or MR images of the brain. Journal of Computer Assisted Tomography 1989;13:20–6.
27. Press WH, Teukolsky SA, Vetterling WT, Flannery BP. Numerical recipes in C: the art of scientific computing 2nd edn. Cambridge: Cambridge University Press, 1992.
28. Faber TL, McColl RW, Opperman RM, Corbett JR, Peshock RM. Spatial and temporal registration of cardiac SPECT and MR images: methods and evaluation. Radiology 1991;179:857–61.
29. Borgefors G. Distance transformations in digital images. Computer Vision, Graphics and Image Processing 1986;34:344–71.
30. Jiang H, Robb RA, Holton KS. New approach to 3-D registration of multimodality medical images by surface matching. SPIE 1992;1808:196–213.
31. Van den Elsen PA, Maintz JBA, Viergever MA. Geometry driven multimodality image matching. Brain Topogr 1992;5:153–8.
32. Van Herk M, Kooy HM. Automated three-dimensional correlation of CT–CT, CT–MRI and CT–SPECT using chamfer matching. Med Phys 1994;21:1163–78.
33. Huang CT, Mitchell OR. A Euclidean distance transform using greyscale morphology decomposition. IEEE Trans Patter Anal Mach Intell 1994;16:443–8.
34. Besl PJ, McKay ND. A method for registration of 3D shapes. IEEE Transactions in Pattern Analysis and Machine Vision. 1992; 14:239–56.
35. Monga O, Benayoun S. Using partial derivatives of 3D images to extract typical surface features. Comput Vision Image Understanding 1995;61:171–89.
36. Gueziec A, Pennec X, Ayache N. Medical image registration using geometric hashing. IEEE Comput Sci Eng 1997;4:29–41.
37. Viola PA. Alignment by maximization of mutual information. PhD Thesis, Massachusetts Institute of Technology, 1995.

38. Lemieux L, Kitchen ND, Hughes S, Thomas DGT. Voxel based localisation in frame-based and frameless stereotaxy and its accuracy. Medical Physics, 1994;21:1301–10.
39. Woods RP, Cherry SR, Mazziotta JC. Rapid automated algorithm for aligning and reslicing PET images. Journal of Comput Assisted Tomogr, 1992;16:620–33.
40. Van den Elsen PA, Pol E-JD, Sumanaweera TS, Hemler PF, Napel S, Adler JR. Grey value correlation techniques used for automatic matching of CT and MR volume images of the head. SPIE 1994; 2359:227–37.
41. Van den Elsen PA, Maintz JBA, Pol E-JD, Viergever MA. Automatic registration of CT and MR brain images using correlation of geometrical features. IEEE Trans Med Imaging 1995;14:384–96.
42. Maintz JBA, van den Elsen PA, Viergever MA. Evaluation of ridge seeking operators for multimodality medical image matching. IEEE Trans Pattern Anal Mach Intell 1996;18:353–65.
43. Woods RP, Mazziotta JC, Cherry SR. MRI PET registration with automated algorithm. Journal of Comput Assisted Tomogr 1993;17:536–46.
44. Netsch T, Roesch P, van Muiseinkel A, Weese J. Towards real-time multi-modality 3-D medical image registration. International Conference on Computer Vision (ICCV'01). Vancouver, pp. 501–8, 2001.
45. Hill DLG, Studholme C, Hawkes DJ. Voxel similarity measures for automated image registration. SPIE 1994; 2359:205–16.
46. Shannon CE. The mathematical theory of communication (parts 1 and 2). Bell Syst Tech J 1948;27:379–423, 623–56. Reprint available from http://www.lucent.com
47. Collignon A, Maes F, Delaere D, Vandermeulen D, Suetens P, Marchal G. Automated multimodality image registration using information theory. In: Bizais Y, Barillot C, Di Paola R eds. Information processing in medical imaging. Dordrecht: Kluwer; 263–74, 1995.
48. Studholme C, Hill DLG, Hawkes DJ. Multiresolution voxel similarity measures for MR-PET registration. In: Bizais Y, Barillot C, Di Paola R eds. Information processing in medical imaging. Dordrecht: Kluwer; 287–98, 1995.
49. Wells WM III, Viola P, Atsumi H, Nakajima S, Kikinis R. Multimodal volume registration by maximisation of mutual information. Med Image Anal 1996;1:35–51.
50. Studholme C, Hill DLG, Hawkes DJ. Automated 3D registration of MR and PET brain images by multi-resolution optimisation of voxel similarity measures. Medical Physics 1997;24:25–36.
51. Studholme C, Hill DLG, Hawkes DJ. An overlap invariant entropy measure of 3D medical image alignment. Pattern Recognition 1999;32:71–86.
52. Edwards PJ, King AP, Maurer Cr, DeCunha DA, Hawkes DJ, Hill DLG et al. Design and evaluation of a system for microscope-assisted guided interventions (MAGI). IEEE Trans Med Imag 2000;19:1082–93.
53. Nijmeh AD, McGurk M, Hawkes D, Odell E, Fenlon M, Edwards P et al. Intraoperative delineation of oral cancer by computer projected 3D PET images. Proc First European Conference of Head and Neck Cancer, Lille, Nov 2001.
54. Studholme C. Measures of 3D image alignment. PhD Thesis: University of London, 1997.
55. Westermann B, Hauser R. Non-invasive 3-D patient registration for image-guided skull base surgery. Comput and Graphics 1996;20:793–9.
56. Chang H, Fitzpatrick JM. A technique for accurate magnetic resonance imaging I the presence of field inhomogeneities. IEEE Trans Med Imaging 1992;11:319–29.
57. Sumanaweera TS, Glover GH, Song SM, Adler JR, Napel S. Quantifying MRI geometric distortion in tissue. Magn Reson Med 1994;31:40–47.
58. Ramsey CR, Oliver AL. Magnetic resonance imaging-based reconstructed radiographs, virtual simulation, and three-dimensional treatment planning for brain neoplasms. Med Phys 1998;25:1928–34.
59. Studholme C, Hill DLG, Hawkes DJ. Incorporating connected region labelling into automated image registration using mutual information. Mathematical methods in bio-medical image analysis. Amini A (ed), 1996, 23–31.
60. Hill DLG, Maurer Jr CR, Studholme C, Maciunas RJ, Fitzpatrick MJ, Hawkes DJ. Correcting scaling errors in tomographic images using a nine degree of freedom registration algorithm. J Comput Assist Tomogr 1998;22:317–23.
61. Fitzpatrick MJ, West J, Maurer Jr C. Predicting error in rigid-body, point-based registration. IEEE Transactions in Medical Imaging 1998;17:694–702.
62. Fitzpatrick JM, Hill DLG, Shyr Y, West J, Studolme C, Maurer Jr CR. Visual assessment of the accuracy of retrospective registration of MR and CT images of the brain. IEEE Trans Med Imaging 1998;17:571–85.
63. Holden M, Hill DLG, Denton ERE, Jarosz JM, Cox TCS, Goodey J et al. Voxel similarity measures for 3D serial image registration. IEEE Trans Med Imaging 2000;19:94–102.

8 Anato-Molecular Imaging: Combining Structure and Function

David W Townsend and Thomas Beyer

Introduction

Historical Perspectives

Non-invasive technologies that image different aspects of disease should really be viewed, in almost all cases, as complementary rather than competing. When the sensitivity and specificity of one imaging technique for diagnosing or staging a specific disease is compared to that of another technique, it is usually to establish the superiority of one of the two techniques. In practice, however, such comparisons are of little real value because anatomical and functional imaging techniques have different physical specifications of spatial, temporal and contrast resolution, and the images even reflect different aspects of the disease process (Fig. 8.1). CT and MRI are used primarily for imaging anatomical changes associated with an underlying pathology, whereas the molecular imaging techniques of PET and SPECT capture functional or metabolic changes associated with that pathology. Historically, CT has been the anatomical imaging modality of choice for the diagnosis and staging of malignant disease and monitoring the effects of therapy. However, more recently, molecular imaging with whole-body PET has begun assuming an increasingly important role in the detection and treatment of cancer [1].

Nevertheless, historically, functional and anatomical imaging modalities have developed somewhat independently, at least from the hardware perspective. For example, until recently, CT developed as a single-slice modality, while PET and SPECT have always essentially been volume imaging modalities even if, for technical reasons, acquisition and reconstruction has been limited to two-dimensional transverse planes. CT detectors integrate the incoming photon flux into an output current whereas PET detectors count individual photons. Some similarity can be found in the data processing as the image reconstruction techniques are

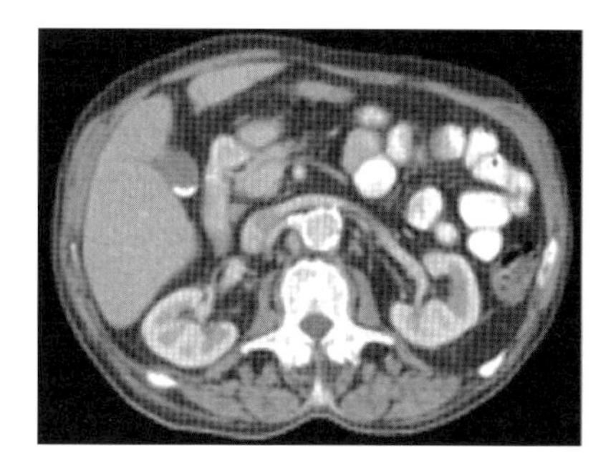

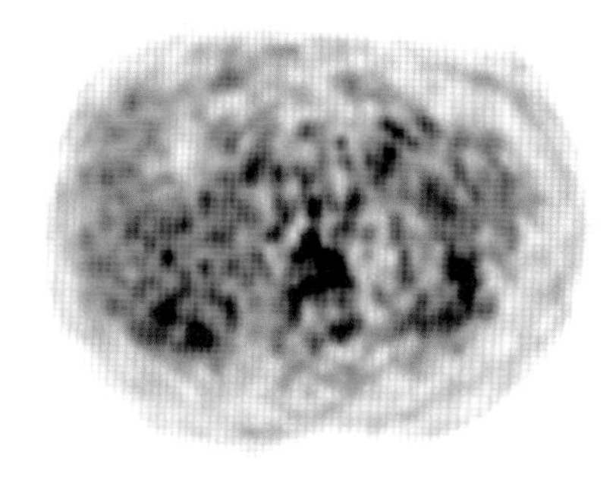

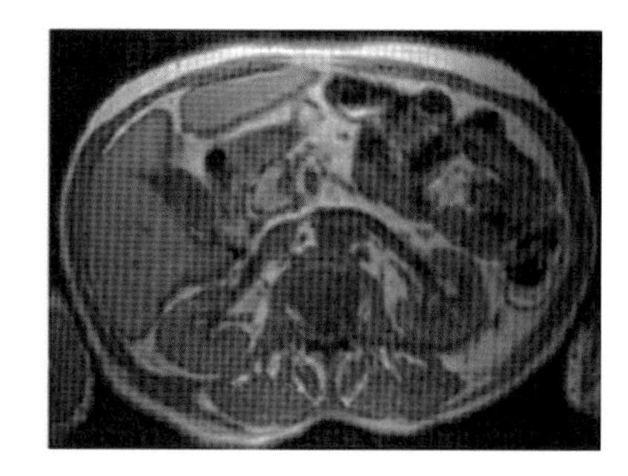

Figure 8.1. Transverse CT (a), PET (b), and MRI (c) image of the abdomen illustrating different aspects of the anatomy and metabolism of the patient as imaged with the different diagnostic imaging techniques.

based on common theoretical principles. The implementation details, however, involve many important differences that set the various modalities apart. Consequently, over the past twenty-five years, the development of anatomical and molecular imaging techniques has followed distinct, but parallel, paths, each supporting its own medical speciality of radiology and nuclear medicine.

Image Fusion: A Hardware Approach

Complementing Anatomy with Function

In clinical practice a PET study, if available, is generally read in conjunction with the corresponding CT scan, acquired on a different scanner and usually on a different day. Adjacent viewing of anatomical and functional images, even without accurate alignment and superposition, can help considerably in the interpretation of the studies. Using the retrospective software-based approaches anatomical and molecular images can be aligned and read as combined, or fused, images. This can be an advantageous procedure because identification of a change in function without knowing accurately where it is localized, or equivalently, knowledge that there is an anatomical change without understanding the nature of the underlying cause, compromises the clinical efficacy of both, the anatomical and functional imaging. More importantly, since a functional change may precede an anatomical change early in the disease process, there may be no identifiable anatomical correlate of the molecular change, although of course a sufficient number of cells must first be affected to produce a macroscopic change that can be imaged with a PET scanner.

Software Approach to Image Registration

While the superimposition of functional and anatomical images was occasionally attempted, it was during the late eighties that the importance of directly combining anatomy and function began to emerge. The first such attempts were software based [2, 3]. These software developments were driven primarily by a demand for accurate localization of cerebral function visualized in PET studies where the low-resolution morphology is, in most cases, insufficient to accurately identify specific cerebral structures. Software fusion techniques were successful for the brain, a rigid organ fixed within the skull, whereas for other parts of the body, image fusion was found to be somewhat problematic [4, 5].

In particular, combining complementary whole-body image data sets retrospectively is not straightforward due to the difference in patient set-up and variable definitions of the axial examination ranges of the two imaging modalities, which are often acquired independently by different medical personal. Further, normal variants of the position and metabolic activity of bowel and intestines at the time of the two scans, as well as dissimilar breathing patterns contribute to additional systematic difference in the two data sets. Although some of the positioning errors may be overcome by non-linear image warping techniques and 3D elastic transformations [6] these registration algorithms are typically limited to a single anatomical region like the thorax [5-9] and are often labour-intensive, thus making them less attractive for routine clinical use in high-throughput situations.

Hardware Approaches to Combined Imaging

An alternative to *post hoc* image fusion by software is, instead, to fuse the hardware from the two imaging modalities. While presenting a significant number of challenges, such an approach overcomes many of the difficulties of the software fusion methods.

In the early nineties, Hasegawa and co-workers at the University of San Francisco developed the first device that could acquire both, anatomical (CT) and functional (SPECT) images, using a single, high-purity germanium detector for both modalities [10, 11]. The CT images were, in addition, used to provide attenuation factors for correction of the SPECT data [12], and operating the device with two different energy windows allowed simultaneous emission-transmission acquisitions to be performed. This pioneering work of Hasegawa, Lang and co-workers is important because it was one of the first to take an alternative, hardware-based approach to image fusion. However, the difficulty of achieving an adequate level of performance for both SPECT and CT with the same detector material and without compromising either modality, led the group to explore a combination of SPECT and CT using different, dedicated imaging systems for each modality – a clinical SPECT camera in tandem with a clinical CT scanner [13]. The CT images are used to correct the SPECT data for photon attenuation, and the device has been used clinically for patient studies since around 1996.

Recognizing the advantages of a hardware solution to combining anatomy and function, Townsend, Nutt and co-workers initiated a design for a combined PET and CT scanner in the early 1990s. The device comprised a clinical CT scanner and a dedicated clinical PET scanner. As with the CT/SPECT scanner of Hasegawa, the CT images are also used to correct the emission data for photon attenuation. The device was completed in early 1998 and underwent an extensive, three-year clinical evaluation programme at the University of Pittsburgh.

A third possibility to have anatomy and function imaged in a single device is to combine PET and MR. Obviously such a combination is technologically more challenging than combining PET with CT in view of the extensive restrictions placed on the imaging environment by the strong magnetic field. Nevertheless, proposals to place PET detectors inside an MR scanner also date back to the mid-nineties [14, 15]. In 1996, an MR-compatible PET scanner was developed at UCLA [16], and then in 1997 a second, larger prototype (5.6 cm diameter ring) was constructed [17] and used in collaboration with researchers at Kings' College London for phantom and animal studies. The studies clearly demonstrated that simultaneous PET and MR images could be acquired using a range of pulse sequences and at different field strengths [18]. The device opened up the possibility of simultaneously imaging [^{18}F]-FDG uptake and measuring MR spectra. Currently a larger, 11.2 cm PET detector ring is being developed, designed to fit inside a 20 cm diameter magnet bore [19]. However, scaling the design up to human dimensions will present many challenges and is still a number of years away.

Design Concept of the Prototype PET/CT Scanner

Design Concept

The design objective for combined PET/CT imaging, as a technically straightforward combination of two complementary imaging modalities, was to provide clinical CT and clinical PET imaging capability within a single, integrated scanner (Table 8.1). The short CT scan duration compared with a typical whole-body PET acquisition time essentially eliminates the requirement for simultaneous CT and PET acquisition. The integration of the two modalities within a single gantry is more straightforward when the simultaneous operation of the CT and PET imaging systems is not required. The PET and CT components are mounted on the same aluminium support with the CT on the front and the PET at the back, as shown schematically in Fig. 8.2. The entire assembly rotates at 30 rpm and is housed within a single gantry of dimensions 170 cm wide and 168 cm high. The patient port is 60 cm in diameter with an overall tunnel length of 110 cm and a 60 cm axial displacement between the center of the CT and the center of the PET imaging fields. A single patient bed is used for both modalities with an axial travel sufficient to cover 100 cm of combined CT and PET imaging. To simplify the prototype development, the acquisition and reconstruction paths are not integrated, with CT and PET scanning controlled from separate consoles, as shown schematically in Fig. 8.2. Once acquired and reconstructed, the CT images are transferred to the

Table 8.1. A comparison of advantages and disadvantages of software- and hardware-based image fusion.

Software fusion	Hardware fusion
Image retrieval from different archives	Images avilable from one device
Repeated patient positioning	Single patient positioning
Different scanner bed profiles	One bed for both scans
Uncontrolled internal organ movement between scans	Consecutive scans with little internal organ movement in between
Disease progression in time between exams	Scans acquired close in time
Limited registration accuracy	Improved registration accuracy
Less convenient for patient (two exams)	Single, integrated exam
Labour-intensive registration algorithms	No further image alignment required

PET computer to provide the attenuation correction factors for the PET emission data. Final PET reconstruction and CT and PET fused image display is performed on the PET computer console.

CT Scanner

The CT components of the prototype PET/CT is a Siemens Somatom AR.SP, a single-slice spiral CT scanner with a 25 kW M-CT 141 tube that produces X-ray spectra of 110 kV_p and 130 kV_p. The detector array comprises 512 xenon gas-filled chambers. A slice thickness of 1, 2, 3, 5 or 10 mm can be selected, with a maximum rotation speed of 1.3 s. For compatibility with the PET components mounted on the same support, the combined assembly is limited to a rotation speed of 30 rpm. The transverse field of view is 45 cm and the patient port is 60 cm. CT data transfer is over mechanical slip rings, as is the power to the X-ray tube and detectors. The tilting capability of the CT is disabled.

PET Scanner

The PET components mounted on the rear of the support are those of a standard Siemens ECAT® ART scanner [20] comprising dual arrays of bismuth germanate (BGO) block detectors. Each array consists of 11 blocks (transverse) by 3 blocks (axial); the blocks are 54 mm × 54 mm × 20 mm in size, cut into 8 × 8 crystals each of dimension 6.75 mm × 6.75 mm × 20 mm. The transverse field-of-view (FOV) is 60 cm and the axial FOV is 16.2 cm, subdivided into 24 (partial) rings of detectors for a plane spacing of 3.375 mm. Shielding from out-of-field activity is provided by arcs of lead, 2.5 cm thick, mounted on both sides of the detector assembly and projecting 8.5 cm into the FOV beyond the front face of the detectors. The PET scanner has no septa and the detector arrays rotate continuously at 30 rpm to collect the full set of projections required for image reconstruction. Power and serial communications to the rotating assembly are transmitted over mechanical slip rings, while high speed digital data transfer is by optical transmission.

Physical performance of the prototype PET/CT

The overall physical performance of the combined scanner is comparable to that of the individual components, the Somatom AR.SP and the ECAT ART. The PET and CT components are mounted on opposite sides of the aluminium support thus minimizing potential interference between the two imaging systems. Although the sensitivity of the PET detectors is temperature-dependent, no significant effect from the operation of the X-ray source has been observed. The PET detectors are never exposed directly to the X-ray flux and the operation of the CT has no residual effect on the photomultiplier tube gains. The PET components can be operated immediately after the acquisition of the CT scan without requiring a recovery time. However, the PET and CT components cannot acquire data simultaneously because of the high flux of scattered CT photons incident on the PET detectors that results in high levels of random coincidences and system dead-time (from pulse pile-up effects). In view of the short time required for the CT, simultaneous operation of both PET and CT scanners is not considered necessary. The PET components are operated with a detector block integration time of 384 ns [21], a coincidence window of 12 ns, and a lower energy threshold of 350 keV. The operation of the Somatom AR.SP is in accordance with standard CT procedures. Complete details of the results of the performance measurements and relevant parameter settings can be found in [22].

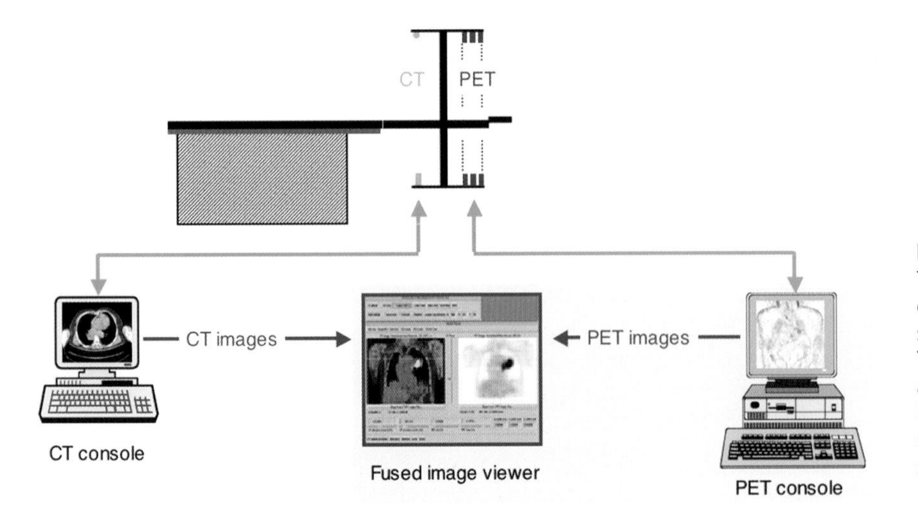

Figure 8.2. Design of the prototype PET/CT. The PET components were mounted on the rear of a common rotating support. The axial separation of the two imaging fields was 60 cm. The entire assembly within the gantry rotated at 30 rpm. The co-scan range for acquiring both PET and CT was 100 cm (maximum). CT and PET scans were acquired and reconstructed on separate consoles but image fusion display was installed on the PET console alone.

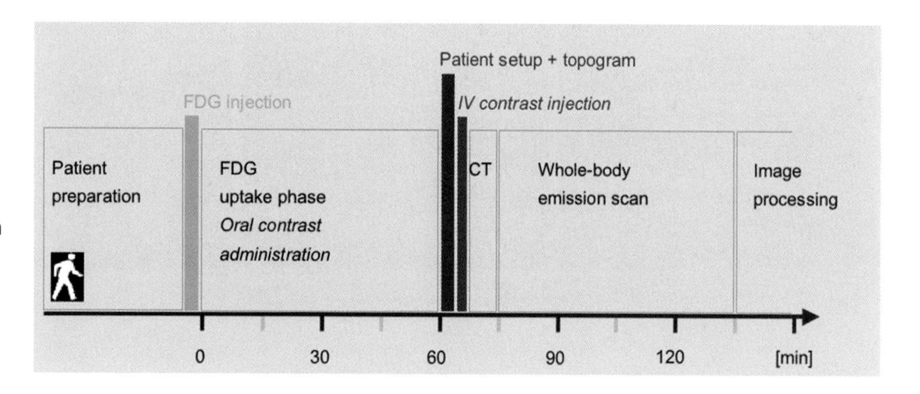

Figure 8.3. Whole-body FDG acquisition protocol for the prototype PET/CT. Note, the standard transmission is replaced with a multi-bed CT scan (tube cooling is needed in case of an extended imaging range) and both, IV and oral CT contrast is given to enhance the diagnostic quality of the transmission images. Attenuation correction factors are calculated on-line but PET images are reconstructed post-acquisition.

Clinical Protocols and Evaluation

When imaging clinically with the PET/CT, a typical acquisition protocol begins with a 260 MBq injection of FDG, followed by a 60 min uptake period. The patient is then positioned in the scanner with the first transverse section to be imaged aligned with the CT field-of-view. An initial scout scan (topogram) is performed to determine the appropriate axial range for the study. The maximum axial extent of a single spiral scan depends on the defined slice-width and pitch. The total axial length to be scanned is subdivided into contiguous, overlapping, 15 cm segments. For the Somatom AR.SP, the spiral scan of each segment typically takes about 40 s, and the 25 kW X-ray tube may sometimes require cooling between segments (Fig. 8.3). Patients are instructed to breathe in a shallow manner during the CT scan. The time for the complete whole-body CT scan is about 5 min. Once the spiral CT covering the required axial length is completed, the patient bed is moved automatically to the start position of the multi-bed PET acquisition, and the PET scan is initiated. An emission scan time of 6-10 min per bed position is selected depending on the number of bed positions, resulting in a total PET scan duration of 45-50 min (Fig. 8.3). An axial overlap of 4 cm is used between bed positions. The CT images are used for attenuation correction as will be described below, and the corrected emission data are reconstructed using Fourier rebinning and attenuation-weighted ordered-subset EM [23]. In this implementation reconstruction takes over one hour to complete.

From July 1998 to July 2001, over 300 patients with a wide variety of different cancers were scanned on the prototype PET/CT [24, 25]. The main indications, most suited to anatomical and functional imaging, are head and neck cancer, and abdominal and pelvic disease, particularly ovarian and cervical cancer. Combined PET/CT in the head and neck is important because normal uptake of FDG in muscles and glands makes interpretation of the studies especially difficult. PET applications in the abdomen and pelvis are complicated by benign, non-specific uptake in the stomach, intestines and bowel that may be difficult to distinguish from malignant disease. Combined imaging for clinical routine allows accurate localization of lesions, the distinction of normal FDG uptake from pathology, and the assessment of response to therapy (Fig. 8.4). Additionally, the use of registered CT and PET images was envisaged for efficient radiation therapy planning, traditionally based on CT alone [26].

CT-based Attenuation Correction

Transforming Attenuation Coefficients

In addition to acquiring co-registered anatomical and functional images, a further advantage of the combined PET/CT scanner is the potential to use the CT images for attenuation correction of the PET emission data, eliminating the need for a separate, lengthy PET transmission scan. The use of the CT scan for attenuation correction not only reduces whole-body scan times by at least 30% [27], but provides essentially noiseless attenuation correction factors compared to those from a standard PET transmission scan. CT-based attenuation values are, however, energy dependent, and hence the correction factors derived from a CT scan at a mean photon energy of 70 keV must be scaled to the PET energy of 511 keV, for which a hybrid scaling algorithm was developed [28]. This scaling approach is based on previous work by La Croix and Tang [29, 30] who have shown that attenuation correction factors for SPECT emission data can be derived from complementary CT transmission data. The single scale-factor approach works well for soft tissues, but serious overestimation of the attenuation properties of cortical bone and ribs is observed, especially with increasing difference of the transmission and emission

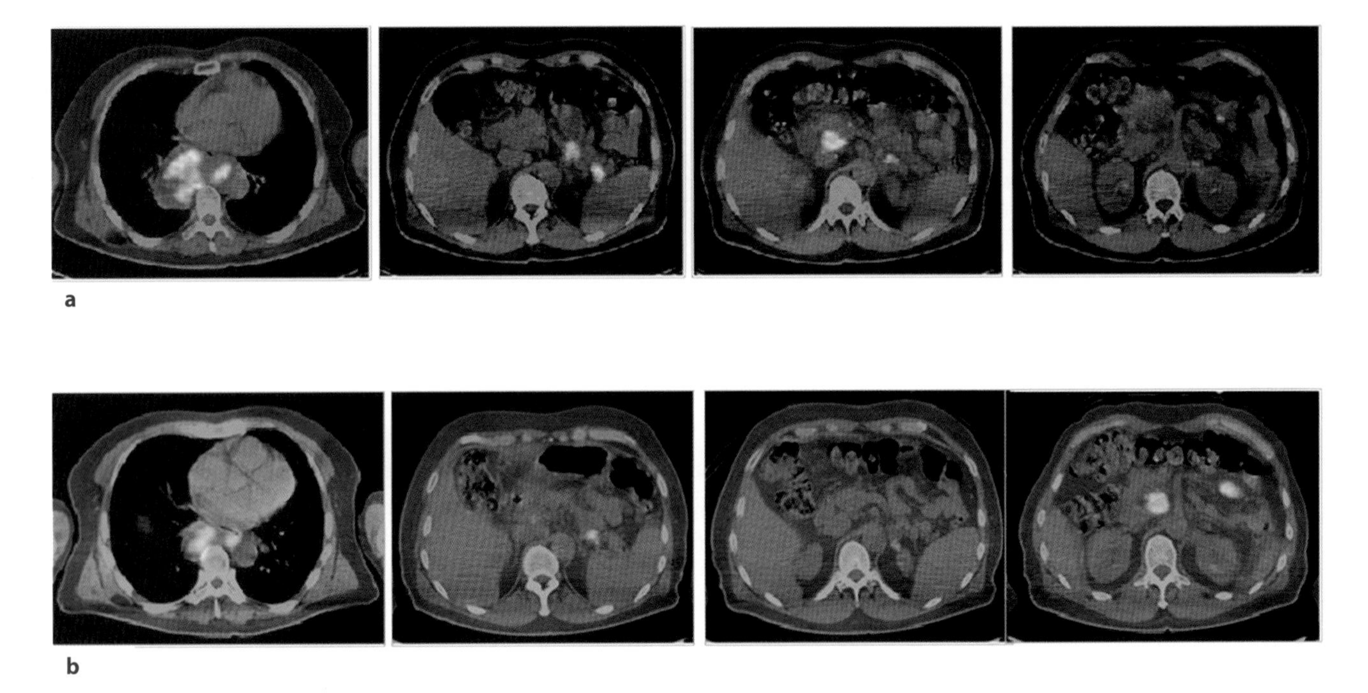

Figure 8.4. Male patient with metastatic melanoma before (a) and after (b) chemotherapy in 11/98 and 12/98, respectively. PET/CT images are from the prototype PET/CT. Multiple lesions are depicted as FDG avid and localized accurately within the anatomy of the patient. Each whole-body scan took about 1 h. PET/CT images before and after therapy are registered by hand and selected axial views are shown to demonstrate multi-variant response to therapy.

photon energies [29]. Nevertheless, the overestimation of attenuation in bone translated into only a minor average overestimation of the tracer uptake in the corrected SPECT images due to the low fraction of voxels containing bone compared to other tissues.

In anticipation of the prototype PET/CT the original scaling approach was extended to CT-based attenuation correction of PET emission data [28]. A bi-linear scaling is employed to account for both the photon energy difference between CT and PET, and the different attenuation properties of low-Z (soft tissues) and high-Z (bone) materials in the range of the lower energy X-ray photons (Fig. 8.5).

The algorithm is based on the observation that for water, lung, fat, muscle and other soft tissues, the mass attenuation coefficient (linear attenuation coefficient divided by density) at CT and PET energies is approximately the same. While the actual value is different at the effective CT energy of 70 keV to that at 511 keV, all the tissues can be scaled with a single factor: the ratio of the mass attenuation coefficient at 511 keV to that at 70 keV. The exception is bone because, at CT energies, the mass attenuation coefficient is somewhat higher than that for the other tissues due to the increased photoelectric contribution from calcium. The ratio of the mass attenuation at 511 keV to that at 70 keV is approximately 0.53 for soft tissues, and 0.44 for bone.

CT-based PET attenuation correction factors are generated in a 4-step procedure [28]:

1. the CT images are divided into regions of pixels classified as either non-bone or bone by simple thresholding. A threshold at 300 Hounsfield Units (HU) separates spongiosa and cortical bone from other tissues,
2. the pixel values in the CT image in HU are converted to attenuation coefficients of tissue (μ_T) at the effective CT energy (~70 keV) using the expression: $\mu_T = \mu_W(HU/1000+1)$; μ_W is the attenuation coefficient for water,
3. the non-bone classified pixel values are then scaled with a single factor of 0.53, and bone classified pixel values are scaled with the smaller scaling factor of 0.44,
4. attenuation correction factors are generated by integrating (forward projecting) along coincidence lines-of-response through the segmented and scaled CT images, with the CT spatial resolution degraded to match that of the PET. Oblique lines-of-response are obtained in the same way by integration through the CT volume.

Today bi-linear scaling methods [28, 31] are widely accepted for clinical PET/CT imaging, and are, with minor modifications, used routinely for CT-based attenuation correction of the PET emission data [32, 33].

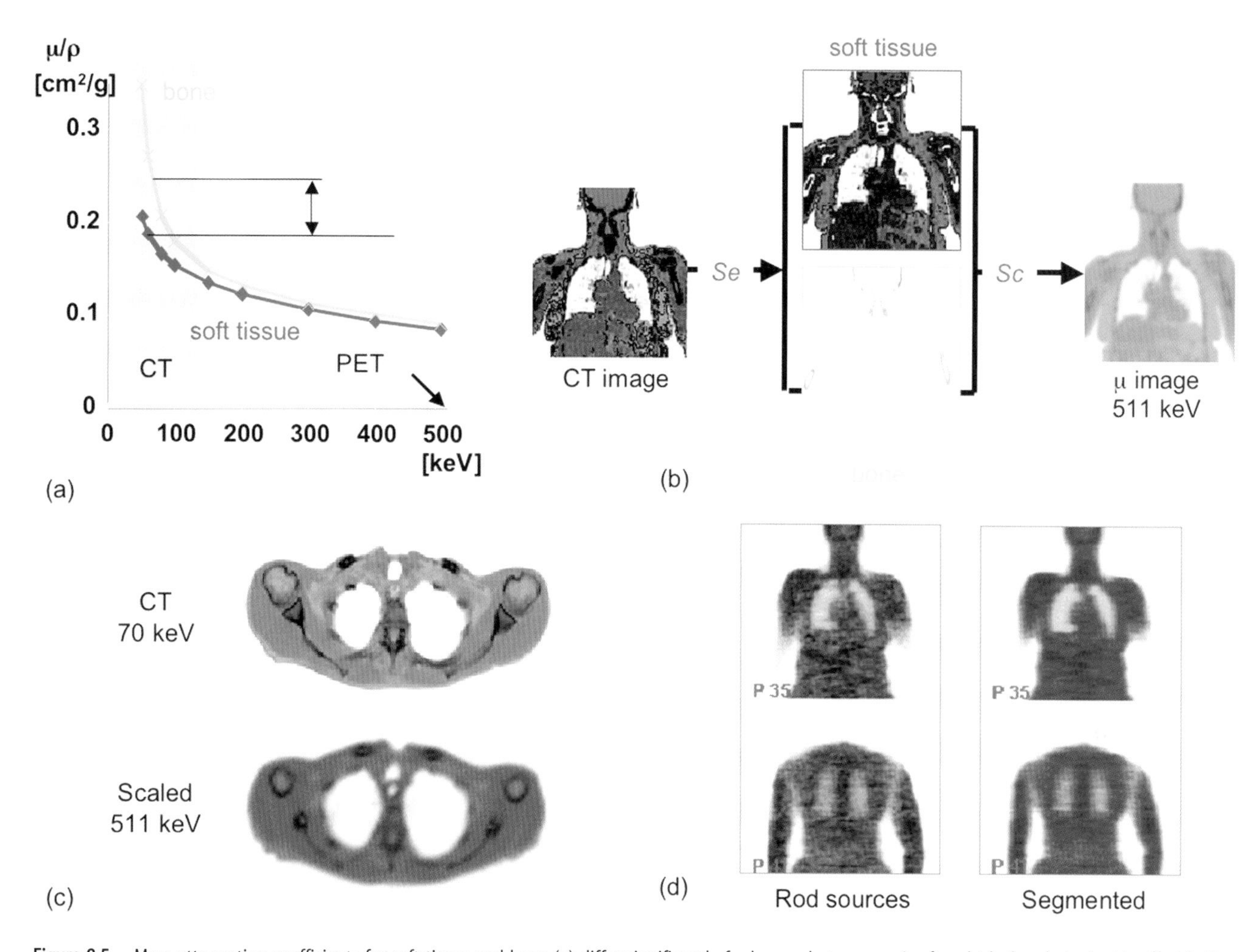

Figure 8.5. Mass attenuation coefficients for soft tissue and bone (a) differ significantly for lower photon energies for which the photoelectric effect is the dominant interaction with matter. The bi-linear scaling and segmentation approach (b) accounts for the different attenuation properties of soft tissue and bone by, first, segmenting (Se) the CT, and, second, by applying a tissue dependent scale factor (Sc) to these pixels. The greatly increased photon flux used in CT results in essentially noiseless attenuation maps (c) and in attenuation correction factors compared with those derived from a standard PET transmission scans (d).

The time for the acquisition of the attenuation data for a whole-body study can be reduced to one minute, or less, by using a fast CT scan instead of a lengthy PET transmission measurement. Furthermore CT transmission data acquired in post-injection scenarios are not noticeably affected by the emission activity inside the patient due to the high X-ray photon flux [22]. Therefore corrective data processing as in post-injection PET transmission imaging [34, 35] is not required.

While, in principle, the CT-based attenuation coefficients are unbiased and essentially noiseless, there are a number of practical limitations. These include respiration effects, truncation of the CT field-of-view when imaging with the arms down, and the effect of using CT contrast.

Patient Respiration

Clinical CT scans of the thorax are normally acquired with breath-hold at full inspiration. The PET image, on the other hand, represents an average over the scan duration of several minutes per bed position, during which the patient breathes normally. Under such a protocol, exact alignment of the CT and PET images, particularly in the lower lungs, is not possible. Typically the movement of the chest wall is suppressed by breath holding during the CT scan, and, with the lungs fully inflated, there is a mismatch between the anterior wall position on CT and the average position in the PET image, as shown in Fig. 8.6. Incorrect attenuation correction factors are then generated by the al-

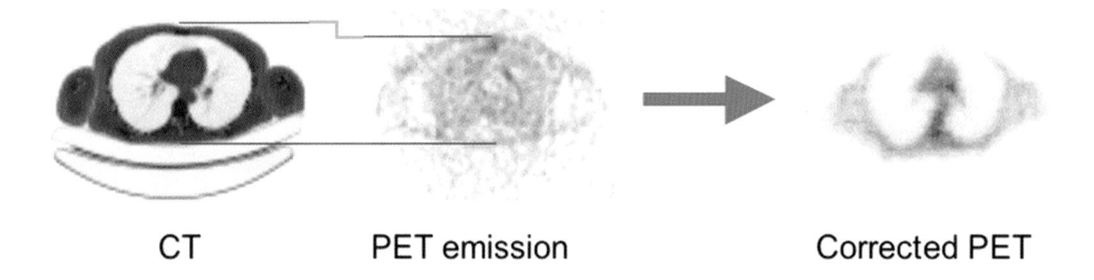

(a)

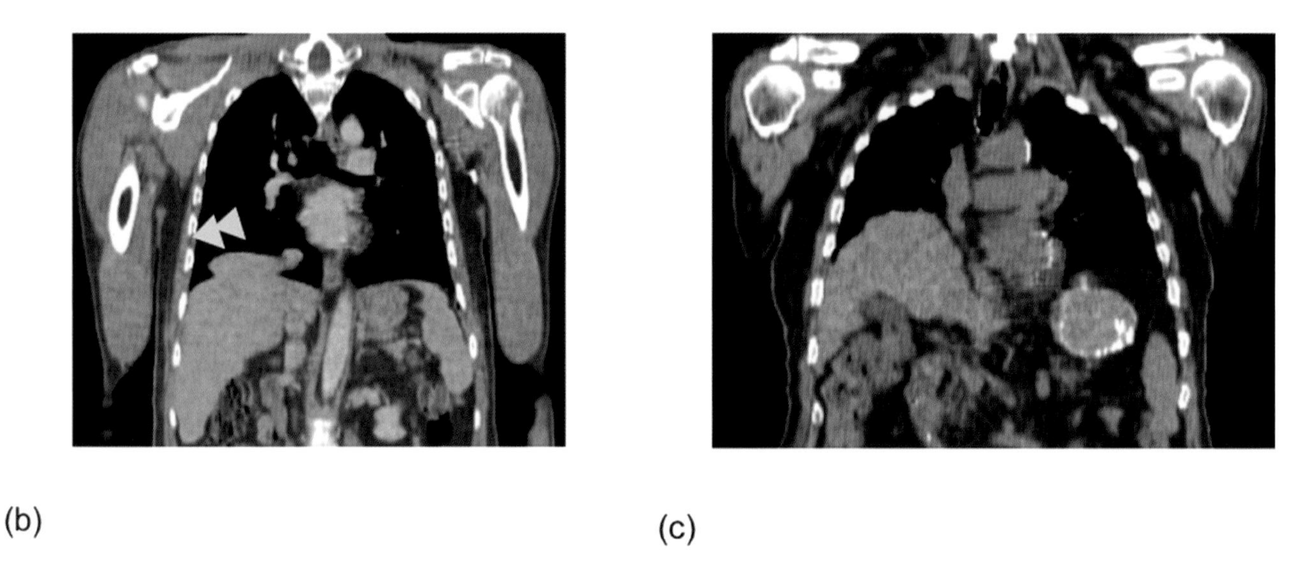

(b) (c)

Figure 8.6. The effect of patient respiration on the CT-based attenuation procedure (a). A CT scan acquired with inspiration breath-hold (left) is not matched with the PET image (right) that is acquired with regular breathing. A CT acquired with normal (b) and shallow (c) breathing; the artifacts near the base of the lung (arrow heads) can be reduced (c).

gorithm described above, since it is based on the assumption that the PET and CT images are accurately co-registered.

When the patient is allowed to breathe normally during both the PET and the CT scans, artefacts appear near the base of the lung and the diaphragm on the CT image (Fig. 8.6b). An alternative protocol is to allow the patient to breathe in a shallow manner (tidal breathing) during both the CT and PET scans, a procedure that minimizes the breathing artefacts on CT (Fig. 8.6c) and the mismatch between the PET and CT images. Nevertheless, a definitive solution to the respiration problem has yet to emerge.

Truncation of the CT Field-of-View

Clinical CT scans of the thorax and abdomen take a few minutes to acquire and can generally be performed with the patient's arms out of the field-of-view. However, in PET/CT imaging total examination time is defined primarily by the time for the emission scan. A typical whole-body PET scan with the prototype PET/CT could last for 1 h, and for scan times of this duration it is difficult for patients to keep their arms comfortably above their head, out of the field-of-view. More recently, with the introduction of full-ring PET components and faster PET detectors into combined PET/CT designs, total examination time for PET/CT is reduced to 30 min [36], or less. Despite the dramatic reduction in total imaging time some patients may still not tolerate having their their arms raised and supported for the duration of the PET/CT scan, and therefore CT and PET imaging must be performed with the arms down and close to the body. However, since the transverse field-of-view of the CT is 50 cm in diameter (45 cm in the prototype), a small angular range of projections around the anterior-posterior direction is, for many patients, truncated (Fig. 8.7). The artefacts caused by truncation affect not only the CT images but also the accuracy of the attenuation correction factors generated from the CT images. The effect is illustrated in Fig. 8.7 for a patient who was imaged on the prototype PET/CT with arms down. The transverse CT field-of-view is limited to 45 cm in diameter. As shown in Fig. 8.7a, truncation leads to ring artefacts around the arms that affect both the accuracy of the CT images and the attenuation correction factors. As a result the tracer distribution in the reconstructed and corrected emission images appears masked near the arms. Figure 8.7b shows a similar patient study from a second generation PET/CT system. Although the transverse CT field-of-view is increased to 50 cm truncation may still occur when imaging large patients. The theory for an effective correction of these truncation artefacts exists today [37, 38], and simplified correction schemes for application in the context of PET/CT imaging are currently being pursued (see Chapter 5).

Effects of CT Contrast Agents

Clinical CT scans are acquired with intravenous contrast and/or oral contrast to enhance the visualization of structures such as the vascular system or the digestive tract. CT contrast media use high atomic number substances such as iodine to increase the attenuation of the vessels or the bowel and intestines above normal, non-enhanced values. Depending on the concentration of the contrast agent, enhancements in CT attenuation values of 1000 HU can be observed. In the presence of contrast agents the routine CT-based attenuation correction algorithm [28] will incorrectly segment and scale the enhanced structures above

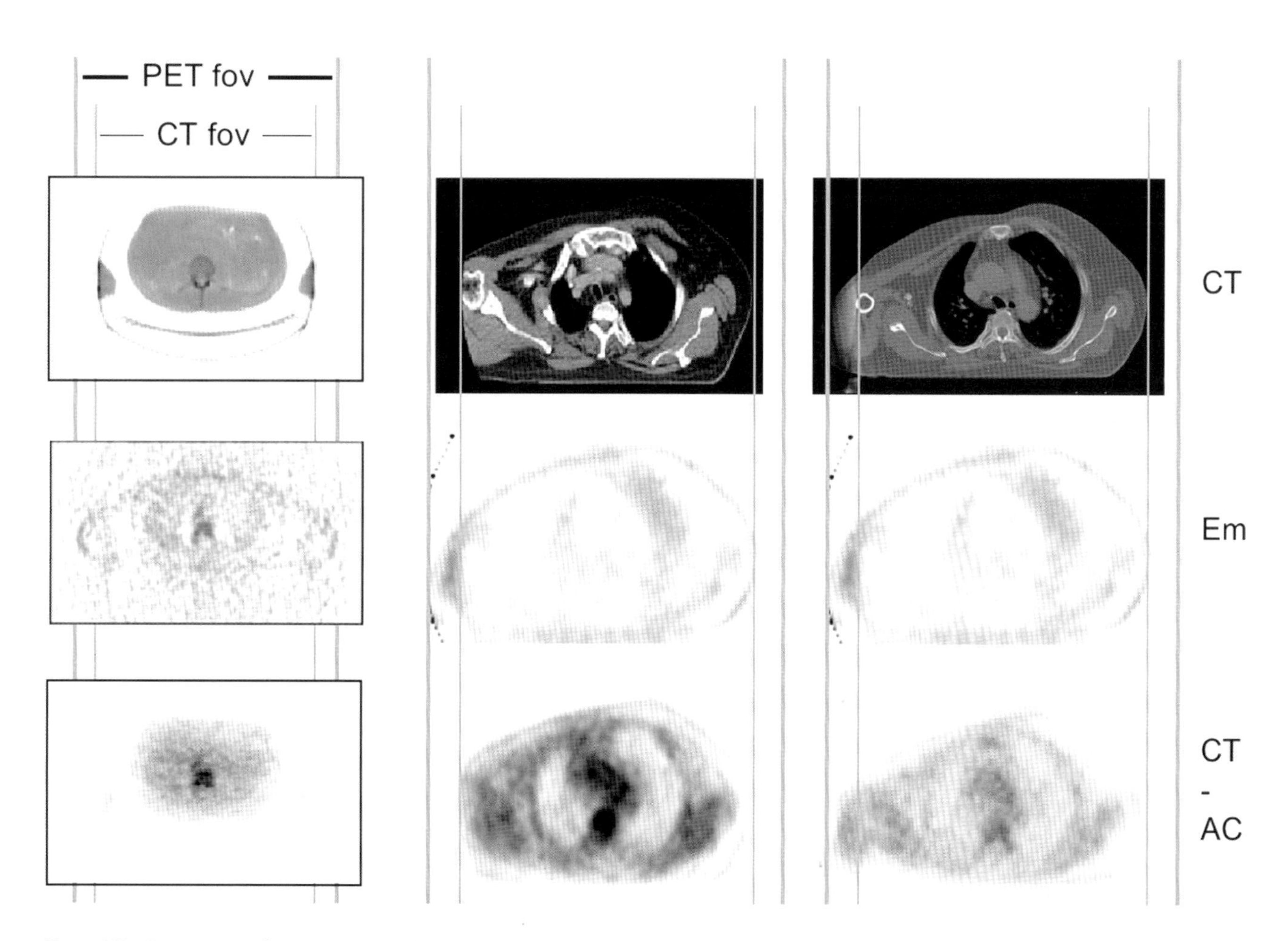

Figure 8.7. Truncation artifacts on CT occur when patients, particularly large patients, are positioned with arms down. The regions outside the maximum transverse field-of-view (45 cm for the prototype PET/CT (a), and 50 cm for second generation PET/CT (b)) are truncated, and the resulting attenuation correction factors for projections traversing the truncated area are underestimated, which yields a "masking effect" of the corrected PET images (CT-AC). The uncorrected emission images (Em) are shown in the middle row. The transverse field-of-view (fov) of the CT and the PET is indicated by the set of the vertical lines. Recently, algorithms have become available that help extrapolate the truncated attenuation information (c), which becomes available for CT-based attenuation correction. (Data processing courtesy of Otto Sembritzki, Siemens Medical Solutions, Forchheim, Germany)

0 HU, resulting in a bias in the attenuation factors. Such biases could potentially generate artefacts in the corrected PET images since the contrast-enhanced CT scan is acquired before the PET scan and then used to generate the attenuation coefficients [39, 40].

Intravenous contrast appears in the vessels as focal regions of elevated attenuation on the CT scan, frequently with attenuation values above 300 HU. Although these small regions are scaled as bone, once the CT resolution has been degraded to match that of the PET (step 4 in the algorithm), there is a significant reduction in image contrast owing to the effect of smoothing. Pixels associated with the vessels nevertheless do have a slightly enhanced value due to the presence of contrast even in the resolution-matched CT images, although the overall effect on the attenuation correction factors is found to be negligible.

Positive oral contrast is potentially more problematic as it collects in larger-volume structures (e.g., intestines) and in a wider range of concentrations. At oral administration, the concentration of the solution will correspond to a pixel value of about 200 HU in the image. However, as water is absorbed from the solution during passage from the stomach and through the intestines, the concentration increases to corresponding CT values of up to 800 HU. Despite these high CT values, phantom studies show only a 2% increase in the linear attenuation coefficient at 511 keV compared to the value for water. Structures containing oral contrast should therefore be transformed, as they would be in the absence of contrast.

Figure 8.8 shows a transverse CT section through the pelvic region containing both regions of bone and positive oral contrast. In Fig. 8.8a, the histogram of CT pixel values exhibits a plateau, or shoulder, due to both bone and oral contrast enhancement. When setting the pixels containing contrast to 0 HU (Fig. 8.8b) a peak at the origin is generated, and a reduced shoulder that repre-

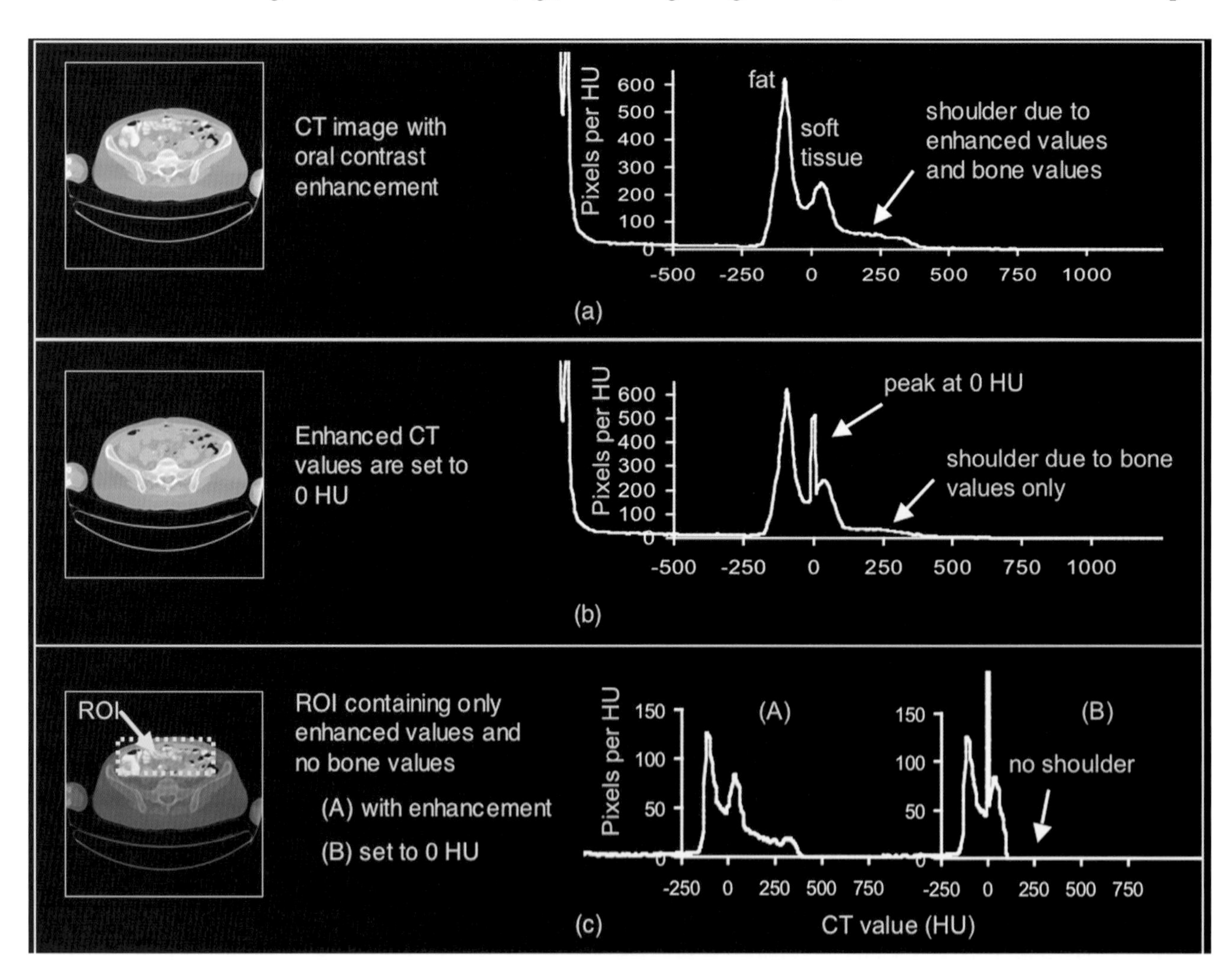

Figure 8.8. Axial CT image in the presence of positive oral contrast (a) and histogram of CT attenuation values (HU). The same image with the contrast enhanced pixel values set to 0 HU is shown in (b), and for the rectangular region-of-interest containing no bone pixels in (c). Selected histograms without (A) and with (B) the enhanced pixels set to 0 HU. (Reproduced from Valk PE, Bailey DL, Townsend DW, Maisey MN. Positron Emission Tomography: Basic Science and Clinical Practice. Springer-Verlag London Ltd 2003, p.206.).

sents bone pixels only. Conversely, within the rectangular region-of-interest indicated (Fig. 8.8c) containing contrast-enhanced pixels and no bone, oral contrast is present in pixels with values above about 150 HU as determined from anatomy-based segmentation of the enhanced colon. Thus, to account for the presence of oral contrast in varying concentrations, pixel values above 150 HU should be set to 0 HU before scaling. Above 300 HU, however, a more complex segmentation procedure than simple thresholding must be used to distinguish contrast enhancement from bone. The presence of both intravenous and oral contrast in the same CT section may further complicate this procedure.

Design Concept of a Production PET/CT Scanner

The somewhat-unanticipated demand for combined PET/CT imaging technology that was created to a large extent by the results from the prototype generated a response from major vendors of medical imaging equipment. However, given the choices for the CT and PET components, a number of decisions had to be made (Table 8.2) that included the appropriate level of CT and PET performance, the extent of hard- and software integration, the potential for upgrades, the targeted users and applications, and of course, the cost [41]. Specifically, the main design questions are:

- what is an appropriate level of CT and PET performance?
- should standard PET transmission sources be provided in addition to CT-based attenuation correction?
- what is the level of hardware integration that can be achieved?
- what co-scan range of PET and CT should the patient handling system offer?
- what level of software integration can be achieved?

In the following sections we describe the design of the first commercial PET/CT tomograph that was developed by CTI PET Systems (CPS Innovations, Knoxville, TN) and presented at the Society of Nuclear Medicine Meeting 2000. Since then several other commercial PET/CT tomographs from other manufacturers have emerged and will be discussed towards the end of this chapter.

The CT and PET Components

The choice of the level of CT and PET performance depends to some extent on the applications envisaged. As with the prototype, the design described here is targeted primarily at PET whole-body oncology, although potential cardiac applications are not excluded.

Since the PET scanner performance is the limiting factor in terms of statistical image quality, spatial resolution, and scan duration, the highest possible PET performance is obviously indicated. This consideration influenced the selection of the ECAT EXACT HR+ [42] as the PET component of choice because of its high sensitivity and high spatial resolution. The scanner has 32 detector rings of crystals of dimensions $4.05 \times 4.39 \times 30$ mm^3, giving an axial plane spacing of 2.43 mm. The detectors cover an axial field-of-view of 15.5 cm. The transmission rod sources are removed and all attenuation correction factors are derived from the CT images, as described in this chapter. The septa are also removed, resulting in a dedicated 3D PET scanner.

The role of the CT is to provide an anatomical infrastructure for the functional images and accurate attenuation correction factors for the PET emission data. However, an objective of this design is also to provide a state-of-the-art CT scan of clinical diagnostic quality, suggesting that a mid to upper range CT scanner can satisfy all three design criteria. For these reasons, the Siemens Somatom Emotion spiral CT scanner (Siemens Medical Solutions, Forchheim, Germany) was selected as the CT component. This CT scanner is available with a single or dual row of Ultra Fast Ceramic (UFC™)

Table 8.2. Design considerations for commercial PET/CT tomographs based on the experiences gained with the prototype PET/CT.

Prototype design	Consequence	PET/CT design goals
ECAT ART	Low-end dedicated PET	Highest PET performance
Somatom AR.SP	Early 90's technology	High-performance CT
30 rpm	Not state-of-art CT	Sub-second rotation
60 cm patient port	Limitation for RTP	Increased port diameter
45 cm CT FOV	CT truncation artifacts	Increased CT FOV
100 cm co-scan	Limitation for whole-body	Whole-body co-scan
No bed support	Bed deflection possible	No bed deflection
Limited access	Service difficulties	Full service access
Dual acquisitions	Operation not integrated	Integrated acquisition

detectors, where each detector row comprises 672 individual elements. The X-ray tube is a 40 kW Siemens DURA 352 with flying spot technology. The full rotation time can be selected from 0.8 s, 1.0 s, and 1.5 s, and slice widths of 1, 2, 3, 5, 8 and 10 mm for the single-slice and 2×1, 2×1.5, 2×2.5, 2×4, 2×5, 8 and 10 mm for the dual-slice. The longest acquisition in a single spiral is 100 s without intermittent tube cooling. The useful diameter of the transverse field-of-view is 50 cm, and the patient port is 70 cm.

Gantry and Patient Handling System

The gantry dimensions of the Somatom Emotion closely match those of the ECAT EXACT HR+, facilitating the mechanical integration of the two units. In comparison to the original prototype design mechanical and thermal isolation is maintained between the two devices for operational and servicing reasons. A schematic of the gantry is shown in Fig. 8.9. The gantry is 188 cm high and 228 cm in width. The overall length is 158 cm, although with the front and rear contouring, the effective tunnel length is only 110 cm. The axial separation of the centres of the CT and PET fields-of-view is 80 cm. The actual 56 cm patient port of the HR+ is increased to 70 cm to match that of the CT by cutting back the side shielding of the PET. The resulting patient port diameter is 70 cm throughout the length of the tunnel, which is essential when positioning most patients from radiation therapy, and which minimizes claustrophobic effects despite the 110 cm tunnel length.

For servicing, the gantries can be separated by moving the PET backwards on rails by about 1 m; access to the rear of the Emotion CT and the front of the EXACT HR+ is then possible. No service procedures on either device have been significantly modified as a consequence of the particular integration into the PET/CT.

The problem of the increasing vertical bed deflection with increasing distance into the scanner encountered with the prototype was resolved by a complete redesign of the patient handling system (PHS). Support of the patient bed throughout the scan range is important to avoid an increasing vertical deflection of the pallet: the pallet deflects downwards with the patient load, a deflection which increases as the bed moves into the tunnel, adversely affecting the CT and PET image registration accuracy. In the prototype (Fig. 8.10a), the pallet is not supported beyond the cantilever point and an approximate correction for the increasing downward deflection is applied in software, based on the patient weight and the pallet position. In the new design, shown in Fig. 8.10b, a carbon fibre pallet is supported at one end by a pedestal that moves horizontally on floor-mounted rails driven by a linear motor. Since the cantilever point does not change, the vertical deflection is limited to a few millimetres once the patient is aligned on the bed, allowing sub-millimeter intrinsic registration accuracy to be achieved between the CT and the PET, independently of the patient weight. A total length, including the head holder, of 145 cm can be scanned with both CT and PET. A flat pallet option is available for use with the PHS when scanning patients undergoing PET/CT for radiation therapy treatment planning.

Software Integration

A key feature of the production PET/CT scanner compared to the prototype is the integration of the CT and PET acquisition and reconstruction software on a single console within a modality-independent software

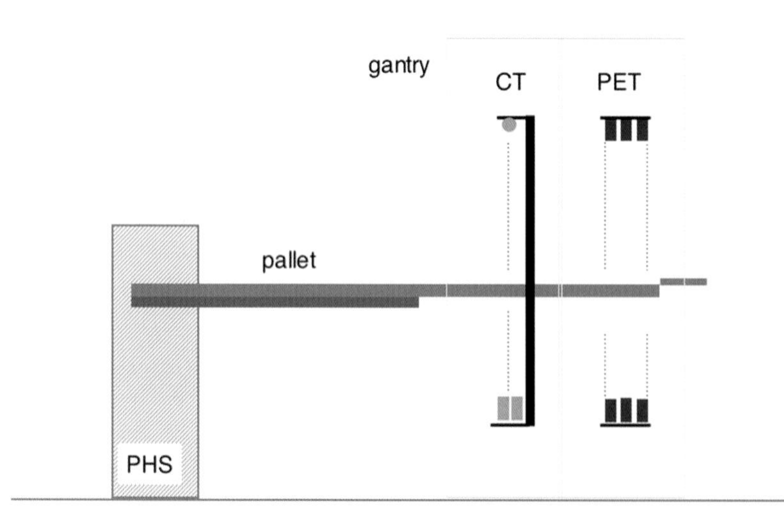

Figure 8.9. A schematic of the biograph PET/CT scanner, as a representative example of the second generation, commercial PET/CT systems. The axial separation of the CT and PET imaging field is 80 cm. A common cantilever design patient handling system (PHS) is mounted to the front of the combined gantry for accurate patient positioning across the 145 cm co-scan range.

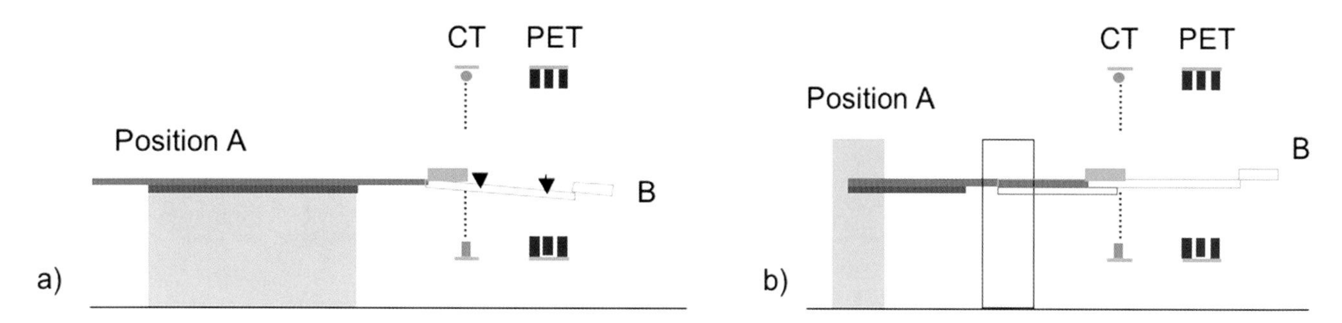

Figure 8.10. A vertical offset (arrows) between the CT and PET data of the same exam is introduced when using a patient handling system with a centre of gravity fixed with respect to the gantry (a). The relative vertical offset between the CT and PET data is eliminated when translating the entire patient handling system, i.e., the patient is moved together with the support structure into the gantry (b).

environment (*syngo*®, Siemens Medical Solutions, Erlangen, Germany). The CT and PET scans can be acquired, reconstructed and viewed on a single console by selecting the appropriate *syngo* task card. An example of the PET/CT examination cards is shown in Fig. 8.11a. The reconstruction software includes CT-based attenuation correction, Fourier rebinning and an attenuation-weighted ordered-subset EM algorithm [43]. The complete whole-body, attenuation-corrected PET images are available within a couple of minutes of the completion of the scan, and all image formats are DICOM compliant to facilitate transfer to PACS or radiation therapy planning systems.

The routine availability of registered CT and PET images highlights the importance of a fused image viewer with a full set of features (Fig. 8.11b). These include transverse, coronal and sagittal displays of CT, PET and fused images using an alpha-blending fusion algorithm. Each modality has the usual set of specific features, for example preset windows and measurement tools for CT, and region-of-interest (ROI) manipulation and SUV calculations for PET. To take full advantage of the registered data sets, enhanced viewing features are required such as linked cursors and common ROI and measurement tools on both CT and PET.

Physical Performance

Other than eliminating the tilt option, no major modifications were required to integrate the Siemens Emotion CT scanner into the PET/CT. The performance characteristics are therefore identical to a standard Emotion CT scanner. The EXACT HR+, however, was modified to accommodate a 70 cm patient port by cutting back the side shielding, and in the absence of septa, all operation is in 3D mode. The increased patient port does not have a significant effect on scanner performance unless there are high levels of activity outside the field-of-view, as can be the case with whole-body imaging. In this situation, with 3D only, the reduced side-shielding results in an increased level of single and scattered photons incident on the detectors, increasing the randoms rate and lowering the peak noise equivalent count rate (NEC). When operated with a 384 ns integration time, the peak NEC for the *biograph* measured with a 70 cm phantom according to the NEMA 2001 standards is 20-30% less than that of a standard EXACT HR+ operated with a 768 ns block integration time. However, if the random coincidences are smoothed prior to subtraction from the prompts (**see Chapter 6**), the peak NEC for the *biograph* is comparable to that of a standard HR+. Hence, by using a shorter integration time, and implementing random coincidence smoothing, the reduced side-shielding in the *biograph* results in no appreciable degradation in the performance of the PET component compared to a standard EXACT HR+.

State-of-the-Art PET/CT Systems

Second Generation

The first commercial scanners (Table 8.3) that followed successful clinical imaging with the prototype PET/CT [24, 25, 44, 45] appeared in early 2001 and consisted of a CT scanner in tandem with a PET scanner, with little or no mechanical integration of the two systems beyond a common gantry cover [46]. This approach, with the CT and PET scanners kept separate, has been a characteristic feature of all commercial designs (Fig. 8.12). The advantage of this design is flexibility in that different levels of CT and PET performance can be combined as required and upgraded independently. An upgrade path from CT or PET to PET/CT can, in some cases, also be envisaged.

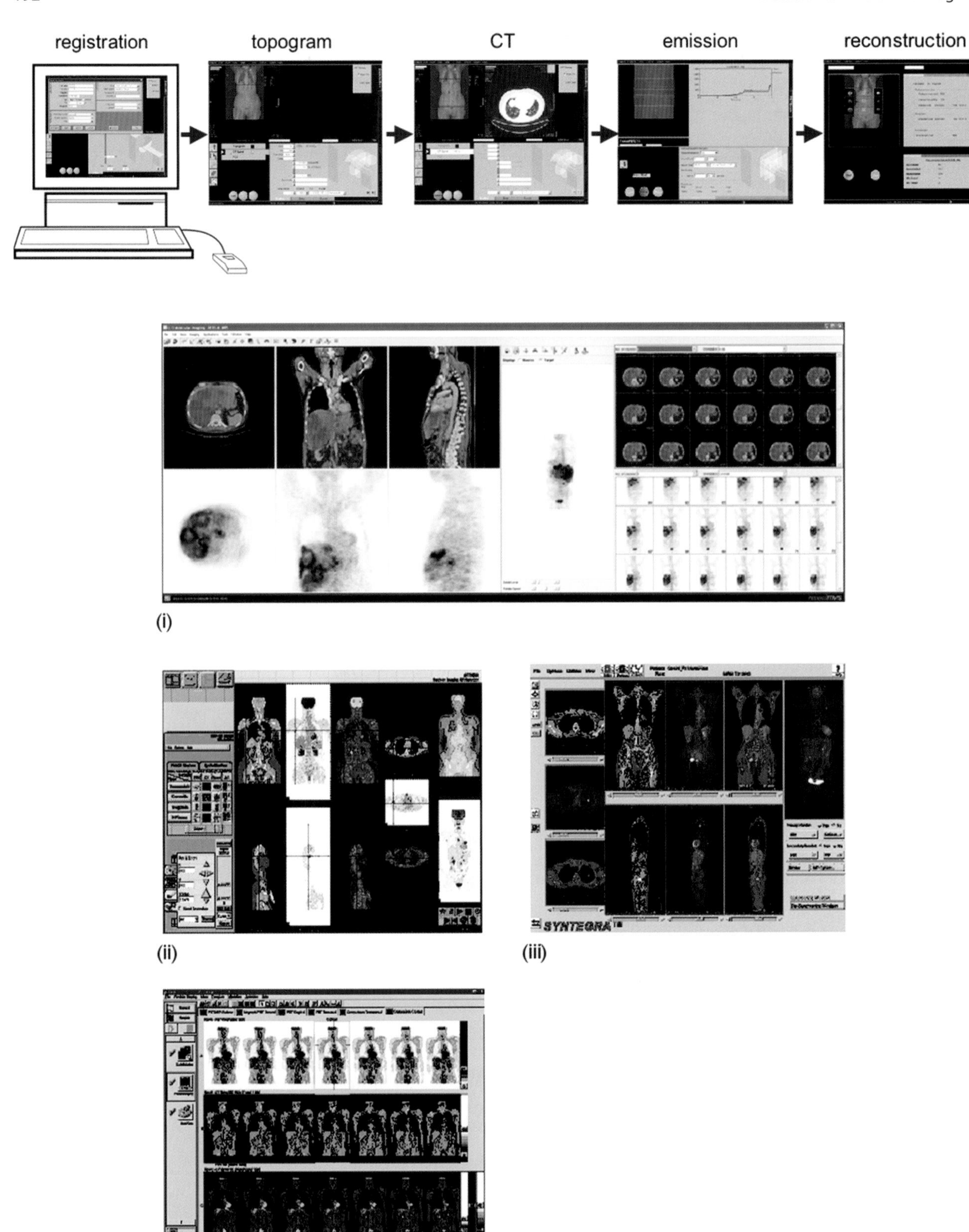

Figure 8.11. (a)State-of-the-art PET/CT tomographs offer joint acquisition and data processing consoles. A single console is used to register patient information, select and acquire a pre-defined acquisition protocol, and reconstruct image sets. (b)Selective screenshots of commercial PET/CT fusion display options: (i) Reveal MVS from CTI Molecular Imaging Inc/Mirada Solutions, (ii) Xeleris by GE Medical Systems, (iii) Syntegra by Philips Medical, and (iv) e.soft 3.0 by Siemens Medical Solutions.

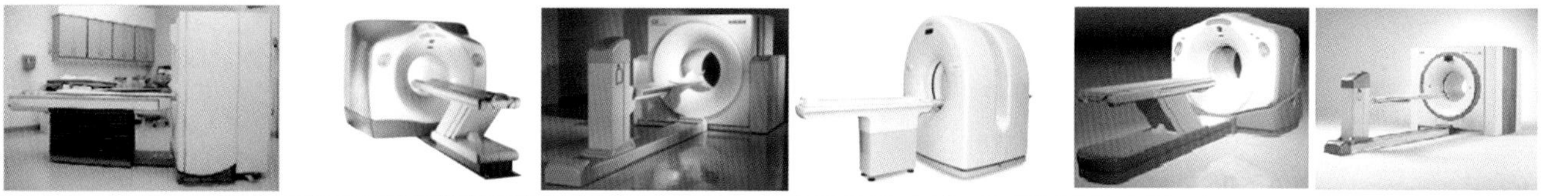

	Prototype	Second generation	Third generation
CT	Single-slice Spiral 80 cm in 240 s	Slices: 2 to 4 Spiral 80 cm in 30 - 80 s	Slices: 4 to 16 Spiral 80 cm in 20 - 30 s
PET	Partial ring BGO - 3D 7 beds in 50 min	Full-ring BGO - 2D/3D. LSO - 3D. GSO - 3D. 7 in <35 min (BGO) or <21 min (LSO)	Full-ring BGO - 2D/3D. LSO - 3D. 7 in <21 min (LSO)
Applications	Whole-body	Whole-body for oncology Radiation therapy planning	Cardiac Therapy planning
	a)	b)	c)

Figure 8.12. Prototype (a), second generation (b), and third generation (c) PET/CT tomographs with major distinguishing features regarding CT and PET technology. Second generation PET/CT employ alternative (LSO- and GSO-based) PET technology for faster whole-body imaging, while ultra-fast and multi-row CT technology in third generation PET/CT opens the applications for cardiac imaging and gated therapy planning. Systems shown from left to right: prototype, Discovery LS (GE Medical Systems), biograph-BGO/LSO (Siemens Medical Solutions), Gemini (Philips Medical Systems), Discovery ST (GEMS), biograph Sensation 16 (SMS).

All designs (Table 8.3) incorporate a common patient couch that is designed to eliminate or minimize vertical deflection due to the weight of the patient and to ensure accurate alignment of the CT and PET images (Fig. 8.10). The actual approach to address this issue, however, varies among the manufacturers of PET/CT systems. In one design vertical deflection of the pallet is eliminated through a pedestal-based patient handling system (Fig. 8.10b), while in another design a standard patient bed is mounted on a rail system in the floor to bridge the distance between the CT and the PET field-of-view. Alternatively, the bed support system is fixed with respect to the gantry, and the pallet moves on rails inside the combined system where it is supported further by a post. The maximum co-axial imaging range with any of these patient handling systems is 145 cm to 200 cm.

With the detector and data acquisition sub-systems kept separate in the second generation commercial PET/CT systems, attempts have been made to integrate the acquisition and data processing more closely. Combined scanners are operated from a single console with application-specific task cards selected for the CT and PET acquisition (see Fig. 8.11). While the early designs involved multiple computers for acquisition, image reconstruction, and image display (Fig. 8.2), progress is being made in combining some of these functions into one computer, thus reducing complexity and increasing reliability. The most recent PET/CT systems are simpler to operate, involve fewer computer systems, and are considerably more reliable. Indeed, poor reliability would be a major concern for the high patient throughput attainable with these scanners.

As mentioned, PET/CT designs have taken advantage of recent advances in both CT and PET technology by maintaining separation of the imaging sub-systems. While the CT scanners in the first PET/CT designs were single or dual-row, more recently 4-, 8- and even 16-row systems have been incorporated into PET/CT scanners. These CT modules offer sequential as well as spiral scanning modes with increased X-ray tube heat capacities to cover large volumes in short scan times

Table 8.3. Design and performance parameters of PET/CT systems. PET Performance parameters were acquired to the NEMA 2001 standard. (CPS Innovations is a joint venture of Siemens Medical Solutions, Inc and CTI Molecular Imaging, Inc).

GE Medical Systems		**Philips Medical**	
PET/CT 2000 **Generation 2**	2002 **Generation 3**	PET/CT 2002 **Generation 2**	
208 / 203 / 205	192 / 230 / 140	206 / 230 / 590	Height/Width/Depth [cm]
142 cm	100	206 with gap (30)	Inner tunnel length [cm]
Tapered, 70 cm–60 cm	Uniform, 70 cm	70 cm CT, 63 cm PET	Patient port diameter
160 cm	160 cm	195 cm	Standard co-scan range
floor-mounted pedestal	floor-mounted pedestal	floor-mounted, dual pallet	PHS
200 kg	200 kg	202	Max patient weight
no	yes	yes	Radiation therapy table attachment
LightSpeed Plus	LightSpeed Range	Mx8000D	System components
yes	yes	yes	Spiral CT
4, 8, 16	4, 8, 16	2, 16	Max number of active detector rings
Solid state–Lumex	Solid state - Lumex	Solid State	Detector material
pixel array	pixel array	2D solid state detector array	Detector design
50 cm	51 cm	50 cm	Measured transverse FOV
0.625 mm–10 mm	0.625 mm–10 mm	0.5, 1, 2.5, 5, 8, 10, 16 mm	Min and Max Slice width [mm]
75 rpm	75 rpm	120 rpm	Max rotation speed
8.5 lp/cm	8.5 lp/cm	22 lp/cm	Maximum spatial Resolution
5 mm / 3 HU / n.a. / 34 mGy / n.a.	6 mm / 3 HU / n.a. / 34 mGy / n.a.	4 mm/ 3 HU/ [n.a.]/ 27mGy/ [n.a.]	Detectability
6.8 mGy (120 kVp)	6.8 mGy (120 kVp)	14–28 mGy (120 kVp)	Centre dose (CTDI_100, body phantom) per 100 mAs
ADVANCE Nxi	unique, not available separately	Allegro	System components
2D and 3D	2D and 3D	3D	PET Acquisition mode
full-ring	full-ring	full-ring	Detector design
yes	yes	no	Septa
BGO	BGO	GSO	Detector material
optional	no	^{137}Cs point (740 MBq)	Transmission sources
55 cm	70 cm	57.6 mm	Transverse FOV
15.2 cm	15.2 cm	18 cm	Axial FOV
35	47	90	Transverse images per bed
4.2 mm	3.2 mm	2 mm	Image plane separation
6.3 mm (2D), 6.4 mm (3D)	5.2 mm (2D), 5.8 mm (3D)	5.4 mm	Axial resolution
4.8 mm (2D), 4.8 mm (3D)	6.2 mm (2D), 6.2 mm (3D)	4.8 mm	Transverse Resolution
1.3 cps/kBq (2D), 6.5 cps/kBq (3D)	1.9 cps/kBq (2D), 9 cps/kBq (3D)	3.8 cps/kBq	Sensititivity
165 kcps at 130 kBq/mL (2D), 42 kcps at 8.5 kBq/mL (3D)	82 kcps at 46 kBq/mL (2D) 62 kcps at 9.5 kBq/mL (3D)	45 kcps at 9 kBq/mL	Peak NEC

[47, 48], a necessary prerequisite for oncology imaging when examining patients with only limited breath-hold capabilities. Alternatively, imaging ranges now can be scanned with finer axial sampling in similar or less time than with a single-row CT. Furthermore, when employing IV contrast agents several organs can be imaged at peak contrast enhancement during a single spiral CT, or repeat CT exams can be acquired for a single IV contrast injection over an individual organ, such as the liver [49]. The use of spiral CT technology in combined PET/CT imaging therefore offers high-quality CT images for a variety of imaging conditions encountered in clinical oncology.

The recent developments in PET scanner technology have been primarily oriented towards the introduction of new scintillators. For over two decades, PET detectors have been based on either thallium-activated sodium iodide (NaI(Tl)) or bismuth germanate (BGO). While NaI(Tl) has high light output, it has low stopping power at PET photon energies (511 keV) and a long decay time. The low stopping power was overcome by BGO, first introduced for PET in 1977 [50] but at the expense of considerably reduced light output compared with NaI(Tl); the decay time for BGO is 30% longer than NaI(Tl). Some physical properties of these scintillators are compared in Table 8.4. The introduc-

Table 8.4. Physical properties of different scintillators for PET

Property	NaI(Tl)	BGO	LSO	GSO
Density [g/mL]	3.67	7.13	7.4	6.7
Effective Z	51	74	66	61
Attenuation length [cm-1]	2.88	1.05	1.16	1.43
Decay time [ns]	230	300	35–45	30–60
Photons/MeV	38,000	8,200	28,000	10,000
Light yield [% NaI]	100	15	75	25
Hygroscopic	Yes	No	No	No

tion of faster scintillators such as gadolinium oxyorthosilicate (GSO) [51] and lutetium oxyorthosilicate (LSO) [52], also compared in Table 8.4, offer enhanced PET scanner performance. LSO in particular outperforms BGO in almost every aspect, especially light output and decay time. The faster scintillators (Table 8.4) have lower dead time and give better count rate performance, particularly at high activity concentrations. This behaviour is confirmed in Fig. 8.13 where the Noise Equivalent Count Rate (NEC) [53] is shown as a function of activity concentration in the NEMA NU-2001 phantom for 2D and 3D BGO scanners compared with a 3D LSO scanner. The curves show the expected behaviour with 3D being superior to 2D and LSO outperforming BGO even at low activity concentrations. The new pico-3D electronics, matched to the physical properties of LSO, show a significant improvement over the older design with peak NEC exceeding 80 kcps. All 3D measurements are for PET/CT scanners, whereas the 2D data are for the standard ECAT EXACT PET scanner (CPS Innovations, Knoxville, TN) with septa extended.

These advances in PET detector technology are reflected in the variety of PET design options in currently available PET/CT technology, which, depending on the manufacturer, employ BGO-, LSO-, or GSO-based detector technology. In the *Discovery LS*, for example, the PET tomograph is based on bismuth germanate (BGO), a scintillator that is the most widely employed detector material in PET. The *ECAT HR+/Somatom Emotion (Duo)* also uses BGO as the PET detector material of choice. CPS Innovations also offers a PET/CT model (*ACCEL/Somatom Emotion (Duo)*) with the PET detector being based on lutetium oxyorthosilicate [54]. Accepting the potential for using faster crystals in PET imaging technology, Philips Medical Systems offer a combination of the GSO-based *Allegro* PET tomograph (GSO: gadolinium oxyorthosilicate) with a state-of-the-art CT scanner. Most commercial PET/CT designs favour the use of 3D-only emission acquisitions by eliminating the septa from the PET components. Assuming proper data processing, 3D PET offers a number of advantages over 2D PET, such as higher sensitivity and higher count rates at lower activity concentrations [55]. Further, the sensitivity advantage of 3D imaging and the fast scintillation properties of either GSO or LSO can be combined to result in high-quality PET images at reduced scan times and increased patient comfort.

Third Generation

By introducing 16-ring CT technology into combined PET/CT designs the advantages of very fast and high-resolution volume coverage by CT are translated directly into the context of anato-metabolic imaging. Durable CT operation at scan speeds of 0.5 s and less help reducing respiration-induced artefacts in extended imaging ranges as well as imaging the anatomy of multiple organs at peak enhancement after intra-

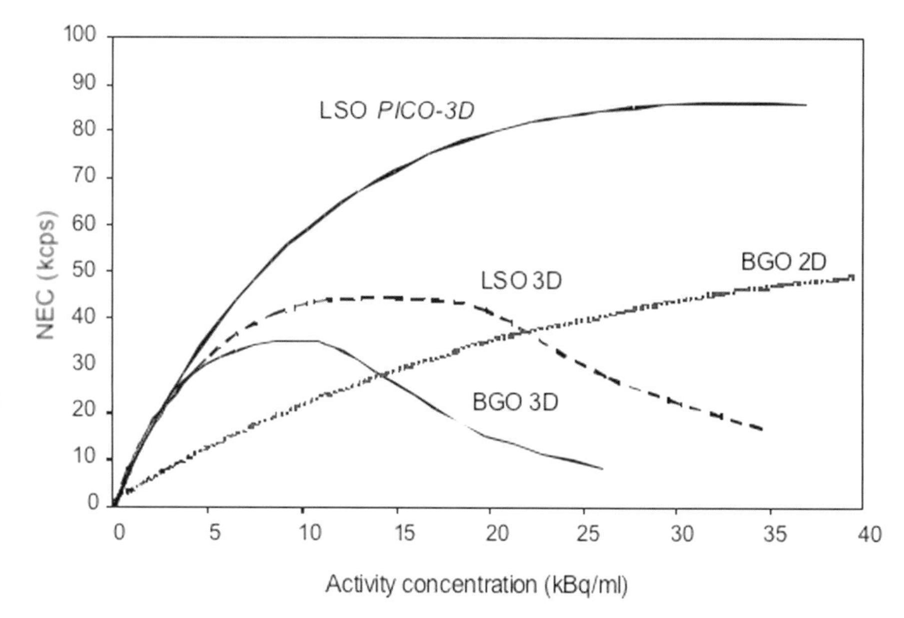

Figure 8.13. Noise Equivalent Count Rate (NEC, k=1) as a function of activity concentration in the NEMA NU-2001, 70 cm phantom. The curves are shown for the standard BGO ECAT EXACT HR+ both, with septa extended (BGO 2D) and with septa retracted (BGO) 3D), a BGO PET/CT scanner (BGO 3D), an LSO PET/CT scanner with standard electronics (LSO 3D), and an LSO PET/CT scanner with the new, high count-rate pico3D electronics (LSO pico3D).

venous contrast injection. As multi-slice CT progresses to a greater number of slices and shorter rotation times, a critical review of the actual requirements of PET/CT for specific applications will be necessary. The benefit of freezing motion during the acquisition of the CT and imaging the anatomy of the patient at a particular point during involuntary periodic motion cycles (cardiac and respiration), for example, is lost without adequate acquisition modes for the PET exam portion. Many of the recent CT improvements are targeted at cardiology, whereas the role of PET/CT in cardiology has yet to be established. Indeed, oncology applications may be adequately addressed with a lower performance CT scanner, such as a 2 or 4-row system [56].

Current developments in improved acquisition and data processing software accompany new hardware developments for updated combined tomograph designs: the *Discovery ST* from GE Medical Systems, and a combined LSO-PET/16-ring CT from CPS Innovations (Table 8.3). The *ST*, for example, is based on a revised full-ring BGO-PET system with 10,080 BGO crystals of $6.3 \times 6.3 \times 30$ mm^3 being arranged in 280 detector blocks (6 by 6) with a somewhat reduced detector ring diameter (88.6 cm) compared to the predecessor (Advance Nxi, 92.7 cm ring diameter) for improved sensitivity in whole-body imaging [57]. The *ST* is available with a choice of 4-, 8-, or 16-ring CT technology. The LSO-PET/16-ring CT by CPS Innovations offers similar-size detector crystals at a reduced coincidence window time (4.5 ns *vs* 11.5 ns) and a light output that is five time that of BGO (Table 8.4). This PET/CT is available exclusively with 16-row CT.

Optimized Protocols for Routine Clinical Procedures

General Considerations for FDG-PET/CT Imaging Protocols

For oncology purposes a standard PET/CT acquisition protocol, in essence, is a modern-day PET oncology imaging protocol, which consists of three steps: (1) patient preparation and positioning, (2) transmission scan, and (3) emission scan. Additional CT scans, such as, e.g., a 3-phase liver CT, or a high-resolution lung scan could be requested by the reviewing physician, but generally these CT scans are not used for attenuation correction. While the clinical acquisition protocols of the PET/CT systems today are similar to those from the prototype [58], demands for high diagnostic image quality increase rapidly, and therefore a number of general and specific considerations apply to current PET/CT imaging protocols (Fig. 8.14). These considerations are described in detail for each of the steps of a combined PET/CT acquisition in [59] and [60]. All PET/CT tomographs offer the use of the available CT transmission images for CT-based attenuation correc-

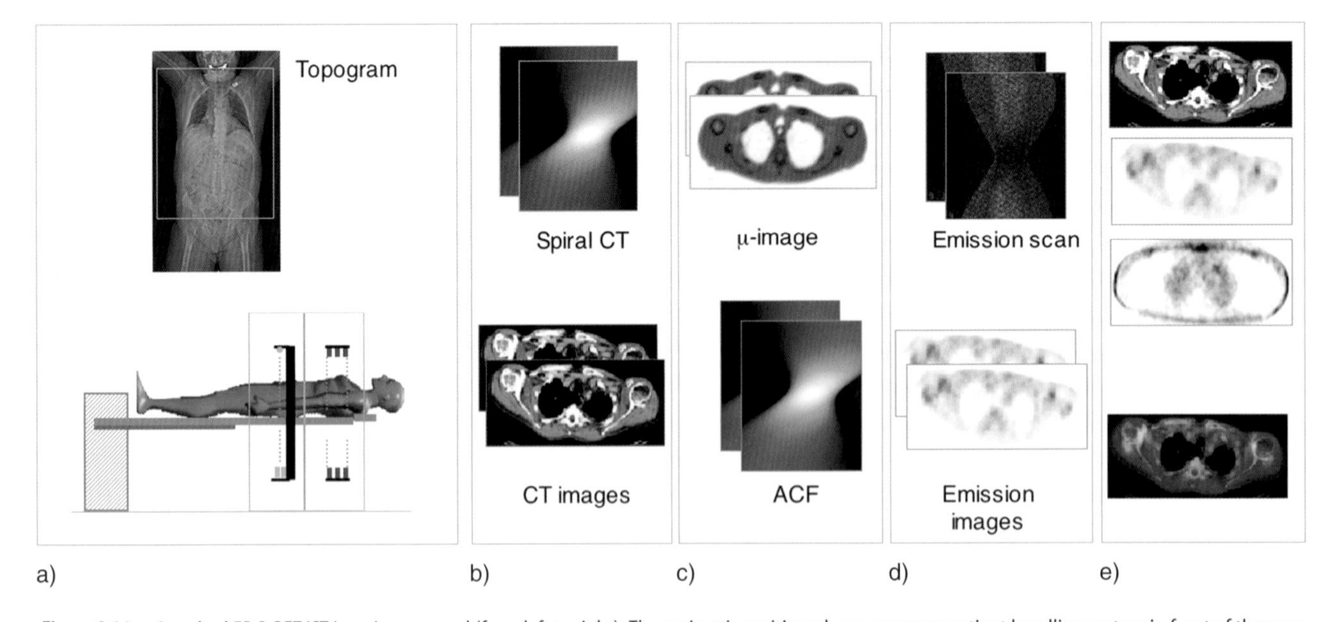

Figure 8.14. Standard FDG-PET/CT imaging protocol (from left to right). The patient is positioned on a common patient handling system in front of the combined gantry (a). First, a topogram is used to define the co-axial imaging range (a). The spiral CT scan (b) precedes the emission scan (d). The CT images are reconstructed on-line and used for the purpose of automatic attenuation correction of the acquired emission data (c). CT, PET with and without attenuation correction, and fused PET/CT images can be used for the clinical image review (e).

tion. To account for the systematic difference in routine acquisition protocols for CT and PET, and to avoid artefacts on the CT transmission data, which may propagate into the attenuation-corrected emission data, particular attention must be given to the preparation of the patient for the CT and to the CT acquisition.

Modifying CT Acquisition Parameters in PET/CT Imaging

Respiratory Motion

Several PET/CT groups have described respiratory motion and the resulting discrepancy of the spatial information from CT and PET as a source of potential artefacts in corrected emission images after CT-based attenuation correction [58, 61, 62]. These artefacts become dominant when standard full-inspiration breath hold techniques are transferred directly from clinical CT to combined PET/CT examination protocols scanning without suitable adaptations (Fig. 8.6a). In the absence of routinely available respiratory gating options the anatomy of the patient captured during the CT scan must be matched to the PET images that are acquired over the course of multiple breathing cycles. Reasonable registration accuracy can be obtained, for example, with the spiral CT scan being acquired during shallow breathing [61, 63, 64]. Alternatively, a limited breath hold protocol can be adopted with either a 1- or a 2-row system, or when dealing with uncooperative patients. Patients are then required to hold their breath in expiration only for the time that the CT takes to cover the lower lung and liver, which is typically around 15 s [65].

Breath hold commands (in normal expiration, for example) can be combined with very fast CT scanning, and therefore may help reduce respiration mismatches over the entire whole-body examination range. With multi-row CT, such as in third generation PET/CT systems (Table 8.3), it is now possible to scan the entire chest at high resolution within a single breath hold. Nevertheless, when respiration commands are not tolerated well and significant respiration-induced artefacts are suspected [66], it is advisable to reconstruct the emission data without attenuation correction and to review the two sets of fused PET/CT images very carefully.

Use of CT Contrast Agents

Clinical CT examinations are almost routinely performed with contrast enhancement to selectively increase the visibility of tissues and organs for an easier and more accurate assessment of disease and existing alterations of the anatomy of the patient. To achieve a diagnostic benefit in contrast-enhanced CT a number of – frequently competing – parameters during the application of the contrast agent must be considered [67]. The unmodified transfer of standard CT contrast protocols into the context of PET/CT has been shown to yield CT and PET images with contrast-related artefacts if attenuation correction is performed based on the acquired CT data [60]. Depending on the contrast concentration CT-based attenuation coefficients were overestimated by 26 % [40] up to 66 % [68]. However, the resulting overestimation of the standardized uptake values (SUV) in the corresponding regions on the corrected PET was only 5 % and thus clinically insignificant assuming the contrast materials were distributed evenly [69]. Nevertheless, the concentration of the oral contrast agent in the colon can vary significantly as reported by Carney *et al.* [70] and may lead to a degradation of the diagnostic accuracy of the corrected PET data. Therefore, threshold-based segmentation algorithms to segment and replace contiguous areas of high-density contrast enhancement on CT images prior to the attenuation correction procedure have been developed [40, 71]. For example, Carney *et al.* have shown that a modification can be made to the original bi-linear scaling algorithm by Kinahan *et al.* [28] to separate contrast-enhanced CT pixels from those of bone, as shown in Fig. 8.15. Pixel enhancement from positive oral contrast is around 200 HU at ingestion through the stomach, increasing to about 800 HU in the lower GI tract as water is absorbed from the contrast solution (Fig. 8.15a). Starting with a contrast enhanced CT scan, cortical bone pixels are identified with a threshold greater than 1500 HU and a region growing algorithm used to identify all contiguous pixels with bone content. The skeleton can then be extracted from the CT images. Contrast-enhanced pixels are identified by applying a simple threshold at, for example, 150 HU, well above any soft tissue value. The CT image pixels identified as oral contrast can be set to a tissue-equivalent value thus ensuring accurate attenuation correction factors for the PET data. Since the presence of contrast material has a negligible effect at 511 keV, the spatial redistribution of the contrast material between the CT and PET scans during the total imaging time does not create a problem; the aim of the modified algorithm is to remove the effect of contrast from the CT images and avoid incorrect scaling of contrast-enhanced regions. The modified algorithm can, to a considerable extent, also reduce artefacts due to catheters and metallic objects in the patient.

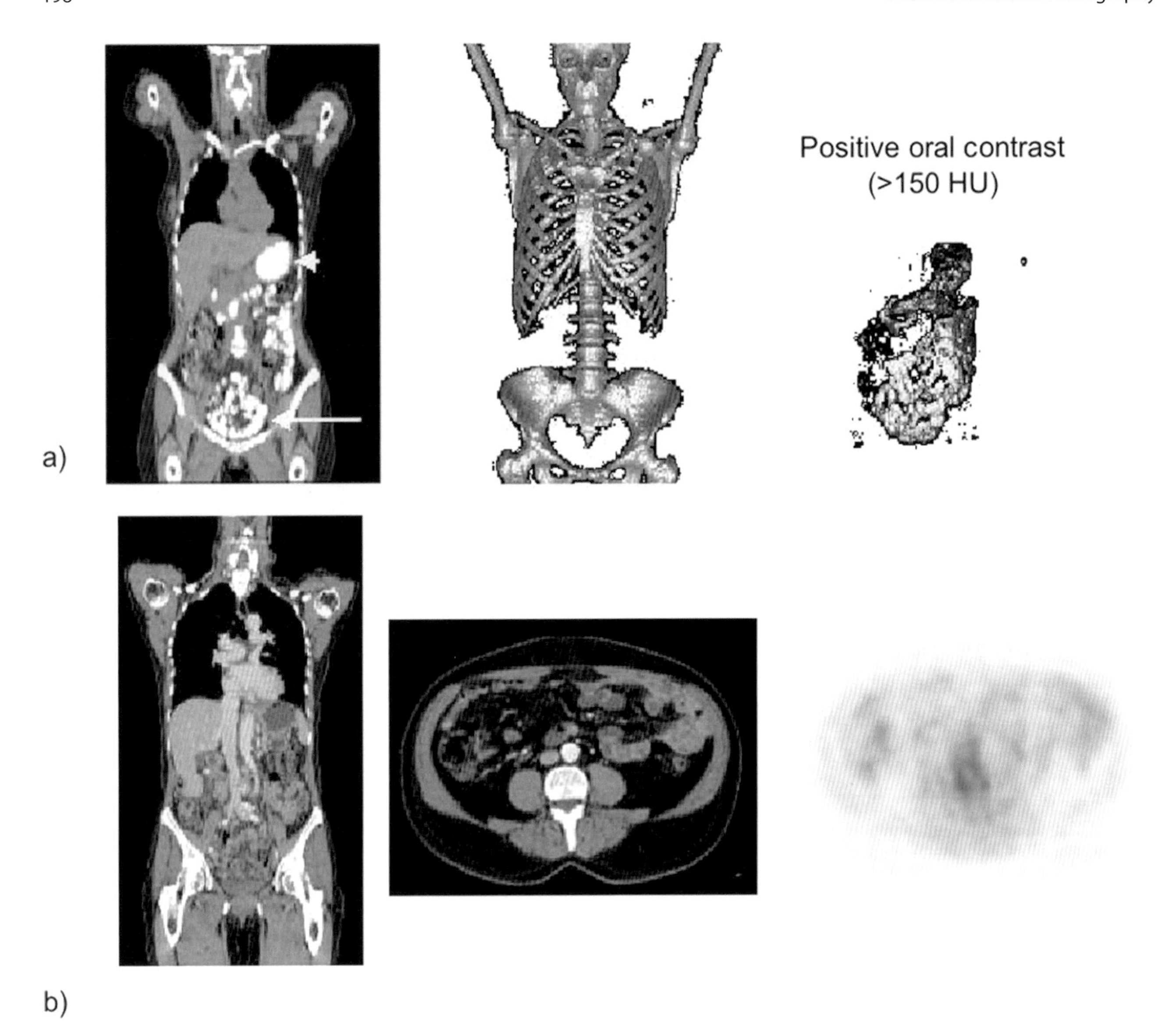

Figure 8.15. Positive oral contrast agents may lead to artifacts on PET/CT images, and thus must be accounted for. For example, CT-based attenuation correction can be modified to account for positive oral contrast agents retrospectively (a). The modified algorithm applies a region-growing technique to the contrast enhanced CT images (left) to extract the skeleton (middle). The contrast-enhanced pixels (right) can then be identified by a simple threshold at 150 HU since after removal of the skeleton the only pixels with values above 150 HU will be those with contrast enhancement. Alternatively, a water-based oral contrast may be used instead of the positive oral contrast (b). Acceptable distention of the bowel and artifact-free fully-diagnostic CT images of the abdomen can be achieved with water-based oral contrast, as seen from the coronal CT (left), transverse CT (middle) and corrected PET (right).

Unlike segmentation techniques that aim at retrospective modifications and corrections of the measured CT-based attenuation map alternative contrast application schemes represent a straightforward approach to avoiding artefacts from high concentrations of positive oral (and IV) contrast agents prospectively. Antoch *et al.* have presented a water-based oral contrast agent, resulting from previous developments for improved MRT contrast enhancement, for PET/CT imaging [72]. This contrast agent is based on a combination of water, 2.5% mannitol, and 0.2% of locust bean gum and allows for good differentiation of bowel loops from surrounding structures (Fig. 8.15b). Unlike iodine, or barium, water-equivalent oral contrast agents do not increase the CT attenuation and thus do not lead to an overestimation of the PET activity in the corrected images.

While alternative contrast materials are inadequate for vascular enhancement, high-density artefacts from the bolus injection of IV contrast agents [39] can be avoided by alternative acquisition protocols [73]. For example, diagnostic quality CT can be achieved and focal contrast enhancement in the thoracic vein can be avoided under the condition of the caudo-cranial (i.e., reverse CT scanning following a somewhat prolonged scan delay after the administration of the IV contrast)

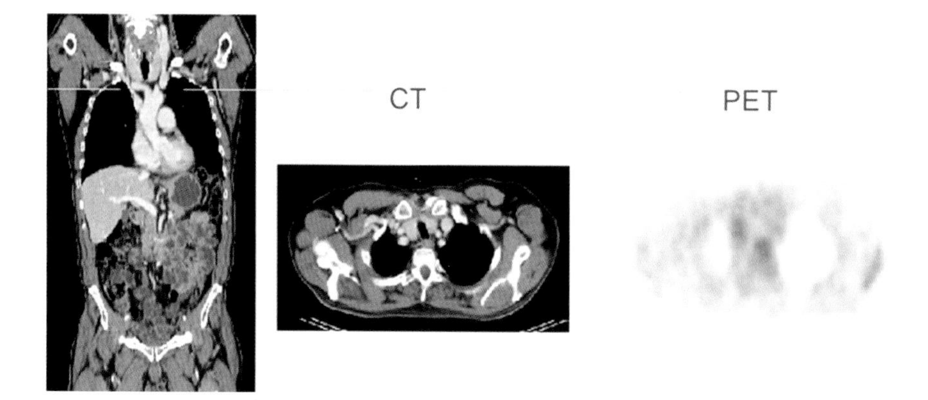

Figure 8.16. Alternative schemes of application of IV contrasts agents are being pursued to avoid contrast-induced artifacts on PET/CT images while providing acceptable image quality to the radiologists.

(Fig. 8.16). A more general solution would be to acquire only a non-enhanced CT for attenuation correction and anatomical labelling. However, additional CT scans with contrast enhancement might be required then for accurate delineation of lesions, thus leading to additional patient exposure and more logistical efforts.

Metal Artefacts

High-density implants, such as dental fillings, pacemakers, prostheses, or chemotherapy infusion ports may lead to serious artefacts in CT images [74, 75]. These CT artefacts have been shown to propagate through CT-based attenuation correction into the corrected PET emission images where artificially increased tracer uptake patterns may then be generated [76-78]. It is therefore recommended that PET images from PET/CT are routinely correlated with the complementary CT, and that these PET data are interpreted with care when lesions are observed in close proximity to artefactual structures on CT. Until robust metal artefact correction algorithms [75, 79] become available routinely in PET/CT the additional evaluation of the emission data without CT-based attenuation correction is also recommended [78].

Truncation Artefacts

Spiral CT technology currently offers a transverse field-of-view of 50 cm, and thus falls short 10 cm less than the corresponding transverse PET field-of-view (Table 8.3). This difference may lead to truncation artefacts in the CT images [80] and to a systematic bias of the recovered tracer distribution when scanning obese patients, or when positioning patients with their arms down (Fig. 8.7). If not corrected for truncation, CT images appear to mask the reconstructed emission data with the tracer distribution being only partially recovered outside the measured CT field of view.

To reduce the amount of truncation on CT and to minimize the frequency of these artefacts, whole-body or thorax patients should be positioned according to CT practice with their arms raised above their head. By keeping the arms outside the field-of-view the amount of scatter [80] and patient exposure are also much reduced. Given the short acquisition times of a PET/CT most patients tolerate to be scanned with their arms raised for the duration of the combined exam.

A number of algorithms have been suggested to extend the truncated CT projections and to recover the unmeasured regions of the attenuation map in cases where truncation is observed. If applied to the CT images prior to CT-based attenuation correction these correction algorithms will help to recover completely the tracer distributions measured with the complementary emission data [81]. Further work is needed, however, to make such algorithms routinely available for clinical diagnostics.

Future Perspectives for PET/CT

The trend of PET/CT scanners is perhaps best illustrated by a design in which a 16-slice CT scanner, the Sensation 16 (Siemens Medical Solutions, Forchheim, Germany) is combined with the recently-announced high-resolution, LSO PET scanner (CPS Innovations, Knoxville, TN). The new PET scanner has unique 13 x 13 LSO block detectors each 4 mm x 4 mm in cross-section (Fig. 8.17). The pico-3D read-out electronics, adapted to the speed and light output of LSO, is operated with a coincidence time window of 4.5 ns and a lower energy threshold of 425 keV. The significance of these high-resolution detectors is illustrated in Fig. 8.17 for a patient with squamous cell carcinoma of the right tonsil. Following treatment that included a right tonsillectomy, radical neck dissection and chemotherapy, the patient was restaged by scanning first on an ECAT EXACT (CPS Innovations, Knoxville, TN) with 6.4 mm x 6.4 mm BGO detectors, and then on

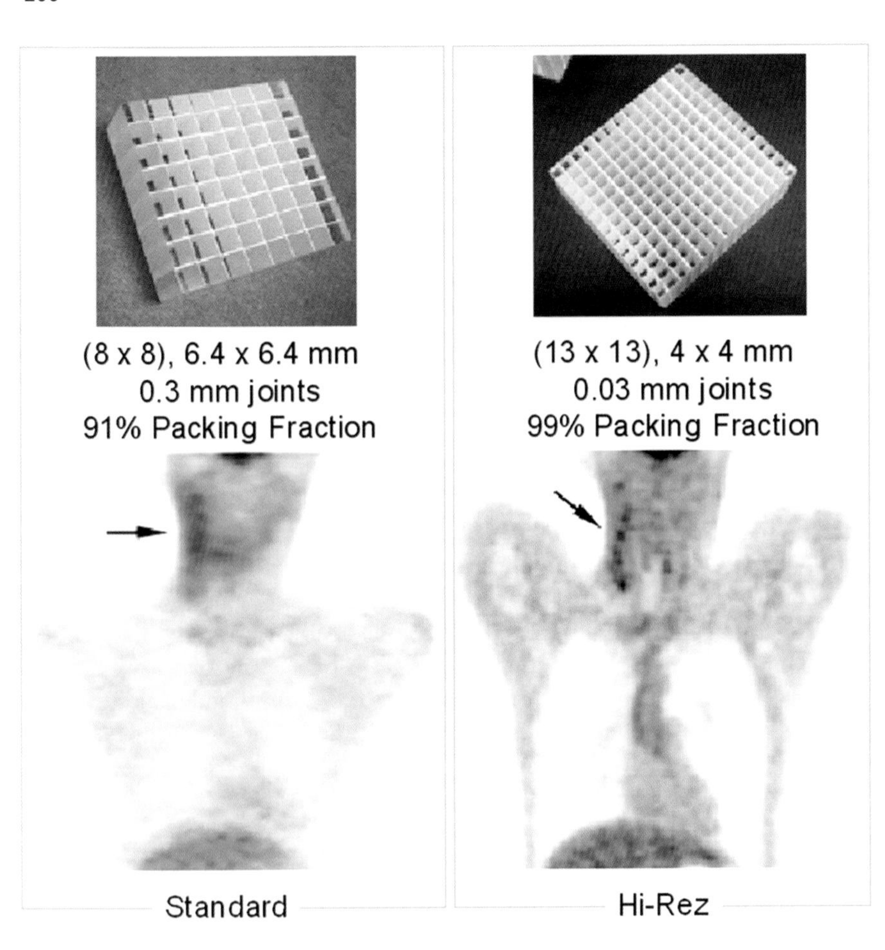

Figure 8.17. High-resolution PET and PET/CT imaging using LSO-based detectors (Hi-Rez). A 52 y/o male patient, 70 kg, diagnosed with squamous cell tonsillar cancer and a 4 cm positive node in the neck. The patient underwent pre-surgical chemotherapy, a right tonsillectomy and a right radical neck dissection for removal of the positive node and 45 additional nodes; all of the additional nodes had negative pathology. The patient suffered post-surgical infectious complications. A follow-up PET scan (Standard) acquired with arms down showed a diffuse band of activity in the right neck (arrow) seen on a coronal section. A PET/CT scan acquired with arms up and with the new high-resolution LSO-based detector blocks (Hi-Rez) clearly resolved this diffuse band of activity into individual, sub clinical lymph nodes (arrow).

the high resolution PET/CT scanner. A coronal section from the PET scan demonstrates a diffuse band of activity in the right neck. The corresponding PET section from the PET/CT scan (Fig. 8.17) resolves this diffuse band into individual nodes in the neck of the patient.

The recent introduction of the fast scintillators LSO and GSO as PET detectors has occurred at just the right moment for PET/CT where a reduction in the lengthy PET imaging time is essential to more closely match that of the CT. These tomographs are aimed primarily at high throughput with whole-body imaging times below 30 min. While it is unlikely that whole-body PET imaging times will be reduced to the 30-60 s that is required for CT scanning, a scan time less than 10 min is feasible with new high-performance LSO area detectors currently under development. Such a design will represent a breakthrough in cancer imaging, eliminating problems of patient movement and truncated CT field-of-view, and substantially reducing artefacts due to respiration. Throughput will increase significantly, as will patient comfort and convenience. New applications, such as dynamic whole-body scans and the use of short-lived radioisotopes (e.g., ^{11}C with a 20 min half-life) will then be within reach.

Future developments in combined PET/CT scanners will be exciting, attaining a higher level of integration of anatomical and functional imaging performance than before. By fulfilling an important role, not only in the diagnosis and staging of cancer, but in designing and monitoring appropriate therapies, the combined PET/CT scanner will undoubtedly have a significant impact on patient care strategies, patient survival and quality of life.

Acknowledgements

The combined PET/CT project involved many people at the University of Pittsburgh, CTI PET Systems in Knoxville, Tennessee and Siemens CT division in Forchheim, Germany over the last few years. In particular, we acknowledge the seminal contribution of Dr

Ron Nutt who led the PET/CT prototype development team at CTI PET Systems, and Dr Charles Watson who led the development of the CPS PET/CT scanner. The PET instrumentation and methodology group at the University of Pittsburgh made a major contribution to this project, in particular Drs Paul Kinahan, Jonathan Carney, David Brasse and Jeffery Yap. Finally we acknowledge the many physicians and technologists in the Pittsburgh PET Facility who contributed to the PET/CT clinical evaluation programme, especially Drs Charron, Meltzer, Blodgett, and McCook, and technologists Denise Ratica, Stacey Mckenzie, Marsha Martinelli and Donna Mason. The PET/CT development project is supported by National Cancer Institute grant CA 65856.

References

1. Weber WA, Avril N, Schwaiger M. Relevance of Positron Emission Tomography (PET) in Oncology. Strahlentherapie und Onkologie 1999;175:356–373.
2. Pelizzari CA, Chen GTY, Spelbring DR, Weichselbaum RR, Chen C-T. Accurate three-dimensional registration of CT, PET, and/or MRI images of the brain. J Comp Assist Tomogr 1989;13:20–26.
3. Woods RP, Mazziotta JC, Cherry SR. MRI-PET registration with automated algorithm. J Comp Assist Tomogr 1993;17:536–46.
4. Pietrzyk U, Herholz K, Heiss W-D. Three-dimensional alignment of functional and morphological tomograms. J Comp Assist Tomogr 1990;14:51–59.
5. Wahl RL, Quint LE, Cieslak RD, Aisen AM, Koeppe RA, Meyer CR. “Anatometabolic” tumor imaging: Fusion of FDG PET with CT or MRI to localize foci of increased activity. J Nucl Med 1993;34:1190–1197.
6. Tai YC, Lin KP, Hoh CK, Huang H, Hoffman EJ. Utilization of 3-D elastic transformation in the registration of chest X-ray CT and Whole Body PET. IEEE Trans Nuc Sci 1997;44:1606–1612.
7. Yu JN, Fahey FH, Harkness BA, Gage HD, Eades CG, Keyes JW. Evaluation of emission-transmission registration in thoracic PET. J Nucl Med 1994;35:1777–1780.
8. Cai J, Chu JCH, Recine D, Sharma M, Nguyen C, Rodebaugh R, et al. CT and PET lung image registration and fusion in radiotherapy treatment planning using the Chamfer-matching method. Int J Rad Onc Bio Phys 1999;43:883–891.
9. Slomka P, Dey D, Przetak C, Baum R. Nonlinear image registration of thoracic FDG PET and CT. J Nucl Med 2001;42:11P.
10. Hasegawa B, Stebler B, Rutt B, Martinez A, Gingold E, Faulkner K, et al. A Prototype High-Purity Germanium Detector System with Fast Photon-Counting Electronics for Medical Imaging. Med Phys 1991;18:900–909.
11. Lang TF, Hasegawa BH, Liew SC, Brown JK, Blankespoor SC, Reilly SM. Description of a prototype emission-transmission computed tomography imaging system. J Nucl Med 1992;33:1881–1887.
12. Hasegawa BH, Lang TF, Brown JK, Gingold EL, Susan M R, Blankespoor SC, et al. Object-specific attenuation correction of SPECT with correlated dual-energy X-ray CT. IEEE Trans Nuc Sci 1993;40:1242–1252.
13. Blankespoor SC, Wu X, Kalki K, Brown JK, Tang HR, Chan CE, et al. Attenuation correction of SPECT using X-ray CT on an emission-transmission CT system: Myocardial perfusion assessment. IEEE Trans Nuc Sci 1996;43:2263–2274.
14. Hammer B, Christensen N, BG BH. Use of a magnetic field to increase the spatial resolution of positron emission tomography. Med Phys 1994;21:1917–1920.
15. Christensen N, Hammer B, Heil B, Fetterly K. Positron emission tomography within a magnetic field using photomultiplier tubes and lightguides. Phys Med Biol 1995;40:691–697.
16. Shao Y, Cherry SR, Farahani K, Slates R, Silverman RW, Meadors K, et al. Development of a PET detector system compatible with MRI/NMR systems. IEEE Transactions on Nuclear Sciences 1997;44:1167–1171.
17. Shao Y, Cherry S, Farahani K, Slates R, Silverman R, K KM. Development of a PET detector system compatible with MRI/NMR systems. IEEE Trans Nucl Sci 1997;44:1167–1171.
18. Slates R, Farahani K, Yiping S, Marsden P, Taylor J, Summers P. A study of artefacts in simultaneous PET and MR imaging using a prototype MR compatible PET scanner. Phys Med Biol 1999;44:2015–2027.
19. Slates R, Cherry S, Boutefnouchet A, Shao Y, Dahlbom M, Farahani K. Design of a small animal MR compatible PET scanner. IEEE Trans Nucl Sci 1999;46:565–570.
20. Bailey DL, Young H, Bloomfield PM, Meikle SR, Glass D, Myers MJ, et al. ECAT ART – A continuously rotating PET camera: Performance characteristics, initial clinical studies and installation considerations in a nuclear medicine department. Eur J Nucl Med 1997;24:6–25.
21. Townsend DW, Beyer T, Jerin J, Watson CC, Young J, Nutt R. The ECAT ART scanner for Positron Emission Tomography: 1. Improvements in performance characteristics. Clin Pos Imag 1999;2:5–15.
22. Beyer T, Townsend DW, Brun T, Kinahan PE, Charron M, Roddy R, et al. A combined PET/CT tomograph for clinical oncology. J Nucl Med 2000;41:1369–1379.
23. Comtat C, Kinahan P, Defrise M, Michel C, Townsend D. Fast reconstruction of 3D PET data with accurate statistical modeling. IEEE Trans Nucl Sci 1998;45:1083–1089.
24. Charron M, Beyer T, Bohnen NN, Kinahan PE, Dachille M, Jerin J, et al. Image analysis in patients with cancer studied with a combined PET and CT scanner. Clinical Nuclear Medicine 2000;25:905–910.
25. Kluetz PG, Meltzer CC, Villemagne MD, Kinahan PE, Chander S, Martinelly MA, et al. Combined PET/CT imaging in oncology: Impact on patient management. Clin Pos Imag 2001;3:1–8.
26. Bradley JD, Perez CA, Dehdashti F, Siegel BA. Implementing Biologic Target Volumes in Radiation Treatment Planning fot Non-Small Cell Lung Cancer. J Nucl Med 2004;45:96S–101S.
27. Schulthess GKv. Cost considerations regarding an integrated CT-PET system. Eur Rad 2000;10:S377–S380.
28. Kinahan PE, Townsend DW, Beyer T, Sashin D. Attenuation correction for a combined 3D PET/CT scanner. Med Phys 1998;25:2046–2053.
29. LaCroix KJ, Tsui BMW, Hasegawa BH, Brown JK. Investigantion of the use of X-ray CT images for attenuation correction in SPECT. IEEE Trans Nuc Sci 1994;41:2793–2799.
30. Tang HR, Brown JK, Silva AJD, Matthay KK, Price D, Huberty JP, et al. Implementation of a combined X-ray CT scintillation camera imaging system for localizing and measuring radionuclide uptake: Experiments in phantoms and patients. IEEE Trans Nuc Sci 1999;46:551–557.
31. Fleming JS. A technique for using CT images in attenuation correction and quantification in SPECT. Nuclear Medicine Communications 1989;10:83–97.
32. Beyer T, Townsend D, Blodgett T. Dual-modality PET/CT tomography for clinical oncology. Quart J Nucl Med 2002;46:24–34.
33. Burger C, Goerres G, Schoenes S, Buck A, Lonn AHR, Schulthess GKv. PET attenuation coefficients from CT images: experimental evaluation of the transformation of CT into PET 511-keV attenuation coefficients. Eur J Nucl Med 2002;29:922–927.
34. Smith RJ, Joel SS K, Muehllehner G, Gualtieri E, Benard F. Singles transmission scans performed post-injection for quantitative whole body PET imaging. IEEE Transactions on Nuclear Sciences 1997;44:1329–1335.
35. Watson CC, Jones WF, Brun T, Veigneur K. Design and performance of a single photon transmission measurement for the ECAT ART. In: IEEE Medical Imaging Conference; 1997; Albuquerque; 1997.
36. Halpern B, Dahlbom M, Waldherr C, Quon A, Schiepers C, Silverman D, et al. A new time-saving whole-body protocol for PET/CT imaging. Molecular Imaging and Biology 2003;5:182.

37. Welch A, Campbell C, Clackdoyle R, Natterer F, Hudson M, Bromiley A, et al. Attenuation correction in PET using consistency information. IEEE Trans Nucl Sci 1998;45:3124–3141.
38. Bromiley A, Welch A, Chilcott F, Waikar S, McCallum S, Dodd M, et al. Attenuation correction in PET using considtency conditions and a three-dimensional template. IEEE Trans Nucl Sci 2001;48:1371–1377.
39. Antoch G, Freudenberg LS, Egelhof T, Stattaus J, Jentzen W, Debatin JF, et al. Focal Tracer Uptake: A Potential Artefact in Contrast-Enhanced Dual-Modality PET/CT Scans. J Nucl Med 2002;43:1339–1342.
40. Dizendorf E, Hany TF, Buck A, von Schulthess GK, Burger C. Cause and Magnitude of the Error Induced by Oral CT Contrast Agent in CT-Based Attenuation Correction of PET Emission Studies. J Nucl Med 2003;44:732–738.
41. Townsend D. A combined PET/CT scanner: The choices. J Nucl Med 2001;43:533–534.
42. Brix G, Zaers J, Adam L-E, Bellemann ME, Ostertag H, Trojan H, et al. Performance evaluation of a whole-body PET scanner using the NEMA protocol. J Nucl Med 1997;38:1614–1623.
43. Comtat C, Kinahan PE, Fessler JA, Beyer T, Townsend DW, Defrise M, et al. Reconstruction of 3D whole-body PET data using blurred anatomical labels. In: IEEE Medical Imaging Conference; 1998; Toronto; 1998.
44. Meltzer C, Martinelli M, Beyer T. Whole-body FDG PET imaging in the abdomen: value of combined PET/CT. J Nucl Med 2001;42:35P.
45. Meltzer C, Snyderman C, Fukui M, Bascom D, Chander S, Johnson J, et al. Combined FDG PET/CT imaging in head and neck cancer: impact on patient management. J Nucl Med 2001;42:36P.
46. Townsend D, Beyer T, Blodgett TM. PET/CT scanners: A hardware approach to image fusion. Semin Nucl Med 2003;XXXIII:193–204.
47. McCollough CH, Zink FE. Performance evaluation of a multi-slice CT system. Med Phys 1999;26:2223–2230.
48. Roos JF, Desbiolles LM, Willmann JK, Weishaupt D, Marincek B, Hilfiker PR. Multidetector-row helical CT: analysis of time management and workflow. Eur Rad 2002;12:680–685.
49. Itoh S, Ikeda M, Achiwa M, Ota T, Satake H, Ishigaki T. Multiphase contrast-enhanced CT of the liver with a multislice CT scanner. Eur Rad 2003;13:1085–1094.
50. Cho Z, Farukhi M. Bismuth germanate as a potential scintillation detector in positron cameras. J Nucl Med 1977;18:840–844.
51. Takagi K, Fukazawa T. Cerium-activated Gd2SiO5 single crystal scintillator. App Phys Lett 1983;42:43–45.
52. Melcher CL, Schweitzer JS. Cerium-doped lutetium oxyorthosilicate: A fast, efficient new scintillator. IEEE Trans Nuc Sci 1992;39:502–504.
53. Strother SC, Casey ME, Hoffman EJ. Measuring PET scanner sensitivity: Relating countrates to image Signal-to-Noise Ratios using Noise Equivalent Counts. IEEE Transactions on Nuclear Sciences 1990;37:763–788.
54. Bruckbauer T, Casey M, Valk PE, Rao J, Finley BR, Farboud B. Optimizing 3D whole body acquisition for oncologic imaging on the ECAT ACCEL LSO PET system. In: EANM; 2001; Naples; 2001. p. OS-405.
55. Bendriem B, Townsend D. The theory and practice of 3D PET. Dordrecht: Kluwer Academic Publishers; 1998.
56. Beyer T, Kuehl H, Stattaus J, DiFilippo F, Schöder H, Roberts F, et al. Respiration artefacts in whole-body studies with 2nd and 3rd generationPET/CT systems employing multi-row CT technology. J Nucl Med 2004;45:P.
57. Bettinardi V, Danna M, Savi A, Lecchi M, Castiglioni I, Gilardi MC, et al. Performance evaluation of the new whole-body PET/CT scanner: Discovery ST. European Journal of Nuclear Medicine and Molecular Imaging 2004;Feb 10.
58. Beyer T, Townsend DW, Nutt R, Charron M, Kinahan PE, Meltzer CC. Combined PET/CT imaging using a single, dual-modality tomograph: a promising approach to clinical oncology of the future. In: Wieler HJ and Coleman RE, editors. PET in Clinical Oncology. Darmstadt: Steinkopff; 2000. p. 101–124.
59. Beyer T, Antoch G, Freudenberg TS, Egelhof T, Müller SP. Considerations on FDG-PET/CT imaging protocols. J Nucl Med 2004;45:25S–35S.
60. Antoch G, Freudenberg LS, Beyer T, Bockisch A, Debatin JF. To enhance or not to enhance? 18F-FDG and CT contrast agents in dual-modality 18F-0FDG PET/CT. J Nucl Med 2004;45:56S–65S.
61. Goerres GW, Kamel E, Heidelberg T-NH, Schwitter MR, Burger C, Schulthess GKv. PET-CT image co-registration in the thorax: influence of respiration. Eur J Nucl Med 2002;29:351–360.
62. Osman MM, Cohade C, Nakamoto Y, Wahl RL. Respiratory motion artefacts on PET emission images obtained using CT attenuation correction on PET-CT. Eur J Nucl Med Mol Imaging 2003;30:603–606.
63. Goerres GW, Kamel E, Seifert B, Burger C, Buck A, Hany TF, et al. Accuracy of image coregistration of pulmonary lesions in patients with non-small cell lung cancer using an integrated PET/CT system. J Nucl Med 2002;43:1469–1475.
64. Goerres GW, Burger C, Schwitter MR, Heidelberg T-NH, Seifert B, Schulthess GKv. PET/CT of the abdomen: optimizing the patient breathing pattern. Eur Rad 2003;13:734–739.
65. Beyer T, Antoch G, Blodgett T, Freudenberg L, Akhurst T, Mueller S. Dual-modality PET/CT imaging: the effect of respiratory motion on combined image quality in clinical oncology. Eur J Nucl Med 2003;30:588–596.
66. Osman MM, Cohade C, Nakamoto Y, Marshall LT, Leal JP, Wahl RL. Clinically Significant Inaccurate Localization of Lesions with PET/CT: Frequency in 300 Patients. J Nucl Med 2003;44:240–243.
67. Beyer T, Antoch G, Muller S, Egelhof T, Freudenberg LS, Debatin J, et al. Acquisition Protocol Considerations for Combined PET/CT Imaging. J Nucl Med 2004;45:25S–35.
68. Nehmeh SA, Erdi YE, Kalaigian H, Kolbert KS, Pan T, Yeung H, et al. Correction for Oral Contrast Artefacts in CT Attenuation-Corrected PET Images Obtained by Combined PET/CT. J Nucl Med 2003;44:1940–1944.
69. Dizendorf EV, Treyer V, von Schulthess GK, Hany TF. Application of Oral Contrast Media in Coregistered Positron Emission Tomography-CT. Am. J. Roentgenol. 2002;179:477–481.
70. Carney J, Beyer T, Brasse D, Yap J, Townsend D. Clinical PET/CT scanning using oral CT contrast agents. J Nucl Med 2002;43:57P.
71. Carney J, Townsend D. CT-based attenuation correction for PET/CT scanners. In: Schultess Gv, editor. Clinical PET, PET/CT and SPECT/CT: Combined Anatomic-Molecular Imaging: Lippincott, Williams and Wilkins; 2002.
72. Antoch B, Kuehl H, Kanja J, Lauenstein T, Schneemann H, Hauth E, et al. Introduction and evaluation of a negative oral contrast agent to avoid contrast-induced artefacts in dual-modality PET/CT imaging. Radiology (to appear) 2004.
73. Beyer T, Antoch G, Rosenbaum S, Freudenberg L, Fehlings T, Stattaus J. Optimized IV contrast administration protocols for diagnostic PET/CT imaging. Eur Rad 2004;14:422–423.
74. Duerinckx AJ, Macovski A. Polychromatic streak artefacts in computed tomography images. J Comp Assist Tomogr 1978;2:481–487.
75. deMan B, Nuyts J, Dupont P, Marchal G, Suetens P. Metal streak artefacts in X-ray Computed Tomography: A simulation study. IEEE Trans Nuc Sci 1999;46:691–696.
76. Bujenovic S, Mannting F, Chakrabarti R, Ladnier D. Artefactual 2-deoxy-2-[18F]fluoro-D-glucose localization surrounding metallic objects in a PET/CT scanner using CT-based attenuation correction. Molecular Imaging and Biology 2003;5:20–22.
77. Kamel EM, Burger C, Buck A, Schluthess GKv, Goerres GW. Impact of metallic dental implants on CT-based attenuation correction in a combined PET/CT scanner. Eur Rad 2002;13:724–728.
78. Goerres GW, Ziegler SI, Burger C, Berthold T, von Schulthess GK, Buck A. Artefacts at PET and PET/CT Caused by Metallic Hip Prosthetic Material. Radiol 2003;226:577–584.
79. Glover GH, Pelc NJ. An algorithm for the reduction of metal clip artefacts in CT reconstructions. Med Phys 1981;8:799–807.
80. Carney JP, Townsend DW, Kinahan PE, Beyer T, Kalender WA, Kachelriess M, et al. CT-based attenuation correction: The effects of imaging with the arms in the field of view. J Nucl Med 2001;42:56–57P.
81. Schaller S, Semrbitzki O, Beyer T, Fuchs T, Kachelriess M, Flohr T. An algorithm for virtual extension of the CT field of measurement for application in combined PET/CT scanners. Radiol 2002;225 (P):497.

9 Radiohalogens for PET Imaging*

N Scott Mason and Chester A Mathis

Introduction

The radiohalogens are particularly attractive to consider as radiolabels for positron emission tomography (PET) radiophamaceuticals. While positron-emitting chlorine has not been utilized, there are several radioiodines and radiobromines and one radiofluorine of great importance to PET (Table 9.1). The chemistry of radioiodine and radiobromine are similar in many respects, but the chemistry of radiofluorine (i.e., fluorine-18) is sufficiently unique to warrant considerable discussion [1]. The emphasis in this chapter is upon fluorine-18 chemistry and ^{18}F-labeled radiopharmaceuticals. This is because ^{18}F, in the form of 2-deoxy-2-[^{18}F]fluoro-D-glucose (FDG), has become the most utilized PET radionuclide. Several positron-emitting radiobromines and radioiodines are not included in Table 9.1, as there is little literature regarding their routine production and use in PET imaging. While ^{75}Br [2, 3] and ^{122}I [4, 5] are included in Table 9.1, they are not discussed further in this chapter. The half-life of ^{75}Br is close to that of ^{18}F, and its production and purification are more complicated. The short half-life of ^{122}I can be an advantage for blood-flow studies, but production constraints have required the use of a high-energy cyclotron and this has limited its application as well.

In general, radioiodines, radiobromines, and fluorine-18 can react as electrophiles or nucleophiles involving species that behave formally as positively charged (X^+) or negatively charged (X^-) ions, respectively (Figs. 9.1 and 9.2). As the names imply, electrophiles are electron-deficient species that seek electron-rich reactants such as carbon atoms with high local electron densities, and nucleophiles are electron-rich species that seek electron-deficient reactants [6]. While free-radical radiohalogen labeling reactions have been utilized, they tend to be disfavored for PET radiolabeling applications as a result of the difficulty in controlling the regioselectivity of this type of reaction. In contrast, electrophilic and nucleophilic radioiodina-

Table 9.1. Cyclotron-produced PET radiohalogens

Radionuclide	Half-life	Decay Modes (%)	Max. β+ Energy (MeV)	Production Reactions
^{18}F	109.8 min	β+ (97) EC (3)	0.635	^{18}O(p,n)^{18}F ^{20}Ne(d,α)^{18}F
^{76}Br	16.1 h	β+ (57) EC (43)	3.98	^{75}As(^{3}He,2n)^{76}Br ^{76}Se(p,n)^{76}Br
^{75}Br	98 min	β+ (76) EC (24)	1.74	^{75}As(^{3}He,3n)^{75}Br ^{78}Kr(p,α)^{75}Br
^{124}I	4.2 d	β+ (25) EC (75)	2.13	^{124}Te(p,n)^{124}I ^{124}Te(d,2n)^{124}I
^{122}I	3.6 min	β+ (77) EC (23)	3.12	^{127}I(p,6n)^{122}Xe/^{122}I

* Chapter reproduced from Valk PE, Bailey DL, Townsend DW, Maisey MN. Positron Emission Tomography: Basic Science and Clinical Practice. Springer-Verlag London Ltd 2003, 217–236.

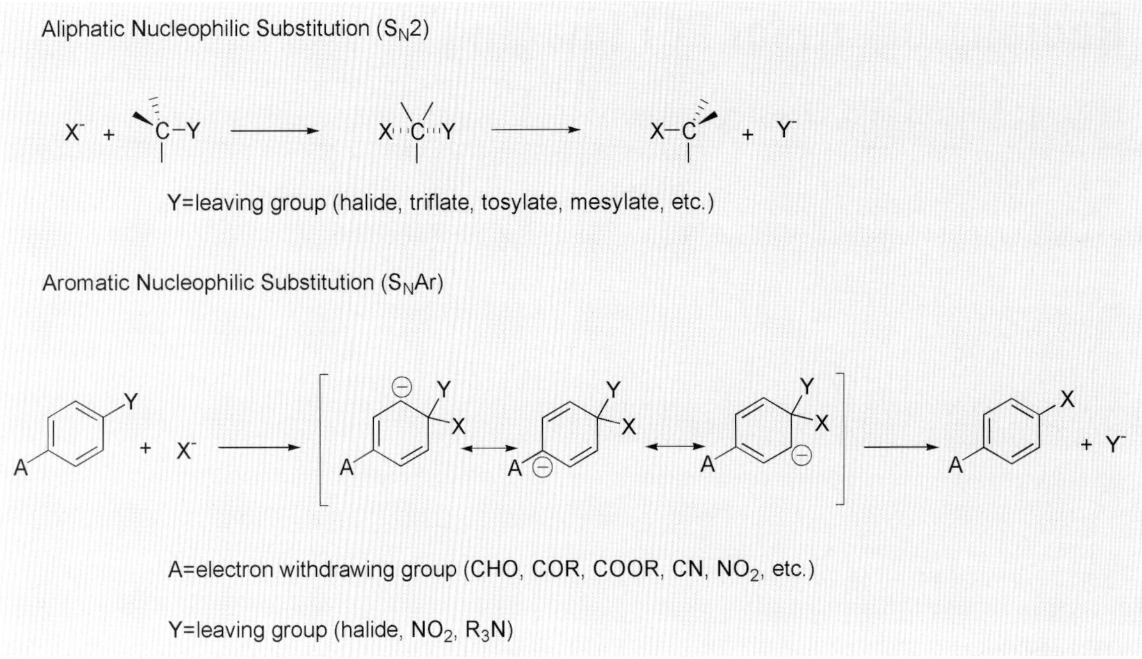

Aliphatic Nucleophilic Substitution (S_N2)

Y=leaving group (halide, triflate, tosylate, mesylate, etc.)

Aromatic Nucleophilic Substitution (S_NAr)

A=electron withdrawing group (CHO, COR, COOR, CN, NO_2, etc.)

Y=leaving group (halide, NO_2, R_3N)

Figure 9.1. Nucleophilic reactions relevant to [^{18}F]fluoride.

tions and radiobrominations are more readily controlled and can be regiospecific in many cases. While nucleophilic reactions involving [^{18}F]fluoride can be regiospecific as well, electrophilic [^{18}F]fluorine is very reactive, and its reactions are more difficult to control. Electrophilic fluorinations require special methods, radiolabeling precursors, and "taming" reagents that are described in this chapter.

Radiohalogen Production

Cyclotrons for PET Radionuclide Production

Cyclotrons have been utilized for the production of medical radionuclides since the 1930s, and cyclotron acceleration of charged particles remains the preferred method to produce short-lived positron-emitting radionuclides. While a variety of different size and energy cyclotrons have been employed for these purposes over the past 70 years, the brief discussion presented here will be limited to an overview of the most common PET cyclotrons in operation (Table 9.2). More complete discussions of cyclotrons that produce radionuclides can be found in several reviews [7–9]. Over the past ten years, most PET radionuclide production sites have installed one of two types of cyclotrons. In many academic medical settings and dedicated ^{18}F-production sites (particularly in the USA), the type of cyclotron most commonly in use is a single-particle cyclotron that produces protons with kinetic energies of about 11 MeV. This type of cyclotron allows access to the four most commonly utilized PET radionuclides (fluorine-18, carbon-11, nitrogen-13, and oxygen-15). Improvements in targetry design and increases in

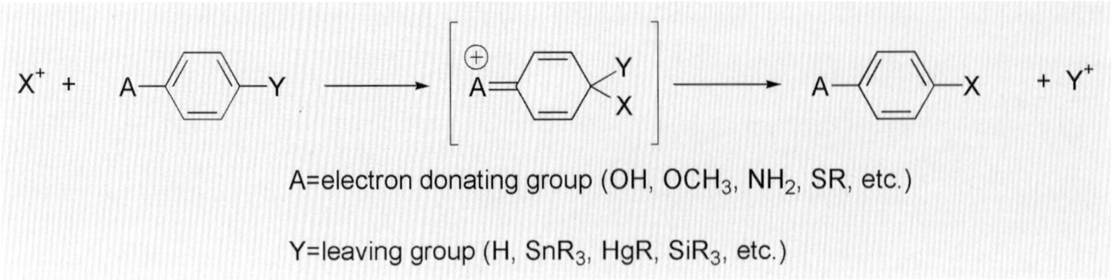

A=electron donating group (OH, OCH_3, NH_2, SR, etc.)

Y=leaving group (H, SnR_3, HgR, SiR_3, etc.)

Figure 9.2. Electrophilic reactions relevant to [^{18}F]fluorine.

Table 9.2. Cyclotron manufacturers and some currently available products

Company	Model	Maximum Energy of Protons(p) and Deuterons (d)	Self-Shielded
CTI	RDS 111	11 MeV (p)	Yes
EBCO	TR19	13-19 MeV (p) 9 MeV (d) optional	Option
GE	PETtrace	16.5 MeV (p) 8.4 MeV (d)	Option
GE	MINItrace	9.6 MeV (p)	Yes
IBA	Cyclone 10/5	10 MeV (p) 5 MeV (d)	No
IBA	Cyclone 18/5	18 MeV (p) 5 MeV (d)	No

available beam currents have made these machines capable of producing multi-Curie (>74 GBq) quantities of the four common PET radionuclides. Low-energy, proton-only cyclotrons are commercially available from several suppliers, and many are available as self-shielded models. This feature helps to simplify site selection in existing building space.

The second most utilized cyclotrons (particularly in larger research-oriented facilities) are dual particle (proton (p) and deuteron (d)) 18 MeV accelerators. These machines can produce usable quantities of other PET radionuclides, such as ^{76}Br, ^{124}I, and ^{64}Cu, in addition to the four common PET radionuclides mentioned above. These cyclotrons also utilize less expensive, naturally abundant [^{14}N]nitrogen to produce [^{15}O]oxygen.

Production of Fluorine-18

The most common method utilized to produce nucleophilic [^{18}F]fluoride is the $^{18}O(p,n)^{18}F$ nuclear reaction (indicating the reaction of an accelerated proton (p) with oxygen-18 to produce a neutron (n) and fluorine-18). The oxygen-18 target material most frequently consists of highly enriched [^{18}O]water [10–12], but [^{18}O]oxygen gas has been used successfully for this purpose as well [13]. Multi-Curie (>74 GBq) quantities of high-specific-activity [^{18}F]fluoride can be produced in a few hours using 11 MeV protons to irradiate [^{18}O]water targets. In addition, the separation and recovery of [^{18}O]water target material from [^{18}F]fluoride is possible [14–16]. While the theoretical specific activity of carrier-free ^{18}F is 1.7×10^6 Ci/mmol (6.3×10^7 GBq/mmol), the no-carrier-added specific activity of [^{18}F]fluoride produced from [^{18}O]water targets has been in the range of about 1×10^4 Ci/mmol (3.7×10^5 GBq/mmol). Other nuclear reaction pathways such as $^{16}O(^3He,p)^{18}F$, $^{16}O(^4He,pn)^{18}F$, and [$^6Li(n,\alpha)^3H$, $^{16}O(^3H,n)^{18}F$] have been utilized in the past [17–21] but are not the current method of choice.

Two processes are currently employed to generate electrophilic [^{18}F]fluorine gas ([$^{18}F]F_2$). One method utilizes deuteron bombardment of neon-20 gas target material to produce ^{18}F by the $^{20}Ne(d,\alpha)^{18}F$ reaction. A passivated nickel target (NiF) is loaded with 0.1% (cold) F_2 in neon-20 and irradiated with 8–18 MeV deuterons. This method produces relatively low specific activity [$^{18}F]F_2$ (~12 Ci/mmol or ~444 GBq/mmol) [22], dependent upon total mass of added carrier F_2. The "double-shoot" method also uses a passivated nickel target; however, the target is loaded with [^{18}O]oxygen gas. Proton irradiation of the target leads to adherence of radioactive ^{18}F species on the target walls. Upon cryogenic removal of the [^{18}O]oxygen from the target, (cold) fluorine (1.0%; 30–70 μmol) diluted in a suitable inert carrier gas (for example, argon) is added to the target. A second short irradiation leads to the interaction of the carrier fluorine and surface-bound ^{18}F to yield recoverable [$^{18}F]F_2$ [23,24]. Both of these methods produce relatively low yields of [$^{18}F]F_2$ (<1 Ci or <37 GBq), and the specific activity of [$^{18}F]F_2$ is low compared to that achievable using [^{18}O]water/[^{18}F]fluoride target technology.

In general, electrophilic ^{18}F-fluorination reactions have resulted in low-specific-activity products (<1 GBq/μmol). A multi-step method used to produce considerably higher-specific-activity [$^{18}F]F_2$ (>50 GBq/μmol) is worthy of note. This method utilized the (p,n) reaction on [^{18}O]water as the starting point, as opposed to deuteron irradiation of neon-20 or proton irradiation of [^{18}O]oxygen gas [24]. Standard proton irradiation of an [^{18}O]water target to yield [^{18}F]fluoride was followed by azeotropic drying of the Kryptofix [2.2.2]®/potassium carbonate/[^{18}F]fluoride mixture. The reactive [^{18}F]fluoride was then reacted with methyl iodide in anhydrous acetonitrile to yield methyl [^{18}F]fluoride, which was isolated by gas chromatography and trapped at liquid nitrogen temperatures. The purified methyl [^{18}F]fluoride was then passed through a discharge chamber operating at 20–30 kV and 280 μA, and small amounts of F_2 (150 nmoles) and neon carrier gases were added to yield [$^{18}F]F_2$ (7.5 GBq [$^{18}F]F_2$ from 37 GBq [^{18}F]fluoride). This [^{18}F]fluoride-to-[^{18}F]fluorine conversion process offers the option of obtaining high-specific-activity [$^{18}F]F_2$ where the use of standard electrophilic [^{18}F]fluorine targetry would lead to unacceptably low specific activity products and/or nucleophilic radiolabeling methods utilizing [^{18}F]fluoride are not practical.

Production of ^{76}Br

Bromine-76 has been produced by the irradiation of natural arsenic with a beam of 30 MeV helium-3 ions via the ^{75}As(^{3}He,2n)^{76}Br reaction [26,27]. Following irradiation, the solid target was dissolved in sulfuric acid and treated with chromic acid. Radioactive bromine was distilled into an ammonia solution using a stream of nitrogen, and the resultant ammonium [^{76}Br]bromide was dried and used as a source of [^{76}Br]bromide for subsequent reactions (either nucleophilic or electrophilic following oxidation). Another method of bromine-76 production utilized [^{76}Se]selenium-enriched (96%) Cu_2Se as the target material [28]. Irradiation with 17 MeV protons produced [^{76}Br]bromine, which was separated from the solid target by thermal diffusion. Lower energy, 11 MeV cyclotrons have also been utilized to produce [^{76}Br]bromide, but to date the yields have been low [29]. Bromine-76 has a more complex decay scheme than fluorine-18, and only about 57% of its transitions result in positron emission (Table 9.1). The positron emitted from ^{76}Br has a considerably higher kinetic energy than that from ^{18}F, resulting in higher patient dose per positron emission and lower imaging resolution.

Production of ^{124}I

Iodine-124 has been produced by the irradiation of 96% enriched [^{124}Te]tellurium (IV) oxide with 15 MeV deuterons via the ^{124}Te(d,2n)^{124}I nuclear reaction [30–32]. Iodine-124 has also been produced by the irradiation of enriched [^{124}Te]tellurium (IV) oxide with 15 MeV protons via the ^{124}Te(p,n)^{124}I nuclear reaction [33]. Some work has been performed utilizing lower energy 11 MeV proton cyclotrons to produce [^{124}I]iodide [29,34,35], but the yields have been relatively low. While ^{124}I has a relatively long half-life (4.2 days), there are problems associated with its use. Like ^{76}Br, ^{124}I has a complex decay scheme with only about 25% of its transitions resulting in positron emission, and the emitted positron has a relatively high energy compared to positrons from ^{18}F (Table 9.1). In addition, several high-energy gamma rays of nuclear origin are emitted along with the positron. Despite these complications, ^{124}I has been used successfully to label PET radiopharmaceuticals because its long half-life provides advantages over ^{18}F for imaging slow pharmacokinetic processes in vivo. In addition, it is possible to achieve relatively high-specific-activity products using electrophilic radiolabeling methods with ^{124}I.

^{18}F Radiochemistry

Nucleophilic reactions with [^{18}F]fluoride (high specific activity)

The majority of PET radiohalogenations reported in the literature are nucleophilic fluorinations utilizing [^{18}F]fluoride. The reasons for this include the availability of high amounts of [^{18}F]fluoride from low- and medium-energy proton-only cyclotrons utilizing [^{18}O]water target material and the generally higher specific activities achievable using [^{18}F]fluoride. A variety of chemical reaction types are amenable to radiolabeling using [^{18}F]fluoride. These can be divided into two principal catagories: SN2-type (substitution nucleophilic bimolecular) reactions with substrates containing leaving groups such as halides or alkyl sulfonate esters, and aromatic nucleophilic substitution reactions utilizing activated aromatic systems with leaving groups such as nitro or trimethylammonium.

Radiofluorination via SN2 Reactions

In SN2-type reactions [6], nucleophiles attack the substrates at 180° opposite to the leaving groups resulting in configurational inversion at the carbon center following substitution (Fig. 9.1). From a kinetic viewpoint, the reaction rates are largely determined by the structures of both the substrates and nucleophiles. Substrates containing bulky substituents near the reaction center or poor leaving groups for substitution generally decrease the reaction rate. Nucleophiles react with substrates as a function of their electron-donating ability, and the fluoride anion is a poor nucleophile. As a result, it has only been within the past 15 years that radiochemists have been able to successfully produce a wide variety of radiofluorinated compounds for PET imaging utilizing this type of reaction.

For SN2-type reactions, the best leaving groups are the weakest bases, which is consistent with the principle that independently stable species make better leaving groups. Of the halides, iodide is the best leaving group and fluoride the worst. The conversion of an alcohol to a sulfonic ester is a useful way to generate a good leaving group. The triflate, tosylate, brosylate, nosylate, and mesylate groups are all better leaving groups than halides. Leaving groups that have been utilized the most for nucleophilic radiofluorination reactions include triflate, tosylate, mesylate, bromide, and iodide [1].

SN2-type radiofluorination reactions generally take place under basic or neutral conditions, as these conditions promote the presence of good nucleophiles (which are often strong bases). The effects of solvents on SN2-type reactions are variable and depend upon the charge dispersal of the reactants and subsequent transition state. An assortment of solvents have supported high-yield radiofluorination reactions, and the solubility of the reactants appears to have played a larger role in solvent choice than their effects on reaction rates. The most common solvent choices (although by no means the only) are dipolar aprotic solvents in which both the alkali metal [^{18}F]fluoride salts (for example, K[^{18}F] or Cs[^{18}F]) and the organic radiolabeling precursors generally show good solubility.

An important point in the general discussion of SN2-type radiofluorinations regards the propensity of fluoride to form tight ion pairs with metal cations. Non-bound or non-coordinated nucleophiles are more reactive. Cryptands and polyaminoethers have been used to coordinate alkali metal cations, for example the potassium ion of K[^{18}F]. This allows the [^{18}F]fluoride anion to be less tightly paired with the cation (termed the "naked ion" effect) and subsequently more reactive [36]. Crown ethers, particularly 18-crown-6, have been used with K[^{18}F] to increase solubility and promote the nucleophilicity of [^{18}F]fluoride. Aminopolyethers (such as Kryptofix [2.2.2]®) have also been used with excellent results in a variety of aliphatic nucleophilic [^{18}F]fluoride substitution reactions [37–39] Cesium and rubidium fluoride salts have been used as radiolabeling sources of reactive [^{18}F]fluoride employing the concept that larger mono-valent cations bind less tightly to [^{18}F]fluoride [40,41]. A variety of tetraalkylammonium [^{18}F]fluoride salts have been used widely in nucleophilic labeling reactions [42]. These salts are very soluble in a variety of organic solvents ranging from nonpolar to dipolar aprotic.

The use of alkyl halides as [^{18}F]fluoride radiolabeling substrates has been viewed as attractive for a variety of compounds, particularly the simple dihalogen substituted alkanes. This approach has been used to produce ^{18}F-radiolabelled synthons such as 1-[^{18}F]fluoro-3-bromopropane (Fig. 9.3), which was then used to radiolabel [^{18}F]β-CFT-FP (radiochemical yield 2–3% decay-corrected to EOB) for use as a dopamine transporter radioligand [43]. Other synthons used in this manner include 1-[^{18}F]fluoro-2-bromoethane and 1-[^{18}F]fluoro-3-iodopropane [44–46].

Another example of the use of a halogen as the leaving group (Fig. 9.4) made use of an iodo group located adjacent to a carbonyl to incorporate [^{18}F]fluoride into 21-[^{18}F]fluoropregnenolone in a radiochemical yield of 20% [47].

Nucleophilic substitution with [^{18}F]fluoride is also possible using benzylic halides as starting materials (Fig. 9.5), as evidenced by the synthesis of 3β-(4-[^{18}F]fluoromethylphenyl)- and 3β-(2-[^{18}F]fluoromethylphenyl) tropane-2β-carboxylic acid methyl esters for use as a dopamine transporter system radioligand [48]. The phenyltropane analog was obtained in 22% radiochemical yield (decay corrected to EOB), with chemical and radiochemical purities >99% of specific activities ranging from 2–5 Ci/μmol.

In those cases where an alkyl alcohol is available, subsequent activation by the formation of the corresponding alkyl sulfonate ester derivative (tosylate, mesylate, triflate, 1,2-cyclic sulfate) has proven to be extremely useful in the preparation of alkyl [^{18}F]fluorides. This reaction is general in scope and has been used to synthesize a variety of complex radioligands containing primary and secondary [^{18}F]fluorides. In a manner analogous to that of the disubstituted haloalkanes, the displacement of mesylate [49,50], tosylate [51], and triflate groups [52] led to the synthesis of substituted ω-[^{18}F]fluoroalkyl synthons.

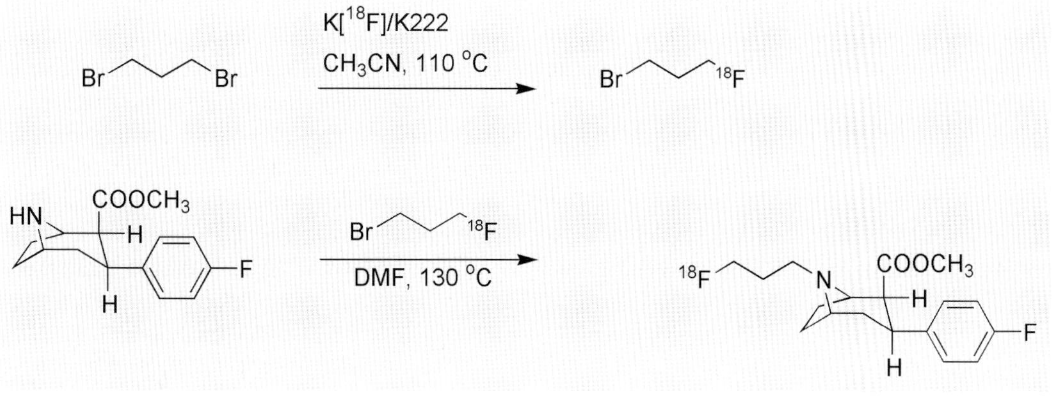

Figure 9.3.

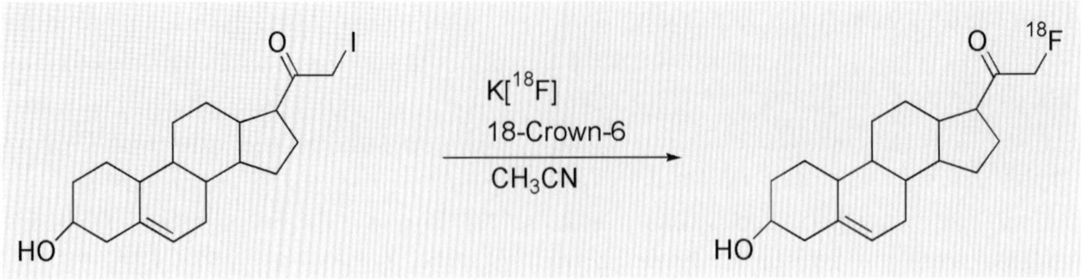

Figure 9.4.

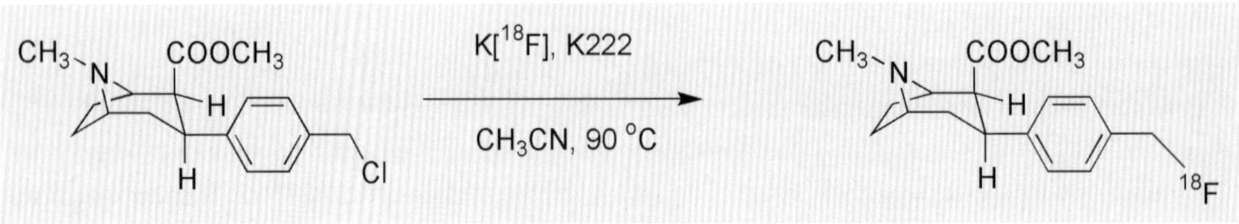

Figure 9.5.

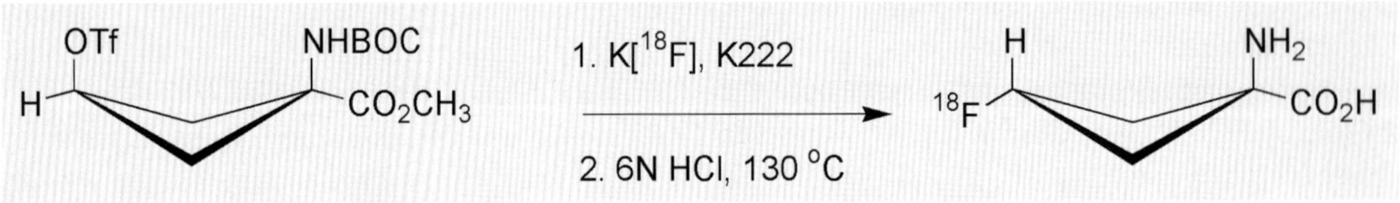

Figure 9.6.

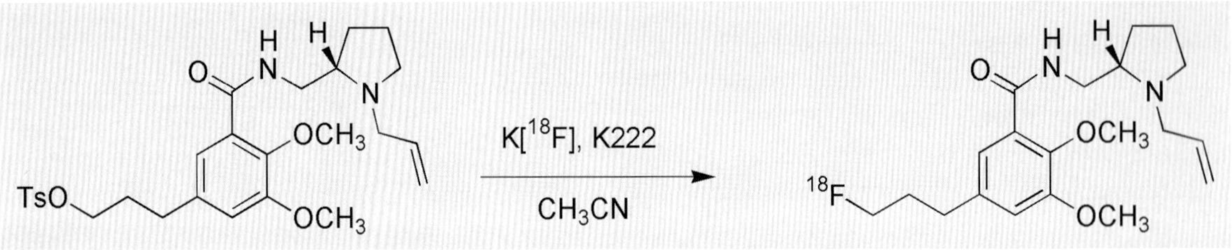

Figure 9.7.

For example, the reaction of 3-bromopropyl triflate with [^{18}F]fluoride produced 1-[^{18}F]fluoro-3-bromopropane [53].

The synthesis of 1-amino-3-[^{18}F]fluorocyclobutane-1-carboxylic acid provides an example of a nucleophilic substitution reaction with [^{18}F]fluoride utilizing a triflate derivative, prepared from the corresponding alcohol, as the radiofluorination substrate [54]. This unnatural, non-metabolized amino acid (Fig. 9.6) was used to visualize malignant tumors and was produced in 12% radiochemical yield with a specific activity >1.5 Ci/μmol (>55 GBq/μmol).

The high-affinity dopamine D_2 receptor antagonist, [^{18}F]fallypride (Fig. 9.7), was synthesized by a nucleophilic [^{18}F]fluoride substitution reaction on the corresponding tosylate in about 20% radiochemical yield at EOS [55].

Another example of the use of the tosylate leaving group to incorporate [^{18}F]fluoride was the synthesis of 9-(4-[^{18}F]fluoro-3-hydroxymethylbutyl)guanine ([^{18}F]FHBG) [56]. [^{18}F]FHBG (Fig. 9.8) was developed as a potential PET imaging agent to assess gene therapy. The masking of other reaction sites on the precursor molecule using trityl protection illustrates the ability to incorporate protecting group methodology into radiosynthetic strategies as a result of both the relatively long half-life of ^{18}F and the chemical stability of the alkyl [^{18}F]fluorides (see also the radiosynthesis of FDG). [^{18}F]FHBG was prepared in 8–22% radiochemical yield (decay corrected to EOB) with specific activities > 450 mCi/μmol (16.7 GBq/μmol).

In some instances, the activated leaving group used to incorporate [^{18}F]fluoride can also act as a protecting group for other functionalities present in the substrate.

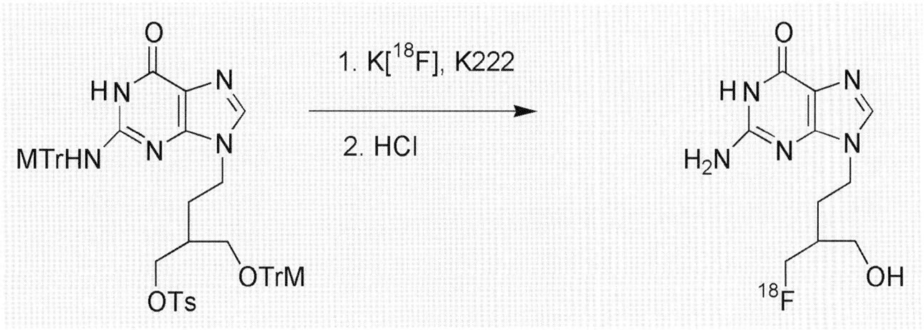

Figure 9.8.

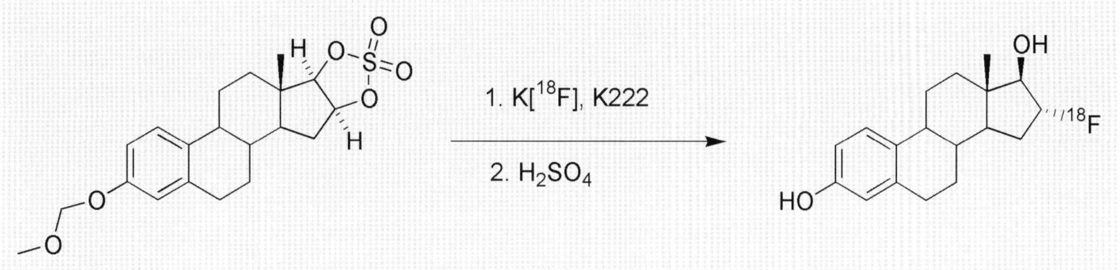

Figure 9.9.

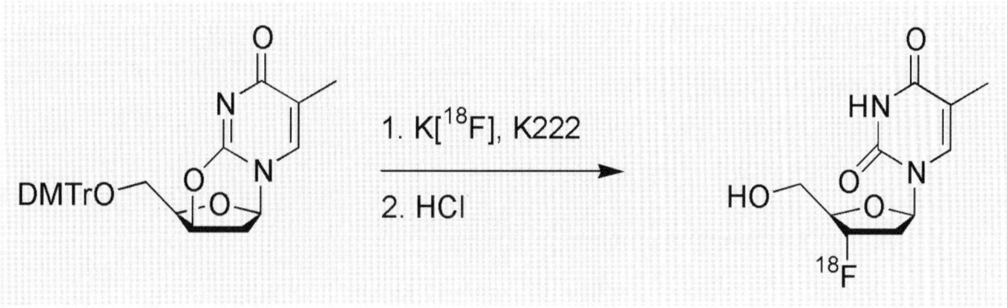

Figure 9.10.

This was the case for 3-O-methoxymethyl-16β,17β-O-epiestriol cyclic sulfone (Fig. 9.9), used as the precursor for the synthesis of 16α-[^{18}F]fluoroestradiol [57]. In this example, the cyclic sulfone acted as a protecting group for the 17β-hydroxyl functionality as well as activating nucleophilic displacement by [^{18}F]fluoride at C-16. The axial methyl group at C-19 prevented attack from the β-face of the D-ring, and there was no evidence of displacement reactions at the C-17 position. 16α-[^{18}F]fluoroestradiol was prepared in 30–45% radiochemical yield (decay corrected to EOB) with specific activities reported to be > 1 Ci/μmol (3.7 GBq/μmol).

Another example of a dual-mode leaving and protecting group can be found in the utilization of 2,3′-anhydro-5′-O-(4,4′-dimethoxytrityl)thymidine to produce 3′-deoxy-3′-[^{18}F]fluorothymidine (FLT) (Fig. 9.10) for use as a cellular proliferation marker in a decay-corrected radiochemical yield of approximately 14% [58,59]. The 2,3′-anhydro structure not only acts as the leaving group for nucleophilic radiofluorination, but also serves as a protecting group for the 3-N-position of the pyrimidine ring.

The most frequently used PET radiopharmaceutical, 2-deoxy-2-[^{18}F]fluoro-D-glucose (FDG), is currently produced utilizing [^{18}F]fluoride-for-alkyl sulfonate ester radiolabeling methodology (Fig. 9.11). The increase in demand for FDG has led to significant effort directed towards the development of routine production methods as well as the design and construction of remote, automated systems dedicated to the synthesis of FDG.

FDG is presently synthesized using modifications of the method developed at the Julich PET Centre [37]. In the original method, aqueous [^{18}F]fluoride was added to a solution consisting of Kryptofix [2.2.2]® and potassium carbonate dissolved in aqueous acetonitrile. The residual water was removed by repeated azeotropic

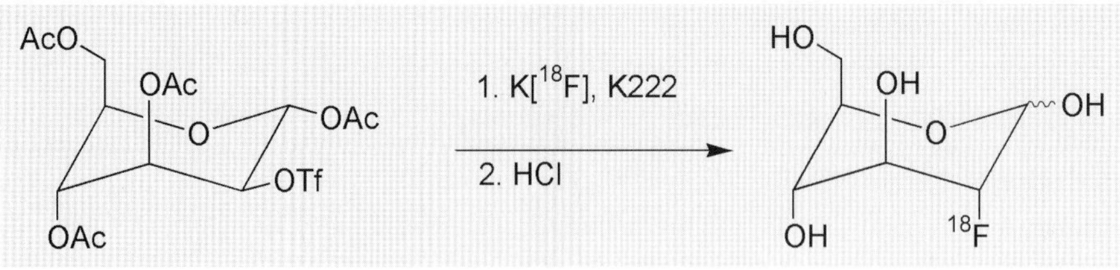

Figure 9.11.

distillations using anhydrous acetonitrile and a stream of nitrogen. The triflate precursor (1,3,4,6-tetra-O-acetyl-2-O-trifluoromethanesulfonyl-β-D-mannopyranose) was dissolved in acetonitrile and added to the dried [^{18}F]fluoride. The reaction mixture was heated to reflux for five minutes. The resultant solution was passed through a C_{18} Sep-Pak® cartridge. The residual aminopolyether was removed by washing the C_{18} Sep-Pak® with 0.1 M hydrochloric acid. The radiolabeled acetylated carbohydrates were eluted into a second reaction vessel using tetrahydrofuran, and the ether was removed. Aqueous hydrochloric acid was added to the acetyl-protected intermediate (2-deoxy-2-[^{18}F]fluoro-1,3,4,6-tetra-O-acetyl-β-D-glucopyrranose), and the solution was heated at 130 °C for 15 minutes. The product was purified by passage through an ion-retardation resin followed by an alumina column. The method was utilized as the basis of a computer-controlled automated synthesizer for the routine production of FDG [60]. Further modifications of the Julich methodology have led to the development of "one-pot" syntheses for the production of FDG. These modifications include the substitution of tetramethylammonium carbonate for Kryptofix [2.2.2]®/potassium carbonate as the phase-transfer reagent and subsequent elimination of the C_{18} Sep-Pak® cartridge-purification step. As a result of these modifications, the acidic hydrolysis was performed in the same reaction vessel. The reported radiochemical yield was 52% at the end-of-synthesis (EOS) with a total synthesis time of 48 minutes [61]. A similar "one-pot" modification was reported that retained Kryptofix [2.2.2]® as the phase-transfer reagent. Several Sep-Pak® cartridges were added to the system to remove unwanted Kryptofix [2.2.2]® and to prevent [^{18}F]fluoride breakthrough. These modifications provided a radiochemical yield of 65–70% decay-corrected to the end-of-bombardment (EOB) in a total synthesis time of approximately 50 minutes [62].

Toxicity concerns associated with Kryptofix [2.2.2]® (LD_{50} 35 mg/kg in rats) have prompted the use of other phase-transfer agents, such as tetrabutylammonium hydroxide or tetrabutylammonium bicarbonate. This modification has been incorporated into a commercially available synthesizer produced by Nuclear Interface, Inc. The Nuclear Interface module is flexible in that it can utilize either tetrabutylammonium bicarbonate or Kryptofix [2.2.2]® as the phase-transfer reagent. In addition, the module can perform the hydrolysis of the radiolabeled intermediate, 2-deoxy-2-[^{18}F]fluoro-1,3,4,6-tetra-O-acetyl-β-D-glucopyrranose, under either acidic or basic (KOH) conditions. The module completes the radiosynthesis in less than thirty minutes with a reported radiochemical yield of approximately 60% at EOS.

Another variation of the FDG radiolabeling scheme used an immobilized quaternary 4-aminopyridinium resin to isolate [^{18}F]fluoride and subsequently incorporate it into the ^{18}F-labelled intermediate [63,64]. The [^{18}F]fluoride solution was passed across the resin column where [^{18}F]fluoride was trapped, and the bulk of the enriched [^{18}O]water was recovered downstream. The resin-bound [^{18}F]fluoride was dried by passing anhydrous acetonitrile across the resin column while heating the column to approximately 100 °C. A solution of the mannose triflate precursor in anhydrous acetonitrile was then passed over the heated resin column in either a slow single-pass or a reciprocating flow across the resin column. The solution containing the radiolabeled intermediate was then transferred to a hydrolysis vessel where the acetonitrile was removed. Following acid hydrolysis, FDG was purified in a manner analogous to the original method described above. The resin methodology formed the basis of a commercially available synthesis unit (PETtrace FDG MicroLab™, GE Medical Systems). This unit utilizes a disposable cassette system for the reaction column as well as disposable transfer and addition lines that facilitate its set-up. Solid-phase support methodology that incorporates basic hydrolysis of the radiolabeled intermediate [65,66] has been implemented in the FDG synthesizer marketed by Coincidence Technologies, Inc. The use of base decreased hydrolysis times to two minutes at room temperature and resulted in no epimerization. In addition, there are commercially available pre-packaged reagent vials and pre-sterilized tubing systems for this automated synthesis module. The module is equipped with a programmable logic controller that regulates the synthesis process. As a result, the system can be operated without a computer controller. The development of high-yield [^{18}O]water targets for the production of [^{18}F]fluoride, the capability of current generation cyclotrons to perform dual-target irradiations at relatively high beam currents, and the availability of efficient automated synthesis modules to produce FDG has made possible the production of multi-Curie (>74 GBq) amounts of FDG in a single cyclotron production run. This capability has significantly increased the utilization of FDG and has led to the growth of regional FDG production facilities that can supply a multitude of off-site users.

Radiofluorination via Aromatic Nucleophilic Substitution Reactions

While alkyl [^{18}F]fluoride derivatives have seen frequent utilization in PET radiopharmaceuticals as noted above, significant effort also has been invested in radiolabeling methods to incorporate [^{18}F]fluoride into aromatic systems. These efforts include radiofluorination methods to incorporate [^{18}F]fluoride directly onto the aromatic ring as well as into prosthetic groups containing aromatic rings. Aromatic nucleophilic substitutions include:

(i) reactions in which the leaving group is activated by the presence of electron-withdrawing groups *ortho* and/or *para* to the leaving group;
(ii) reactions catalyzed by strong bases that proceed through an aryne (triple bond) intermediate; and
(iii) reactions in which the nitrogen of a diazonium salt is replaced by the nucleophile.

The first two examples can be classified as SNAr reactions, but the third example is an SN1-type reaction. The first class of reaction is by far the most commonly used for aromatic radiofluorinations, wherein a leaving group is activated by the presence of *ortho* and/or *para* electron-withdrawing groups on the aromatic ring. An approximate ranking of common substituents in order of decreasing activation ability includes: NO_2>CF_3>CN>CHO>COR>COOR>COOH>Br>I>F>Me>NMe_2>OH>NH_2 [1,67]. The leaving group will also have an effect on the reaction rate. The following list is an approximate order of leaving group ability in aromatic nucleophilic substitution reactions: NMe_3^+>NO_2>CN>F>Cl,Br,I>OAr>OR>SR>NH_2. There have been attempts to correlate the radiochemical yields in nucleophilic radiofluorination reactions with the C-13 NMR chemical shifts of the corresponding fluoro-, nitro-, and trialkylammonium-substituted aryl aldehydes, ketones, and nitriles. While good agreement was found for the displacement of the substituted fluoro and nitro groups, the trialkylammonium group did not show the same correlation pattern [68,69].

The radiosynthesis of [^{18}F]altanserin, a serotonin 5-HT_{2A} receptor ligand, utilized the corresponding aromatic nitro precursor in the nucleophilic substitution reaction with potassium [^{18}F]fluoride in the presence of potassium carbonate and Kryptofix [2.2.2]® and illustrates activation of the nitro group by a carbonyl group situated *para* to the leaving group (Fig. 9.12). The decay-corrected radiochemical yield was reported to be 20% [70].

As a result of the relatively long half-life of ^{18}F, there are several examples of multi-step radiosynthetic pathways where the radionuclide is incorporated very early in the process. An example of a multi-step radiosynthetic pathway is the no-carrier added synthesis of 6-[^{18}F]fluoro-L-DOPA [71]. In this case, [^{18}F]fluoride was used as the radiofluorinating agent in the preparation of 3,4-dimethoxy-2-[^{18}F]fluorobenzaldehyde from the corresponding nitro-substituted compound (Fig. 9.13). The resultant ^{18}F-labeled product was then reacted in a enantiomerically pure variant of the 2-phenyl-5-oxazolone procedure to yield ^{18}F-labeled α,β-didehydro derivatives. Following enatioselective reduction and deprotection, 6-[^{18}F]fluoro-L-DOPA was isolated in 3% decay-corrected radiochemical yield with an enantiomeric excess >90%.

There are many examples of the use of aryl trialkylammonium salts as alternatives to nitro-substituted

K[^{18}F], K222
DMSO

Figure 9.12.

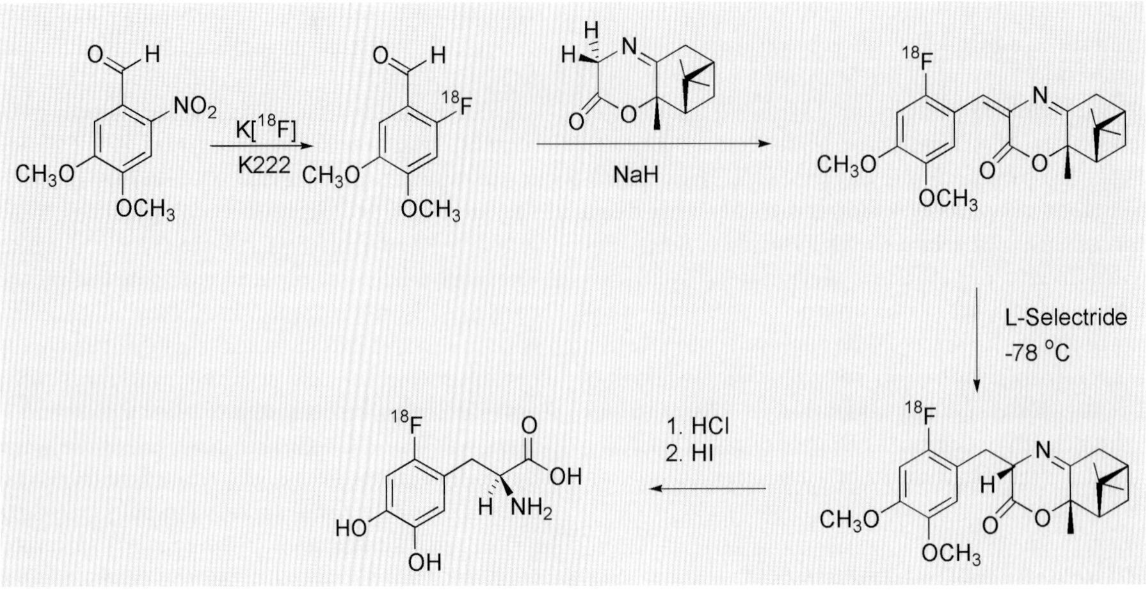

Figure 9.13.

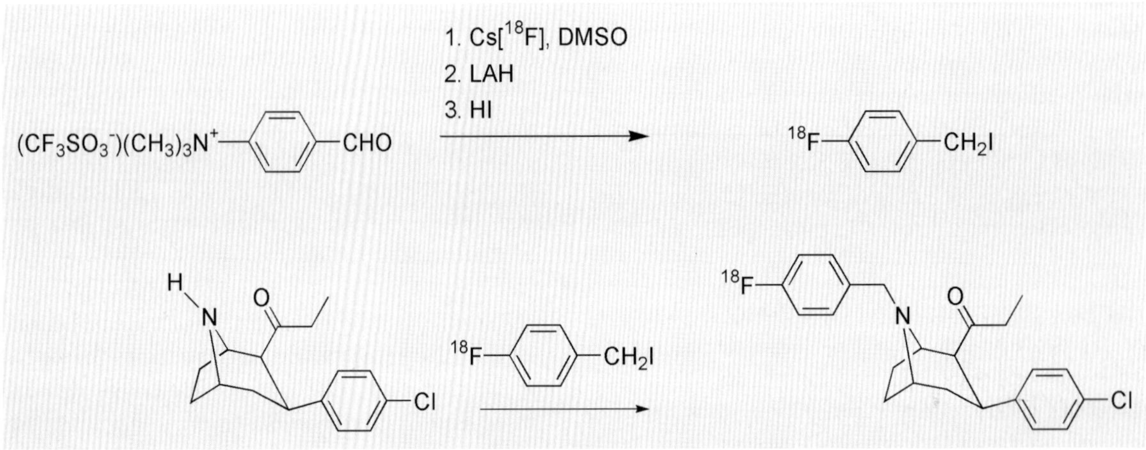

Figure 9.14.

aryl precursors. Aryl trialkylammonium salts tend to be more reactive and require milder conditions for [^{18}F]fluoride incorporation. An example of this type of S_NAr reaction is the radiosynthesis of 4-[^{18}F]fluorobenzyl iodide (Fig. 9.14) [41]. 4-Trimethylammoniumbenzaldehyde trifluoromethanesulfonate in aqueous dimethyl sulfoxide was reacted with Cs[^{18}F]. The resultant substituted [^{18}F]fluorobenzaldehyde was reduced to the benzyl alcohol followed by treatment with hydriodic acid to yield 4-[^{18}F]fluorobenzyl iodide in approximately 25% yield (EOS). This prosthetic group is amenable to incorporation into a variety of radiopharmaceuticals, including (+)-N-(4-[^{18}F]fluorobenzyl)-2β-propanoyl-3β-(4-chlorophenyl)tropane [72].

The synthesis of [^{18}F]norchlorofluoroepibatidine provides an example of the use of [^{18}F]fluoride with trialkylammonium salts as the leaving group from a heteroaromatic ring [73]. Using the tert-BOC-protected epibatidine derivative with trimethylammonium iodide as the leaving group, the desired ^{18}F-labeled compound was prepared in 70% radiochemical yield (Fig. 9.15). Subsequent deprotection and N-methylation afforded overall radiochemical yields of 45–55% for [^{18}F]N-methyl-norchlorofluoroepibatidine with a specific activity of 2–6 Ci/μmol at EOS.

An ^{18}F-radiolabeled prosthetic group approach similar to the use of [^{18}F]fluorobenzyl iodide has been used to label oligonucleotides [74]. In this case, the desired prosthetic group, N-(4-[^{18}F]fluorobenzyl)-2-

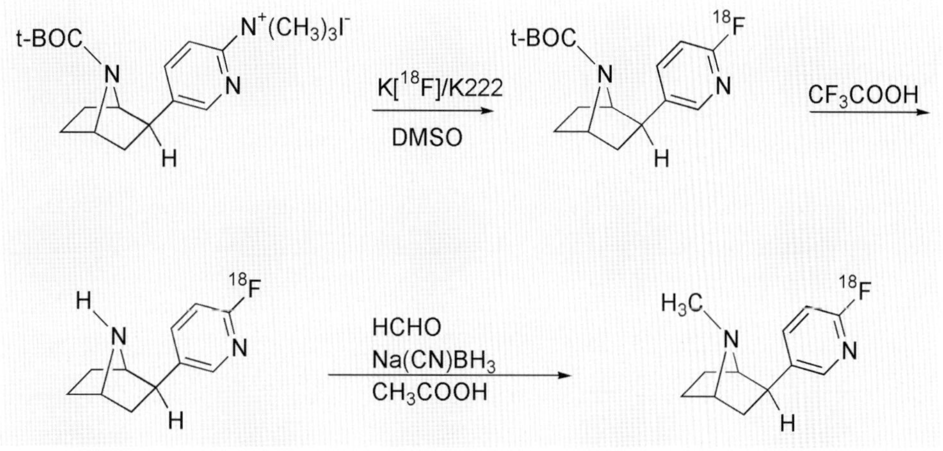

Figure 9.15.

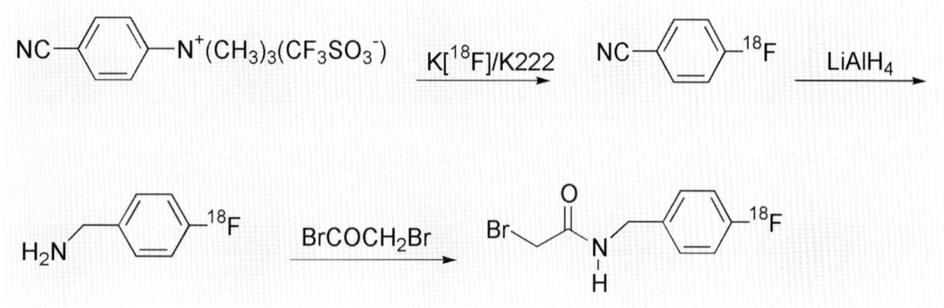

Figure 9.16.

bromoacetamide, was produced in a three-step synthesis that began with the nucleophilic aromatic radiofluorination of trimethylammonium benzonitrile triflate. The resultant 4-[^{18}F]benzonitrile was reduced to provide the radiolabeled benzyl amine, which was allowed to react with bromoacetyl bromide to yield the desired prosthetic group (Fig. 9.16). The radiochemical yield was approximately 12% and the specific activity of the resultant labeled oligonucleotide was 1 Ci/μmol (3.7 GBq/μmol) at EOS. Similar approaches have been reported by other groups utilizing N-succinimidyl-4-[^{18}F]fluorobenzoate [75] as well as a solid-phase based approach using 4-[^{18}F]fluorobenzoic acid [76].

Other Nucleophilic [^{18}F]Fluorination Reactions

A classic method for the synthesis of aromatic fluorides, known as the Balz–Schiemann reaction, involves the thermal decomposition of aryl diazonium tetrafluoroborate salts (Fig. 9.17). While this methodology has been applied to the production of aryl [^{18}F]fluorides, it suffers several drawbacks [1]. The use of [^{18}F]BF_4^- as the counter anion results in low radiochemical yields as well as low specific activity products.

There has been little utilization of this methodology for the synthesis of more complex radiofluorinated compounds. It should be noted that this reaction could allow the incorporation of [^{18}F]fluoride into an aromatic ring that is not activated to nucleophilic substitution reactions.

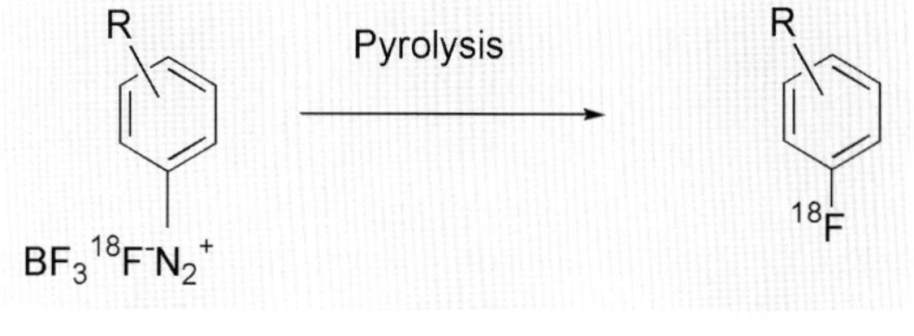

Figure 9.17.

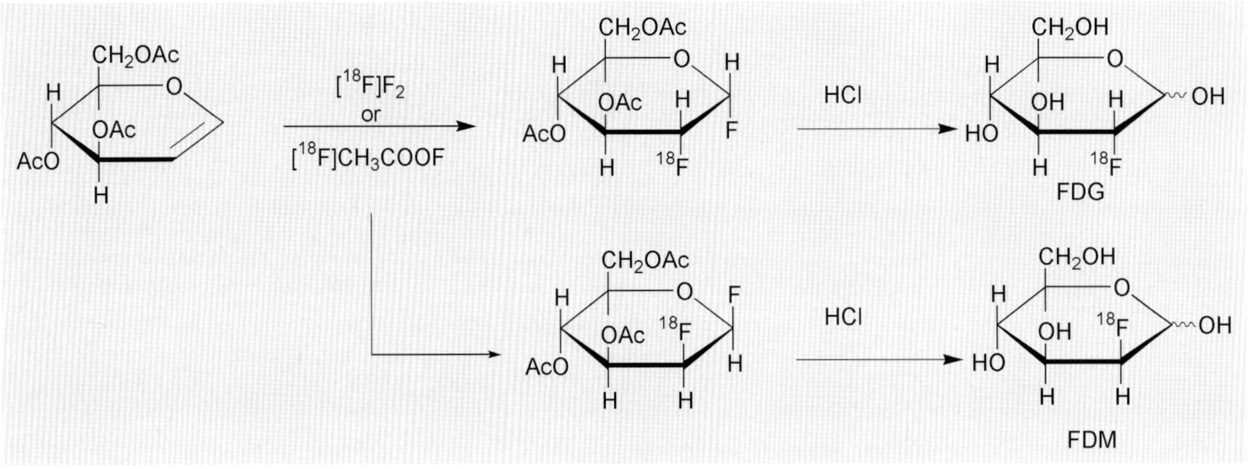

Figure 9.18.

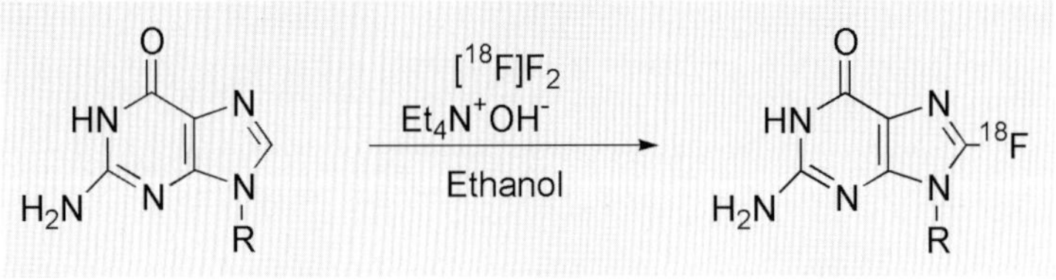

Figure 9.19.

Electrophilic Reactions with $^{18}F^{+}$ (Low Specific Activity)

Fluorine is the most electronegative of all the elements. Fluorine exists as a colorless to pale yellow corrosive gas (F_2) that reacts with many organic and inorganic substances. Fluorine is a powerful oxidizing agent and attacks both quartz and glass, making its handling problematic. The use of fluorine gas as a carrier in the production of $[^{18}F]F_2$ leads to several orders of magnitude lower-specific-activity reaction products compared to methods using no-carrier added $[^{18}F]$fluoride obtained from $[^{18}O]$water targets.

Early synthetic methods for the production of FDG were based on electrophilic radiofluorination chemistry (Fig. 9.18). The reaction of $[^{18}F]F_2$ with 3,4,6-tri-O-acetyl glucal in fluorotrichloromethane (Freon-11) was the first synthetic method employed for this important PET radiopharmaceutical [77,78]. One disadvantage of this method, due to the highly reactive nature of F_2, was the production of the protected fluoromannopyrranosyl leading to the undesired 2-deoxy-2-$[^{18}F]$fluoro-D-mannose (FDM) derivative (Fig. 9.18). The use of an alternative fluorinating agent, acetyl $[^{18}F]$hypofluorite, was proposed as a method for the routine production of FDG [79,80]. The use of acetyl $[^{18}F]$hypofluorite resulted in the regioselective (95%) synthesis of the desired glucose configuration under optimal reaction conditions. However, non-optimal conditions led to the production of undesired FDM in high yield [81]. In addition to this problem, electrophilic methods using $[^{18}F]F_2$ to produce mono-fluorinated products also suffer from the loss of 50% of the radioactivity. The development of high-yield $[^{18}F]$fluoride targets and stereospecific nucleophilic radiofluorination chemistries have replaced electrophilic methods for the production of FDG.

One example of a direct electrophilic radiofluorination of an activated aromatic substrate is the synthesis of a series of purine derivatives that have shown promise for PET imaging of a reporter system to assess viral gene therapy [82]. Earlier work demonstrated the capability of direct incorporation of fluorine into the C-8 position of a series of substituted purine derivatives [83,84]. The analogous radiofluorinations (Fig. 9.19) using $[^{18}F]F_2$ led to the production of 8-$[^{18}F]$fluoroganciclovir, 8-$[^{18}F]$fluoropenciclovir, 8-$[^{18}F]$fluoroacyclovir, and 8-$[^{18}F]$fluoroguanosine. While this method does not require the use of protecting groups, the radiochemical yields were low (0.9–1.2% decay-corrected to EOB). Nevertheless, sufficient material was produced to allow for the utilization of these $[^{18}F]$fluoropurine derivatives in animal studies [85].

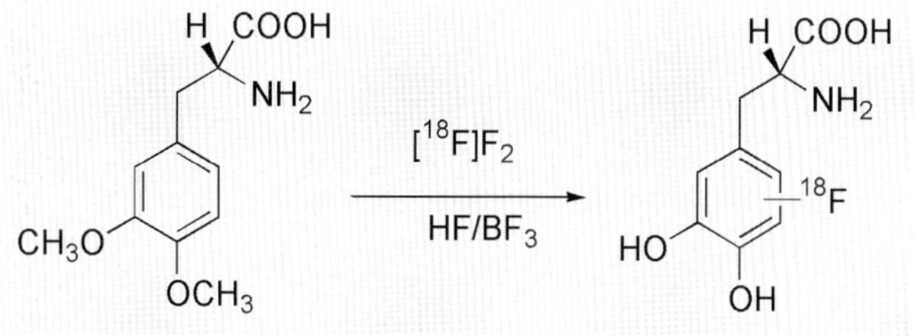

Figure 9.20.

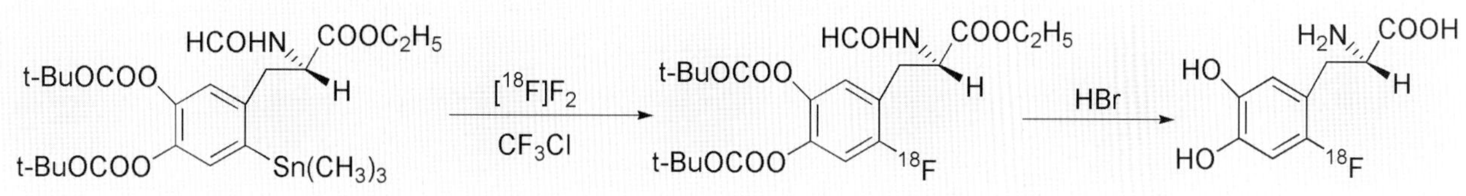

Figure 9.21.

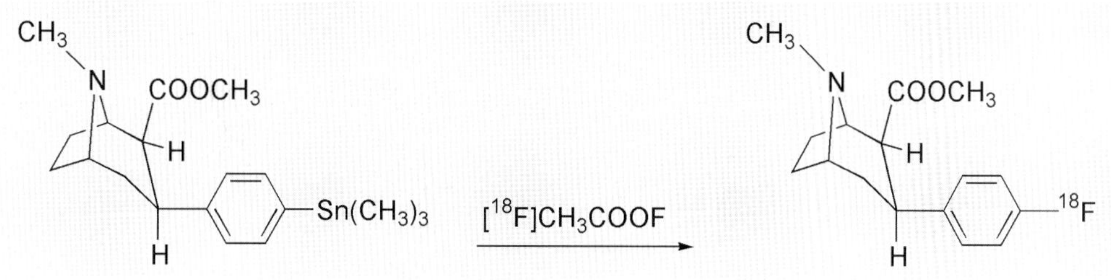

Figure 9.22.

The direct electrophilic fluorination of aromatic rings has been accomplished using a variety of ^{18}F-radiolabeled electrophilic fluorinating agents including $[^{18}F]F_2$, xenon $[^{18}F]$difluoride, and acetyl $[^{18}F]$hypofluorite. These types of radiofluorinations are not generally regioselective. As an example, the reaction of 3,4-dihydroxyphenylalanine with $[^{18}F]F_2$ using liquid hydrogen fluoride as a solvent (Fig. 9.20) yielded 2-, 5-, and 6-$[^{18}F]$fluoro-L-DOPA in the ratio of 35:5:59, respectively [86].

The lack of regiospecificity in direct electrophilic radiofluorination reactions has resulted in the increased use of demetallation reactions. Regioselective demetallations have been used to great advantage with $[^{18}F]F_2$. Examples can be found in the literature for the use of aryl tin, mercury, silicon, selenium, and germanium compounds as substrates for radiofluorinations using $[^{18}F]F_2$. This type of reaction has been utilized to produce 6-$[^{18}F]$fluoro-L-DOPA [87,88]. The most common precursor is an aryl substituted trialkyl tin derivative (Fig. 9.21). The decay-corrected radiochemical yield was reported to be 29–37% using a protected trialkyl tin derivative with trifluorochloromethane as the reaction solvent, followed by acidic removal of the phenol and amino acid protecting groups. This methodology has been utilized for the automated production of 6-$[^{18}F]$fluoro-L-DOPA using a commercially available computer-controlled synthesis apparatus (Nuclear Interface, Inc.).

Another example of electrophilic regioselective fluorodemetallation is the radiosynthesis of $[^{18}F]$WIN 35,428 ($[^{18}F]\beta$-CFT) [89]. In this case, $[^{18}F]\beta$-CFT was prepared by the electrophilic radiofluorination of 2β-carbomethoxy-3β-(4-trimethylstannylphenyl)tropane using acetyl $[^{18}F]$hypofluorite as the fluorination agent (Fig. 9.22) with a specific activity of 18–25 GBq/mmol and a yield of 0.9–2.0%.

In a manner analogous to the other halogens, fluorine will add to double bonds (Fig. 9.23). The radiosynthesis of $[^{18}F]$2-(2-nitro-1[H]-imidazol-1-yl)-N-(2,2,3,3,3-pentafluoropropyl)-acetamide ($[^{18}F]$EF5), a radiotracer used to assess tissue hypoxia, took advantage of this reactivity by using $[^{18}F]F_2$ in trifluoroacetic acid to radiofluorinate the perfluoro alkene in 10–15% radiochemical yield [90].

The synthesis of 5-$[^{18}F]$fluoro-2′-deoxyuridine (Fig. 9.24) is another example of the addition of fluorine to double bonds. In this case, the formal addition of

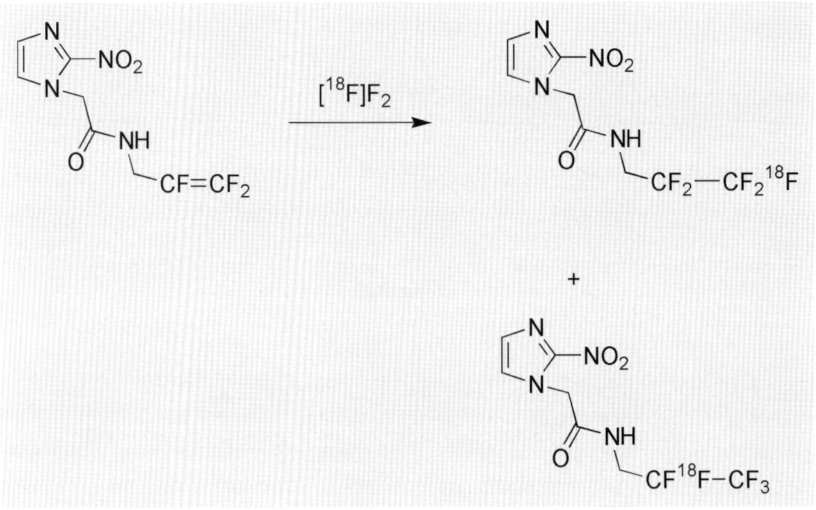

Figure 9.23.

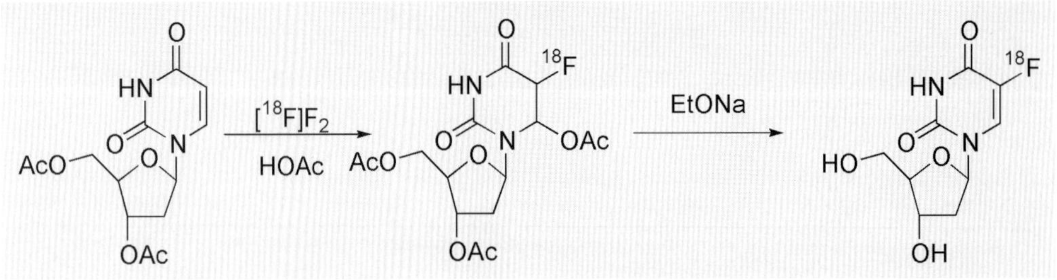

Figure 9.24.

"fluorine acetate" to the 5,6-double bond led to an intermediate, which upon base catalyzed elimination of acetic acid yielded the target molecule in a radiochemical yield of 15–25% [91].

Electrophilic fluorination of carbanions using fluorinating agents such as perchloryl[^{18}F]fluoride, N-[^{18}F]fluoro-N-alkylsulfonamides, N-[^{18}F]fluoropyridinium triflate has seen some utilization. A series of N-[^{18}F]fluoro-N-alkyl sulfonamides have been synthesized using [^{18}F]F_2 and shown to be suitable for use in radiolabeling a variety of structurally simple aryl lithium and aryl Grignard reagents [92]. However, the methodology has not been widely utilized to radiolabel more complicated target compounds.

Radiobromine and Radioiodine for PET

Fluorine-18 is not the only radiohalogen that has shown utility in the synthesis of PET radiopharmaceuticals. While there are no useful positron-emitting radionuclides of chlorine or astatine, there are useful positron-emitting radionuclides of both bromine (^{76}Br) and iodine (^{124}I). These radionuclides have suitable half-lives (^{76}Br $t_{\frac{1}{2}}$ = 16.1 hours and ^{124}I $t_{\frac{1}{2}}$ = 4.2 days) for use in PET studies and can be produced in sufficient quantities to foster the development of a variety of radiopharmaceuticals.

Over the past 40 years, a great deal of radiochemistry effort has been devoted to methods for attaching single photon- and beta-emitting radioiodines, such as ^{125}I, ^{131}I, and ^{123}I, onto large and small molecules for in vitro and in vivo experimental uses [94–98]. These radiolabeling methods are also applicable to positron-emitting radioiodines, such as ^{124}I, and to a large extent to radiobromines as well [99–102].

As with most halogen chemistry, [^{76}Br]bromide can be used either as a source of nucleophilic bromide or as a source of electrophilic bromine upon oxidation. [^{76}Br]β-CBT, a dopamine transporter radioligand (Fig. 9.25), has been radiolabeled using both nucleophilic and electrophilic radiobromination chemistry [103].

A variety of dopamine receptor ligands have been labeled using bromine-76 including [^{76}Br]FLB 457, [^{76}Br]FLB 463, [^{76}Br]bromolisuride, [^{76}Br]bromospiper-

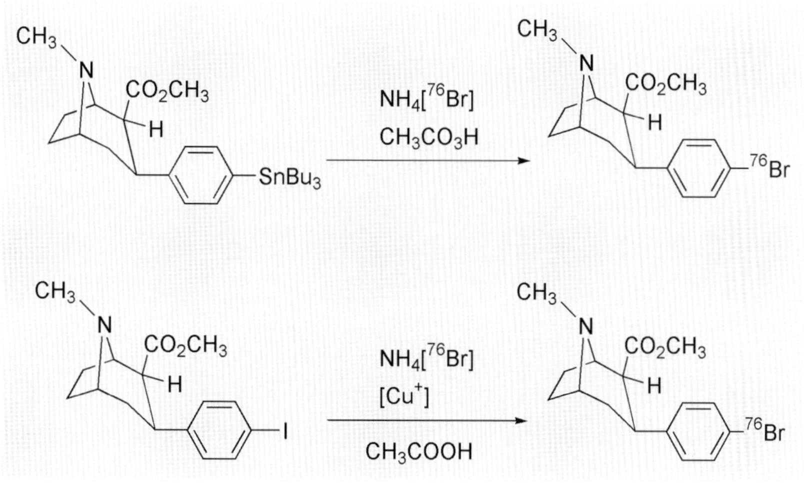

Figure 9.25.

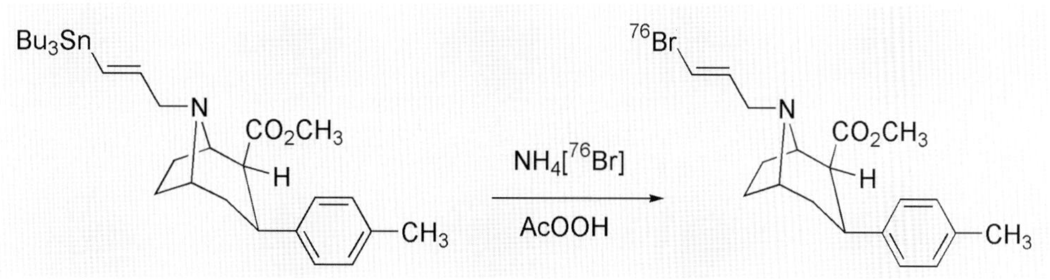

Figure 9.26.

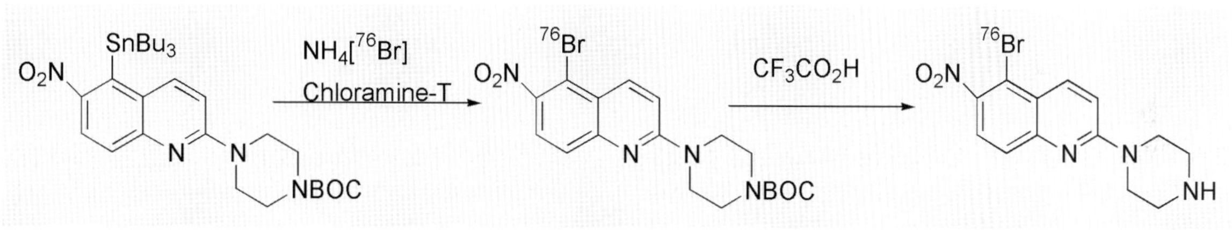

Figure 9.27.

one, and [^{76}Br]PE2Br [104–107]. The radiolabeling of [^{76}Br]PE2Br, (E)-N-(3-bromoprop-2-enyl)-2β-carbomethoxy-3β-(4'-tolyl)nortropane illustrates the capability of incorporating the radiobromine and radioiodine into vinylic positions (Fig. 9.26). This moiety offers increased stability for the halides compared to alkyl-substituted analogs. [^{76}Br]PE2Br was synthesized in a radiochemical yield of approximately 80% using NH_4[^{76}Br] and peracetic acid with the vinyl tri-n-butylstannane substituted tropane analog.

The serotonin transporter ligand 5-[^{76}Br]bromo-6-nitoquipazine also has been synthesized (Fig. 9.27) [108], as well as a norepinephrine transporter agent [^{76}Br]MBBG [109,110].

The electrophilic radiobromination of metaraminol (Fig. 9.28) yielded a mixture of the 4- and 6-substituted bromometaraminols (17% and 38% non-decay corrected radiochemical yields, respectively), which were separable by HPLC. These compounds have shown promise as radiotracers for the myocardial norepinephrine reuptake system [111].

Bromine-76 has also been utilized to radiolabel intact monoclonal antibodies where its longer half-life allows for longer clearance times [112,113]. The synthesis of a radiobrominated thymidine analog ([^{76}Br]FBAU) for use as a cellular proliferation marker for PET has also been reported [114]. Ammonium [^{76}Br]bromide was used in an electrophilic destannyla-

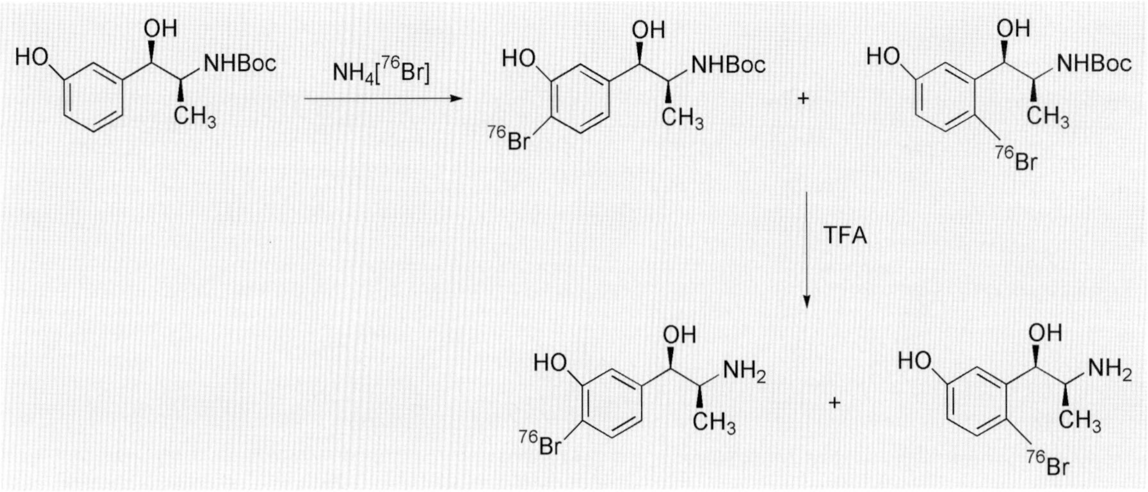

Figure 9.28.

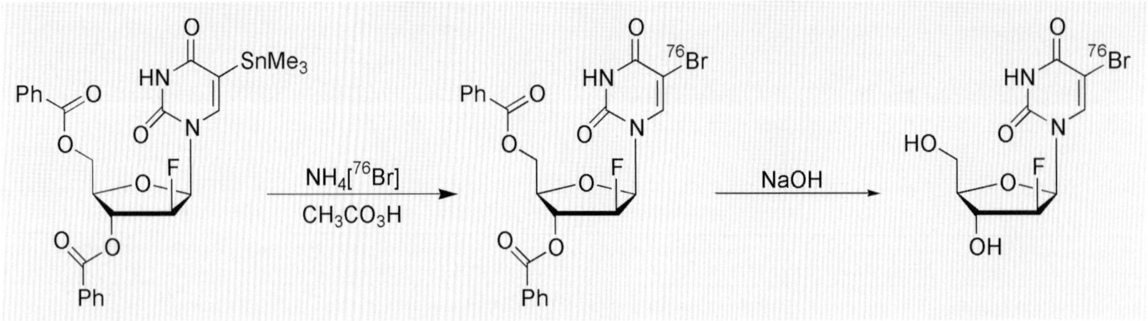

Figure 9.29.

tion reaction to prepare the 3′,5′-dibenzoyl protected analog, which yielded the desired [^{76}Br]FBAU following base hydrolysis (Fig. 9.29).

Iodine-124 has been used to radiolabel intact monoclonal antibodies, where its longer half-life allows for longer metabolic clearance times [115–117]. In a manner analogous to the radiobromination chemistry discussed above, iodine-124 also has been used in both nucleophilic and electrophilic radiolabeling reactions. An example of this was the radiosynthesis of [^{124}I]β-CIT (Fig. 9.30), which was labeled using both electrophilic iododestannylation and nucleophilic substitution via iodo-for-bromo exchange [118].

In addition, insulin has been labeled with ^{124}I on the fourteenth amino acid residue (tyrosine) using electrophilic radiolabeling conditions [119]. The nucleoside analog 2′-fluoro-2′-deoxy-1β-D-arabinofuranosyl-5-iodouracil (FIAU) has also been labeled using Na[^{124}I] and a stannylated uracil derivative as the precursor (Fig. 9.31) [120–122]. This radioiodinated derivative has been reported to result in significantly higher specific accumulation of radioactivity compared to [^{18}F]FHPG in tumor-bearing BALB/c mice [123].

Conclusions

Fluorine-18 is currently the most utilized PET radiohalogen as a result of its relatively facile production in large quantities, its convenient half-life, and its nearly optimal decay properties. Efficient incorporation of the ^{18}F-radiolabel into a variety of radiopharmaceuticals is possible using either nucleophilic routes with high specific activity [^{18}F]fluoride or electrophilic routes with lower specific activity [^{18}F]fluorine. The longer half-lives of ^{76}Br and ^{124}I can provide advantages over ^{18}F to image slower physiological processes, but the production of these radiohalogens is more demanding and their decay properties are more complex than those of ^{18}F. Thus, ^{18}F has become the radiohalogen of choice for a variety of PET imaging applications.

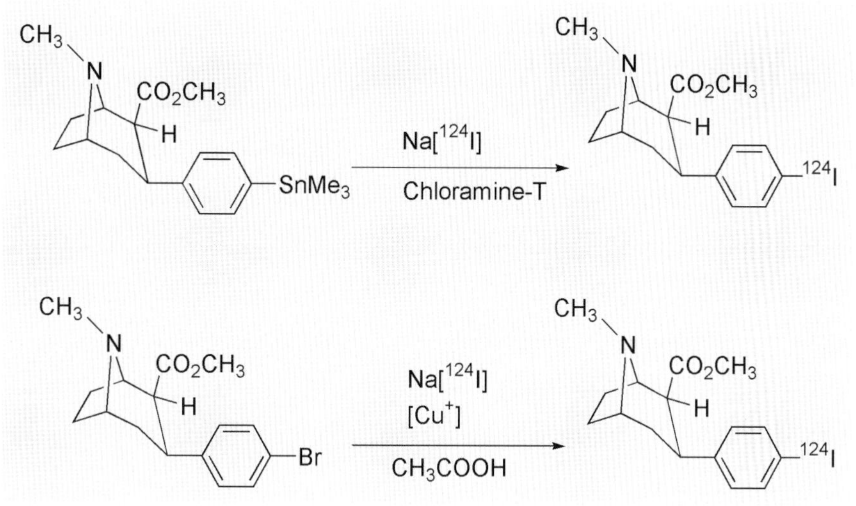

Figure 9.30.

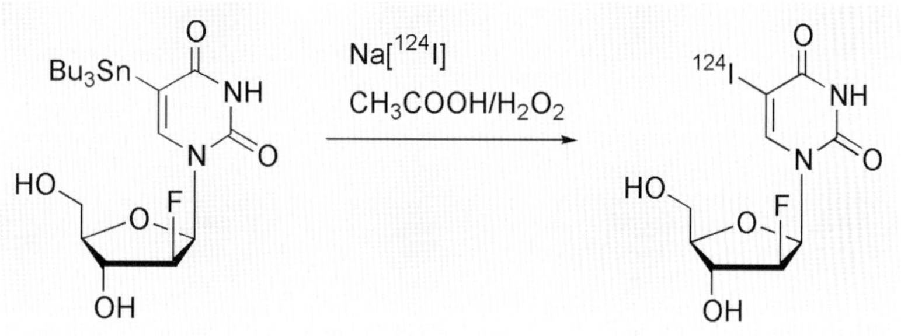

Figure 9.31.

References

1. Kilbourn MR. Fluorine-18 labeling of radiopharmaceuticals, Nuclear Science Series NAS-NS-3203, Washington DC: National Academy Press,1990.
2. Stocklin G. Molecules labeled with positron emitting halogens. Int J Rad Appl Instrum B 1986; 13:109–18.
3. Moerlein SM, Laufer P, Stocklin G, Pawlik G, Wienhard K, Heiss WD. Evaluation of ^{75}Br-labeled butyrophenone neuroleptics for imaging cerebral dopaminergic receptor areas using positron emission tomography. Eur J Nucl Med 1986; 12:211–16.
4. Mathis CA, Sargent III T, Shulgin AT. Iodine-122-labeled amphetamine derivative with potential for PET brain blood-flow studies J Nucl Med 1985; 26:1295–301.
5. Mathis CA, Lagunas-Solar MC, Sargent III T, Yano Y, Vuletich A, Harris LJ. A ^{122}Xe-^{122}I generator for remote radio-iodinations. Appl Radiat Isot 1986; 37:258–60.
6. March J. Advanced organic chemistry reactions, mechanisms, and structure, 4th edn. New York: John Wiley & Sons, 1992.
7. Fowler JS, Wolf AP. Positron emitter-labeled compounds: priorities and problems. In: Phelps ME, Mazziotta JC, Schelbert HR (eds). Positron emission tomography and autoradiography principles and applications for the brain and heart. New York: Raven Press, 1986, p 391–450.
8. McCarthy TJ, Welch MJ. The state of positron emitting radionuclide production in 1997. Sem Nucl Med 1998; 28:235–46.
9. Welch MJ, Redvanly CS (eds). Handbook of radiopharmaceuticals: radiochemistry and applications. Sussex: John Wiley & Sons, 2002.
10. Kilbourn MR, Hood JT, Welch MJ. A simple ^{18}O water target for ^{18}F production. Appl Radiat Isot 1984; 35:599–602.
11. Kilbourn MR, Jerabek PA, Welch MJ. An improved [^{18}O]water target for [^{18}F]fluoride production. Int J Appl Radiat Isot 1985; 36:327–8.
12. Wieland BW, Hendry GO, Schmidt DG, Bida G, Ruth TJ. Efficient small-volume ^{18}O-water targets for producing ^{18}F-fluoride with low energy protons. J Label Compds Radiopharm 1986; 23:1205–7.
13. Ruth TJ, Buckle KR, Chun KS, Hurtado ET, Jivan S, Zeisler S. A proof of principle for targetry to produce ultra high quantities of ^{18}F-fluoride. Appl Radiat Isot 2001; 55:457–61.
14. Schlyer DJ, Bastos M, Wolf AP. A quantitative separation of fluorine-18 fluoride from oxygen-18 water. J Nucl Med 1987;28:764.
15. Schlyer DJ, Bastos MA, Wolf AP. Separation of [^{18}F]fluoride from [^{18}O]water using anion exchange resin. Int J Rad Appl Instrument [A] 1990;41:531–3.
16. Jewett DM, Toorongian SA, Mulholland GK, Watkins GL, Kilbourn MR. Multiphase extraction: rapid phase-transfer of [^{18}F]fluoride ion for nucleophilic radiolabeling reactions. Appl Radiat Isot 1988;39:1109–11.
17. Chan PKH, Firnau G, Garnett ES. An improved method for the production of fluorine-18 in a reactor. Radiochem Radioanal Lett 1974;19:237–42.

18. Thomas CC, Sondel JA, Kerns RC. Production of carrier-free fluorine-18. Int J Appl Radiat Isot 1965;16:71–4.
19. Clark JC, Silvester DJ. A cyclotron method for the production of fluorine-18. Int J Appl Radiat Isot 1966;17:151–4.
20. Hinn GM, Nelp WB, Weitkamp WG. A high-pressure non-catalytic method for cyclotron production of fluorine-18. Int J Appl Radiat Isot 1971;22:699–701.
21. Knust EJ, Machulla H-J. High yield production in a water target via the $^{16}O(^{3}He,p)^{18}F$ reaction. Int J Appl Radiat Isot 1983;34:1627–8.
22. Casella V, Ido T, Wolf AP, Fowler JS, MacGregor RR, Ruth TJ. Anhydrous F-18 labeled elemental fluorine for radiopharmaceutical preparation. J Nucl Med 1980;21:750–7.
23. Chirakal R, Adams RM, Firnau G, Schrobilegen GJ, Coates G, Garnett ES. Electrophilic ^{18}F from a Siemens 11 MeV proton-only cyclotron Nucl Med Biol 1995;22:111–6.
24. Nickles RJ, Daube ME, Ruth TJ. An O_2 target for the production of $[^{18}F]F_2$. Appl Radiat Isot 1984;35:117–22.
25. Bergman J, Solin O. Fluorine-18-labeled fluorine gas for synthesis of tracer molecules. Nucl Med Biol 1997;24:677–83.
26. Maziere B, Loc'h C. Radiopharmaceuticals labelled with bromine isotopes Appl Radiat Isot 1986;37:703–13.
27. Qaim SM. Recent developments in the production of ^{18}F, $^{75,76,77}Br$, and ^{123}I Appl Radiat Isotop 1986;37:803–10.
28. Tolmachev V, Lövqvist A, Einarsson L, Schultz J, Lundqvist H. Production of ^{76}Br by a low-energy cyclotron. Appl Radiat Isot 1998;49:1537–40.
29. Nickles RJ. Production of a broad range of radionucldies with an 11 MeV proton cyclotron. J Label Compd Radiopharm 1991;30:120–21.
30. Knust EJ, Dutschka K, Weinreich R. Preparation of ^{124}I solutions after thermodistillation of irradiated $^{124}TeO_2$ targets. Appl Radiat Isot 2000;52:181–4.
31. Lambrecht RM, Sajjad M, Syed RH, Meyer W. Target preparation and recovery of enriched isotopes for medical radionuclide production. Nucl Instr Meth Phys Res 1989;282:296–300.
32. Weinreich R, Knust EJ. Quality assurance of iodine-124 produced via the nuclear reaction $^{124}Te(d,2n)^{124}I$. J Radioanal Nucl Chem Lett 1986;213:253–61.
33. Sheh Y, Koziorowski J, Balatoni J, Lom C, Dahl JR, Finn RD. Low energy cyclotron production and chemical separation of "no carrier added" iodine-124 from a reusable, enriched tellurium-124 dioxide/aluminum oxide solid target. Radiochimica Acta 2000;88:169–73.
34. Qaim SM, Blessing G, Tarkanyi F, Lavi N, Bräutigam W, Scholten B et al. In: Cornell JC (ed). Proceedings of the 14th International Conference on Cyclotrons and their Applications, World Scientific, Singapore, p 541,1996.
35. Qaim SM, Hohn A, Nortier FM, Blessing G, Schroeder IW, Scholten B et al. (eds). Proceedings of the Eighth International Workshop on Targetry and Targetry Chemistry , St. Louis, USA, p 131, 2000.
36. Guibe F, Bram G. Réactivité SN_2 des formes dissociée et associée aux cations alcalins des nucléophiles anioniques. Bull Soc Chim Fr 1975;3:933–48.
37. Hamacher K, Coenen HH, Stocklin G. Efficient stereospecific synthesis of no-carrier-added 2-$[^{18}F]$-fluoro-2-deoxy-D-glucose using aminopolyether-supported nucleophilic substitution. J Nucl Med 1986;27:235–8.
38. Block D, Klatte B, Knochel A, Beckman R, Holm U. NCA $[^{18}F]$-labeling of aliphatic compounds in high yields via aminoether-supported nucleophilic substitution. J Label Compd Radiopharm 1986;23:467–77.
39. Coenen HH, Klatte B, Knochel A, Schuller M, Stöcklin G. Preparation of NCA $[17\text{-}^{18}F]$fluoroheptadencanoic acid in high yields via aminopolyether-supported, nucleophilic fluorination. J Label Compd Radiopharm 1986;23:455–66.
40. Shiue C-Y, Fowler JS, Wolf AP, Watanabe M, Arnett CD. Syntheses and specific activity determinations of no-carrier-added (NCA) F-18-labeled butyrophenone neuroleptics-benperidol, haloperidol, spiroperidol, and pipamperone. J Nucl Med 1985;26:181–6.
41. Mach RH, Elder ST, Morton TE, Nowak PA, Evora PH, Scipko JG et al. The use of $[^{18}F]$4-Fluorobenzyl iodide (FBI) in PET radiotracer synthesis: model alkylation studies and its application in the design of dopamine D_1 and D_2 receptor-based imaging agents. Nucl Med Biol 1993;20:777–94.
42. Kiesewetter DO, Eckleman WC, Cohen RM, Finn RD, Larson SM. Syntheses and D_2 receptor affinities of derivatives of spiperone containing aliphatic halogens. Appl Radiat Isot 1986;37:1181–8.
43. Kämäräinen E-L, Kyllönen T, Airaksinen A, Lundkvist C, Yu M, Na[o]gren K et al. Preparation of $[^{18}F]\beta$CFT-FP and $[^{11}C]\beta$CFT-FP, selective radioligands for visualisation of the dopamine transporter using positron emission tomography (PET). J Label Compd Radiopharm 2000;43:1235–44.
44. Block D, Coenen HH, Stocklin G. The NCA nucleophilic ^{18}F-fluorination of 1,N-disubstituted alkanes as fluoralkylation agents. J Label Compd Radiopharm 1987;24:1029–42.
45. Chi DY, Kilbourn MR, Katznellenbogen JA, Welch MJ. A rapid and efficient method for the fluoroalkylation of amines and amides. Development of a method suitable for the incorporation of the short-lived positron emitting radionuclide fluorine-18. J Org Chem 1987;52:658–64.
46. Shiue C-Y, Bai L-Q, Teng R-R, Wolf AP. Syntheses of no-carrier-added (NCA) $[^{18}F]$fluoroalkyl halides and their application in the syntheses of $[^{18}F]$fluoroalkyl derivatives of neurotransmitter receptor active compounds. J Label Compd Radiopharm 1987;24:53–64.
47. Eng RR, Spitznagle LA, Trager WF. Preparation of radiolabeled pregnenolone analogs 21-fluoropregnenolone-21-^{18}F, 21-fluoropregnenolone-3-acetate-21-^{18}F, 21-fluoropregnenolone-7-^{3}H, and 21-fluoropregnenolone-3-acetate-^{3}H. J Label Compd Radiopharm 1983;20:63–72.
48. Petric A, Barrio JR, Namavari M, Huang S-C, Satyamurthy N. Synthesis of 3β-(4-$[^{18}F]$fluoromethylphenyl)- and 3β-(2-$[^{18}F]$fluormethylphenyl)tropane 2β-carboxylic acid methyl esters: new ligands for mapping brain dopamine transporter with positron emission tomography. Nucl Med Biol 1999;26:529–35.
49. Kim SH, Jonson SD, Welch MJ, Katzenellenbogen JA. Fluorine-substituted ligands for the peroxisome proliferator-activated receptor gamma (PPAR?): potential imaging agents for metastatic tumors. J Label Compd Radiopharm 2001;44:S316.
50. de Groot T, Van Oosterwijck G, Verbruggen A, Bormans G. $[^{18}F]$Fluoroethyl β-D-glucoside and methyl β-D-3-$[^{18}F]$fluoro-3-deoxyglucoside as ligands for the in vivo assessment of SGLT1 receptors. J Label Compd Radiopharm 2001;44:S301.
51. Schirrmacher R, Hamkens W, Piel M, Schmitt U, Lüddens H, Hiemke C et al. Radiosynthesis of (±)-(2-((4-(2-$[^{18}F]$fluoroethoxy)phenyl)bis(4-methoxy-phenyl)methoxy)ethyl-piperidine-3-carboxylic acid: a potential GAT-3 PET ligand to study GABAergic neuro-transmission in vivo. J Label Compd Radiopharm 2001;44:627–42.
52. Al-Qahtani MH, Hostetler ED, McCarthy TJ, Welch MR. Improved labeling procedure of $[^{18}F]$FFNP for in vivo imaging of progesterone receptors. J Label Compd Radiopharm 2001;44:S305.
53. Oh S-J, Choe YS, Chi DY, Kim SE, Choi Y, Lee KH et al. Re-evaluation of 3-bromopropyl triflate as the precursor in the preparation of 3-$[^{18}F]$fluoropropyl bromide. Appl Radiat Isot 1999;51:293–7.
54. Shoup TM, Goodman MM. Synthesis of [F-18]-1-amino-3-fluorocyclobutane-1-carboxylic acid (FACBC): a PET tracer for tumor delineation. J Label Compd Radiopharm 1999;42:215–25.
55. Mukherjee J, Yang Z-Y, Das MK, Brown T. Fluorinated benzamide neuroleptics-III. Development of (S)-N-[(1-allyl-2-pyrrolidinyl)methyl]-5-(3-$[^{18}F]$fluoropropyl)-2,3-dimethoxybenzamide as an improved dopamine D-2 receptor tracer. Nucl Med Biol 1995;22:283–96.
56. Alauddin MM, Conti PS. Synthesis and preliminary evaluation of 9-(4-$[^{18}F]$-fluoro-3-hydroxymethylbutyl)guanine ($[^{18}F]$FHBG): a new potential imaging agent for viral infection and gene therapy using PET. Nucl Med Biol 1998;25:175–80.
57. Lim JL, Zheng L, Berridge MS, Tewson TJ. The use of 3-methoxymethyl-16β,17β-epiestriol-O-cyclic sulfone as the precursor in the synthesis of F-18 16α-fluoroestradiol. Nucl Med Biol 1996;23:911–15.
58. Wodarski C, Eisenbarth J, Weber K, Henze M, Heberkorn U, Eisenhut M. Synthesis of 3′-deoxy-3′-$[^{18}F]$fluorothymidine with

2,3′-anhydro-5′-O-(4,4′-dimthoxytrityl)thymidine. J Label Compd Radiopharm 2000;43:1211–18.
59. Machulla H-J, Blocher A, Kuntzsch M, Piert M, Wei R, Grierson JR. Simplified labeling approach for synthesizing 3′-deoxy-3′-[^{18}F]fluorothymidine ([^{18}F]FLT). J Radioanal Nucl Chem 2000;243:843–6.
60. Padgett HC, Schmidt DG, Luxen A, Bida GT, Satyamurthy N, Barrio JR. Computer-controlled radiochemical synthesis: a chemistry process control unit for the automated production of radiochemicals. Appl Radiat Isot 1989;40:433–45.
61. Mock BH, Vavrek MT, Mulholland GK. Back-to-back "one-pot" [^{18}F]FDG syntheses in a single Siemens-CTI chemistry process control unit. Nucl Med Biol 1996;23:497–501.
62. Padgett H, Wilson D, Clanton J, Zigler S. Two for the price of one: single-vessel FDG syntheses using the CPCU- the PETNet Experience. RDS Users' Meeting, San Francisco, California, April 2–4 (1998).
63. Toorongian SA, Mulholland GK, Jewett DM, Bachelor MA, Kilbourn MR. Routine production of 2-deoxy-2-[^{18}F]fluoro-D-glucose by direct nucleophilic exchange on a quaternary 4-aminopyridinium resin. Int J Rad Appl Instrum [B] 1990;17:273–9.
64. Mulholland GK, Mangner TJ, Jewett DM, Kilbourn MR. Polymer-supported nucleophilic radiolabeling reactions with [^{18}F]fluoride and [^{11}C]cyanide ion on quaternary ammonium resins. J Label Compd Radiopharm 1989;26:378–80.
65. Mosdzianowski C, Lemaire C, Lauricella B, Aerts J, Morelle J-L, Gobert F et al. Routine and multi-Curie level productions of [^{18}F]FDG using an alkaline hydrolysis on solid support. J Label Compd Radiopharm 1999;42:S515–16.
66. Lemaire C, Damhaut P, Lauricella B, Mosdzianowski C, Morelle J-L, Monclus M et al. Fast [^{18}F]FDG synthesis by alkaline hydrolysis on a low polarity solid phase support. J Label Compd Radiopharm 2002;45:435–47.
67. Angelini G, Speranza M, Wolf AP, Shiue C-Y. Nucleophilic aromatic substitution of activated cationic groups by ^{18}F-labeled fluoride. A useful route to no-carrier-added (nca) ^{18}F-labeled aryl fluorides. J Fluor Chem 1985;27:177–91.
68. Haka MS, Kilbourn MR, Watkins GL, Toorongian SA. Aryltrimethylammonium trifluoromethanesulfonates as precursors to aryl [^{18}F]fluorides: improved synthesis of [^{18}F]GBR-13119. J Label Compd Radiopharm 1989;27:823–33.
69. Rengan R, Chakraborty PK, Kilbourn MR. Can we predict reactivity for aromatic nucleophilic substitution with [^{18}F]fluoride ion? J Label Compd Radiopharm 1993;33:563–72.
70. Lemaire C, Cantineau MG, Plenevaux A, Christiaens L. Fluorine-18-altanserin: a radioligand for the study of serotonin receptors with PET: radiolabeling and in vivo biologic behavior in rats. J Nucl Med 1991;32:2266–72.
71. Horti A, Redmond Jr. DE, Soufer R. No-carrier-added (nca) synthesis of 6-[^{18}F]fluoro-L-DOPA using 3,5,6,7,8,8a-hexahydro-7,7,8a-trimethyl-[6S-(6α,8α,8αβ)]-6,8-methano-2H-1,4-benzoxazin-2-one. J Label Compd Radiopharm 1995;36:409–23.
72. Mach RH, Nader MA, Ehrenkaufer RL, Gage HD, Childers SR, Hodges LM et al. Fluorine-18-labeled tropane analogs for PET imaging studies of the dopamine transporter. Synapse 2000;37:109–17.
73. Ding Y-S, Liang F, Fowler JS, Kuhar MJ, Carroll FI. Synthesis of [^{18}F]norchlorofluoroepibatidine and its N-methyl derivative: new PET ligands for mapping nicotinic acetylcholine receptors. J Label Compd Radiopharm 1997;39:827–32.
74. Kuhnast B, Dolle F, Tavitian B. Fluorine-18 labeling of peptide nucleic acids. J Label Compd Radiopharm 2002;45:1–11.
75. Fredriksson A, Johnström P, Stone-Elander S, Jonasson P, Nygren P-Å, Ekberg K et al. Labeling of human C-peptide by conjugation with N-succinimidyl-4-[^{18}F]fluorobenzoate. J Label Compd Radiopharm 2001;44:509–19.
76. Sutcliffe-Goulden JL, O'Doherty MJ, Bansal SS. Solid phase synthesis of [^{18}F]labelled peptides for positron emission tomography. Bioorg Med Chem Lett 2000;10:1501–3.
77. Ido T, Wan C-N, Casella V, Fowler JS, Wolf AP, Reivich M et al. Labeled 2-deoxy-D-glucose analogs. ^{18}F-labeled 2-deoxy-2-fluoro-D-glucose, 2-deoxy-2-fluoro-D-mannose and ^{14}C-2-deoxy-2-fluoro-D-glucose. J Label Compd Radiopharm 1978;14:175–83.
78. Fowler JS, MacGregor RR, Wolf AP, Farrell AA, Karlstrom KI, Ruth TJ. A shielded synthesis system for the production of 2-deoxy-2-[^{18}F]fluoro-D-glucose. J Nucl Med 1981;22:376–80.
79. Shiue C-Y, Salvadori PA, Wolf AP, Fowler JS, MacGregor RR. A new improved synthesis of 2-deoxy-2-[^{18}F]fluoro-D-glucose from ^{18}F-labeled acetyl hypofluorite. J Nucl Med 1982;23:899–903.
80. Ehrenkaufer RE, Potocki JF, Jewett DM. Simple synthesis of F-18-labeled 2-fluoro-2-deoxy-D-glucose: concise communication. J Nucl Med 1984;25:333–7.
81. Coenen HH, Pike VW, Stöcklin G, Wagner R. Recommendation for a practical production of 2-[^{18}F]Fluoro-2-Deoxy-D-Glucose. Appl Radiat Isot 1987;38:605–10.
82. Namavari M, Barrio JR, Toyokuni T, Gambhir SS, Cherry SR, Herschmann HR et al. Synthesis of 8-[^{18}F]fluoroguanine derivatives: in vivo probes for imaging gene expression with positron emission tomography. Nucl Med Biol 2000;27:157–62.
83. Barrio JR, Namavari M, Phelps ME, Satyamurthy N. Elemental fluorine to 8-fluoropurines in one step. J A C S 1996;118:10408–11.
84. Barrio JR, Namavari M, Phelps ME, Satyamurthy N. Regioselective fluorination of substituted guanines with dilute F_2: A facile entry to 8-fluoroguanine derivatives. J Org Chem 1996;61:6084–5.
85. Iyer M, Barrio JR, Namavari M, Bauer E, Satyamurthy N, Nguyen K et al. 8-[^{18}F]Fluoropenciclovir: An improved reporter probe for imaging HSV1-tk reporter gene expression in vivo using PET. J Nucl Med 2001;42:96–105.
86. Firnau G, Chirakal R, Garnett ES. Aromatic radiofluorination with [F-18]F_2 in anhydrous hydrogen fluoride. J Label Compd Radiopharm 1986;23:1106–8.
87. deVries EFJ, Luurtsema G, Brüsser M, Elsinga PH, Vaalburg W. Fully automated synthesis module for the high yield one-pot preparation of 6-[^{18}F]fluoro-L-DOPA. Appl Radiat Isot 1999;51:389–94.
88. Namavari M, Bishop A, Satyamurthy N, Bida G, Barrio JR. Regioselective radiofluorodestannylation with [^{18}F]F_2 and [^{18}F]CH_3COOF: A high yield synthesis of 6-[^{18}F]Fluoro-L-dopa. Int J Rad Appl Instrum [A] 1992;43:989–96.
89. Haaparanta M, Bergmann J, Laakso A, Hietala J, Solin O. [^{18}F]CFT ([^{18}F]WIN 35,428), a radioligand to study the dopamine transporter with PET: Biodistribution in rats. Synapse 1996;23:321–7.
90. Dolbier Jr. WR, Li A-R, Koch CJ, Shiue C-Y, Kachur AV. [^{18}F]-EF5, a marker for PET detection of hypoxia: Synthesis of precursor and a new fluorination procedure. Appl Radiat Isot 2001;54:73–80.
91. Ishiwata K, Monma M, Iwata R, Ido T. Automated synthesis of [^{18}F]-5-fluoro-2'-deoxyuridine. J Label Compd Radiopharm 1984;21:1231–3.
92. Satyamurthy N, Bida GT, Phelps ME, Barrio JR. N-[^{18}F]fluoro-N-alkylsulfonamides: Novel reagents for mild and regioselective radiofluorination. Int J Rad Appl Instrum [A] 1990;41:733–8.
93. Seevers RH, Counsell RE. Radioiodination techniques for small organic molecules. Chem Rev 1982;82:575–90.
94. Goodman MM, Kung MP, Kabalka GW, Kung HF, Switzer R. Synthesis and characterization of radioiodinated N-(3-iodopropen-1-yl)-2 beta-carbomethoxy-3 beta-(4-chlorophenyl)tropanes: Potential dopamine reuptake site imaging agents. J Med Chem 1994;37:1535–42.
95. Laruelle M, Baldwin RM, Innis RB. SPECT imaging of dopamine and serotonin transporters in nonhuman primate brain. NIDA Research Monograph 1994;138:131–59.
96. Stöcklin G. Bromine-77 and iodine-123 radiopharmaceuticals. Int J Appl Radiat Isot 1977;28:131–47.
97. Moerlein SM, Parkinson D, Welch MJ. Radiosynthesis of high effective specific-activity [^{123}I]SCH 23982 for dopamine D-1 receptor-based SPECT imaging. Appl Radiat Isot 1990;41:381–5.
98. Bolton R. Radiohalogen incorporation into organic systems. J Label Compd Radiopharm 2002;45:485–528.
99. Stöcklin G. Molecules labeled with positron emitting halogens. Int J Rad Appl Instrum [B] 1986;13:109–18.
100. Moerlein SM, Hwang DR, Welch MJ. No-carrier-added radiobromination via cuprous chloride-assisted nucleophilic aromatic bromodeiodination. Appl Radiat Isot 1988;39:369–72.

101. McElvany KD, Carlson KE, Welch MJ, Senderoff SG, Katznellenbogen JA. In vivo comparison of 16-alpha[^{77}Br] bromoestradiol-17-beta and 16-alpha-[^{125}I]iodoestradiol-17-beta. J Nucl Med 1982;23:420–4.
102. Coenen HH, Harmond MF, Kloster G, Stöcklin G. 15-(p-[^{75}Br] bromophenyl)pentadecanoic acid: pharmacokinetics and potential as heart agent. J Nucl Med 1981;22:891–6.
103. Maziere B, Loc'h C, Müller L, Halldin CH. ^{76}Br-beta-CBT, a PET tracer for investigating dopamine neuronal uptake. Nucl Med Biol 1995;22:993–7.
104. Loc'h C, Halldin C, Bottlaender M, Swahn C-G, Moresco R-M, Maziere M et al. Preparation of [^{76}Br]FLB 457 and [^{76}Br]FLB 463 for examination of striatal and extrastriatal dopamine D-2 receptors with PET. Nucl Med Biol 1996;23:813–19.
105. Maziere B, Loc'h C, Stulzaft O, Hantraye P, Ottaviani M, Comar D et al. [^{76}Br]Bromolisuride: a new tool for quantitative in vivo imaging of D-2 dopamine receptors. Eur J Pharm 1986;127:239–47.
106. Maziere B, Loc'h C, Baron J-C, Sgouropoulos P, Duquesnoy N, D'Antona R et al. In vivo quantitative imaging of dopamine receptors in human brain using positron emission tomography and [^{76}Br]bromospiperone. Eur J Pharm 1985;114:267–72.
107. Helfenbien J, Emond P, Loc'h C, Bottlaender M, Ottaviani M, Guilloteau D et al. Synthesis of (E)-N-(3-bromoprop-2-enyl)-2β-carbomethoxy-3β-(4′-tolyl) nortropane (PE2Br) and radiolabelling of [^{76}Br]PE2Br: A potential ligand for exploration of the dopamine transporter by PET. J Label Compd Radiopharm 1999;42:581–8.
108. Lundkvist C, Loc'h C, Halldin C, Bottlaender M, Ottaviani M, Coulon C et al. Characterization of bromine-76-labelled 5-bromo-6-nitroquipazine for PET studies of the serotonin transporter. Nucl Med Biol 1999;26:501–7.
109. Maziere B, Valette C, Loc'h C. ^{76}Br-MBBG, a PET radiotracer to investigate the norepinephrine neurological and vesicular transporters in the heart. Nucl Med Biol 1995;22:1049–52.
110. Valette H, Loc'h C, Mardon K, Bendriem B, Merlet P, Fuseau C et al. Bromine-76-metabromobenzylguanidine: a PET radiotracer for mapping sympathetic nerves of the heart. J Nucl Med 1993;34:1739–44.
111. Langer O, Dolle F, Loc'h C, Halldin C, Vaufrey F, Coulon C et al. Preparation of 4- and 6-[^{76}Br]bromometaraminol, two potential radiotracers for the study of the myocardial norepinephrine neuronal reuptake system with PET. J Label Compd Radiopharm 1997;39:803–16.
112. Lövqvist A, Sundin A, Roberto A, Ahlström H, Carlsson J, Lundqvist H. Comparative PET imaging of experimental tumors with bromine-76-labeled antibodies, fluorine-18-fluorodeoxyglucose, and carbon-11-methionine. J Nucl Med 1997;38:1029–35.
113. Lövqvist A, Sundin A, Ahlström H, Carlsson J, Lundqvist H. Pharmacokinetics and experimental PET imaging of a bromine-76-labeled monoclonal anti-CEA antibody. J Nucl Med 1997;38:395–401.
114. Kao C-HK, Sassaman MB, Szajek LP, Ma Y, Waki A, Eckelman WC. The sequential syntheses of [^{76}Br]FBAU 3′,5′-dibenzoate and [^{76}Br]FBAU. J Label Compd Radiopharm 2001;44:889–98.
115. Larson SM, Pentlow KS, Volkow ND, Wolf AP, Finn RD, Lambrecht RM et al. PET scanning of iodine-124-3F9 as an approach to tumor dosimetry during treatment planning for radioimmunotherapy in a child with neuroblastoma. J Nucl Med 1992;33:2020–3.
116. Bakir MA, Eccles SA, Babich JW, Aftab N, Styles JM, Dean CJ et al. C-erbB2 protein overexpression in breast cancer as a target for PET using iodine-124-labeled monoclonal antibodies. J Nucl Med 1992;33:2154–60.
117. Daghighian F, Pentlow KS, Larson SM, Graham MC, Diresta GR, Yeh SDJ et al. Development of a method to measure kinetics of radiolabelled monoclonal antibody in human tumour with applications to microdosimetry: positron emission tomography studies of iodine-124 labelled 3F8 monoclonal antibody in glioma. Eur J Nucl Med 1993;20:402–9.
118. Coenen HH, Dutschka K, Müller SP, Geworski L, Farahati J, Reiners C. NCA radiosynthesis of [123,124I]beta-CIT, plasma analysis and pharmacokinetic studies with SPECT and PET. Nucl Med Biol 1995;22:977–84.
119. Glaser M, Brown DJ, Law MP, Iozzo P, Waters SL, Poole K et al. Preparation of no-carrier-added [^{124}I]A_{14}-iodoinsulin as a radiotracer for positron emission tomography. J Label Compd Radiopharm 2001;44:465–80.
120. Jacobs A, Bräunlich I, Graf R, Lercher M, Sakaki T, Voges J et al. Quantitative kinetics of [^{124}I]FIAU in cat and man. J Nucl Med 2001;42:467–75.
121. Jacobs A, Tjuvajev JG, Dubrovin M, Akhurst T, Balatoni J, Beattie B et al. Positron emission tomography-based imaging of transgene expression mediated by replication-conditional, oncolytic herpes simplex virus type 1 mutant vectors in vivo. Cancer Res 2001;61:2983–95.
122. Doubrovin M, Ponomarev V, Beresten T, Balatoni J, Bornmann W, Finn R et al. Imaging transcriptional regulation of p53-dependent genes with positron emission tomography in vivo. PNAS 2001;98:9300–5.
123. Brust P, Haubner R, Friedrich A, Scheunemann M, Anton M, Koufaki O-N et al. Comparison of [^{18}F]FHPG and [$^{124/125}$I]FIAU for imaging herpes simplex virus type 1 thymidine kinase gene expression. Eur J Nucl Med 2001;28:721–9.

10 Progress in ^{11}C Radiochemistry*

Gunnar Antoni and Bengt Långström

Introduction

The development of detector systems for in vivo imaging of compounds labeled with the accelerator-produced short-lived β^+-emitting radionuclides ^{11}C, ^{15}O, ^{13}N, and ^{18}F, applicable in clinical diagnosis, has been an incentive for the development of new tracer molecules. The sensitivity of the positron emission tomography (PET) technique and the possibility of performing non-invasive studies have thus opened up new ways of studying in vivo biochemistry and pharmacology in man.

In the past several years, commercial networks for the delivery of tracers such as 2-[^{18}F]fluorodeoxyglucose (FDG) and other ^{18}F-labeled compounds have increased the clinical usage of PET. However, it is clear that the PET technology has a wider potential, and that additional ^{18}F-labeled tracers need to be developed and to be complemented by compounds labeled with other radionuclides. Carbon-11 especially, with a half-life short enough to allow repeated PET investigations on the same subject within short time intervals, but long enough to perform multi-step synthesis, has proven to be a useful alternative. There are, however, limitations for the development of PET technology related to tracer production with the short-lived ^{11}C, ^{15}N and ^{15}O for clinical applications. Today, tracers containing these radionuclides can be used only when there is access to in-house production facilities, and such sites benefit from the experience of a research-oriented background. There is thus a potential for further development of tracers and technology applicable in the clinical setting.

There is no doubt in our minds that the great potential of PET technology lies very much in the development of ^{11}C-labeled tracer molecules for routine applications because of the synthetic versatility of carbon. In this chapter we will illustrate some approaches to labeling synthesis and give examples of ^{11}C tracers which have been applied in clinical PET studies. Many of the ^{11}C compounds used in clinical research have not been evaluated in clinical trials, and the future of the clinical use of PET technology will be dependent to some extent on the development of organizational structures where such trials can be performed routinely in an efficient way.

^{11}C Labeling Strategies

Biological Considerations

In order to address a given biological, pharmacological or medical question, the design of labeled tracer molecules need special consideration and there are a few points which need to be addressed:

(i) The labeling position must be considered since the metabolic pathway of the compound might have an impact on the interpretation of the PET data,
(ii) Labeling in different positions in the molecule may give additional information. An illustration of this is ^{11}C-labeled L-DOPA and 5-hydroxy-L-tryptophan where different tissue kinetics are obtained if the tracers are labeled either in the carboxylic- or β-positions (Fig. 10.1). With the label in the β-position, the products obtained after enzymatic

* Chapter reproduced from Valk PE, Bailey DL, Townsend DW, Maisey MN. Positron Emission Tomography: Basic Science and Clinical Practice. Springer-Verlag London Ltd 2003, 237–250.

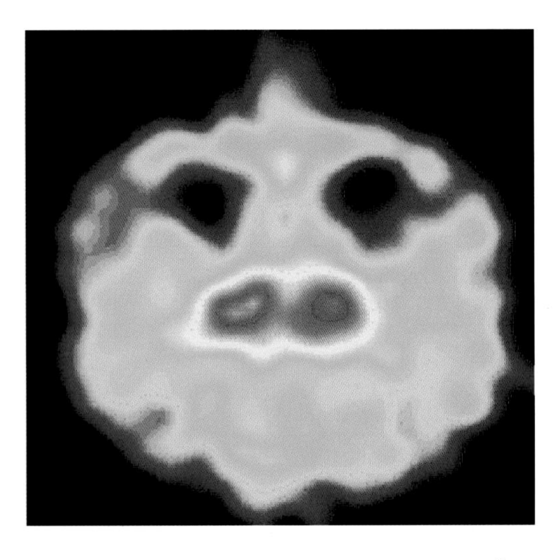
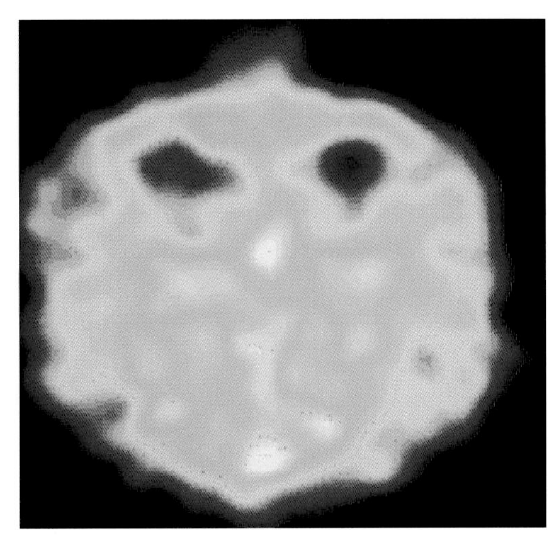

Figure 10.1. Summation images showing the different fate of ^{11}C-labeled L-DOPA depending on the position of the label in monkey brain. Left image: ^{11}C-label in the β-position gives labeled dopamine as the primary metabolite, which is trapped in the neurons. Right image: same monkey two hours later but now using DOPA labeled in the carboxylic position yielding [^{11}C]carbon dioxide which is not stored in the neurons.

decarboxylation are the labeled neurotransmitters serotonin or dopamine. Labeling in the carboxylic position generates [^{11}C]carbon dioxide [1],

(iii) The prodrug concept: This can be used when targeting specific delivery of a drug or labeled tracer to the target organ is restricted. The labeling of the corresponding methyl ester of a prostacyclin receptor ligand either in the methyl ester group or at a metabolic stable methyl group on the phenyl ring is an example of that [2]. The penetration of the prostacyclin over the blood–brain barrier (BBB) is very low but the corresponding methyl ester is, however, transported through the BBB and the prostacyclin receptor ligand produced inside the brain by the action of esterases. With the label in the methyl ester position, [^{11}C]methanol is formed, which gives a uniform brain distribution, whereas with the label in the methyl group on the phenyl ring, a selective uptake in certain brain areas is obtained. This can be used as an indication for the formation of the [^{11}C]prostacyclin receptor ligand within the brain [3],

(iv) Addition or substitution of structural elements (functional groups) or atoms in target molecules can be used to fine-tune the molecular properties of the tracer,

(v) Kinetic Isotope Effect (KIE): Here, the design perspectives should be based on experiences in medicinal chemistry. This means that replacing or adding a small structural element can change properties such as lipophilicity (logP) or pKa, which might have significant impact on the biological behavior of the tracer. Another type of fine-tuning molecular properties is exemplified by substituting hydrogen with deuterium, exemplified by L-deprenyl, a selective monoamine oxidase B (MAO-B) inhibitor used for quantification of regional brain MAO-B activity. The interpretation of the PET data obtained with L-deprenyl was difficult because tracer delivery was highly flow-dependent. A less flow-dependent tracer was obtained by substituting the hydrogens in the propargyl group in L-deprenyl with deuterium. The deuterated molecule's interaction with the enzyme changed (KIE) since the rate-limiting step included the abstraction of the protons/deuterons in the propargyl group. The rate of reaction between MAO-B and the deuterated deprenyl was reduced by a factor of three as compared to the protium compound and the enzyme activity could be measured [4],

(vi) In some applications the physical half-life of the radiotracers need to be adjusted to the biological equivalence.

This is exemplified by the selection of ^{15}O ($t_{\frac{1}{2}}$ = 2.03 min) as radionuclide for blood flow measurements using [^{15}O]-water while ^{18}F-labelled tracers ($t_{\frac{1}{2}}$ = 110 min) are preferred in studies of slower biological processes like protein synthesis and cell proliferation.

Synthetic Considerations

Several aspects apart from those in conventional synthesis have to be considered when planning syntheses of compounds labeled with short-lived β^+-emitting nuclides. For example, the time factor, radiation protection, labeling position and specific radioactivity are points which need consideration. Furthermore, in production of tracers for in vivo human applications the final product has to be sterile, endotoxin-free and dissolved in an appropriate physiological vehicle. The whole procedure has to be achieved within the time frame set by the physical half-life of the radionuclide used, and a rule of thumb is three half-lives. For ^{11}C this is approximately 60 min. As a consequence of the time constraint, synthetic methods are often modified when applied in tracer production.

The development of methods and techniques for rapid tracer synthesis is of special importance when working with short-lived radionuclides such as ^{11}C. The demand for high specific radioactivity introduces further constraint on the quality of reagents and techniques used in terms of reducing isotopic dilution. In ^{11}C-labelling synthesis, the time and the concentrations of reactants become essential factors to recognize [5]. Access to labeled precursors, available for routine preparation, is one important feature in the development of labeling synthesis. Other aspects to consider are related to the production of short-lived radionuclides with high specific radioactivity, allowing studies of high-affinity receptors present in very low concentrations. Important aspects to consider in this context are:

(i) the importance of introducing the radionuclide as late as possible in the synthetic sequence,
(ii) minimizing the synthesis time will increase both the radiochemical yield and the specific radioactivity.

Due to the short reaction time, drastic reaction conditions can be used in the synthesis. Provided that the increase in reaction rate is larger that the decomposition rate, a favorable ratio between product formation and decomposition is achieved. The choice of protective groups and the type of techniques used for synthesis and work-up are all factors that might influence the time optimization. Examples are the use of one-pot procedures, ultrasound and microwave technology [6], in order to reduce production time by simplifying handling and/or increasing reaction rates.

The stoichiometric ratio between substrate and labeled reagent may be in the order 10^3 to 1, due to the small amounts of labeled reagents. A consequence of this is that the labeled reagent is consumed quickly by pseudo first-order reaction kinetics. The small amounts of substance may also be advantageous from a technical point of view by simplifying technical handling. The convenient application of semi-preparative high-performance liquid chromatography (HPLC) and the possibilities of miniaturization of the equipment in order to facilitate automation and to speed up the production of a tracer are illustrative examples. It is relevant to state that the work on increasing specific radioactivity and the advent of new precursors and synthetic methods are very much related to technological development. The possibility of using [^{11}C]carbon monoxide as a labeled precursor has, for example, significantly increased after recent technical improvements. The use of supercritical ammonia in ^{11}C-labelling synthesis is another example where technological improvement was of crucial importance for the development of the methodology [7].

The factors discussed above, combined with aspects on radiation safety, have pushed the need for development of synthetic technology that can meet the demands of routine pharmaceutical production. Therefore, processor-controlled automated synthetic devices have been developed [8] and are routinely applied. This technology is, furthermore, mandatory in order to meet the increasing demands related to Good Laboratory Practice (GLP) and Good Manufacturing Practice (GMP).

Tracer Production with ^{11}C

Radionuclide Production

Two nuclear reactions used to produce ^{11}C are presented in Table 10.1. The most commonly applied production method is the $^{14}N(p,\alpha)^{11}C$ nuclear reaction. This nuclear reaction can be performed with low-energy particles and ^{11}C is obtained with high specific radioactivity. The recovery of ^{11}C-radioactivity from the target in the form of [^{11}C]carbon dioxide or [^{11}C]methane is achieved by automated systems.

Precursor Production

The development of new precursors [9] is important for the development of new labeled substances. A number of precursors more or less routinely available from target-produced [^{11}C]carbon dioxide are shown in Fig. 10.2.

The most frequently employed precursor is [^{11}C]methyl iodide [10]. There are two synthetic methods available: converting [^{11}C]carbon dioxide to [^{11}C]methoxide followed by reaction with hydroiodic acid, or by a gas phase reaction where [^{11}C]methane is reacted with iodine. Methyl iodide is a useful alkylating

Table 10.1.

Nuclear reaction	Threshold energy (MeV)	Target produced precursor of practical interest
$^{14}N(p,\alpha)^{11}C$	3.1	$[^{11}C]CO_2$, $[^{11}C]CH_4$
$^{11}B(p,n)^{11}C$	3.0	$[^{11}C]CO_2$

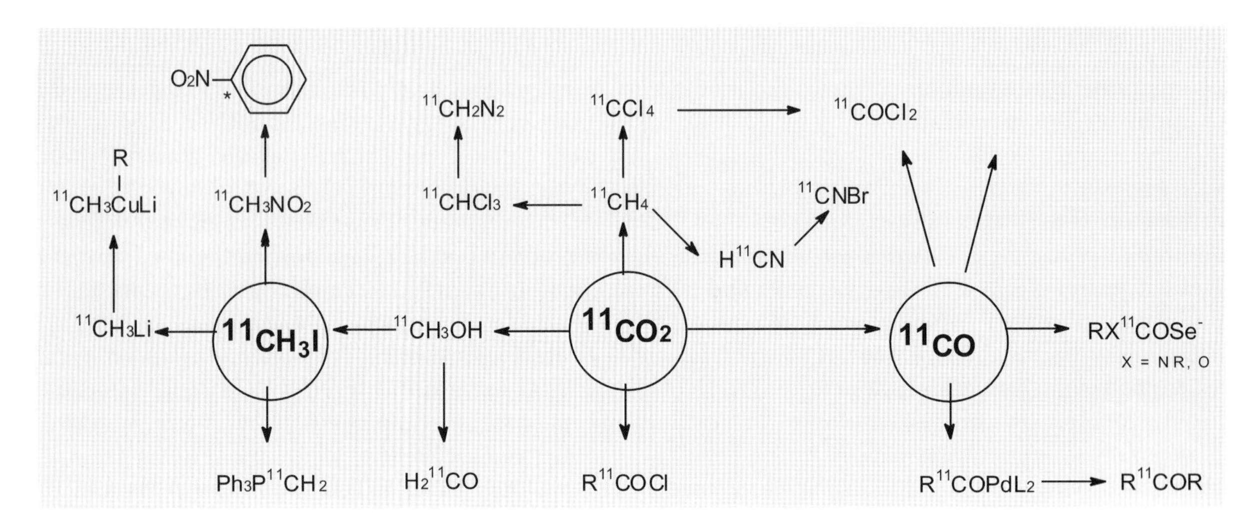

Figure 10.2. Examples of precursors available from [^{11}C]carbon dioxide.

agent for nucleophiles such as carbanions and nucleophilic hetero atoms. It can also be used for preparation of several other valuable precursors such as [^{11}C]methyl triflate [11], [^{11}C]methyl lithium [12], [^{11}C]methylcuprates [13], [^{11}C]nitro methane [14] and [^{11}C]methyltriphenylphosphorane [15].

Another precursor, routinely applied in labeling synthesis, is hydrogen [^{11}C]cyanide obtained on-line from [^{11}C]carbon dioxide or [^{11}C]methane. The labeled nitriles obtained from substitution reactions with [^{11}C]cyanide can be converted to amines, amides and carboxylic acids.

Recently [^{11}C]carbon monoxide has proved useful for synthesis of labeled carbonyl compounds. Some of these labeled carbonyl compounds (i.e., aldehydes and ketones) might themselves be valuable precursors. It is likely that [^{11}C]carbon monoxide in the future will be as important as [^{11}C]methyl iodide in the routine production of PET tracers.

Synthesis of Compounds Labeled with ^{11}C

The labeling synthesis can be divided into two areas of chemistry, ^{11}C-hetereo (*N*, *O*, and *S*) and ^{11}C–C bond-forming reactions. The first application of [^{11}C]methyl iodide was an alkylation on a sulfur nucleophile in the synthesis of ^{11}L-[^{11}C]methionine [16] as is presented in Fig. 10.3. Later, the general utilization of [^{11}C]methyl iodide in alkylation reactions with *N*-, *O*- and *S*-nucleophiles such as amines, amides, phenolates, carboxylates and thiolates became the most common way of introducing ^{11}C in a molecule. A large number of receptor ligands and enzyme substrates have been ^{11}C-labeled using *N*- or *O*-nucleophiles.

Although a substantial number of the compounds used as pharmaceuticals today contain an *N*-methyl group, and may thus potentially be labeled by [^{11}C]methyl iodide, this is not always the preferred position due to metabolic cleavage. The need for synthetic strategies that give access to other labeling positions is obvious. The ability to build up key structural units for use in further coupling reactions is also important.

In tracer synthesis, the following ^{11}C–C bond-forming reactions have been applied:

(i) Alkylation on stabilized carbanions using ^{11}C-labeled alkyl halides [17], as exemplified in Fig. 10.4, by an asymmetric synthesis of ^{11}C-amino acids [18],
(ii) Cuprate-mediated coupling reactions using an ^{11}C-labeled alkyl iodide or an [^{11}C]methyl cuprate [19],
(iii) Reactions with the anion of an [^{11}C]nitroalkane [20],
(iv) Alkene synthesis using [^{11}C]methylenetriphenylphosphorane,
(v) Reactions of [^{11}C]cyanide with electrophilic carbons [21],

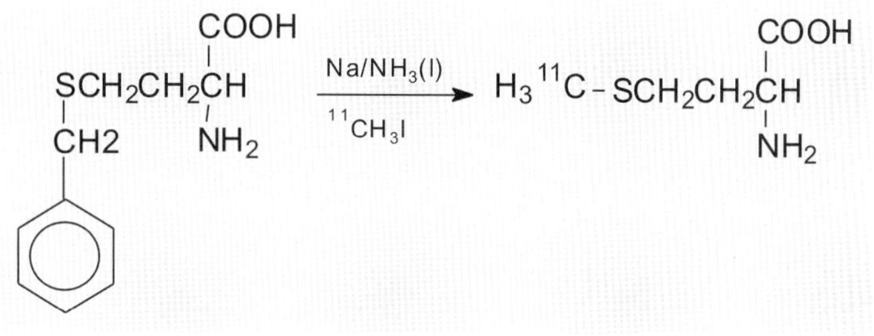

Figure 10.3. Synthesis of L-[^{11}C]methionine.

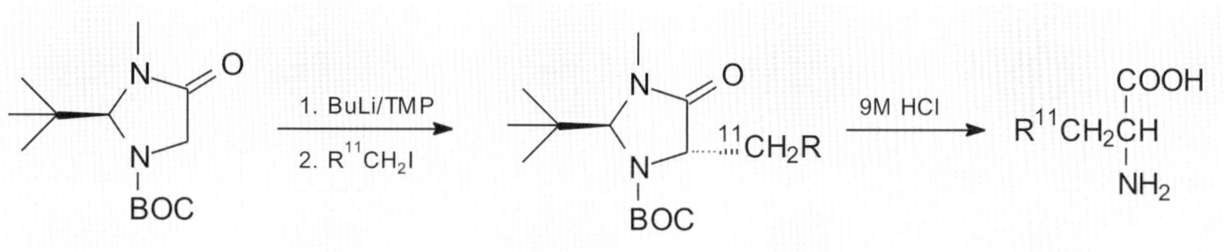

Figure 10.4. Asymmetric synthesis of some ^{11}C-amino acids.

(vi) Carbonation of organometallic reagents with [^{11}C]carbon dioxide.

Palladium has been used in metal-mediated ^{11}C–C bond-forming reactions. Aromatic and aliphatic cyanations, Stille and Suzuki cross-coupling reactions [22] and Heck reactions [23] are important examples of useful ^{11}C–C bond-forming reactions. Of special importance are palladium-mediated carbonylation reactions utilizing [^{11}C]carbon monoxide. These reactions will be discussed in more detail further on.

Enzyme Catalyzed Reactions

Enzyme catalysis has been utilized in labeling synthesis, especially for the preparation of endogenous compounds. Enzyme catalysis has proved to efficiently prepare biomedically interesting compounds in high chemo-, regio-, and stereo-selectivity. The application in labeling synthesis with short-lived radionuclides is particularly rewarding since the small amounts of labeled substance makes it possible to achieve high yields and short reaction times with low enzyme concentrations. Enzyme catalysis may thus give access to compounds that otherwise are difficult to prepare within the time limitations caused by the short half-life. The use of enzymes in labeling synthesis has, so far, mostly been focused on the development of reliable methods for the production of ^{11}C-amino acids. In several of the methods, the same-labeled precursors and similar reaction sequences are applied. This simplifies automation of the tracer production, which is important when high levels of radioactivity are used.

Enzymes can be used either free in solution or immobilized on a solid support. The use of immobilized enzymes has several advantages, such as reducing the risk for contamination of the product with trace amounts of biological materials, and allows repeated use of the same enzymes. Automation is simplified, facilitating the implementation of GMP in tracer production, an important aspect with respect to quality assurance.

Amino acids constitute a class of compounds important as precursors for proteins and certain neurotransmitters. Amino acids labeled with ^{11}C have thus great interest as tracers. Enzymes have at least two functions in amino acid synthesis:

(i) Functional group conversion, for example, as resolving agents where the undesired enantiomer is converted to another compound [24].
(ii) As a catalyst to create new bonds, for example, C–N or C–C.

Several amino acids have been labeled with ^{11}C by enzyme-catalyzed reaction routes [25]. The combination of a chemical and enzymatic reaction route that gives the possibility of labeling several aromatic amino acids in two different positions is shown in Fig. 10.5.

The labeling of L-DOPA and 5-hydroxy-L-tryptophan in the carboxylic- or β-position allows the in vivo measurement of neurotransmitter synthesis rate, i.e., formation of dopamine and serotonin, respectively.

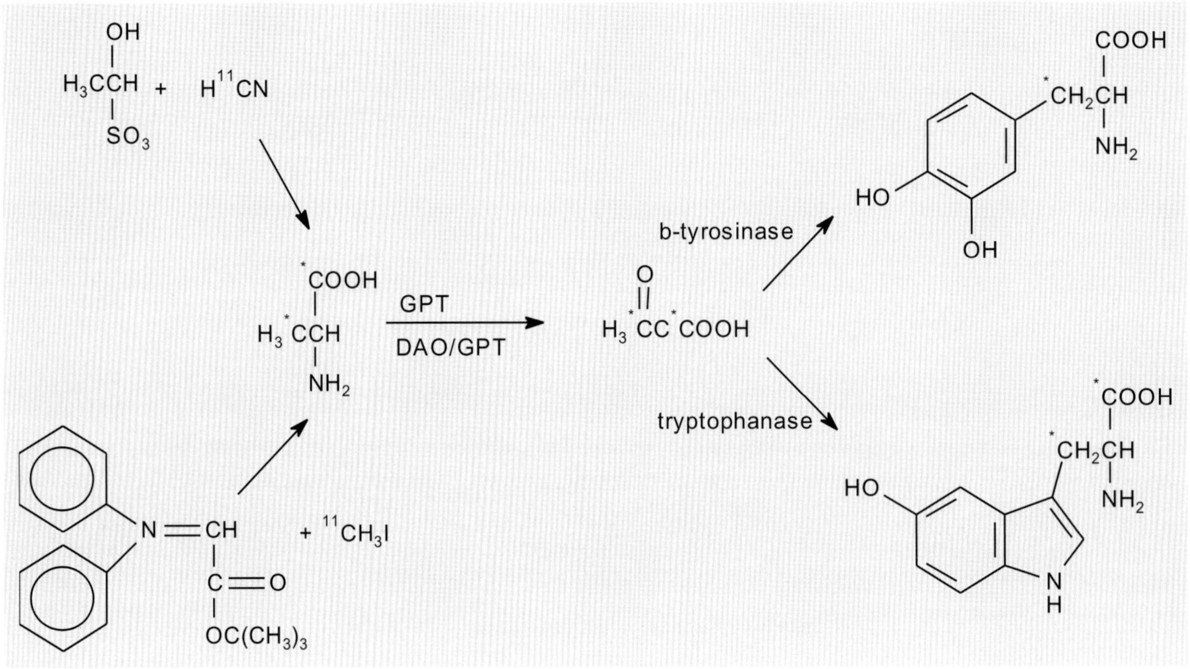

Figure 10.5. Synthesis of aromatic amino acids labeled in two positions.

Metal-Mediated Reactions:

Cyanation Reactions Using Hydrogen [^{11}C]cyanide

Palladium-mediated coupling reactions have allowed formation of various carbon–carbon bonds that were previously difficult to make. One example is the synthesis of ^{11}C-labeled aromatic [26] and vinylic [27] nitriles via the palladium-mediated reaction of [^{11}C]cyanide and an aromatic or vinyl halide (Fig. 10.6).

Cross-Couplings Using [^{11}C]Methyl Iodide

Palladium-mediated cross-coupling reactions are important in organic synthesis. The Stille reaction is a cross coupling of an organotin reagent with an organohalide and the related Suzuki reaction is a coupling of an organoboron compound with an organohalide [28]. Since many functional groups are tolerated during these reactions, protective groups are usually not needed. The Stille reaction has been selected for ^{11}C-labeling of a wide array of different substances including some prostacyclin analogues (Fig. 10.7) [2, 29].

Organocuprates

Coupling reactions with organocuprates were studied with the prime objective to develop methods for ^{11}C-labeling of fatty acids in selected positions. Bis-Grignard reagents were used in coupling reactions with ^{11}C-labeled alkyl iodides in the syntheses of a broad range of saturated fatty acids so that it was also possible to label the polyunsaturated fatty acid, arachidonic acid [30]. The lithium [^{11}C]methyl(2-thienyl)cuprates were, for example, used in the syntheses of [^{11}C]octane, [^{11}C]acetophenone [31], [21-^{11}C]progesterone [32] and [1α-methyl-^{11}C]mesterolone [33] (Fig. 10.8).

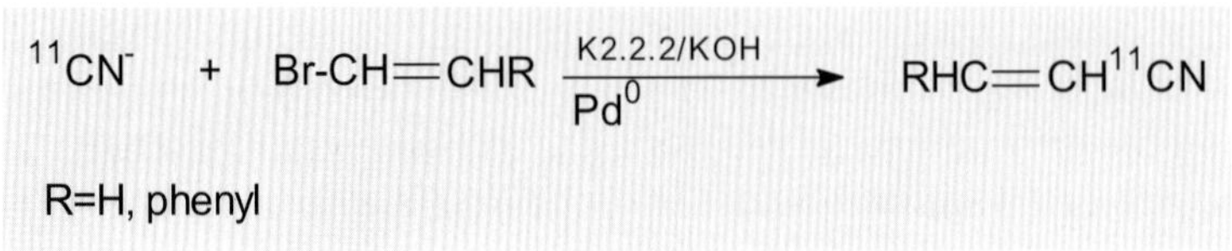

Figure 10.6. Palladium-mediated cyanations.

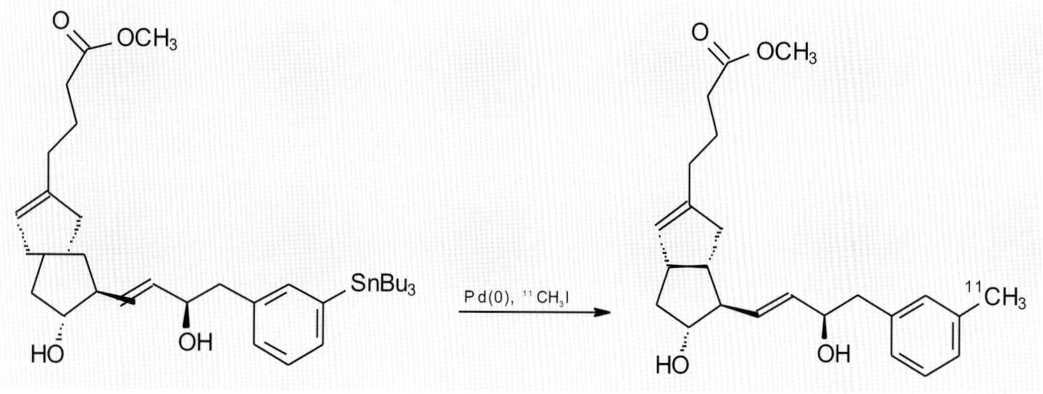

Figure 10.7. Synthesis of a [^{11}C] prostacyclin receptor analog by a Stille coupling.

^{11}C-Labelling Reactions Using [^{11}C]Carbon Monoxide

Due to recent technical developments it is now possible to use [^{11}C]carbon monoxide to synthesize various types of carbonyl compounds [34]. Since carbonyl groups are common in biologically interesting compounds the potential of [^{11}C]carbon monoxide as a precursor in tracer synthesis is significant. This is further reinforced by the fact that carbonyl compounds are useful substrates for many chemical group transformations. Compounds where the ^{11}C-carbonyl group is bound to one or two carbon atoms have been produced using palladium-mediated reactions and compounds where the carbonyl group is bound to two heteroatoms (for example, ureas, carbamates and carbonates) by selenium-mediated reactions (Fig. 10.9). Protective groups are usually not needed and the syntheses can, in most cases, be carried out by a one-pot procedure [35].

Further work on the scope and limitation of using [^{11}C]carbon monoxide in labeling synthesis is currently in progress in our laboratory.

PET as a Tracer Method in Drug Development

The development of new drugs is a time consuming and expensive process. The costs are markedly increased the closer the drug proceeds towards market approval. This is especially clear when the process has reached the phase of clinical trials. During drug development a number of decisions have to be taken. At each point, adequate information must be available to give an optimal base for decision. The relevance of pre-

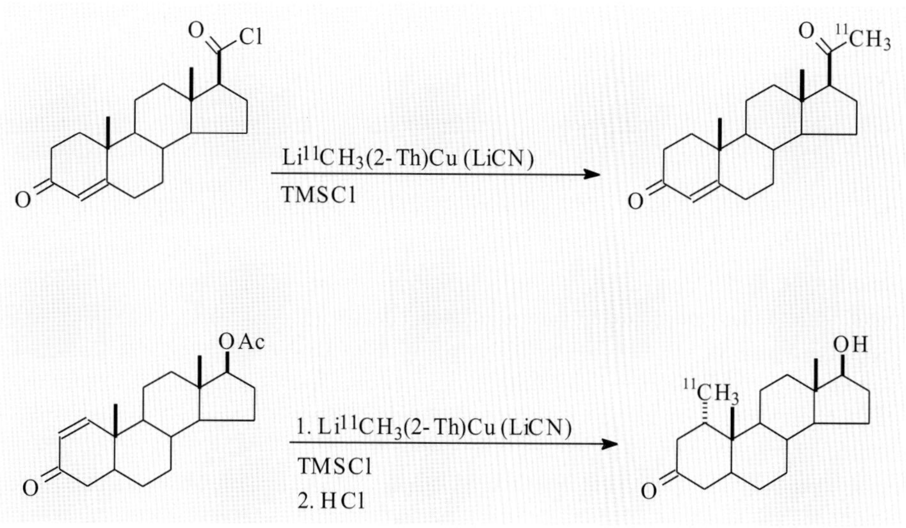

Figure 10.8. Synthesis of [21-^{11}C]-progesterone and [1α-methyl-^{11}C]-mesterolone.

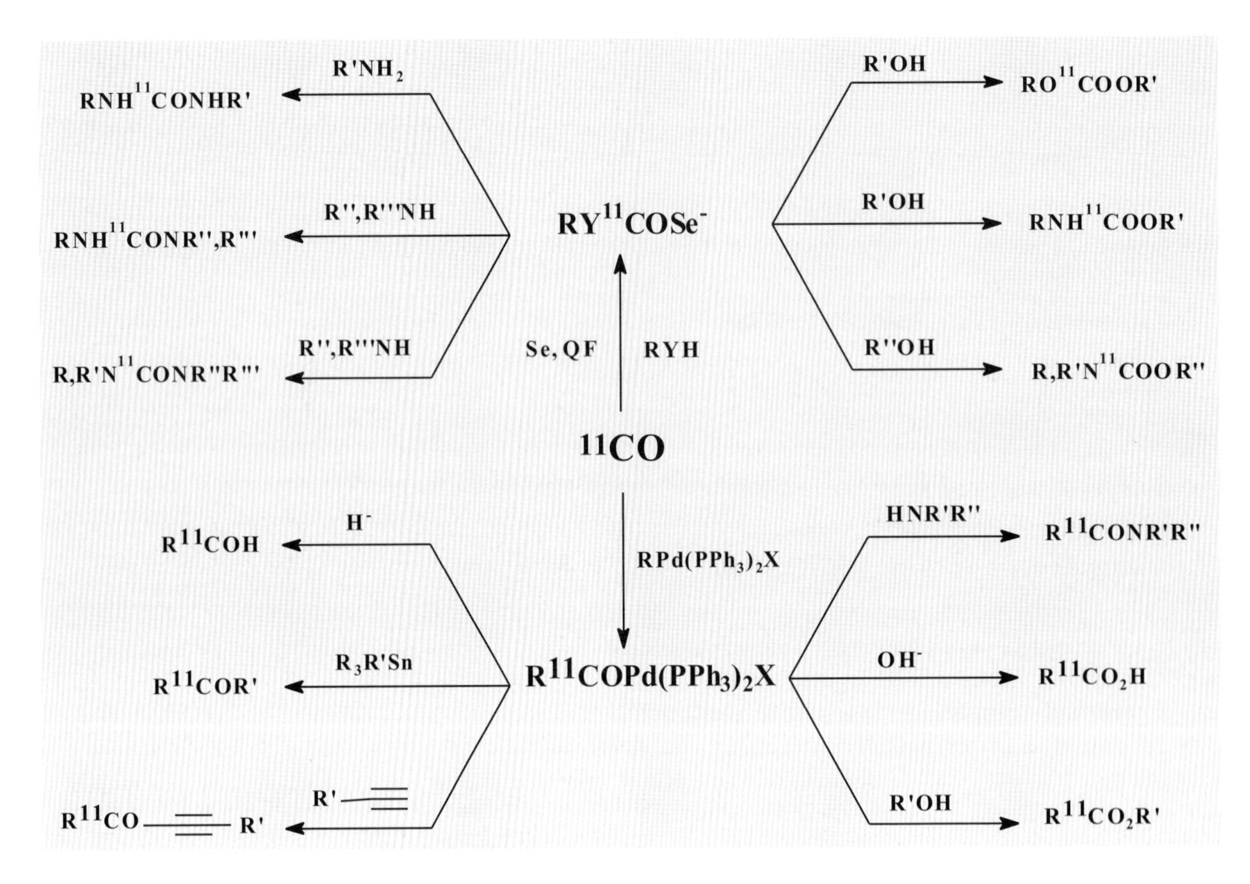

Figure 10.9. Palladium- and selenium-mediated carbonylations.

clinical studies (for example, in genetically modified cells or small animals) for the drug behavior in man is a key issues in the phase I and phase II clinical trials. PET combined with in vitro studies may bridge the gap between specific biology in cell and tissue and the complicated, integrated biology in man. PET is thus increasingly applied in different areas of drug development for these reasons. The possibility to label drug candidates with ^{11}C for preclinical and early human PET studies (for example, phase I) may have significant impact on the choice of lead compound as well as on selection of dose and dose level in subsequent clinical trial.

Drug Distribution

Knowledge of the drug distribution in the body is important for assessment of possible desired or undesired interactions in various tissues. This type of information is not readily attained in humans, and except for PET studies, has to be derived indirectly from mathematical models with input from plasma pharmacokinetics and extrapolation from animal data. These extrapolations are mostly uncertain because factors like distribution, plasma protein binding, metabolism in different organs and binding to receptors and enzymes are often markedly different between species as well as between different individuals of the same species. Significant changes in all modulating factors on drug distribution may occur in the disease or after pharmacological treatment with drugs. PET gives opportunities to measure tissue radioactivity and correlate it with drug concentration, allowing the observation of temporal changes with high accuracy and precision.

Studies of the deposition of inhaled compounds in humans have so far mainly been made with single-photon scintigraphy. A strength of this method is that the majority of the respiratory system can be monitored simultaneously due to the large field of view of the gamma camera. There are a number of limitations in this method which makes PET a strong alternative. (see also chapter 22) PET, using the labeled active component, allows its deposition and disposition to be monitored [36]. Some compounds experience a rapid disposition, for example, from the oral cavity to the stomach or from lung parenchyma to blood. These studies are performed with the administration of very low amounts of radioactivity, 5–20 MBq, thereby reducing the radiation dose.

The use of ^{11}C-labeled compounds in this context is advantageous since several PET studies can be performed on the same subject on the same day; for example, before and after a drug challenge.

Interaction with Biochemical Targets

A number of PET tracers have been developed and validated allowing the in vivo monitoring of interaction or binding to specific receptors and enzyme systems. This gives the opportunity of measuring the number of free receptors in a certain anatomical structure on one occasion, and evaluating changes in this receptor population induced during drug treatment. For drugs where a defined biochemical target is assumed, this gives possibilities to verify that the target system is indeed affected and to assess the degree of interaction. Biochemical systems other than the primary target might also be monitored for assessment of potential side effects.

Dosing Based on Receptor Occupancy

The use of PET for the evaluation of degree and duration of receptor occupancy in relation to dose has become standard for new antipsychotic and antidepressive drugs. Important information derived from such studies include the following:

(i) drug interaction with the assumed biochemical target and hence information that the drug reaches the target tissue, for example, passes through the blood–brain barrier,
(ii) degree of receptor occupancy reached for a given dose, i.e., the dose–occupancy relations,
(iii) duration of action on the target system and its relation to conventional plasma kinetics,
(iv) variability in a patient population with respect to receptor occupancy, and
(v) degree of receptor occupancy affected by other drugs.

Some of the above factors are essential for a decision on dose and dose interval in a clinical trial [37].

PET in Clinical Practice and Medical Research

PET should be regarded as a general tracer method for various types of biological applications. Its role in the recording of biochemistry and physiology in vivo is highlighted by the fact that some tracers are used routinely in clinics for diagnosis and as an important contributor in decisions on patient treatment strategy.

PET has proven to be particularly cost effective in oncology by reducing unnecessary surgery when tumor spread is already present, whilst giving stronger indications for focal treatment when such spread is not demonstrated . Some tracers with clinical potential such as metomidate [38] (11–β-hydroxylase inhibitor), [^{11}C]acetate [39], [^{11}C]choline [40] and, as already mentioned, L-[^{11}C]methionine [41] (tumor metabolism, 5-hydroxy-L-[^{11}C]tryptophan [42], L-[β- ^{11}C]DOPA [43] (endocrine tumors) are shown in Fig. 10.10.

Methionine is a useful tracer for the delineation and measurement of the metabolic activity of brain tumors. This can be used to follow up treatment and obtain information about the response to the treatment before any reduction in the tumor size can be seen with other techniques such as CT or MRI (Fig. 10.11) [44].

[^{11}C]metomidate, a 11β-hydroxylase inhibitor, can be used to study lesions in the adrenal glands as shown in Fig. 10.12.

In neurology, metabolic tracers are used for monitoring of the brain's residual function in stroke patients and for differential diagnosis and treatment monitoring in Alzheimer's disease [45]. Specific tracers for the dopaminergic pathways are used in Parkinson's disease [46] and other movement disorders as well as in schizophrenia [47]. Enzymatic activity in vivo (i.e., aromatic amino acid decarboxylase) can be assessed via position-specific labeling of L-DOPA, as shown in Fig. 10.13.

In CNS-related PET work the study of different receptor systems by selective receptor ligands is important both in clinical research and drug development. Some examples of useful receptor ligands can be found in Fig. 10.14 [48].

Another important use of PET is in cardiology, where heart viability after ischemic insults are evaluated with metabolic tracers such as [^{11}C]acetate, [^{11}C]pyruvate, and [^{11}C]lactate (Fig. 10.15). Pyruvate and lactate can be labeled either in the carboxylic position or in the 3-position and acetate either in the carboxylic or 2-position [49].

The development of tracers for the study of cell proliferation has been an important area of research. The labeling of thymidine in different positions is an example [50]. The methyl position is not optimal since it produces interfering labeled metabolites. Labeling in any of the carbonyl positions seems advantageous since [^{11}C]carbon dioxide is then the most prominent labeled metabolite. This has been achieved by using

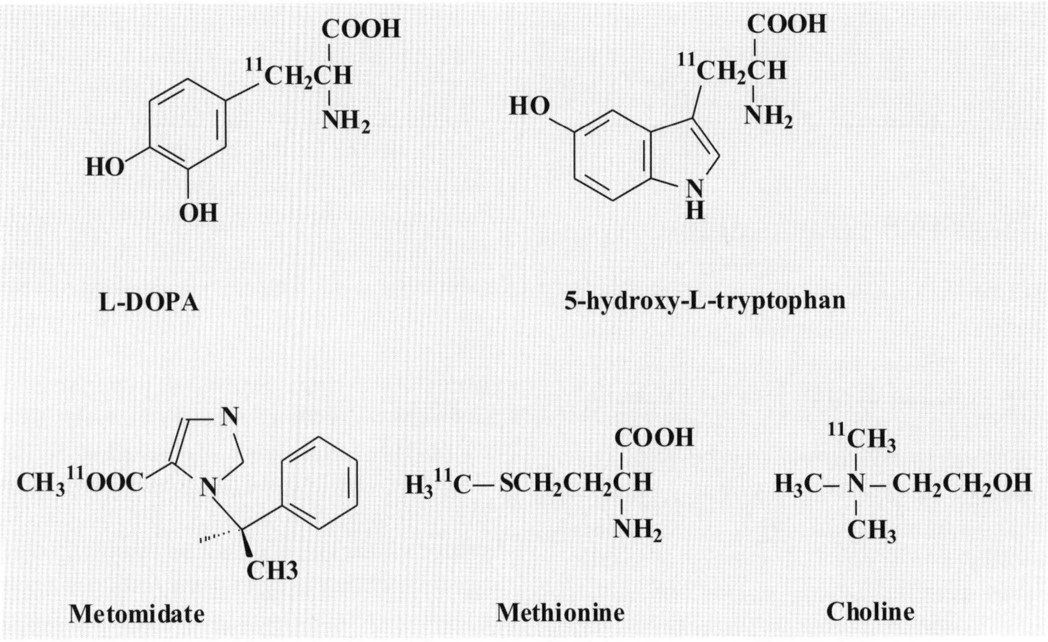

Figure 10.10. [^{11}C]choline, [^{11}C]methomidate, L-[^{11}C]methionine, L- [^{11}C]Dopa, 5-hydroxy-L-[^{11}C]tryptophan.

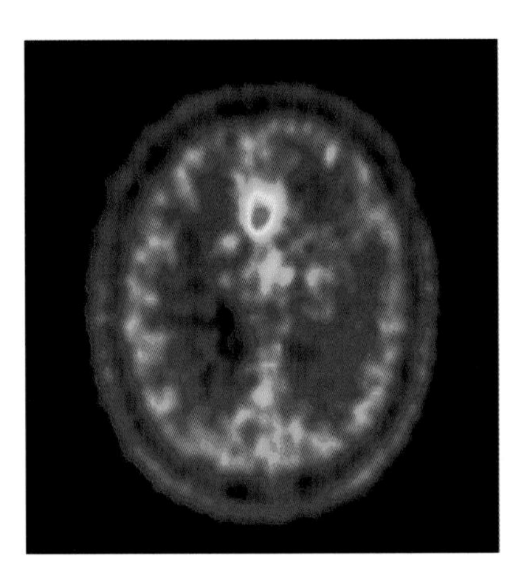

Figure 10.11. Diagnosis of primary malignant brain tumor using [^{11}C]methionine. Methionine gives a measure of the metabolic activity of the tumor.

[^{11}C]urea or [^{11}C]phosgene in ring-closure reactions [51]. Other potentially useful tracers for the study of cell proliferation are, for example, the uracil analogue [^{11}C]FMAU [52] and [methyl-^{11}C]methyl-2′-deoxyuridine [53].

Conclusions

With this contribution focusing on developments in labeling chemistry using ^{11}C- as the radionuclide, we hope that this might inspire further development of using of short-lived β^+-emitting tracers, especially ^{11}C in medicine. There is no doubt that from the synthetic perspective the potential of ^{11}C as a radionuclide in future tracer development, even for clinical applications, is outstanding. However, there are still severe problems which need to be addressed, and these include developing technology so that the tracer production can be improved to such an extent that the existing infrastructure and technicians will be capable of operating with the sophistication of the chemistry which is needed. As this field of imaging science is in an early phase of its development, there is still a strong need for continuing investigations on the scope and limitation of new synthetic methods, technology and its integration in applications in biological systems.

The list of specific clinical questions in which PET may contribute is constantly growing with the increased availability of new tracers and the exploration of their potential to reveal biological function and physiology. Most of the applications of PET in medicine today, and probably also in the future, are in medical research of either basic nature or in explorative clinical trials.

Adrenocortical adenomas

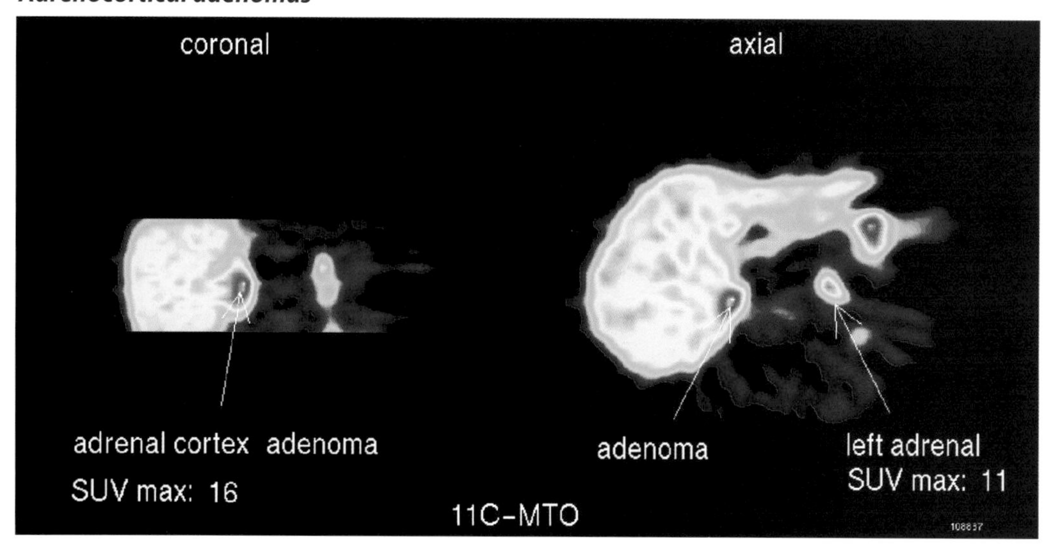

Adrenocortical carcinoma

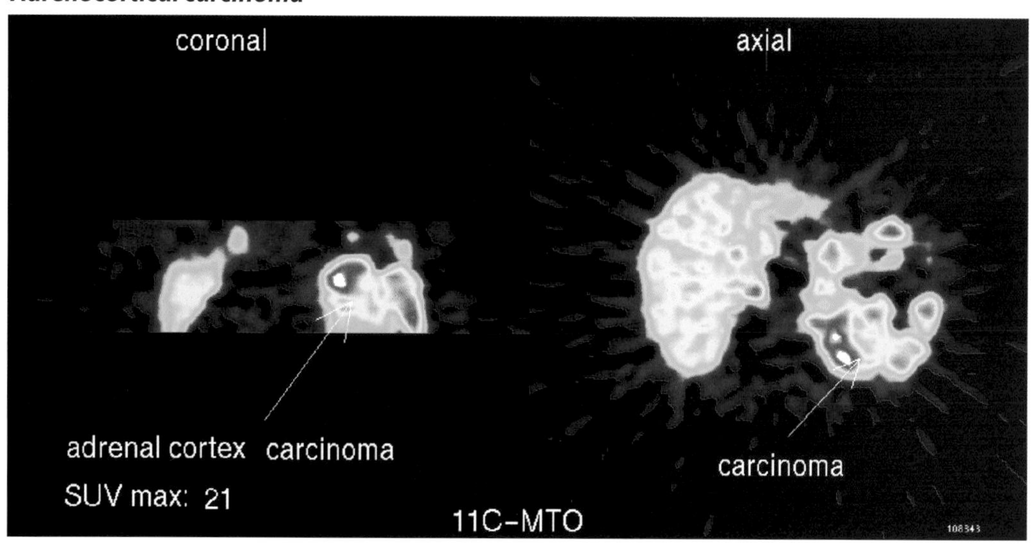

Figure 10.12. The use of [^{11}C]metomidate and [^{18}F]FDG for the differential diagnosis of lesions in the adrenal glands. Adrenocortical adenomas can be distinguished from tumors.

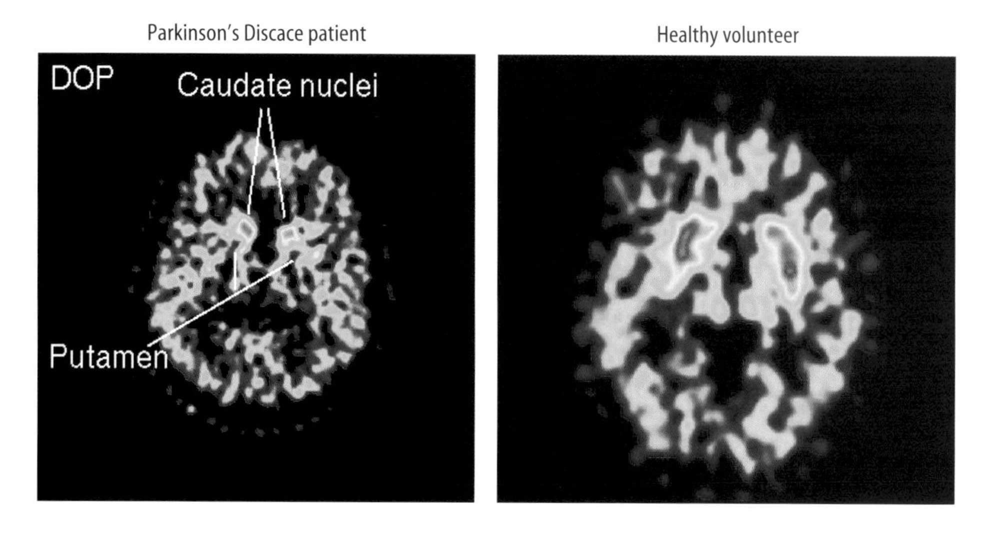

Figure 10.13. Aromatic amino acid decarboxylase activity in striatum measured with L-[β-^{11}C]DOPA. Lower enzyme activity and formation of [^{11}C]dopamine was found in patients with Parkinson's disease as compared with healthy age-matched volunteers.

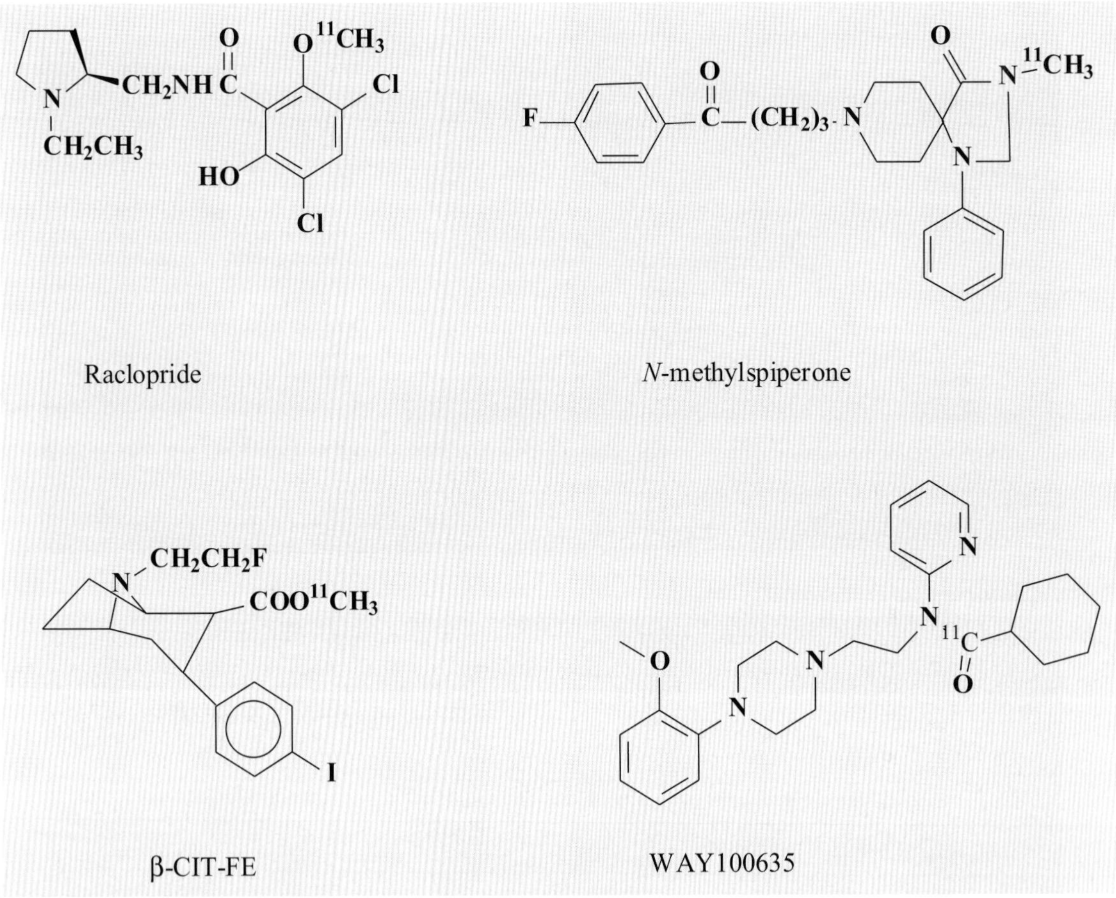

Raclopride

N-methylspiperone

β-CIT-FE

WAY100635

Figure 10.14. Examples of receptor ligands for the serotonergic and dopaminergic systems. Raclopride is a dopamine D_2 antagonist, β-Cl-FE a dopamine transporter antagonist, N-methylspiperone a 5-HT_{2A} antagonist, and WAY100635 a $5HT_{1A}$ antagonist.

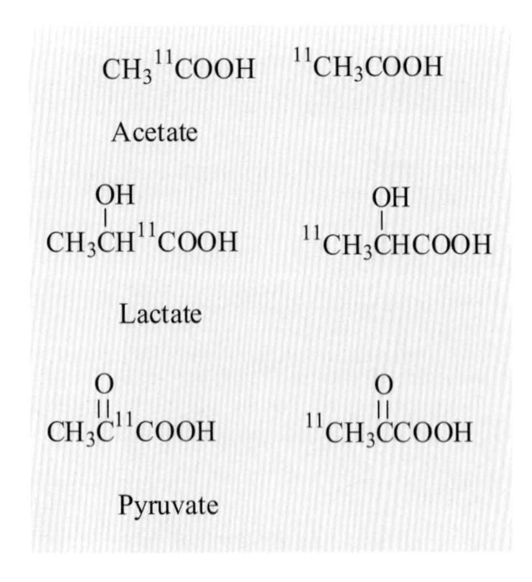

Figure 10.15. Tracers for cardiology, [^{11}C]acetate, [^{11}C]pyruvate, [^{11}C]lactate.

Suggested Further Reading

Fowler JS, Wolf AP. (1986) In: M. Phelps, J. Mazzioatta and H. Schelbert, eds. Positron emission tomography and autoradiography: principles and applications for the brain and heart. New York: Raven Press, 1986.

Långström B, Dannals RF. Carbon-11 compounds. In: Wagner HN, Szabo Z, Buchanan JW, eds. Principles of nuclear medicine. W.B. Saunders, 1995;166–78.

Långström B, Kihlberg T et al. Compounds labelled with short-lived β^+-emitting radionuclides and some applications in life sciences. The importance of time as a parameter. Acta Chem Scand 1999;53: 651–69.

References

1. Tedroff J, Aquilonius SM et al. Estimation of regional cerebral utilization of ß-[^{11}C]-L-dihydroxyphenylalnine (DOPA) in the primate by positron emission tomography. Acta Neurol Scand 1992;85:166–73.

2. Björkman M, Andersson Y, Doi H, Kato K, Suzuki M, Noyori R et al. Synthesis of $^{11}C/^{13}C$-labelled prostacyclins. Acta Chem Scand 1998;52:635–40.
3. (a) Watanabe Y, Kobayashi M et al. Developmental brain and mind on the biochemical machinery. Science Frontier Symposium "Brain Science", Tsukuba 1999. (b) Watanabe Y, Matsumura K et al. A novel subtype of prostacyclin receptor in the central nervous system. J Neurochem 1999;72:2583–92.
4. Pappas N, Alexoff DL, Patlak C, Wolf AP, Selective reduction of radiotracer trapping by deuterium substitution: comparison of carbon-11 L-deprenyl and carbon-11 deprenyl-D2 for MAO B mapping. J Nucl Med 1995;36:1255–62.
5. (a) Långström B, Bergson G. The determination of optimal yields and reaction times in syntheses with short-lived radionuclides of high specific radioactivity. Radiochem Radioanal Lett 1980;43:47–54. (b) Långström B, Obenius U, Sjöberg S, Bergson G. Kinetic aspects of the syntheses using short-lived radionuclides. J Radionnal Chem 1981;64:273–80.
6. (a) Niisawa K, Ogawa K, Saio J, Taki K, Karazawa T, Nozaki T. Production of no-carrier-added ^{11}C-carbon disulfide and ^{11}C-hydrogen cyanide by microwave discharge. Int J Appl Radiat Isot 1984;35:29–33, (b) Thorell JO, Stone-Elander S, Elander N. Use of a microwave cavity to reduce reaction times in radiolabelling with [^{11}C]cyanide. J Labelled Compd Radiopharm 1992;31:207–18.
7. Jacobson G, Markides K, Långström B. Supercritical fluid synthesis in the preparation of $\beta^{?}$-emitting labelled compounds. Acta Chem Scand 1994;48:428–33.
8. Bjurling P, Reineck R, Westerberg G, Schultz J, Gee A, Suthcliffe J et al. Synthia, a compact radiochemistry system for automated production of radiopharmaceuticals. Sixth Workshop on Targetry and Target Chemistry. Vancouver, Canada 1995, p. 282.
9. Långström B, Dannals RF In: Wagner Jr HN, Szabo Z, Buchanan JW, eds. Principles of nuclear medicine, 2nd ed. Philadelphia: WB Saunders,1995.
10. (a) Långström B, Antoni G, Gullberg P, Halldin C, Malmborg P, Någren K et al. Synthesis of L- and D-[methyl-^{11}C]methionine. J Nucl Med 1987;28:1037–40. (b) Larsen P, Ulin J, Dahlström K, Jensen M. Synthesis of [^{11}C]iodomethane by iodination of [^{11}C]methane. Appl Radiat Isot 1997;48:153–7. (c) Bolton R. Isotopic methylation. J Labelled Compd Radiopharm 2001;44:701–36.
11. Jewett DM. A simple synthesis of [^{11}C]methyl triflate. Appl Radiat Isot 1991;43:1383–5.
12. Reiffers S, Vaalburg W, Weigman T, Wynberg H, Woldring MG. Carbon-11-labeled methyllithium as methylating agent: the addition to 17-keto steroids. Int J Appl Radiat Isot 1980;31:535–9.
13. Kihlberg T, Neu H, Långström B. [^{11}C]Methyl(2-thienyl)cuprates, New 11C-precursors used in the syntheses of ^{11}C-labelled methyl ketones and octane. Acta Chem Scand 1997;51:791–6.
14. Schoeps KO, Långström B, Stone-Elander S, Halldin C. Synthesis of [1-^{11}C]D-glucose and [1-^{11}C]mannose from on-line produced [^{11}C]nitromethane. Appl Radiat Isot 1991;42:877–82.
15. Kihlberg T, Gullberg P, Långström B. ^{11}C-Methylenetriphenylphosphorane, a new ^{11}C-precursor used in a one-pot Wittig synthesis of ß-^{11}C-styrene. J Lab Comp Radiopharm 1990;28:1115–20.
16. Långström B, Lundqvist H. The preparation of [^{11}C]methyl iodide and its use in the synthesis of [methyl-^{11}C]-L-methionine. Int J Appl Radiat Isot 1976;27:357–63.
17. (a) Kilbourn MR, Dischino DD, Welch MJ. Synthesis of DL-[3-^{11}C]phenylalanine. Int J Appl Radiat Isot 1984;35:603. (b) Antoni G, Långström B. Synthesis of 3-^{11}C-labelled alanine, 2.aminobutyric acid, norvaline, norleucine, leucine and phenylalanine and preparation of L-[3-^{11}C]alanine and L-[3-^{11}C]phenylalanine. J Labelled Compd Radiopharm 1987;24:125–43.
18. Fasth KJ, Hörnfeldt K, Långström B. Asymmetric synthesis of ^{11}C-labelled L- and D-amino acids by alkylation of imidazolidinone derivatives. Acta Chem Scand 1995;49:301–4.
19. Kihlberg T, Långström B. Cuprate-mediated ^{11}C–C coupling reactions using Grignard reagents and ^{11}C-alkyl iodides. Acta Chem Scand 1994;48:570–7.
20. Schoeps KO, Halldin C. Synthesis of racemic [α-^{11}C]amphetamine and [α-^{11}C]phenethylamine from [^{11}C]nitroalkanes. J Labelled Compd Radiopharm 1992;31:891–902.
21. Hörnfeldt K, Långström B. Synthesis of [^{11}C]cyanoalkyltriphenylphosphoranes and their use in synthesis of ^{11}C-olefins. Acta Chem Scand 1994;48:665–9.
22. Andersson Y, Cheng A, Långström B. Palladium-promoted coupling reactions of ^{11}C-methyl iodide with organotin and organoboron compounds. Acta Chem Scand 1995;49:683–8.
23. Björkman M, Långström B. Functionalisation of ^{11}C-labelled olefins via a Heck coupling reaction. J Chem Soc 2000:3031–4.
24. Antoni G, Långström B. Synthesis of DL-[3-^{11}C]valine using [2-^{11}C]isopropyl iodie, and preparation of L-[3-^{11}C]valine by treatment with D-amino acid oxidase. Appl Radiat Isot 1987;38:655–9.
25. (a) Bjurling P, Watanabe Y, Tokushige M, Oda T, Långström B. Syntheses of β-^{11}C-labelled L-tryptophan and 5-hydroxytryptophan by using a multi-enzymatic route. J Chem Soc Perkin Trans1989;1331–4. (b) Bjurling P, Antoni G, Watanabe Y, Långström B. Enzymatic synthesis of carboxy-^{11}C-labelled L-tyrosine, L-DOPA, L-tryptophan and 5-hydroxy-L-tryptophan. Acta Chem Scand 1990;44:178–82. (c) Antoni G, Omura H, Ikemoto M, Moulder R, Watanabe Y, Långström B. Enzyme-catalysed synthesis of L-[4-^{11}C]asparate and L-[5-^{11}C]glutamate. J Labelled Compd Radiopharm 2001;44:285–94.
26. (a) Andersson Y, Långström B. Transition metal-mediated reactions using [^{11}C]cyanide in the synthesis of ^{11}C-labelled aromatic compounds. J Chem Soc Perkin Transact 1994:1395–400. (b) Andersson Y, Tyrefors N, Sihver S, Onoe H, Watanabe Y, Tsukada H et al. Synthesis of a ^{11}C-labelled derivative of the N-methyl-D-aspartate receptor antagonist MK-801. J Labelled Compd Radiopharm1998; 49:567–76.
27. Antoni G, Långström B. Synthesis of ^{11}C-labelled ,α,β-unsaturated nitriles. Appl Radiat Isot 1992;43:903–5.
28. Andersson Y, Långström B. ^{11}C-Methyl iodide and ^{11}C-Carbon monoxide in palladium-promoted coupling reactions. J Lab Comp Radiopharm.1995;37:84–7.
29. Björkman M, Doi H, Resul B, Suzuzki M, Noyori R, Watanabe Y et al. Synthesis of a ^{11}C-labelled prostaglandin $F_{2\alpha}$ analogue using an improved method for Stille reactions with [^{11}C]methyl iodide. J Labelled Compd Radiopharm 2000;43:1327–34.
30. Kihlberg T, Långström B. Synthesis of [19-^{11}C]arachidonic acid. J Lab Comp Radiopharm 1994;34:617–26.
31. Neu H, Kihlberg T, Långström B. [^{11}C]Methyl(2-thienyl)cuprates, new ^{11}C-precursors in the syntheses of methyl ^{11}C-labelled fatty acids and ketones. J Labelled Compd Radiopharm 1995;37:357–9.
32. Lidström P, Neu H, Långström B. Syntheses of [21-^{11}C] and (21-^{13}C) progesterone. J Labelled Compd Radiopharm 1997;39:695–701.
33. Neu H, Bonasera T, Långström B. Lithium [^{11}C]methyl (2-thienyl)cuprate LiCN in 1,4-additions to •••-unsaturated ketones. $^{11}C/^{13}C$-labelling of the androgen mesterolone. J Labelled Compd Radiopharm 1998;41:227–35.
34. Andersson Y, Långström B. Synthesis of ^{11}C-labelled ketones via carbonylative coupling reactions using [^{11}C]carbon monoxide. J Chem Soc Perkin Trans 1995;1:287–9.
35. (a) Kihlberg T, Karimi F, Långström B. [^{11}C]Carbon monoxide in the synthesis of carbonyl compounds using palladium- or selenium-mediated reactions. J Labelled Compd Radiopharm 1999;42 suppl. 1:86–8. (b) Kihlberg T, Långström B. Biological active ^{11}C-labelled amides using palladium-mediated reactions with aryl halides and [^{11}C]carbon monoxide. J Org Chem 1999;64:9201–5.
36. (a) Bergström M, Nordberg A et al. Regional deposition of inhaled ^{11}C-nicotine vapour as visualized by positron emission tomography. Clin Pharm Therapeut 1995;57:309–17. (b) Bergström M, Cass L et al. Deposition and disposition of ^{11}C-Zanamivir follwing administration as an intranasal spray – evaluation with positron emission tomography. Clinical Pharmacokinetics 1998;36 Suppl:33–9.
37. Bergström M, Westerberg G. et al. MAO-A inhibition in brain after dosing with esuprone, moclobemide and placebo in healthy volunteers: in vivo studies with positron emission tomography. Eur J Clin Pharmacol 1997;52:121–8.

38. Bergström M, Juhlin C et al. PET imaging of adrenal cortical tumors with the 11b-hydroxylase tracer ^{11}C-metomidate. J Nucl Med 2000;41:275–82.
39. Pike VW, Eakins MN, Allan RM, Selwyn AP. Preparation of [1-^{11}C]acetate – an agent for the study of myocardial metabolism by positron emission tomography. Int J Appl Radiat Isot 1982;33:505–12.
40. Hara T, Kosaka N, Shinoura N, Kond T. PET imaging of brain tumours with [methyl-^{11}C]choline. J Nucl Med 1997;38:842–7.
41. Lilja A, Bergström K et al. Dynamic study of supratentorial gliomas with methyl-^{11}C-L-methionine and positron emission tomography. A J N R 1985;6:505–14.
42. Eriksson B, Lilja A, Ahlström H, Bjurling PMB, Lindner KJ, Långström B et al. In: Wiedenmann B, Kvols LKAR, E-O R, eds. Molecular and cell biological aspects of astroenteropancreatic neuroendocrine tumor disease. 1994, 446–52.
43. Bergström M, Eriksson B et al. In vivo demonstration of enzyme activity in endocrine pancreatic tumors – decarboxylation of ^{11}C-DOPA to dopamine. J Nucl Med 1996;37:32–7.
44. Bergström M, Muhr C et al. Rapid decrease in amino acid metabolism in prolactin-secreting pituitary adenomas after bromocriptine treatment – a positron emission tomography study. J Comput Ass Tomograph 1987;11:815–19.
45. (a) Nordberg A, Lundqvist H, Hartvig P, Andersson J, Johansson M, Hellström-Lindahl E et al. Dementia and geriatric cognitive disorders 1997;8:78. (b) Nordberg A, Lundqvist H, Hartvig P, Lilja A, Långström B. Alzheimer´s disease and assoc. diosorders, 1995;9: 21.
46. (a) Tedroff J, Aquilonius S-M et al. Functional positron emission tomographic studies of striatal dopaminergicm activity. Changes induced by drugs and nigrostriatal degeneration. Advances in Neurology 1996;69:443–8. (b) Tedroff J, Ekesbo A et al. Regulation of dopaminergic activity in early Parkinson's disease. Ann Neurology 1999;46:359–65.
47. Hagberg G, Gefvert O, Bergström M, Wieselgren I-M, Lindström L, Wiesel F-A et al. Psychiatry research: Neuroimaging, 1998;82:147.
48. (a) Mathis CA, Simpson NR, Mahmod K, Kinahan PE, Mintun MA. [^{11}C]WAY100635: a radioligand for imaging 5-HT1A receptors with positron emission tomography. Life sciences 55,1994;PL:403–7. (b) Ehrin E, Gawell L, Högberg T, de Paulis T, Ström P. Synthesis of [methoxy-^{3}H]- and [methoxy-^{11}C]labelled raclopride. Specific dopamine-D2 receptor ligand. J Labelled Compd Radiopharm 1986;24:931–40, (c) Halldin C, Farde L, Lundkvist C, Ginovart N, Nakashima Y, Karlsson P et al. [^{11}C]β-CIT-FE, a radioligand for quantitation of the dopamine transporter in the living brain using positron emission tomography. Synapse 1996;22:386–90, (d) Wagner HN, Burns HD, Dannals RF, Wonf DF, Långström B, Duelfers T et al. Imaging dopamine receptors in the human brain by positron emission tomography. Science 1983;221:1264–6.
49. (a) Bjurling P, Långström B. Synthesis of 1- and 3-^{11}C-labelled L-lactic acid using multi-enzymatic catalysis. J Labelled Compd Radiopharm 1989;38:427–32. (b) Bjurling P, Watanabe Y, Långström B. The synthesis of [3-^{11}C]pyruvic acid, a useful synthon, via an enzymatic route. Appl Radiat Isot 1988;39:627–30. (c) Kihlberg T, Valind S, Långström B. Synthesis of [1-^{11}C], [2-^{11}C], [1-^{11}C]($^{2}H_3$) and [2-^{11}C] ($^{2}H_3$)acetate for in vivo studies of myocardium using PET. Nucl Med Biol 1994;21:1067–72.
50. (a) Goethals P, Sambre J, Coene M, Casteleyn K, Poupeye E. A remotely controlled production system for routine preparation of [methyl-^{11}C]thymidine. App Radiat Isot 1992;43:952–4. (b) Alauddin MM, Ravert HT, Musachio JL, Mathews WB, Dannals RF, Conti P. Selective alkylation of pyrimidyl dianions III: no-carrier-added synthesis of [^{11}C-methyl]-thymidine. Nucl Med Biol 1995;22:791–5.
51. (a) Labar D, Vander Borght T. Total synthesis of [2-^{11}C]thymidine from [^{11}C]urea: a tracer of choice for measurement of cellular proliferation using PET. J Labelled Compd Radiopharm 1991;30:342. (b) Steel CJ, Brown GD, Dowsett K, Turton DR, Luthra SK, Tochan-danguy H et al. Synthesis of 2-[^{11}C]thymine from [^{11}C]phosgene: a precursor for 2-[^{11}C]thymidine. J Labelled Compd Radiopharm 1993;32:178–9.
52. Conti PS, Alauddin MM, Fissekis JR, Schmall B, Watanabe KA. Synthesis of 2´-fluoro-5-[^{11}C]-methyl-1-•-D-arabinofuranosyluracil ([^{11}C]-FMAU): a potential nucleoside analog for in vivo study of cellular proliferation with PET. Nucl Med Biol 1995;22:783–9.
53. Goethals P, Volders F, van der Eycken. Synthesis of 6-methyl[^{11}C]-2′-deoxyuridine and evaluation of its in vivo distribution in wistar rats. Nucl Med Biol 1997;24:713–18.

11 Metal Radionuclides for PET Imaging*

Paul McQuade, Deborah W McCarthy and Michael J Welch

Introduction

Although the majority of PET radiopharmaceuticals in clinical and research use are labeled with the four common PET radionuclides, ^{15}O, ^{13}N, ^{11}C and ^{18}F, a number of metal radionuclides have been studied. ^{68}Ga, produced from a $^{68}Ge/^{68}Ga$ generator, was initially used for brain imaging in 1964 [1, 2]. The $^{82}Sr/^{82}Rb$ generator was originally developed by researchers at the Squibb Institute for Medical Research [3]. This generator, now marketed by Bracco, is the only industry-approved PET radiopharmaceutical with a Food and Drug Administration New Drug Application (NDA) in the United States. Over the past several years, there has been increasing interest in other metal-based radionuclides, particularly nuclides of copper, ^{66}Ga and ^{86}Y. Several of these nuclides can be distributed from a central production site and thereby have the potential for use in PET centers without a cyclotron.

In this chapter the available metal PET radionuclides will be summarized and those with the greatest potential for widespread use will be discussed in detail.

Cyclotron PET Radionuclide Production

In general, the majority of radionuclides used in PET imaging are produced by cyclotrons, either on-site or at a site near the scanner. Table 11.1 lists many of the metal PET isotopes produced by a cyclotron. Isotopes that have a suitable half-life (^{64}Cu, ^{66}Ga, ^{86}Y) can potentially be transported over a long distance from the cyclotron facility.

The following section contains a brief description of the more commonly used cyclotron-produced metal PET isotopes.

Copper Radionuclides

A number of copper PET radionuclides can be produced on a biomedical cyclotron including: ^{60}Cu, ^{61}Cu, ^{64}Cu with half-lives of 23.4 min, 3.32 h, and 12.8 h respectively. ^{64}Cu, ^{61}Cu and ^{60}Cu are produced at Washington University using a specially designed solid target holder [6, 7]. The isotopically enriched nickel targets are electroplated for irradiation, separated using ion exchange chromatography, and can then be recycled [7–9]. Large quantities of these Cu isotopes have been produced (yields of up to 33 GBq (~900 mCi) of ^{60}Cu, 17 GBq (~150 mCi) of ^{61}Cu, and 40 GBq (~1 Ci) of ^{64}Cu) [6, 7]. At present, ^{64}Cu is produced routinely at Washington University. As a result of its 12.8-hour half-life, it can be distributed to investigators across the country. ^{64}Cu is also produced as a by-product of the $^{68}Zn(p,2n)^{67}Ga$ reaction, where it is separated from the ^{67}Ga [10]. However, the specific activity of ^{64}Cu produced in this manner is much lower (30 TBq/mol (~860 Ci/mmol)) than that achieved using the $^{64}Ni(p,n)^{64}Cu$ reaction (>370 TBq/mol (10,000 Ci/mmol)) [7]. ^{64}Cu complexed to pharmaceuticals are used for PET imaging, biodistribution studies, and therapy studies. The ^{64}Cu metal ion is used for studies

* Chapter reproduced from Valk PE, Bailey DL, Townsend DW, Maisey MN. Positron Emission Tomography: Basic Science and Clinical Practice. Springer-Verlag London Ltd 2003, 251–264.

Table 11.1. Cyclotron-produced metal PET isotopes

Isotope	Half-life	Decay modes (%)	Max β^+ energy (MeV)	Reaction	Natural abundance of target isotope (%)
^{60}Cu	23.4 m	β^+(93), EC(7)	3.92	$^{60}Ni(p,n)^{60}Cu$ [4]	26.16
^{45}Ti	3.09 h	β^+(86), EC(14)	1.04	$^{45}Sc(p,n)^{45}Ti$ [4]	100
^{61}Cu	3.32 h	β^+(62), EC(38)	1.22	$^{61}Ni(p,n)^{61}Cu$ [4]	1.25
				$^{60}Ni(d,n)^{61}Cu$ [4]	26.16
^{66}Ga	9.45 h	β^+(57), EC(43)	4.153	$^{66}Zn(p,n)^{66}Ga$ [4]	27.81
^{64}Cu	12.8 h	β^+(19), EC(41), β^-(40)	0.656	$^{64}Ni(p,n)^{64}Cu$ [4]	1.16
^{86}Y	14.74 h	β^+(34), EC(66)	3.15	$^{86}Sr(p,n)^{86}Y$ [4]	9.86
^{94m}Tc	53 m	β^+(72), EC(28)	2.47	$^{94}Mo(p,n)^{94m}Tc$ [4]	9.12
^{55}Co	17.5 h	β^+(77), EC(23)	1.50	$^{54}Fe(d,n)^{55}Co$ [4]	5.84
^{89}Zr	78.4 h	β^+(22), EC(78)	0.90	$^{89}Y(p,n)^{89}Zr$ [4]	100
				$^{89}Y(d,2n)^{89}Zr$ [4]	100
^{83}Sr	32.4 h	β^+(24), EC(76)	1.15	$^{85}Rb(p,3n)^{83}Sr$ [4]	72.15
^{38}K	7.71 m	β^+(100)	2.68	$^{38}Ar(p,n)^{38}K$ [5]	0.063
				$^{40}Ca(d,\alpha)^{38}K$ [4]	96.97
^{52}Mn	5.6 d	β^+(28), EC(72)	0.575	$^{52}Cr(p,n)^{52}Mn$ [4]	83.76
				$^{52}Cr(d,2n)^{52}Mn$ [4]	83.76
^{72}As	26 h	β^+(77), EC(23)	3.34	$^{72}Ge(p,n)^{72}As$ [4]	27.37
^{82m}Rb	6.3 h	β^+(26), EC(74)	0.78	$^{82}Kr(d,2,)^{82m}Rb$ [4]	11.56
^{52}Fe	8.2 h	β^+(57), EC(43)	0.80	$^{50}Cr(^{4}He,2n)^{52}Fe$ [4]	4.31
^{51}Mn	46.2 m	β^+(97), EC(3)	2.17	$^{50}Cr(d,n)^{51}Mn$ [4]	4.31

involving copper metabolism (Menkes' syndrome and Wilson's disease), nutrition, and copper transport [11]. In addition, ^{60}Cu and ^{64}Cu have clinical applications in PET [12].

Gallium-66

Gallium-66 can be produced on a biomedical cyclotron via the $^{66}Zn(p,n)^{66}Ga$ nuclear reaction [13]. ^{66}Ga is produced at Washington University using isotopically enriched ^{66}Zn. The Zn can then be separated from the ^{66}Ga either by cation exchange [14] or by solvent extraction techniques [15]. ^{66}Ga is of interest because it is a medium half-life isotope ($t_{\frac{1}{2}}$ = 9.45 h) with potential for both imaging and therapy as a result of its high-energy positron (4.1 MeV) and other high-energy gamma rays [16].

Technetium-94m

^{94m}Tc can be produced via the $^{94}Mo(p,n)^{94m}Tc$ nuclear reaction [17]. Using enriched ^{94}Mo provides a good yield with relatively low levels of impurities [18]. Thermo-chromatographic separation of Mo from ^{94m}Tc is achieved using a steam distillation method [19]. ^{94m}Tc is an attractive PET isotope ($t_{\frac{1}{2}}$ = 52 min) as it can potentially be used as a substitute in the typical ^{99m}Tc-labeled single-photon-emitting radiopharmaceuticals.

Yttrium-86

^{86}Y can be produced on a biomedical cyclotron via the $^{86}Sr(p,n)^{86}Y$ nuclear reaction using isotopically enriched ^{86}Sr foil or carbonate pellet [20]. The separation of ^{86}Sr from the ^{86}Y is carried out by dissolving the carbonate in an acidic solution and then co-precipitating the Sr with lanthanum. The precipitate is then dissolved and separated using anion exchange [20]. ^{86}Y is appealing as an isotope ($t_{\frac{1}{2}}$ = 14.74 h) because of its potential use for evaluating dosimetry prior to ^{90}Y radiotherapy.

With the current interest in small-animal PET imaging [21], a word should be said about the application of these metal radionuclides in such devices. One of the biggest constraints for these scanners is the positron range of the nuclide used. Many of the nuclides described in this chapter have β^+_{max} energies in excess of 1 MeV. Above this threshold, significant reduction in resolution is seen. One isotope, ^{64}Cu, has demonstrated particular utility and promise in small-animal imaging. Its β^+_{max} energy of 646 keV puts it in the range of ^{18}F, and the resolution of the images is comparable [22–24].

Table 11.2. Generator-produced metal PET isotopes [4].

Daughter Isotope	Daughter half-life	Daughter decay mode (%)	Max β^+ energy (MeV)	Parent Isotope	Parent half-life	Parent decay mode (%)
^{128}Cs	3.8 m	β^+(61), EC(39)	2.90	^{128}Ba	2.43 d	EC(100)
^{44}Sc	3.92 h	β^+(95), EC(5)	1.47	^{44}Ti	48 y	EC(100)
^{62}Cu	9.76 m	β^+(98), EC(2)	2.91	^{62}Zn	9.13 h	β^+(7), EC(93)
^{82}Rb	1.25 m	β^+(96), EC(4)	3.15	^{82}Sr	25 d	EC(100)
^{68}Ga	68.3 m	β^+(90), EC(10)	1.90	^{68}Ge	275 d	EC(100)
^{52m}Mn	21.1 m	β^+(98), EC(2)	1.63	^{52}Fe	8.2 h	β^+(57), EC(43)
^{110}In	66 m	β^+(71), EC(29)	2.25	^{110}Sn	4.0 h	EC(100)
^{118}Sb	3.5 m	β^+(77.5), EC(22.5)	2.67	^{118}Te	6.00 d	EC(100)

Generator PET Radionuclide Production

There are many short-lived radionuclides available from radionuclide generators. Generator systems consist of a long-lived parent radionuclide, which decay to a short-lived daughter radionuclide. These generator systems offer a continuous supply of relatively short-lived daughter radionuclides. It has been suggested that the use of generators for PET is very important because they make PET studies possible at centers that are remote from a cyclotron facility. Some of the generator metal PET isotopes are listed in Table 11.2.

Following is a brief description of the more commonly used generator-produced metal PET isotopes.

$^{62}Cu/^{62}Zn$ Generator

^{62}Cu is the daughter isotope of ^{62}Zn and is obtained by eluting the $^{62}Cu/^{62}Zn$ generator [25, 26]. Due to the short half-life of ^{62}Cu ($t_{\frac{1}{2}}$ = 9.76 m), its production is well suited to a generator. The relatively short half-life of the parent (^{62}Zn, $t_{\frac{1}{2}}$ = 9.13 h) means that these generators must be replaced every few days. An automated $^{62}Cu/^{62}Zn$ generator that produces the perfusion agent [^{62}Cu]pyruvaldehyde-bis(n^4-methylthiosemicarbazone) (PTSM) has been used in clinical trials [27].

$^{82}Rb/^{82}Sr$ Generator

^{82}Rb ($t_{\frac{1}{2}}$ = 76 s) is the daughter isotope of ^{82}Sr. The generator is a microprocessor-controlled, self-contained infusion system [3]. ^{82}Rb is mainly used in myocardial perfusion studies in which the $^{82}Rb^+$ acts as an analog of potassium [28]. In addition, it has been utilized to evaluate blood–brain barrier changes in patients with Alzheimer's-type dementia [29].

$^{68}Ga/^{68}Ge$ Generator

The $^{68}Ga/^{68}Ge$ generator produces ^{68}Ga as either [^{68}Ga]ethylenediaminetetraacetic-acid ([^{68}Ga]EDTA) or [^{68}Ga]Cl_3 in 1M HCl [30]. ^{68}Ga has been used to label blood constituents, proteins, peptides, and antibodies (see section below on Gallium). There have been a limited number of patient studies using the $^{68}Ga/^{68}Ge$ generator. The most common use of ^{68}Ga is in the form of [^{68}Ga]citrate, which upon administration produces [^{68}Ga]transferrin. This approach has been used to measure pulmonary transcapillary escape rate in various disease states [31–35].

Metal-based Radiopharmaceuticals

The next section of this chapter will deal with the labeling studies that have been carried out with these PET radionuclides.

Gallium

Introduction

In 1871 Dmitri Mendeleev predicted the existence of gallium (which he named eka aluminum) on the basis of the periodic table. It was not until 1875 that this prediction was confirmed, when gallium was discovered spectroscopically and obtained as the free metal by Lecoq de Boisbaudian. Naturally occurring gallium has two isotopes, ^{69}Ga (60.1% natural abundance) and ^{71}Ga (39.9% natural abundance). Three radioisotopes can be produced, all of which are useful for incorporation into radiopharmaceuticals. Two of these, ^{66}Ga ($t_{\frac{1}{2}}$ = 9.45 h) and ^{68}Ga ($t_{\frac{1}{2}}$ = 68 min), decay by β^+-emission and are used for PET imaging, and ^{67}Ga ($t_{\frac{1}{2}}$ = 78 h) decays by γ-emission and is used for single photon imaging [4].

Chemistry of Gallium

Gallium is a group 13 element and exists most commonly in the +3 oxidation state. While lower oxidation states have been observed, all relevant radiopharmaceuticals occur in this oxidation state. Similar to other group 13 elements (B, Al, and In), Ga^{3+} is classified as a "hard acid", bonding strongly to highly ionic, non-polarizable Lewis bases. As a result, gallium chemistry is dominated by ligands containing oxygen and nitrogen donor atoms [36]. There are two requirements for using gallium complexes as radiopharmaceuticals. The first is that gallium complexes must resist hydrolysis at physiological pH. $Ga(OH)_3$, the primary product formed at physiological pH, is insoluble. It is not until the pH is increased above 9.6 that the soluble species $[Ga(OH)_4]^-$ is formed. The second concern is that Ga^{3+} has an electronic configuration of $3d^{10}$, and this is similar to that of high spin Fe^{3+}, which has a half-filled 3d shell. As a result, their ionic radii, ionization potential, and coordination environment are similar. Thus, gallium radiopharmaceuticals must be stable enough to avoid trans-chelation of Ga^{3+} to various iron-binding proteins, particularly transferrin. Transferrin has two binding sites, for which the gallium binding constants are 20.3 and 19.3 [37]. In practice, these requirements necessitate the coordination of Ga^{3+} by a polydentate ligand, typically forming gallium species that are six-coordinate. However, several Ga(III) complexes with coordination numbers four or five are stable in vivo.

Gallium-citrate/transferrin

[^{67}Ga]citrate has been used as a tumor imaging agent for over 30 years [38]. It was subsequently discovered that trans-chelation of gallium to the iron-binding protein transferrin was the actual tumor imaging agent [39]. Further work demonstrated that gallium is completely bound to transferrin as soon as 15 minutes after administration of [^{67}Ga]citrate [40]. The effectiveness of this radiopharmaceutical is such that it remains in use today in the clinical diagnosis of certain types of neoplasia, lung cancer, non-Hodgkin's disease, lymphoma, malignant melanoma, and leukemia. To date, the mechanism by which gallium–transferrin enters tumors is unknown.

[^{68}Ga]citrate also has been used in diagnostic imaging, but due to its shorter half-life, the disease states studied are different. Upon injection, Ga-citrate/transferrin is immediately taken up by the lungs and has been used to study pulmonary vascular permeability (PVP) using PET imaging techniques that are not possible with ^{67}Ga and single photon imaging, due to improved quantification [32, 41–43].

Only a limited number of studies have been carried out with [^{66}Ga]transferrin. These show that with the longer half-life of ^{66}Ga compared to ^{68}Ga and with the higher sensitivity of PET versus single photon imaging, [^{66}Ga]transferrin may provide an alternative to [^{67}Ga]transferrin [44].

Somatostatin Analogs

Somatostatin is a cyclic 14-amino acid peptide that was initially found in the hypothalamus, which has an inhibiting effect on growth hormone secretion [45]. Receptors for somatostatin can be found in the brain, pituitary gland, gastrointestinal tract, endocrine and exocrine pancreas, and the thyroid [46, 47]. Somatostatin receptors have also been found in a large number of human tumors [48]. Unfortunately, somatostatin has a short plasma half-life, therefore analogs such as octreotide and DOTATOC have been developed. The structural formulas of somatostatin, octreotide, and DOTATOC are shown in Fig. 11.1.

Octreotide is an 8-amino acid cyclic peptide that can be labeled with ^{68}Ga using the bifunctional chelator (BFC) DTPA. However, in vitro studies have shown that trans-chelation of gallium takes place [49]. An in vitro and in vivo stable conjugate of octreotide was synthesized using the bifunctional chelator desferrioxoamine B (DFO) [49, 50], and it was rapidly cleared from the circulation via the kidneys and showed high tumor uptake [50].

Similar to octreotide, the radiopharmaceutical DOTATOC (DOTA-(D)Phe^1-Tyr^3-octreotide) was designed. This can be labeled with a variety of metals in the +3 oxidation state such as Ga, In, B, and Y, and has shown high stability in human serum and has a high affinity to somatostatin receptors. Studies have shown that DOTATOC labeled with gallium is one of the best somatostatin analogs developed to date due to its high tumor-to-blood ratio [51]. DOTADOC has been labeled with both ^{66}Ga and ^{68}Ga [51–53]. ^{66}Ga with its longer half-life may be preferred over ^{68}Ga for PET diagnosis, and, due to other aspects of its decay properties, may have potential for tumor radiotherapy.

Serum Albumin Microspheres

The classical radionuclide technique for measurement of organ perfusion involves vascular injection of

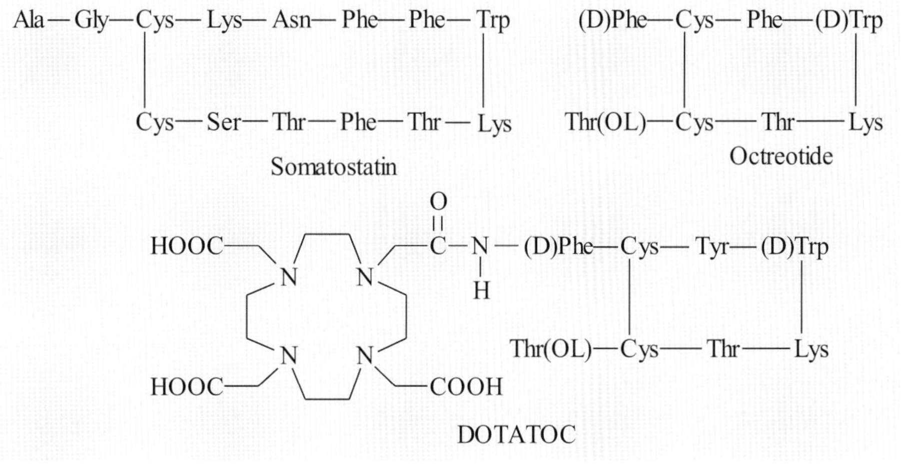

Figure 11.1. Structural formulas of somatostatin, octreotide, and DOTATOC.

labeled microspheres of such a size that can be trapped by the first capillary bed they encounter [36]. For this purpose, commercial kits of biodegradable albumin colloids are available for ^{99m}Tc labeling. As mentioned earlier, however, the labeling with PET radionuclides offers a distinct advantage over single-photon radionuclides. ^{66}Ga- and ^{68}Ga-labeled albumin microspheres have been studied. The microspheres can be labeled directly with gallium, or through a bifunctional chelator that is covalently bound to the microspheres. ^{68}Ga-labeled microspheres have been used as regional pulmonary and cerebral blood flow agents [54–57]. However, for studying slow dynamic processes such as lymphatic transport, ^{66}Ga-labeled albumin colloids have been proposed [58].

Gallium Labeling of Antibodies and Proteins

The monoclonal antibody antimyosin has been labeled with both ^{66}Ga and ^{68}Ga, via the bifunctional chelator DTPA, for the imaging of acute myocardial infarction [59]. The slow antigen–antibody reaction during the diffusion of antimyosin in necrotic myocardium requires several hours to equilibrate, therefore the longer-lived ^{66}Ga would be more appropriate [59]. Twenty-nine hours after administrating [^{66}Ga]DTPA-antimyosin, the normal-to-infarcted myocardial ratio was 2.7, and this agent can be viewed as a valid tracer for myocardial necrosis via PET imaging.

DTPA has also been used to label ^{68}Ga to the monoclonal antibody fragment BB5-G [60], which is specific for a human parathyroid surface antigen. This study showed that ^{68}Ga-labeled antibodies are potential candidates for imaging with PET. However, the slow antibody clearance along with the slow antigen–antibody uptake may require the use of a longer-lived radionuclide, such as ^{66}Ga, to improve target-to-non-target ratios.

Elevated plasma concentrations of low-density lipoprotein (LDL) have been linked to atherogenesis and coronary heart disease [61]. To study this, LDL was labeled with ^{68}Ga via the bifunctional chelator DTPA. Studies involving [^{68}Ga]DTPA-LDL showed significant uptake in LDL receptor-rich tissue, and this gallium radiopharmaceutical has potential to evaluate lipoprotein metabolism non-invasively [61]. A similar study involved the labeling of platelets with ^{68}Ga [62], since platelet deposition occurs at the site of endothelial injury and this can be caused by LDL. ^{68}Ga-labeled platelets have potential for the imaging of platelet accumulation at the site of vascular lesions.

The examples given thus far have all utilized DTPA as the bifunctional chelator agent, but other ligands can be used for this purpose as well. Other examples include desferrioxamine-B (DFO), which through three hydroxamate groups can co-ordinate the gallium and can be attached to the C- or N-terminus of the biomolecule, and also 1,4,7-triazacyclononane-1-succinic acid 4,7-diacetic acid (NODASA) [63], shown in Fig. 11.2. Studies have also been undertaken to synthesize new bifunctional chelators containing a more stable and lipophilic metal center to increase clearance of the labeled biomolecule from the liver [64].

Ga-labeled Myocardial Imaging Agents

Due to the convenient half-life of ^{68}Ga, considerable work has been done in the development of ^{68}Ga-labeled myocardial agents. Gallium radiopharmaceuticals that localize in the heart are lipophilic in nature and can be either neutral or cationic.

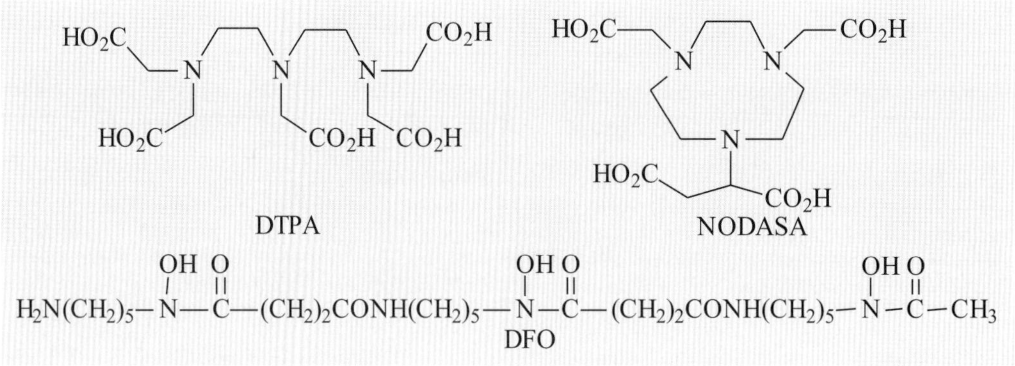

Figure 11.2. Bifunctional chelators for labeling with $^{66/68}Ga$.

A series of uncharged lipophilic tripodal tris(salicylaldimine) ligands (Fig. 11.3) have been investigated with limited success [65]. This work was continued, and analogous species were prepared in which alkoxy substituents were placed on the ethane backbone, resulting in more lipophilic ligands [66]. These ligands provided increased uptake in the heart and higher heart-to-blood ratios, but their increased lipophilicity resulted in higher accumulation in the liver. Along a similar vein, the cationic species $[^{68}Ga(4,6\text{-}MeO_2sal)_2BAPEN]^+$ showed significant myocardial uptake and retention, with a heart-to-blood ratio of 45.6 ± 4.0 achieved two hours post injection [67]. This represents an improvement over the analogous neutral salicylaldimine ligands.

Other ligands examined include the neutral complexes of the type 1-aryl-3-hydroxy-2-methyl-4-pyridones, which were found to have heart uptake in animal models [68]. Unfortunately, these species had a short plasma half-life and were only stable long enough for a first-pass extraction by the heart.

The ligands described so far have the gallium in a hexadentate environment, but four co-ordinate ligands have also been examined. These include ligands of the type N_2S_2 (BAT-TECH) [69] and *tris*(2-mercaptobenzyl)amine (S_3N), which showed both brain and myocardial uptake [70] (Fig. 11.4).

Ga-labeled Brain Imaging Agents

The work on gallium radiopharmaceuticals that can cross the blood–brain barrier (BBB) has been conducted for some time with only limited success. Species such as $[Ga]THM_2BED$ show low brain uptake immediately after injection, but have a very fast washout [71]. [Ga]EDTA has also been used to show BBB defects at the site of brain tumors and multiple sclerosis plaques [72–75]. The most promising gallium brain imaging radiopharmaceutical developed to date is that complexed with the small lipophilic S_3N species [70]. This agent has a gradual increase in brain uptake, followed by a slow washout. This was demonstrated by a blood-to-brain ratio of 3.5 at 15 minutes after administration that increased to 5.2 after one hour.

Copper

Introduction

Copper has two natural occurring isotopes, ^{63}Cu (69.2%) and ^{65}Cu (30.8%). Several radionuclides can be produced for use in PET imaging, ^{60}Cu, ^{61}Cu, ^{62}Cu and ^{64}Cu, with ^{67}Cu and ^{64}Cu used for therapeutic purposes.

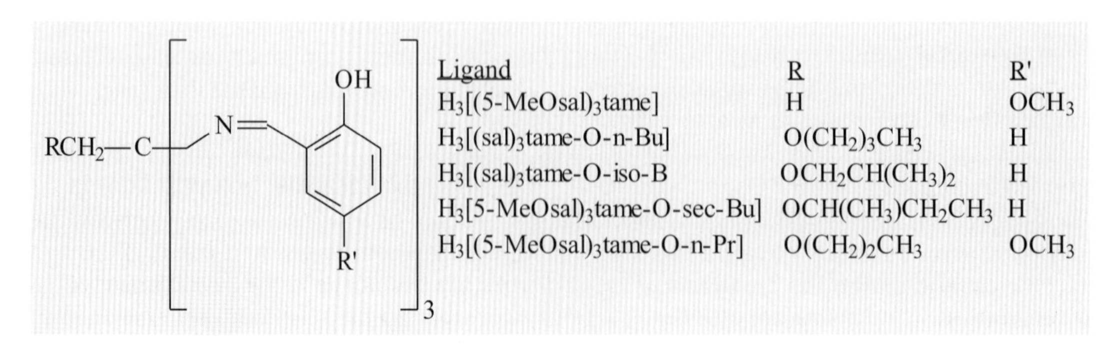

Ligand	R	R'
$H_3[(5\text{-}MeOsal)_3tame]$	H	OCH_3
$H_3[(sal)_3tame\text{-}O\text{-}n\text{-}Bu]$	$O(CH_2)_3CH_3$	H
$H_3[(sal)_3tame\text{-}O\text{-}iso\text{-}B$	$OCH_2CH(CH_3)_2$	H
$H_3[5\text{-}MeOsal)_3tame\text{-}O\text{-}sec\text{-}Bu]$	$OCH(CH_3)CH_2CH_3$	H
$H_3[(5\text{-}MeOsal)_3tame\text{-}O\text{-}n\text{-}Pr]$	$O(CH_2)_2CH_3$	OCH_3

Figure 11.3. Neutral salicyladimine ligands.

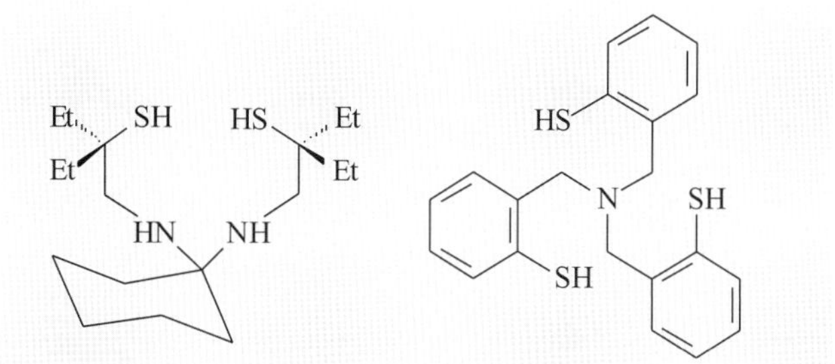

BAT-TECH

tris(2-mercatobenzyl)amine (S_3N)

Figure 11.4. ^{68}Ga-labelled myocardial and brain imaging agents.

For a brief description of the radioactive properties and production see the section on copper radionuclides (above).

Chemistry of Copper

Copper is situated in the first row of the transition elements and has an electron configuration of $[Ar]4s^1 3d^{10}$. The chemistry of copper is dominated by two oxidation states, I and II, although Cu(III) complexes have been reported.

Copper (I) has the electron configuration $[Ar]3d^{10}$, so its complexes tend to be colorless and diamagnetic. It prefers "soft" Lewis bases such as thioethers, phosphines, and nitriles, and usually forms tetradentate complexes that adopt a tetrahedral environment. If Cu(I) is bound by weakly coordinating ligands, it disproportionates in solution to give Cu(0) and Cu(II).

Cu(II) has an electron configuration of $[Ar]3d^9$, so all mononuclear Cu(II) complexes are paramagnetic. Cu(II) can be termed an "intermediate" Lewis acid and as such is bound most strongly by nitrogen- and sulfur-containing ligands. The co-ordination number can vary from four to six, with tetradentate complexes preferring a square–pyramidal arrangement and hexadentate complexes adopting a distorted octahedral environment. The distorted octahedral environment is caused by the partially filled d-orbital that causes tetragonal elongation along the z-axis.

Cu(III) complexes are rare and require strong π-donating ligands for stability. The copper in these species have a $[Ar]3d^8$ electronic configuration similar to that for Ni(II) and form predominately square planar complexes. Cu(III) complexes are powerful oxidants and in solution require a high pH to remain stable; because of this they cannot be utilized in biological systems.

Biochemistry of Copper

The human body contains about 100 mg of copper, the third most abundant trace metal after iron and zinc, and it is distributed mainly in the muscle, bone, liver, and blood. Although the amount of copper is small, it is an essential component in several enzymatic systems [76].

Copper *bis* (thiosemicarbazones)

One of the most commonly used Cu(II)-bis(thiosemicarbazone) is Cu(II)-pyruvaldehyde-bis(N^4-methylthiosemicarbazone) ([Cu]PTSM), (Fig. 11.5). [$^{62/64}$Cu]PTSM is a highly lipophilic complex, and is used as a blood flow tracer. Following administration of [$^{62/64}$Cu]PTSM, tissue uptake is rapid, and the tracer is trapped in most major tissues such as the brain [77–80], heart [77, 78, 81], and tumors [82–85]. Tissue retention of [$^{62/64}$Cu]PTSM and other *bis*(thiosemicarbazone) derivatives is believed to result when the lipophilic complex diffuses across the cell membrane. The Cu(II) is then reduced to Cu(I), at which time the copper complex decomposes. Cu(I) is trapped by binding to intracellular macromolecules [86]. Unlike [$^{62/64}$Cu]PTSM, studies involving [$^{60/62/64}$Cu(II)]diacetyl-*bis*([N^4]methylthiosemicarbazone) ([$^{60/62/64}$Cu]ATSM) (Fig. 11.5) have shown that [$^{60/62/64}$Cu]ATSM has selective uptake in hypoxic tissue [6, 87, 88]. Hypoxia is a condition in which the oxygen demands exceed the oxygen available. Hypoxic tissue can be found in myocardial infarction, brain injury, and tumors. The selectivity of hypoxic imaging agents is a balance between its redox potential and cell uptake. The cell uptake is important because hypoxic tissue has poor blood supply, and therefore the ability to deliver the tracer to the tissue is reduced. The difference in uptake of [$^{62/64}$Cu]PTSM and [$^{60/62/64}$Cu]ATSM is believed to result from the additional methyl group that results in a lower redox potential for [$^{60/62/64}$Cu]ATSM. Research is on-going to develop new mixed *bis*(thiosemicarbazone) ligands in an attempt to develop superior hypoxic imaging agents [89–92].

Design of Copper Celates

The design of copper complexes that are inert poses a significant challenge due to the lability of Cu(II). The choice of ligand can greatly influence the biokinetics, biodistribution, and metabolism of the radiopharmaceutical, thus affecting its usefulness. In an attempt to reduce Cu(II) lability, sterically encumbered ligands and macrocyclic ligands based on Cyclan have been utilized [94–97]. Macrocyclic

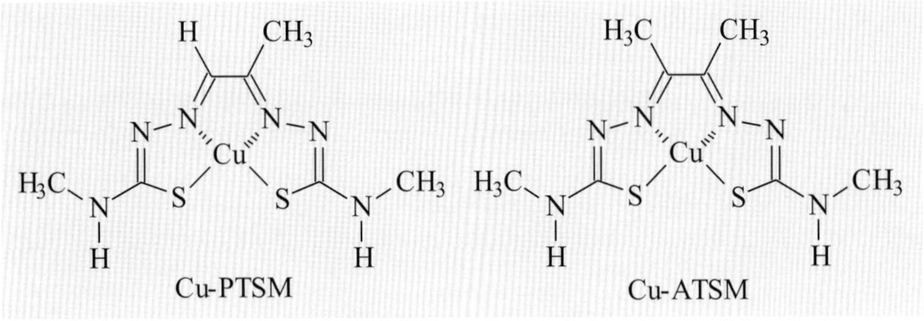

Figure 11.5. Cu(II) *bis*(thiosemicarbazone) complexes, [Cu]PTSM and [Cu]ATSM.

ligands are also used as bifunctional chelators to couple copper to monoclonal antibodies and other biomolecules. Studies of these macrocyclic ligands have shown that the charge carried and the number of functional groups attached to the nitrogens affect the stability and clearance properties of these radiopharmaceuticals [94, 96]. Other ligands examined (Fig. 11.7) include the cationic copper(I) *bis*(diphosphines), which act as substrates for P-glycoprotein (Pgp) [98, 99], and Schiff base complexes of Cu(II) as possible perfusion tracers for the brain, heart, and kidneys [100–102].

Copper-labeled Antibodies, Proteins and Peptides

^{64}Cu has labeled monoclonal antibodies (MAbs), proteins, and peptides for PET imaging. Copper is most commonly attached via macrocyclic ligands such as TETA (1,4,8,11-tetraazacyclotetradecane-N,N′,N″,N‴-tetraacetic acid), BAD (bromoacetamidobenzyl-1,4,8,11-tetraazacyclotetradecane- N,N′,N″,N‴-tetraacetic acid), CPTA (1,4,8,11-tetraazazcyclotetradecane-1-(α-1,4-toluic acid)) and DOTA (1,4,7,10-tetraazacyclododecane-N,N′,N″,N‴-tetraacetic acid). Schematic representations of these ligands are shown in Fig. 11.7. Direct labeling of Cu(I) to antibodies has also been attempted. Stannous tartrate was used to reduce disulfide groups, which then bound Cu(I) [103].

Examples of biomolecules labeled with copper include: the anti-colorectal monoclonal antibody (Mab) 1A3 for PET imaging of primary and metastatic colorectal tumors [104–106]; human serum albumin for perfusion and blood pool imaging [107–109]; and somatostatin analogs, such as octreotide and Y3-TATE, for imaging of somatostatin receptors (SSR) [12, 110–112].

Technetium

Introduction

Technetium was predicted on the basis of the periodic table and was incorrectly reported to be discovered in

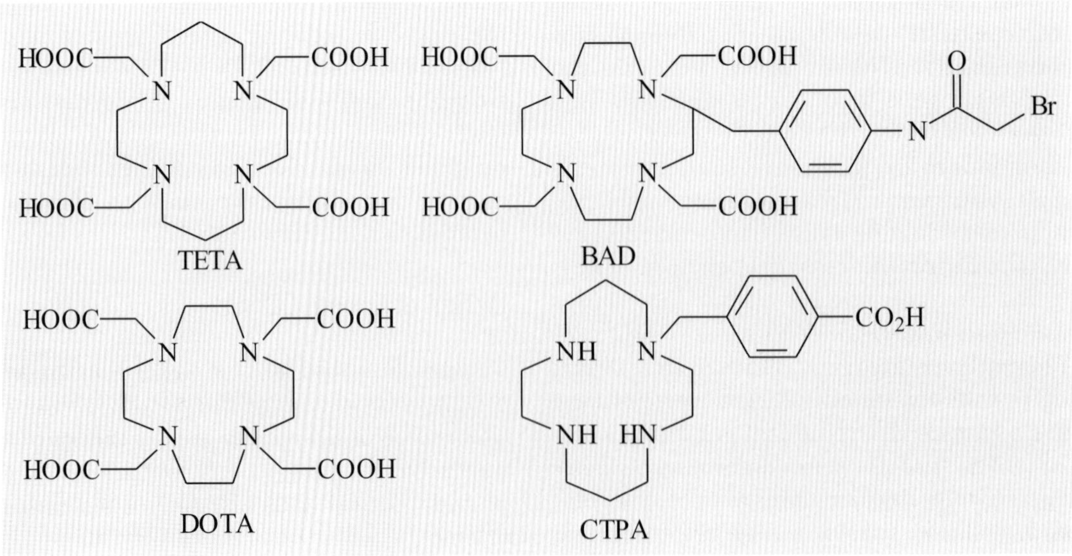

Figure 11.6. Cu(I) and Cu(II) chelates.

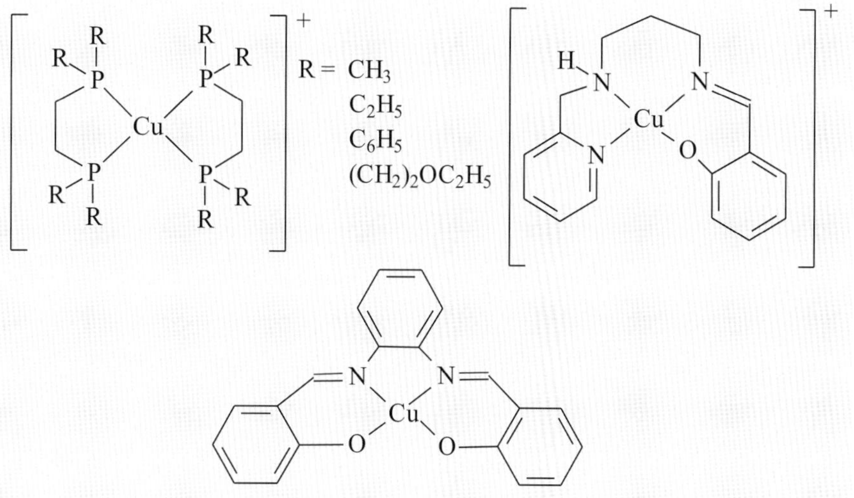

Figure 11.7. Macrocyclic chelates for Cu(II).

1925, at which time it was named masurium [113]. Technetium-99 was first produced by Perrier and Segre in 1937 and was the first artificially produced element [114]. Naturally occurring technetium-99 was isolated from African pitchblende (a uranium-rich ore) in 1961 [115]. Twenty-two isotopes of technetium with masses ranging from 90 to 111 have been reported; all of them are radioactive. Technetium is one of only two elements with mass number less than 83 that have no stable isotopes, the other is promethium with a mass number of 61. There are three long-lived radioactive isotopes of technetium: ^{97}Tc ($t_{\frac{1}{2}}$ =2.6 x 10^6 years), ^{98}Tc ($t_{\frac{1}{2}}$ = 4.2 x 10^6 years) and ^{99}Tc ($t_{\frac{1}{2}}$= 2.1 x 10^5 years). The most-used isotope of technetium is ^{99m}Tc ($t_{\frac{1}{2}}$ = 6.01 hours), which accounts for about 80% of the radiopharmaceuticals in use. It is used frequently as a result of its short half-life, its low-energy gamma ray, and its ability to be chemically bound to many biologically active molecules as well as the availability and long shelf life of ^{99m}Tc generators. More recently, ^{94m}Tc ($t_{\frac{1}{2}}$ = 52.5 min) has been examined. ^{94m}Tc can be produced on a biomedical cyclotron and decays by β^+-emission and therefore can be used in PET imaging.

Chemistry of Technetium

Technetium is a second-row group-VII transition metal, with an electron configuration of [Kr]$5s^2 4d^5$. The chemistry exhibited by Tc is vast as a result of the fact that it can occupy several different oxidation states (−1 to +7, although only +1 through +7 can exist in solution) and the large number of coordination geometries it can adopt [116]. These two factors mean that Tc can bind to a large number of donor ligands. However, this flexibility can cause problems in the design of Tc radiopharmaceuticals, as well as the in vivo biodistribution of Tc coordination compounds. Technetium is a group-VII congener of rhenium, and as a result their chemistries are similar. The only major difference is that the redox potential of analogous complexes can differ significantly, with Tc complexes being more easily reduced. Due to their chemical similarities, rhenium has often been used as an alternative to technetium in preliminary radiopharmaceutical investigations.

Radiolabeling Studies Using ^{94m}Tc

The first technetium complex to give consistently high uptake and slow washout in myocardial tissue was the technetium(I) cationic complex $[Tc(tbi)_6]^+$ [117]. In this complex, the Tc(I) oxidation state is stabilized by the back-bonding isonitrile ligands, which adopt an octahedral geometry around the technetium. In an effort to improve the heart-to-non-target organ ratios, the isonitrile ligand was functionalized. This led to the development of the complex technetium-methoxyisobutylisonitrile (Tc-sestamibi, Fig. 11.8) [117], which is now a commercially available instant kit for ^{99m}Tc labeling. Studies have been undertaken in which ^{94m}Tc was used to label sestamibi, in the hope that the increased resolution of PET imaging over SPECT would be useful in detecting myocardial perfusion [118, 119]. Another approach to imaging myocardial perfusion is via the neutral, lipophilic, seven-coordinate Tc(III) complex Tc-teboroxime. Tc-teboroxime can be produced in 90% yield from the reaction of TcO_4^-, dioxime, methyl boronic acid, and $SnCl_2$ under acidic conditions [120]. [^{94m}Tc]teboroxime was prepared by using $^{94m}TcO_4^-$ and a commercially available kit for teboroxime [121]. The results obtained were consistent

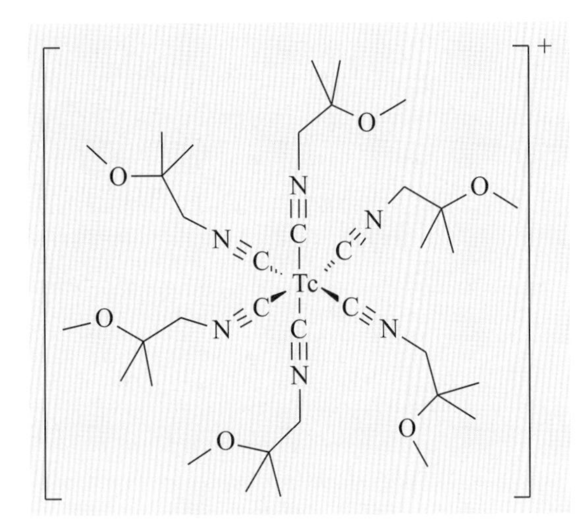

Figure 11.8. [^{94m}Tc]-sestamibi.

with those reported for [^{99m}Tc]teboroxime [121], and demonstrate that ^{94m}Tc provides an opportunity to study many of the new technetium compounds with PET. Another study involving instant labeling kits available for ^{99m}Tc was the anti-carcinoembryonic antibody Fab' fragment (CEA-Scan) [122]. This monoclonal Fab' fragment when labeled with ^{99m}Tc is used for imaging colorectal cancer [123]. The labeling with ^{94m}Tc was as straightforward as those observed for ^{99m}Tc, with labeling yields typically around 99%.

These studies have shown that ^{94m}Tc labeling can be used as an alternative to ^{99m}Tc and can directly replace ^{99m}Tc in instant labeling kits. By utilizing PET imaging, ^{94m}Tc should provide image resolution superior to that of the corresponding ^{99m}Tc-labeled agent obtained with single photon imaging. Therefore, ^{94m}Tc labeled compounds would make superior radiopharmaceuticals relative to the analogous ^{99m}Tc compounds.

Other PET Metaloradiopharmaceuticals

Rubidium-82

As mentioned previously, ^{82}Rb is a short-lived ($t_{\frac{1}{2}}$ = 1.27 min) positron-emitting radionuclide that is commercially available as a generator system from the long-lived ^{82}Sr ($t_{\frac{1}{2}}$ = 25.6 days) [3]. This generator system is the most widely used in PET imaging, with its in vivo applications based on the ability of rubidium to mimic K^+ [28]. Hence, ^{82}Rb has been used to study myocardial perfusion and blood flow [124–127] as well as the integrity of the BBB [128–131].

Potassium-38

^{38}K decays 100% by β^+-emission and has a half-life ($t_{\frac{1}{2}}$ = 7.71 min) that is short enough to allow multiple studies in the same patient [4]. Similar to ^{82}Rb, PET imaging has focused on $^{38}K^+$ as a tracer for myocardial perfusion [132–134] and as a cerebral tracer [134, 135].

Yttrium-86

^{86}Y has a 14.7-hour half life and decays 33% by β^+-emission [4]. The long half life means that PET images can be acquired several days after administrating the radiopharmaceutical. [^{86}Y]citrate and [^{86}Y]EDTMP ([^{86}Y]ethylenediamine-tetramethylene-phosphonate) have been used to study bone metastases [136–138]. ^{86}Y has also labeled somatostatin analogs for imaging of somatostatin receptors [139–141].

Manganese-51, -52 and -52m

Both ^{51}Mn and ^{52}Mn are cyclotron-produced positron-emitting radionuclides with half-lives of 46.2 min and 5.6 days, respectively, while ^{52m}Mn is generator produced with a half-life of 21.1 min [4]. $^{51/52}$Mn have been suggested for use in the diagnosis and treatment of blood diseases [142] and as cationic perfusion tracers [143]. ^{52m}Mn has been used to study the myocardial and cerebral perfusion [144, 145].

Cobalt-55

Finally, ^{55}Co has a half life of 18.2 hours and decays 81% by β^+-emission [4]. It has been used in PET imaging as [^{55}Co]Cl_2, where $^{55}Co^{2+}$ has been used as a marker for calcium uptake in degenerating brain tissue [146–150] and complexed to molecules such as oxine and MPO (mercaptopyridine-N-oxide) for platelet labeling [151] and to EDTA for renal function assessments [152]. It has also been chelated to biomolecules including bleomycin for studies of lung cancer and brain metastases [153–155] and the MAb LS-174T [156].

Conclusions

There are many possible metal-based positron-emitting radionuclides that can be utilized in positron tomography. At the present time, the most promising nuclides are copper, gallium, and yttrium. It is highly likely that radiopharmaceuticals labeled with these nuclides will be generally utilized in clinical practice over the next several years.

References

1. Anger HO, Gottschalk A. Localization of brain tumors with the positron scintillation camera. J Nucl Med 1964;5:326.
2. Shealy CN, Aronow S, Brownell GL. Gallium-68 as a scanning agent for intracranial lesions. J Nucl Med 1964:161–7.
3. Gennaro GP, Neirinckx RD, Bergner B, Muller WR, Waranis A, Haney TA, et al. A radionuclide generator and infusion system for pharmaceutical-quality rubidium-82. ACS Symp Ser 1984:135–50.
4. Lederer CM, Shirley VS. Table of Isotopes. 7th Ed, 1978:1632 pp.
5. Guillaume M, De Landsheere C, Rigo P, Czichosz R. Automated production of potassium-38 for the study of myocardial perfusion using positron emission tomography. Appl Radiat Isot 1988:97–107.
6. McCarthy DW. High-purity production and potential applications of copper-60 and copper-61. Nucl Med Biol 1999;26(4):351–8.
7. McCarthy DW, Shefer RE, Klinkowstein RE, Bass LA, Margeneau WH, Cutler CS, et al. Efficient production of high-specific-activity 64Cu using a biomedical cyclotron. Nucl Med Biol 1997:35–43.
8. Szelecsenyi F, Blessing G, Qaim SM. Excitation functions of proton-induced nuclear reactions on enriched nickel-61 and nickel-64: possibility of production of no-carrier-added copper-61 and copper-64 at a small cyclotron. Appl Radiat Isot 1993:575–80.
9. Zweit J, Smith AM, Downey S, Sharma HL. Excitation functions for deuteron-induced reactions in natural nickel: production of no-carrier-added copper-64 from enriched nickel-64 targets for positron emission tomography. Appl Radiat Isot 1991:193–7.
10. Smith SV, Waters DJ, Di Bartolo N. Separation of 64Cu from 67Ga waste products using anion exchange and low acid aqueous/organic mixtures. Radiochim Acta 1996:65–8.
11. Payne AS, Kelly EJ, Gitlin JD. Functional expression of the Wilson disease protein reveals mislocalization and impaired copper-dependent trafficking of the common H1069Q mutation. PNAS 1998;95(18):10854–9.
12. Anderson CJ, Dehdashti F, Cutler PD, Schwarz SW, Laforest R, Bass LA, et al. 64Cu-TETA-octreotide as a PET imaging agent for patients with neuroendocrine tumors. J Nucl Med 2001;42(2):213–21.
13. Szelecsenyi F, Boothe TE, Tavano T, Plitnikas ME, Tarkanyi F. Compilation of cross sections/thick target yields for 66Ga, 67Ga and 68Ga production using Zn targets up to 30 MeV proton energy. Appl Radiat Isot 1994:473–500.
14. Tarkanyi F, Szelecsenyi F, Kovacs Z, Sudar S. Excitation functions of proton induced nuclear reactions on enriched zinc-66, zinc-67 and zinc-68. Production of gallium-67 and gallium-66. Radiochim Acta 1990:19–26.
15. Brown LC. Chemical processing of cyclotron-produced gallium-67. Int J Appl Radiat Isotop 1971:710–13.
16. Zweit J, Sharma H, Downey S. Production of gallium-66, a short-lived, positron emitting radionuclide. Int J Rad Appl Instrum A 1987:499–501.
17. Roesch F, Qaim SM. Nuclear data relevant to the production of the positron emitting technetium isotope 94mTc via the 94Mo(p,n)-reaction. Radiochim Acta 1993:115–21.
18. Qaim SM. Production of high-purity 94mTc for positron emission tomography studies. Nucl Med Biol 2000:323–8.
19. Roesch F, Novgorodov AF, Qaim SM. Thermochromatographic separation of 94mTc from enriched molybdenum targets and its large-scale production for nuclear medical application. Radiochim Acta 1994:113–20.
20. Roesch F, Qaim SM, Soecklin G. Nuclear data relevant to the production of the positron emitting radioisotope yttrium-86 via the 86Sr(p,n)- and natRb(3He,xn)-processes. Radiochim Acta 1993:1–8.
21. Cherry SR, Gambhir SS. Use of positron emission tomography in animal research. Ilar J 2001:219–32.
22. Wipke BT, Wang Z, Kim J, McCarthy TJ, Allen PM. Dynamic visualization of a joint-specific autoimmune response through positron emission tomography. Nat Immunol 2002:366–72.
23. Wu AM, Yazaki PJ, Tsai S-w, Nguyen K, Anderson A-L, McCarthy DW, et al. High-resolution microPET imaging of carcinoembryonic antigen-positive xenografts by using a copper-64-labelled engineered antibody fragment. PNAS 2000;97(15):8495–500.
24. Adonai N, Nguyen KN, Walsh J, Iyer M, Toyokuni T, Phelps ME, et al. Ex vivo cell labelling with 64Cu-pyruvaldehyde-bis(N4-methylthiosemicarbazone) for imaging cell trafficking in mice with positron-emission tomography. PNAS 2002;99(5):3030–5.
25. Ramamoorthy N, Pao PJ, Watson IA. Preparation of a zinc-62–copper-62 generator and of copper-61 following alpha particle irradiation of a nickel target. Radiochem Radioanal Lett 1981:371–9.
26. Fujibayashi Y, Matsumoto K, Yonekura Y, Konishi J, Yokoyama A. A new zinc-62/copper-62 generator as a copper-62 source for PET radiopharmaceuticals. J Nucl Med 1989:1838–42.
27. Haynes NG, Lacy JL, Nayak N, Martin CS, Dai D, Mathias CJ, et al. Performance of a 62Zn/62Cu generator in clinical trials of PET perfusion agent 62Cu-PTSM. J Nucl Med 2000:309–14.
28. Sapirstein LA. Regional blood flow by fractional distribution of indicators. Am J. Physiol 1958:161–8.
29. Yano Y, Budinger TF, Cahoon JL, Huesman RH. An automated microprocessor-controlled rubidium-82 generator for positron emission tomography studies. ACS Symp Ser 1984:97–122.
30. McElvany KD, Hopkins KT, Welch MJ. Comparison of germanium-68/gallium-68 generator systems for radiopharmaceutical production. Int J Appl Radiat Isot 1984:521–4.
31. Kaplan JD, Calandrino FS, Schuster DP. Effect of smoking on pulmonary vascular permeability. A positron emission tomography study. Am Rev Res Dis 1992;145(3):712–5.
32. Kaplan JD. Pulmonary vascular permeability in interstitial lung disease. A positron emission tomographic study. Am Rev Res Dis 1992;145(6):1495–8.
33. Kaplan JD, Trulock EP, Cooper JD, Schuster DP. Pulmonary vascular permeability after lung transplantation. A positron emission tomographic study. Am Rev Res Dis 1992:954–7.
34. Palazzo R, Hamvas A, Shuman T, Kaiser L, Cooper J, Schuster DP. Injury in nonischemic lung after unilateral pulmonary ischemia with reperfusion. J App Physio 1992:612–20.
35. Hamvas A, Park CK, Palazzo R, Liptay M, Cooper J, Schuster DP. Modifying pulmonary ischemia-reperfusion injury by altering ventilatory strategies during ischemia. J App Physio 1992:2112–9.
36. Green MA. Gallium radiopharmaceutical chemistry. Int J Rad Appl Instrum B 1989;16(5):435–48.
37. Harris WR. Thermodynamic binding constants for gallium transferrin. Biochem 1983;22(2):292–9.
38. Edwards CL. Tumor scanning with 67Ga citrate. J Nucl Med 1969;10(2):103–5.
39. Gunasekera SW. The behaviour of tracer gallium-67 towards serum proteins. Clinica Chimica Acta 1972;39(2):40–-6.
40. Vallabhajosula SR. The mechanism of tumor localization of gallium-67 citrate: role of transferrin binding and effect of tumor pH. Int J Nucl Med Biol 1981;8(4):363–70.
41. Brunetti A. Gallium-transferrin as a macromolecular tracer of vascular permeability. Int J Rad Appl Instrum B 1988;15(6):665–72.
42. Mintun MA. Measurements of pulmonary vascular permeabil ity with PET and gallium-68 transferrin. J Nucl Med 1987;28(11):1704–16.
43. Calandrino FS, Jr., Anderson DJ, Mintun MA, Schuster DP. Pulmonary vascular permeability during the adult respiratory distress syndrome: a positron emission tomographic study. Am Rev Resp Dis 1988;138(2):421–8.
44. Graham MC, Pentlow KS, Mawlawi O, Finn RD, Daghighian F, Larson SM. An investigation of the physical characteristics of

66Ga as an isotope for PET imaging and quantification. Med Phys 1997;24(2):317–26.

45. Brazeau P, Vale W, Burgus R, Ling N, Butcher M, Rivier J, et al. Hypothalamic polypeptide that inhibits the secretion of immunoreactive pituitary growth hormone. Science 1973;179(68): 77–9.
46. Szabo S, Reichlin S. Somatostatin depletion of the gut and pancreas induced by cysteamine is not prevented by vagotomy or by dopamine agonists. Regulatory Peptides 1983;6(1):43–9.
47. Robbins RJ, Reichlin S. Somatostatin biosynthesis by cerebral cortical cells in monolayer culture. Endocrinology 1983;113(2): 574–81.
48. Reubi JC, Kvols L, Krenning E, Lamberts SW. Distribution of somatostatin receptors in normal and tumor tissue. Metabolism: Clinical & Experimental 1990;39(9 Suppl 2):78–81.
49. Smith-Jones PM, Stolz B, Bruns C, Albert R, Reist HW, Fridrich R, et al. Gallium-67/gallium-68-[DFO]-octreotide – a potential radiopharmaceutical for PET imaging of somatostatin receptor-positive tumors: synthesis and radiolabelling in vitro and preliminary in vivo studies. J Nucl Med 1994;35(2):317–25.
50. Stolz B, Smith-Jones PM, Albert R, Reist H, Macke H, Bruns C. Biological characterisation of [67Ga] or [68Ga] labelled DFO-octreotide (SDZ 216-927) for PET studies of somatostatin receptor positive tumors. Horm Meta Res 1994;26(10):453–9.
51. Henze M, Schuhmacher J, Hipp P, Kowalski J, Becker DW, Doll J, et al. PET imaging of somatostatin receptors. J Nucl Med 2001;42(7):1053–6.
52. Hofmann M, Maecke H, Borner R, Weckesser E, Schoffski P, Oei L, et al. Biokinetics and imaging with the somatostatin receptor PET radioligand (68)Ga-DOTATOC: preliminary data. Eur J Nucl Med 2001;28(12):1751–7.
53. Ugur O, Kothari PJ, Finn RD, Zanzonico P, Ruan S, Guenther I, et al. Ga-66 labelled somatostatin analogue DOTA-DPhe1-Tyr3-octreotide as a potential agent for positron emission tomography imaging and receptor-mediated internal radiotherapy of somatostatin receptor-positive tumors. Nucl Med Biol 2002:147–57.
54. Maziere B. Stable labelling of serum albumin microspheres with gallium-68. Int J Rad Appl Instrum A 1986;37(4):360–1.
55. Even GA. Gallium-68-labelled macroaggregated human serum albumin, 68Ga-MAA. Int J Rad Appl Instrum B 1989;16(3): 319–21.
56. Mintun MA, Ter-Pogossian MM, Green MA, Lich LL, Schuster DP. Quantitative measurement of regional pulmonary blood flow with positron emission tomography. J Appl Physio 1986:31–26.
57. Steinling M, Baron JC, Maziere B, Lasjaunias P, Loc'h C, Cabanis EA, et al. Tomographic measurement of cerebral blood flow by the 68Ga-labelled-microsphere and continuous-C15O2-inhalation methods. Eur J Nucl Med 1985:29–32.
58. Goethals P, Coene M, Slegers G, Agon P, Deman J, Schelstraete K. Cyclotron production of carrier-free gallium-66 as a positron-emitting label of albumin colloids for clinical use. Eur J Nucl Med 1988:152–4.
59. Goethals P, Coene M, Slegers G, Vogelaers D, Everaert J, Lemahieu I, et al. Production of carrier-free gallium-66 and labelling of antimyosin antibody for positron imaging of acute myocardial infarction. Eur J Nucl Med 1990:237–40.
60. Otsuka FL. Antibody fragments labelled with fluorine-18 and gallium-68: in vivo comparison with indium-111 and iodine-125-labelled fragments. Int J Rad Appl Instrum B 1991;18(7):813–6.
61. Moerlein SM. Metabolic imaging with gallium-68- and indium-111-labelled low-density lipoprotein. J Nucl Med 1991;32(2): 300–7.
62. Yano Y. Gallium-68 lipophilic complexes for labelling platelets. J Nucl Med 1985;26(12):1429–37.
63. Heppeler A, Froidevaux S, Eberle AN, Maecke HR. Receptor targeting for tumor localisation and therapy with radiopeptides. Curr Med Chem 2000:971–94.
64. Mathias CJ. Targeting radiopharmaceuticals: comparative biodistribution studies of gallium and indium complexes of multidentate ligands Int J Rad Appl Instrum B 1988;15(1):69–81.
65. Green MA. Potential gallium-68 tracers for imaging the heart with PET: evaluation of four gallium complexes with functionalized tripodal tris(salicylaldimine) ligands. J Nucl Med 1993;34(2):228–33.
66. Green MA. Gallium-68 1,1,1-tris (5-methoxysalicylaldiminomethyl) ethane: a potential tracer for evaluation of regional myocardial blood flow. J Nucl Med 1985;26(2):170–80.
67. Tsang BW. A gallium-68 radiopharmaceutical that is retained in myocardium: 68Ga[(4,6-MeO2sal)2BAPEN]+. J Nucl Med 1993;34(7):1127–31.
68. Zhang Z, Lyster DM, Webb GA, Orvig C. Potential 67Ga radiopharmaceuticals for myocardial imaging: tris(1-aryl-3-hydroxy-2-methyl-4-pyridinonato)gallium(III) complexes. Int J Rad Appl Instrum B 1992:327–35.
69. Kung HF. A new myocardial imaging agent: synthesis, characterization, and biodistribution of gallium-68-BAT-TECH. J Nucl Med 1990;31(10):1635–40.
70. Cutler CS. Evaluation of gallium-68 tris(2-mercaptobenzyl)amine: a complex with brain and myocardial uptake. Nucl Med Biol 1999;26(3):305–16.
71. Madsen SL. ^{68}GaTHM2BED: a potential generator-produced tracer of myocardial perfusion for positron emission tomography. Int J Rad Appl Instrum B 1992;19(4):431–44.
72. Pozzilli C, Bernardi S, Mansi L, Picozzi P, Iannotti F, Alfano B, et al. Quantitative assessment of blood–brain barrier permeability in multiple sclerosis using 68-Ga-EDTA and positron emission tomography. J Neurol Neuros Psych 1988:1058–62.
73. Ericson K, Lilja A, Bergstrom M, Collins VP, Eriksson L, Ehrin E, et al. Positron emission tomography with ([11C]methyl)-L-methionine, [11C]D-glucose, and [68Ga]EDTA in supratentorial tumors. J Comp Ass Tomog 1985:683–9.
74. Mosskin M, von Holst H, Ericson K, Noren G. The blood tumour barrier in intracranial tumours studied with X-ray computed tomography and positron emission tomography using 68-Ga-EDTA. Neurorad 1986:259–63.
75. Madar I, Anderson JH, Szabo Z, Scheffel U, Kao PF, Ravert HT, et al. Enhanced uptake of [11C]TPMP in canine brain tumor: a PET study. J Nucl Med 1999:1180–5.
76. O'Dell BL. Biochemistry of copper. Medical Clinics of North America 1976;60(4):687–703.
77. Green MA. A potential copper radiopharmaceutical for imag ing the heart and brain: copper-labelled pyruvaldehyde bis(N4-methylthiosemicarbazone). Int J Rad Appl Instrum B 1987;14(1):59–61.
78. Green MA, Klippenstein DL, Tennison JR. Copper(II) bis(thiosemicarbazone) complexes as potential tracers for evaluation of cerebral and myocardial blood flow with PET. J Nucl Med 1988;29(9):1549–57.
79. Mathias CJ, Welch MJ, Raichle ME, Mintun MA, Lich LL, McGuire AH, et al. Evaluation of a potential generator-produced PET tracer for cerebral perfusion imaging: single-pass cerebral extraction measurements and imaging with radiolabelled Cu-PTSM. J Nucl Med 1990;31(3):351–9.
80. Taniuchi H, Fujibayashi Y, Okazawa H, Yonekura Y, Konishi J, Yokoyama A. Cu-pyruvaldehyde-bis(N4-methylthiosemicarbazone) (Cu-PTSM), a metal complex with selective NADH-dependent reduction by complex I in brain mitochondria: a potential radiopharmaceutical for mitochondria-functional imaging with positron emission tomography (PET). Biol Pharm Bull 1995;18(8):1126–9.
81. Wada K, Fujibayashi Y, Taniuchi H, Tajima N, Tamaki N, Konishi J, et al. Effects of ischemia-reperfusion injury on myocardial single pass extraction and retention of Cu-PTSM in perfused rat hearts: comparison with 201T1 and 14C-iodoantipyrine. Nucl Med Biol 1994;21(4):613–7.
82. Flower MA. 62Cu-PTSM and PET used for the assessment of angiotensin II-induced blood flow changes in patients with colorectal liver metastases. Eur J Nucl Med 2001;28(1):99–103.
83. Mathias CJ. Investigation of copper-PTSM as a PET tracer for tumor blood flow. Int J Rad Appl Instrum B 1991;18(7):807–11.
84. Minkel DT, Saryan LA, Petering DH. Structure–function correlations in the reaction of bis(thiosemicarbazonato) copper(II) complexes with Ehrlich ascites tumor cells. Cancer Res 1978;38(1):124–9.
85. Mathias CJ, Green MA, Morrison WB, Knapp DW. Evaluation of Cu-PTSM as a tracer of tumor perfusion: comparison with labelled microspheres in spontaneous canine neoplasms. Nucl Med Biol 1994;21(1):83–7.

86. Young H, Carnochan P, Zweit J, Babich J, Cherry S, Ott R. Evaluation of copper(II)-pyruvaldehyde bis-(N-4-methylthiosemicarbazone) for tissue blood flow measurement using a trapped tracer model. Eur J Nucl Med 1994:336–41.
87. Fujibayashi Y, Taniuchi H, Yonekura Y, Ohtani H, Konishi J, Yokoyama A. Copper-62-ATSM: a new hypoxia imaging agent with high membrane permeability and low redox potential. J Nucl Med 1997;38(7):1155–60.
88. Lewis JS. Evaluation of 64Cu-ATSM in vitro and in vivo in a hypoxic tumor model. J Nucl Med 1999;40(1):177–83.
89. Horiuchi K, Tsukamoto T, Saito M, Nakayama M, Fujibayashi Y, Saji H. The development of (99m)Tc-analog of Cu-DTS as an agent for imaging hypoxia. Nucl Med Biol 2000;27(4):391–9.
90. Dearling JLJ, Blower PJ. Redox-active metal complexes for imaging hypoxic tissues: structure-activity relationships in copper(II) bis(thiosemicarbazone) complexes. Chem Commun 1998:2531–2.
91. Ackerman LJ, West DX, Mathias CJ, Green MA. Synthesis and evaluation of copper radiopharmaceuticals with mixed bis(thiosemicarbazone) ligands. Nucl Med Biol 1999;26(5):551–4.
92. Dearling JL, Lewis JS, Mullen GE, Rae MT, Zweit J, Blower PJ. Design of hypoxia-targeting radiopharmaceuticals: selective uptake of copper-64 complexes in hypoxic cells in vitro. Eur J Nucl Med 1998;25(7):788–92.
93. Park G, Dadachova E, Przyborowska A, Lai S-j, Ma D, Broker G, et al. Synthesis of novel 1,3,5-cis,cis-triaminocyclohexane ligand-based Cu(II) complexes as potential radiopharmaceuticals and correlation of structure and serum stability. Polyhedron 2001:3155–63.
94. Cutler CS. Labeling and in vivo evaluation of novel copper(II) dioxotetraazamacrocyclic complexes. Nucl Med Biol 2000;27(4):375–80.
95. Sun X, Wuest M, Weisman GR, Wong EH, Reed DP, Boswell CA, et al. Radiolabelling and in vivo behavior of copper-64-labelled cross-bridged cyclam ligands. J Med Chem 2002;45(2):469–77.
96. Motekaitis RJ, Rogers BE, Reichert DE, Martell AE, Welch MJ. Stability and structure of activated macrocycles. Ligands with biological applications. Inorg Chem 1996:3821–7.
97. Jones-Wilson TM. The in vivo behavior of copper-64-labelled azamacrocyclic complexes. Nucl Med Biol 1998;25(6):523–30.
98. Lewis JS, Dearling JL, Sosabowski JK, Zweit J, Carnochan P, Kelland LR, et al. Copper bis(diphosphine) complexes: radiopharmaceuticals for the detection of multi-drug resistance in tumours by PET. Eur J Nucl Med 2000;27(6):638–46.
99. Lewis JS, Zweit J, Dearling JLJ, Rooney BC, Blower PJ. Copper(I) bis(diphosphine) complexes as a basis for radiopharmaceuticals for positron emission tomography and targeted radiotherapy. Chem Commun 1996:1093–4.
100. John EK, Bott AJ, Green MA. Preparation and biodistribution of copper-67 complexes with tetradentate Schiff-base ligands. J Pharm Sci 1994;83(4):587–90.
101. Luo H, Fanwick PE, Green MA. Synthesis and structure of a novel Cu(II) complex with a monoprotic tetradentate Schiff base ligand. Inorg Chem 1998:1127–30.
102. Sri-Aran M, Mathias CJ, Lim JK, Green MA. Synthesis and evaluation of a monocationic copper(II) radiopharmaceutical derived from N-(2-pyridylmethyl)-N'-(salicylaldimino)-1,3-propanediamine. Nucl Med Biol 1998;25(2):107–10.
103. Blower PJ, Lewis JS, Zweit J. Copper radionuclides and radiopharmaceuticals in nuclear medicine. Nucl Med Biol 1996;23(8):957–80.
104. Cutler PD. Dosimetry of copper-64-labelled monoclonal antibody 1A3 as determined by PET imaging of the torso. J Nucl Med 1995;36(12):2363–71.
105. Philpott GW. RadioimmunoPET: detection of colorectal carcinoma with positron-emitting copper-64-labelled monoclonal antibody. J Nucl Med 1995;36(10):1818–24.
106. Anderson CJ. Preparation, biodistribution and dosimetry of copper-64-labelled anti-colorectal carcinoma monoclonal antibody fragments 1A3-F(ab')2. J Nucl Med 1995;36(5):850–8.
107. Okazawa H, Fujibayashi Y, Yonekura Y, Tamaki N, Nishizawa S, Magata Y, et al. Clinical application of 62Zn/62Cu positron generator: perfusion and plasma pool images in normal subjects. Ann Nucl Med 1995;9(2):81–7.
108. Mathias CJ. In vivo comparison of copper blood-pool agents: potential radiopharmaceuticals for use with copper-62. J Nucl Med 1991;32(3):475–80.
109. Anderson CJ, Rocque PA, Weinheimer CJ, Welch MJ. Evaluation of copper-labelled bifunctional chelate–albumin conjugates for blood pool imaging. Nucl Med Biol 1993;20(4):461–7.
110. Lewis JS, Lewis MR, Srinivasan A, Schmidt MA, Wang J, Anderson CJ. Comparison of four 64Cu-labelled somatostatin analogues in vitro and in a tumor-bearing rat model: evaluation of new derivatives for positron emission tomography imaging and targeted radiotherapy. J Med Chem 1999;42(8): 1341–7.
111. Lewis JS, Lewis MR, Cutler PD, Srinivasan A, Schmidt MA, Schwarz SW, et al. Radiotherapy and dosimetry of 64Cu-TETA-Tyr3-octreotate in a somatostatin receptor-positive, tumor-bearing rat model. Clin Cancer Res 1999;5(11):3608–16.
112. Anderson CJ. Radiotherapy, toxicity and dosimetry of copper -64-TETA-octreotide in tumor-bearing rats. J Nucl Med 1998;39(11):1944–51.
113. Berg O, Tacke I. Two new elements of the manganese group. II. Roentgen spectroscopy. Sitzb Preuss Akad Wissenschaften 1925:405–9.
114. Perrier C, Segre E. Chemical properties of element 43. Atti accad Lincei Classe sci fis mat nat 1937:723–30.
115. Kenna BT, Kuroda PK. Isolation of naturally occurring technetium. J Inorg Nucl Chem 1961:142–4.
116. Liu S, Edwards DS, Barrett JA. 99mTc labelling of highly potent small peptides. Bioconj Chem 1997;8(5):621–36.
117. Packard AB, Kronauge JF, Brechbiel MW. Metalloradiopharmaceuticals. Top Biol Inorg Chem 1999:45–115.
118. Smith MF, Daube-Witherspoon ME, Plascjak PS, Szajek LP, Carson RE, Everett JR, et al. Device-dependent activity estimation and decay correction of radionuclide mixtures with application to Tc-94m PET studies. Med Phys 2001;28(1):36–45.
119. Stone CK, Christian BT, Nickles RJ, Perlman SB. Technetium 94m-labelled methoxyisobutyl isonitrile: dosimetry and resting cardiac imaging with positron emission tomography. J Nucl Cardio 1994;1(5 Pt 1):425–33.
120. Stewart RE, Schwaiger M, Hutchins GD, Chiao PC, Gallagher KP, Nguyen N, et al. Myocardial clearance kinetics of technetium-99m-SQ30217: a marker of regional myocardial blood flow. J Nucl Med 1990;31(7):1183–90.
121. Nickles RJ, Nunn AD, Stone CK, Christian BT. Technetium-94m-teboroxime: synthesis, dosimetry and initial PET imaging studies. J Nucl Med 1993:1058–66.
122. Griffiths GL, Goldenberg DM, Roesch F, Hansen HJ. Radiolabelling of an anti-carcinoembryonic antigen antibody Fab' fragment (CEA-Scan) with the positron-emitting radionuclide Tc-94m. Clin Cancer Res 1999:3001s–3s.
123. Verhaar-Langereis MJ, Bongers V, de Klerk JM, van Dijk A, Blijham GH, Zonnenberg BA. Interferon-alpha-induced changes in CEA expression in patients with CEA-producing tumours. Eur J Nucl Med 2000;27(2):209–13.
124. Coxson PG, Brennan KM, Huesman RH, Lim S, Budinger TF. Variability and reproducibility of rubidium-82 kinetic parameters in the myocardium of the anesthetized canine. J Nucl Med 1995:287–96.
125. Herrero P, Markham J, Shelton ME, Bergmann SR. Implementation and evaluation of a two-compartment model for quantification of myocardial perfusion with rubidium-82 and positron emission tomography. Circ Res 1992:496–507.
126. Gould KL, Yoshida K, Hess MJ, Haynie M, Mullani N, Smalling RW. Myocardial metabolism of fluorodeoxyglucose compared to cell membrane integrity for the potassium analog rubidium-82 for assessing infarct size in man by PET. J Nucl Med 1991:1–9.
127. Rigo P, de Landsheere C, Raets D, Delfiore G, Quaglia L, Lemaire C, et al. Myocardial blood flow and glucose uptake after myocardial infarction. Eur J Nucl Med 1986:S59–61.
128. Zunkeler B, Carson RE, Olson J, Blasberg RG, Girton M, Bacher J, et al. Hyperosmolar blood-brain barrier disruption in baboons: an in vivo study using positron emission tomography and rubidium-82. J Neuros 1996:494–502.
129. Roelcke U, Radue EW, von Ammon K, Hausmann O, Maguire RP, Leenders KL. Alteration of blood-brain barrier in human

brain tumors: Comparison of [18F]fluorodeoxyglucose, [11C]methionine and rubidium-82 using PET. J Neurol Sci 1995:20–7.
130. Roelcke U, Radue EW, Leenders KL. [11C]methionine and rubidium-82 uptake in human brain tumors: comparison of carrier-dependent blood–brain barrier transport. Dev Nucl Med 1993:197–9.
131. Dhawan V, Poltorak A, Moeller JR, Jarden JO, Strother SC, Thaler H, et al. Positron emission tomographic measurement of blood-to-brain and blood-to-tumor transport of rubidium-82. I: Error analysis and computer simulations. Phys Med Biol 1989:1773–84.
132. Melon PG, Brihaye C, Degueldre C, Guillaume M, Czichosz R, Rigo P, et al. Myocardial kinetics of potassium-38 in humans and comparison with copper-62-PTSM. J Nucl Med 1994:1116–22.
133. De Landsheere C, Mannheimer C, Habets A, Guillaume M, Bourgeois I, Augustinsson LE, et al. Effect of spinal cord stimulation on regional myocardial perfusion assessed by positron emission tomography. Am J Cardio 1992:1143–9.
134. Takami A, Yoshida K, Tadokoro H, Kitsukawa S, Shimada K, Sato M, et al. Uptakes and images of 38K in rabbit heart, kidney, and brain. J Nucl Med 2000:763–9.
135. Duncan CC, Lambrecht RM, Bennett GW, Rescigno A, Ment LR. Observations of the dynamics of ionic potassium-38 in brain. Stroke 1984:145–8.
136. Herzog H, Rosch F, Stocklin G, Lueders C, Qaim SM, Feinendegen LE. Measurement of pharmacokinetics of yttrium-86 radiopharmaceuticals with PET and radiation dose calculation of analogous yttrium-90 radiotherapeutics. J Nucl Med 1993:2222–6.
137. Rosch F, Herzog H, Neumaier B, Muller-Gartner HW, Stocklin G. Quantitative whole-body pharmacokinetics of yttrium-86 complexes with PET and radiation dose calculation of analogous yttrium-90 radiotherapeuticals. Int Radiopharm Dosimetry Symp Proc Conf 6th, Gatlinburg, Tenn., May 7–10, 1996, 1999:101–10.
138. Rosch F, Herzog H, Plag C, Neumaier B, Braun U, Muller-Gartner HW, et al. Radiation doses of yttrium-90 citrate and yttrium-90 EDTMP as determined via analogous yttrium-86 complexes and positron emission tomography. Eur J Nucl Med 1996;23(8):958–66.
139. Foerster GJ, Engelbach M, Brockmann J, Reber H, Buchholz H-G, Maecke HR, et al. Preliminary data on biodistribution and dosimetry for therapy planning of somatostatin receptor positive tumours: comparison of 86Y-DOTATOC and 111In-DTPA-octreotide. Eur J Nucl Med 2001:1743–50.
140. Wester H-J, Brockmann J, Roesch F, Wutz W, Herzog H, Smith-Jones P, et al. PET-pharmacokinetics of 18F-octreotide: a comparison with 67Ga-DFO- and 86Y-DTPA-octreotide. Nucl Med Biol 1997:275–86.
141. Rosch F, Herzog H, Stolz B, Brockmann J, Kohle M, Muhlensiepen H, et al. Uptake kinetics of the somatostatin receptor ligand [86Y]DOTA-dPhe1-Tyr3-octreotide ([86Y]SMT487) using positron emission tomography in non-human primates and calculation of radiation doses of the 90Y-labelled analog. Eur J Nucl Med 1999:358–66.
142. Sastri CS, Petri H, Kueppers G, Erdtmann G. Production of manganese-52 of high isotopic purity by helium-3 activation of vanadium. Int J Appl Radiat Isot 1981:246–7.
143. Daube ME, Nickles RJ. Development of myocardial perfusion tracers for positron emission tomography. Inter J Nucl Med Biol 1985:303–14.
144. Buck A, Nguyen N, Burger C, Ziegler S, Frey L, Weigand G, et al. Quantitative evaluation of manganese-52m as a myocardial perfusion tracer in pigs using positron emission tomography. Eur J Nucl Med 1996;23(12):1619–27.
145. Calonder C, Wurtenberger PI, Maguire RP, Pellikka R, Leenders KL. Kinetic modeling of 52Fe/52mMn-citrate at the blood–brain barrier by positron emission tomography. J Neurochem 1999;73(5):2047–55.
146. Gramsbergen JBP, Veenma von der Duin L, Loopuijt L, Paans AMJ, Vaalburg W, Korf J. Imaging of the degeneration of neurons and their processes in rat or cat brain by calcium-45 chloride autoradiography or cobalt-55(2+) chloride positron emission tomography. J Neurochem 1988:1798–807.
147. Jansen HM, Pruim J, vd Vliet AM, Paans AM, Hew JM, Franssen EJ, et al. Visualization of damaged brain tissue after ischemic stroke with cobalt-55 positron emission tomography. J Nucl Med 1994:456–60.
148. Jansen HM, Willemsen AT, Sinnige LG, Paans AM, Hew JM, Franssen EJ, et al. Cobalt-55 positron emission tomography in relapsing-progressive multiple sclerosis. J Neurol Sci 1995:139–45.
149. Jansen HM, van der Naalt J, van Zomeren AH, Paans AM, Veenma-van der Duin L, Hew JM, et al. Cobalt-55 positron emission tomography in traumatic brain injury: a pilot study. J Neurol Neuros Psych 1996:221–4.
150. Jansen HM, Dierckx RA, Hew JM, Paans AM, Minderhoud JM, Korf J. Positron emission tomography in primary brain tumours using cobalt-55. Nucl Med Commun 1997:734–40.
151. Schmaljohann J, Karanikas G, Sinzinger H. Synthesis of cobalt-55/57 complexes for radiolabelling of platelets as a potential PET imaging agent. J Lab Comp Radiopharm 2001:395–403.
152. Goethals P, Volkaert A, Vandewielle C, Dierckx R, Lameire N. 55Co-EDTA for renal imaging using positron emission tomography (PET): a feasibility study. Nucl Med Biol 2000:77–81.
153. Neirinckx RD. Cyclotron production of nickel-57 and cobalt-55 and synthesis of their bleomycin complexes. Int J Appl Radiat Isot 1977:561–2.
154. Paans AMJ, Wiegman T, De Graaf EJ, Kuilman T, Nieweg OE, Vaalburg W, et al. The production and imaging of cobalt-55-labelled bleomycin. Int Conf Cyclotrons Their Appl [Proc.], 9th, 1982:699–701.
155. Nieweg OE, Beekhuis H, Paans AM, Piers DA, Vaalburg W, Welleweerd J, et al. Detection of lung cancer with 55Co-bleomycin using a positron camera. A comparison with 57Co-bleomycin and 55Co-bleomycin single photon scintigraphy. Eur J Nucl Med 1982:104–7.
156. Srivastava SC, Mausner LF, Kolski KL, Mease RC, Joshi V, Meinken GE, et al. Production and use of cobalt-55 as an antibody label for PET imaging. J Label Comp Radiopharm 1994;35:389–91.

12 Radiation Dosimetry and Protection in PET*

Jocelyn EC Towson

Introduction

Positron emission tomography (PET), after having spent 20 years or more being developed within the research environment, has now emerged as a clinical diagnostic tool. This means that PET facilities are now being located in imaging departments of nuclear medicine and radiology within the hospital environment. In most cases, these facilities were not designed for the higher-energy annihilation radiation of PET tracers. In addition, many of the approaches already employed in the nuclear medicine departments to reduce radiation exposure need to be re-thought to implement with PET, again due to the higher-energy nature of the radiation.

This chapter will discuss many of the radiation safety issues that have arisen from the transition of PET from the research/university environment into the clinical environment.

Impact of PET Radionuclide Decay Schemes

The short half-lives of clinical PET radionuclides limit the internal radiation dose to patients and the external radiation dose to persons who come in contact with the patient some time after the PET scan. However, they confer no particular benefit on PET staff who must contend with high dose rates from patients and many patients to be scanned each day. The various aspects of "exposure to radiation" need to be described in specific dosimetric terms.

The radiation *absorbed dose*, D, is the energy deposited per unit mass of an absorbing material, including biological tissue. In SI units, absorbed dose is expressed in grays (Gy): 1 Gy is 1 joule per kilogram. Absorbed doses from natural background radiation are of the order of 2 mGy per year, absorbed doses in medicine typically range up to a few tens of mGy from diagnostic procedures and tens of Gy to tissues targeted in therapeutic applications. Two derived dose quantities are invoked to regulate the exposure of persons at work and the public at large [1]: *equivalent dose*, H, and *effective dose*, E, both of which are expressed in sieverts (Sv). Equivalent dose is absorbed dose weighted for the type of radiation and averaged over the whole organ or tissue (except in the case of skin). Fortunately, for simplicity in most medical applications, the radiation weighting factor for electrons, positrons, X- and gamma rays is one and therefore the equivalent dose in Sv is numerically equal to the absorbed dose in Gy. Effective dose is also a mathematical construct: a weighted sum of the equivalent doses to the individual organs and tissues of the body. The tissue weighting factors take account of the relative susceptibility of different tissues to radiation damage. Effective dose represents the long-term risk of harm from low-level exposure, essentially the risk of radiogenic cancer.

Most countries have adopted the dose limits recommended by the International Commission on Radiological Protection (ICRP) as shown in Table 12.1. Medical exposures are not included in the system of recommended limits. The limit on effective dose for occupational exposure is associated with an acceptable long-term risk compared to most other occupational hazards; the limit for members of the public is considered to be acceptable because it is comparable to varia-

* Chapter reproduced from Valk PE, Bailey DL, Townsend DW, Maisey MN. Positron Emission Tomography: Basic Science and Clinical Practice. Springer-Verlag London Ltd 2003, 265–279.

Table 12.1. Dose limits recommended by the ICRP [1]

	Occupational	Public
Effective dose	20 mSv y^{-1}, averaged over 5 y and not more than 50 mSv in any 1 y	1 mSv y^{-1}
Equivalent dose		
Lens of the eye	150 mSv y^{-1}	15 mSv y^{-1}
Skin	500 mSv y^{-1}	50 mSv y^{-1}
Hands and feet	500 mSv y^{-1}	–

tions in natural background radiation [1]. The effective dose limits are supplemented by limits on the equivalent dose to the tissues most likely to receive a high exposure at work – the skin, eyes and the hands and feet ("extremities") – to avoid damage to skin and formation of cataracts in the lens of the eye. For the purposes of monitoring a person's exposure to an external source of radiation, the *ambient dose equivalent* at a depth of 10 mm in tissue, H*(10), also called the deep dose equivalent (DDE), may be taken as the effective dose from a uniform whole-body exposure. The *directional dose equivalent* at a depth of 0.07 mm in tissue, H'(0.07), also called the shallow dose equivalent (SDE), can be taken as the equivalent dose at the average depth – 70 μm–of the basal cell layer in skin [2].

In terms of energy deposition in tissue, PET radionuclides have more in common with the radionuclides used for therapy than those used for diagnostic imaging. The amount of energy deposited locally or at a distance from disintegrating atoms in an infinite medium is indicated by the equilibrium absorbed dose constant, Δ, as shown in Table 12.2 for a selection of radionuclides used for diagnosis and therapy [3, 4]. Positrons, being non-penetrating charged particles, deposit their energy locally and account for most of the dose to the organs and tissues of PET patients. The annihilation photons are penetrating and account for the exposure of persons nearby. The influence of half-life on the energy available from the total decay of a source is also evident in Table 12.2.

External exposure is the most significant pathway for occupational exposure in PET facilities. The high dose rates from PET radionuclides relative to other radionuclides used for diagnostic imaging are due to their high photon energy (511 keV) and abundance (197–200% for the PET radionuclides shown in Table 12.2 as there are two photons for each positron emitted). Other potential pathways are

(i) a skin dose from surface contamination,
(ii) a deep dose from bremsstrahlung generated in lead or other shielding material of high atomic number,
(iii) a superficial dose from positrons emitted from the surface of uncovered sources,
(iv) an immersion and inhalation dose from a release of radioactive gas into the room air.

The starting point when planning protection against external exposure from a radioactive source is a knowledge of the dose rate from the radionuclide in question. However, it is not always a straightforward exercise to find the appropriate value from published data. Variables include the physical quantity and absorbing medium (for example, exposure, absorbed dose, kerma in air; kerma or equivalent dose in tissue), distance from the source (for example, 1 cm, 30 cm, 1 meter), source configurations (for example, point source, vial), lower bound on photon energy (for example, 10 keV, 20 keV) and, of course, units (SI or old system).

Dose rates in air were traditionally calculated using the specific gamma ray constant (m^2 R mCi^{-1} h^{-1} in old units) for the exposure rate at 1 meter from the nuclide in question. The conversion factor from exposure in air (roentgens, R) to absorbed dose in tissue was close to unity and was generally ignored.[1] With the introduction of SI units, the International Commission on Radiation Units and Measurements (ICRU) recommended that the specific gamma ray dose constant should be phased out and replaced by the air kerma rate constant [5].[2] The conversion factor from air kerma to ambient dose equivalent is not close to unity. It takes account of scattering and attenuation in tissue and depends on the photon energy [6]. The dose rates in air and tissue at a distance of 1 meter from a 1GBq "point" source of commonly used radionuclides are given in Table 12.3 [6, 7].

There is good agreement between the data in Table 12.3 for the ambient and deep dose equivalent rates from photon emissions. The rate constants can be used to check the response of survey meters, whether displayed in air kerma or ambient dose equivalent, to a reference source of known activity. Dose rates from

[1] For an approximate conversion of exposure in roentgens to absorbed dose in rads, multiply by 0.87 for air, or 0.97 for tissue.
[2] "Kerma" stands for **k**inetic **e**nergy **r**eleased in unit **ma**ss and is expressed in the same units as absorbed dose. It is the sum of the initial kinetic energies of all the charged particles produced by photons incident on the unit mass. The kerma value may be slightly lower than the absorbed dose if some of the charged particle energy is deposited elsewhere (for example, after conversion to bremsstrahlung) [47].

Table 12.2. Energy available from the decay of nuclear medicine radionuclides

	Equilibrium absorbed dose constant, Δ [a] g Gy MBq^{-1} h^{-1} Non-penetrating Δn–p	Penetrating Δp	Total Δ	$T_{\frac{1}{2}}$	Energy from total decay of 1MBq μJ
^{11}C	0.227	0.588	0.815	20.3 m	397
^{13}N	0.281	0.589	0.870	10.0 m	209
^{15}O	0.415	0.589	1.004	2.07 m	50
^{18}F	0.139	0.570	0.709	1.83 h	1868
^{90}Y	0.539	–	0.539	2.7 d	50295
^{99m}Tc	0.010	0.072	0.082	6.0 h	708
^{131}I	0.109	0.219	0.328	8.05 d	91250

[a] derived from data in [3, 4]. The unit g is mass in grams.

small-volume sources such as vials, syringes, or capsules containing typical "unit dosage" activities administered to a patient are illustrated in Fig. 12.1.

The superficial dose rates given for betas and electrons only do not allow for absorption in the source and walls of the container, and may substantially overestimate actual dose rates, however, they do indicate that skin and eye doses from open PET sources could be reduced significantly by interposing a barrier as thick as the maximum beta range (see Table 12.4).

Medical Exposures: Internal Dosimetry

Despite the high energy of the annihilation photons, the radiation dose to patients from PET procedures is comparable with many diagnostic nuclear medicine procedures utilizing radionuclides with single photon emissions. The absorbed dose is limited by a short physical half-life. The dose may also be limited by the maximum amount of activity that can be administered to the patient without taxing the response of the detector. For example, the maximum amount of [^{18}F]-FDG administered to a "standard" 70 kg adult patient for a whole-body oncology study with a sodium iodide (NaI) coincidence gamma camera is about 200 MBq. For a whole-body study with a bismuth germanate (BGO) camera in 3D mode, the administered activity should be less than 500 MBq. The count rate capability of newer cameras with fast LSO detectors is not as restrictive.

Organ and Tissue Dosimetry

Dose estimates for diagnostic nuclear medicine procedures are usually generic rather than patient specific, and are calculated using the methodology developed

Table 12.3. External dose rates from radionuclides used for diagnostic imaging

	Air kerma rate constant [a] m^2 μGy GBq^{-1} h^{-1}	Ambient dose equivalent H*(10) rate constant [a] m^2 μSv GBq^{-1} h^{-1}	Deep tissue dose rate at 1m [b] 1GBq point source μSv h^{-1}	Superficial tissue dose rate at 1m [c] 1GBq point source μSv h^{-1}
^{11}C	140	170	170	11,700
^{13}N	140	170	170	10,800
^{15}O	140	170	170	10,800
^{18}F	140	170	160	10,800
^{67}Ga	19	27	25	0
^{99m}Tc	14	21	23	0
^{111}In	75	88	89	8
^{123}I	36	44	47	0
^{131}I	53	66	66	7,700
^{201}Tl	10	17	18	0

[a] photons >20 keV [6], [b] photons, and [c] electrons, derived from data in [7]

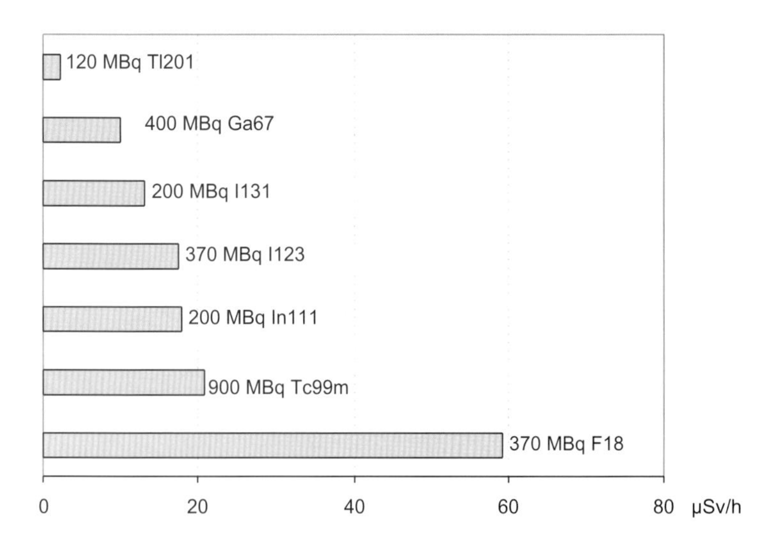

Figure 12.1. Equivalent dose rates in tissue at a distance of one meter in air from small-volume sources. Derived from data in [7].

by the Medical Internal Radiation Dose (MIRD) Committee of the Society of Nuclear Medicine [8] and software is available for this purpose [9]. The MIRD method requires an estimate of the spatial and temporal distribution of radioactivity in the body. The dose to a "target" organ from radioactivity in a "source" organ (D_{st}) is calculated as the product of two factors:

$$D_{st} = \tilde{A}_s \text{ x } S_{st} \qquad (1)$$

$\tilde{A}_s$ is the cumulated activity in a source organ. It represents the total number of disintegrations which occur in the source organ and therefore depends on the physiological behavior of the radiopharmaceutical. The cumulated activity per unit-administered activity, $\tilde{A}_s/A_0$, is called the residence time (τ) for the radioactivity in the organ. $\tilde{A}_s$ is obtained from biokinetic data, being the area under the activity–time curve for the organ. For example, the cumulated activity in an organ which has an initial rapid uptake of a fraction F_s of an administered activity A_0 (MBq), followed by an inverse exponential clearance with an effective half-life of $t_{1/2}$ (eff) (hours), would be

$$\tilde{A}_s = F_s \times A_0 \times 1.44 \times t_{\frac{1}{2}}(\text{eff}) \qquad (\text{MBq h}) \qquad (2)$$

The effective half-life combines biological clearance and radioactive decay, in this simple case as

$$1/t_{\frac{1}{2}}(\text{eff}) = 1/t_{\frac{1}{2}}(\text{biol}) + 1/t_{\frac{1}{2}}(\text{phys}) \qquad (3)$$

Standard biokinetic models developed by the ICRP are used for the movement of activity through complex anatomical and physiological compartments including the gastrointestinal tract, the cerebrospinal fluid space and the kidneys and bladder.

S_{st} is the S factor, the absorbed dose in a target organ per unit cumulated activity in the source organ (for example, in mGy MBq^{-1} h^{-1}). The S factor is determined by the physical properties of the radionuclide;

Table 12.4. Positron characteristics relevant to radiation protection

	β+ Emax MeV	Range in glass [a] mm	Range in plastic [a] mm	% of incident β+ energy converted to bremsstrahlung in lead	in tungsten	Skin dose rate: 1 kBq in 0.05 mL droplet [a] mSv h^{-1}
^{11}C	0.960	1.6	3.0	2.8	2.5	1.1
^{13}N	1.199	2.1	4.0	3.4	3.1	1.2
^{15}O	1.732	3.4	6.4	5.0	4.5	1.4
^{18}F	0.634	0.9	1.7	1.8	1.6	0.8

[a] [7]

tabulated S values for pairs of standardized source and target organs are available for many radionuclides [10]. The total dose to a particular target organ or tissue is then obtained by summing the contributions from all the identified source organs:

$$D_t = \{\tilde{A}_s \times S_{st}\} \qquad \text{(mGy)} \qquad (4)$$

alternatively

$$D_t = \{\tau_s \times S_{st}\} \qquad \text{(mGy per MBq administered)} \qquad (5)$$

Absorbed doses to selected organs and tissues for many radiopharmaceuticals have been compiled by the ICRP [11]. The ICRP has noted the difficulties of applying conventional dosimetry methods to very short-lived PET tracers [12]. For example, the radioactivity may not last long enough to allow true equilibration of the tracer in body compartments, or the highest dose may be received by organs or tissues – such as the trachea or walls of major blood vessels – which are not included in ICRP-listed sources and targets. The ICRP foreshadowed the development of novel ad hoc methods of dose estimation. The dose estimates for injected [^{15}O]-water and inhaled [^{15}O]-gases are cases in point, as are the estimates of reduced bladder dose for ^{18}F compounds by strategies such as hydration and frequent voiding [13–17]. Absorbed doses to maximally exposed organs or tissues and the gonads and uterus are shown in Table 12.5 for selected PET radiopharmaceuticals.

The Lactating Patient

PET studies with [^{18}F]-FDG for oncology or epilepsy investigations are infrequently requested for a woman who is breast feeding an infant. Avid uptake of [^{18}F]-FDG in lactating breast tissue has been reported in a small series of patients [18]. The uptake of [^{18}F]-FDG appears to be mediated by the GLUT-1 transporter, which is activated by suckling, not by prolactin. By imaging the breast before and after the expression of milk and counting the activity in milk samples, it was confirmed that [^{18}F]-FDG, being metabolically inert, is not secreted in milk to any significant amount but is retained in glandular tissue. The dose to glandular tissue will be higher than the value for the non-lactating breast, which is 0.0117 mGy per MBq injected [19]. The ^{18}F concentration in milk, measured in samples from one patient, was approximately 10 Bq mL^{-1} at one hour and 5 Bq mL^{-1} at three hours after injection. It was postulated that the ^{18}F activity was associated with the cellular elements in milk, mainly lymphocytes. Using the standard model of breast feeding with the first feed at three hours after injection [20], it was estimated that the dose to the infant from ingested milk would not exceed 85 μSv following an injection of up to 160 MBq of [^{18}F]-FDG. In addition, the infant would receive an external dose while being held while feeding from the breast or bottle. The dose rate against the chest was about 60 μSv h^{-1}, 2.5 hours after injection of 160 MBq of [^{18}F]-FDG. Breast feeding and cuddling of the infant should be postponed for several hours after an [^{18}F]-FDG study, particularly if a large amount of activity has been administered, if the infant's dose is to be kept below a dose constraint of 0.3 mSv for a single event.

The Pregnant Patient

The exposure of an embryo or fetus will depend on the biodistribution of the radiopharmaceutical. In the earliest stages of pregnancy, it is usual to take the dose to the embryo as being the same as the dose to the uterus. After about 12 weeks, when trophoblastic nutrition has been replaced by placental nutrition, the fetal dose will depend on whether the radiopharmaceutical or any of its metabolites accumulates in or is transferred across the placenta as well as on the distribution of activity in the mother. Where placental transfer of radioactivity occurs, the activity is generally assumed for dosimetry purposes to be distributed uniformly in the fetus. Fetal dose at various stages of gestation has been calculated for a range of radiopharmaceuticals using the MIRD methodology and an anatomical model of a pregnant female at three, six, and nine months' gestation [21]. In the absence of documented evidence of placental uptake or transfer of a particular radiopharmaceutical, the fetal dose was calculated from the maternal radioactivity only, which was the case for the three PET tracers in the report as shown in Table 12.6. For example, the fetal dose estimate following the administration of 500 MBq of [^{18}F]-FDG to the mother would range from 13.5 mSv in early pregnancy to 4 mSv at term. However, the second and third trimester doses may have been underestimated. Fetal uptake of [^{18}F]-FDG in early pregnancy has subsequently been demonstrated on a PET scan [22]. Iodide is also known to cross the placenta. The fetal thyroid begins to concentrate iodine from about the 13th week of pregnancy and reaches a maximum concentration at about the 5th to 6th month [23]. Fetal thyroid dose estimates for [^{124}I]-NaI are included in Table 12.6.

Table 12.5. Radiation dose estimates for selected PET radiopharmaceuticals

	[Ref.]	Organ (1)	Absorbed Dose mGy MBq^{-1} Organ (2)	Organ (3)	Uterus	Effective Dose mSv MBq^{-1}
[^{11}C]-CO single inhale, 20″ hold*	11, 12	0.02 heart	0.014 lungs	0.013 spleen	0.0023	0.0048
[^{11}C]-CO_2 single inhale, 20″ hold*	11, 12	0.0024 lungs	0.0018 adrenals	0.0017 intestine	0.0016	0.0016
[^{11}C]-diprenorphine	¶	0.015 testes	0.012 sm.intestine	0.0097 kidney	0.0034	0.009
[^{11}C]-methionine	52	0.027 bladder	0.019 pancreas	0.018 liver	–	0.0052
[^{11}C]-methyl-thymidine	11	0.032 liver	0.031 kidneys	0.0055 gall bladder	0.0015	0.0035
[^{11}C]-raclopride	53, ¶	0.011 ovaries	0.0083 kidneys	0.0063 liver	–	0.0053 (EDE)
[^{11}C]-spiperone	11	0.0038 adrenals	0.0035 pancreas	0.0031 kidneys	0.0022	0.0053
[^{11}C]-thymidine	11	0.011 kidneys	0.0052 liver	0.0034 heart	0.0024	0.0027
[^{13}N]-ammonia	11,12	0.0081 bladder	0.0046 kidneys	0.0042 brain	0.0019	0.002
[^{15}O]-O_2 single inhale, 20″ hold*	11, 12	0.0024 lungs	0.00033 heart	0.00022 spleen	0.000057	0.00037
[^{15}O]-water	11	0.0019 heart	0.0017 kidney	0.0016 liver, lung	0.00035	0.00093
[^{15}O]-butanol	54	0.0015 liver	0.0011 kidney	0.0011 spleen	0.00036	0.00036
[^{15}O]-CO single inhale, 20″ hold*	11,12	0.0034 lungs	0.0033 heart	0.0021 spleen	0.0003	0.00081
[^{15}O]-CO_2 single inhale, 20″ hold*	11,12	0.0012 lungs	0.00049 adrenals	0.00048 breast	0.00043	0.00051
[^{18}F]-FDG	11	0.16 bladder	0.062 heart	0.028 brain	0.021	0.019
[^{18}F]–FDOPA	17	0.16 bladder	0.011 intestine	0.010 gonads	0.016	0.018
[^{18}F]–FDOPA carbidopa pre-dose	55	0.15 bladder	0.027 kidneys	0.02 pancreas	0.019	0.02
[^{18}Fluoride]	11,12	0.22 bladder	0.04 bone	0.04 marrow	0.019	0.024
[^{18}F]-5-fluorouracil (5FU)	¶	0.12 bladder	0.06 kidneys	0.041 thyroid	0.025	0.022
[^{18}F]-FMISO	56	0.021 bladder	0.0185 heart wall	0.0183 liver	0.0183	0.0139
[^{18}F]-fluorethyl-flumazenil	†	0.0187 bladder	0.0046 thyroid	0.0036 gall bladder	0.0213	0.07

¶ Page BC, Medical Research Council (UK) Cyclotron Unit, Hammersmith Hospital (Personal communication).
† Bartenstein P, Klinik und Poliklinik für Nuklearmedizin, Klinikum der Johannes Gutenberg-Univesität, Mainz, DE (Personal communication).
* [11] also contains dose estimates for continuous, as opposed to single, inhalation.

Table 12.6. Fetal absorbed dose (excluding contribution from any radioactivity which crosses the placenta) from PET radiopharmaceuticals administered to the mother

	Fetal absorbed dose per unit activity administered to mother mGy MBq^{-1}			
	Early	3 months	6 months	9 months
[^{18}F]-FDG [a]	0.027	0.017	0.0094	0.0081
[^{18}F]-NaF [a]	0.022	0.017	0.0075	0.0068
[^{124}I]-NaI [a]	0.14	0.1	0.059	0.046
" fetal thyroid [b]	–	130	680	300

[a] [19], [b] [23]

Effective Dose

Of more general interest for diagnostic medical procedures is the concept of effective dose, which allows a comparison of the relative risk of non-uniform exposures [1]. Effective dose in adult patients from a PET scan can be estimated from the effective dose coefficients in Table 12.5. Values are generally in the range of 5–10 mSv, which is comparable to the dose from many nuclear medicine procedures, and also to the dose from CT examinations for similar diagnostic purposes [26].

PET procedures often include a transmission scan for attenuation correction of image data. The transmission scan is usually obtained with a coincidence source of germanium-68/gallium-68 or a single photon source of caesium-137. The short duration of the exposure and the collimation of the transmission source are such that the dose to the patient is insignificant [24]. A recent development in PET technology is the incorporation of a CT X-ray unit in the PET camera to facilitate the fusion of functional and anatomical image information. The additional radiation dose would be similar to that for the equivalent CT scan conducted separately from the PET study, assuming that the scan parameters are the same. The CT component of the hybrid scanner may be a multi-slice unit. It has been estimated that the average effective dose from CT with a multi-slice scanner increased by 30% for a scan of the head and 150% for scans of the chest and abdomen, to 1.2, 10.5 and 7.7 mSv respectively, compared to a conventional single-section scanner [25].

The activity of a radiopharmaceutical administered for pediatric studies is usually calculated by scaling down the adult dosage by the child's body weight or surface area, subject to a minimum acceptable amount for very small children and infants [27]. The effective dose as a function of age for oncologic imaging is illustrated in Fig. 12.2 for both scaling methods applied

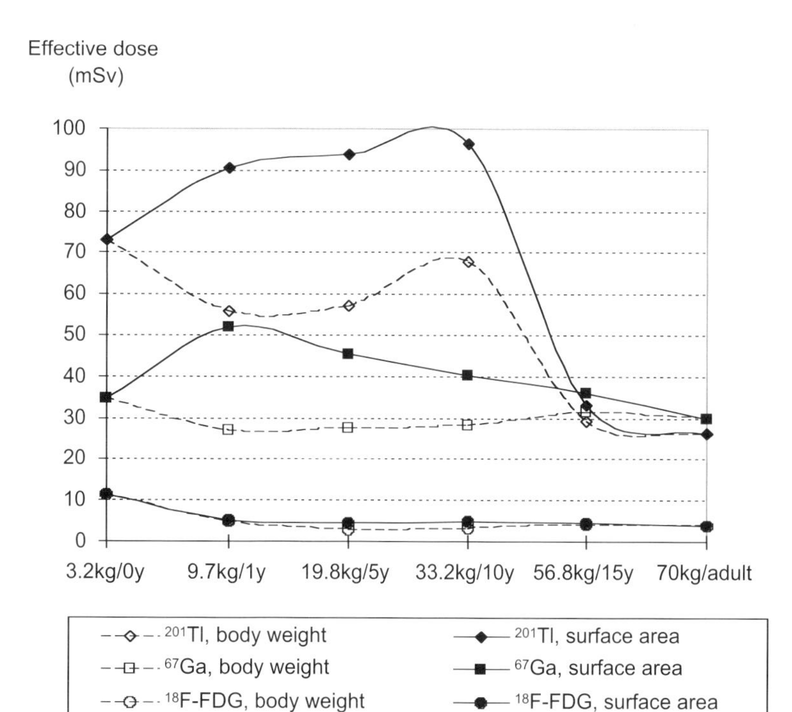

Figure 12.2. Dosimetry of radionuclide imaging in pediatric oncology: Administered activity scaled according to body weight or surface area between a minimum and maximum activity of 30–300 MBq of [^{67}Ga]-citrate, 20–120 MBq of [^{201}Tl]-chloride, and 50–200 MBq of [^{18}F]-FDG.

to an activity range of 50–200 MBq of [^{18}F]-FDG, 20–120 MBq of [^{201}Tl]-chloride, and 30–300 MBq of [^{67}Ga]-citrate [28]. Although the radiation dose from diagnostic imaging is not a prime concern for these patients, [^{18}F]-FDG dosimetry is lower, particularly when repeated studies are required.

One area in which the effective dose should be considered is the recruitment of volunteers to participate in research studies. Regulatory authorities in many countries have adopted the recommendations of the ICRP [29]:

(i) an exposure for research purposes is treated on the same basis as a medical exposure - and is therefore not subject to a specific dose limit,
(ii) the dose should be commensurate with the potential benefit of the research findings,
(iii) the study protocol should be approved by a properly constituted ethics committee.

A dose constraint may apply where the participant is not expected to benefit from the exposure, as in the exposure of "normal" subjects or patients enrolled in a clinical trial which involves additional or different exposures to what would be considered necessary for clinical management. There is limited scope for many PET studies to meet such dose constraints, for example, in determining the maximum number of injections of [^{15}O]-water for studies of cognitive function.

Table 12.8. Dose rate measurements near [^{18}F]-FDG patients: 95th percentiles

Patient position, measurement location	Distance m	Normalized dose rate μSv h^{-1} per MBq injected	
		Post injection [a]	2 h post injection [b]
Standing, at anterior chest	0.5	0.60	0.20
	2.0	0.10	0.03
Supine, at side	0.5	0.85	–
	2.0	0.11	–
Supine, at head	0.5	0.36	–
	2.0	0.075	–
Supine, at feet	0.5	0.078	–
	2.0	0.023	–

[a] [33], [b] [38]

Occupational and Public Exposures

Healthcare Workers Within and Outside the PET Facility

The radiation dose to a technologist performing a PET study is generally higher than for conventional nuclear medicine imaging [30–32]. Comparison of staff doses between PET facilities is not very informative because of the variability in clinical workload and scan procedures. With attention to shielding and protocols, it should be possible to maintain occupational exposure below approximately 6 mSv y^{-1} in most circumstances [33]. Hand doses may also be higher but are very variable because the radiation fields when manipulating partially shielded syringes and vials are highly directional. Substantial shielding of syringes, vials, transmission and quality control sources is standard practice in PET facilities. With inanimate sources effectively shielded, attention has turned to minimizing the exposure to staff from patients. Education of staff on the importance of distance and time is a key factor in dose control [33–36].

Dose rates near the [^{18}F]-FDG patient have been measured in a number of studies, at different orientations to the patient and at different times after injection [31–33, 35–38]. Representative values are shown in Table 12.8. The data could be adapted for other PET nuclides after correcting for the branching ratio and radioactive decay.

High dose rates require close attention to strategies that shorten, eliminate, or postpone close contact with the patient. Task-specific monitoring can be used to identify actions that contribute most to staff exposure and to suggest areas for improvement [39–42]. The dose from individual events, for example, dispensing and injecting [^{18}F]-FDG, or positioning and scanning the patient, typically ranges from 1 to 4 μSv as shown in Fig. 12.3. One of the most important factors is the duration of close contact with the patient. As vials and syringes can be shielded but dose rates within 0.5 m of the patient can be of the order of 4 to 8 μSv min^{-1} following an injection of [^{18}F]-FDG, this is to be expected.

The quantitative measurement of cerebral glucose metabolism originally required a number of blood samples to be taken over a period of 30–40 minutes following the injection of [^{18}F]-FDG, resulting in significant operator exposure [39]. A two-sample method has been developed which allows an 84% reduction in staff dose, from approximately 10 μSv to 1.5 μSv per study [43]. Other strategies to reduce staff exposure include setting up an intravenous line and a urinary catheter (if required) prior to administering the dose, postponing the removal of lines and catheters

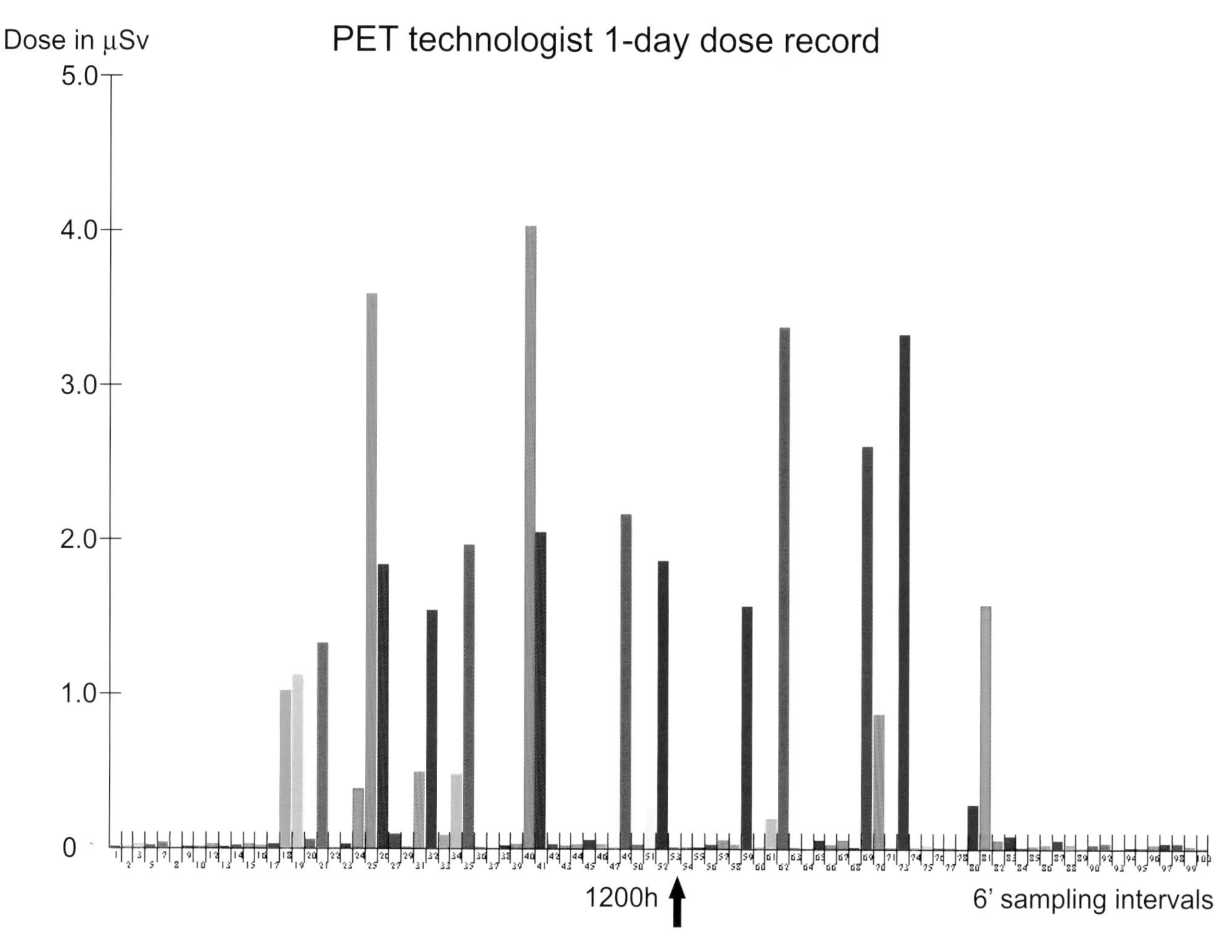

Figure 12.3. Daily pattern of exposure to PET technologist: personal dosimeter readings in six-minute intervals (Eurisys Dosicard). Peaks (1–4 μSv) correspond to dispensing and injecting multiple [^{18}F]-FDG doses, fitting a head mask and taking blood samples for neurological studies, escorting patients to and from the scanner for whole-body and neurological studies. Note low background at all other times.

until the conclusion of scanning, using a tourniquet or asking the patient to maintain pressure on the puncture site after removal of a line, using a wheelchair to move frail patients to and from the scanner as quickly as possible, and enlisting other persons to assist with patient handling. Nurses working within hospital PET facilities which scan many high-dependency patients may be the "critical group" as far as staff exposure is concerned.

The duties and previous personal dosimetry history of a staff member who declares that she is pregnant and wishes to continue at work during the pregnancy should be reviewed. The ICRP recommends that the equivalent dose to the surface of the mother's abdomen should not exceed 2 mSv for the remainder of the pregnancy [1]. It also notes that the use of source-related dose constraints – i.e., shielding and/or isolation of sources – will usually provide an adequate guarantee of compliance with this guideline without the need for specific restrictions on the employment of pregnant women. However, restrictions may need to be placed on the number or duration of close contacts a pregnant staff member should have with PET patients.

Following a PET scan, the patient may come into close contact with other health professionals. Dose rate measurements at various distances from the patient on leaving the PET facility, combined with modeling of potential patterns of close contact, indicated that nurses on a ward that regularly sends patients for PET scans are unlikely to receive more than 24 μSv per day [38]. The exposure rate to a sonographer working at 0.5 m from a patient who had received 400 MBq of [^{18}F]-FDG two hours previously would be about 40 μSv h^{-1} [44]. In circumstances where a staff member may have frequent contact with PET patients, for example, nursing staff or porters of an oncology ward, personal dosimeters can be used to establish the level of exposure for informed guidance on policies and procedures.

Table 12.7. Effective dose values in adult patients for PET and comparable diagnostic imaging procedures

Application		Protocol	Effective dose mSv
Oncology	[^{18}F]-FDG	370 MBq	7.0
	[^{11}C]-methionine	400 MBq	2.1
	[^{67}Ga]-citrate	400 MBq	40.0
	[^{201}Tl]-chloride	120 MBq	26.4
	[^{99m}Tc]-mibi	1 GBq, resting	9.0
Brain	[^{15}O]-water	1 GBq	0.93
	[^{18}F]-FDG	250 MBq	4.8
	[^{99m}Tc]-HMPAO	800 MBq	7.4
Myocardium	[^{13}N]-ammonia	550 MBq	1.1
	[^{18}F]-FDG	250 MBq	4.8
	[^{99m}Tc]-mibi	1.3 GBq; 1 day rest/stress protocol	10.6
	[^{201}Tl]-chloride	140 MBq; stress/reinjection protocol	30.8
Bone	[^{18}F]-NaF	250 MBq	6.0
	[^{99m}Tc]-MDP	800 MBq	4.6
CT [a]	head	average 1.63 scans per examination	2.6
	chest	average 1.40 scans per examination	10.4
	abdomen	average 1.72 scans per examination	16.7
	pelvis	average 1.50 scans per examination	11.0

[a] National survey of CT practice in Australia, 1994–1995 [26]

Family Members, Carers and Members of the Public

A dose limit of 1 mSv y^{-1} for members of the public is recommended by the ICRP and has been widely adopted [1]. This limit is accepted as a criterion when discharging radionuclide therapy patients from hospital, although a dose constraint of 0.3 mSv from any single event has subsequently been proposed (by the European Union) to allow for several exposures occurring during the course of a year [45]. As with the exposure of staff, the dose to persons near the patient will depend on the activity in the patient and excretion (if any), and the pattern of close contact effectively within a distance of about two meters or less. Only ^{18}F sources need be considered. Based on measured dose rates at various close-contact distances and times of leaving the PET facility (Table 12.7), the dose to persons near a patient has been modeled for a number of scenarios [38]. Results show that there is no need to restrict the activities of patients for the remainder of the day of their PET scan in order to satisfy a dose constraint of 0.3 mSv per event for a member of the public.

Persons "knowingly and willingly assisting with the support...of the patient" are regarded by the ICRP as carers, and are not subject to the dose limit for members of the public [1]. A dose constraint of 5 mSv per event has been proposed for carers [45]. Other family members – especially children – should be subject to the same dose constraint as members of the public, as it is quite possible that the patient will undergo more than one radionuclide imaging procedure within a year. Not all the accompanying persons in a common waiting area of a PET facility may qualify as "carers". For them, the 0.3 mSv dose constraint may be exceeded if they are seated next to one or more patients who have been injected with [^{18}F]-FDG and are waiting to be scanned. Patients should be advised at the time of booking that they should not be accompanied by pregnant women, infants, or children when attending for the scan. If this cannot be arranged, patients should remain in a single room or dedicated post-injection waiting area during the [^{18}F]-FDG uptake phase.

Facilities and Equipment

Instrumentation

The dose calibrator used in general nuclear medicine applications is adequate for PET in a clinical setting, whereas in a PET production laboratory a high-ranging chamber may be required if measuring very high activities. The chamber should be provided with additional shielding, up to 50 mm of lead, to protect the operator during PET nuclide measurements. At 511 keV, no corrections should be required for the geometry of the source container (for example, syringe, vial) or volume, with the manufacturer's settings possibly overestimating the activity of ^{18}F by 3–6% depending on the geometry [46].

Radiation instrumentation should include a survey meter, preferably a dual-purpose instrument for measurement of dose rate and surface contamination. Geiger-Mueller (GM) detectors have good sensitivity to PET radiations, and their energy response is fairly uniform over the photon energy range of a few hundred keV. No energy response correction should be necessary for a GM meter which has been calibrated at 660 keV with a ^{137}Cs source. An electronic personal dosimeter is a valuable means of monitoring staff who are in training or performing specific tasks where dose rates are high or there is prolonged close contact with a source.

Shielding

With the advent of commercial PET scanners in nuclear medicine departments more than a decade ago, it rapidly became apparent that the shields for the nuclides used in conventional gamma camera imaging were inadequate at 511 keV. PET radionuclides present more of a challenge because of their higher photon energy and hence smaller cross section for photoelectric absorption. Adequate shielding of small sources at all times is essential. The major component of the dose to PET staff now comes from close contact with patients.

The aim of shielding design is to achieve a desired transmission, K, the ratio of dose rates – or doses integrated over a specified interval – with and without the shield in place. Containers for sources in the workplace or during transport should attenuate the maximum expected intensity (I_0) to an acceptable value (I), for example, 10 μSv h^{-1}, at a specified close distance. Barriers such as walls or floors between a source and an occupied area should attenuate the dose (D_0) for the maximum anticipated workload during a specified interval, for example, one week, to an acceptable value (D, the design limit).

$$K = I/I_0 \text{ or } K = D/D_0 \qquad (6)$$

Dose rates in the vicinity of unshielded PET sources can be calculated using the dose rate constants given in Table 12.3. Vials and syringes can be treated as point sources, with the dose rate being inversely related to the square of the distance from the source. The dose rate near a line, planar, or cylindrical source can be calculated by the appropriate formula [47]. The dose rate near an [^{18}F]-FDG patient has been measured at various distances and intervals after injection [33, 38, 48]. Not surprisingly, given the biodistribution of FDG, the dose rate is higher at the head than the feet, a fact which can be employed to advantage when siting a camera and operator console [37]. A very conservative approach is to treat the activity in the patient as a point source located at the midpoint of the trunk with no self-attenuation or excretion losses (which would overestimate the dose rate by about 40% at two meters and 30% at five meters [48]), and disregard attenuation in the scanner gantry or other hardware. The determination of K also requires an estimation of the maximum anticipated workload at each source location for the specified design interval, say in GBq h in one week. The highest such source term is likely to be in the room set aside for an [^{18}F]-FDG patient in the uptake phase after injection, when the activity is maximal, the room is fully utilized and the room dimensions – and possibly distance to the nearest occupied areas – are small.

The value of D depends on who has access to the area in question, and for how long. Regulatory authorities should be consulted for local requirements. The design limit for areas to which the public has access [32] may be 20 μSv per week or a lower dose constraint, with an adjustment for partial occupancy where appropriate. The design limit for areas used by occupationally exposed persons only is taken to be less than the occupational dose limit pro rata for the specified interval, for example, 100 μSv per week compared to 400 μSv per week, although generally greater than the design limit for areas to which the public has access [32]. Full-time occupancy is assumed if not known. This approach is adequate if the source is a type which can be turned off or retracted into shielding before the operator enters the room, but may not be sufficiently conservative when the source is a radioactive patient with whom intermittent close contact without a barrier is unavoidable. One survey has found that the average time spent by the technologist at a distance of one meter or less from the patient was 32 minutes per day [33]. For the remaining seven hours or so of the working day, the technologist is usually at the PET console. If distance does not provide sufficient protection, barriers between the PET console or control room and "hot" patients should be designed to reduce the dose to a low level (for example,. 20 μSv per week) to optimize staff protection (see Fig. 12.2). Depending on the layout, the construction materials and the dimensions of the facility, supplementary shielding in walls, or a mobile barrier may be required [37, 48].

Table 12.9. Attenuation of 511 keV photons under broad beam geometry

	Atomic number Z	Density ρ g cm^{-3}	HVL cm	TVL [a] cm
concrete	–	2.2	11[a] 6.4[c]	24[a] 22[c]
iron	26	7.87	1.6[c]	5.5[c]
tungsten	74	19.30	0.32[c]	1.1[c]
lead	82	11.35	0.5[a] 0.6[b] 0.56[c]	1.6[a] 1.7[b] 2.0[c]

Derived from data in: [a] [47], [b] [7], [c] [51]

Under narrow-beam conditions, with scatter excluded by collimating the source and detector, the attenuation of a beam of radiation through an absorbing medium is described by

$$I = I_0 e^{-\mu x} \quad (7)$$

where I and I_0 are the dose rates at the same location with and without the shield in place and μ is the linear attenuation coefficient of the shielding material. The thickness, x, of the shield can readily be determined from tabulated values of μ or of the total mass attenuation coefficient μ/ρ and the density ρ [47,49]. However, the effect of broad-beam geometry should be considered when shielding extended sources of energy more than a few hundred keV with material of low atomic number, in which Compton scattering is the predominant interaction. Under broad-beam conditions, the equation with the narrow beam μ value can be modified to

$$I = I_0 B e^{-\mu x} \quad (8)$$

where B is a build-up factor to account for transmitted radiation which has been scattered within the barrier. An iterative approach can be used to determine shield thickness for the desired K using tabulated values of B for the number of relaxation lengths (μx) in various materials [47, 49]. Attenuation of 511 keV photons under broad-beam conditions can also be estimated using point kernel modeling software [50], as shown in Fig. 12.4. Values obtained with this method suggest that narrow-beam analysis may underestimate the transmission of 511 keV photons through 10 cm of concrete by about 40% and through 20 cm by about 15% (JC Courtney, 2001, personal communication). In practice, tabulated Half Value Layers (HVL, K = 0.5) and Tenth Value Layers (TVL, K = 0.1) under broad-beam conditions, as shown in Table 12.9, are convenient for simple shielding assessments [7, 50, 51].

Typical thicknesses of lead are 50 mm for bench shields and storage caves for waste and PET camera quality control sources, 30 mm for vial containers located behind a bench shield, and 15 mm for syringe shields. Vial and syringe shields of these dimensions are too heavy to manipulate with safety, hence mechanical supports are desirable when dispensing and injecting PET radiopharmaceuticals. Tungsten may be preferable to lead for small PET source containers. For example, a cylindrical pot which achieves 1% trans-

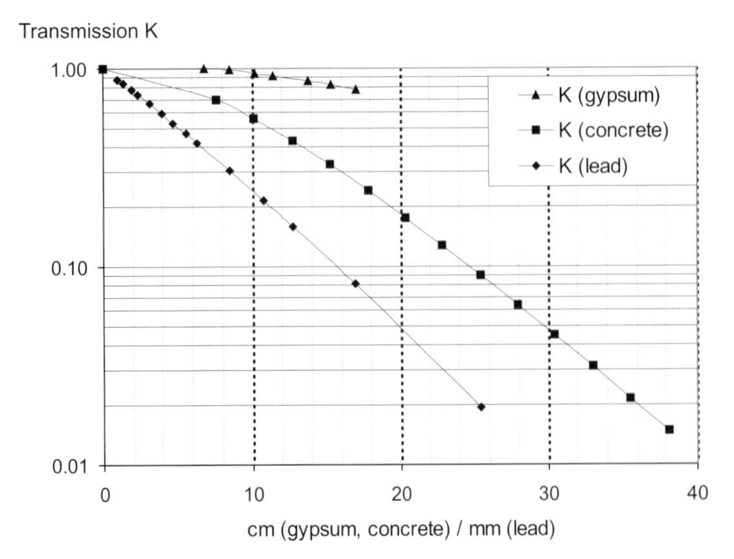

Figure 12.4. Transmission factors for some common materials used for shielding at 511 keV.

mission (thickness 2TVL) around a source cavity 5 cm high and 2 cm in diameter would be approximately 25% heavier and 30% wider if fabricated in lead rather than tungsten. Similarly, cost factors apart, lead has the advantage over concrete for walls because it requires less space and reduces the floor loading by roughly half.

Perspex or plastic liners may be used within lead or tungsten vial and syringe shields to absorb all positrons, although most positrons from ^{18}F would be absorbed in the vial or syringe wall. ^{15}O positrons, if absorbed directly in lead with negligible loss in the walls of the vial or syringe, could generate bremsstrahlung (X-ray) photons up to their maximum energy of 1.7 MeV. However, the amount of energy converted to bremsstrahlung radiation, being a small fraction of the average positron energy incident on the shield as shown in Table 12.4, is trivial compared to the 1.02 MeV of the two annihilation photons. The main practical value of perspex syringe shield inserts is to increase the distance of the fingers from the source and possibly screen the fingers from longer range positrons.

When installing a PET scanner that incorporates a CT unit, no additional shielding in the walls, floor, or ceiling should be necessary beyond what would be required to shield the annihilation photons. The primary X-ray beam is tightly collimated and mostly absorbed in the patient and scanner. The secondary radiation (leakage from the X-ray tube housing and scattered radiation, mainly from the patient) is of short duration and substantially lower energy than 511 keV.

Other Design Considerations

Spills are an uncommon event and usually result from mishaps with intravenous lines or urinary catheters. Strategically located glove dispensers can facilitate the use of disposable gloves when dealing with doses, patients, and waste. The importance of personal protective equipment can be seen from the dose rates for skin contamination in Table 12.4: a droplet from an [^{18}F]-FDG solution with a concentration of 100 MBq/mL could deliver 500 mSv, the annual dose limit for the skin, in six minutes.

A separate waiting area should be provided for patients during the "uptake phase" after the injection of [^{18}F]-FDG. Care needs to be taken at the design stage to avoid situations that prolong the period of close contact by staff with frail patients who require support or assistance, for example, in escorting to and from the scanner, toileting, or dressing. Toilets, change cubicles, and injection rooms should be generously sized with ready access to the scanner room. The PET scanner design and location should allow a patient to get on and off the scanner bed easily, with minimal assistance from staff.

In facilities located near a cyclotron, PET gas tracers may be used. The gas supply and return lines from the radiochemistry laboratory to the scanner room will require shielding. In occupied areas and the PET scanner room itself, a thickness of 20–25 mm of lead around the lines may be sufficient. A mask over the patient's head should effectively contain the administered gas and scavenge the exhaled gas for venting via a stack to the atmosphere. The air of the PET scanner room should be continuously monitored during a gas study. Because of the high background radiation level in the room, an air sampler is required to pass the air through a sensitive detector in a remote low-background area. The scanner room should be kept at negative pressure to the adjacent areas. The room air should not be recirculated, but vented direct to atmosphere.

Given the high costs of construction and possible space limitations when planning or modifying space for a clinical PET facility, the impact of the layout and the projected workload on radiation safety needs to be kept under review. In most facilities, it is unlikely that administered doses of ^{18}F or longer-lived PET nuclides will exceed 800 MBq. However, in those facilities fortunate enough to acquire cameras with fast detector systems, both the administered activity and the number of patients scanned in a day may increase in the future.

References

1. International Commission on Radiological Protection. 1990 Recommendations of the International Commission on Radiological Protection. ICRP Publication 60. Oxford: Pergamon Press, 1991.
2. International Commission on Radiation Units and Measurements. Determination of Dose Equivalents Resulting from External Radiation Sources. ICRU Report No. 39. Bethesda, MD: ICRU, 1985.
3. Dillman LT. Radionuclide decay schemes and nuclear parameters for use in radiation-dose estimation. MIRD Pamphlet No. 4. J Nucl Med 1969;10 (Supplement No.2).
4. Dillman LT. Radionuclide decay schemes and nuclear parameters for use in radiation-dose estimation, Part 2. MIRD Pamphlet No. 6. J Nucl Med 1970;11 (Supplement No.4).
5. International Commission on Radiation Units and Measurements. Radiation quantities and units. ICRU Report No. 33. Bethesda, MD: ICRU, 1980.
6. Groenewald W, Wasserman HJ. Constants for calculating ambient and directional dose equivalents from radionuclide point sources. Health Phys 1990; 58:655–8.

7. Delacroix D, Guerre JP, Leblanc P, Hickman C. Radionuclide and Radiation Protection Handbook 1998. Rad Prot Dosim 1998;76.
8. Loevinger R, Budinger TF, Watson EE. MIRD Primer for Absorbed Dose Calculations. New York: Society of Nuclear Medicine, 1988.
9. Stabin MG. MIRDOSE: Personal computer software for internal dose assessment in nuclear medicine. J Nucl Med 1996; 37:538–46. (The MIRDOSE program is available from the Radiation Internal Dose Information Center, Oak Ridge Institute for Science and Education, PO Box 117, MS51, Oak Ridge, TN 37831-01117, USA.)
10. Snyder WS, Ford MR, Warner GG, Watson SB. Absorbed dose per unit cumulated activity for selected nuclides and organs. MIRD Pamphlet No.11. New York: Society of Nuclear Medicine, 1975.
11. International Commission on Radiological Protection. Radiation Dose to Patients from Radiopharmaceuticals Addendum to ICRP53. ICRP Publication 80. Oxford: Pergamon Press, 1998.
12. International Commission on Radiological Protection. Radiation Dose to Patients from Radiopharmaceuticals. ICRP Publication 53. Oxford: Pergamon Press, 1987.
13. Brihaye C, Depresseux JC, Comar D. Radiation dosimetry for bolus administration of oxygen-15 water. J Nucl Med 1995; 36:651–6.
14. Smith T, Tong C, Lammertsma AA et al. Dosimetry of intravenously administered oxygen-15 labelled water in man: a model based on experimental human data from 21 subjects. Eur J Nucl Med 1994; 21:1126–34.
15. Deloar HM, Watabe H, Nakamura T et al. Internal dose estimation including the nasal cavity and major airway for continuous inhalation of $C^{15}O_2$, $^{15}O_2$ and $C^{15}O$ using the thermoluminescent method. J Nucl Med 1997; 38:1603–13.
16. Dowd MT, Chen C-T, Wendel MJ, Faulhaber PJ, Cooper MD. Radiation dose to the bladder wall from 2-[^{18}F] fluoro-2-deoxy-D-glucose in adult humans. J Nucl Med 1991; 32:707–12.
17. Dhawan V, Belakhlef A, Robeson W, Ishikawa T, Margouleff C, Takikawa S, et al. Bladder wall radiation dose in humans from fluorine-18-FDOPA. J Nucl Med 1996;37(11):1850–2.
18. Hicks RJ, Binns D, Stabin MG. Pattern of uptake and excretion of [^{18}F]-FDG in the lactating breast. J Nucl Med 2001; 42:1238–42.
19. Stabin MG. Health concerns related to radiation exposure of the female nuclear medicine patient. Environ Health Perspect 1997; 105 (suppl 6):1403–9. Also at www.orau.gov/ehsd/ridic.htm.
20. Stabin MG, Breitz HB. Breast milk excretion of radiopharmaceuticals: mechanisms, findings and radiation dosimetry. J Nucl Med 2000; 41:863–73.
21. Stabin MG, Watson EE, Cristy M et al. Mathematical models of the adult female at various stages of pregnancy. ORNL Report No. ORNL/TM-12907. Oak Ridge, TN: 1995.
22. Bohuslavizki KH, Kroger S, Klutmann S, Greiss-Tonshoff M, Clausen M. Pregnancy testing before high-dose radioiodine treatment: a case report. J Nucl Med Technol 1999; 27:220–1.
23. Watson EE. Radiation absorbed dose to the human foetal thyroid. In: Fifth International Radiopharmaceutical Dosimetry Symposium. Oak Ridge, TN: Oak Ridge Associated Universities, pp 179–87, 1992.
24. Almeida P, Bendriem B, de Dreuille O, Peltier A, Perrot C, Brulon V. Dosimetry of transmission measurements in nuclear medicine: a study using anthropomorphic phantoms and thermoluminescent dosimeters. Eur J Nucl Med 1998; 25:1435–41.
25. Huda W, Mergo PJ. How will the introduction of multi-slice CT affect patient doses? In: IAEA International Conference, Malaga 2001 Radiological Protection of Patients in Diagnostic and Interventional Radiology, Nuclear Medicine and Radiotherapy. Vienna: IAEA, 2001.
26. Thomson J, Tingey D. Radiation doses from computed tomography in Australia. Australian Radiation Laboratory Report ARL/TR123. Canberra: Australian Government Department of Health and Aged Care, 1997.
27. Paediatric Task Group of European Association of Nuclear Medicine. A radiopharmaceuticals schedule for imaging in paediatrics. Eur J Nucl Med 1990; 17:127–9.
28. Towson J, Smart R. Radiopharmaceutical activities administered for paediatric nuclear medicine procedures in Australia. Radiation Protection in Australasia 2000; 17:110–20.
29. International Commission on Radiological Protection. Radiological Protection in Biomedical Research. ICRP Publication 62. Oxford: Pergamon Press, 1991.
30. Bloe F, Williams A. Personnel monitoring observations. J Nucl Med Technol 1995; 23:82–6.
31. Chiesa C, De Sanctis V, Crippa F et al. Radiation dose to technicians per nuclear medicine procedure: comparison between technetium-99m, gallium-67 and iodine-131 radiotracers and fluorine-18 fluorodeoxyglucose. Eur J Nucl Med 1997; 24:1380–9.
32. Kearfott KJ, Carey JE, Clemenshaw MN, Faulkner DB. Radiation protection design for a clinical positron emission tomography imaging suite. Health Phys 1992; 63:581–9.
33. Benatar NA, Cronin BF, O'Doherty M. Radiation dose rates from patients undergoing PET: implications for technologists and waiting areas. Eur J Nucl Med 2000; 27:583–9.
34. Bixler A, Springer G, Lovas R. Practical aspects of radiation safety for using fluorine-18. J Nucl Med Technol 1999; 27:14–16.
35. Brown TF, Yasillo NJ. Radiation safety considerations for PET centers. J Nucl Med Technol 1997; 25:98–102.
36. Dell MA. Radiation safety review for 511-keV emitters in nuclear medicine. J Nucl Med Technol 1997; 25:12–17.
37. Bailey DL, Young H, Bloomfield PM et al. ECAT ART – a continuously rotating PET camera: performance characteristics, initial clinical studies, and installation considerations in a nuclear medicine department. Eur J Nucl Med 1997; 24:6–15.
38. Cronin B, Marsden PK, O'Doherty MJ. Are restrictions to behaviour of patients required following fluorine-18 fluorodeoxyglucose positron emission tomographic studies? Eur J Nucl Med 1999; 26:121–8.
39. McCormick VA, Miklos JA. Radiation dose to positron emission tomography technologists during quantitative versus qualitative studies. J Nucl Med 1993; 34:769–72.
40. McElroy NL. Worker dose analysis based on real time dosimetry. Health Phys 1998; 74:608–9.
41. Bird NJ, Barber RW, Turner KB, Meara S. Radiation doses to staff during gamma camera PET [abst]. Nucl Med Commun 1999; 20:471.
42. Towson J, Brackenreg J, Kenny P, Constable C, Silver K, Fulham M. Analysis of external exposure to PET technologists [abstr]. Nucl Med Commun 2000; 21:497.
43. Eberl SE, Anayat AA, Fulton RR, Hooper PK, Fulham MJ. Evaluation of two population-based input functions for quantitative neurological FDG PET studies. Eur J Nucl Med 1997; 24:299–304.
44. Griff M, Berthold T, Buck A. Radiation exposure to sonographers from fluorine-18-FDG PET patients. J Nucl Med Technol 2000; 28:186–7.
45. Council of the European Union. Council Directive 96/29/Euratom on basic safety standards for the protection of the health of workers and the general public. Official J Eur Commun 1996; L159:1–114.
46. Zimmerman BE, Kubicek GJ, Cessna JT, Plascjak PS, Eckelman WC. Radioassays and experimental evaluation of dose calibrator settings for ^{18}F. Applied Radiation & Isotopes 2001; 54:113–22.
47. Cember H. Introduction to Health Physics (3rd ed). New York: McGraw-Hill, 1996.
48. Courtney J, Mendez P, Hidalgo-Salvatierra O, Bujenovic S. Photon shielding for a positron emission tomography suite. Health Phys 2001;81(Supplement):S24–8.
49. Schleien B (ed). The Health Physics and Radiological Health Handbook, 2nd edn. Silver Spring MD: Scinta Inc, 1992.
50. Negin C, Worku G. Microshield version 4.0: A microcomputer code for shielding analysis and dose assessment. Rockville MD: Grove Engineering, Inc.; 1992.
51. Wachsmann F, Drexler G. Graphs and Tables for Use in Radiology. Berlin: Springer-Verlag, 1976.
52. Deloar H, Fujiwara T, Nakamura T, Itoh M, Imai D, Miyake M et al. Estimation of internal absorbed dose of L-[methyl-^{11}C] methionine using whole-body positron emission tomography. Eur J Nucl Med 1998; 25:629–33.
53. Wrobel MC, Carey JE, Sherman PS, Kilbourn MR. Simplifying the dosimetry of carbon-11-labelled radiopharmaceuticals. J Nucl Med 1997;38(4):654–60.

54. Herzog H, Seitz R, Tellmann L et al. Pharmacokinetics and radiation dose of oxygen-15 labelled butanol in rCBF studies in humans. Eur J Nucl Med 1994; 1:138–43.
55. Brown WD, Oakes TR, DeJesus OT, Taylor MD, Roberts AD, Nickles RJ, et al. Fluorine-18-Fluoro-L-DOPA dosimetry with carbidopa pretreatment. J Nucl Med 1998;39(11):1884–91.
56. Graham MM, Peterson LM, Link JM, Evans ML, Rasey JS, Koh W-J, et al. Fluorine-18-fluoromisonidazole radiation dosimetry in imaging studies. J Nucl Med 1997;38(10):1631–6.

13 Whole-Body PET Imaging Methods

Paul D Shreve

PET instrumentation has been available for over 25 years, yet the clinical applications of PET largely languished until the past decade. As the impediments to clinical PET [1] have fallen, the clinical value of the technology has become widely recognized and PET is now emerging as a mainstream diagnostic imaging modality. Over the past five years it has become clear PET using the glucose metabolism tracer [^{18}F]-fluorodeoxyglucose (FDG) will have a major role in the management of patients, particularly oncology patients. This has shaped the recent development of commercial PET tomographs and the evolution of clinical imaging protocols to accommodate rapid imaging of the whole-body (*c.f.*, torso) in a clinical setting.

Historical Development

Coincidence detection of positron radiotracers was accomplished as early as the mid 1960s using opposed detectors. Modern PET imaging systems are based on principles established by Phelps, Ter-Pogossian, and Hoffman at Washington University, St. Louis, in the early 1970s [2, 3]. The first commercial PET scanner, the ECAT II, introduced in the late 1970s, was capable of brain imaging and could accommodate the torso of a narrow patient [4]. Few of these were sold, as positron radiotracers were available only in research institutions with a cyclotron and appropriate radiochemistry facilities. The detection efficiency of these early tomographs was limited by the sodium iodide scintillator which has limited stopping power for the 511 keV annihilation photons. Scintillators with higher density and hence greater stopping power were investigated, including bismuth germanate oxyorthosilicate (BGO), gadolinium orthosilicate (GSO), and barium fluoride (BF), among others. In the late 1970s tomographs using the high density scintillators with attendant improvements in sensitivity and count rate capability were built in commercial and academic settings in limited quantities. These were dedicated brain PET tomographs, however, in part reflecting the exclusive research applications of PET at the time. A commercial PET brain tomograph using a high density scintillator (BGO), with at least a tentative intended clinical market niche, was introduced in 1978 [5].

Early success in the brain PET research encouraged some investigators and clinicians to contemplate analogous studies of tissue blood flow and metabolism in the heart and other organs of the body. Tomographs capable of accommodating the whole-body with more than just a few axial tomographic planes presented substantial cost barriers. The volume of scintillator and number of electronic channels required the development of block detectors, essentially very small Anger cameras, to limit the number of expensive photomultiplyer tubes and associated electronics [6]. In the early 1980s three commercial vendors introduced full ring BGO PET tomographs with gantry diameters capable of accommodating an adult torso. These scanners were capable of continuous data sampling of approximately 10 cm axial extent of the body in 15 or more contiguous transaxial planes. Transmission scanning using germanium-68 sources was incorporated in to the tomograph design to allow attenuation correction of the emission data. Each tomograph typically contained over 500 small photomultiplyer tubes, tungston axial septa, and substantial data processing hardware to reconstruct the thousands of coincidence lines. Due to the small commercial market at the time, such tomographs were among the most expensive medical diagnostic imaging devices.

The commercial availability of whole-body PET tomographs in the 1980s allowed investigators with clinical interests to explore applications of PET to the heart, other major organs, and extra-cranial neoplasms. The theoretical potential of quantitative tracer kinetics, however, still remained strong, particularly among brain researchers. Hence, refinements in tomograph design in the mid 1980s retained the capacity for high count rate dynamic imaging of a limited axial field of view, accommodating the brain or heart, and transmission scanning for attenuation correction of the subject prior to injection of tracer. By the late 1980s, FDG imaging of extra-cranial neoplasms was emerging as a potentially more broadly applicable use of PET in clinical practice than imaging of the brain or heart. Initially, oncology imaging involved specific problem solving such as pulmonary nodules or masses identified on CT scans [7]. It soon became clear that a key advantage of FDG PET was detection of regional and distant metastatic disease not detected on conventional anatomical imaging [8]. While enthusiasm for the potential of quantitative kinetic analysis of imaging data applied to tumor imaging broadly, but most particularly to tumor response and drug evaluation remained [9], kinetic rate constants derived from quantitative dynamic imaging appeared to have no clear advantage over semi-quantitative or simple qualitative scan interpretation of FDG PET studies applied to simple diagnosis [10]. Consequently, for clinical diagnosis, FDG PET scan protocols emphasized static imaging of as much of the body as practical beginning roughly one hour after FDG administration.

With 10 cm axial field of view, as many as 7 bed positions would be required for a whole-body scan, and at 10 minutes emission acquisition per bed position, a whole-body scan required over one hour. Attenuation correction of the emission data required comparable time for the emission scans acquisitions, and since the transmission scanning had to be performed prior to tracer administration, an additional 50 minutes for FDG uptake and distribution prior to emission acquisitions was added for a total scanner time of 3 hours. As such, much of the body imaging in the 1980s was limited to non-attenuation corrected whole torso or attenuation corrected imaging limited to 2 or 3 bed positions, a procedure which generally could be accomplished in between 1 to 2 hours of scanner time. These protocols provided the early evidence of the importance of whole-body PET imaging for cancer staging [11]. The notion that clinical PET would involve primarily static imaging of a large axial extent (*i.e.*, the entire torso) and the need to reduce the expense of a clinical PET tomograph led to the development of innovative tomographs by the late 1980s employing conventional sodium iodide Anger cameras with 2.5 cm thick crystals in a hexagonal array [12]. To compensate for reduced sensitivity, such tomographs employed septa-less or 3D emission acquisition architecture with an extended axial field of view of 25 cm. Sealed point sources for singles transmission scanning were later developed, allowing the transmission scans to be performed even in the presence of tracer in the patient [13]. Whole-body attenuation scans could be performed in about an hour, although image contrast was degraded by the high fraction of scatter and random coincidence events consequent to the fully 3D emission acquisition. Partial ring rotating tomographs based on BGO detectors were also developed in an effort to reduce cost [14]. Again, 3D emission acquisition architecture to compensate for the reduced sensitivity was used along with sealed point sources for transmission scanning.

In the early 1990s, a new generation of full ring BGO commercial tomographs was introduced [15-17]. These tomographs used full ring BGO block detector design with a 15 cm axial field of view and removable axial septa such that they could be operated in both 2D and 3D acquisition modes [16–17]. Germanium-68 rod sources were used which allowed transmission scanning in the presence of tracer in the patient. Such "post injection transmission scans" made whole-body attenuation corrected imaging possible in roughly one hour on the BGO ring scanners (Fig. 13.1). Throughout the 1990s whole-body FDG PET imaging performed on the ring BGO and sodium iodide scanners provided the clinical experience and scientific evidence supporting government and private payment for clinical PET exams. Growing whole-body clinical applications increased the need to reduce imaging time while improving image quality. Much of the progress in improving image quality and reducing overall scan time involved improvements in image reconstruction algorithms (Fig. 13.2). Conventional filtered back projection algorithms were supplanted with statistical reconstruction algorithms using segmentation methods on the transmission scan data to reduce overall scan time on both emission and transmission acquisitions and yet still improve image quality on the whole-body scans [18, 19].

The need for attenuation correction on whole-body imaging remained somewhat controversial in the 1990s [20]. Transmission scan time added to overall imaging time, and noise from the transmission scan propagated into the final attenuation corrected emission scan, reducing lesion contrast. Further, many PET centers were performing whole-body non-attenuation corrected scans routinely with diagnostic results comparable to

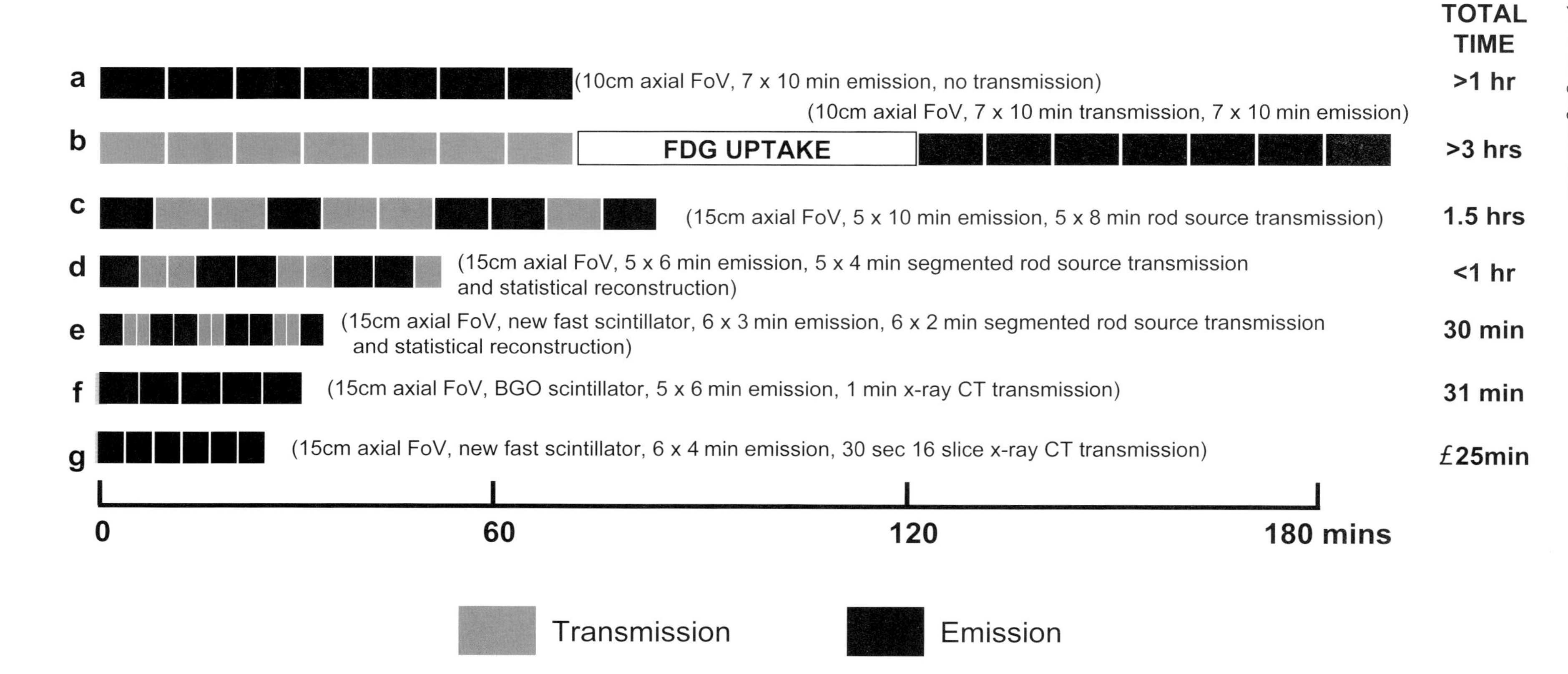

Figure 13.1. Whole body imaging protocols. Comparison of image acquisition protocols and scanner time required to complete a whole body acquisition. Ring tomographs in the 1980s were limited by a 10 centimetre axial field of view and transmission scanning they could only be used prior to tracer administration. A whole-body scan of roughly 70 cm axial length thus required more than one hour without attenuation correction (**a**), and an impractical three hours with attenuation correction (**b**). By the early 1990s, tomographs with longer axial fields of view (15 cm BGO and 25 centimetre NaI(Tl)) and the capability to perform transmission acquisition after tracer injection became available. These could complete a 70 cm examination with attenuation correction in about 1.5 hours (**c**). Refinements in image reconstruction permitted reduction in emission and transmission acquisition times, allowing for whole body image acquisition time to be reduced to less than one hour by the late 1990s (**d**). New detector technology permitting 3D image acquisition following injection of 400–550 mBq of tracer now allows whole body emission and transmission image acquisition of 70 centimetre axial length in about 30 minutes (**e**). Likewise combined PET/CT tomographs reduce whole body imaging time by markedly reducing transmission scan time, allowing whole body attenuation corrected studies in 30 minutes or less (**f** and **g**). Note that the actual imaging time is slightly greater than acquisition time alone due to additional time needed for scanner bed and transmission source movement. There is greater field of view overlapping in 3D acquisition mode than 2D mode, hence the extra bed position in (**e**).

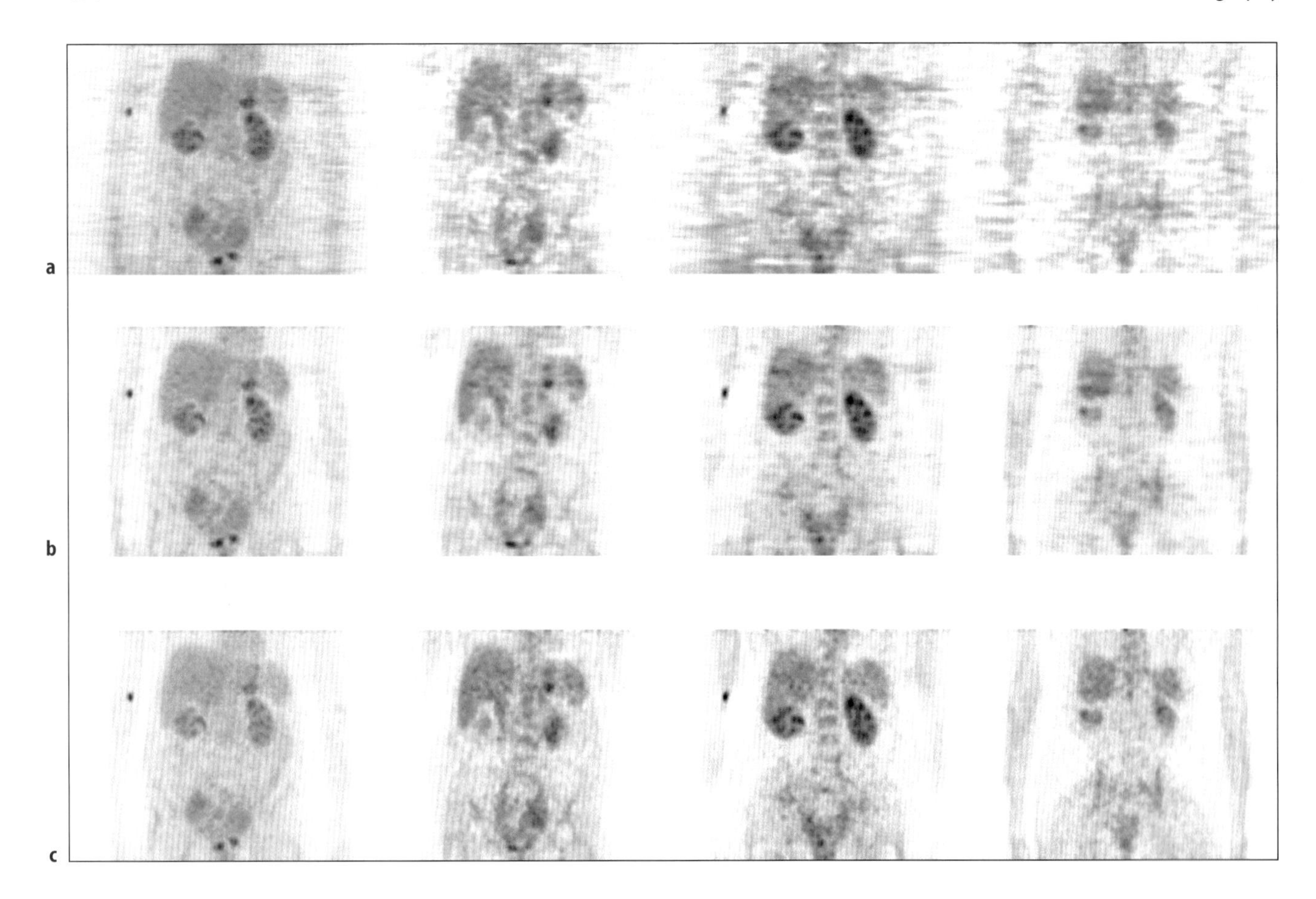

Figure 13.2. Comparison of image reconstruction methods. Substantial improvements in whole-body FDG PET image quality have occurred due to refinements in image reconstruction algorithms. Shown above are anterior projection images and coronal image sections of the abdomen and pelvis, with acquisition beginning about one hour after FDG administration. Two-dimensional acquisition was used (six minutes per bed position) and transmission scanning was performed with rotating Ge-68 rod sources (four minutes per bed position). The same emission and transmission data are shown after reconstruction by filtered back-projection (**a**), by statistical or iterative reconstruction using ordered subset expectation maximization (OSEM) (**b**), and by attenuation-weighted statistical reconstruction (**c**). [Courtesy of Technical University of Munich and CPS Innovations Inc., Knoxville, Tennessee.]

centers using attenuation corrected protocols. Never the less, the need for anatomical correlation and the diminishing time and noise penalty for attenuation correction have resulted in most centers now performing attenuation correction routinely on whole-body protocols. Further, refinements in attenuation weighted statistical reconstruction algorithms which allow for further improvements in image quality (Fig. 13.2) require the anatomical map of a transmission scan.

Early commercial PET tomograph design relied on axial septa to reduce random and scatter coincidences and constrain image reconstruction to a series of 2D tasks which could be solved in a reasonable time using existing computational hardware and software. By the late 1980s and throughout the 1990s efforts to move towards 3D imaging in the body accelerated as image reconstruction computational hardware and software advanced [21, 22]. By expanding the number of accepted out of plane coincidence events, tomograph sensitivity increases dramatically. Removing the axial septa altogether also reduces cost and permits a wider gantry aperture, improving patient comfort. Unfortunately random and scatter coincidence events increase even more so than the true coincident events, and degrade image contrast unless successfully corrected. The full potential in increased sensitivity is also not fully realized as detectors with slow light decay scintillators such as BGO and sodium iodide cannot accommodate the high photon flux associated with patients given FDG in excess of approximately 350 MBq. Due to the relative limited scatter medium and out of field source of random coincidences, 3D acquisitions of the brain are readily accomplished and have allowed for reduced imaging time and improved image quality. 3D acquisitions in the body however have yielded poorer image contrast than 2D acquisitions, particu-

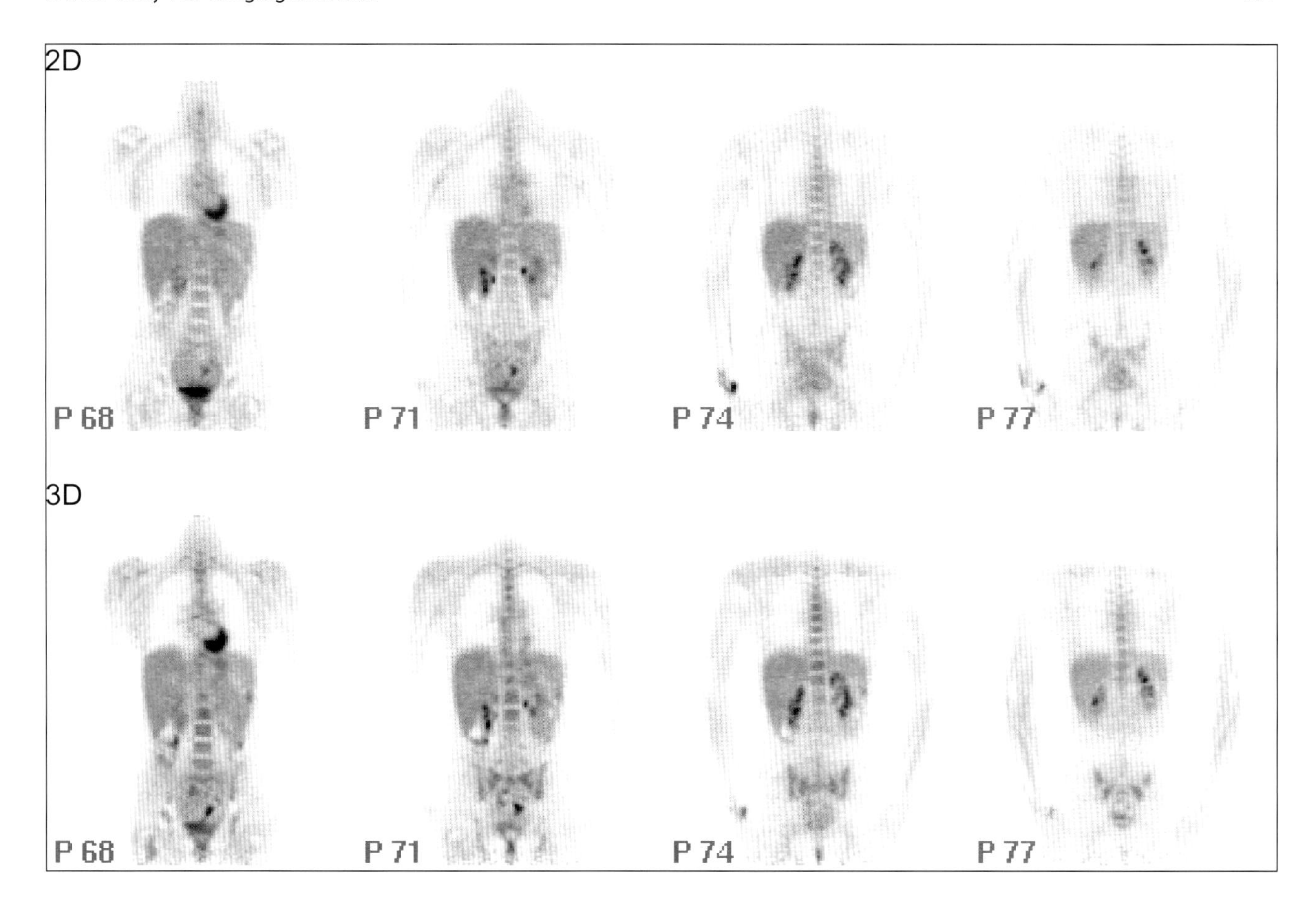

Figure 13.3. Whole-body 2D vs 3D image acquisition. Emission acquisitions can be performed in a 2D mode in which axial septa are employed to permit only a narrow angle of photon acceptance, or a 3D mode in which axial septa are absent. The 3D mode allows higher overall sensitivity but requires image reconstruction algorithms that are capable of correcting for the much higher contributions of random coincidences and scattered photons. Image reconstruction algorithms now allow reconstruction of images acquired in 3D mode which are comparable to 2D images in quality, but require a lower injected dose or shorter acquisition time due to the higher sensitivity. Shown above is a comparison of FDG PET whole-body coronal images of the same patient obtained on a BGO tomograph, first in 2D mode and then in 3D mode . Total image acquisition times were comparable, since the 3D image was obtained after a longer tracer decay time. [Courtesy of Kettering Memorial Hospital, Kettering, Ohio, and CPS Innovations Inc., Knoxville, Tennessee.]

larly evident in larger patients. Refinements in randoms and scatter correction have shown improvements in whole-body 3D acquisitions such that they are now becoming comparable to high quality 2D studies performed on full ring BGO tomographs (Fig. 13.3).

Current Trends in Whole-Body Tomographs

Body oncology applications of clinical PET are driving scanner technology to further improve image quality for small lesion detection and reduce scanning time for faster patient throughput in the clinic. Such improvements will require higher sensitivity for shorter emission acquisition times and shorter transmission scan time. High sensitivity can be achieved in 3D mode if a tomograph can accommodate the markedly increased detector event rates for higher true coincident count rates while the scatter and random coincident contributions to the final reconstructed images are minimized. Scintillators with faster light decay times such as lutetium oxyorthosilicate (LSO) or gadolinium oxyorthosilicate (GSO) and improved detector system energy resolution will allow for further increments in performance. Faster light decay times permit much higher detector count rate capability permitting full 3D body emission image acquisitions at patient tracer doses limited by tracer dosimetry rather than detector count rate capability (Fig. 13.4). For example, a full ring LSO tomograph based on conventional BGO design, has equivalent sensitivity to the BGO tomograph, but much higher count rate capability, allowing 3D body

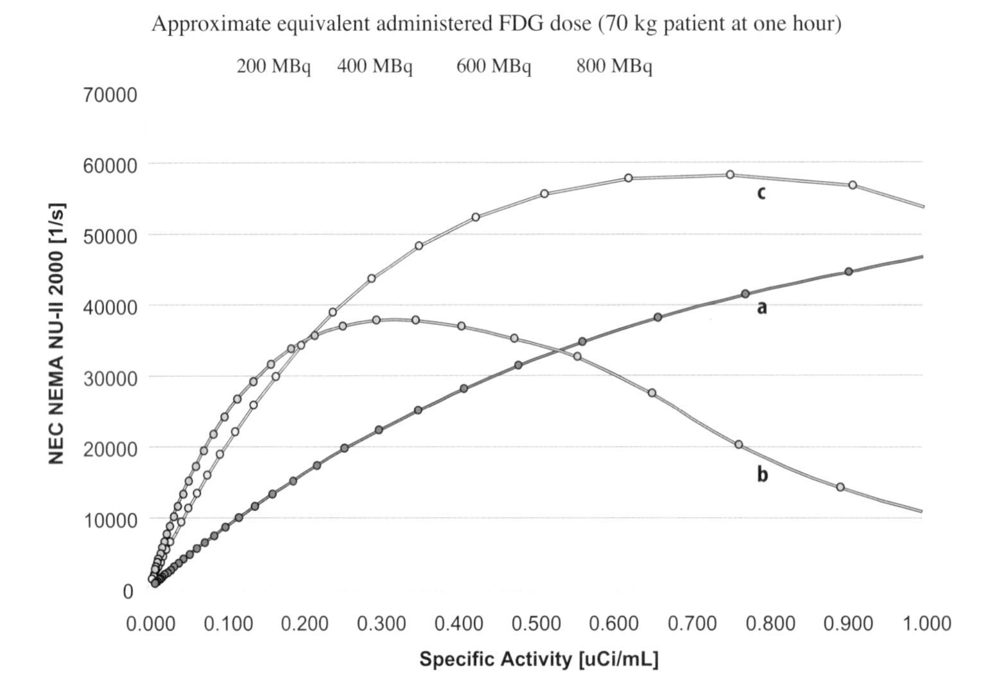

Figure 13.4. Count rate performance of BGO vs LSO detector materials. A useful method of evaluating and comparing PET tomograph performance is the noise equivalent count rate curve. A standard cylindrical phantom containing tracer in water is imaged, and the true, random and scattered coincidence counts as measured by the tomograph are recorded. Using a standard formula, the noise equivalent count rate (NEC) is computed. The test is performed over a range of tracer activity in the phantom, and a curve is generated showing the increasing NEC with increasing tracer activity. The NEC is generally limited by the tomograph count-rate capability and the exponential increase in random coincidence rate with increasing dose. For these reasons, the NEC curve reaches a maximum and then declines. The NEC is an approximate measure of useful coincidence events. Shown above are NEC curves for (**a**) 30 mm-deep BGO detectors in 2D mode and (**b**) 3D mode, compared to (**c**) 25 mm-deep LSO detectors in 3D mode. The BGO detectors in 2D mode do not reach a maximum at these levels of activity, while the same detectors in 3D mode quickly reach a maximum NEC and then decline due to increasing random coincidences and count rate limitations. With BGO, the 3D mode does not allow higher count rates and shorter imaging times but does permit use of lower tracer doses. Because of faster light decay and a narrower coincidence time window, the LSO detectors in 3D mode permit higher count rates with half the relative random coincidence contribution of the "slower" BGO. Consequently, a higher maximum NEC is achieved with the LSO detectors in 3D mode, even though the BGO detectors in 3D mode have slightly higher sensitivity. The approximate range of tracer concentration in the body of a 70 kg adult, one hour after administration of 200 to 800 MBq FDG is noted for comparison. [Courtesy of CPS Innovations, Knoxville, Tennessee.] (Reproduced from Valk PE, Bailey DL, Townsend DW, Maisey MN. Positron Emission Tomography: Basic Science and Clinical Practice. Springer-Verlag London LTD, 2003, p. 486.)

image acquisitions with patient does of FDG up to 700 MBq. Random coincidence contributions can be reduced by the very narrow coincidence timing window (4–6 nanoseconds) and scatter coincidences are at least partially corrected using anatomy based algorithms. Contribution of scatter coincidences to image degradation can be reduced by improvements in effective energy resolution both due to properties of scintillation crystals and sampling and analysis of scintillator light output [23], in addition to attenuation weighted scatter correction algorithms [24]. The high count rate capability of fast light decay scintillators accommodates higher activity transmission scan sources allowing for emission image acquisitions as short as 1 minute per bed position. Thus with emission scan acquisitions shortened to 3 , or even 2 minutes per bed position and high count rate transmission scans shortened to 1 minute per bed position, high image quality whole-body exams can be performed in under 20 minutes using a 15 cm axial field of view tomograph (Fig. 13.5). The shortened acquisition times in turn further improve image quality by minimizing patient movement during scan acquisition.

The need for anatomical correlation in both interpretation and therapy planning is also driving the merging of PET and CT tomograph technology (25, 26). The X-ray CT attenuation map, which can be scaled to a 511 keV transmission map [26] provides a very rapid transmission scan with minimal noise. Effectively the transmission scan contribution to overall scanning time is reduced to less than one minute for a whole-body exam. A conventional full ring BGO detector of 15 cm axial field of view in the PET/CT configuration can acquire a whole-body diagnostic CT and attenuation corrected PET in about 30 minutes (Fig. 13.6). With fast crystal detector PET tomographs operated in full 3D mode in a PET/CT configuration, whole torso have been reported in under

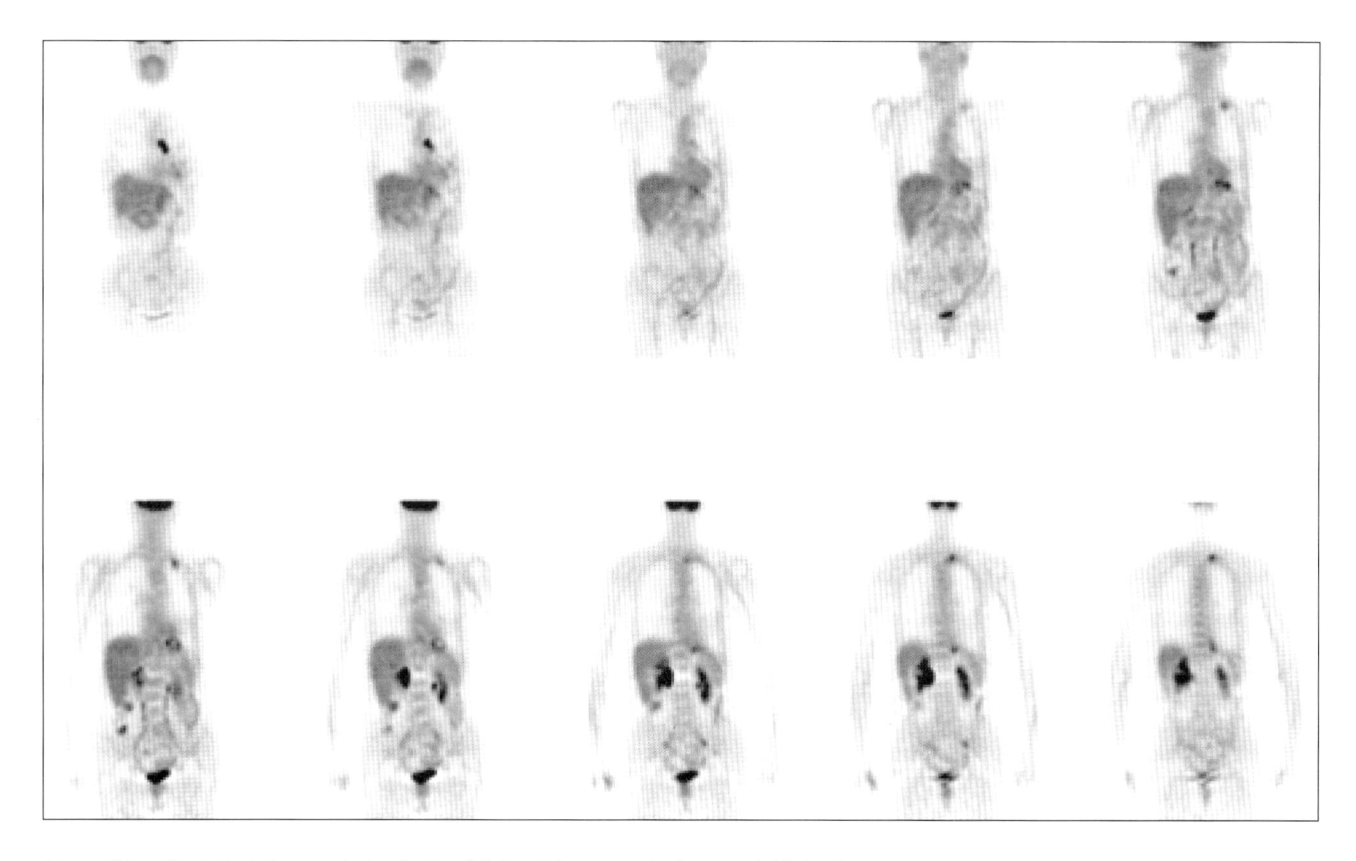

Figure 13.5. Whole-body images obtained with a full-ring LSO tomograph of 15 cm axial field-of-view, operating in 3D mode. Attenuation-corrected whole-body images were obtained in less than 25 minutes, using a dose of 500 MBq FDG. [Courtesy of Northern California PET Imaging Center, Sacramento, California.] (Reproduced from Valk PE, Bailey DL, Townsend DW, Maisey MN. Positron Emission Tomography: Basic Science and Clinical Practice. Springer-Verlag London LTD, 2003, p. 487.)

10 minutes total scan acquisition time [27]. The advantages of accurately registered and aligned PET metabolic and CT anatomical images for both diagnosis and radiation therapy planning are already becoming clear [28–31]. As predicted [26], the grafting of the PET metabolic images on the familiar CT anatomical images is speeding acceptance of PET as a mainstream imaging modality among radiologists, oncologists and surgeons. Already the majority of PET scanners sold are in the configuration of a PET/CT scanner, and it appears the standard modality for body oncology imaging will be the combined PET/CT. Perhaps the most important consequence of this evolution of body PET imaging to PET/CT is the merger of metabolic and anatomical diagnosis into one imaging procedure and one overall medical imaging interpretation.

Whole-Body Imaging in Oncology

Overview

Presently whole-body FDG imaging protocols for oncology applications takes full advantage of the recent developments in tomographs and image reconstruction. For most extra-cranial malignancies, the goal is to image the entire torso, from skull base to pelvis within one hour or less of scanner time, and in fact with recent developments in PET tomograph detector technology and combined PET/CT tomographs, imaging of the entire torso in under 30 minutes is becoming routine. The brain is usually not included in the axial field-of-view for imaging extra-cranial malignancies as

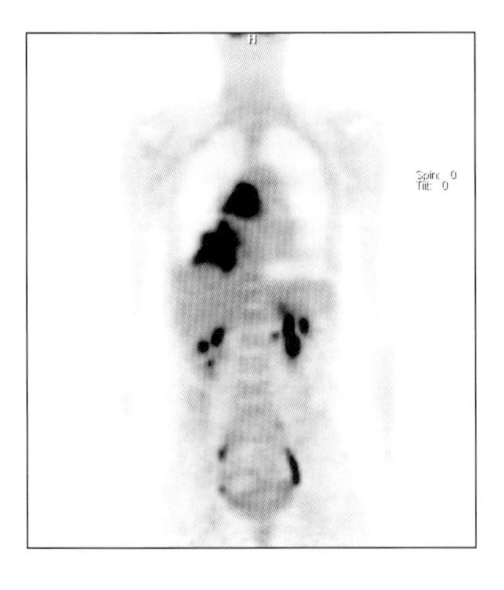

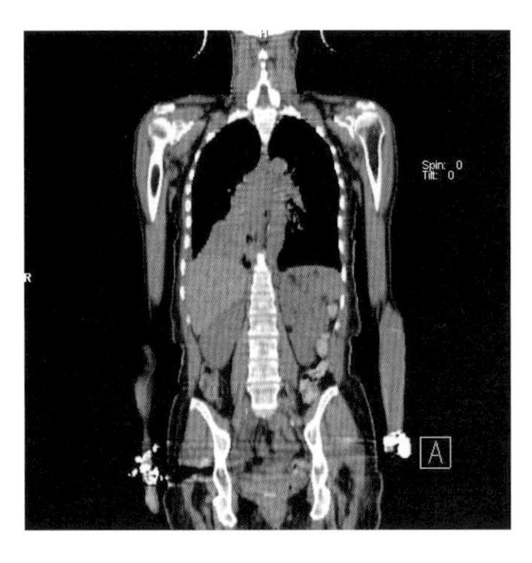

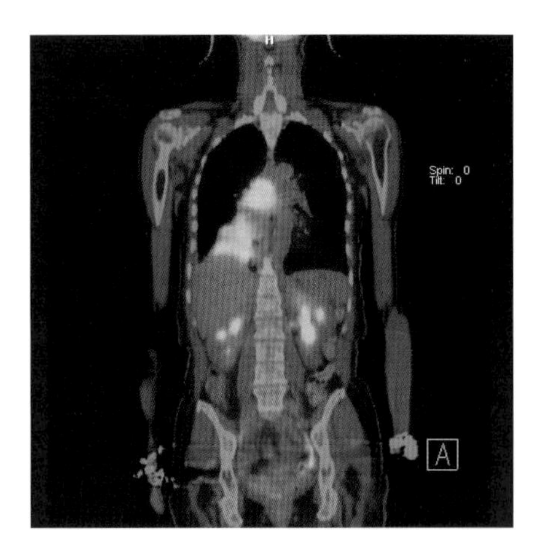

a FDG PET **b** CT **c** Fused : 58% PET/ 42% CT

Figure 13.6. Whole-body images obtained with a combined PET/CT tomograph. Whole-body images obtained with a full-ring BGO tomograph with 15 cm axial field of view, operating in 3D mode and using CT image data for attenuation correction. (**a**) Attenuation-corrected coronal whole-body PET image, and (**b**) coronal reconstruction of diagnostic CT data, obtained sequentially in 30 minutes total, producing highly registered metabolic and anatomic images. (**c**) fused PET and CT images. [Courtesy of Memorial Sloan-Kettering Cancer Center.]

FDG uptake in the brain is of a level found in most malignant neoplasms, hence detection of brain metastases is still best accomplished by contrast enhanced CT or MR. One notable exception is melanoma, a neoplasm with such intense FDG uptake and propensity for widespread and unpredictable metastatic sites, that brain metastases can be detected on FDG PET, and therefore many centers include the entire head in whole-body imaging of melanoma patients. For lung cancer and oesophageal patients, the top of the axial field-of-view should include the base of the neck, for head and neck cancer and lymphoma patients up to the skull base. Patients with cancers originating in the gastrointestinal and genito-urinary tract require body positioning to insure the entire pelvis, to below the level of the pubic symphysis, in addition to the abdomen and chest is included. In lymphoma patients the caudal extent of axial field-of-view typically includes the upper thighs to insure inguinal lymph nodes are included. Again, with melanoma patients, some or all of the lower extremities will be included at some centers, often using shorter imaging acquisition times for the lower extremities. Hence for most patients 70–80 cm axial extent imaged will comprise a whole-body study, for lymphoma and melanoma patients 90–120 cm.

Patient Preparation

Patient preparation for whole-body FDG PET examinations is an essential part of the procedure both to optimize image quality and to minimize physiological variants and artefacts [26]. Patients should be fasted a minimum of 4 hours to insure serum glucose and endogenous serum insulin levels are low at the time of FDG administration. Typically patients are fasted overnight with no breakfast for morning appointments, and can have a light breakfast, but no subsequent lunch or snakes for afternoon appointments. Glucose competes with FDG for cellular uptake, and there is some evidence elevated serum glucose will lower observed FDG uptake in malignant neoplasms [33]. Equally significant, elevated serum insulin promotes FDG uptake in the liver and muscle. Hence a recent carbohydrate containing meal (even a snack) or administration of exogenous insulin in attempt to lower blood glucose levels prior to FDG administration can yield extensive muscle uptake (Fig. 13.7). Such muscle uptake will not preclude evaluation of centrally located abnormalities such as lung nodules or mediastinal lymph nodes, but can potentially reduce conspicuity of osseous and peripheral lymph node basin involvement and reduce available circulating FDG for tumor uptake. Myocardial FDG uptake will be absent given a sufficiently long fast (18–24 hours) due to the shift to fatty acids as an energy source. With shorter fasts (which patients can tolerate such as overnight) myocardial FDG uptake will vary from uniform intense, to irregular, to absent in largely unpredictable patterns among patients. Hence the goal of fasting is not ordinarily to eliminate the myocardial FDG uptake.

In general, a serum glucose level under 150 mg/dL at the time of FDG administration is preferred, with less than 200 mg/dL acceptable. With serum glucose levels above 200 mg/dL, noticeable degradation in image quality due to reduced tissue uptake of FDG and sustained blood pool tracer activity can occur. It is relatively easy to measure serum glucose prior to FDG administration, and this is routine in many centers. Use of exogenous insulin to reduce serum glucose immediately prior to FDG administration is not generally indicated, as this will result in accelerated FDG uptake in muscle and the liver. It is much preferred to manage known diabetic patients such that at the time of FDG administration, serum glucose levels are under roughly 150 mg/dL. This must be arranged in consultation with the patient and the physician treating their diabetes, as the strategy used will depend on the patient's treatment regimen and history of serum glucose control. For example, patients who are non-insulin requiring may present with an acceptable serum glucose levels with an overnight fast, while insulin requiring patients may need a fraction of their usual morning dose of short acting insulin in addition to the overnight fast. For patients with poor serum glucose control, the goal is not entirely normal serum glucose, simply less than roughly 150 mg/dl, as there is a risk of hypoglycemia. If a patient is found to be hypoglycemic at the time of FDG administration, and is not symptomatic, it should be noted that approximately 30 minutes after FDG administration, much of the FDG uptake has occurred, and serum glucose can be subsequently by normalized without compromising the examination. Simple ingestion of a sweet juice drink is often sufficient to insure adequate glucose levels for the duration of the examination. Patients without a diagnosis of diabetes mellitus will occasionally present for an FDG PET scan with abnormal elevated fasting serum glucose, but this rarely exceeds 150 mg/dL, much less 200 mg/dL. When patients do present with and abnormally diffuse increased muscle and liver uptake, it may well reflect a recent snack (Fig. 13.7); hence it is important to emphasize the meaning of fasting for the exam at the time the patient is scheduled.

Hydration prior to FDG administration will, as with any tracer cleared by urinary excretion, facilitate tracer clearance from blood pool and the urinary tract. Patients should be encouraged to drink plenty of water, but only water (no sugar containing beverage), prior to FDG administration. After FDG administration

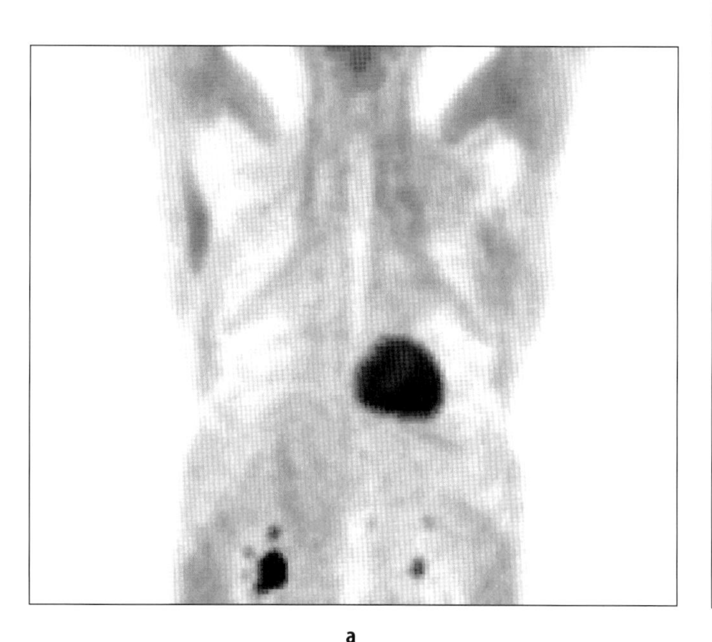

a

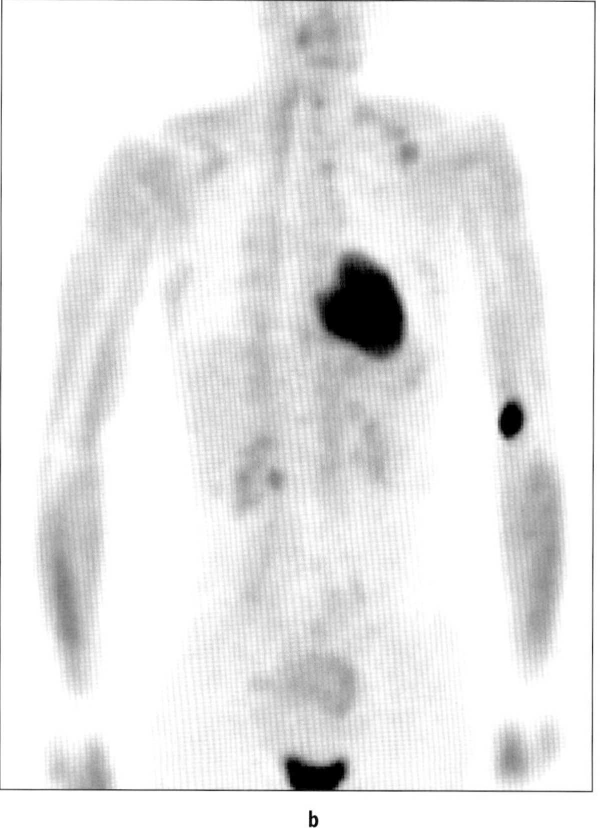

b

Figure 13.7. Effect of exogenous and endogenous insulin. Whole-body anterior projection images of (**a**) a patient given 10 units of regular insulin intravenously prior to FDG administration, attempting to normalize a serum glucose level of 180 mg/dL, and (**b**) a patient who fasted overnight but ate an apple and half a granola bar prior to FDG administration. In both cases there is extensive skeletal muscle uptake, uniform and symmetrical, due to the action of insulin.

drinking, chewing or even talking must be eschewed for at least approximately 30 minutes to avoid muscle FDG uptake. The presence of urinary tracer in the upper urinary tract and bladder can be confounding [33]. Pooling of urinary tracer in the renal calyces can mimic a renal or adrenal mass, while urinary tracer in an extra-renal pelvis or isolated down the course of the ureters can mimic FDG avid retro-peritoneal lymph nodes. Such mimics involving urinary tracer activity are less problematic with properly registered and aligned PET and CT images such as generated on combined PET/CT scanners. Intense urinary tracer activity in the bladder can result in reconstruction artefacts limiting evaluation of adjacent structures, including uterus, adnexa, prostate, rectum and obturator nodal groups.

It is possible to minimize or largely eliminate urinary tracer activity by hydration and use of intravenous diuretics [34], and such maneuvers have been advocated, particularly in patients with GI, GU malignancies and abdominal lymphoma. The use of diuretics typically mandates bladder catheterization when scan table times approach an hour, as the patient typically will need to void during the scan without urinary bladder drainage. With scan table times well under 30 minutes, this is less problematic. When bladder catheterization is used, best results are obtained using a multi-lumen catheter such that the urinary bladder undergoes continuous lavage [35]. This insures a relatively full bladder with only dilute urinary tracer. Some centers report with aggressive hydration and use of diuretics, very little urinary tracer is present in the bladder at the time of imaging. In any case, the use of an indwelling urinary bladder catheter adds an invasive component to the exam, and requires skilled personnel time, and this must be weighted against the advantages of reducing or eliminating bladder urinary tracer activity.

Bowel FDG physiological tracer activity is commonly observed on whole-body FDG PET scans and can be confounding due to the inconsistent and unpredictable patterns [31, 35]. While the aetiology of this uptake is not entirely understood, maneuvers have been advocated to reduce or eliminate bowel tracer activity, including isosmotic bowel preparations the evening prior to the exam [34], or bowel smooth muscle relaxants prior to FDG administration [36]. These maneuvers must again be viewed in the context of patient compliance and increasing the complexity of the examination. Isosmotic bowel cleansing is demanding on the patient to say the least, and further, bowel FDG activity is usually readily identified on high quality PET images. In certain patients, however, such as ovarian carcinoma, abdominal lymphoma and colon cancer, where mesenteric or bowel serosal implants are possible, such preparation to eliminate or minimize bowel physiological tracer activity may well be indicated.

Muscle FDG uptake is also common physiological variant which can be confounding, particularly in the neck [31, 35]. As noted above, generalized muscle uptake due to insulin action can be eliminated by sufficient fasting and avoidance of exogenous insulin prior to FDG administration. Skeletal muscle contraction during the uptake phase of FDG (principally 30 minutes following intravenous administration) can result in fairly intense FDG accumulation, hence patients should be seated or recumbent following FDG administration, and not engaged in any physical activity for 20–30 minutes. Talking, chewing or swallowing should be avoided during this period, as this can result in FDG uptake in the tongue, muscles of mastication, and larynx [37]. Intense muscle activity prior to administration of FDG, even hours prior, can result in elevated muscle uptake as well, and hence patients are generally advised to avoid strenuous physical activity prior to undergoing an FDG PET scan. FDG uptake in major muscles of the neck including the sterno-cleido-mastoid and scalene muscles can be seen also be seen, and particularly with patients undergoing studies for malignancies involving the neck, a recumbent position with head support during the FDG uptake phase is commonly advocated. More confounding in the neck and supraclavicular region is brown adipose tissue (BAT) FDG uptake, originally thought to reflect physiological muscle uptake. Brown adipose fat is involved in thermoregulation, and on adrenergic stimulation, can accumulate FDG avidly [38]. Recognition of this source of confounding physiological uptake occurred with the advent of combined PET/CT scanners [39, 40]. Most typically seen in the lower neck, medial shoulders and paraspinal fat between muscle groups, brown adipose tissue uptake can be seen in mediastinal fat and subdiaphragmatic and perirenal fat [41]. Typically seen in younger and slimmer patients, particularly when cold, brown adipose tissue FDG uptake is often quite focal, and can be very difficult to distinguish from lymph node metastases without accurately registered and aligned images such as provided by combined PET/CT scanners. Anxiolytics have been shown to be effective in eliminating neck muscle and brown adipose fat uptake [42], presumably due to reduced muscle tension and diminished adrenergic stimulative output, respectively. Some centers keep patients warm with blankets or elevated ambient temperature as a strategy to reduce brown adipose fat uptake.

Anxiolytics, including intravenous short acting benzodiazepines such as midazolam, routinely in adult patients to improve patient comfort and compliance are used at some centers. With the substantially shorter whole-body imaging times of contemporary tomographs, however, the importance of such light conscious sedation has diminished. Certainly for patients with back pain or other impediments to maintaining motionless supine position during the scan acquisition, sedation and pain management entirely analogous to that used for MRI imaging is indicated. In general, there is far less difficulty with patient claustrophobia in whole-body PET acquisitions relative to closed magnet MRI acquisitions. Oral alprazolam, 05–1.5 mg, depending on body weight, an hour before scan acquisition, is very effective in managing patients with claustrophobia.

Image Acquisition

As with any clinical medical imaging procedure, the goal is to obtain the highest quality image in a limited image acquisition time in order to minimize patient movement and maximize scanner throughput. As noted above, in the past whole-body FDG PET images had been obtained without attenuation correction, less the total scan acquisition time exceed two hours. Contemporary PET scanners, and particularly combined PET/CT tomographs, allow whole-body attenuation corrected images routinely in as little as 30 minutes. As patient movement between the transmission and emission image acquisitions will result in artefacts on the final attenuation corrected emission images, the emission and transmission image acquisitions should be temporally as close as possible when sealed source transmission scans are used (Fig. 13.1). The patient's arms are a common source of movement artefact. Most patients can comfortably keep their arms up out of the torso field of view when properly supported for 30 minutes or less. For whole-body scan acquisitions on fast crystal (LSO or GSO) tomographs and combined PET/CT tomographs, where such short whole-body acquisition times are feasible, an arms up supine configuration is optimal for chest, abdomen and pelvis examinations. Head and neck image acquisitions are, as with CT, optimally performed with arms down.

Image acquisition times and FDG dose are related, but not in an entirely inverse fashion of single photon radiotracer medical imaging. Regarding sealed source transmission scans, with image segmentation, acquisition time per bed position can be shorted to 3 or even 2 minutes in conventional BGO whole ring tomographs, while sealed point sources using with fully 3D tomographs and higher activity rod sources used in fast crystal 2D tomographs allow for reduction in acquisition time to 2 minutes or less per bed position. Such short transmission acquisition times can result in segmentation errors, particularly associated with the diaphragm. CT based attenuation correction allows for the entire body transmission scan without noise or segmentation errors [43] in less than 30 seconds with multi-detector helical CT. Since the PET emission acquisitions are acquired after the single CT scan of the entire torso, the transmission image acquisition of the last portion of the body to undergo emission image acquisition may be separated by up to 30 minutes, hence absence of patient movement is essential. Also, shallow relaxed breathing will help minimize image registration errors when X-ray CT is used for the transmission image sinogram because even when the CT acquisition is performed during free breathing, the temporal relation about the diaphragm over seconds captured by CT will be different than the PET emission acquisition captured over a few to several minutes. This appears to be less of an issue with increasing number of detector channels with helical CT such as up to 16, as the increasing speed of CT scan acquisition reduces diaphragm motion artefact [44]. Alternatively, abdominal binders can be employed to constrain diaphragmatic excursion to minimize such errors. Mid exhalatory breath hold provides improved quality CT images and can closely match average diaphragmatic position on the PET images [45]. Patients short of breath due to pulmonary disease may benefit from supplemental oxygen to reduce lung tidal volume, and use of abdominal binders can aid in insuring shallow breathing.

Due to the nature of contaminating scatter and random coincidence events detected, the relationship between FDG dose and useable image counting statistics is neither direct nor linear, and depends on the geometry of the tomograph, type of detector crystal, size of the patient, and the reconstruction algorithm used. The general rule of single photon medical radiotracer imaging, that a larger radiopharmaceutical dose results in more useable counts, does not uniformly apply in whole-body PET imaging. Because random coincidences increase exponentially (to the second power) with tracer activity while true coincidences increase linearly with the tracer activity, useable count rates are eventually limited by randoms coincidence contribution. In general, ring tomographs in 2D mode with thick axial septa will increase useable true coincidences with increasing patient tracer activity (administered dose) out to the upper range of dosimetry limited [46] administered FDG (about 700 MBq).

Hence increasing administered dose can be used to reduce emission image acquisition times, from, for example, 8 minutes to 4 minutes per bed position. Tomographs with greater axial cross-plain acceptance and finer septa, and especially tomographs operating in fully 3D mode (no septa), may reach randoms limiting count rate contributions with administered doses as low as 300 MBq or less. Hence, increasing the dose of administered FDG above such levels does not allow for decreased scan acquisition times, indeed image degradation can be observed with increasing dose using the same scan acquisition times, as predicted by NEC curves [Fig. 13.4]. Likewise, larger patients provide an expanded source of photon emissions to the detector ring, resulting in higher randoms, especially in tomographs operating in 3D mode. Hence in a large patient a lower FDG dose with an extended image acquisition time can improve final image quality in tomographs operated in fully 3D mode. The 3D mode does result in substantial gains in sensitivity, however, such that useable true coincidence count rates are acquired with lower administered FDG in the range of 200–300 MBq. The fast crystal tomographs (LSO and GSO) allow for considerably narrower coincidence acceptance windows (4–6 nsec vs 8–12 nsec) than BGO or NaI(Tl) based tomographs, reducing relative random contributions by roughly half. The rapid light decay of such fast crystals also permits very high detector count rates, permitting 3D mode acquisitions using patient FDG doses well above 500 MBq (Fig. 13.4). Such tomographs in 3D mode yield useable increases in true coincidence count rates with administered doses of FDG in excess of 500 MBq, allowing emission image acquisitions to be shortened to 2–3 minutes per bed position (Fig. 13.5).

Photon scatter contributes to background noise in the reconstructed images and degrades image contrast in PET, just as with single photon radionuclide imaging. Larger patients will thus present greater image degradation due to a larger scatter component emanating from the patient. Again, 2D tomographs with thick septa will be less degraded by the increased scatter contribution than tomographs with more limited, or no, axial collimation, as with a fully 3D operation. Advances in scatter correction algorithms have greatly improved image quality, particularly in 3D mode (Figs. 13.2 and 13.3), by correcting for scatter coincidence contributions. In addition, refinements in energy resolution (raising the lower threshold of energy window from 350 keV to as high as 425 keV) has resulted in further improvements in reduce scatter contribution with attendant improvements in whole-body image quality in a 3D emission acquisition mode, even in large patients (Fig. 13.6).

A clear consensus of the optimal time for whole-body image acquisition following FDG administration has not yet emerged. Typically image acquisition for body imaging has commenced 40–60 minutes following FDG administration. This delay is based in part on the time required for a majority of blood pool activity to clear and the majority of the tumor accumulation of tracer to occur, and in part on the historical need to minimize the time between pre-injection transmission scans and the commencement of emission image acquisition. With whole-body imaging times up to and exceeding an hour, emission acquisition of the last portion of the body imaged can occur over two hours post tracer administration. There is continued accumulation of FDG in malignant neoplasms and other FDG avid tissues such as bone marrow beyond one hour, with continued clearance of blood pool [47]. Hence, a longer delay in the commencement of image acquisition has been advocated to enhance tumor conspicuity and allow for more complete clearance of upper urinary tract tracer activity. Upper urinary tract tracer activity in the absence of aggressive hydration and use of diuretics is not assured even at 2 hour post FDG administration [48], and an increase in tumor to background is offset by the physical decay of tracer, lowering counting statistics. For tomographs that are count rate limited or encounter dominating random coincidence contributions with FDG doses exceeding 300 MBq, a longer delay, such as 90–120 minutes, with a corresponding higher administered FDG dose may provide optimal whole-body imaging. For tomographs operated in 2D with relatively heavy septa or in 3D with fast scintillation crystals, the optimal dose of FDG and delay time has not yet been fully explored, but likely is between one and two hours. Increasingly centers are moving to 90 minute uptake times for routine body oncology imaging, and in some instances advocating uptake times of 2 to 3 hours for cancers with modest average FDG accumulation such as breast or pancreatic cancer.

Image Display and Interpretation

Whole-body FDG PET images routinely are displayed as a combination of a series of orthogonal tomographic images in the transaxial, coronal, and sagittal planes, and a whole-body rotating projection image. The rotating projection image provides an invaluable rapid assessment of the overall status of FDG avid malignancy in the body, and can be very helpful in discerning the three dimensional relationships of abnormalities to normal structures. Interpretation of whole-body images is thus best accomplished using both the rotating whole-body projection image and the serial tomographic images. As noted above, emission only (non-attenuation corrected) images are being largely supplanted by attenuation corrected images, however as image reconstruction artefacts can occur due to patient movement between the emission and transmission image acquisitions, the non-attenuation corrected images can serve as a useful fall back set of images. Additionally, iterative image reconstruction methods can constrain some aspects of movement artefacts and noise into discrete focal abnormalities potentially mistaken for clinically significant abnormalities such as metastatic deposits. Consequently viewing non-attenuation corrected images reconstructed with conventional filtered back projection has been advocated as a routine adjunct or at least fall back set of images to the attenuation corrected images reconstructed with statistical reconstruction algorithms. There remains controversy over the use of semi-quantitative measures of FDG uptake in the setting of routine diagnostic FDG PET applications in oncology, with some centers using semi-quantitative measures such as the Standardized Uptake Value (SUV) routinely, while others rely entirely on visual interpretation. SUVs should be used with caution as an absolute criteria for malignancy, not only because the degree of FDG uptake implies a probability of malignancy, not an absolute threshold, but more importantly because SUVs reported in the literature are generally insufficiently standardized amongst different PET imaging laboratories to be universally applied [48]. When a patient undergoes serial PET imaging on the same tomograph a the same institution using the same imaging protocol to assess change in FDG uptake such as in the setting of therapy monitoring, SUV or similar semi-quantitative measurements may well be a very useful adjunct to visual interpretation, although current data regarding therapy monitoring with FDG PET suggests it is the complete resolution of abnormal FDG uptake (essentially a qualitative interpretation) that is most predictive of progression free survival [49].

Interpretation of PET and CT images generated by combine PET/CT scanners or by registered and aligned images of PET and CT images acquired on separate tomographs requires workstations capable of displaying both PET and CT images in full resolution (512 × 512 matrix for CT) and rapid stacked image display at full image fidelity for rapid and full interpretation of the PET and CT images, along with the rotating whole-body projection images. It should be noted that CT images produced by combined PET/CT scanners are fully diagnostic CT images, and hence interpretation of PET/CT involves both anatomical diagnosis based on CT images and metabolic diagnosis based on the PET images. The CT images, while used for anatomical reference and localization of abnormalities on the PET images to aid in PET interpretation, contain essential independent anatomical diagnostic information as well, and hence display of CT images must be displayed in full fidelity with rapid access to window and level settings as well as image reconstruction algorithms (soft tissue vs lung/bone). So-called fusion images, in which the CT and PET images are superimposed using gray scale for the CT and a color scale for the PET are of limited utility as the PET color image obscures the CT image and subtle findings on PET are obscured by the CT images. Side by side registered and aligned images in gray scale with coordinated cursors provides rapid access to all information on both images and permits efficient complete interpretation of both PET and CT image sets.

Conclusions

In as little as a decade whole-body PET imaging has emerged as an essential component of medical imaging in oncology. Rather than being a competitor of CT based anatomical diagnosis in body oncology imaging, the complimentary value of metabolic diagnosis provided by PET and anatomical diagnosis provided by CT is now manifest in combined PET/CT scanners, which likely will quickly become the standard for body oncology medical diagnosis.

References

1. Shreve PD. Status of clinical PET in the USA and the role and activities of the institute for clinical PET. In: Positron Emission Tomography: A Critical Assessment of Recent Trends. [Bulyas B and Muller-Gartner HW, Eds] Dordrecht: Kluwer 1998:33–42.
2. Phelps ME, Hoffman EJ, Mullani NA, Ter-Pogossian MM. Application of annihilation coincidence detection to transaxail reconstruction tomography. J Nucl Med 1975;16:210–224.
3. Ter-Pogossian MM, Phelps ME, Hoffman EJ, et. al. A positron-emission transaxial tomograph for nuclear medicine imaging (PETT). Radiology 1975;114:89–98.
4. Willimas, Crabtree, Burgiss. Design and performance characteristics of a positron emission computed axial tomograph-ECAT-II. IEEE Trans Nucl Sci 1979;26.
5. Hoffman EJ, Phelps ME, Huang S-C. Perfomance evaluation of a positron tomograph designed for brain imaging. J Nucl Med 1983;24:245–257.
6. Casey ME, Nutt R. A multicrystal two-dimensional BGO detector system for postiron emission tomography. IEEE Trans. Nucl Sci 1986;NS33:460–463.

7. Gupta NC, Frank AR, Dewan NA, et. al. Solitary pulmonary nodules: detection of malignancy with PET with 2-[F-18]-fluoro-2Deoxy-D-glucose. Radiology 1992;184:441–444.
8. Glaspy JA, Hawkins R, Hoh CK, Phelps ME. Use of positron emission tomography in oncology. Oncology 1993;7:41–50.
9. Price P. Is there a future for PET in oncology. Eur J Nucl Med 1997;24:587–589.
10. Wahl RL, Zasadny K, Helvie M, Hutchins GD, Weber B, Cody R. Metabolic monitoring of breast cancer chemohormonotherapy using positron emission tomography: initial evaluation. J Clin Onc 1993;11:2101–2111.
11. Rigo P, Paulus P, Kaschten BJ, et. al. Oncologic applications of positron emission tomography with fluorine-18 fluorodeoxyglucose. Eur J Nucl Med 1996;23:1641–1674.
12. Muehllehner G, Karp JS, Mankoff DA, Beerbohm, Ordonez CE. Design and performance of a new positron emission tomograph. IEEE Trans Nucl Sci 1988;35:670–674.
13. Karp JS, Muehllehner G, Qu H, Yan X-H. Singles transmission in volume-imaging PET with a ^{137}Cs source. Phys Med Biol 1995;40:929–944.
14. Townsend DL, Wensveen M, Byars LG, et. al. A rotating PET scanner using BGO block detectors: design, performance, and applications. J Nucl Med 1993;34:1367–1376.
15. Mullani NA, Gould KL, Hitchens RE, et. al. Design and performance of POSICAM 6.5 BGO positron camera. J Nucl Med 1990;31:610–616.
16. Wienhard K, Eriksson L, Grootoonk S, Casey M, Pietrzyk U, Heiss W. Performance evaluation of the positron scanner ECAT EXACT. JCAT 1992;16:804–813.
17. DeGrado TR, Turkington TG, Williams JJ, Stearns CW, Hoffman J, Coleman RE. Performance characteristics of a whole-body PET scanner. J Nucl Med 1994;35:1398–1406.
18. Hudson HM, Larkin RS. Accelerated image reconstruction using ordered subsets of projection data. IEEE Tans Med Imaging. 1994;13:601–609.
19. Xu M, Cutler PD, Luk WK. Adaptive, segmented attenuation correction for whole-body PET imaging. IEEE Trans Nucl Sci. 1996;43:331–336.
20. Wahl RW. To AC or not to AC: that is the question. J Nucl Med 1999;40:2025–2028.
21. Kinahan PE, Rogers JG. Analytic 3D image reconstruction using all detected events. IEEE Tans. Nucl. Sci. 1989;36:964–968.
22. Townsend DW, Geissbuhler A, Defrise M, et. al. Fully three-dimensional reconstruction for a PET camera with retractable septa. IEEE Tans. Med. Imaging 1991;MI-10:505–512.
23. Muehllehner G. Design considerations for PET scanners. Quarterly Journal of Nuclear Medicine. 2002;45: 16–23.
24. Watson CC. New, faster, image-based scatter correction for 3D PET. IEEE 2000;47:1587–1594.
25. Beyer T, Townsend DW, Brun T, et. al. A combined PET/CT scanner for clinical oncology. J Nucl Med 2000;41:1369–1379.
26. Shreve PD. Adding structure to function. J Nucl Med 2000;41:1380–1382.
27. Kinahan PE. Townsend DW, Beyer T, Sashin D. Attenuation correction for a combined 3D PET/CT scanner. Med. Phys. 1998;25:2046–2053.
28. Halpern B, Dahlbom M, Vranjesevic D, et. al. LSO-PET/CT whole-body imaging in 7 minutes: is it feasible? J Nucl Med 2003;44:380–381.
29. Bar-Shalom R, Yefremov N, Guralnik L, et. al. Clinical performance of PET/CT in evaluation of cancer: Additional value for diagnostic imaging and patient management. J Nucl Med 2003;44:1200–1209.
30. Cohade C, Osman M, Leal J, Wahl RL. Direct comparison of 18F-FDG and PET/CT in patients with colorectal carcinoma. J Nucl Med 2003;44:1797–1803.
31. Ollenberger GP, Weder W, von Schulthess GK, Steinert HC. Staging of lung cancer with integrated PET-CT. N Engl J Med 2004;350: 86–87.
32. Shreve, PD, Anzai Y, Wahl RW. Pitfalls in oncologic diagnosis with FDG PET imaging: Physiologic and benign variants. Radiographics 1999;19:61–67.
33. Lindholm P, Minn H, Leskinen-Kallio S, Bergman J, Ruotsalainen U, Joensuu H. Influence of the blood glucose concentration of FDG uptake in cancer: a PET study. J Nucl Med 1993;34:1–6.
34. Vesselle HJ, Miraldi FD. FDG PET of the retroperitoneum: normal anatomy, variants, pathological conditions, and strategies to avoid diagnostic pitfalls. RadioGraphics 1998;18:805–823.
35. Brigid GA, Flanagan FL, Dehdashti F. Whole-body positron emission tomography: normal variations, pitfalls, and technical considerations. AJR 1997;169:1675–1680.
36. Miraldi F, Vesselle H, Faulhaber PF, Adler LP, Leisure GP. Elimination of artifactual accumulation of FDG in PET imaging of colorectal cancer. Clin Nucl Med 1998;23:3–7.
37. Stahl A, Weber W, Avril N, Schwaiger M. The effect of N-butylscopolamine on intestinal uptake of F-18 fluorodeoxyglucose in PET imaging of the abdomen. Eur J Nucl Med 1999;26(P):1017.
38. Kostakoglu L, Wong JCH, barrington SF, Cronin BF, Dynes AM, Maisey MN. Speech-related visualization of laryngeal muscles with fluorine-18 FDG. J Nucl Med 1996;37:1771–1773.
39. Hany TF, Gharelpapagh E, Kamel E, Buch A, Himms-Hagen J, von Schulthess G. Brown adipose tissue: a factor to consider in symetrical tracer uptake in the neck and upper chest region. Eur J Nucl Med Mol Imaging 2002;29:1393–1398.
40. Cohade C, Osman M, Pannu HK, Wahl RL. Uptake in supraclavicular area fat ("USA-Fat"): Description on 18F-FDG PET/CT. J Nucl Med 2003;44:170–176.
41. Yeung HWD, Grewal RK, Gonen M, Schoder H, Larson SM. Patterns of 18-F FDG uptake in adipose tissue and muscle: A potential source of false-positives for PET. J Nucl Med 2003; 44:1789–1796.
42. Barrington SF, Maisey MN. Skeletal muscle uptake of fluorine-18-FDG: effect on oral diazepam. J Nucl Med 1996;37:1127–1129.
43. Beyer T. Personnel communication.
44. Beyer T, Antoch G, Muller S, Egelhof T, Freudenberg LS, Debatin J, Bockisch A. Acquisition protocol considerations for combined PET/CT imaging. J Nucl Med 2004;45:25S–35S.
45. Jones SC, Alavi A, Christman D, Montanez I, Wolf AP, Reivich M. The radiation dosimetry of 2-[F-18]fluoro-2Deoxy-D-glucose in man. J Nucl Med 1982;23:613–617.
46. Hamberg LM, Hunter GJ, Alpert NM, Choi NC, Babich JW, Fischman AJ. The dose uptake ratio as an index of glucose metabolism: useful parameter or oversimplification? J Nucl Med 1994; 35:1308–1312.
47. Lowe V. Personnel communication.
48. Keyes JW Jr. SUV: standard uptake value or silly useless value? J Nucl Med 1995;36:1836–1839.
49. Kostakoglu L, Goldsmith SJ. 18F FDG PET evaluation of the response to therapy for lymphoma and for breast, lung and colorectal carcinoma. J Nucl Med 2003;44:224–239.

14 Artefacts and Normal Variants in Whole-Body PET and PET/CT Imaging

Gary JR Cook

Introduction

The number of clinical applications for PET continues to increase, particularly in the field of oncology. In parallel with this is growth in the number of centres that are able to provide a clinical PET or PET/CT service. As with any imaging technique, including radiography, ultrasound, computed tomography, magnetic resonance imaging and conventional single photon nuclear medicine imaging, there are a large number of normal variants, imaging artefacts and causes of false positive results that need to be recognised in order to avoid misinterpretation. It is particularly important to be aware of potential pitfalls while PET is establishing its place in medical imaging so that the confidence of clinical colleagues and patients is maintained. In addition, the advent of combined PET/CT scanners in clinical imaging practice has brought its own specific pitfalls and artefacts.

The most commonly used PET radiopharmaceutical in clinical practice is ^{18}F-fluorodeoxyglucose (^{18}FDG). As it has a half-life of nearly 2 hours, it can be transported to sites without a cyclotron, and in view of this and the fact that there is a wealth of clinical data and experience with this compound, it is likely to remain the mainstay of clinical PET for the immediate future.

Mechanisms of Uptake of ^{18}F-fluorodeoxyglucose

^{18}FDG, as an analogue of glucose, is a tracer of energy substrate metabolism, and although it has been known for many years that malignant tumours show increased glycolysis compared to normal tissues, its accumulation is not specific to malignant tissue. ^{18}FDG is transported into tumour cells by a number of membrane transporter proteins that may be overexpressed in many tumours. ^{18}FDG is converted to ^{18}FDG-6-phosphate intracellularly by hexokinase, but unlike glucose does not undergo significant enzymatic reactions. In addition, because of its negative charge, remains effectively trapped in tissue. Glucose-6-phosphatase mediated dephosphorylation of ^{18}FDG occurs only slowly in most tumours, normal myocardium and brain, and hence the uptake of this tracer is proportional to glycolytic rate. Rarely, tumours may have higher glucose-6-phosphatase activity resulting in relatively low uptake, a feature that has been described in hepatocellular carcinoma [1]. Similarly, some tissues have relatively high glucose-6-phosphatase activity, including liver, kidney, intestine and resting skeletal muscle, and show only low uptake. Conversely, hypoxia, a feature common in malignant tumours, is a factor that may increase ^{18}FDG uptake, probably through activation of the glycolytic pathway [2].

Hyperglycaemia may impair tumour uptake of ^{18}FDG because of competition with glucose [3], although it appears that chronic hyperglycaemia, as seen in diabetic patients, only minimally reduces tumour uptake [4]. To optimise tumour uptake, patients are usually asked to fast for four to six hours prior to injection to minimise insulin levels. This has also been shown to reduce uptake of ^{18}FDG into background tissues including bowel, skeletal muscle and myocardium [5]. In contrast, insulin induced hypoglycaemia may actually impair tumour identification by reducing tumor uptake and increasing background muscle and fat activity [6].

In addition to malignant tissue, ^{18}FDG uptake may be seen in activated inflammatory cells [7,8], and its use has even been advocated in the detection of inflammation [9]. An area where benign inflammatory uptake of ^{18}FDG may limit specificity is in the assessment of response to radiotherapy [10]. Here uptake of ^{18}FDG has been reported in rectal tumours and in the brain in relation to macrophage and inflammatory cell activity [11–13]. This may make it difficult to differentiate persistent tumour from inflammatory activity for a number of months following radiotherapy in some tumors. Non-specific, inflammatory and reactive uptake has also been recorded following chemotherapy in some tumours [14, 15], and there is no clear consensus on the optimum time to study patients following this form of therapy.

Normal Distribution of ^{18}FDG

The normal distribution of ^{18}FDG is summarised in Table 14.1. The brain typically shows high uptake of ^{18}FDG in the cortex, thalamus and basal ganglia. Cortical activity may be reduced in patients who require sedation or a general anaesthetic, a feature that might limit the sensitivity of detection of areas of reduced uptake as in the investigation of epilepsy. It is not usually possible to differentiate low-grade uptake of ^{18}FDG in white matter from the adjacent ventricular system (Fig. 14.1).

In the neck, it is common to see moderate symmetrical activity in tonsillar tissue. This may be more difficult to recognise as normal tissue if there has been previous surgery or radiotherapy that may distort the anatomy, resulting in asymmetric activity or even unilateral uptake on the unaffected side. Adenoidal tissue is not usually noticeable in adults but may show marked uptake in children. Another area of lymphoid activity that is commonly seen in children is the thymus. This usually has a characteristic shape (an inverted V) and is therefore not usually mistaken for anterior mediastinal tumour (Fig. 14.2). Clinical reports vary as to the incidence of diffuse uptake of ^{18}FDG in the thyroid [16–18]. This may be a geographical phenomenon, because its presence is more likely in women and has been correlated with the presence of thyroid autoantibodies and chronic thyroiditis [18].

In the chest, there is variation in regional lung activity, this being greater in the inferior and posterior segments, and it has been suggested that this might reduce sensitivity in lesion detection in these regions [19]. In the abdomen, homogeneous, low-grade accumulation is seen in the liver and to a lesser extent, the spleen. Small and large bowel activity is quite variable, and unlike glucose, ^{18}FDG is excreted in the urine, leading to variable appearances of the urinary tract, both of which are discussed further below. Resting skeletal muscle is usually associated with low-grade activity, but active skeletal muscle may show marked uptake of ^{18}FDG in a variety of patterns that are discussed later in this chapter.

Myocardial activity may also be quite variable. Normal myocardial metabolism depends on both glucose and free fatty acids (FFA). For oncologic scans, it is usual to try to reduce activity in the myocardium, so as to obtain clear images of the mediastinum and

Table 14.1. Normal distribution of ^{18}FDG.

Organ/system	Pattern
Central nervous system	High uptake in cortex, basal ganglia, thalami, cerebellum, brainstem. Low uptake into white matter and cerebrospinal fluid.
Cardiovascular system	Variable but homogeneous uptake into left ventricular myocardium. Usually no discernible activity in right ventricle and atria.
Gastrointestinal system	Variable uptake into stomach, small intestine, colon and rectum.
Reticuloendothelial and lymphatic	Liver and spleen show low grade diffuse activity. No uptake in normal lymph nodes but moderate activity seen in tonsillar tissue. Age related uptake is seen in thymic and adenoidal tissue.
Genitourinary system	Urinary excretion can cause variable appearances of the urinary tract. Age related testicular uptake is seen.
Skeletal muscle	Low activity at rest
Bone marrow	Normal marrow shows uptake that is usually less than liver.
Lung	Low activity (regional variation)

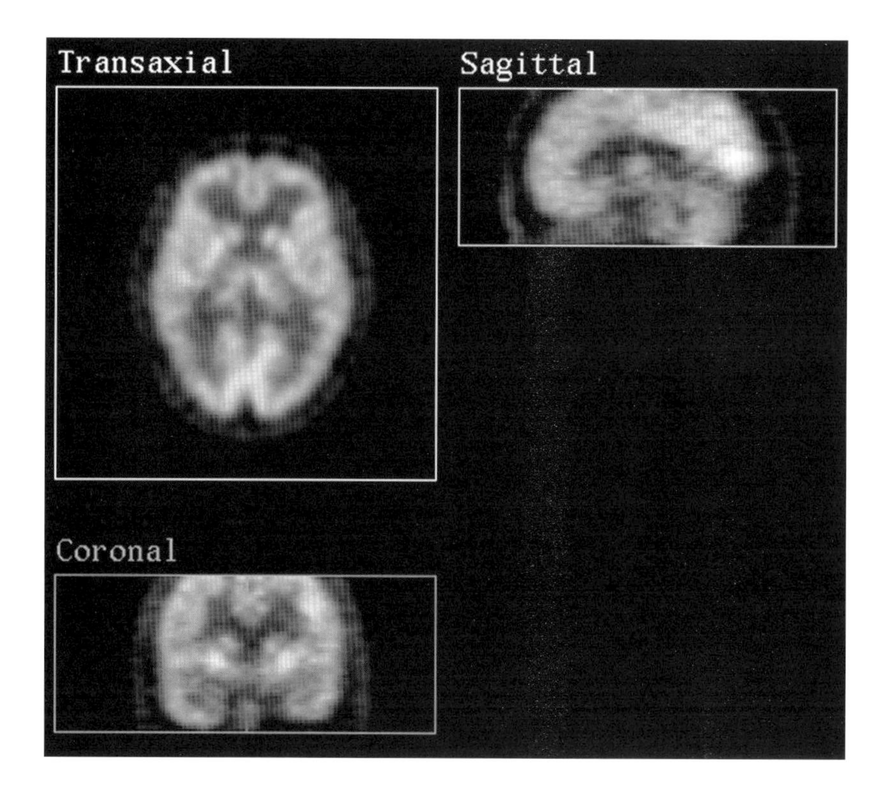

Figure 14.1. Normal ^{18}FDG brain scan. The transaxial image is taken at the level of the basal ganglia and thalami.

Figure 14.2. Transaxial (above) and coronal (below) ^{18}FDG images in a child showing normal thymic activity.

adjacent lung. Although most centres fast patients for at least 4 to 6 hours before ^{18}FDG injection, reducing insulin levels and encouraging FFA acid metabolism in preference to glucose, myocardial activity may still be quite marked and varies among patients. Another possible intervention that has not been quantified or validated as yet is to administer caffeine to the patient to encourage FFA metabolism.

For cardiac viability studies, it is necessary to achieve high uptake of ^{18}FDG into the myocardium. Patients may receive a glucose load to encourage glucose (and hence ^{18}FDG) rather than FFA metabolism, and it may also be necessary to administer insulin to enhance myocardial uptake, particularly in diabetic patients [20–22]. The hyperinsulinaemic euglycaemic clamping method may further improve myocardial uptake but is technically more difficult [23–26]. This allows maximum insulin administration without rendering the patient hypoglycaemic. An alternative method is to encourage myocardial glucose metabolism by reducing FFA levels pharmacologically. Improved cardiac uptake of ^{18}FDG has been described following oral nicotinic acid derivatives such as acipimox, a simple and safe measure that may also be effective in diabetic patients [27].

Variants That May Mimic or Obscure Pathology

A number of physiological variations in uptake of ^{18}FDG have been recognised, some of which may mimic pathology [16, 28–30], and are summarised in Table 14.2.

Skeletal muscle uptake is probably the most common cause of interpretative difficulty. Increased aerobic glycolysis associated with muscle activation, either after

Table 14.2. Variants that may mimic or obscure pathology.

Organ/system	Variant
Skeletal muscle	High uptake after exercise or due to tension, including eye movement, vocalisation, swallowing, chewing gum, hyperventilation.
Adipose tissue	Uptake in brown fat may be seen particularly in winter months in patients with low body mass index.
Myocardium	Variable (may depend on or be manipulated by diet and drugs).
Endocrine	Testes, breast (cyclical, lactation, HRT), follicular ovarian cysts, thyroid
Gastrointestinal	Bowel activity is variable and may simulate tumour activity
Genitourinary	Small areas of ureteric stasis may simulate paraaortic or pelvic lymphadenopathy

exercise or because of involuntary tension, leads to increased accumulation of ^{18}FDG that may mimic or obscure pathology. Exercise should be prohibited before injection of ^{18}FDG and during the uptake period to minimise muscle uptake.

A pattern of symmetrical activity commonly encountered in the neck, supraclavicular and paraspinal regions (Fig. 14.3) was initially assumed to be the result of involuntary muscle tension but with the advent of PET/CT it has become obvious that this activity originates in brown fat, a vestigial organ of thermogenesis that is sympathetically innervated and driven. To support this hypothesis it has been noted that this pattern is commoner in winter months and in patients with lower body mass index [31]. It appears that benzodiazepines are able to reduce the incidence of this potentially confusing appearance, possibly the result of a generalised reduction in sympathetic drive.

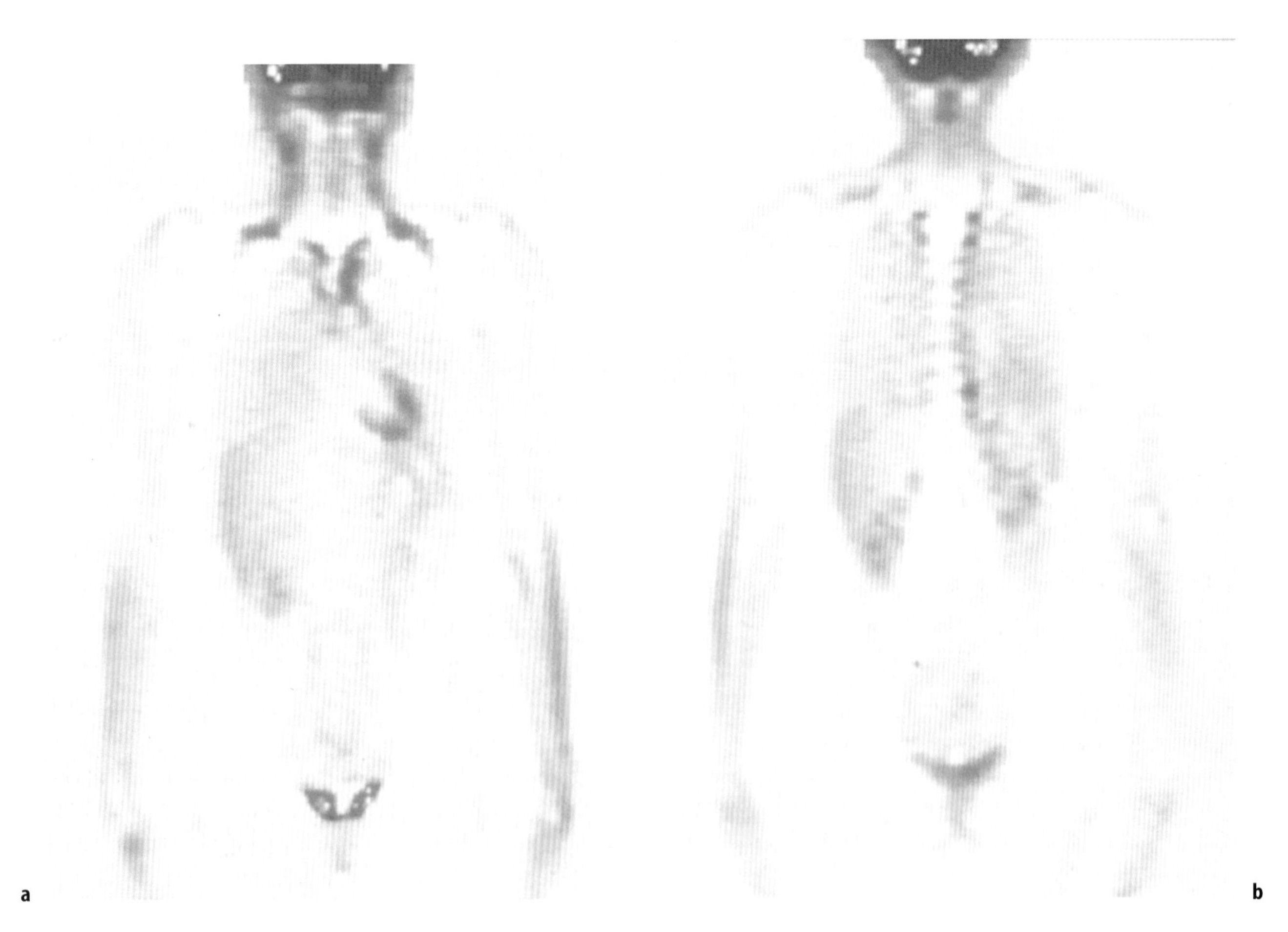

Figure 14.3. Coronal sections from a ^{18}FDG study. Symmetrical brown fat activity is seen in the neck (a) and paraspinal (b) regions. Although this is a recognisable pattern it can be appreciated that metastatic lymphadenopathy may be obscured, especially in the neck.

Even apparently innocent activities such as talking or chewing gum may lead to muscle uptake that simulates malignant tissue (Figs. 14.4 and 14.5). In patients being assessed for head and neck malignancies, it is therefore important that they maintain silence and refrain from chewing during the uptake period. In addition, anxious or breathless patients may hyperventilate, producing increased intercostal and diaphragmatic activity, and involuntary muscle spasm such as that seen with torticollis may lead to a pattern that is recognisable but may obscure diseased lymph nodes.

The symmetrical nature of most muscle uptake usually alerts the interpreter to the most likely cause, but occasionally unilateral muscle uptake may be seen when there is a nerve palsy on the contralateral side and may be mistaken for an abnormal tumour focus. This has been described in recurrent laryngeal nerve palsy and in VIth cranial nerve palsy [30]. Diffusely increased uptake of ^{18}FDG may also be seen in dermatomyositis complicating malignancy, a factor that may reduce image contrast and tumour detectability.

Uptake in the gastrointestinal system is quite variable and is most commonly seen in the stomach (Fig. 14.6) and large bowel (Fig. 14.7) and to a lesser extent in loops of small bowel. It is probable that activity in bowel is related to smooth muscle uptake as well as activity in intralumenal contents [32, 33]. If it is important to reduce intestinal physiological activity,

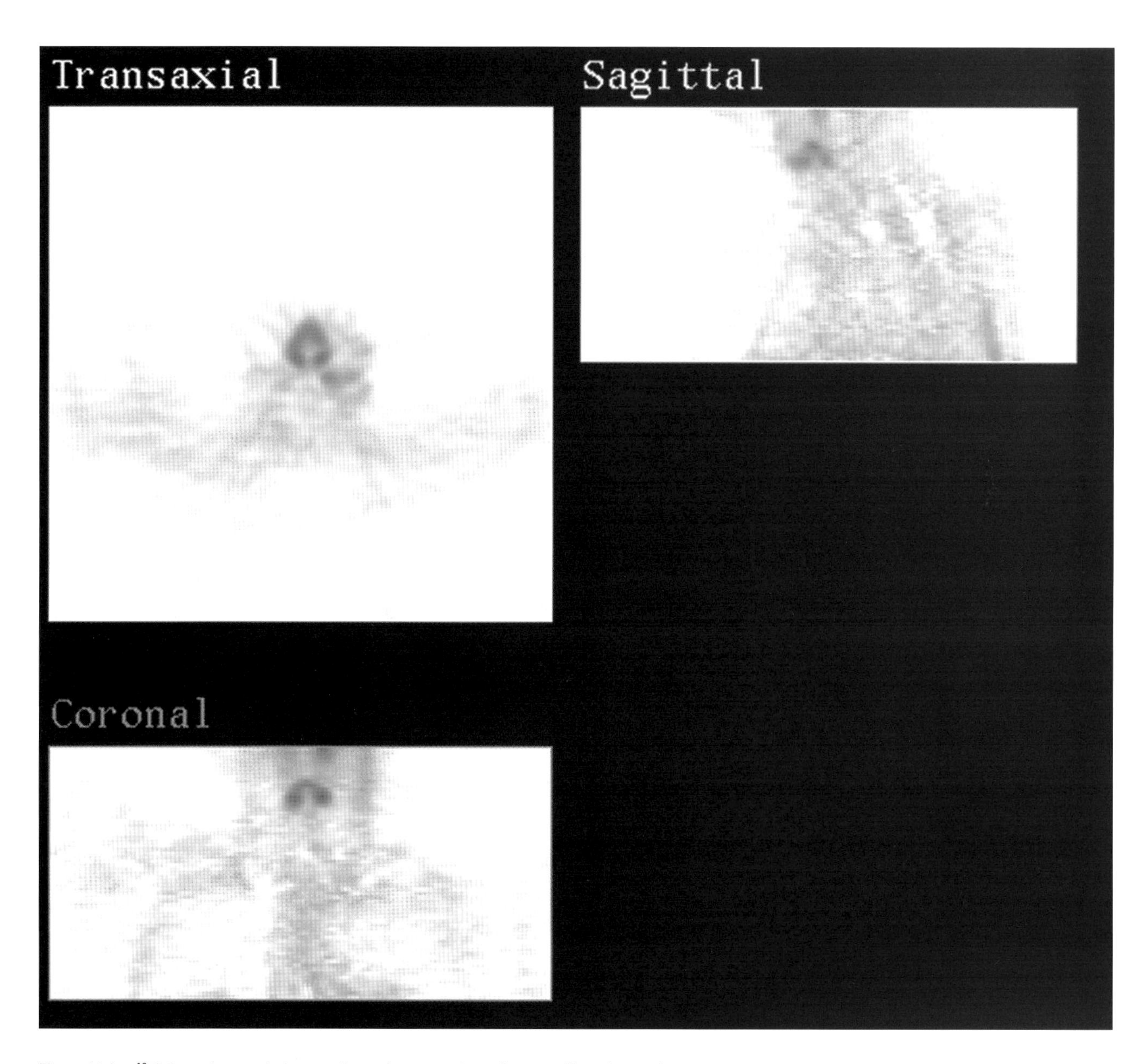

Figure 14.4. ^{18}FDG uptake seen in laryngeal muscles in a patient who was talking during the uptake period.

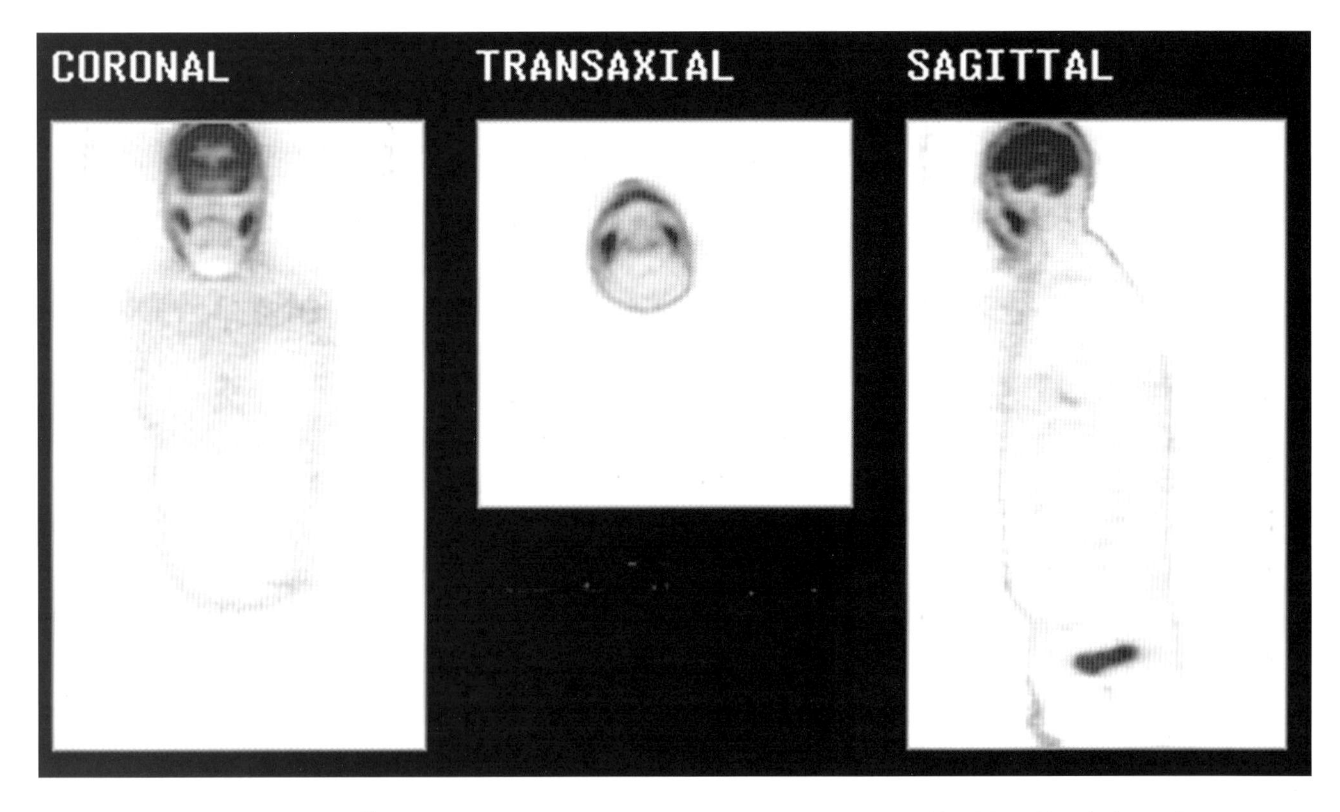

Figure 14.5. Symmetrical ^{18}FDG uptake in the masseter muscles in a patient chewing gum, resembling bilateral lymphadenopathy.

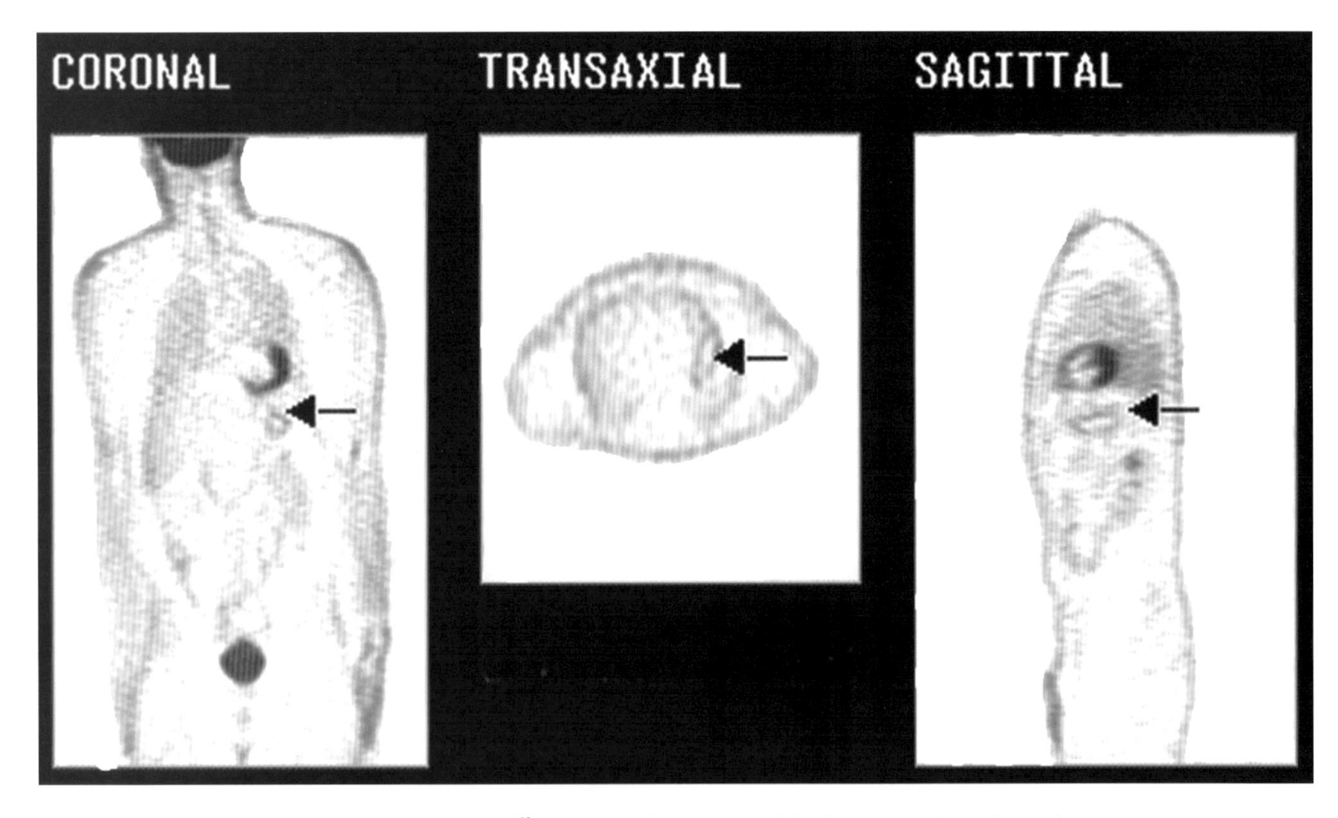

Figure 14.6. Physiological uptake of ^{18}FDG is seen in the stomach wall. Moderate myocardial activity is also seen.

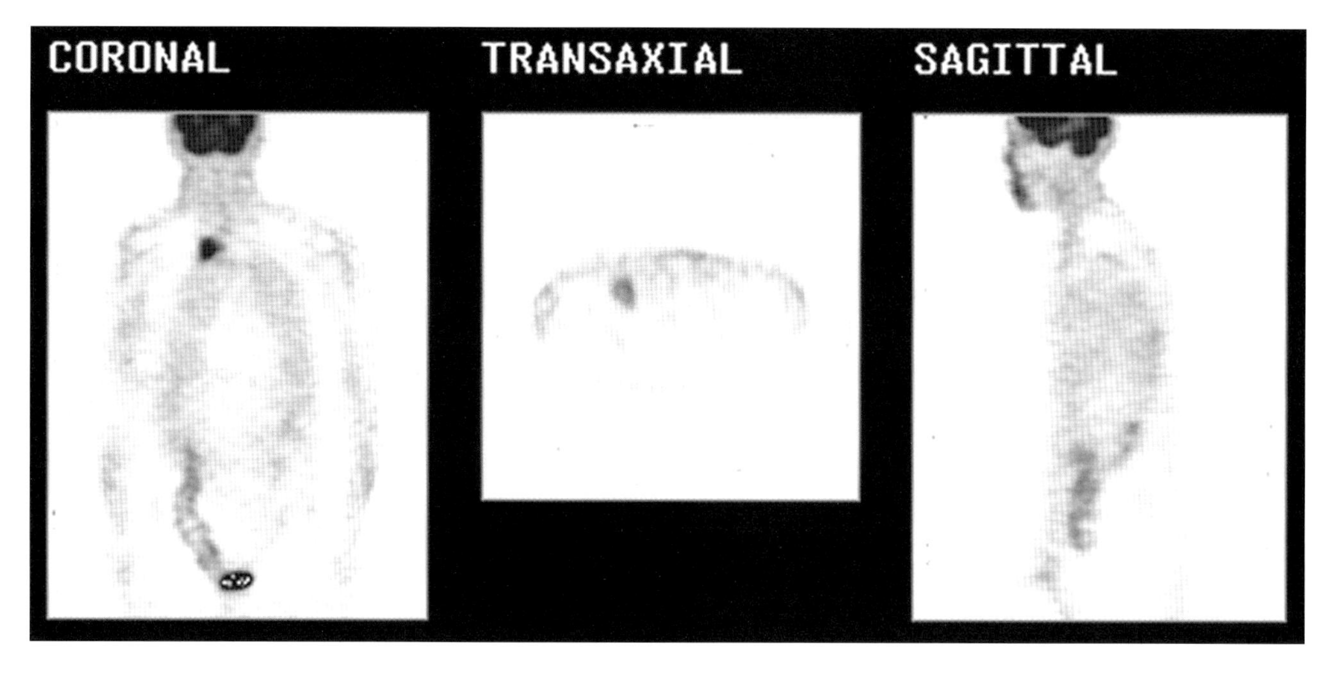

Figure 14.7. Marked physiological uptake is seen in the region of the caecum and ascending colon in a patient with a primary lung cancer that can also be seen on these images at the right lung apex (coronal section).

pharmacological methods to reduce peristalsis as well as bowel lavage could be useful. This is too invasive and is unnecessary for routine patient preparation, and in most situations it is possible to differentiate physiological uptake within bowel from abdominal tumour foci by the pattern of uptake, the former usually being curvilinear and the latter being focal (Fig. 14.8). Some centres use a mild laxative as a routine in any patient requiring abdominal imaging, but improvement in interpretation has not been demonstrated.

Unlike glucose, ^{18}FDG is not totally reabsorbed in the renal tubules, and urinary activity is seen in all patients and may be present in all parts of the urinary tract. This may interfere with a study of renal or pelvic tumours, either by obscuring local tumours or by causing reconstruction artefacts that reduce the visibility of abnormalities adjacent to areas of high urinary activity. Using iterative reconstruction algorithms rather than filtered back projection can reduce this problem. Catheterisation and drainage of

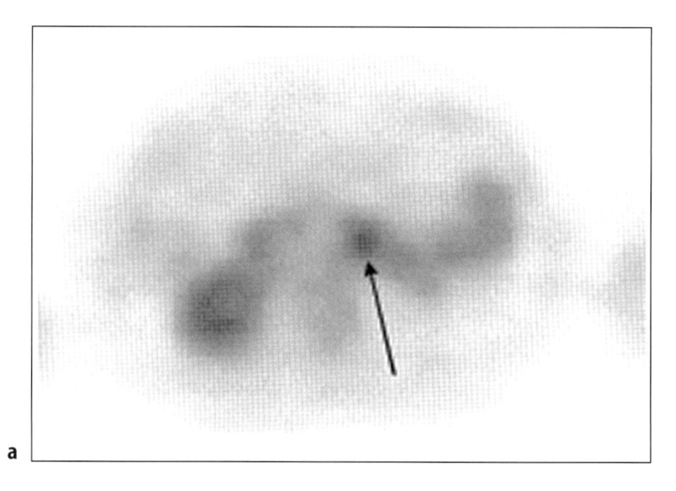

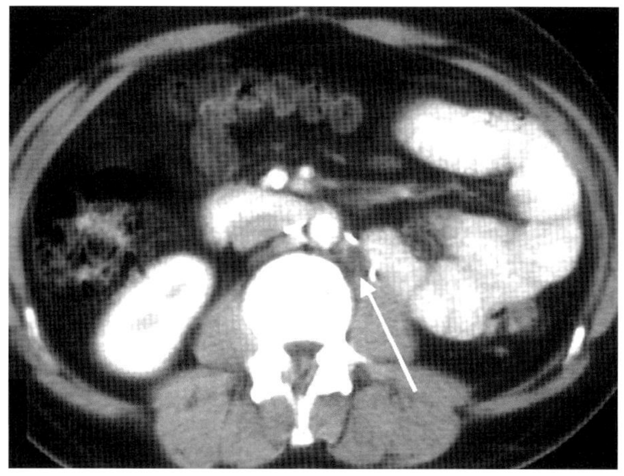

Figure 14.8. (a) Transaxial ^{18}FDG slice through the upper abdomen and (b) corresponding CT slice in a patient with a history of seminoma and previous para-aortic lymph node dissection but rising tumour markers. The linear area of low grade ^{18}FDG activity can be seen to correspond to a barium filled loop of bowel but the more focal area of high uptake (arrow) corresponds to a small density located adjacent to the previous surgical clips indicating recurrent disease at this site. The case demonstrates how normal bowel activity can be differentiated from tumour foci.

urinary activity may reduce bladder activity that may obscure perivesical or intravesical tumours. However, this may still leave small pockets of concentrated activity that may resemble lymphadenopathy, causing even greater problems in interpretation. Bladder irrigation may help to some extent, but is associated with increased radiation dose to staff and may introduce infection.

We have found it beneficial to hydrate the patient and administer a diuretic. This approach leads to a full bladder with *dilute* urine, making it easier to differentiate normal urinary activity from perivesical tumour activity and allowing the bladder to be used as an anatomical landmark. By diluting vesical ^{18}FDG activity, reconstruction artefacts from filtered back projection algorithms are also reduced. It is often helpful to perform image registration with either CT or MRI in the pelvis. Here it may be helpful to administer a small amount of ^{18}F-fluoride ion in addition to ^{18}FDG, to allow easy identification of bony landmarks for registration purposes. Although excreted ^{18}FDG may be seen in any part of the urinary tract, it is important to gain a history of any previous urinary diversion procedures, since these may cause areas of high activity outside the normal renal tract and may result in errors of interpretation unless this is appreciated.

Glandular breast tissue often demonstrates moderate ^{18}FDG activity in premenopausal women and postmenopausal women taking oestrogens for hormone replacement therapy. The pattern of uptake is usually symmetrical and easily identified as being physiological, but there is the potential for lesions to be obscured by this normal activity. Breast feeding mothers show intense uptake of ^{18}FDG bilaterally (Fig. 14.9). Similarly in males, uptake of ^{18}FDG may be seen in normal testes and appears to be greater in young men than in old [34].

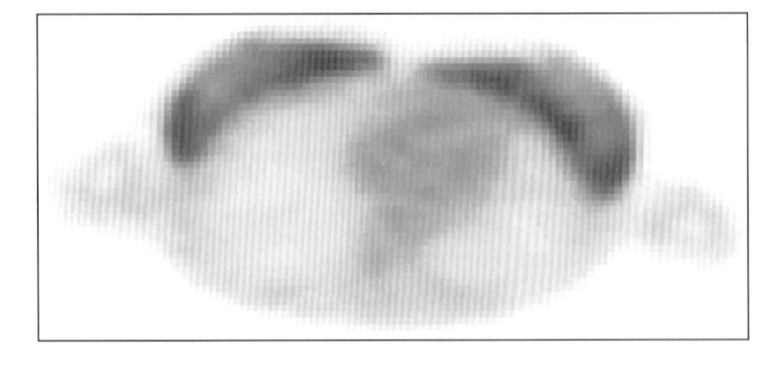

Figure 14.9. Transaxial ^{18}FDG scan of a breast-feeding mother in whom intense symmetrical breast activity can be seen. (Reproduced from Valk PE, Bailey DL, Townsend DW, Maisey MN. Positron Emission Tomography: Basic Science and Clinical Practice. Springer-Verlag London LTD, 2003, p. 502.)

Table 14.3. Artefacts

Attenuation correction related	Apparent superficial increase in activity and lung activity if no correction applied.
Injection related	Lymph node uptake following tissued injection. Reconstruction artefacts due to tissued activity. Inaccuracies in SUV calculation
Attenuating material	Coins, medallions, prostheses
Patient movement	Poor image quality. Artefacts on applying attenuation correction.

Artefacts

Image reconstruction of PET images without attenuation correction may lead to higher apparent activity in superficial structures, that may obscure lesions e.g. cutaneous melanoma metastases [28]. A common artefact arising from this phenomenon is caused by the axillary skin fold, where lymphadenopathy may be mimicked in coronal image sections. However, the linear distribution of activity can be appreciated on transaxial or sagittal slices and should prevent misinterpretation. Another major difference between attenuation corrected and non-corrected images is an apparent increase in lung activity in the latter due to relatively low attenuation by the air-containing lung.

Filtered back projection reconstruction leads to streak artefacts and may obscure lesions adjacent to areas of high activity. Many of these artefacts can be overcome by using iterative reconstruction techniques (Fig. 14.10).

Patient movement may compromise image quality. In brain imaging it is possible to split the acquisition into a number of frames, so that if movement occurs in one frame then this can be discarded before summation of the data [35]. When performing whole body scans, unusual appearances may result if the patient moves between bed scan positions. This most commonly occurs when the upper part of the arm is visible in higher scanning positions, but the lower part disappears when moved out of the field of view on lower subsequent scanning positions.

Special care is required in injecting ^{18}FDG since soft-tissue injection may cause reconstruction artefacts across the trunk, and may even cause a low-count study or inaccuracies in standardised uptake value (SUV) measurements. Axillary lymph nodes, draining the region of tracer extravasation, may also accumulate activity following extravasated injections. The site of

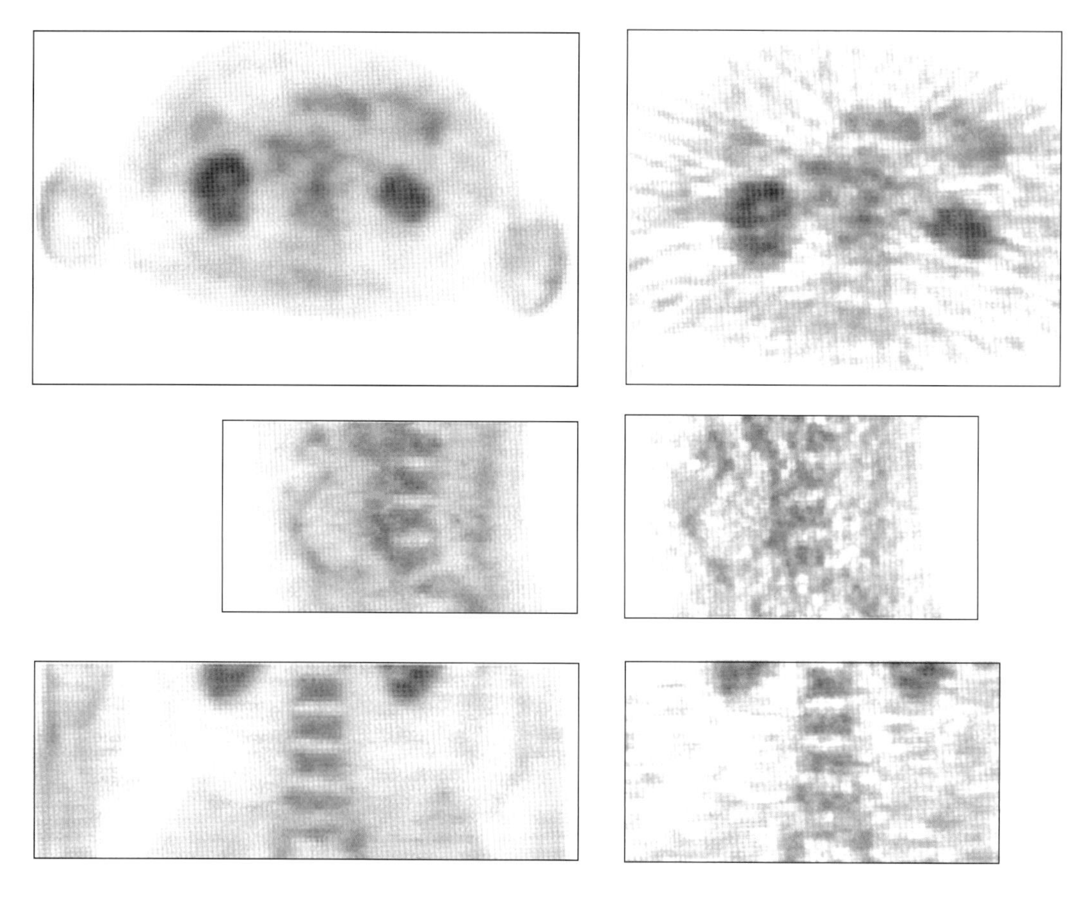

Figure 14.10. Transaxial, sagittal and coronal abdominal ^{18}FDG images from iterative reconstruction (left) and filtered back projection (right) demonstrating the improved image quality and reduction in streak artefacts possible with the former.

administration should be chosen carefully, so as to minimise the risk of false positive interpretation should extravasation occur.

Artefacts caused by prostheses are usually readily recognisable. Photon deficient regions may result from metallic joint prostheses or other metallic objects carried by the patient. Ring artefacts may occur if there is misregistration between transmission and emission scans due to patient movement, and are particularly apparent at borders where there are sudden changes in activity concentrations (e.g., at a metal prosthesis). Misregistration artefacts between emission and transmission scans have become less frequent now that interleaved or even simultaneous emission/transmission scans are being performed.

Benign Causes of ^{18}FDG Uptake

Uptake of ^{18}FDG is not specific to malignant tissue, and it is well recognised that inflammation may lead to accumulation in macrophages and other activated inflammatory cells [7, 8]. In oncological imaging, this inflammatory uptake may lead to decrease in specificity. For example, it may be difficult to differentiate benign postradiotherapy changes from recurrent tumour in the brain, unless the study is optimally timed or unless alternative tracers such as ^{11}C methionine are used. Apical lung activity may be seen following radiotherapy for breast cancer, and moderate uptake may follow radiotherapy for lung cancer [36]. It may also be difficult to differentiate radiation changes from recurrent tumour in patients who have undergone radiotherapy for rectal cancer within six months of the study [12].

Pancreatic imaging with ^{18}FDG may be problematic. In some cases, uptake into mass-forming pancreatitis may be comparable in degree to uptake in pancreatic cancer. Conversely, false negative results have been described in diabetic patients with pancreatic cancer. However, if diabetic patients and those with raised inflammatory markers are excluded, then ^{18}FDG PET may still be an accurate test to differentiate benign from malignant pancreatic masses [37].

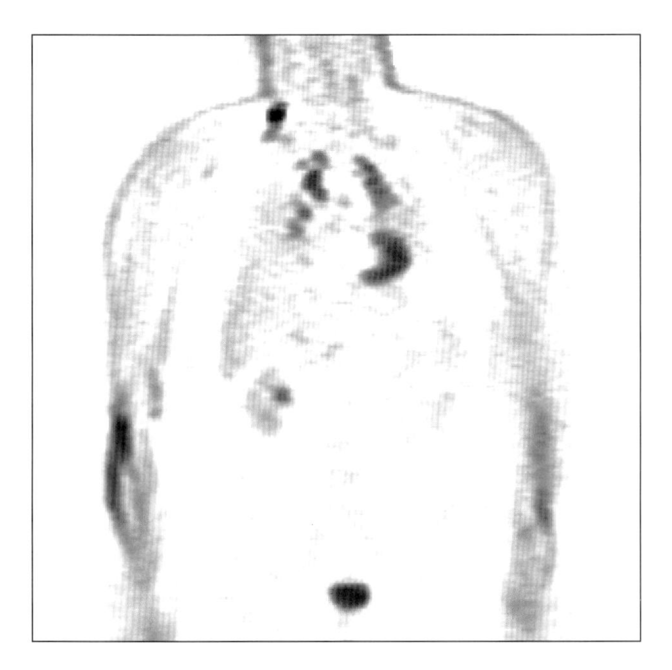

Figure 14.11. Coronal ^{18}FDG scan demonstrating high uptake in lymph nodes in a patient with sarcoidosis.

Table 14.4. Benign causes of ^{18}FDG uptake

Organ/Type	Disease
Brain	Postradiotherapy uptake.
Pulmonary	Tuberculosis, sarcoidosis, histoplasmosis, atypical mycobacteria, pneumoconiosis, radiotherapy.
Myocardium	Heterogeneous left ventricular activity possible after myocardial infarction, increased right ventricular activity in right heart failure
Bone/bone marrow	Paget's disease, osteomyelitis, hyperplastic bone marrow.
Inflammation	Wound healing, pyogenic infection, organising haematoma, oesophagitis, inflammatory bowel disease, lymphadenopathy associated with granulomatous disorders, viral and atypical infections, chronic pancreatitis, retroperitoneal fibrosis, radiation fibrosis (early), bursitis.
Endocrine	Graves' disease and chronic thyroiditis, adrenal hyperplasia.

A number of granulomatous disorders have been described as leading to increased uptake of ^{18}FDG, including tuberculosis [38], and sarcoidosis [39] (Fig. 14.11). It is often necessary to be cautious in ascribing ^{18}FDG lesions to cancer in patients who are known to be immunocompromised. It is these patients who often have the unusual infections that may lead to uptake that cannot be differentiated from malignancy. PET remains useful in these patients despite a lower specificity, as it is often able to locate areas of disease that have not been identified by other means and that may be more amenable to biopsy [40].

A more comprehensive list of benign causes of abnormal ^{18}FDG uptake is displayed in Table 14.4.

Specific Problems Related to PET/CT

One of the most exciting technological advances in recent years is the clinical application of combined PET/CT scanners. However, this new technology has come with its own particular set of artefacts and pitfalls.

One of the biggest problems with PET/CT imaging in a dedicated combined scanner is related to differ-

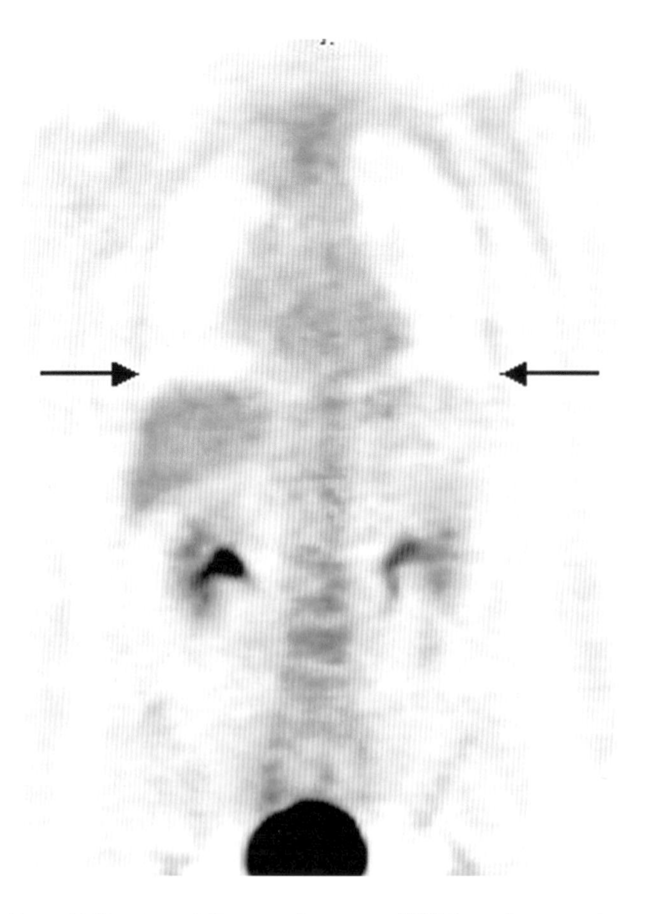

Figure 14.12. Coronal CT attenuation corrected ^{18}FDG scan demonstrating an apparent loss of activity at the level of the diaphragm (arrows) due to differences in breathing patterns between the CT and PET scans.

ences in breathing patterns between the CT and the PET acquisitions. CT scans can be acquired during a breath hold but PET acquisitions are taken during tidal breathing and represent an average position of the thoracic cage over 30 minutes or more. This may result in mis-registration of pulmonary nodules between the two modalities particularly in the peripheries and at the bases of the lungs where differences in position may approach 15 mm [41]. Mis-registration may be reduced by performing the CT scan while the breath is held in normal expiration [42, 43]. It has been noted that deep inspiration during the CT acquisition can lead to deterioration of the CT-attenuation corrected PET image with the appearance of cold artefacts (Fig. 14.12) and can even lead to the mis-positioning of abdominal activity into the thorax [44]. CT acquisition during normal expiration minimises the incidence of such artefacts and also optimises co-registration of abdominal organs.

High-density contrast agents, *e.g.* oral contrast, or metallic objects (Fig. 14.13) can lead to an artefactual overestimation of activity if CT data are used for attenuation correction [45–51]. Such artefacts may be recognised by studying the uncorrected image data. Low-density oral contrast agents can be used without significant artefact [52, 53] or the problem may be avoided by using water as a negative bowel contrast agent. Algorithms have been developed to account for the overestimation of activity when using CT-based attenuation correction that may minimise these effects in the future [53].

The use of intravenous contrast during the CT acquisition may be a more difficult problem. Similarly the concentrated bolus of contrast in the large vessels may lead to over correction for attenuation, particularly in view of the fact that the concentrated column of contrast has largely dissipated by the time the PET emission scan is acquired. Artefactual hot spots in the attenuation corrected image [48] or quantitative overestimation of ^{18}FDG activity may result. When intravenous contrast is considered essential for a study then the diagnostic aspect of the CT scan is best performed as a third study with the patient in the same position, after first, a low current CT scan for attenuation correction purposes and second, the PET emission scan.

While many centres have found low current CT acquisitions to be adequate for attenuation correction and image fusion [54], it may be necessary to increase CT tube current in larger patients to minimise beam-hardening artefacts on the CT scan that may translate through to incorrect attenuation correction of the PET emission data [49]. This effect can be caused by the

a

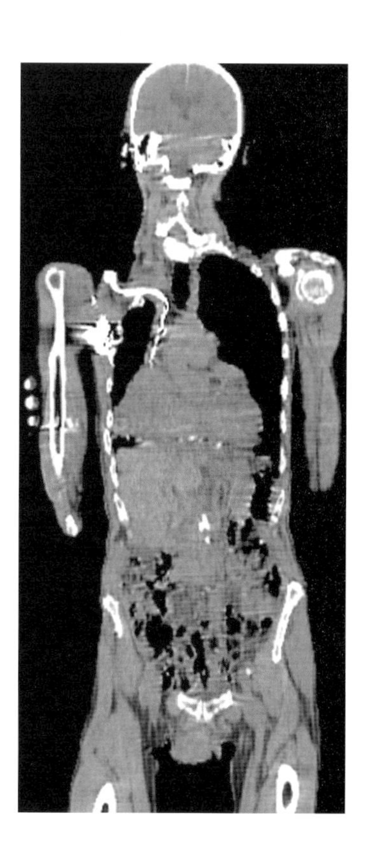
b

c

Figure 14.13. Coronal ^{18}FDG scan with CT attenuation correction (a), CT alone (b), uncorrected ^{18}FDG (c), of a patient with a metallic pacemaker placed over the right upper chest demonstrating artefactual increased uptake on the corrected images.

patient's arms being in the field of view and may be minimised by placing arms above the head for imaging. Differences in the field of view diameter between the larger PET and smaller CT parts of combined scanners can lead to truncation artefacts at the edge of the CT image but these are generally small and can be minimised by the use of iterative image reconstruction methods [53].

Although some new artefacts are introduced by combined PET/CT imaging, it is likely that many pitfalls caused by normal variant uptake may be avoided by the ability to correctly attribute ^{18}FDG activity to a structurally normal organ on the CT scan. This may be particularly evident in the abdomen when physiological bowel activity or ureteric activity can otherwise cause interpretative difficulties. PET/CT also has the potential to limit false negative interpretations in tumours that are not very ^{18}FDG avid by recognising uptake as being related to structurally abnormal tissue and increasing the diagnostic confidence in tumour recognition by the use of the combined structural and functional data. Similarly, it may be possible to detect small lung metastases of a few millimetres on CT lung windows that are beyond the resolution of ^{18}FDG PET. The full use of the combined data, including the corrected and non-corrected PET emission data, and the inspection of soft-tissue, lung and bone windows on the CT data, may also allow the description and correct diagnosis of pertinent ^{18}FDG negative lesions, e.g. liver cysts, and incidental ^{18}FDG negative CT abnormalities, *e.g.* abdominal aortic aneurysm, to provide an integrated interpretation of all the available data resulting from this technology.

References

1. Torizuka T, Tamaki N, Inokuma T et al. In vivo assessment of glucose metabolism in hepatocellular carcinoma with FDG-PET. J Nucl Med 1995;36(10):1811–1817.
2. Minn H, Clavo AC, Wahl RL. Influence of hypoxia on tracer accumulation in squamous cell carcinoma: in vitro evaluation for PET imaging. Nucl Med Biol 1996;23:941–946.
3. Wahl RL, Henry CA, Ethier SP. Serum glucose: effects on tumour and normal tissue accumulation of 2-[F-18]-fluoro-2-deoxy-D-glucose in rodents with mammary carcinoma. Radiology 1992; 183:643–647.
4. Torizuka T, Clavo AC, Wahl RL. Effect of hyperglycaemia on in vitro tumour uptake of tritiated FDG, thymidine, L-methionine and L-leucine. J Nucl Med 1997;38:382–386.
5. Yasuda S, Kajihara M, Fujii H, Takahashi W, Ide M, Shohtsu A. Factors influencing high FDG uptake in the intestine, skeletal muscle and myocardium. J Nucl Med 1999;40:140P.
6. Torizuka T, Fisher SJ, Wahl RL. Insulin induced hypoglycaemia decreases uptake of 2-[F-18]fluoro-2-deoxy-D-glucose into experimental mammary carcinoma. Radiology 1997;203:169–172.
7. Yamada S, Kubota K, Kubota R et al. High accumulation of fluorine-18-fluorodeoxyglucose in turpentine-induced inflammatory tissue. J Nucl Med 1995;36:1301–1306.
8. Kubota R, Kubota K, Yamada S et al. Methionine uptake by tumor tissue: a microautoradiographic comparison with FDG. J Nucl Med 1995; 36:484–492.
9. Sugawara Y, Braun DK, Kison PV, et al. Rapid detection of human infections with fluorine-18 fluorodeoxyglucose and positron emission tomography: preliminary results. Eur J Nucl Med 1998; 25:1238–1243.
10. Reinhardt MJ, Kubota K, Yamada S, Iwata R, Yaegashi H. Assessment of cancer recurrence in residual tumors after fractionated radiotherapy: a comparison of fluorodeoxyglucose, L-methionine and thymidine. J Nucl Med 1997;38:280–287.
11. Strauss LG. Fluorine-18 deoxyglucose and false-positive results: a major problem in the diagnostics of oncological patients. Eur J Nucl Med 1996;23:1409–1415.
12. Haberkorn U, Strauss LG, Dimitrakopoulou A et al. PET studies of fluorodeoxyglucose metabolism in patients with recurrent colorectal tumors receiving radiotherapy. J Nucl Med 1991;32:1485–1490.
13. Kubota R, Kubota K, Yamada S et al. Methionine uptake by tumour tissue: a microautoradiographic comparison with ^{18}FDG. J Nucl Med 1995;36:484–492.
14. Nuutinen JM, Leskinen S, Elomaa I et al. Detection of residual tumours in postchemotherapy testicular cancer by FDG-PET. Eur J Cancer. 1997;33:1234–1241.
15. Jones DN, McCowage GB, Sostman HD et al. Monitoring of neoadjuvant therapy response of soft-tissue and musculoskeletal sarcoma using fluorine-18-FDG PET. J Nucl Med 1996;37:1438–1444.
16. Shreve PD, Anzai Y, Wahl RL. Pitfalls in oncologic diagnosis with FDG PET imaging: physiologic and benign variants. Radiographics 1999;19:61–77.
17. Kato T, Tsukamoto E, Suginami Y et al. Visualization of normal organs in whole-body FDG-PET imaging. Jpn J Nucl Med 1999; 36:971–977.
18. Yasuda S, Shohtsu A, Ide M et al. Chronic thyroiditis: diffuse uptake of FDG at PET. Radiology 1998;207:775–778.
19. Miyauchi T. Wahl RL. Regional 2-[18F]fluoro-2-deoxy-D-glucose uptake varies in normal lung. Eur J Nucl Med 1996; 23:517–523.
20. Kubota K, Kubota R, Yamada S,Tada M, Takahashi T, Iwata R. Reevaluation of myocardial FDG uptake in hyperglycaemia. J Nucl Med 1996;37:1713–1717.
21. Knuuti MJ, Maki M, Yki-Jarvinen et al. The effect of insulin and FFA on myocardial glucose uptake. J Mol Cell Cardiol 1995; 27:1359–1367.
22. Choi Y, Brunken RC, Hawkins RA et al. Factors affecting myocardial 2-[F-18]fluoro-2-deoxy-D-glucose uptake in positron emission tomography studies of normal humans. Eur J Nucl Med 1993; 20:308–318.
23. Bax JJ, Visser FC, Raymakers PG et al. Cardiac 18F-FDG-SPET studies in patients with non-insulin dependent diabetes mellitus during hyperinsulinaemic euglycaemic clamping. Nucl Med Commun 1997;18:200–206.
24. Huitink JM, Visser FC, van Leeuwen GR et al. Influence of high and low plasma insulin levels on the uptake of fluorine-18 fluorodeoxyglucose in myocardium and femoral muscle assessed by planar imaging. Eur J Nucl Med 1995;22:1141–1148.
25. Locher JT, Frey LD, Seybold K, Jenzer H. Myocardial 18F-FDG-PET. Experiences with the euglycaemic hyperinsulinaemic clamp technique. Angiology 1995;46:313–320.
26. Ohtake T, Yokoyama I, Watanabe T et al. Myocardial glucose metabolism in noninsulin dependent diabetes mellitus patients evaluated by FDG-PET. J Nucl Med 1995;36:456–463.
27. Bax JJ, Veening MA, Visser FC et al. Optimal metabolic conditions during fluorine-18 fluorodeoxyglucose imaging: a comparative study using different protocols. Eur J Nucl Med 1997;24:35–41.
28. Engel H, Steinert H, Buck A et al. Whole body PET: physiological and artifactual fluorodeoxyglucose accumulations. J Nucl Med 1996;37:441–446.
29. Cook GJR, Fogelman I, Maisey M. Normal physiological and benign pathological variants of 18-fluoro-2-deoxyglucose positron emission tomography scanning: potential for error in interpretation. Semin Nucl Med 1996;24:308–314.

30. Cook GJR, Maisey MN, Fogelman I. Normal variants, artefacts and interpretative pitfalls in PET imaging with 18-fluoro-2-deoxyglucose and carbon-11 methionine. Eur J Nucl Med 1999;26:1363–1378.
31. Hany TF, Gharehpapagh E, Kamel EM et al. Brown adipose tissue: a factor to consider in symmetrical tracer uptake in the neck and upper chest region. Eur J Nucl Med 2002;29:1393–1398.
32. Bischof Delalove A, Wahl RL. How high a level of FDG abdominal activity is considered normal? J Nucl Med 1995;36:106P.
33. Nakada K, Fisher SJ, Brown RS, Wahl RL. FDG uptake in the gastrointestinal tract : can it be reduced? J Nucl Med 1999;40:22P–23P.
34. Kosuda S, Fisher S, Kison PV, Wahl RL, Grossman HB. Uptake of 2-deoxy-2-[18F]fluoro-D-glucose in the normal testis: retrospective PET study and animal experiment. Ann Nucl Med 1997;11:195–199.
35. Picard Y, Thompson CJ. Motion correction of PET images using multiple acquisition frames. IEEE Trans Med Imaging 1997;16:137–144.
36. Nunez RF, Yeung HW, Macapinlac HA, Larson SM. Does post-radiation therapy changes in the lung affect the accuracy of FDG PET in the evaluation of tumour recurrence in lung cancer. J Nucl Med 1999;40:234P.
37. Diederichs CG, Staib L, Vogel J et al. Values and limitations of 18F-fluorodeoxyglucose-positron-emission tomography with preoperative evaluation of patients with pancreatic masses. Pancreas 2000;20:109–116.
38. Knopp MV, Bischoff HG. Evaluation of pulmonary lesions with positron emission tomography. Radiologe 1994;34:588–591.
39. Lewis PJ, Salama A. Uptake of Fluorine-18-Fluorodeoxyglucose in sarcoidosis. J Nucl Med 1994;35:1–3.
40. O'Doherty MJ, Barrington SF, Campbell M, et al; PET scanning and the human immunodeficiency virus-positive patient. J Nucl Med 1997;38:1575–1583.
41. Goerres GW, Kamel E, Seifert B et al. Accuracy of image coregistration of pulmonary lesions in patients with non-small cell lung cancer using an integrated PET/CT system. J Nucl Med 2002;43:1469–1475.
42. Goerres GW, Kamel E, Heidelberg TN et al. PET-CT image co-registration in the thorax: influence of respiration. Eur J Nucl Med 2002;29:351–360.
43. Goerres GW, Burger C, Schwitter MR et al. PET/CT of the abdomen: optimizing the patient breathing pattern. Eur Radiol 2003;13:734–739.
44. Osman MM, Cohade C, Nakamoto Y et al. Clinically significant inaccurate localization of lesions with PET/CT: frequency in 300 patients. J Nucl Med 2003;44:240–243.
45. Dizendorf E, Hany TF, Buck A et al. Cause and magnitude of the error induced by oral CT contrast agent in CT-based attenuation correction of PET emission studies. J Nucl Med 2003;44:732–738.
46. Goerres GW, Hany TF, Kamel E et al. Head and neck imaging with PET and PET/CT: artefacts from dental metallic implants. Eur J Nucl Med 2002;29:367–370.
47. Kamel EM. Burger C. Buck A. von Schulthess GK. Goerres GW. Impact of metallic dental implants on CT-based attenuation correction in a combined PET/CT scanner. Eur Radiol 2003;13:724–728.
48. Antoch G, Freudenberg LS, Egelhof T et al. Focal tracer uptake: a potential artifact in contrast-enhanced dual-modality PET/CT scans. J Nucl Med 2002;43:1339–1342.
49. Cohade C, Wahl RL. Applications of PET/CT image fusion in clinical PET – Clinical use, interpretation methods, diagnostic improvements. Semin Nucl Med 2003;33:228–237.
50. Goerres GW, Ziegler SI, Burger C et al. Artifacts at PET and PET/CT caused by metallic hip prosthetic material. Radiology 2003;226:577–584.
51. Kinahan PE, Hasegawa BH, Beyer T. X-ray based attenuation correction for PET/CT scanners. Semin Nucl Med 2003;33:166–179.
52. Cohade C, Osman M, Nakamoto Y et al. Initial experience with oral contrast in PET/CT: phantom and clinical studies. J Nucl Med 2003;44:412–416.
53. Dizendorf EV, Treyer V, Von Schulthess Gk et al. Application of oral contrast media in coregistered positron emission tomography-CT. AJR 2002;179:477–481.
54. Hany TF, Steinert HC, Goerres GW et al. PET diagnostic accuracy: improvement with in-line PET-CT system: initial results. Radiology 2002;225:575–581.

15 The Technologist's Perspective

Bernadette F Cronin

Introduction

From the technologist's point of view, positron emission tomography (PET) combines the interest derived from the 3D imaging of X-ray CT and MRI with the functional and physiological information of Nuclear Medicine. Until recently PET imaging was performed on "PET only" scanners and any direct comparison between the PET image and an anatomical one was either done by eye or required sophisticated hardware and software as well as extra human power to create a registered image. Now all the key manufacturers have developed and are marketing combined PET/CT scanners that allow patients to have both a PET scan and a CT scan without getting off the scanning couch. This creates an exciting imaging modality that offers the technologist the chance to develop a new range of skills.

Although clinical PET has been developing over the last 12 to 15 years, it is still relatively new and the high cost of introducing such a service has limited access for staff wishing to enter the field. As a result it is still the case that few staff entering the field will have previous experience. However, technologists with experience in other modalities will bring with them useful knowledge and, although PET in most cases is developing as part of nuclear medicine, recruitment does not need to be confined to nuclear medicine technologists alone. In the UK, training for PET technologists is performed on site with staff learning and gaining experience while working full time. Most recognized undergraduate and post-graduate courses cover PET in a limited capacity, providing theoretical information on a subject that is intensely practical. As with all new techniques there will, for a while, be a discrepancy between the number of centres opening around the country and the availability of trained staff. It is important for all concerned, and for the viability of PET itself, that training is formalized to create high national standards which can be easily monitored and maintained.

A common view is to assume that a PET scanner can be sited in an existing nuclear medicine department and that, once commissioned, will function with little extra input. Although there are obvious similarities between PET and nuclear medicine, there are also quite subtle differences and it is not unusual for experienced nuclear medicine staff to find PET quite bewildering. The advent of PET/CT adds an extra dimension as many Nuclear Medicine Technologists will not have any CT training or background. This too must be addressed in a structured way to ensure that the technique develops optimally. Currently, it is probably the least predictable imaging modality and staff working in a PET facility need to possess certain characteristics if the unit is to be successful. PET is multi-disciplinary, requiring input from various professional groups, and it is important that all these groups work together towards the same end. Teamwork is an essential component of a successful PET unit and without this, departments can easily flounder. More people are gaining the required expertise to work in these units; however, the numbers of people with the necessary expertise are still small and may not always be available to staff newly emerging departments. In the event that departments are unable to recruit people with existing PET experience, it is advisable to send staff to comparable PET centers to obtain initial training.

Setting up a PET Service

When setting up a PET service, several factors have to be considered. First, and most important, what is the unit to be used for? If the only requirement is for clinical oncology work, then a stand-alone scanner with the tracer being supplied by a remote site may well be sufficient. For the technologist this option is in some ways the least attractive as it will always leave the unit vulnerable to problems over which it has no control, and will provide the least diverse workload. At the moment, most of the work will be done using fluorine-18-labeled tracers (mainly 2-[^{18}F]-fluoro-2-deoxy-D-glucose ([^{18}F]-FDG)) and although the half-life is reasonably long (109.7 minutes) it does not provide much margin for error and therefore scheduling the patients can present added difficulties. If you are fortunate enough to be involved with a unit which has both scanner(s) and a cyclotron with chemistry facilities, the opportunities expand enormously. Scheduling, although not necessarily easier, can be more flexible, enabling better use of the equipment at your disposal. In addition, the variety of scans that can be performed will increase allowing you to provide a full clinical oncology, neurology, and cardiology service as well as the flexibility required for research work. Another factor to be considered is what service you will offer if you are purchasing a PET/CT scanner. All commercially available PET/CT scanners combine state-of-the-art PET scanners with diagnostic CT scanners. The specification of the CT component varies between manufacturers, but all are capable of high quality diagnostic work. Opinions will vary as to whether the CT scanner should be used only for attenuation correction and image fusion or whether the patient should have both their PET scan and their diagnostic CT scan performed on the one machine during a single hospital/clinic visit. The concept of "one-stop shops" has also been considered, where patients will have their PET, CT and radiotherapy planning all performed in a single visit.

Staffing Requirements and Training

Staffing levels will reflect the aims of the unit. If the intention is to have a stand-alone scanner with no on-site cyclotron and chemistry facilities, then, assuming a throughput of ~800 patients per annum, the department will need a minimum of two technologists. Two are needed to ensure that there will always be cover for annual leave and sickness and also to share the radiation dose. This number of staff assumes that there will be some additional scientific and clerical support staff. However, with a scanner plus cyclotron and chemistry facilities, the annual patient throughput is likely to increase due to greater availability of radiopharmaceuticals for clinical studies and, possibly, research work. Assuming an annual workload of ~1200 patients, there would need to be at least three technologists with a corresponding increase in support staff. Provision of [^{18}F]-FDG for 8:00 am injection, allowing scanning to start at 9:00 am, requires cyclotron and chemistry staff to begin work very early. This has obvious revenue consequences as it will necessitate out-of-hours or enhanced payment and may make recruitment difficult. For this reason it is possible that in many centers the [^{18}F]-FDG will not be available for injection much before 9:30–10:00 am, with the first scan not starting until nearly 11:00 am. To maximize the resources available it will probably be necessary to extend the working day into the early evening by working a split shift system. Once the PET technologists have been selected, consideration must be given to how they will be trained. There is little point sending a technologist who is going to be working in a clinical PET center to train in a center that undertakes research work only. The two units will operate in very different ways with different priorities, and are likely to be using a different range of tracers. To gain an in-depth working knowledge of clinical PET, technologists should spend at least four and preferably eight weeks at their "training" center. The amount of training required will depend on the existing knowledge of the technologist. It must be remembered that in order to function successfully in a PET unit, the technologist should be experienced in patient care, handling of unsealed sources, intravenous cannulation, and administration of radiopharmaceuticals as well as the acquisition and subsequent reconstruction of the data.

A good basic understanding of physics and computing will also help, particularly when it comes to troubleshooting problems. CT training is also likely to be an advantage as time goes on and there are several short courses available that will provide a good grounding in basic CT. If the work is likely to include full diagnostic CT, there are staffing implications with regards to state registration and the operation of CT scanners which cannot be ignored.

Planning the PET Service

Clinical PET is divided into three main categories comprising oncology, neurology, and cardiology. Most clinical PET centers have found that the workload is split among these three areas in a similar way, with oncology taking about 75 to 85% and the balance being shared between the other two.

When setting up a clinical PET service it is important to consider the types of scans that will be offered. When deciding which protocols are going to be used, the most important thing to consider is what question is being asked of PET. This may not always be clear in the initial referral, and it is important that this is established prior to the study so that PET is not used inappropriately. Other things to consider are whether quantification (for example, a semi-quantitative standardized uptake value (SUV) [1, 2]) of the data will be required as this can only be performed on attenuation corrected studies. It is also important to know whether the patient is a new patient or is attending for a PET study as part of their follow-up. If they are for a follow-up scan it is essential to know what treatment they have had and, critically, when this treatment finished or was last given. Following the completion of chemotherapy, approximately 4–6 weeks should elapse before the patient is scanned, and after radiotherapy it is ideal to wait longer, up to 3–6 months, although clinically this may not be practical. If patients have their PET scan midway through a chemotherapy regime, the aim should always be to scan the patient as close to the next cycle of chemotherapy as possible, allowing at least 10 days to elapse since the previous cycle. Use should also be made of other diagnostics tests which the patient may have undergone, and so for the PET center it is valuable to have details about previous scans (CT, MRI, *etc*) and preferably to have the scan reports or images available before the appointment is made.

For PET, patient compliance must also be considered. Patients are accustomed to the idea that diagnostic tests on the whole are becoming quicker. However, with PET, the scanning times are significantly longer than other scanning techniques. Scans can take anything from 20 minutes up to three hours for some of the more complex studies, and it is important to establish whether the patient is going to be able to tolerate this. The design of PET systems, although not as confined as MRI, can mean that patients who are claustrophobic may not be able to complete the scan without help and may need to be given some mild sedation in order to undergo this type of investigation. Recently developed PET/CT dual-modality systems will exacerbate this even more due to their increased axial length. Some patients, particularly younger subjects, will not be able to keep still for the length of time required for the scan and may well need heavier sedation or perhaps a general anaesthetic.

The other key area that must be considered when setting up a PET service is which tracers are available. In centers with a cyclotron on site, a full range of clinical radiopharmaceuticals should be available. However, when operating a PET scanner at a site remote from a cyclotron, there will be much greater limitation as to what tracers a re available (essentially, ^{18}F-labeled tracers or those labeled with longer-lived radioisotopes) and that in turn will limit the types of studies possible. The half-lives of these tracers can be very short, which, again, will influence the types of scan possible as well as the way the work is scheduled. Production times for the various tracers range from just a few minutes up to two or three hours and so scans must be booked accordingly. As with all imaging units, it is important to ensure that the best and most efficient use of the equipment and personnel available is made. PET scans are costly both financially and in terms of the time required, and a single scanner may only be able to accommodate 7–10 scans per day. The way the schedule is arranged should ensure that the scanner is in continuous use throughout the day to maximize throughput. Currently in the UK there are few centers producing [^{18}F]-FDG for distribution, and those centers which do cannot produce it in vast quantities. It is essential that tracer is used in a way that ensures that little, if any, is wasted, and this requires a lot of thought on the part of the people booking the scans.

Information that should be given to the patient obviously includes the time, date, and location of where the examination will take place, along with a clear explanation of what the scan involves and any special dietary requirements including special instructions for diabetics. An information booklet may be of help but it may need to be tailored to specific scan types and the working practice of individual departments. It is a good idea to ask the patient to provide confirmation of their intention to attend for their appointment. A patient not attending at the prearranged time can cause a huge disruption and waste of both tracer and scanner time, which is difficult to fill at short notice.

Patient Preparation and Scanning Protocols

Different centers will have different methods for preparing and performing the tests, and these variations will be determined by several factors, including the type of PET system available to them. However, regardless of the type of PET camera being used, there will be some similarities in technique with the greatest overlap in the area of patient preparation and management of the patient once they are in the department in order to obtain the best diagnostic data.

PET Scanning in Oncology

Today, [^{18}F]-FDG PET scanning is regarded as useful in many oncological conditions. The most frequent applications are in staging of disease, the assessment and monitoring of treatment response, and in the evaluation of tumor recurrence, particularly when morphological imaging techniques are equivocal or difficult to assess for technical reasons.

Patient preparation for [^{18}F]-FDG PET scans in oncology is fairly simple. Patient referrals are divided into two categories, those who are insulin-dependant diabetics (IDD) and all others. Insulin-dependant diabetics are asked to drink plenty of water in the six hours leading up to their scan appointment, but they are not asked to fast because it is both inappropriate and unnecessary to disrupt their blood glucose levels. All other patients are asked to fast for at least six hours prior to their scan but are again encouraged to drink plenty of water. On arrival in the department, the patient's personal details should be checked and a clear explanation of what the scan involves should be given. It is important to give the patient the opportunity to ask questions at this stage so as to ensure that they are completely relaxed. For this reason, the appointment time should include a pre-injection period of about 15 minutes. For [^{18}F]-FDG PET scanning it is advantageous to have the patient lying down for both the injection and the uptake period as this will hopefully aid relaxation and reduce unwanted muscle and brown fat uptake. In order to get the best out of the [^{18}F]-FDG PET scan it is important that certain key areas are given due consideration. The site of the body chosen for injection of the [^{18}F]-FDG is important. Any extravasation at the injection site is unsatisfactory. A small amount of tissued radiotracer at the injection site will result in a local radiation dose caused primarily by the positrons themselves due to their short path length and high linear energy transfer. In addition to this, the extravasated tracer will be cleared from the injection site via the lymphatic system and may result in uptake in more proximal lymph nodes that could be confused with uptake due to lymph node disease. Even small amounts of extravasated tracer can cause devastating artefacts on the 3D reconstruction of the data and, in extremis, can render the scan non-diagnostic if it is adjacent to an area of interest (see example in Fig. 15.1). Iterative reconstruction methods will help by

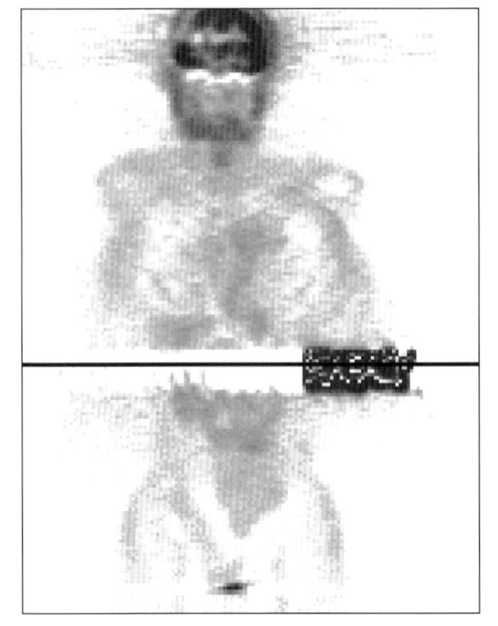

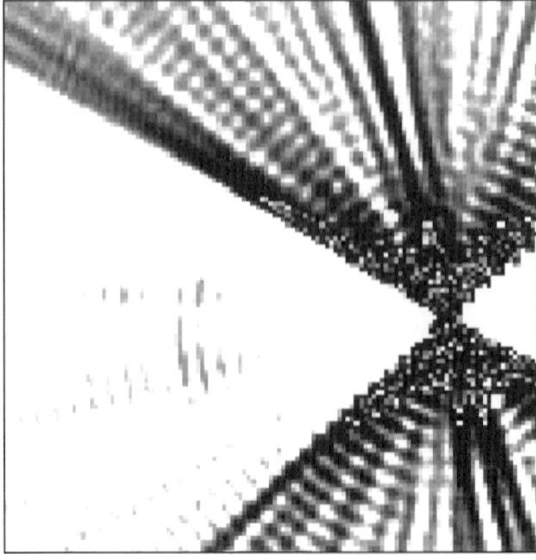

Figure 15.1. Reconstruction artifacts caused by extravasated tracer at the injection site can leave the final image difficult or even impossible to interpret.

reducing these artefacts but may not remove them altogether.

Many of the patients attending for these scans will have already undergone chemotherapy regimes that may have made their veins extremely difficult to access. It is therefore essential that the technologist giving the injections is well trained and well practiced in venepuncture. A good, precise injection technique is essential with the emphasis on accuracy. Unfortunately, there will occasionally be mistakes, but with careful thought it is possible to ensure that these mistakes do not make the scan completely worthless. It is always advisable to keep the injection site as far away as possible from the area of interest. If the abdomen is under investigation it is best to avoid the ante-cubital fossae and likewise the hands or wrists when the pelvic area is significant. For many patients the arms may not be at all appropriate for the injection; for example, patients with a history of carcinoma of the breast, patients with bilateral axillary, neck, or supra-clavicular disease, and patients whose primary lesion is on their arm. In any of these instances it is advisable to inject the patient in their foot so that if there is a problem it will not affect the scan to such a degree that it may cause confusion. A note should always be made of the site of the injection so that the reporting doctor can refer to this if clarification is needed. A note should also be made if an unsuccessful attempt at venepuncture has been made prior to the successful one, as leakage of injected tracer can occasionally be seen at this initial site. If it is suspected that tracer has extravasated it is advisable to perform a quick test acquisition at that site to establish if this is the case, as it is not helpful to remove the injection site from the field of view without establishing whether it is necessary. If it is, then remedial action can be taken; for example, lifting the arm above the patient's head and out of the field of view. With PET/CT it is likely that patients will be routinely imaged with their arms raised above their head. This is done to prevent beam hardening artefacts in the CT image. However, an image where the injection site (the arm for example) is not seen but which does include axillary uptake will be difficult to assess unless the reporting clinician can be sure that the injection was without incident. For this reason a quick acquisition of the injection site will avoid any subsequent confusion.

Normal physiological muscle and brown fat uptake of [^{18}F]-FDG can also give rise to confusion when reporting PET scans (Fig. 15.2) [3, 19]. The causes of this increased uptake are not fully understood, however, it is clear that physical exertion and cold will inevitably give rise to increased levels of [^{18}F]-FDG in these areas. In addition to this, stress and nervous tension also play their part.

Unfortunately it is not always manifestly clear which patients are going to produce high levels of muscle/brown fat uptake, and the calmest of patients can produce a very "tense" – looking scan. When uptake of this type occurs it is only the distribution that may give a clue as to the origins of the pattern of uptake, and it is almost impossible to distinguish this physiological response from any underlying pathological cause. In order to reduce this uptake it is suggested that patients be given 5–10 mg of diazepam orally one hour prior to the [^{18}F]-FDG injection. Clearly this is going to increase the time they spend in the department, as well as influencing their homeward journey, as they should not drive following this. It is not something that can be done at short notice as patients will have to be brought in for their appointment an hour

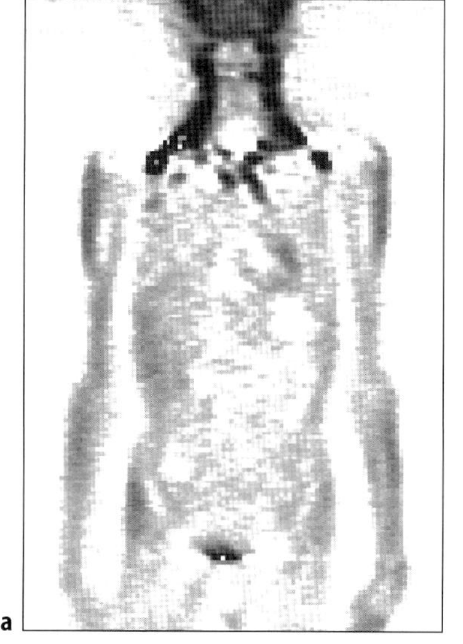

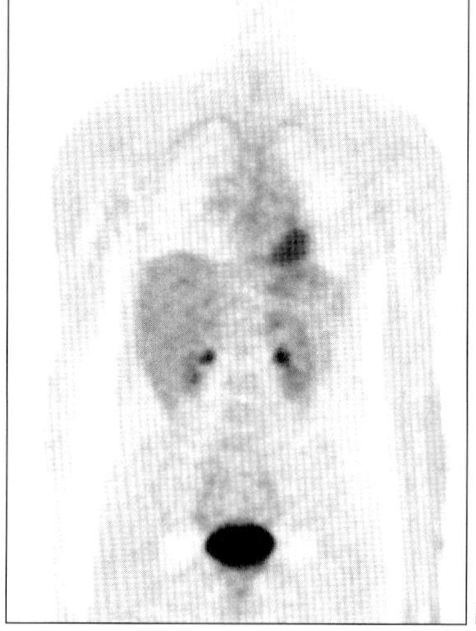

Figure 15.2. FDG uptake into tense muscle can mask or simulate underlying pathology (**a**). The use of diazepam prior to FDG injection can significantly reduce this effect (**b**).

earlier than usual and an area will have to be provided where they can rest comfortably following the administration of the diazepam. Although the dosage suggested is relatively low, patients' reactions range from barely negligible to sleeping throughout most of their visit to the department. The unpredictability of this reaction means that it is generally advisable to suggest that the patient bring someone along with them for their scan to ensure that they are accompanied once they leave the department. Too low a temperature may also increase the incidence of this physiological uptake so it is also important that, on arrival, patients are injected and allowed to rest in a warm comfortable area. Although the likelihood of physiologically increased muscle/brown fat uptake is difficult to predict, there are certain categories of patients where diazepam should be given prophylactically. Obviously, any patient who has previously demonstrated physiological uptake with [^{18}F]-FDG should be given diazepam on all subsequent visits. For patients attending for their first PET scan, those with lymphoma where the axillae, neck, and supraclavicular area is of particular interest may well be given diazepam, and patients with carcinoma of the breast are also likely to be considered. Adolescents also give rise to significantly increased amounts of physiological muscle/brown fat uptake and they too are routinely given diazepam, although in this group of patients it is less successful than with adults. In addition to the diazepam, every effort should be made to ensure that the total environment and experience is as relaxing as possible for the patient and that they feel comfortable and well informed about the procedure. It is difficult to evaluate the true effect of the oral diazepam because for each patient there is no control study. However, it does appear to reduce the incidence of this type of normal uptake and is therefore used frequently by many centers.

Uptake of [^{18}F]-FDG into normal myocardium is both unpredictable and difficult to manage. When investigating the thorax of a patient, particularly when looking for small-volume disease close to the myocardial wall, high levels of myocardial uptake can cause such severe artefacts that this can be almost impossible to assess (Fig. 15.3).

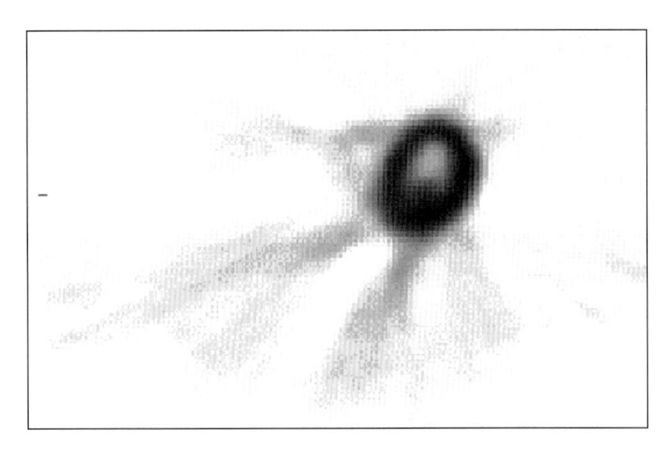

Figure 15.3. High myocardial uptake in a scan looking for small-volume disease in the chest can render the scan uninterpretable.

The use of iterative reconstruction methods will help decrease the magnitude of the artefact, but ideally it is desirable to minimize the myocardial uptake of the [^{18}F]-FDG when necessary. The physiology of uptake of [^{18}F]-FDG into the myocardium is complex and there are, as yet, no straightforward methods of reducing it. One suggestion is that by increasing the free fatty acid levels in the patient the heart may be persuaded to utilize this for its primary energy source and ignore the available glucose and [^{18}F]-FDG. One way of doing this is to ensure that the patients have a high caffeine intake prior to, and immediately after, the [^{18}F]-FDG injection. This can be achieved simply by encouraging the patient to drink unsweetened black coffee instead of water in the hours leading up to the scan and during the uptake period or, more palatably, a diet cola drink as this contains no sugar but high levels of caffeine.

Unlike glucose, [^{18}F]-FDG is excreted through the kidneys into the urine, and artefacts can be produced if the radioactivity concentration in the urine is high. These artefacts can be seen around the renal pelvis and the bladder and can make it very difficult to assess these areas with confidence. In patients where the upper abdomen or pelvis is in question, it is advisable to give 20mg intravenous frusemide at the same time as their [^{18}F]-FDG injection. The frusemide should be injected slowly and the patient should be encouraged to drink plenty of water to ensure that they do not become dehydrated. Patients should be given an opportunity to void immediately prior to the start of the [^{18}F]-FDG scan and if the pelvis is to be included in the scan, then the patient should be scanned from the bladder up. Obviously, it is not easy for patients to be given a diuretic and then have limited access to a toilet, but in the main, with a good explanation, patients tolerate this quite well.

Finally, in patients where the laryngeal and neck area is being investigated, normal physiological muscle uptake into the laryngeal muscles can cause confusion. Patients who speak before, during and after injection are likely to demonstrate this type of uptake, whereas patients who are silent for this period do not [4]. For this reason it is recommended that when this area is of clinical significance, the patient should be asked to remain silent for about 20 minutes before injection, during injection and for the majority of the uptake period (the first half hour being the most important time after injection). Although this is not as easy as it sounds, most patients, if they understand why it is necessary, are sufficiently well motivated to comply.

PET Scanning in Neurology

For clinical PET in neurology, the most frequently used tracer is [^{18}F]-FDG. It features in the general work-up for patients who have intractable epilepsy for whom the next option would be surgery, and PET is one of the many investigations these patients will undergo to confirm the exact site of the epileptogenic focus. PET is also used in the investigation of primary brain tumors, mainly in patients who have had tumors previously treated by surgery, radiotherapy, or a combination of both. In these patients, standard anatomical techniques – for example, CT and MRI – are often difficult to interpret because of the anatomical disruption caused by the treatment. [^{18}F]-FDG PET scanning may also be helpful when investigating patients with dementia-like symptoms, as it can provide a differential diagnosis for the referring clinician.

When investigating partial epilepsy, [^{11}C]-flumazenil may also be used. This is a radiopharmaceutical that looks at the distribution of benzodiazepine receptors and it can be useful in more accurately localizing the epileptogenic focus [5] than [^{18}F]-FDG alone. Likewise, in the investigation of primary brain tumors, the use of [^{18}F]-FDG alone may not always be sufficient to answer the clinical question. Normal cortical brain has a high uptake of [^{18}F]-FDG and so makes visualization of tumor tissue over and above this background level quite difficult. The grade of the brain tumor is correlated with the degree of uptake into the tumor [6],with low-grade tumors having a lower than normal uptake, intermediate-grade tumors having a similar uptake to normal brain, and high-grade tumors having a higher than normal uptake. For this reason it is common for patients who are being assessed for recurrent tumor to have a [^{11}C]-L-Methionine scan in addition to their [^{18}F]-FDG scan. Methionine is an amino acid that demonstrates protein synthesis rates, so, whilst the [^{18}F]-FDG gives information about the grade of the tumor, the methionine will provide information about the extent of any tumor present. To use any ^{11}C-labeled tracer obviously requires that the scanner be sited close to the cyclotron and radiochemistry unit where the tracer will be produced, as the 20-minute half-life along with the low quantities currently produced precludes transport to a distant site.

The exact protocol for brain scanning will vary from center to center but again there are certain key points that can be taken from one site to another. For all of these scans, patients may have to keep still for quite a significant period of time. When dealing with a patient suffering from epilepsy, particularly if they are very young, it may be that the risk of them having a seizure during the scan is quite high. To ensure that this causes the minimum amount of disruption to the total study, it is advisable to acquire the data as a series of sequential dynamic time frames rather than as one long, single, static acquisition. This means that if the patient does move during the scan it is possible to exclude that frame from the total scan data. If the patient is clearly marked at various points corresponding to the positioning laser lights, they can be repositioned so that only the affected frame is lost. However, any repositioning needs to be done with extreme accuracy to ensure co-alignment of the transmission data for attenuation correction, which is often recorded prior to the start of the emission scan. Patient movement is a particularly important issue when dealing with the shorter-lived ^{11}C tracers. If a patient moves during a single static acquisition, not only is the scan going to be corrupted, but the short half-life of the tracer will make it very difficult to repeat the scan as the resultant scan quality will be very poor. An alternative which is becoming increasingly popular on PET scanners with the ability to operate in 3D mode, is to acquire these as high-count, short-duration 3D scans only. Emission scan times in 3D for a brain may be of the order of only 5–10 minutes.

Preparation of patients for brain scans is again concerned with ensuring that the injected [^{18}F]-FDG does not have to compete with high levels of normal glucose in the patient. So, as for oncology scans, the patients are fasted and asked to drink water only in the few hours preceding the study, but because of the high avidity of uptake in the brain this fasting period is restricted to three hours only. As with all other areas of clinical PET, insulin-dependent diabetics should be told to maintain their normal diet and insulin intake and should not be fasted. Following injection it is important that the patient remains quiet and undisturbed for the uptake period. External stimuli should be avoided, as these may cause "activation" within different areas of the brain, which, in extreme cases, might be evident in the final image produced. Ideally, it is best to ask the patient to remain in a darkened room with their eyes closed and no distractions. This applies primarily to [^{18}F]-FDG scanning but should also be adhered to when using other tracers so that the preparation is standard and reproducible. Many patients being investigated for epilepsy may well be young children, and may also have concurrent behavioural problems. As a result it can be difficult to keep these patients still for the length of time required for the scan, and it is prudent therefore to consider scanning them with some form of sedation. In general, it is preferable to scan under a full general anaesthetic rather than a light sedation. This is because the latter can be unpredictable and difficult to manage. However, a general anaesthetic requires good cooperation with the anaesthetics department and an anesthetist who has a reasonable understanding of PET and the time constraints associated with the technique. One option is to have a regular general anaesthetic time booked, the frequency of which can be governed by the individual demands of the PET center. The anaesthetic will unfortunately result in a globally reduced [^{18}F]-FDG uptake, which will give a slightly substandard image, and the best results will be achieved if the patient is awake for their [^{18}F]-FDG uptake period but under anaesthetic for the scan. This is not always possible, particularly when patients are having a dual-study protocol with [^{18}F]-FDG scans in conjunction with an ^{11}C-labelled tracer.

Sequential Dual Tracer Neurological Studies

In general, the technique for performing dual-tracer studies is usually for the scan using the shorter halflife tracer to be performed first, immediately followed by the [^{18}F]-FDG scan. When the patient is sedated or under a general anaesthetic, the uptake period for the [^{18}F]-FDG will inevitably be whilst the patient is anaesthetized. However, the advantages of having the patient remaining still for the study usually outweigh the slightly reduced quality of the study due to lower global uptake. In children under 16 years who are being investigated for epilepsy, it is advisable to perform an EEG recording for the whole duration of the uptake period.

[^{18}F]-FDG PET brain scans for the investigation of epilepsy are always performed inter-ictally. The short half-life of the tracer makes ictal scanning technically very challenging and the comparatively long uptake period of [^{18}F]-FDG into the brain could render ictal scanning unreliable. However, if a patient does enter status epilepticus during the uptake period it may have an effect on the final image, and so it is worthwhile noting brainwave activity during this period. [^{18}F]-FDG scans in the inter-ictal state in epilepsy are usually employed to locate regions of diminished uptake. Most adults can tolerate [^{18}F]-FDG brain scanning with or without an ^{11}C scan fairly well and it is rare to require either general anaesthetic or sedation for these patients. Once the acquisition is completed the individual frames can be assessed for patient movement and if there has been no movement they can be summed and reconstructed as a single frame study. Alternatively, it is possible, if there have been small amounts of movement between frames, to register sequential frames to each other before summing and reconstruction, although great care must be taken to ensure that the frames are co-aligned with the transmission scan data used for attenuation correction. Where there has been significant movement on a frame this will probably have to be excluded from the study and the remaining frames summed and reconstructed.

The production of ^{11}C radiopharmaceuticals tends to be is less reliable than [^{18}F]-FDG and the yields tend to be lower. In addition to this, the half-life of ^{11}C is only 20.4 minutes, which means that there is a limited period of time during which there is enough radioactivity to achieve a good diagnostic scan. Radiopharmaceuticals labeled with ^{11}C are produced on demand and can take approximately one hour to make. As a result, it is common practice to book patients so that they arrive in the PET scanning department about one hour before their scan appointment time. This means that the production need not be started until the patient presents for the study, thus removing the possibility of having tracer ready to inject but no patient. It is important that patients understand the reason for this so that they are aware that, on arrival in the department, there will be a delay before their scan is started.

PET Scanning in Cardiology

Aside from some special research studies, PET cardiology studies are usually performed to assess hibernating myocardium or viability in patients with left ventricular (LV) dysfunction. This can involve assessing myocardial perfusion using ammonia ([^{13}N]-NH3) or [^{15}15O]-H_2O and glucose metabolism with [^{18}F]-FDG. Hibernating myocardium may be thought of in terms of myocardium with reduced function due to an adaptive response to chronic hypoperfusion and which, if revascularized, may recover function. Patients with poor left ventricular function and ischaemic heart disease represent only a small percentage of surgical candidates but are an important group. They have increased mortality and morbidity compared to patients with normal LV function, but stand to gain most from revascularization in terms of increased survival. Separating those who may benefit from surgery (in whom the associated operative risks are worth taking) from those whom surgery cannot hope to benefit is vital.

Rest/stress flow studies can be used in the investigation of reversible ischaemia in cases where [^{99m}Tc]-labeled perfusion agents or ^{201}Tl scanning is equivocal, difficult to interpret, or does not match the clinical findings. ^{13}N has a 9.97-minute half-life, which means that this tracer can only be used if the scanner is situated close to the production facility. [^{13}N]-NH_3 is a highly diffusible tracer which, following intravenous injection, is efficiently extracted into perfused tissue where it remains for a considerable period [7, 8]. Thus, it acts as a flow tracer with the uptake being proportional to perfusion. Dynamic [^{13}N]-NH_3 scanning allows quantfication of blood flow to the myocardium in mls/min/gm of tissue as well as giving qualitative information [9]. Likewise, [^{15}O]-H_2O is a freely diffusible tracer that distributes in the myocardium according to perfusion. However, the short half-life and rapid wash-out from myocardium makes it more difficult to use than [^{13}N]-NH_3.

[^{18}F]-FDG uptake gives information about the viability of the heart muscle. When the heart is normally perfused it relies on the oxidation of free fatty acids in the fasted state and aerobic glucose metabolism post prandially. When myocardial perfusion is reduced, the heart switches from free fatty acid metabolism almost entirely to glucose metabolism and, therefore, preferential uptake of [^{18}F]-FDG should occur in ischaemic regions (Fig. 15.4). As myocardial perfusion and oxygen consumption continue to reduce, myocardial contractility worsens and then stops whilst glucose consumption by the myocardium increases. However, in the fasted state, uptake of glucose can be very variable in both normal and ischaemic myocardium, making some studies difficult to interpret. Following a glucose load (50g orally), glucose becomes the primary energy source, with both normal and ischaemic myocardium taking up [^{18}F]-FDG. To minimize any inhomogeneity of [^{18}F]-FDG uptake it is best to glucose-load the patient prior to scanning. Uptake of [^{18}F]-FDG into the myocytes can be enhanced by insulin, however, the use of a feedback-controlled hyperinsulinemic clamp is technically time-consuming. We have found a reduced number of uninterpretable studies using a sliding scale of insulin, based on blood glucose prior to [^{18}F]-FDG injection [10]. In order to get acceptable [^{18}F]-FDG uptake into the myocardium it is vital that the patient is prepared correctly and managed in a controlled way throughout the PET scan. Uptake of [^{18}F]-FDG into the heart muscle is dependent on achieving the correct insulin/glucose balance in the patient prior to [^{18}F]-FDG injection, and so the

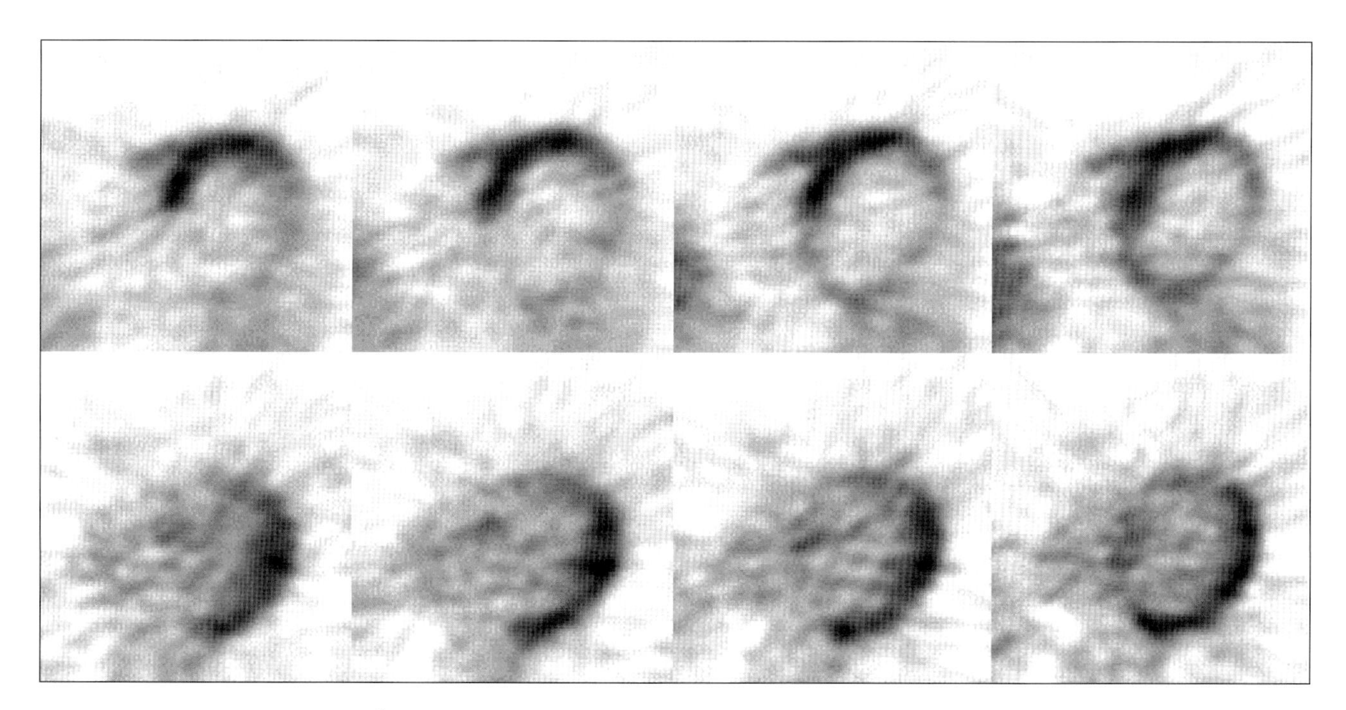

Figure 15.4. Myocardial perfusion with [^{13}N]-NH_3 (top row) and [^{18}F]-FDG (bottom row) demonstrates almost complete mismatch. The areas of low myocardial perfusion on the [^{13}N]-NH_3 scan demonstrate high [^{18}F]-FDG uptake, indicating hypoperfused, viable myocardium.

management of these patients is focused on this aspect.

Cardiac PET scanning is perhaps the most complex and time-consuming of the three main clinical groups. The patients are prepared slightly differently depending on whether they are non-diabetic, insulin dependent diabetic (IDD) or non-insulin-dependent diabetic (NIDDM). All categories of patients should be asked to refrain from caffeine-containing products for at least 24 hours prior to their appointment time. This minimizes both the inherent stress effect of caffeine and also lowers the free fatty acid level that might otherwise compete with the [^{18}F]-FDG for provision of the myocardium's energy source. Non-diabetic patients should be asked to fast (water only) for six hours prior to their appointment time. The non-insulin dependent diabetics should fast for six hours but if their appointment is booked for the afternoon they are encouraged to have breakfast and take their normal morning oral hypoglycemic as usual. The insulin-dependent diabetics are instructed to eat and take their insulin as normal with no modifications to their normal routine. Many patients attending for cardiac PET scanning will be on medication and it is not necessary for them to stop any of this prior to their scan.

After arrival in the department, the procedure should be explained to the patient and a blood sample taken to measure the baseline blood glucose level. If the blood glucose is below 7 mmol/L then the patient will have to be glucose loaded. The glucose can be given in the form of dextrose monohydrate powder dissolved in water and the aim is to give this drink approximately one hour before the [^{18}F]-FDG is due to be given. If the patient is not a diabetic they are given 50 g of dextrose monohydrate and if they are diabetic (IDD or NIDDM) they receive 25 g. About 10 minutes before the [^{18}F]-FDG is due to be injected another blood glucose measurement should be made, and depending on this result the patient may or may not be given insulin prior to [^{18}F]-FDG injection. The amount of insulin given will increase on a sliding scale and an example of one such scale is shown below:

Table 15.1. Sliding scale of insulin given dependent on blood glucose levels in cardiac PET [18F]-FDG scanning

Glucose Concentration (mmol/L)	Action
<5.0	no insulin
5.0–8.0	3 units of insulin
8.0–12.0	4.5 units of insulin
>12.0	(at clinician's discretion)

In the above example, the insulin used is Human Actrapid and this is diluted in about 1ml normal saline and given intravenously to the patient about five minutes before the [^{18}F]-FDG injection. The key to achieving a good [^{18}F]-FDG scan is to push the glucose level up high enough so that either the patient's own regulatory system will produce insulin in response to this, or the patient's glucose level will be high enough that exogenous insulin can be administered safely. Obviously great care must be taken when giving insulin to a patient. It is important that intravenous 50% glucose is on hand if required in case the patient becomes hypoglycemic, and it is also essential that the patient is not allowed to leave the department until they are safe and well to do so. If patients have been given insulin they should be infused with about 50ml of 20% glucose in order to counteract any effects from the insulin during the last 10–15 minutes of the [^{18}F]-FDG scan. It is also sensible to make sure the patient has something to eat and drink, and to check the patient's blood glucose level before they leave.

If the blood glucose level is checked after the patient has been injected with the [^{18}F]-FDG, it is important to ensure that the blood sample is not taken from the same line through which the [^{18}F]-FDG was given. About 20 minutes after the [^{18}F]-FDG has been administered, a short test acquisition should be performed to assess whether [^{18}F]-FDG is getting into the heart muscle. Even though the uptake may still be quite low at this stage and, because of the short acquisition time the counting statistics will be poor, an experienced eye should be able to judge whether or not there is uptake. If uptake is not seen at this stage it may be worth waiting another 5 or 10 minutes and then repeating the test acquisition. If there is still no uptake after this time a second administration of insulin (as prescribed by the doctor on site) may be enough to push the remaining circulating [^{18}F]-FDG into the myocardium. Despite the fact that by this stage it may be more than 30 minutes after the [^{18}F]-FDG injection it should not be assumed that the [^{18}F]-FDG scan is going to fail. This glucose loading protocol for cardiac imaging seems to work very successfully with a very low (<5%) failure rate.

Further information for assessing the viability of the myocardium may be obtained by gating the [^{18}F]-FDG scans. Gating of [^{18}F]-FDG scans is done in the same way as gating of nuclear medicine studies. The patient preparation is identical to that already described as the tracer used will still be [^{18}F]-FDG. The patient's electrocardiograph is monitored for a short time prior to the start of acquisition and the average R–R interval is divided into a selected number of equal segments.

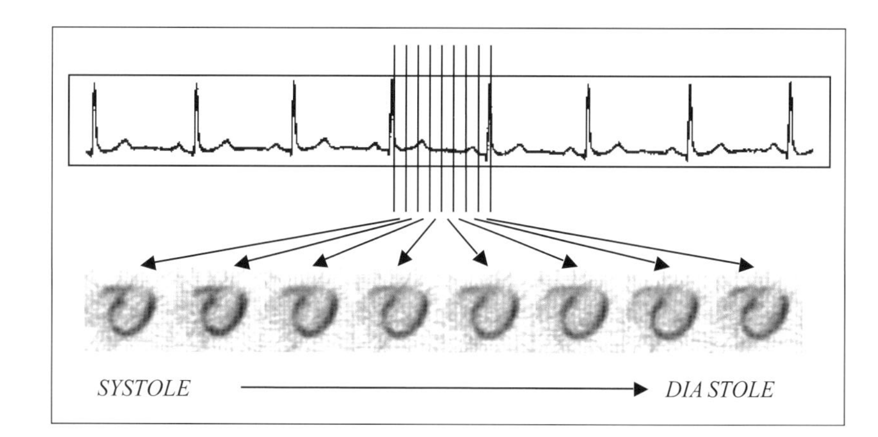

Figure 15.5. Gating of the cardiac cycle produces a series of images of the heart that can be viewed as a cine.

During the acquisition the same segment from each cardiac cycle acquired is stored in the same location of computer memory (Fig. 15.5).

Thus, once the data have been reconstructed, a ciné image of the beating heart can be viewed tomographically. Due to the relatively large volume of data created in PET scanning, fewer temporal bins (typically eight) are used than with nuclear medicine single-photon planar gamma camera studies. List mode acquisition of the data can be used to overcome this limitation, but is not available on all PET systems. Patients who have gated [^{18}F]-FDG cardiac scans will be unlikely to have [^{13}N]-NH_3 scans at the same visit. The acquisition time for the [^{18}F]-FDG scan should be slightly longer than for static [^{18}F]-FDG cardiac scans to compensate for the poorer counting statistics in each individual gate. Gated [^{18}F]-FDG scans appear to be slowly replacing the need for combined rest [^{13}N]-NH_3 / [^{18}F]-FDG scans and because only [^{18}F]-FDG is used this makes it much more accessible as centers remote from a cyclotron will be able to undertake clinical cardiac work more easily.

Radiation Protection for the PET Technologist

Over the last decade there has been a significant increase in interest in the clinical use of positron-emitting tracers. [^{18}F]-FDG with its 109.8-minute half-life lends itself to use not only at sites close to a cyclotron and chemistry facility, but also allows transportation to scanners at sites remote from the cyclotron. It is the most commonly used tracer in clinical PET at present and the combination of these two factors means that hospitals are looking more and more at the possibility of introducing this radiopharmaceutical for use within existing departments. However, clinical units usually require high throughput (compared with research facilities) and this inevitably means increased numbers of scans and an increased radiation burden in the department.

Shielding

For most hospitals the natural assumption will be that PET scanning, either on a dedicated ring system or a dual-headed coincidence gamma camera PET (GC-PET) system, will be performed in the Nuclear Medicine Department. It should be remembered that although there are similarities between nuclear medicine and PET, the latter makes use of radiotracers with a much higher photon energy than those usually encountered in nuclear medicine (511 keV compared to 140 keV). Local legislation in the United Kingdom [11, 12] sets the annual whole-body dose limit for unclassified radiation workers at 6 mSv. This is a significant reduction from the previous limit of 15 mSv, and consequently places a greater burden of responsibility on departments to ensure that staff members are provided with the means to keep their doses below this level. Doses to staff include not only the whole-body dose but also the dose to extremities. When handling unsealed sources the key extremity dose is that to the hands. The photon energy of 511 keV means that conventional amounts of shielding with lead or tungsten will not be sufficient and more emphasis on the inverse square law, as a means of reducing dose, may need to be employed.

Currently, [^{18}F]-FDG is not produced in enormous quantities, and on arrival the activity in the multi-dose vial will range from 7 to 20 GBq. These multi-dose vials should be shielded within a lead pot with a minimum thickness of 18 mm, although thicker pots (34 mm) are being introduced by some departments particularly in the light of increasing tracer yields. The design of the dose preparation area should be similar to that used in nuclear medicine. An L-shield with a lead glass insert and a lead base should be used. The thickness of the base and the shield should be of the order of 6 cm and the lead glass window should have a 5.5 cm lead glass equivalent. Measurements of instantaneous dose rate taken from a shielded vial of [^{18}F]-FDG (7–8 GBq) placed behind the L-shield range from 20 μSv/hr immediately in front of the L-shield to 2 μSv/hr at 2.0 m from the shield. Clearly, the dose rate immediately in front of the shield is quite high and, as this is the place where the technologist will be standing when drawing up the tracer, it is essential that a good, reliable and fast technique is developed so that the dose received is reduced to a minimum.

Dispensing

The suggested technique for drawing up the tracer differs from that commonly used in nuclear medicine, in that it is inadvisable to invert the lead pot and vial in order to withdraw the tracer. A better technique involves the use of a long flexible needle that will reach to the bottom of the vial coupled with a filter needle for venting. The vial can be held firmly in place in the lead pot with a pair of long-handled forceps, and the syringe, once attached to the long needle, can be angled away from the top of the vial. Thus, both hands are kept out of the direct beam of the photons being emitted from the unshielded top of the vial. In this way dose to the hands can be kept to a minimum and there is no risk of the vial sliding out from its shielding.

Various other methods have been used to try to reduce the hand dose further. For example, in some departments the needle within the [^{18}F]-FDG vial is connected to a long length of flexible tubing which itself is shielded and enables the technologist to put an even greater distance between their hands and the vial of activity. In addition, commercially available purpose-built units are available which provide a significantly higher degree of physical shielding. Each unit will need to be tested for their ease of use, especially in departments where more than one radiopharmaceutical is likely to be used, as this will require the technologist swapping vials between the unit and their own lead pot.

Opinions vary slightly from department to department, but in the majority of PET centers the technologists draw up into unshielded syringes. Once the tracer has been drawn up the syringe can be placed into a syringe shield. Lead and tungsten syringe shields tend not to be used as the thickness of metal required to produce effective shielding would render the shield unmanageable. Instead, many centers make use of 1 cm thick perspex (lucite) to manufacture the syringe shield. The effect of the perspex is to act as a barrier to any stray positrons and to increase the distance between the syringe and the technologist's hands thus making use of the inverse square law to reduce the dose received. The shielded syringe can be transported to the patient for injection in an 18 mm (minimum) lead carry pot, which will mean that the technologist need only be further exposed to the radiation during the injection itself.

Finger doses vary from center to center depending on the throughput. At the Guy's and St. Thomas' Hospital, a relatively busy clinical PET center in London, the monthly hand doses were compared to the activity handled by the technologists. Each technologist was injecting approximately 1GBq of radioactivity per day, although this activity was being drawn up from vials containing much higher levels of activity. The average hand dose for these technologists was between 4 and 7 μSv per month which, over a twelve month period, means that their extremity dose is comfortably within the current United Kingdom annual limit for unclassified workers (150 μSv). However, as [^{18}F]-FDG production and workload increases, technologists will find themselves handling more and more radioactivity and so they must not become complacent, but should continue to find ways to keep these doses to a minimum. The average hand doses given were with technologists handling [^{18}F]-FDG only, however, some of the shorter-lived tracers require higher injected activities to get a good diagnostic result. In centers where [$^{15}15_{O}$]-H_2O is used, a manual injection for a single scan (2.3 GBq) can result in a dose to each hand of nearly 1 mSv. It is important to ensure that these higher doses are shared between staff members and that centers do not end up with only one person performing this type of work.

Whole-body Dose

The whole-body doses to staff in PET are slightly higher than those typically seen in nuclear medicine.

Benatar *et al* looked at the whole-body doses received by staff in a dedicated clinical PET centre and then used measured data to estimate the doses which might be received by staff in departments using [^{18}F]-FDG, but not originally designed to do so [13]. The results showed that in a dedicated unit the average whole-body dose received by each technologist was around 5.5 μSv per patient study and approximately 14.0 μSv per day. It should be noted that the new annual dose limit for unclassified workers of 6 μSv equates to an average daily dose of 24 μSv . With the exception of same-day rest/stress [^{99m}Tc]-Myoview studies [14], the average dose per patient study in nuclear medicine is only 1.5 μSv [15], which is a lot lower than for PET studies. However, the throughput per technologist in a PET unit is likely to be lower than that in nuclear medicine and therefore, the daily whole-body dose received is comparable. This work was carried out in a dedicated purpose-built PET unit where the technologists are shielded from any radioactive sources for the majority of the working day and only had significant contact during injection and positioning of the patient. In this unit the control room is separate to, and shielded from, the scanning area and the room where the patients wait after injection prior to the scan. This contrasts with the set-up commonly seen in nuclear medicine where the control console is often sited adjacent to the gamma camera with no shielding between the two. In order to make estimates of the doses likely to be received by technologists using [^{18}F]-FDG in an existing department, Benatar *et al* measured the instantaneous dose rates at 0.1 m, 0.5 m, 1.0 m, and 2.0 m from the anterior chest wall in patients immediately following their [^{18}F]-FDG injection. The average dose rates per MBq injected at each of the four distances were used to estimate the whole-body dose that could be received by technologists. Figure 15.6 shows the whole-body dose received during a one-hour scan commencing one hour after injection based on an injected activity of between 100 and 200 MBq. These injected activities were used to try to reproduce the likely usage on a gamma camera PET system. As can be seen, at a distance of 2 m with an average injected activity of 200 MBq the whole-body dose received is of the order of 12 μSv. Although in isolation this dose may not seem particularly high, it must be remembered that this is the dose from each individual PET patient whilst each technologist is likely to scan many more than this in a year, as well as a number of nuclear medicine patients. Five hundred PET patients in a year would take the technologist close to the 6 μSv annual limit. Therefore, if PET is going to be introduced into existing departments, some thought needs to be given to the positioning of the control console relative to the scanner itself. Figure 15.7 shows the instantaneous dose rates measured from a patient one hour after being injected with [^{18}F]-FDG.

The dose rate immediately next to the bed is quite high (120 μSv.hr^{-1}). At one meter from the bedside it drops to 15 μSv.hr^{-1} but significantly, at the foot end of the bed the dose rate is only 4.0 μSv.hr^{-1}. Obviously, the dose received by a technologist can be reduced by careful planning even within a pre-existing unit. An alternative, if space is a constraint and the operator's console must be included in the scanning room, is to use a mobile lead shield. A practical shield, on wheels, has been reported as reducing the exposure rate by 90% from approximately 20 μSv.hr^{-1} at an operator's console roughly one meter from a patient who had been injected with 370 MBq of [^{18}F]-FDG to 2 μSv.hr^{-1} [16]. The comparable exposure rate from a typical [^{99m}Tc]-MDP bone scan patient at a similar distance is 6 μSv.hr^{-1}. Consideration must also be given where a patient who has been injected with a positron-emitting tracer waits during the uptake period. With [^{18}F]-FDG it is important that the patient is able to rest, preferably

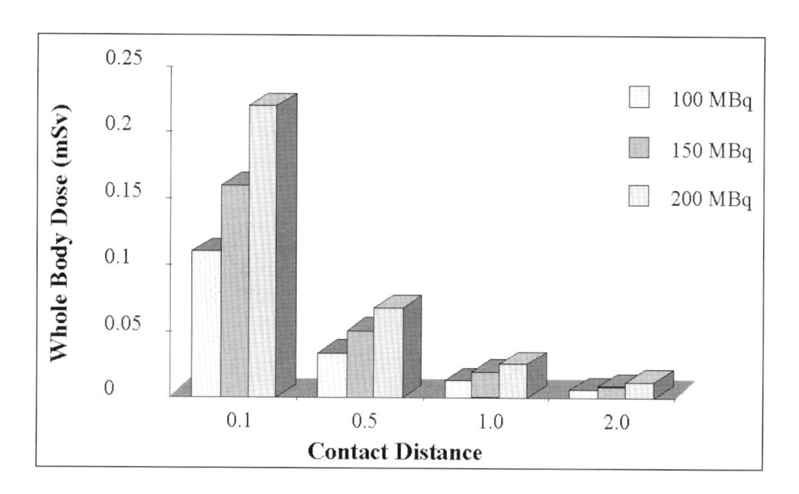

Figure 45.6. Estimated whole body doses likely to be received by technologists during 1 hour scan beginning 60 minutes after [^{18}F]-FDG injection.

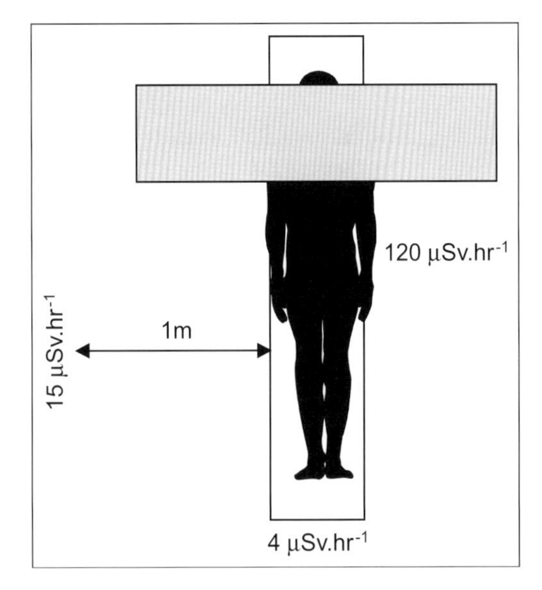

Figure 15.7. Dose rates measured from a patient in the scanner 55 minutes after injection of 334MBq of [18F-FDG. (Reproduced from Valk PE, Bailey DL, Townsend DW, Maisey MN. Positron Emission Tomography: Basic Science and Clinical Practice. Springer-Verlag London LTD, 2003, p. 789.)

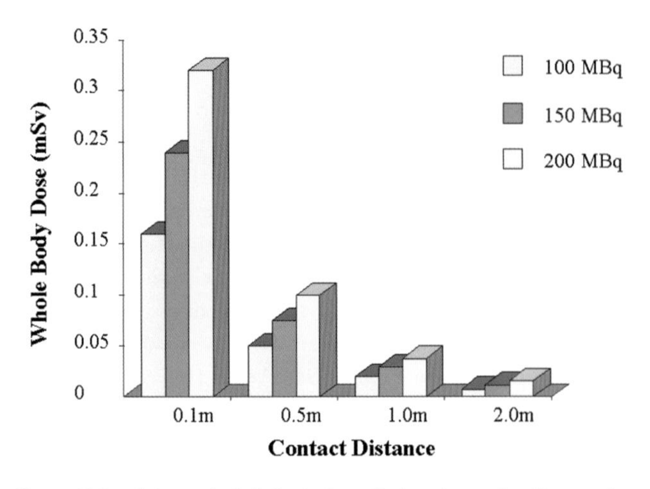

Figure 15.8. Estimated whole body doses likely to be received by member of the public / accompanying person during the 1 hour immediately following injection of [18F]-FDG

lying down, and should not be allowed to leave the department during this time. Ideally a separate room should be provided for this but in the event that this is not available and that they will have to share a room it should be remembered that the dose rates from patients immediately following injection can be quite high and may present a radiation hazard. Benatar *et al* [13] used the dose rate data to estimate the doses which might be received by people coming into contact with injected patients during the one-hour uptake period. Figure 15.8 shows that, following administration of [^{18}F]-FDG, the dose received during the following hour can range from just over 0.3 μSv if sitting right next to the patient down to 0.16 μSv at 2 m.

Risk to the General Public

The United Kingdom annual dose limit for members of the general public is 1mSv and there is a recommendation that at any one exposure the dose should not exceed 30% of this [17]. Clearly, in a general waiting area there will not be just one PET patient, and an escort or carer could receive in excess of 0.3 μSv when taking into account the contribution from all the patients present. Ideally it is best to have a "hot" waiting room separate to the "cold" waiting room, and the planning arrangements in the "hot" waiting room should ensure adequate separation of seating.

The majority of clinical PET radiotracers have very short half-lives and therefore patients pose little radiation risk to members of the public once they have completed their scan and are ready to leave the department. [^{18}F]-FDG, with its longer half life, will not have completely decayed away before the patient leaves and so consideration must be given to the severity of this risk. The critical group is likely to be health care workers who may have contact with several patients who have undergone PET investigations or with an individual patient who may be having other radiopharmaceutical studies in addition to their PET scan. United Kingdom guidelines on when patients can leave hospital following administration of a radiopharmaceutical are currently based on the activity retained within the patient on exit rather than the total doses which would be received by others coming into contact with them. A study published in 1999 measured the instantaneous dose rates from patients just prior to leaving the department [18]. The integrated doses were calculated for various contact times at differing distances. The results showed that for positron-emitting tracers with a half-life of 110 minutes or less, the use of the activity-energy product is not a useful concept for restricting critical groups to an infrequent exposure. Instead, greater emphasis should be placed on the true estimate of actual radiation dose which could be received, or the predicted doses. With longer-lived PET tracers, for example 124I (half-life 4.2 days), the recommendations would need to be reviewed. The recommendations given as a result of the work suggested that following an [^{18}F]-FDG PET study using an injected activity of 350 MBq there was no need to prevent contact with a patient's partner following a scan. There was no need to restrict travel on public transport following the scan, despite the fact that the activity-energy product may exceed the local rules. Clearly, young children of patients should never accompany their parents to the PET or nuclear medicine department. However, these data showed there was no need for restrictive advice following an [^{18}F]-FDG scan once the patient has left the department, although most centers recommend that close contact is kept to a minimum for the rest of the day. For a single nurse caring for a patient following an [^{18}F]-FDG scan there is virtually no risk; however, if this nurse is working on a ward where there is a high throughput of either PET and/or nuclear medicine patients, the annual dose received may become significant. It is unlikely, though, that the 6 μSv annual dose limit would be reached by any member of the nursing staff. Overall, there is a slightly higher risk with positron emitting tracers than there is with conventional nuclear medicine tracers because of the higher photon energy involved. However, with care and good working practice it is possible to keep the staff doses to an acceptable level, although the issue of waiting room space will have to be addressed by departments wishing to undertake PET scanning, especially if space is at a premium.

Conclusion

The technologist provides a vital interface between the patient, the clinician, the nursing and the scientific staff. The information obtained from the patients can be structured to give additional information to the clinicians interpreting the scans if recorded methodically. Technologists are aware of the finer technical aspects of the scans and the procedures that the patients will need to follow, and can both reassure and answer most of the questions asked. Technologists also have to be able to adapt to changing situations when there are failures of tracer production and the combination of their practical experience and theoretical knowledge means that they are well placed to do this effectively. The interaction with other cross-sectional imaging modalities, particularly in the light of an increasing use of image registration ("fusion") techniques, further brings together skills from radiographic and nuclear medicine technology backgrounds. This interaction should encourage individuals from varied backgrounds to consider PET as a dynamic, evolving, imaging specialty that has a lot to offer.

References

1. Zasadny KR, Wahl RL. Standardized uptake values of normal tissues at PET with 2-[fluorine-18]-fluoro-2-deoxy-D-glucose: variations with body weight and a method for correction. Radiology 1993;189:847–50.
2. Keyes JR. SUV: standard uptake or silly useless value? J Nucl Med 1995;36:1836–9.
3. Barrington SF, Maisey MN. Skeletal muscle uptake of F-18 FDG: Effect of oral diazapam. J Nucl Med 1996;37:1127–9.
4. Kostagoglu L, Wong JCH, Barrington SF, Cronin BF, Dynes AM, Maisey MN. Speech-related visualisation of laryngeal muscles with F-18 FDG. J Nucl Med 1996;37:1771–3.
5. Richardson MP, Koepp MJ, Brooks DJ, Fish DR, Duncan JS. Benzodiazepine receptors in focal epilepsy with cortical dysgenesis: an [^{11}C]-flumazenil PET study. Ann Neurol 1996;40(2):188–98.
6. Di Chiro G, De La Paz RL, Brooks RA, Sokoloff L, Kornblith PL, Smith PH, et al. Glucose utilization of cerebral gliomas measured by ^{18}F-flourodeoxyglucose and PET. Neurology 1982;32:1323–9.
7. Harper PV, Lathrop KA, Krizek H, Lembares N, Stark V, Hoffer PB. Clinical feasibility of myocardial imaging with ^{13}NH3. J Nucl Med 1972;13:278–80.
8. Phelps ME, Hoffman EJ, Coleman RE, Welch MJ, Raichle ME, Weiss ES, et al. Tomographic images of blood pool and perfusion in brain and heart. J Nucl Med 1976;17:603–12.
9. Schelbert HR, Phelps ME, Huang S-C, MacDonald NS, Hansen H, Selin C, et al. N-13 ammonia as an indicator of myocardial blood flow. Circulation 1981;63:1259–72.
10. Lewis P, Nunan T, Dynes A, Maisey MN. The use of low-dose intravenous insulin in clinical myocardial F-18 [^{18}F]-FDG PET scanning. Clinical Nucl Med 1996;21:15–18.
11. HMSO. The Ionising Radiations Regulations (1985). London: Her Majesty's Stationary Office; 1985. Report No.: S.I.1985 No.1333.
12. HMSO. The Ionising Radiations (Protection of Persons undergoing Medical Examination or Treatment) Regulations. London: Her Majesty's Stationary Office; (In Preparation).
13. Benatar NA, Cronin BF, O'Doherty MJ. Radiation dose rates from patients undergoing PET: implications for technologists and waiting areas. Eur J Nucl Med 2000;27(5):583–9.
14. Clarke EA, Notghi A, Harding LK. Are MIBI/tetrofosmin heart studies a potential radiation hazard to technologists? Nucl Med Commun 1997;18:574–7.
15. Clarke EA, Thomson WH, Notghi A, Harding LK. Radiation doses from nuclear medicine patients to an imaging technologist: relation to ICRP recommendations for pregnant workers. Nucl Med Commun 1992;13:795–8.
16. Bailey DL, Young HE, Bloomfield PM, Meikle SR, Glass DE, Myers MJ, et al. ECAT ART – A continuously rotating PET camera: performance characteristics, comparison with a full ring system, initial clinical studies, and installation considerations in a nuclear medicine department. Eur J Nucl Med 1997;24(1):6–15.
17. Council Directive 96/29/Euratom of 13 May 1966 laying down basic safety standards for the protection of the health of workers and the general public against dangers arising from ionising radiation. Official Journal of the European Communities: Council of the European Union; 1996. Report No. L159.
18. Cronin BF, Marsden PK, O'Doherty MJ. Are restrictions to behaviour of patients required following [^{18}F]-FDG PET studies? Eur J Nucl Med 1999;26:121–8.
19. Hany TF, Gharehpapagh E, Kamel EM, Buck A, Himms-Hagen J, von Schulthess GK. Brown adipose tissue: a factor to consider in symmetrical tracer uptake in the neck and upper chest region. Eur J Nucl Med Mol Imaging 2002 Oct;29(10):1393–8.

16 PET Imaging in Oncology

Andrew M Scott

Introduction

Positron emission tomography (PET) is an imaging technique that provides *in vivo* measurements in absolute units of a radioactive tracer. One of the attractive aspects of PET is that the radioactive tracer can be labelled with short-lived radioisotopes of the natural elements of the biochemical constituents of the body. This provides PET with a unique ability to detect and quantify physiologic and receptor processes in the body, particularly in cancer cells, that is not possible by any other imaging technique.

The clinical role of PET has evolved considerably over the last decade. From its first applications principally in neurology and cardiology, the evaluation of oncology patients has become a pre-eminent clinical role for PET worldwide. Oncology PET studies now represent almost 90% of clinical studies performed in clinical PET Centres worldwide[1–5]. The dramatic rise in the number of PET oncology studies performed is related both to recent reimbursement approvals (particularly in the USA), as well as the increasing evidence for the role of PET in the staging, monitoring treatment response and biologic characterisation of tumours.

PET Radionuclides in Oncology

The short-lived radionuclides (radioisotopes) required for PET are produced in cyclotrons. In PET oncology clinical applications, the most commonly used positron-emitting tracer is ^{18}F-fluoro-2-deoxy-glucose, or [^{18}F]-FDG [6]. The unique versatility of PET lies in the ability to study numerous physiologic and biochemical processes *in vivo*. The measurement of tissue blood flow, oxygen metabolism, glucose metabolism, amino acid and protein synthesis and nucleic acid metabolism have all been demonstrated in PET oncology clinical studies [6–9]. To exploit these physiologic and molecular targets, there are a number of positron emitting radiopharmaceuticals that have been used in clinical oncology studies to date (Table 16.1). Labelling of a large array of other compounds including hypoxic markers, amino acids, DNA proliferation markers and chemotherapy drugs with ^{11}C and ^{18}F have also been studied in clinical trials [3, 7, 8, 10].

The Evidence for Clinical Use of PET in Oncology

The experience over the last decade is that the most important clinical role of [^{18}F]-FDG PET is in oncology. In many cancers, [^{18}F]-FDG PET has been shown to be the most accurate non-invasive method to detect and

Table 16.1. Positron-emitting radionuclides used in oncology clinical studies (see appendix for further radionuclide information).

Radionuclide	Half-Life
^{15}O	122 seconds
^{13}N	9.97 minutes
^{11}C	20.4 minutes
^{18}F	109.8 minutes
^{124}I	4.17 days
^{86}Y	14.7 hours
^{64}Cu	12.8 hours

stage tumours [1–5, 7–9, 11–21]. This has major implications in terms of improving the planning of treatment and avoiding unnecessary treatment and its associated morbidity and cost. Evaluation of the evidence for [^{18}F]-FDG PET in clinical oncology practice has been complicated by the inherent diagnostic nature of this imaging technique. While standard evidence-based approaches to treatment require randomised controlled trials to establish the appropriate outcome or efficacy measures for assessment, imaging techniques provide information which is commonly used as only a part of the management paradigm of most patients. As such, the practical and ethical issues surrounding this make randomised controlled trials for PET extremely difficult to perform or inappropriate in the majority of clinical scenarios [5, 22]. The establishment of diagnostic accuracy, and impact on patient management (including cost), are therefore the most appropriate levels of evidence that can be accurately obtained for PET in clinical practice [5, 22].

Brain Tumours

Brain tumours are a common and often devastating malignancy that impacts on both paediatric and adult populations. In adults, brain tumours are the leading cause of death for males aged 15 to 34 years, and are the fourth commonest cause of cancer death in females of this age group. Paediatric brain tumours are the second commonest cancer, and the second leading cause of death from cancer, in that age group [23, 24]. The evaluation of brain tumours with [^{18}F]-FDG PET is the longest established oncologic application of PET. Tumour grade can be assessed accurately and non-invasively by [^{18}F]-FDG PET, as the rate of glucose utilisation is directly proportional to the degree of malignancy [25]. This can be used in the planning of biopsies, and in monitoring high grade recurrence, particularly in patients with low grade glioma. Increased [^{18}F]-FDG uptake is seen in high grade glial tumours, as well as in primary cerebral lymphomas, pilocytic astrocytomas, and some unusual tumours (e.g., pleomorphic xanthoastrocytoma. Low grade gliomas) (Fig. 16.1) and other primary brain tumours (e.g., meningiomas) do not usually show increased [^{18}F]-FDG uptake except in more aggressive tumours and in post-radiation meningiomas.

Cerebral metastases occur in 20 to 40% of systemic malignancies, and may be the initial presentation of malignancy in 16 to 35% of cases. [^{18}F]-FDG PET has been extensively studied in patients with cerebral metastases, and has been shown to have a sensitivity ranging from 68 to 79% [26]. The principal issue with

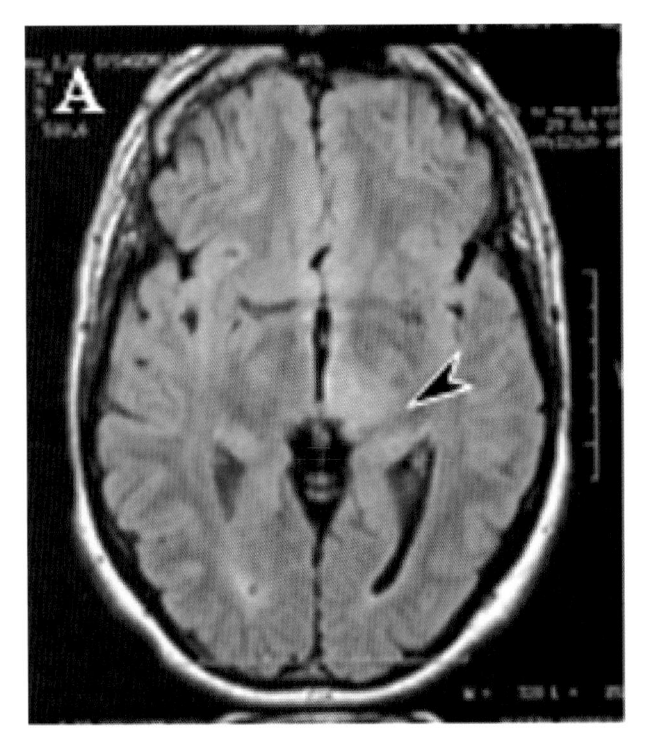

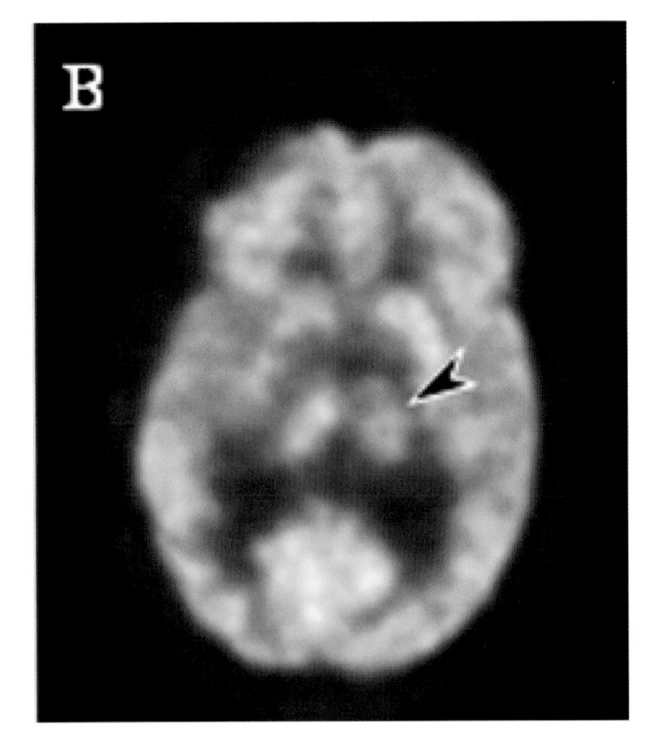

Figure 16.1. (A) MRI scan, and (B) corresponding transaxial [^{18}F]-FDG PET scan of a 53-year-old patient with bilateral hearing loss. A lesion in the left thalamus (arrow) is hypometabolic on [^{18}F]-FDG PET scan. This was subsequently demonstrated to be a low grade glioma.

FDG PET in this setting is the frequent hypometabolic nature of cerebral metastases, and in addition, metastatic lesions are often small (<1 cm in size), and because metastases most often occur at the interface between grey and white matter, identification of lesions can be problematic.

An important role of [^{18}F]-FDG PET is in the assessment of tumour recurrence compared to radiation necrosis, factors critical to the management of these patients and often impossible to determine accurately by CT or MRI (27). The uptake of [^{18}F]-FDG in normal grey matter may make evaluation of tumour recurrence difficult in some cases, and care is required in order to interpret PET scans in these patients.

Patient prognosis may also be evaluated by the presence and degree of uptake of [^{18}F]-FDG [28]. While the evaluation of disease recurrence may also be evaluated with other techniques including ^{201}Tl SPECT, which is more readily available, the problems of blood-brain barrier disruption and mixed grade tumour assessment with ^{201}Tl SPECT may be more accurately assessed with [^{18}F]-FDG PET. Incorporating [^{18}F]-FDG PET into radiotherapy treatment planning, and in monitoring response to therapy, has also emerged as important applications of PET in brain tumours.

A range of other PET tracers have been studied in brain tumours, examining DNA proliferation, protein expression, hypoxia and even gene reporter expression [29, 30]. The most commonly studied tracer is [^{11}C]-methionine PET, which has some advantages in detecting low grade gliomas, although the ability to discriminate between high grade and low grade tumours may be less accurate with [^{11}C]-methionine PET compared to [^{18}F]-FDG PET [31]. In summary, PET studies are an established part of the management of patients with brain tumours in major neuro-oncology centres.

Lung Carcinoma

Solitary Pulmonary Nodules

There have been numerous studies examining the accuracy of [^{18}F]-FDG PET in evaluating solitary pulmonary nodules [32, 33]. Analysis of the published data has shown a high sensitivity (average 96%) and accuracy (average 94%) for determining malignancy (Figs. 16.2 and 16.3 [2, 8, 34]. The specificity is also high but the variation is slightly greater and is dependent on the local prevalence of the known causes of false positive cases, particularly granulomatous diseases such as tuberculosis and histoplasmosis. False negative results may arise where small lesions are present (<0.6 cm), due to the resolution limitations of PET scanners and respiratory motion over the acquisition period, and also with certain types of lung cancer such as broncho-alveolar and carcinoid. The use of [^{18}F]-FDG PET for solitary pulmonary nodules is often reserved for those cases where fine needle aspiration is technically difficult in view of the superior diagnostic yield of this technique.

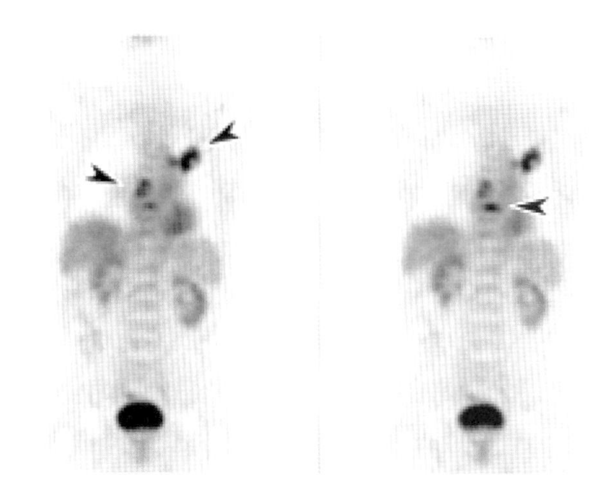

Figure 16.2. Coronal sections of a [^{18}F]-FDG PET scan performed in a 72-year-old man with a left lung carcinoma. Uptake of [^{18}F]-FDG in the primary lung cancer (arrow) and lymph nodes (arrow) is seen, as well as an unsuspected metastasis in a thoracic vertebrae (arrow).

Staging of Non-Small Cell Lung Carcinoma

In patients with known non-small cell lung carcinoma (NSCLC), the results of staging both within and outside the thorax are key in determining operability. Demonstration of unilateral hilar adenopathy is not a contra-indication to surgery if the nodes can be resected with the primary tumour. Conversely, extensive mediastinal involvement, involvement of contralateral lymph nodes, as well as pleural or distant metastases should contra-indicate surgery given the high surgical morbidity and the poor prognosis. The relatively sub-optimal sensitivity, specificity and accuracy of conventional imaging techniques, including CT and MRI, for staging of lung carcinoma has been demonstrated repeatedly [35].

Numerous studies have evaluated the role of [^{18}F]-FDG PET for staging NSCLC [11, 17, 18, 35–38]. The reported sensitivity for lymph node staging in non-small

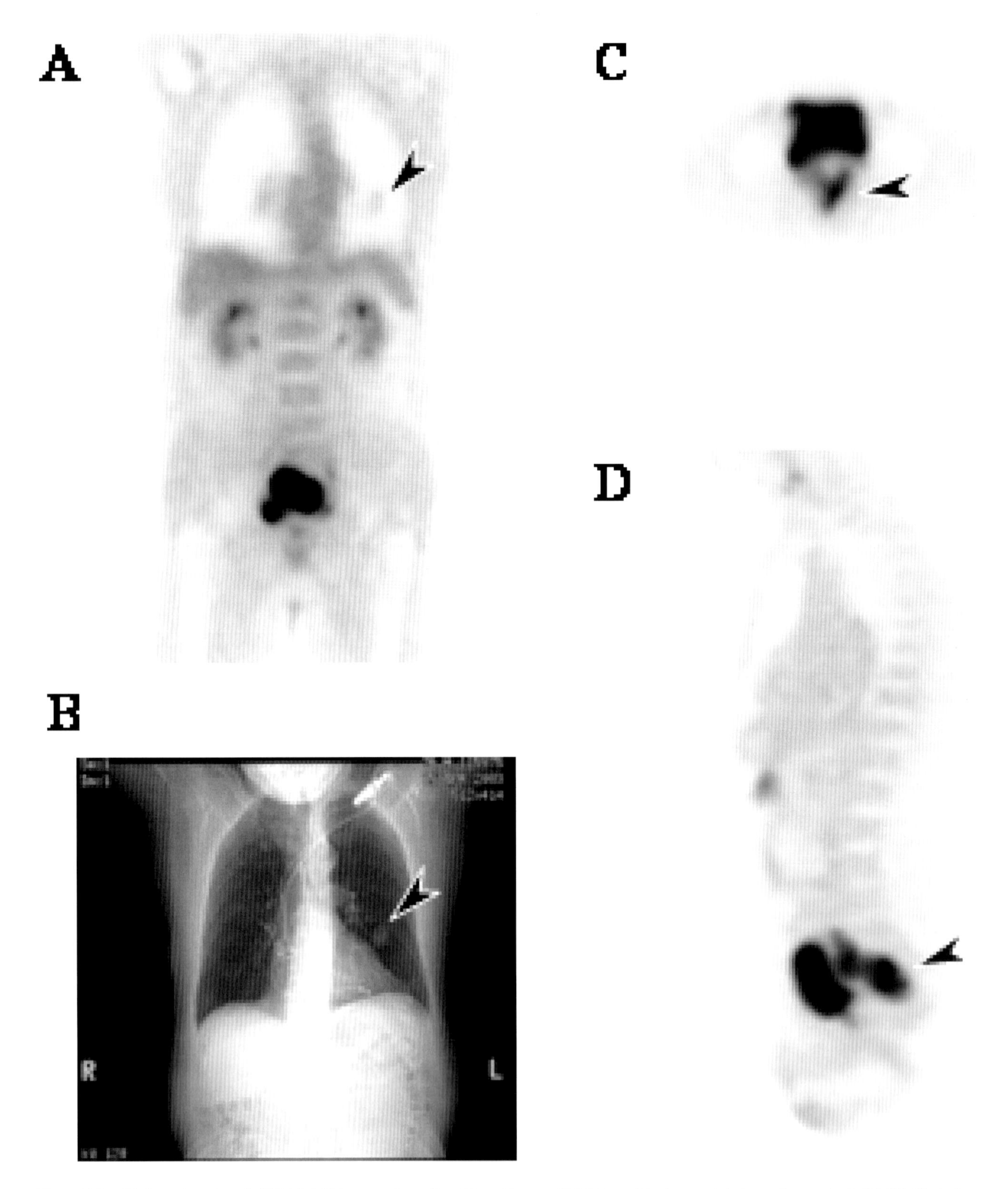

Figure 16.3. (A) Coronal section of a [^{18}F]-FDG PET scan performed in a 76 yr old man with left lower lobe lung nodule (arrow), seen also in (B) CXR. The lesion was hypometabolic on [^{18}F]-FDG PET scan, and was non-malignant. An incidental finding of increased [^{18}F]-FDG uptake in the rectum, seen on (C) transaxial and (D) sagittal images (arrow) was due to a non-diagnosed rectal cancer.

cell lung carcinoma varies from 82 to 100% and the specificity from 73 to 100% [11, 16, 36, 37, 39, 40]. In all series, [^{18}F]-FDG PET has been shown to outperform CT in staging lymph node spread of disease, and has been shown to correctly stage disease over CT scan results in up to 24% of patients (Fig. 16.2) (37). In a recent randomised controlled trial of [^{18}F]-FDG PET in staging patients with newly diagnosed NSCLC, the inclusion of [^{18}F]-FDG PET in diagnostic work up reduced futile thoracotomies by half, indicating the important role of [^{18}F]-FDG PET in this staging process (18). The negative predictive value of [^{18}F]-FDG PET is sufficiently high that a negative mediastinum on [^{18}F]-FDG PET scanning may preclude mediastinoscopy in patient work-up, and a positive mediastinum on [^{18}F]-FDG PET should always be further assessed to exclude false positive results (e.g., sarcoid) [5, 38]. In addition, studies utilising whole body scanning have reported unsuspected metastatic disease in up to 15% of patients (Fig. 16.2) [39]. This is particularly relevant for adrenal lesions, where benign enlargement is common, and [^{18}F]-FDG PET is highly accurate in identifying metastatic involvement [16]. Through correct staging by [^{18}F]-FDG PET, therapy has been shown to be changed in $>$40% of patients, mainly by obviating unnecessary surgery [40]. [^{18}F]-FDG PET has emerged as a standard pre-operative assessment test in patients with NSCLC.

Recurrent Lung Carcinoma and Response to Therapy Assessment

The ability to accurately evaluate residual masses following surgery or radiotherapy for lung cancer is essential in many patients. Post treatment fibrosis and scarring is common, and [^{18}F]-FDG PET has been shown in a number of small series to be accurate in detecting residual tumour, which allows treatment planning decisions to be reliably made [15]. The potential role of [^{18}F]-FDG PET in treatment planning for radiotherapy of unresectable lung cancer has also been explored, and may provide improved treatment with a reduced incidence of relapse outside the radiotherapy field [41]. In addition, in patients with NSCLC undergoing radical radiotherapy or chemoradiotherapy, [^{18}F]-FDG PET performed soon after therapy has been shown to be a superior predictor of survival than CT response, stage, or pre-treatment performance status [42].

Colorectal Carcinoma

Staging of Primary Colorectal Cancer

The diagnosis of colorectal cancer is principally based on colonoscopy and biopsy, with imaging being performed primarily to assist in initial surgical planning. There have been a number of studies examining the utility of PET for staging primary colorectal carcinoma. In one study of 16 patients with known or suspected primary or recurrent colon and rectal cancer studied with [^{18}F]-FDG PET and CT scans, PET detected all 12 sites of disease in bowel, whereas CT only detected 6 [43]. Other small studies have confirmed these results, although lymphatic spread of tumour was poorly detected with [^{18}F]-FDG PET due to the small size of involved lymph nodes in many cases [44]. Primary colorectal cancers occasionally present as an incidental finding on [^{18}F]-FDG PET (Fig. 16.3), and [^{18}F]-FDG uptake has been reported in adenomatous polyps, a pre-cursor for colon cancer [45]. However, the presence of physiological gut uptake of FDG combined with false positive uptake in inflammatory disease along with low sensitivity to lesions less than 1 cm precludes a significant role for FDG PET in primary diagnosis or screening [46]. The role of PET in primary colon cancer remains limited, and should be reserved for clinical situations where resection of metastatic disease requires accurate staging of distant spread. This is in contrast to advanced rectal cancer, where [^{18}F]-FDG PET has been shown to have a significant impact on management in up to one third of patients planned for preoperative adjuvant treatment (chemoradiation), indicating the potential role of [^{18}F]-FDG PET in this clinical setting [47].

Staging of Metastatic Colon Cancer

Hepatic Metastases

Clinical studies have shown [^{18}F]-FDG PET to be a highly sensitive technique for the detection of hepatic metastases of colorectal cancer. In published series the accuracy of [^{18}F]-FDG PET in identifying metastatic colorectal carcinoma in the liver has ranged from 90 to 98% [8]. The role of PET in this clinical setting appears to be complementary to CT scan and CT portography [48–50]. Mucinous adenocarcinoma may have a lower sensitivity for detection [51]. Importantly, [^{18}F]-FDG

PET has been shown to identify previously unsuspected extrahepatic disease in patients with liver metastases in up to 20% of patients, and can change management in reported series in up to 35% of patients [12, 52].

Extrahepatic Metastases

The detection of extrahepatic metastases of colorectal carcinoma remains a difficult clinical problem. While CT scans are sensitive for hepatic metastases, they are less sensitive for detecting extrahepatic disease [12, 53]. MRI, while as accurate as CT for detecting hepatic metastases of colorectal carcinoma, remains less sensitive in extrahepatic intra-abdominal disease [54]. This is of particular importance in patients considered for surgery for metastatic disease, in view of the high prevalence of undiagnosed sites of extrahepatic disease which results from conventional techniques preoperatively.

Studies examining the accuracy of PET in extrahepatic metastases of colorectal carcinoma have demonstrated higher accuracy than conventional scanning techniques (including CT scan) [12, 52]. Extrahepatic disease has been detected with [^{18}F]-FDG PET with an accuracy of 92 to 93% in recent series. In

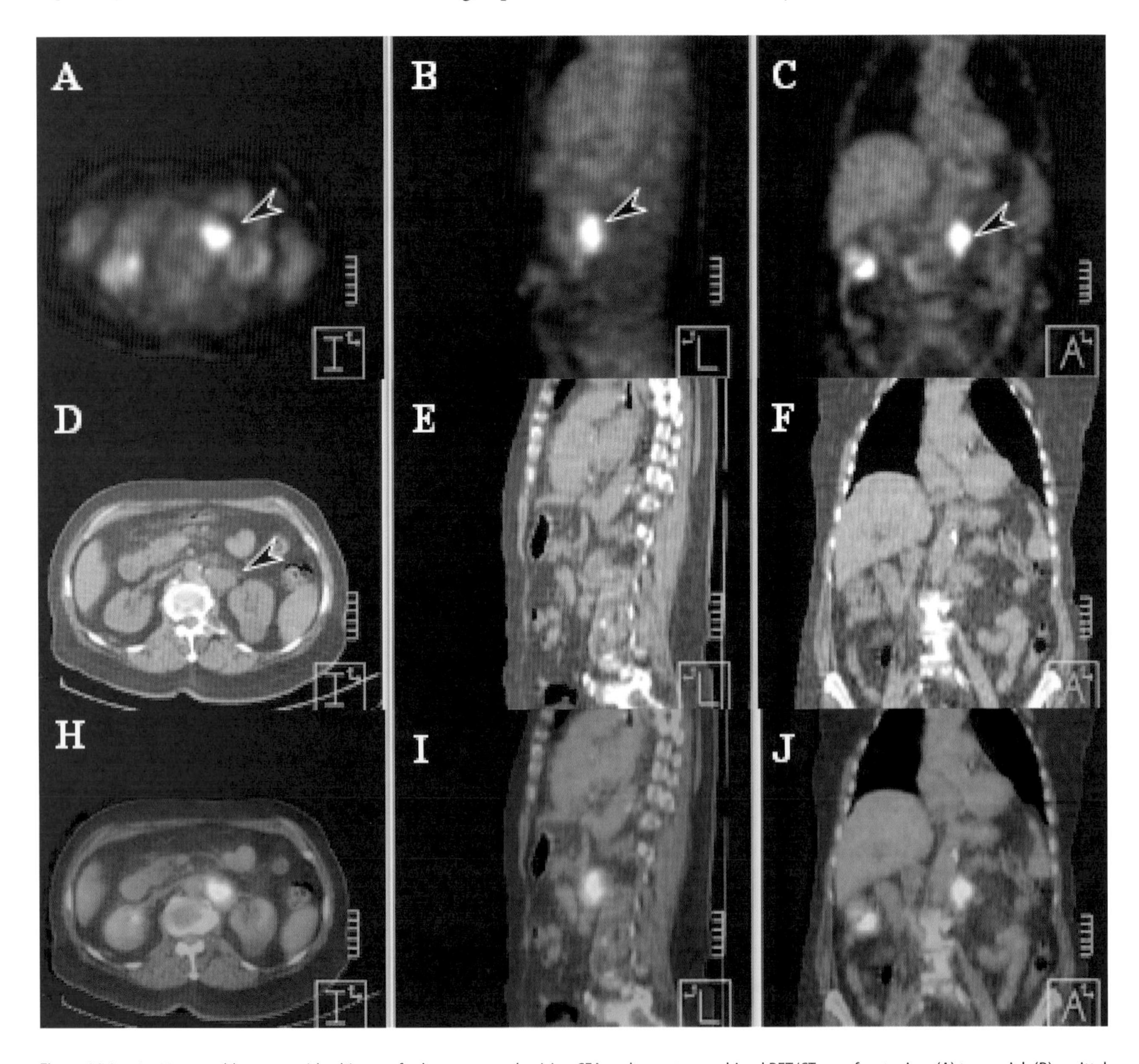

Figure 16.4. An 83-year-old woman with a history of colon cancer and a rising CEA underwent a combined PET/CT scan for staging. (A) transaxial, (B) sagittal and (C) coronal images show a focal area of abnormal [^{18}F]-FDG uptake (arrow), corresponding to a retroperitoneal lymph node region on corresponding CT slices [(D), (E) and (F)]. Co-registered [^{18}F]-FDG PET/CT images [(G), (H) and (I)] show precise localisation of the recurrent colon cancer in the retroperitoneum.

patients with elevated serum CEA markers, occult disease (often extrahepatic recurrence) has been identified accurately with [^{18}F]-FDG PET (Fig. 16.4) [12, 52]. These results have been confirmed in a recent meta-analysis of whole-body [^{18}F]-FDG PET studies in patients with colorectal carcinoma, where the sensitivity and specificity of [^{18}F]-FDG PET in detecting tumour was 97% and 76% respectively, and the change in management was calculated to be 29%, both by upstaging and downstaging disease [55].

Rectal Carcinoma Recurrence

Recurrent colorectal (pelvic) cancer has been reported to occur in 20 to 40% of patients within two years after potentially curative surgery [56]. Both CT and MRI, however, have significant difficulties in reliably distinguishing local spread of disease, and recurrence of rectal cancer from post surgical change [53]. There have been a number of extremely promising clinical studies with [^{18}F]-FDG PET in the evaluation of possible rectal carcinoma recurrence. In one study of 37 patients with suspected rectal carcinoma recurrence, 32/32 patients with recurrent rectal carcinoma were correctly identified with [^{18}F]-FDG PET [57]. The accuracy of [^{18}F]-FDG PET in this setting has also been demonstrated in other published clinical studies [8].

Management Impact of PET and Response to Therapy Assessment

The management impact of [^{18}F]-FDG PET in recurrent colorectal carcinoma has been clearly demonstrated [58, 59]. Studies of [^{18}F]-FDG PET post therapy (chemo/radiotherapy) have also demonstrated a potential role of [^{18}F]-FDG PET in this setting [13, 60, 61]. Based on available evidence, [^{18}F]-FDG PET should be used in the detection of advanced or metastatic colorectal cancer where management will be altered by disease presence and extent.

Lymphoma

The ability of whole body [^{18}F]-FDG PET to accurately stage lymphoma has emerged as an important role of PET in patient management. In the staging of Hodgkins Disease and non-Hodgkins lymphoma (NHL) the sensitivity and specificity of [^{18}F]-FDG PET in detecting sites of disease has been reported as 86–90% and 93–96% respectively, and superior to CT scan [2, 8, 62]. Compared to conventional imaging (e.g., CT scans) [^{18}F]-FDG PET is more sensitive in identifying extra-nodal sites of disease. [^{18}F]-FDG PET has been shown to change management in up to 40% of patients undergoing staging at initial diagnosis [21, 63–65]. In comparison to ^{67}Ga scans, [^{18}F]-FDG PET has been shown to have a greater sensitivity for disease detection (particularly spleen), and in view of the potential advantage of a same day procedure has supplanted ^{67}Ga scans in many oncology centres [66]. In some cases of low grade lymphoma [^{18}F]-FDG uptake may not be high, however this is dependent on histologic subtype, and high sensitivity for disease sites has been reported for follicular and marginal zone low grade lymphoma, and with management change reported in up to 34% of patients (Fig. 16.5) [64, 67].

[^{18}F]-FDG PET has also been shown to have superior accuracy compared to conventional imaging in the restaging of lymphoma patients, particularly where residual masses are present [68, 69]. In comparison to ^{67}Gallium scans, [^{18}F]-FDG PET has been shown to be more accurate in identifying active disease in residual masses (66, 70). [^{18}F]-FDG PET performed as part of the assessment of treatment response for NHL has also been shown to be superior to conventional imaging and to be a strong prognostic indicator of response and progression free survival [68, 71]. [^{18}F]-FDG PET therefore has a major role in both the initial staging, and restaging/therapy response assessment of patients with lymphoma.

Melanoma

The role of [^{18}F]-FDG PET in the initial staging of low-risk (<1 mm thickness) and intermediate-risk (1–3 mm thickness) melanoma is limited, due to the low prevalence of metastatic disease (particularly in low-risk melanoma), and the often small size of metastatic deposits in nodes involved with metastatic spread [72, 73].

Malignant melanoma can spread widely and unpredictably throughout the body, and median survival after the appearance of distant metastases is approximately six months [74]. The accuracy of [^{18}F]-FDG PET in detecting metastatic melanoma has been reported to range from 81 to 100%, and in one series of 100 patients demonstrated a sensitivity of 93% [14]. [^{18}F]-FDG PET has been shown to be particularly sensitive in detecting subcutaneous and visceral metastases

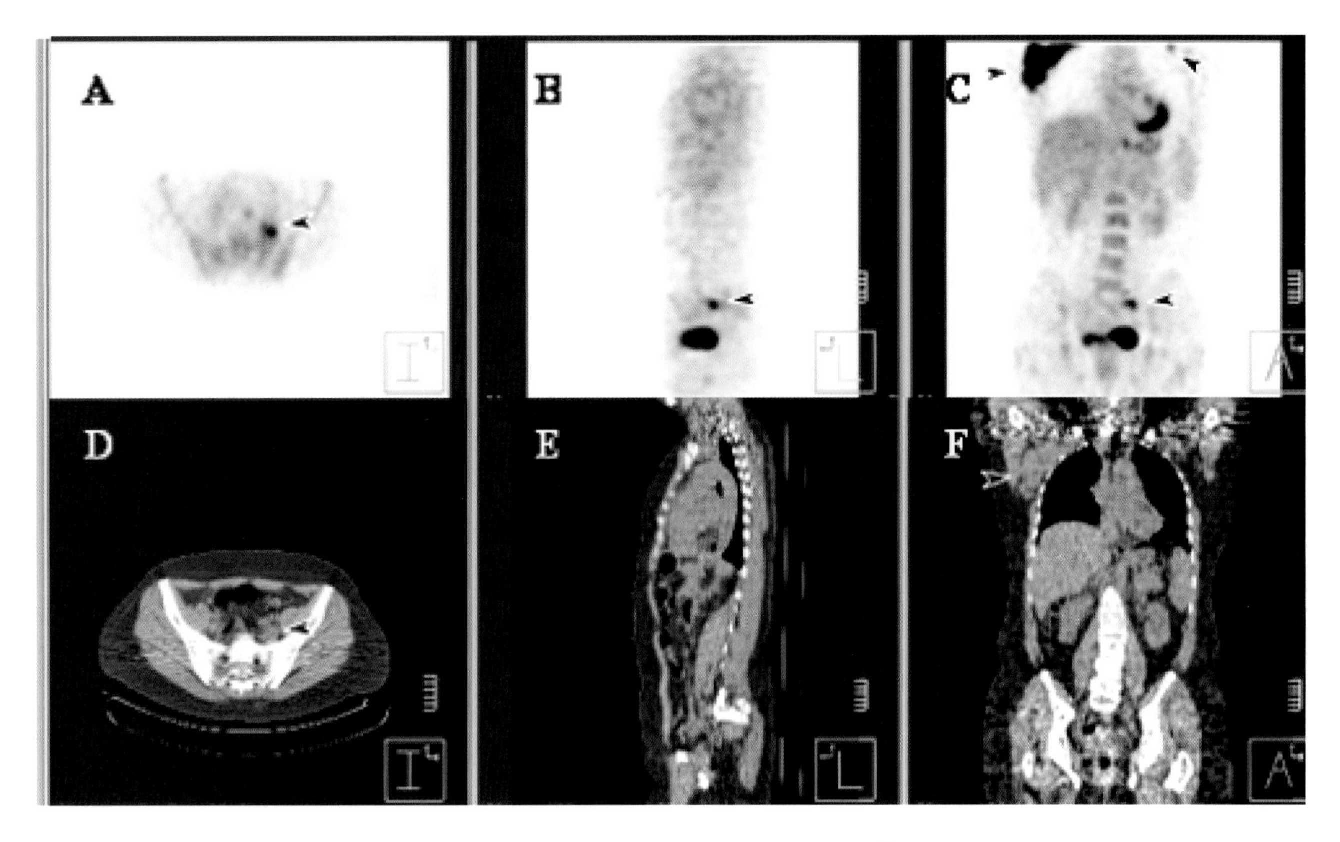

Figure 16.5. A 48-year-old woman with Follicular lymphoma of the right axilla underwent a staging [^{18}F]-FDG PET/CT scan prior to planned radiotherapy. (A) transaxial, (B) sagittal and (C) coronal whole body staging [^{18}F]-FDG PET images show lymphoma in right axillary lymph nodes (arrow), as well as additional lymphoma sites in the left axilla and left illiac nodes (arrows). Corresponding CT scan images [(D), (E) and (F)] show right axillary node enlargement only. The [^{18}F]-FDG PET results upstaged the patient from Stage I to Stage III.

(Fig. 16.6). In published studies and meta-analyses of the literature, [^{18}F]-FDG PET has been demonstrated to detect disease up to six months earlier than conventional techniques, and alter management in 22 to 32% of patients, principally by altering plans for surgical resection of metastatic disease [14, 75–77]. As approximately one quarter of all melanoma patients with metastatic disease are potential candidates for surgical resection, and long disease-free intervals are possible in patients rendered clinically disease free by the surgery, [^{18}F]-FDG PET has a clear role in this clinical setting [5, 78].

[^{18}F]-FDG PET has also been compared to other staging investigations for metastatic melanoma. In a series of 121 patients with metastatic melanoma where [^{18}F]-FDG PET was compared to ^{67}Ga scintigraphy, PET was more accurate and provided incremental and clinically important information in 10% of patients, and at a lower cost [79]. The role of [^{18}F]-FDG PET in melanoma is therefore principally in the evaluation of extent of metastatic disease, the accurate assessment of which can alter patient management particularly where surgery is planned.

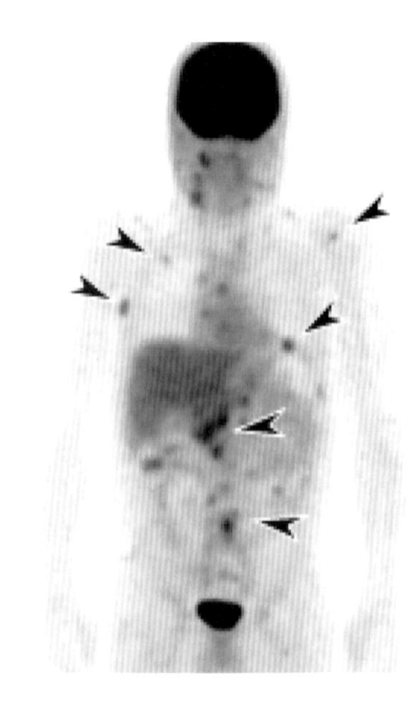

Figure 16.6. A 45-year-old woman with a history of right postauricular melanoma and previous resection of metastases in the right parotid and ipsilateral cervical lymph nodes, was referred for a restaging [^{18}F]-FDG PET scan prior to possible radiotherapy to the operative sites. A coronal whole body image shows multiple metastatic lesions in subcutaneous, lymph node, lung and visceral sites (arrows) throughout the body.

Head and Neck Tumours

The presence of lymph node spread of head and neck tumours is associated with substantially worse prognosis, and clinical examination and imaging techniques (CT and MRI) detect fewer than 50% of involved lymph nodes, which may result in unnecessary neck surgery. In patients with head and neck tumours studied prior to initial surgery, the sensitivity and specificity of [^{18}F]-FDG PET in detecting nodal metastases has been reported ranging from 71 to 91%, and 88-100%, respectively (Fig. 16.7) [8, 80–83]. Metastatic disease outside the neck can also be identified with [^{18}F]-FDG PET scans [84]. Primary tumours can also be detected with a similar sensitivity to CT/MRI. In patients studied after initial treatment of metastatic nodal disease with radiotherapy, [^{18}F]-FDG PET is often accurate only after a three month period [20, 85–87]. This has been shown to be due to the effects of stunning of tumour cells, where the metabolic rate and proliferation of cancer cells may be suppressed after radiotherapy, as well as the possible presence of microscopic disease only after radiotherapy treatment which is below the resolution of the PET camera. In both patient groups, the accuracy of [^{18}F]-FDG PET has the potential to direct surgeons to otherwise unknown sites of metastatic disease, as well as avoiding surgery in areas of the neck where [^{18}F]-FDG PET scans show negative results. [^{18}F]-FDG PET has also been shown to be a prognostic factor for radiotherapy response [88]. In patients with advanced loco-regional head and neck cancer, radiotherapy treatment planning incorporating [^{18}F]-FDG PET information has emerged as an exciting new method of potentially improving response rates.

Breast Carcinoma

The use of [^{18}F]-FDG PET in breast carcinoma has been evaluated in the initial assessment and staging of disease, and in monitoring response to therapy. In

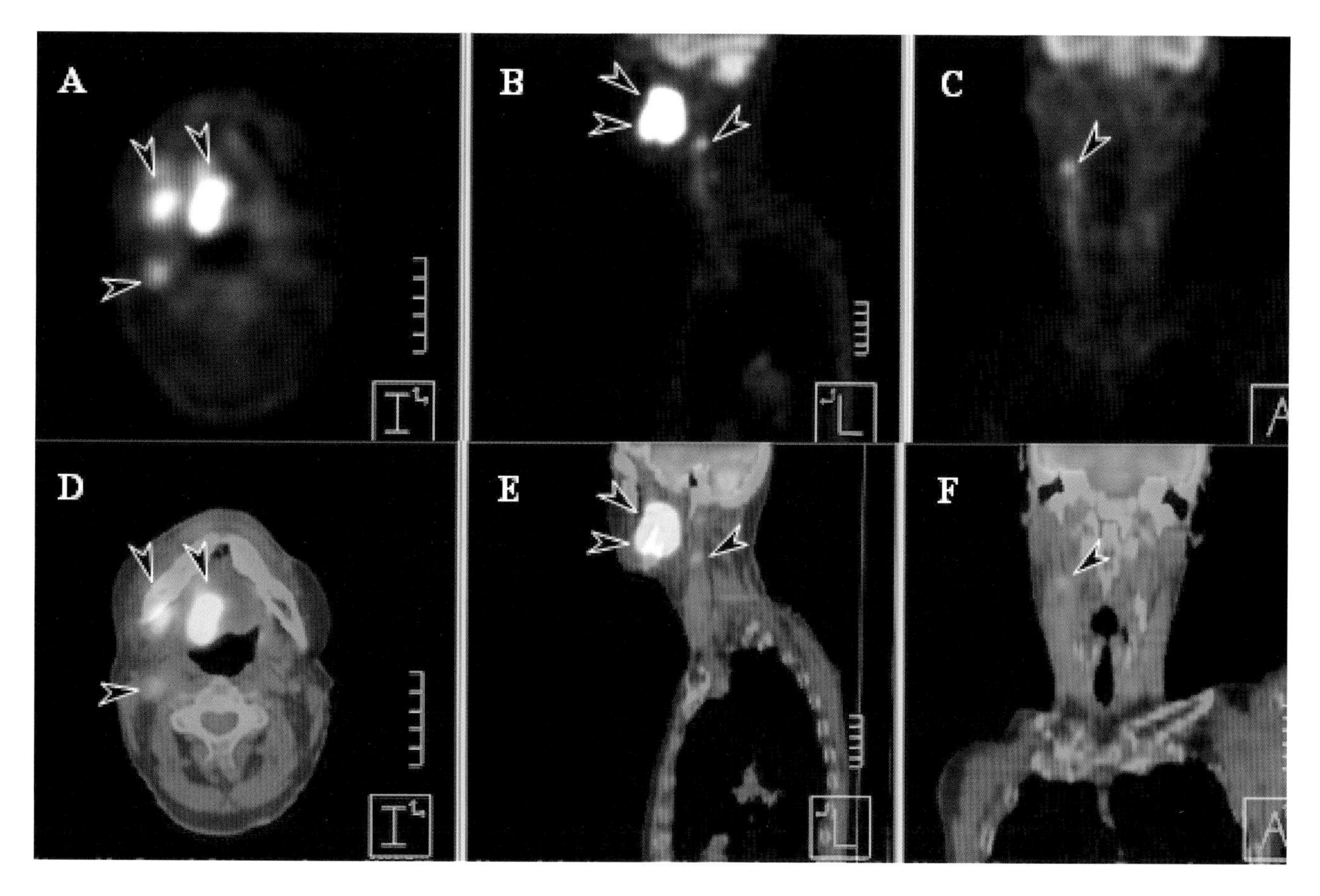

Figure 16.7. (A) transaxial, (B) sagittal and (C) coronal [^{18}F]-FDG PET scan images of a 60 yr old man with carcinoma of the right tonsil and base of tongue (arrows). A previously undiagnosed lymph node metastasis is also seen (arrow). Co-registered [^{18}F]-FDG PET/CT images on corresponding images [(D), (E) and (F)] show the precise localisation of sites of tumour.

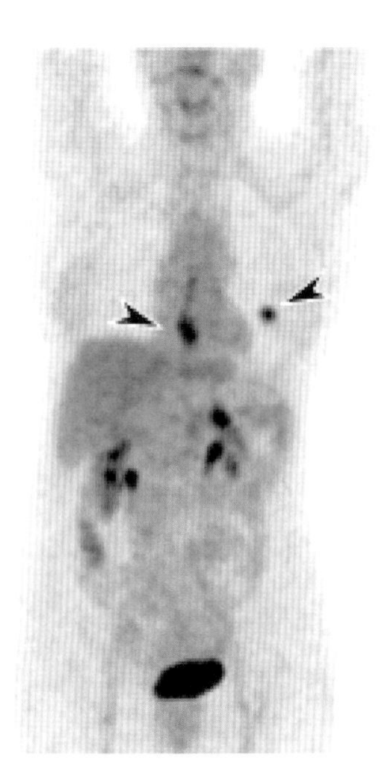

Figure 16.8. Whole body 3D coronal [^{18}F]-FDG PET image of an 81-year-old woman with an oesophageal carcinoma (arrow). A previously undiagnosed left breast carcinoma was also identified (arrow), subsequently confirmed on biopsy. Normal excretion of [^{18}F]-FDG from kidneys, and in bladder, and normal uptake of [^{18}F]-FDG in ascending colon, is also evident.

primary breast tumours, [^{18}F]-FDG PET has been shown to have a mean sensitivity and specificity for tumour detection of 88 and 79% respectively in a recent meta-analysis (Fig. 16.8) [89]. Axillary nodal involvement is a critical issue in the management of patients with breast carcinoma, and [^{18}F]-FDG PET has been shown to have a sensitivity ranging from 57 to 100% and specificity of 66 to 100% across reported series [8]. In a recent prospective study of 360 patients, [^{18}F]-FDG PET was shown to have moderate accuracy in detecting axillary metastases, but often missed small or few nodal metastases [90]. [^{18}F]-FDG PET is therefore not routinely recommended for the axillary staging of breast cancer patients.

One potential area where [^{18}F]-FDG PET has shown great promise is in whole body staging of metastatic breast cancer, where the accuracy of [^{18}F]-FDG PET has been shown to be higher than conventional staging techniques [91, 92]. [^{18}F]-FDG PET has also been evaluated for assessment of tumour response to therapy [61, 93]. Further evaluation in monitoring response to chemotherapy is warranted before the true role of [^{18}F]-FDG PET in this setting can be determined.

Gastric and Oesophageal Tumours

Although the detection of primary gastric and oesophageal tumours with [^{18}F]-FDG PET has been reported to be excellent, the identification of regional nodal metastases has been restricted by the presence of small volume disease in some lymph nodes (Fig. 16.8) [32]. Unsuspected distant disease may be detected by [^{18}F]-FDG PET in up to 20% of cases, and recurrent disease may also be evaluated more accurately than CT scan, which may represent the more appropriate clinical utility of this technique [32].

[^{18}F]-FDG PET has been shown to significantly improve detection of haematogenous and distant lymphatic metastasis in carcinoma of the oesophagus and gastro-oesophageal junction (GEJ) [32, 94–96]. There is no difference in accuracy in detecting squamous cell carcinoma or adenocarcinoma of the oesophagus with [^{18}F]-FDG PET. [^{18}F]-FDG PET may not be as accurate as EUS or CT in determining wall invasion or close lymph node spread of disease, however the diagnostic specificity of lymph node involvement is greatly improved with [^{18}F]-FDG PET [95]. [^{18}F]-FDG PET is more accurate in detecting distant disease, and is also highly accurate in the diagnosis of recurrent disease [94–97]. In assessing response of induction therapy (chemotherapy – radiotherapy) in locally advanced disease, [^{18}F]-FDG PET also appears to be of high value in predicting response [98].

Ovarian Carcinoma

Ovarian carcinoma is the leading cause of death among gynaecological tumours [99]. The treatment of ovarian carcinoma primarily consists of surgical resection followed by chemotherapy and/or radiotherapy. Accurate staging is essential, particularly in the restaging of patients with elevated serum markers (CA-125). [^{18}F]-FDG PET has been shown to have high accuracy in detecting in ovarian carcinoma lesions greater than 1 cm in size, but the detection of micrometastatic disease (one of the most important issues in this disease) has been difficult [100–102]. In several series, [^{18}F]-FDG PET has been shown to be accurate in restaging patients with ovarian cancer, and may be better than CA-125 in this clinical setting [100, 101]. The role of [^{18}F]-FDG PET in the management of ovarian carcinoma remains to be clearly defined, but may principally be in the evaluation of recurrent masses following therapy, or distant disease.

Germ Cell, Renal Cell and Other Tumours

Studies of [^{18}F]-FDG PET in germ cell tumours and in bone and soft tissue sarcomas have shown high accuracy in detecting metastatic disease and monitoring response to therapy, although reported patient numbers are small [102, 103]. Renal cell carcinoma may also be accurately staged with [^{18}F]-FDG PET, and is particularly useful for the detection of metastatic disease, while false negative primary tumours are occasionally observed [19, 104, 105]. The role of [^{18}F]-FDG PET in tumours such as hepatoma has been limited, due to the low grade nature of these tumours and the poor uptake of [^{18}F]-FDG in clinical studies [2]. In cervical carcinoma, recent reports have shown that [^{18}F]-FDG PET has a high positive predictive value for the detection of metastatic nodes in the pelvis and para-aortic regions, which may assist with surgical planning or possibly indicate cases where chemoradiotherapy may be the most appropriate therapy (Fig. 16.9) [106]. An important emerging role of [^{18}F]-FDG PET is in the detection of metastatic thyroid carcinoma (particularly where ^{131}I scan is negative and serum thyroglobulin levels are elevated), and evaluation of medullary thyroid carcinoma. The application of [^{18}F]-FDG PET in detecting sites of disease for many other less common malignant tumours is the subject of continuing clinical review.

Monitoring Response of Tumour to Therapy

An emerging area of clinical utility of PET is in the monitoring of tumour response to therapy, principally with [^{18}F]-FDG. Accurate evaluation of response to both chemotherapy and radiation therapy, often prior to CT scan changes, have been reported in glioma, colorectal, NSCLC, lymphoma, head and neck tumours, and soft tissue sarcomas [2, 8, 16, 17, 42, 61, 107]. The timing and reliability of [^{18}F]-FDG PET studies in predicting tumour response is the subject of numerous prospective studies. The implications of this approach are significant in terms of optimising treatments, minimising unnecessary morbidity, and reducing costs.

Perhaps the most potentially important clinical use of PET can be found in the ability to label with positron emitters compounds that are either physiologic targets or therapeutics used in cancer therapy. This enables the development of novel compounds, either therapy compounds themselves (e.g., ^{18}F-Fluorouracil), or tracers that measure physiologic events (e.g. ^{18}F-fluoromisonidazole for hypoxia, and $H_2{}^{15}O$ for perfusion), that can predict the success of a therapeutic approach [3, 5, 108, 109]. The pharmacokinetics and pharmacodynamics of therapeutics, measurements of biologic change (e.g., DNA proliferation, signalling events, oxygen metabolism, protein synthesis), or success of treatment *e.g.*, gene therapy, can be accurately evaluated by PET, and may enable more scientific decisions to be made as to treatment efficacy, and how improvements in effect can be achieved [5, 29, 110–114]. These "surrogate markers" for tumour biology provide a unique link between the molecular

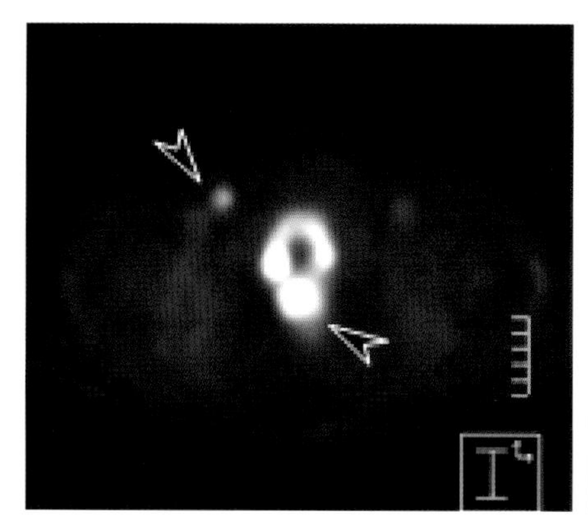

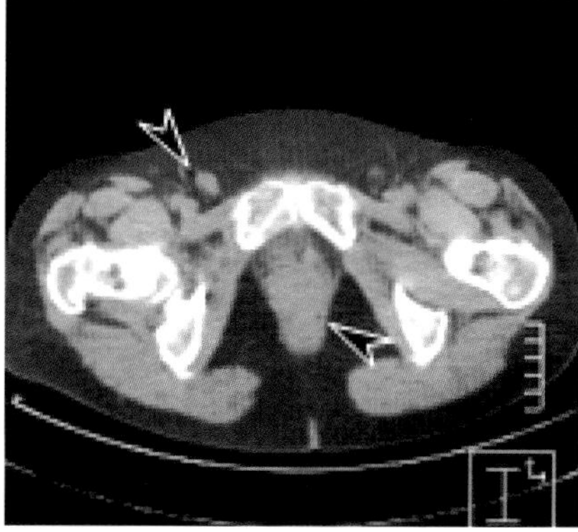

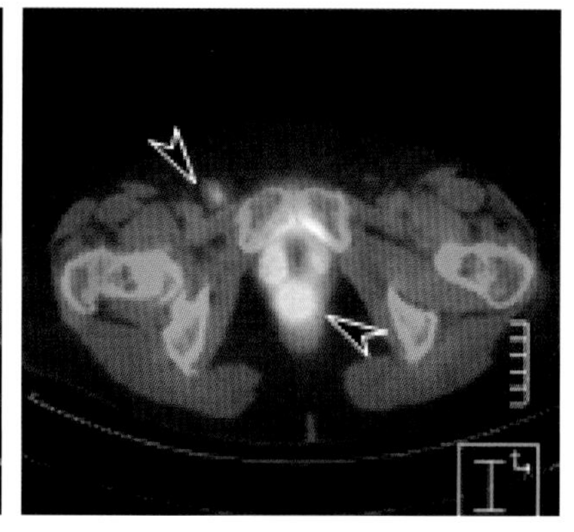

Figure 16.9. A 71-year-old woman with a cervical carcinoma underwent a PET/CT scan for staging. Transaxial (A) [^{18}F]-FDG PET image shows a large primary tumour (arrow) posterior to the bladder, as well as a right groin lymph node (arrow). (B) CT image, and (C) co-registered [^{18}F]-FDG PET/CT scan show the location of the primary and metastatic tumour.

events associated with cancer to the biology of tumour cells, and will dramatically assist in the development of innovative approaches to cancer therapy .

PET/CT scanners in Oncology

The development of combined PET/CT scanners has dramatically changed the approach to PET image interpretation, as the seamless integration of anatomic and PET images allows more accurate determination of abnormal sites and potential clinical relevance [115]. The superior accuracy of this approach in the assessment of many cancers including NSCLC, colorectal cancer, lymphoma and melanoma has been reported [116, 117]. Further assessment of this technology, including the ability to integrate PET/CT data in radiotherapy treatment planning, is an area of significant importance for the management of oncology patients in the future.

Conclusions

PET has emerged as a powerful diagnostic tool in the management of patients with cancer. The published literature has provided evidence of the superior utility over conventional imaging methods of the principal PET tracer [^{18}F]-FDG in the staging of a range of cancers, in monitoring disease recurrence, and in changing patient management to more appropriate therapy. Emerging new PET tracers that can quantitate non-invasively biologic processes within tumours, and the introduction of PET/CT into routine clinical practice, are further enhancing the role of PET in oncology. PET has the potential to dramatically improve our ability to manage patients with cancer, and is making major contributions to our understanding of cancer biology and in developing new therapies.

References

1. Strauss LG, Conti PS. The applications of PET in clinical oncology. J Nucl Med 1991;32(4):623–648; discussion 649–650.
2. Bar-Shalom R, Valdivia AY, Blaufox MD. PET imaging in oncology. Semin Nucl Med 2000;30(3):150–185.
3. Tochon-Danguy HJ, Sachinidis JI, Egan GF, Chan JG, Berlangieri SU, McKay WJ, et al. Positron emission tomography: radioisotope and radiopharmaceutical production. Australas Phys Eng Sci Med 1999;22(4):136–144.
4. Hicks RJ, Binns DS, Fawcett ME, Ware RE, Kalff V, McKenzie AF, et al. Positron emission tomography (PET): experience with a large-field-of-view three-dimensional PET scanner. Med J Aust 1999;171(10):529–532.
5. Scott AM. Current status of positron emission tomography in oncology. Australas Radiol 2002;46(2):154–162.
6. Som P, Atkins HL, Bandoypadhyay D, Fowler JS, MacGregor RR, Matsui K, et al. A fluorinated glucose analog, 2-fluoro-2-deoxy-D-glucose (F-18): nontoxic tracer for rapid tumor detection. J Nucl Med 1980;21(7):670–675.
7. Scott AM, Larson S. Tumor Imaging and Therapy. In: Coleman E, editor. Nuclear Medicine. Radiologic Clinics of North America. New York: W.B. Saunders Inc.; 1993. pp. 859–880.
8. Conti PS, Lilien DL, Hawley K, Keppler J, Grafton ST, Bading JR. PET and [18F]-FDG in oncology: a clinical update. Nucl Med Biol 1996;23(6):717–735.
9. Reutens DC, Bittar RG, Tochon-Danguy H, Scott AM. Clinical Applications of [(15)O] H(2)O PET Activation Studies. Clin Positron Imaging 1999;2(3):145–152.
10. Foo SS, Khaw P, Mitchell PL, al e. Evaluation of tumour hypoxia in non-small cell lung cancer (NSCLC) using 18F-Fluoromisonidazole Positron EmissionTomography (18F-FMISO PET) studies. In: 10th World Conference of Lung Cancer-IASLC.; 2003; Vancouver, Canada; 2003.
11. Berlangieri SU, Scott AM, Knight SR, Fitt GJ, Hennessy OF, Tochon-Danguy HJ, et al. F-18 fluorodeoxyglucose positron emission tomography in the non-invasive staging of non-small cell lung cancer. Eur J Cardiothorac Surg 1999;16 Suppl 1:S25–30.
12. Scott AM, Berlangieri SU. Positron emission tomography in colorectal carcinoma. Diagnostic Oncology 1995;4:123–129.
13. Findlay M, Young H, Cunningham D, Iveson A, Cronin B, Hickish T, et al. Noninvasive monitoring of tumor metabolism using fluorodeoxyglucose and positron emission tomography in colorectal cancer liver metastases: correlation with tumor response to fluorouracil. J Clin Oncol 1996;14(3):700–708.
14. Damian DL, Fulham MJ, Thompson E, Thompson JF. Positron emission tomography in the detection and management of metastatic melanoma. Melanoma Res 1996;6(4):325–329.
15. Kiffer JD, Berlangieri SU, Scott AM, Quong G, Feigen M, Schumer W, et al. The contribution of 18F-fluoro-2-deoxy-glucose positron emission tomographic imaging to radiotherapy planning in lung cancer. Lung Cancer 1998;19(3):167–177.
16. Berlangieri SU, Scott AM. Metabolic staging of lung cancer. N Engl J Med 2000;343(4):290–292.
17. Shon IH, O'Doherty M J, Maisey MN. Positron emission tomography in lung cancer. Semin Nucl Med 2002;32(4):240–271.
18. van Tinteren H, Hoekstra OS, Smit EF, van den Bergh JH, Schreurs AJ, Stallaert RA, et al. Effectiveness of positron emission tomography in the preoperative assessment of patients with suspected non-small-cell lung cancer: the PLUS multicentre randomised trial. Lancet 2002;359(9315):1388–1393.
19. Hain SF, Maisey MN. Positron emission tomography for urological tumours. BJU Int 2003;92(2):159–164.
20. Lowe VJ, Boyd JH, Dunphy FR, Kim H, Dunleavy T, Collins BT, et al. Surveillance for recurrent head and neck cancer using positron emission tomography. J Clin Oncol 2000;18(3):651–658.
21. Naumann R, Beuthien-Baumann B, Reiss A, Schulze J, Hanel A, Bredow J, et al. Substantial impact of FDG PET imaging on the therapy decision in patients with early-stage Hodgkin's lymphoma. Br J Cancer 2004;90(3):620–625.
22. Valk PE. Randomized controlled trials are not appropriate for imaging technology evaluation. J Nucl Med 2000; 41(7):1125–1126.
23. Surawicz TS, Davis F, Freels S, Laws ER, Jr., Menck HR. Brain tumor survival: results from the National Cancer Data Base. J Neurooncol 1998;40(2):151–160.
24. Black WC. Increasing incidence of childhood primary malignant brain tumors–enigma or no-brainer? J Natl Cancer Inst 1998; 90(17):1249–1251.
25. Di Chiro G, Brooks RA. PET-FDG of untreated and treated cerebral gliomas. J Nucl Med 1988;29(3):421–423.

26. Jeong HJ, Chung JK, Kim YK, Kim CY, Kim DG, Jeong JM, et al. Usefulness of whole-body (18)F-FDG PET in patients with suspected metastatic brain tumors. J Nucl Med 2002;43(11): 1432–1437.
27. Di Chiro G, Oldfield E, Wright DC, De Michele D, Katz DA, Patronas NJ, et al. Cerebral necrosis after radiotherapy and/or intraarterial chemotherapy for brain tumors: PET and neuropathologic studies. AJR Am J Roentgenol 1988;150(1):189–197.
28. Patronas NJ, Di Chiro G, Kufta C, Bairamian D, Kornblith PL, Simon R, et al. Prediction of survival in glioma patients by means of positron emission tomography. J Neurosurg 1985;62(6): 816–822.
29. Gambhir SS, Bauer E, Black ME, Liang Q, Kokoris MS, Barrio JR, et al. A mutant herpes simplex virus type 1 thymidine kinase reporter gene shows improved sensitivity for imaging reporter gene expression with positron emission tomography. Proc Natl Acad Sci U S A 2000;97(6):2785–2790.
30. Scott AM, Ramdave S, Hannah A, al e. Correlation of hypoxic cell fraction with glucose metabolic rate in gliomas with 18F-fluoromisonidazole (FMISO) and 18F-fluorodeoxyglucose (FDG) positron emission tomography. J Nucl Med 2001;42:678.
31. Chung JK, Kim YK, Kim SK, Lee YJ, Paek S, Yeo JS, et al. Usefulness of 11C-methionine PET in the evaluation of brain lesions that are hypo- or isometabolic on [18F]-FDG PET. Eur J Nucl Med Mol Imaging 2002;29(2):176–182.
32. Gupta N, Bradfield H. Role of positron emission tomography scanning in evaluating gastrointestinal neoplasms. Semin Nucl Med 1996;26(1):65–73.
33. Lowe VJ, Fletcher JW, Gobar L, Lawson M, Kirchner P, Valk P, et al. Prospective investigation of positron emission tomography in lung nodules. J Clin Oncol 1998;16(3):1075–1084.
34. Gould MK, Maclean CC, Kuschner WG, Rydzak CE, Owens DK. Accuracy of positron emission tomography for diagnosis of pulmonary nodules and mass lesions: a meta-analysis. Jama 2001;285(7):914–924.
35. Webb WR, Gatsonis C, Zerhouni EA, Heelan RT, Glazer GM, Francis IR, et al. CT and MR imaging in staging non-small cell bronchogenic carcinoma: report of the Radiologic Diagnostic Oncology Group. Radiology 1991;178(3):705–713.
36. Vansteenkiste JF, Stroobants SG, De Leyn PR, Dupont PJ, Verschakelen JA, Nackaerts KL, et al. Mediastinal lymph node staging with FDG-PET scan in patients with potentially operable non-small cell lung cancer: a prospective analysis of 50 cases. Leuven Lung Cancer Group. Chest 1997;112(6):1480–1486.
37. Valk PE, Pounds TR, Hopkins DM, Haseman MK, Hofer GA, Greiss HB, et al. Staging non-small cell lung cancer by whole-body positron emission tomographic imaging. Ann Thorac Surg 1995;60(6):1573–81; discussion 1581–1582.
38. Reed CE, Harpole DH, Posther KE, Woolson SL, Downey RJ, Meyers BF, et al. Results of the American College of Surgeons Oncology Group Z0050 trial: the utility of positron emission tomography in staging potentially operable non-small cell lung cancer. J Thorac Cardiovasc Surg 2003;126(6):1943–1951.
39. Bury T, Dowlati A, Paulus P, Corhay JL, Hustinx R, Ghaye B, et al. Whole-body 18FDG positron emission tomography in the staging of non-small cell lung cancer. Eur Respir J 1997;10(11): 2529–2534.
40. Gambhir SS, Hoh CK, Phelps ME, Madar I, Maddahi J. Decision tree sensitivity analysis for cost-effectiveness of FDG-PET in the staging and management of non-small-cell lung carcinoma. J Nucl Med 1996;37(9):1428–1436.
41. Ling CC, Humm J, Larson S, Amols H, Fuks Z, Leibel S, et al. Towards multidimensional radiotherapy (MD-CRT): biological imaging and biological conformality. Int J Radiat Oncol Biol Phys 2000;47(3):551–560.
42. Mac Manus MP, Hicks RJ, Matthews JP, McKenzie A, Rischin D, Salminen EK, et al. Positron emission tomography is superior to computed tomography scanning for response-assessment after radical radiotherapy or chemoradiotherapy in patients with non-small-cell lung cancer. J Clin Oncol 2003;21(7):1285–1292.
43. Falk PM, Gupta NC, Thorson AG, Frick MP, Boman BM, Christensen MA, et al. Positron emission tomography for preoperative staging of colorectal carcinoma. Dis Colon Rectum 1994;37(2):153–156.
44. Abdel-Nabi H, Doerr RJ, Lamonica DM, Cronin VR, Galantowicz PJ, Carbone GM, et al. Staging of primary colorectal carcinomas with fluorine-18 fluorodeoxyglucose whole-body PET: correlation with histopathologic and CT findings. Radiology 1998; 206(3):755–760.
45. Yasuda S, Fujii H, Nakahara T, Nishiumi N, Takahashi W, Ide M, et al. [18F]-FDG PET detection of colonic adenomas. J Nucl Med 2001;42(7):989–992.
46. Arulampalam TH, Costa DC, Bomanji JB, Ell PJ. The clinical application of positron emission tomography to colorectal cancer management. Q J Nucl Med 2001;45(3):215–230.
47. Heriot AG, Hicks RJ, Drummond EG, Keck J, Mackay J, Chen F, et al. Does Positron Emission Tomography Change Management in Primary Rectal Cancer? A Prospective Assessment. Dis Colon Rectum 2004.
48. Akhurst T, Larson SM. Positron emission tomography imaging of colorectal cancer. Semin Oncol 1999;26(5):577–583.
49. Lai DT, Fulham M, Stephen MS, Chu KM, Solomon M, Thompson JF, et al. The role of whole-body positron emission tomography with [18F]fluorodeoxyglucose in identifying operable colorectal cancer metastases to the liver. Arch Surg 1996; 131(7):703–707.
50. Valk PE, Abella-Columna E, Haseman MK, Pounds TR, Tesar RD, Myers RW, et al. Whole-body PET imaging with [18F]fluorodeoxyglucose in management of recurrent colorectal cancer. Arch Surg 1999;134(5):503–11; discussion 511–513.
51. Whiteford MH, Whiteford HM, Yee LF, Ogunbiyi OA, Dehdashti F, Siegel BA, et al. Usefulness of FDG-PET scan in the assessment of suspected metastatic or recurrent adenocarcinoma of the colon and rectum. Dis Colon Rectum 2000;43(6):759–67; discussion 767–770.
52. Schiepers C, Penninckx F, De Vadder N, Merckx E, Mortelmans L, Bormans G, et al. Contribution of PET in the diagnosis of recurrent colorectal cancer: comparison with conventional imaging. Eur J Surg Oncol 1995;21(5):517–522.
53. Krestin GP, Steinbrich W, Friedmann G. Recurrent rectal cancer: diagnosis with MR imaging versus CT. Radiology 1988; 168(2):307–311.
54. de Lange EE, Fechner RE, Wanebo HJ. Suspected recurrent rectosigmoid carcinoma after abdominoperineal resection: MR imaging and histopathologic findings. Radiology 1989;170(2): 323–328.
55. Huebner RH, Park KC, Shepherd JE, Schwimmer J, Czernin J, Phelps ME, et al. A meta-analysis of the literature for whole-body FDG PET detection of recurrent colorectal cancer. J Nucl Med 2000;41(7):1177–1189.
56. Haberkorn U, Strauss LG, Dimitrakopoulou A, Engenhart R, Oberdorfer F, Ostertag H, et al. PET studies of fluorodeoxyglucose metabolism in patients with recurrent colorectal tumors receiving radiotherapy. J Nucl Med 1991;32(8):1485–1490.
57. Ito K, Kato T, Ohta T, Tadokoro M, et al. F-18-FDG-PET in recurrent rectal cancer: relation to tumor size. J Nucl Med 1994;35:119P.
58. Kalff V, Hicks RJ, Ware RE, Hogg A, Binns D, McKenzie AF. The clinical impact of (18)F-FDG PET in patients with suspected or confirmed recurrence of colorectal cancer: a prospective study. J Nucl Med 2002;43(4):492–499.
59. Meta J, Seltzer M, Schiepers C, Silverman DH, Ariannejad M, Gambhir SS, et al. Impact of [18F]-FDG PET on managing patients with colorectal cancer: the referring physician's perspective. J Nucl Med 2001;42(4):586–590.
60. Guillem JG, Puig-La Calle J, Jr., Akhurst T, Tickoo S, Ruo L, Minsky BD, et al. Prospective assessment of primary rectal cancer response to preoperative radiation and chemotherapy using 18-fluorodeoxyglucose positron emission tomography. Dis Colon Rectum 2000;43(1):18–24.
61. Kostakoglu L, Goldsmith SJ. [18F]-FDG PET evaluation of the response to therapy for lymphoma and for breast, lung, and colorectal carcinoma. J Nucl Med 2003;44(2):224–239.
62. Gambhir SS, Czernin J, Schwimmer J, Silverman DH, Coleman RE, Phelps ME. A tabulated summary of the FDG PET literature. J Nucl Med 2001;42(5 Suppl):1S–93S.
63. Delbeke D, Martin WH, Morgan DS, Kinney MC, Feurer I, Kovalsky E, et al. 2-deoxy-2-[F-18]fluoro-D-glucose imaging with

positron emission tomography for initial staging of Hodgkin's disease and lymphoma. Mol Imaging Biol 2002;4(1):105–114.
64. Jerusalem G, Beguin Y, Najjar F, Hustinx R, Fassotte MF, Rigo P, et al. Positron emission tomography (PET) with 18F-fluorodeoxyglucose ([18F]-FDG) for the staging of low-grade non-Hodgkin's lymphoma (NHL). Ann Oncol 2001;12(6):825–830.
65. Schoder H, Meta J, Yap C, Ariannejad M, Rao J, Phelps ME, et al. Effect of whole-body (18)F-FDG PET imaging on clinical staging and management of patients with malignant lymphoma. J Nucl Med 2001;42(8):1139–1143.
66. Foo SS, Mitchell PL, Berlangieri SU, Smith C, Scott AM. F-18 Fluorodeoxyglucose positron emission tomography (PET) provides additional staging information in patients with Hodgkin's and non-Hodgkin's lymphoma compared to conventional imaging. Int Med J 2004;in press.
67. Blum RH, Seymour JF, Wirth A, MacManus M, Hicks RJ. Frequent impact of [18F]fluorodeoxyglucose positron emission tomography on the staging and management of patients with indolent non-Hodgkin's lymphoma. Clin Lymphoma 2003;4(1): 43–49.
68. Mikhaeel NG, Timothy AR, Hain SF, O'Doherty MJ. 18-FDG-PET for the assessment of residual masses on CT following treatment of lymphomas. Ann Oncol 2000;11 Suppl 1:147–150.
69. de Wit M, Bumann D, Beyer W, Herbst K, Clausen M, Hossfeld DK. Whole-body positron emission tomography (PET) for diagnosis of residual mass in patients with lymphoma. Ann Oncol 1997;8 Suppl 1:57–60.
70. Shen YY, Kao A, Yen RF. Comparison of 18F-fluoro-2-deoxyglucose positron emission tomography and gallium-67 citrate scintigraphy for detecting malignant lymphoma. Oncol Rep 2002;9(2):321–325.
71. Spaepen K, Stroobants S, Dupont P, Van Steenweghen S, Thomas J, Vandenberghe P, et al. Prognostic value of positron emission tomography (PET) with fluorine-18 fluorodeoxyglucose ([18F]FDG) after first-line chemotherapy in non-Hodgkin's lymphoma: is [18F]FDG-PET a valid alternative to conventional diagnostic methods? J Clin Oncol 2001;19(2):414–419.
72. Acland KM, Healy C, Calonje E, O'Doherty M, Nunan T, Page C, et al. Comparison of positron emission tomography scanning and sentinel node biopsy in the detection of micrometastases of primary cutaneous malignant melanoma. J Clin Oncol 2001; 19(10):2674–2678.
73. Macfarlane DJ, Sondak V, Johnson T, Wahl RL. Prospective evaluation of 2-[18F]-2-deoxy-D-glucose positron emission tomography in staging of regional lymph nodes in patients with cutaneous malignant melanoma. J Clin Oncol 1998;16(5): 1770–1776.
74. Balch CM, Soong SJ, Gershenwald JE, Thompson JF, Reintgen DS, Cascinelli N, et al. Prognostic factors analysis of 17,600 melanoma patients: validation of the American Joint Committee on Cancer melanoma staging system. J Clin Oncol 2001;19(16): 3622–3634.
75. Harris MT, Berlangieri SU, Cebon J, Davis ID, Scott AM. 18F-Fluorodeoxyglucose positron emission tomography (FDG-PET) in patients with advanced melanoma. In: ASCO 38th Annual Meeting; 2002; Orlando, Florida; 2002. p. 1390.
76. Mijnhout GS, Hoekstra OS, van Tulder MW, Teule GJ, Deville WL. Systematic review of the diagnostic accuracy of (18)F-fluorodeoxyglucose positron emission tomography in melanoma patients. Cancer 2001;91(8):1530–1542.
77. Schwimmer J, Essner R, Patel A, Jahan SA, Shepherd JE, Park K, et al. A review of the literature for whole-body FDG PET in the management of patients with melanoma. Q J Nucl Med 2000; 44(2):153–167.
78. Karakousis CP, Velez A, Driscoll DL, Takita H. Metastasectomy in malignant melanoma. Surgery 1994;115(3):295–302.
79. Kalff V, Hicks RJ, Ware RE, Greer B, Binns DS, Hogg A. Evaluation of high-risk melanoma: comparison of [18F]FDG PET and high-dose 67Ga SPET. Eur J Nucl Med Mol Imaging 2002; 29(4):506–515.
80. Adams S, Baum RP, Stuckensen T, Bitter K, Hor G. Prospective comparison of [18F]-FDG PET with conventional imaging modalities (CT, MRI, US) in lymph node staging of head and neck cancer. Eur J Nucl Med 1998;25(9):1255–1260.
81. Hannah A, Scott AM, Tochon-Danguy H, Chan JG, Akhurst T, Berlangieri S, et al. Evaluation of 18 F-fluorodeoxyglucose positron emission tomography and computed tomography with histopathologic correlation in the initial staging of head and neck cancer. Ann Surg 2002;236(2):208–217.
82. Paulus P, Sambon A, Vivegnis D, Hustinx R, Moreau P, Collignon J, et al. 18FDG-PET for the assessment of primary head and neck tumors: clinical, computed tomography, and histopathological correlation in 38 patients. Laryngoscope 1998;108(10):1578–1583.
83. Wong WL, Chevretton EB, McGurk M, Hussain K, Davis J, Beaney R, et al. A prospective study of PET-FDG imaging for the assessment of head and neck squamous cell carcinoma. Clin Otolaryngol 1997;22(3):209–214.
84. McGuirt WF, Greven K, Williams D, 3rd, Keyes JW, Jr., Watson N, Cappellari JO, et al. PET scanning in head and neck oncology: a review. Head Neck 1998;20(3):208–215.
85. Keyes JW, Jr., Watson NE, Jr., Williams DW, 3rd, Greven KM, McGuirt WF. FDG PET in head and neck cancer. AJR Am J Roentgenol 1997;169(6):1663–1669.
86. Greven KM, Keyes JW, Jr., Williams DW, 3rd, McGuirt WF, Joyce WT, 3rd. Occult primary tumors of the head and neck: lack of benefit from positron emission tomography imaging with 2-[F-18]fluoro-2-deoxy-D-glucose. Cancer 1999;86(1):114–118.
87. Farber LA, Benard F, Machtay M, Smith RJ, Weber RS, Weinstein GS, et al. Detection of recurrent head and neck squamous cell carcinomas after radiation therapy with 2-18F-fluoro-2-deoxy-D-glucose positron emission tomography. Laryngoscope 1999;109(6):970–975.
88. Rege S, Safa AA, Chaiken L, Hoh C, Juillard G, Withers HR. Positron emission tomography: an independent indicator of radiocurability in head and neck carcinomas. Am J Clin Oncol 2000;23(2):164–169.
89. Samson D, Flamm CR, Aronson N. FDG positron emission tomography for evaluation of breast cancer. Blue Cross and Blue Shield Association. 2001:1–93.
90. Wahl RL, Siegel BA, Coleman RE, Gatsonis CG. Prospective multicenter study of axillary nodal staging by positron emission tomography in breast cancer: a report of the staging breast cancer with PET Study Group. J Clin Oncol 2004;22(2):277–285.
91. Moon DH, Maddahi J, Silverman DH, Glaspy JA, Phelps ME, Hoh CK. Accuracy of whole-body fluorine-18-FDG PET for the detection of recurrent or metastatic breast carcinoma. J Nucl Med 1998;39(3):431–435.
92. Bender H, Kirst J, Palmedo H, Schomburg A, Wagner U, Ruhlmann J, et al. Value of 18fluoro-deoxyglucose positron emission tomography in the staging of recurrent breast carcinoma. Anticancer Res 1997;17(3B):1687–1692.
93. Schelling M, Avril N, Nahrig J, Kuhn W, Romer W, Sattler D, et al. Positron emission tomography using [(18)F]Fluorodeoxyglucose for monitoring primary chemotherapy in breast cancer. J Clin Oncol 2000;18(8):1689–1695.
94. Luketich JD, Friedman DM, Weigel TL, Meehan MA, Keenan RJ, Townsend DW, et al. Evaluation of distant metastases in esophageal cancer: 100 consecutive positron emission tomography scans. Ann Thorac Surg 1999;68(4):1133–1136; discussion 1136–1137.
95. Lerut T, Flamen P. Role of FDG-PET scan in staging of cancer of the esophagus and gastroesophageal junction. Minerva Chir 2002;57(6):837–845.
96. Meltzer CC, Luketich JD, Friedman D, Charron M, Strollo D, Meehan M, et al. Whole-body FDG positron emission tomographic imaging for staging esophageal cancer comparison with computed tomography. Clin Nucl Med 2000;25(11):882–887.
97. Fukunaga T, Okazumi S, Koide Y, Isono K, Imazeki K. Evaluation of esophageal cancers using fluorine-18-fluorodeoxyglucose PET. J Nucl Med 1998;39(6):1002–1007.
98. Flamen P, Van Cutsem E, Lerut A, Cambier JP, Haustermans K, Bormans G, et al. Positron emission tomography for assessment of the response to induction radiochemotherapy in locally advanced oesophageal cancer. Ann Oncol 2002;13(3):361–368.
99. Einhorn N, Nilsson B, Sjovall K. Factors influencing survival in carcinoma of the ovary. Study from a well-defined Swedish population. Cancer 1985;55(9):2019–2025.
100. Karlan BY, Hawkins R, Hoh C, Lee M, Tse N, Cane P, et al. Whole-body positron emission tomography with 2-[18F]-fluoro-

2-deoxy-D-glucose can detect recurrent ovarian carcinoma. Gynecol Oncol 1993;51(2):175–181.
101. Hubner KF, McDonald TW, Niethammer JG, Smith GT, Gould HR, Buonocore E. Assessment of primary and metastatic ovarian cancer by positron emission tomography (PET) using 2-[18F]deoxyglucose (2-[18F]FDG). Gynecol Oncol 1993;51(2):197–204.
102. Stephens AW, Gonin R, Hutchins GD, Einhorn LH. Positron emission tomography evaluation of residual radiographic abnormalities in postchemotherapy germ cell tumor patients. J Clin Oncol 1996;14(5):1637–1641.
103. Miraldi F, Adler LP, Faulhaber P. PET imaging in soft tissue sarcomas. Cancer Treat Res 1997;91:51–64.
104. Hoh CK, Seltzer MA, Franklin J, deKernion JB, Phelps ME, Belldegrun A. Positron emission tomography in urological oncology. J Urol 1998;159(2):347–356.
105. Ramdave S, Thomas GW, Berlangieri SU, Bolton DM, Davis I, Danguy HT, et al. Clinical role of F-18 fluorodeoxyglucose positron emission tomography for detection and management of renal cell carcinoma. J Urol 2001;166(3):825–830.
106. Narayan K, Hicks RJ, Jobling T, Bernshaw D, McKenzie AF. A comparison of MRI and PET scanning in surgically staged locoregionally advanced cervical cancer: potential impact on treatment. Int J Gynecol Cancer 2001;11(4):263–271.
107. Thomas DM, Mitchell PL, Berlangieri SU, Tochon-Danguy H, Knight S, Clarke CP, et al. Positron emission tomography in assessing response to neoadjuvant chemotherapy for non-small-cell lung cancer. Med J Aust 1998;169(4):227.
108. Herbst RS, Mullani NA, Davis DW, Hess KR, McConkey DJ, Charnsangavej C, et al. Development of biologic markers of response and assessment of antiangiogenic activity in a clinical trial of human recombinant endostatin. J Clin Oncol 2002;20(18): 3804–3814.
109. Foo SS, Abbott DF, Lawrentschuk N, Scott AM. Functional imaging of intratumoural hypoxia. J Mol Imag Biol 2004;in press.
110. Phelps ME. PET: the merging of biology and imaging into molecular imaging. J Nucl Med 2000;41(4):661–681.
111. Lee FT, Rigopoulos A, Hall C, Clarke K, Cody SH, Smyth FE, et al. Specific localization, gamma camera imaging, and intracellular trafficking of radiolabelled chimeric anti-G(D3) ganglioside monoclonal antibody KM871 in SK-MEL-28 melanoma xenografts. Cancer Res 2001;61(11):4474–4482.
112. Ackermann U, Tochon-Danguy H, Young K, Sachinidis JI, Chan JG, Scott AM. C-11 labelling of AG957 – a potential tyrphostin radiotracer for PET. J Label Compd Radiopharm 2002;45: 157–165.
113. Francis DL, Visvikis D, Costa DC, Arulampalam TH, Townsend C, Luthra SK, et al. Potential impact of [18F]3'-deoxy-3'-fluorothymidine versus [18F]fluoro-2-deoxy-D-glucose in positron emission tomography for colorectal cancer. Eur J Nucl Med Mol Imaging 2003;30(7):988–994.
114. Tjuvajev JG, Doubrovin M, Akhurst T, Cai S, Balatoni J, Alauddin MM, et al. Comparison of radiolabeled nucleoside probes (FIAU, FHBG, and FHPG) for PET imaging of HSV1-tk gene expression. J Nucl Med 2002;43(8):1072–1083.
115. Townsend DW, Carney JP, Yap JT, Hall NC. PET/CT today and tomorrow. J Nucl Med 2004;45 Suppl 1:4S–14S.
116. Schoder H, Larson SM, Yeung HW. PET/CT in oncology: integration into clinical management of lymphoma, melanoma, and gastrointestinal malignancies. J Nucl Med 2004;45 Suppl 1:72S–81S.
117. Lardinois D, Weder W, Hany TF, Kamel EM, Korom S, Seifert B, et al. Staging of non-small-cell lung cancer with integrated positron-emission tomography and computed tomography. N Engl J Med 2003;348(25):2500–2507.

17 The Use of Positron Emission Tomography in Drug Discovery and Development

William C Eckelman

Introduction

To accelerate the drug discovery process and to increase the annual number of approved drugs, pharmaceutical chemists have developed new procedures, such as combinatorial chemistry and parallel synthesis, and biologists have developed high-throughput screening methods that test various properties of the drug candidate in vitro (Table 17.1). Since the time line for drug discovery and development can be as long as 16 years, these techniques are necessary to expedite the drug discovery process. Currently, discovery and preclinical testing can take up to 6 years. Phase I and Phase II studies usually average 2 years each. Phase III takes longer and the final U.S. Food and Drug Administration (FDA) approval process takes one year [1]. The goal of this review is to show that studies carried out in vivo using external imaging of all kinds,

Table 17.1. High throughput in vitro screening

Physical Properties
pKa, solubility, dissolution rate, crystallinity and chemical stability
Drug Metabolism, Pharmacokinetics (PK), and Pharmacodynamics (PD)
Affinity (interaction with membranes, binding proteins, transporters and enzymes)
Organ blood flow, glomerular filtration rate, and tissue uptake
Absorption potential (Caco-2 cell permeability)
Blood Brain Barrier penetration (bovine brain microvessel endothelial cells)
Metabolism (cytochrome P450)
Metabolic stability (hepatocyte metabolites analysed by liquid chromatography/mass spectrometry)

Source: Ref. 55.

but especially Positron Emission Tomography (PET), offer great promise for accelerating the process from pre-clinical discovery to Phase III studies. Although high-throughput, in vitro screening has been useful, the results must be extrapolated to *in vivo* use. Even though *in vivo* imaging is first carried out in animals, the drug candidate is being tested in an intact species and this should yield more complete information. Imaging is especially important in paradigms where animals are studied before and after treatment with the drug candidate. As paired statistics can be carried out in the same animal, fewer animals are needed. Therefore, *in vivo* imaging has an advantage over the autoradiographic approaches that are carried out *in vivo* but require the sacrifice of the animal after each study.

The development of radiolabelled biochemicals and imaging devices to detect the radioactivity by external imaging has expanded the use of nuclear medicine studies in drug development. The common radionuclides for Single Photon Emission Computed Tomography (SPECT) are ^{99m}Tc (half life = 6 hr) and ^{123}I (half life = 13 hr). For PET, the radionuclides emit a positron that annihilates to give two 511 keV gamma rays at approximately 180 degrees. The positron-emitting radionuclides (with their half lives) used most frequently are ^{15}O (2.07 min), ^{11}C (20.4 min), ^{13}N (10 min) and ^{18}F (109.7 min). The specific activities (radioactivity per mass unit, usually measured in Ci/mM or TBq/mM) of these radionuclides are high because they are made through a nuclear transformation; that is, one element is converted into another so that, except for trace contaminants, they are carrier free. The actual specific activities for the most-used PET radionuclides, ^{18}F and ^{11}C, are of the order of 37–185 TBq/mM (1000–5000 Ci/mM) at the end of the cyclotron bombardment. Therefore, these radioactive probes are injected at tracer levels (~nM injected). The uniqueness of the nuclear medicine technique based on this tracer principle is in measuring biochemistry *in vivo*, especially the biochemistry of easily saturated sites such as receptors, by external imaging.

The Tracer Principle

George Charles de Hevesy is usually considered the first to identify the tracer principle [2]. In 1923, he used ^{212}Pb (half life = 10.6 hr) to study the uptake of solutions in bean plants. Although lead is generally considered toxic, he was able to use small, non-toxic amounts because of the sensitivity of the radioactivity techniques. The following year he carried out the first experiment in animals using ^{210}Bi to label and follow the circulation of bismuth in rabbits after intra-muscular injection of bismuth-containing antisyphilitic drugs. Martin D. Kamen, in the three editions of his book titled *Isotopic Tracers in Biology,* chronicled the rapid progress in applying the newly discovered tracer principle to clinical research [3]. Even today it is the only method of monitoring low concentrations of either receptor or enzymes using external imaging.

Compared with other external imaging modalities, the radiotracer approach requires injection of the least amount of mass to obtain the necessary biochemical information. The high specific activity of the radioligands allows detection of very low density sites. For example, for ^{11}C or ^{18}F-labelled radiopharmaceuticals, the specific activity at the time of injection is usually greater than 37TBq/mM (1000 Ci/mM). Since the radiopharmaceutical is injected in amounts approximating 370MBq, the associated mass injected is 10 nM in a 70 kg human. This amount is minimal compared with Magnetic Resonance Imaging (MRI), which requires concentrations of 10 to100 μM/kg (~700 to 7000 μM) for iodinated contrast media where a concentration of >100 μM (~7 mM) is required. The advantage of using smaller amounts can be seen from the Scatchard transformation of the equilibrium binding constant. At high specific activity, the bound-to-free ratio (B/F) reaches a maximum in that $(B_{max} - B) / K_D$ approaches B_{max}/K_D where B_{max} represents the free receptor concentration and K_D is the dissociation equilibrium constant. At lower specific activity, the ratio is decreased, or in the extreme case the receptor will be saturated [3]. In general, these saturating concentrations using radioligands will still be orders of magnitude less concentrated than those required for MRI and CT. However, the higher instrument resolution of MRI and CT is a driving force to apply these tracer principles to MRI contrast media and iodinated contrast media. An example of the effect of injected dose can be seen from data showing binding of the muscarinic M2 subtype-specific ligand FP-TZTP [(1,2,5-thiadiazol-4-yl)-tetrahydro-1-methylpyridine]. As the dose of injected material increases, the percent injected dose (%ID) decreases (Fig. 17.1). A saturating amount of drug can be estimated from such studies. In this particular example, the receptor is easily saturated at 140 μM/kg, a much lower amount than the 700 to 7000 μM/kg required for standard MRI studies.

Case Method Analysis of the Drug Discovery Process Using External Imaging

The Society of Noninvasive Imaging in Drug Development (SNIDD) was formed in 1990 by a small

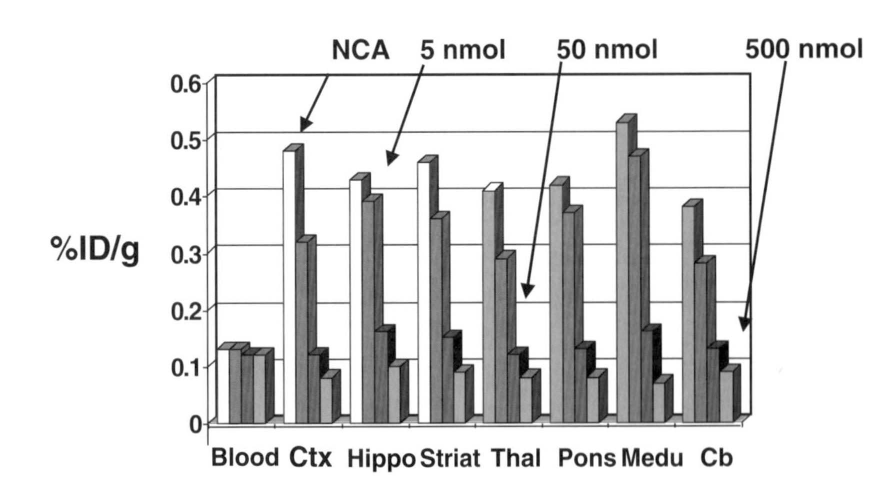

Figure 17.1. The percent injected dose per gram (%ID/g) in rat brain for [^{18}F]-FP-TZTP as a function of the co-injected dose of P-TZTP. Tissues assayed are cortex (Ctx), hippocampus (Hippo), striatum (Striat), thalamus (Thal), pons, medulla (Medu) and cerebrum (Cb).

group of researchers from academia, government laboratories and the pharmaceutical industry. This group believed that non-invasive imaging technologies such as PET and SPECT can be powerful research tools for the drug industry and that these tools can help reduce the time and cost of discovering and bringing new drugs to the marketplace by improving the efficiency of both pre-clinical and clinical research on new drug candidates and identifying drug candidates that will ultimately fail. SNIDD has held two meetings to illustrate the use of imaging in drug development. The first of these symposia was held on September 11–13, 1998 at the National Institutes of Health (NIH). The symposium was organized using the case method technique, popularized by the Harvard Business School as a framework for analyzing examples of actual business decisions and, from that, obtaining insight into the process [4]. Several conferences have addressed the use of imaging in drug research and development, but none, in our experience, had analyzed decisions made during the drug development process based on collaborative imaging studies. Nuclear Medicine imaging techniques were used in the eleven case method presentations. The special issue reporting on this conference contains descriptions of pre-clinical studies in knockout mice, pre-clinical PET studies in non-human primates, a series of Phase I studies (dosing studies, pharmacokinetics [unique delivery systems], dose, dose interval, drug interactions), Phase I/II (dose response, response duration, pharmacokinetic/pharmacodynamic

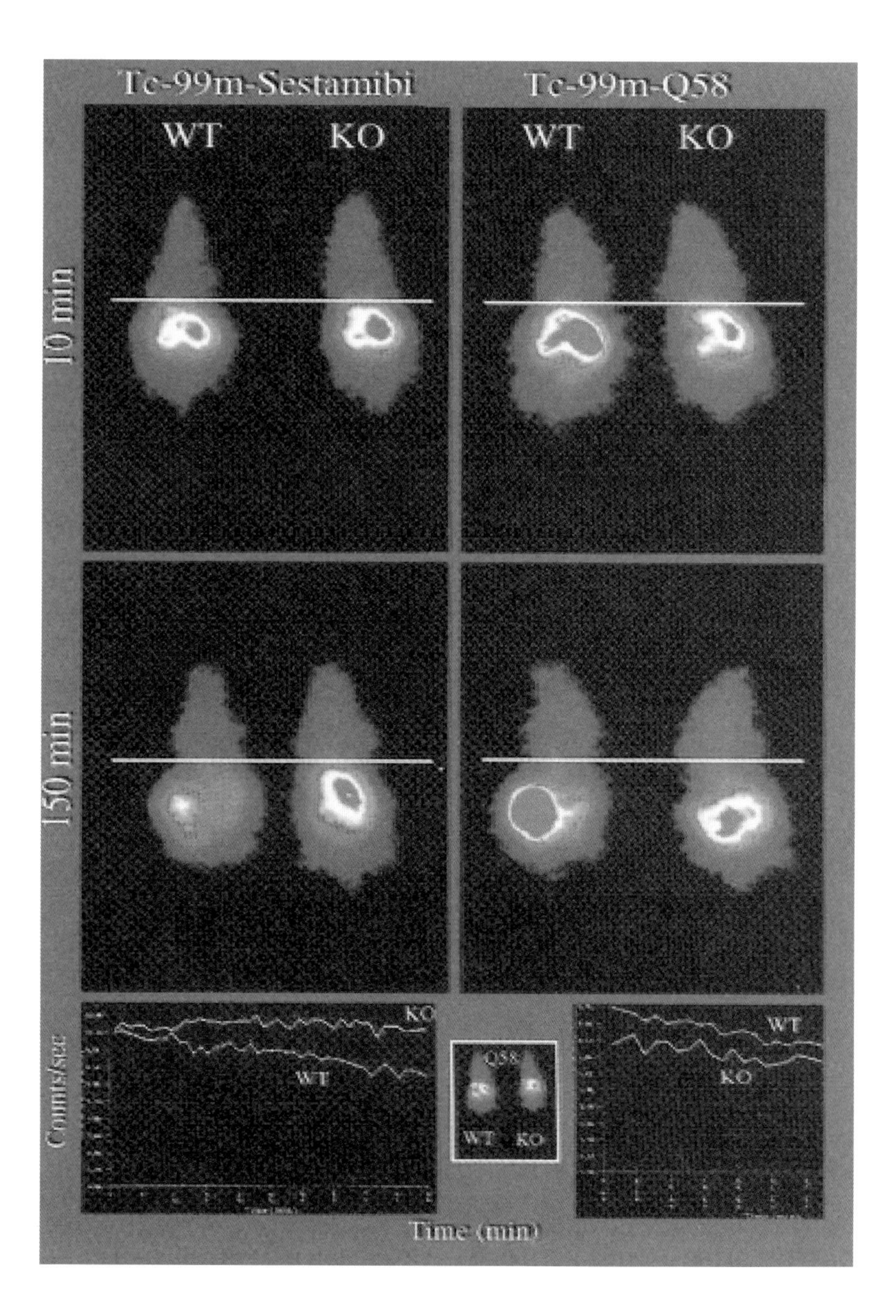

Figure 17.2. Representative planar images of wild-type and knockout mice obtained 10 minutes (upper panels) and 150 minutes (lower panels) post-injection. Both [^{99m}Tc]-sestamibi and [^{99m}Tc]-Q58 show Pgp-mediated hepatobiliary clearance in wild-type mice. However, in knockout mice, Tc-99m sestamibi lacked hepatobiliary clearance but [^{99m}Tc]-Q58 did not.

[PK/PD] relationship), Phase III "surrogate markers" or mechanistic intermediates [5], and marketing studies.

Use of Knockout Mice in Drug Discovery

Imaging can save considerable time in pre-clinical studies. Multi-drug resistance (MDR) in cancer chemotherapy is mediated by P-glycoprotein. Monitoring the effectiveness of modulators is an important area of imaging. Piwnica-Worms and Marmion presented a case study that used knockout mice to decide whether [^{99m}Tc]-Q complexes would be effective as probes for P-glycoprotein. The [^{99m}Tc]-Q complexes tested in MDR 1a/ab knockout mice showed enhancement of initial brain uptake but no significant delay in liver clearance (Fig. 17.2). These agents were deemed to have cross-reactivity for another organic cation transporter expressed in mouse liver and were not advanced to clinical studies [6].

Drug Distribution Studies

Two studies were carried out to evaluate drug distribution in the gut and in the lung. Producing a true tracer situation is important to these studies. In the analysis of modified-release formulations, drugs such as diltiazem are formulated with small amounts of stable [^{152}Sm]-samarium oxide. The tablets can then be activated by neutron activation to produce radioactive ^{153}Sm and followed *in vivo* using planar imaging. This Phase I study allowed quantitative distribution of the tablet in the gastrointestinal tract [7]. In another example, the lung distribution of the corticosteroid triamcinolone acetonide was studied using the dispenser Azmacort, a pressurized aerosol metered-dose inhaler formulation. The steroid was radiolabelled with ^{11}C and introduced into the dispenser. The time-activity curve obtained using PET showed a significant increase of the steroid in the lung and a significant decrease in the mouth (Fig. 17.3). These data were submitted to the FDA in support of the Azmacort inhaler's superiority [8].

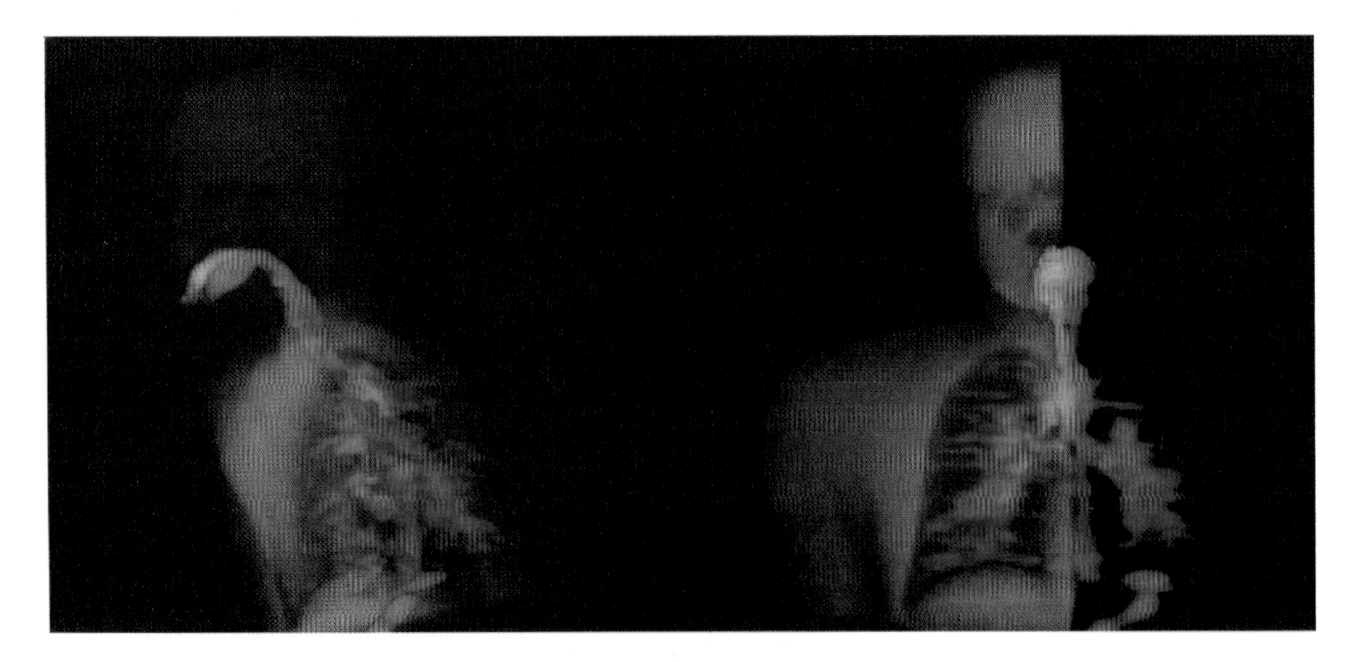

Figure 17.3. Azmacort distribution in airway, lung, GI tract after administration. 3-D translucent, low-resolution CT (flesh tone) overlaid with the PET image of drug concentration (green).

Drug Discovery in Oncology

The conventional paradigm of demanding proof of an anti-cancer drug's potency by demonstrating tumour shrinkage using conventional anatomical imaging is not consistent with the newer therapeutics, especially the chemostatic agents where the size of the tumour does not necessarily decrease in successful treatment. The use of proliferation marker agents such as ^{18}F labelled fluorothymidine (FLT) and PET may show an earlier response [9]. Pharmaceutical companies now have the option of using radiolabelled proliferation agents in Phase I or Phase II to choose which compounds to advance to the next phase.

Drug Discovery in Neuroscience

PET is frequently used to evaluate neuroreceptor ligands. Salazar and Fischman evaluated BMS 181101, a drug with agonist and antagonist activity at various sites in the serotonin system [10]. The ^{11}C labelled form of the drug was used to show that the residence time in the brain was short and, as a result, specific binding could not be determined by external imaging. These studies showed that the drug may have a narrow therapeutic index and may not be suitable for once or twice daily dosage. In this case the drug itself was radiolabelled, but the remaining examples used an established radiotracer with varying concentrations of the potential drug. The main goal of these studies was to measure saturable binding site occupancy. Fowler [11] showed techniques for measuring occupancy for the dopamine transporter (DAT) and for monoamine oxidase B (MAO-B). In the case of dopamine transporter , the occupancy of the dopamine transporter by cocaine was faster than that found by methylphenidate (Ritalin), but both reached the same saturation level at 0.6 mg/kg. For the reversible MAO-B inhibitor lazabemide, 90% saturation was demonstrated at a dose of 50mg and was reversible. On the other hand, deprenyl occupancy at MAO-B was long lasting as measured by PET, demonstrating the effectiveness of doses lower than those currently used in the clinic. The putative anti-psychotic drug M100907 was studied indirectly using [^{11}C]-spiperone, which binds to both the dopamine D_2 receptor and the 5-HT_{2A} serotonin receptor [12]. The therapeutic index of M100907 was defined in Phase I single and multiple dose tolerability studies. PET was then used to confirm the mechanism of action of M100907 in humans and to define an appropriate dose range and regimen of 20 mg per day.

General Approaches Using Radiolabelled Tracers in Drug Discovery

Most studies used in drug development apply the tracer principle to specific examples involving receptor/ transporter binding drugs. Receptor binding studies generally use one of the following methods:

1. To determine the interaction of the drug with a desired binding site (e.g., receptor or enzyme):
   a. Radiolabelled the potential drug in such a way as to not disturb the biochemical parameter to be measured.
   b. Use a radioligand with the desired properties and study potential drug candidate binding by competition. Usually co-injection is used, but pre-or post-injection may be required depending on the relative pharmacokinetics. Measure neurotransmitter concentration changes with reversible receptor radioligand indirectly after administering the potential drug, whose putative mode of action is through neurotransmitter release. One example of this is the use of [^{11}C]-raclopride, a dopamine receptor antagonist, to indirectly measure increases in the neurotransmitter dopamine as a function of the pharmacologic action of an amphetamine-like potential drug candidate [13], or,
2. Measure enzyme inhibition indirectly by measuring neurotransmitter concentration. An example is the measurement of the effect of cholinesterase inhibitors on the increase in acetylcholine concentration. This increase is measured indirectly by using a radiolabelled muscarinic receptor agonist that is bound reversibly [14].

Radiolabelling the potential drug was not the method of choice in this symposium and other such symposia. The approaches described in 1b and 2 above offer the advantage of quick answers because the radiotracer is already characterized, a clear advantage given the cost of delaying new drug development [15]. In this era of increased interest in surrogate or mechanistic markers to accelerate the drug development process, nuclear medicine imaging is an important technique and an unique approach when low-density binding sites are the drug targets. The one negative implication of the approach outlined in 1b or 2 is that the drug being developed is in an area that is well enough characterized such that radioligands have already been developed for immediate use. In general, new targets are the goal of pharmaceutical discovery, which implies that a new radioligand will have to be developed, which could slow the drug development in this new therapeutic area.

Accelerating Drug Discovery

The second SNIDD symposium, which was held in October 2000 [16], discussed whether the use of PET can be implemented fast enough to compensate for the increased time and expense of these studies by decreasing the number of studies that must be completed. The goal of the second SNIDD conference was to discuss how the drug discovery process could be accelerated using well-established radiopharmaceuticals, such as ^{15}O water for blood flow and 2-[^{18}F]-fluoro-2-deoxyglucose ([^{18}F]-FDG) for glucose metabolism. The advantages of these radiotracers are their availability and their potential to measure indirectly the effect of drugs on specific neurotransmitter systems. Another approach that was outlined in the first symposium is the use of well-established radiolabelled ligands that can measure indirectly the effect of drugs on specific neurotransmitter systems but require a different ligand for each application. The issue of speed depends on whether the binding site has been characterized and radioligands are already developed as discussed in the previous section.

Radiolabelled ^{15}O water and [^{18}F]-FDG are general probes that have been used in drug discovery. Most university PET centers have access to both ^{15}O water and [^{18}F]-FDG, while other centers have access to ^{15}O water only if there is a cyclotron on site because of the 2 minute physical half life of ^{15}O. On the other hand, ^{18}F labelled compounds such as FDG can be shipped from regional cyclotrons because of the 109.7 minute half life of ^{18}F.

Herscovitch has recently reviewed the uses of radioactive water in drug development [17]. Regional cerebral blood flow (rCBF) is thought to be coupled to local neuronal activity and metabolism. The kinds of studies that can be performed are similar to those performed with a radioligand in that dosing studies and target saturation studies can be measured indirectly through the observed coupling between rCBF and neuronal activity and metabolism. These studies were carried out with drugs that bound to neurotransmitter receptors such as lorazepam binding to the benzodiazepine receptor, D_1 and D_2 dopamine agonists binding to the dopamine receptors, and physostigmine increases in acetylcholine binding to muscarinic receptors. Often, the site of maximal CBF change, and presumably neuronal activity, is not correlated with the maximum known receptor density indicating that the changes occurred downstream from sites of highest receptor density, at distinct neuronal projections from the sites of greatest drug binding [17]. There are many studies where the effects of drugs on blood flow have been studied primarily in the heart, but in other organs as well. [^{18}F]-FDG has also been used to look at drug responses. There is a complete listing of studies in rodents, which was published in the early 1980s [18]. The glucose metabolic consequences of manipulating central neurotransmitter systems was demonstrated primarily for the dopamine system using apomorphine, d-amphetamine and haldol. The most striking effect was the blunting of the effect of apomorphine in choral-hydrate anaesthetised animals versus awake animals. This has ramifications for screening techniques using anaesthetised animals in PET imaging. Other neurotransmitter systems were studied as well. Administration of alpha blocks generally increased glucose utilization, whereas beta receptor antagonists reduced cerebral glucose utilization. The effects of muscarinic antagonists, opiate peptides, adrenocorticotrophic hormone, thyrotropin releasing hormone and GABA-ergic agonists were also evaluated.

Lammertsma presented the measurement of tumour response using [^{18}F]-FDG and PET in human tumours [19]. Various analytical methods have been used ranging from visual inspection of the images to a two-compartment model. The European Organisation for Research and Treatment of Cancer (EORTC) has made recommendations on both analysis and reporting results. This is an active area of discussion by the National Cancer Institute as well, which recently sponsored a similar analysis of the use of [^{18}F]-FDG in determining the effectiveness of chemotherapeutics [20].

Krohn et al. compared the ability of FDG and thymidine analogs (proliferation agents) to measure response to therapy [21]. For example, radiolabelled proliferation agents have been used to measure new anticancer agents such as thymidylate synthetase inhibitors. Both [^{18}F]-FDG and radiolabelled thymidine have also been used to study the effect of Interleuken IL-2 treatment. The information obtained from [^{18}F]-FDG and radiolabelled thymidine images differed in half the cases, indicating that thymidine provided different biochemical information (Fig. 17.4).

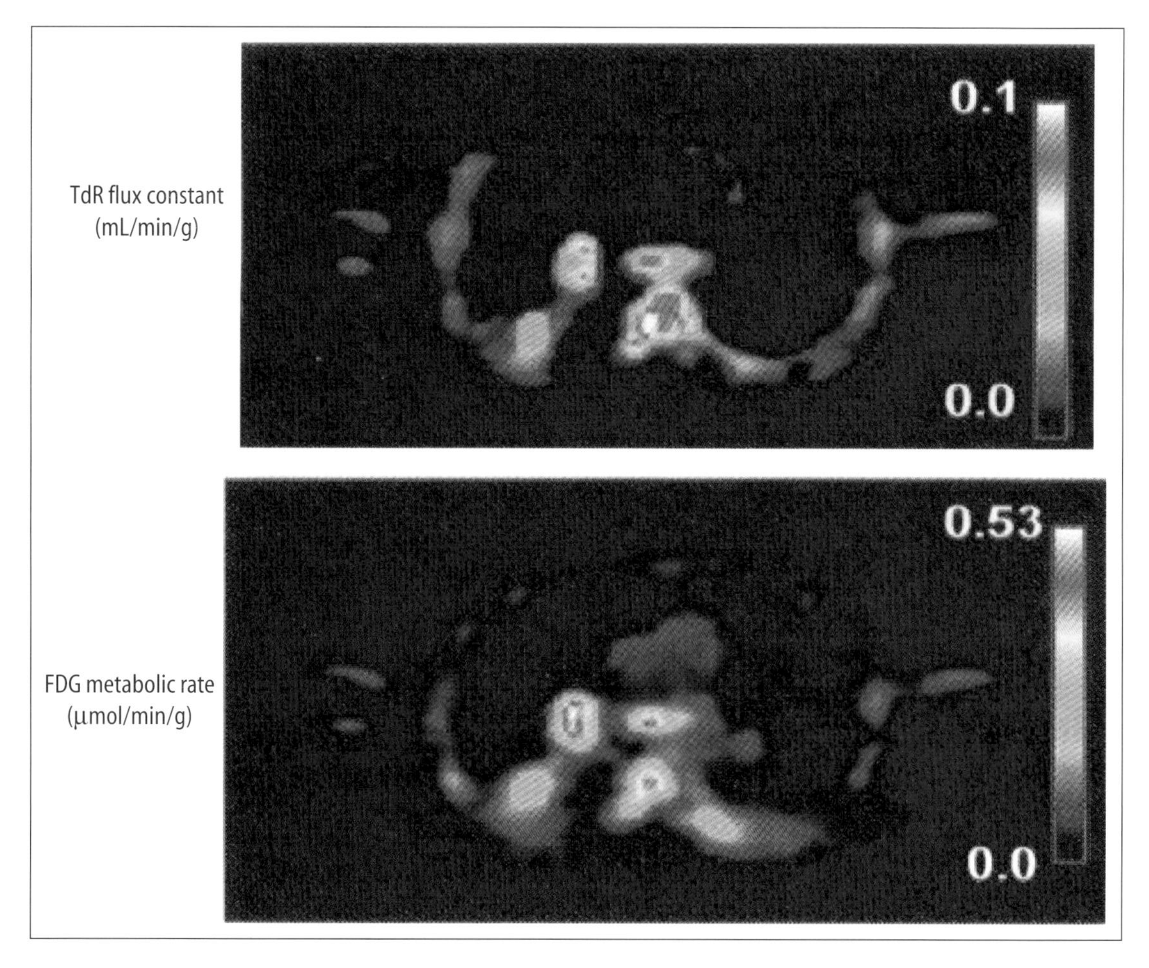

Figure 17.4. Transaxial mixture analysis images of a patient with small-cell cancer showing pixel-wise estimates of thymidine flux (top) and glucose metabolism.

Testing Multidrug Resistance in chemotherapeutics

Three additional presentations, which built on the information reported at the first meeting, described the testing of drugs to reverse MDR using primarily [^{99m}Tc]-sestamibi as the radiolabelled probe. These presentations addressed the development of modulators for multi-drug resistance [22–24]. Multi-drug resistance has emerged as a major obstacle to successful chemotherapy because many chemotherapeutics are removed from the tumour by energy dependent efflux pumps. Various radiopharmaceuticals are being developed to measure the ability of modulators to block the efflux of chemotherapeutics, thereby improving the efficacy of the treatment. These radiopharmaceuticals can also be used to screen patients for resistance to a particular chemotherapeutic agent. The primary result of these studies is the increased uptake in the brain of [^{99m}Tc]-sestamibi in multi-drug resistance 1a/1b knockout mice (i.e., mice lacking the p-glycoprotein pump). This has major implications for drug delivery to the brain.

Monitoring Biochemical Changes in the Brain

The first symposium emphasised monitoring biochemical changes in the brain using PET and SPECT. This topic played a major role in the second symposium as well. Doudet reviewed the advantages and disadvantages of using [^{18}F]-FDG, a dopamine precursor ([^{18}F]-DOPA), a tyrosine derivative (6-fluoro-m-tyrosine,) and a number of radiolabelled substrates for the dopamine transporter to measure changes in Parkinson's disease [25]. She also reviewed the use of a radioligand for the vesicular monoamine transporter system (VMAT2), a measure of the number of dopamine terminals. Many of the clinical assessments are qualitative and are not specific for dopamine dysfunction, which makes evaluating these radioligands against these "gold standards" difficult. The recommendation is to combine several radiotracers with the clinical parameters to increase the sensitivity to small changes in function. Kapur has proposed a number of paradigms for making decisions on the unprecedented number of potential therapeutic agents that have been made available through molecular biology and combinatorial chemistry [26]. Although the cause of many neuropsychiatric disorders is not known, molecular alterations have been identified and can be targets for neuroimaging. The altered brain function leads to system pathophysiology. Functional neuroimaging can be indexed by a change in blood flow, oxygen utilization or glucose metabolism. As stated earlier, the radiopharmaceuticals necessary for these measurements are readily available and ideal for measurements of system pathophysiology. Neurochemical neuroimaging requires a radiolabelled ligand for the molecular alteration involved, which means radiolabelling the drug or using a radiolabelled drug in the same class to determine the changes in biochemistry caused by the disease. Kapur also suggests several strategies for answering drug-related questions, including determining brain access distribution, functional effects in the brain using blood flow, [^{18}F]-FDG, or MRI and measuring pharmacology at the target site. Kapur's report is a comprehensive review of an approach to using imaging in neuropsychiatric disorders, but its principles can be applied to drug development in general. Finally, Halldin has reviewed the current status of development and standardization of radioligands to monitor changes in the serotonin 5-HT_{1A} receptor density and occupancy [27].

Chemotherapeutics and Sentinel Lymph Node Imaging

The use of radiotracers in the analysis of the immunological and molecular properties of the sentinel lymph node in cancer is a new area that is attracting attention in both diagnosis and therapy. This effort promises to yield diagnostic approaches that will result in rapid screening of new chemotherapeutics and biotherapeutics [28]. Finally, the use of microPET for the development of neural repair therapeutics shows promise in that complex rodent models can be studied in a serial fashion before and after therapy [29]. A dedicated PET camera with higher resolution and sensitivity in small animals will be important in the future drug development paradigm.

An Example of Radiolabelling a New Class of Drug Candidates

What is the best approach to developing a new drug and a new radioligand? Some in vitro screening is necessary because of the large number of drug candidates being produced. Imaging can certainly play a part in the screening of the final candidates. If the target system is new, then each drug candidate must be radiolabelled. For example, Novo Nordisk prepared a series of muscarinic agonists, which were tetrahydropyridine thiadiazoyl derivatives. We radiolabelled a number of these with ^{18}F and tested them *in vitro* and *in vivo* for M2 selectivity. The primary focus of our work has been the development of M2 subtype-selective cholinergic ligands, based on the observation that this subtype is lost in the cerebral cortex in Alzheimer's disease [30–32]. Post-mortem quantification of muscarinic subtypes indicated a selective loss of M2 subtype in the hippocampus and a trend toward a decrease in cortical regions while there was an increase of the M2 subtype in the striatum as reported by Rodriquez-Puertas et al. Thus, an M2 selective ligand labelled with a positron-emitting radionuclide would allow determination of M2 subtype concentrations in the living human brain, which could provide information on the progression of Alzheimer's dementia, early diagnosis and non-invasive monitoring of drug therapies.

The Pharmacologic Definition of a Receptor Binding Radiotracer

The muscarinic acetylcholine receptor (mAChR) was one of the earliest receptors studied with *in vivo* techniques. As early as 1973, Farrow and O'Brien [33] described the use of [^{3}H]-atropine to define the mAChR.

With the development of higher affinity and higher specific activity compounds such as [^{3}H]-QNB, receptor distribution was mapped in isolated tissue [34, 35]. The first ligand to map mAChR in humans *in vivo* was the radioiodinated form of QNB, 3-R-quinuclidinyl 4-S-iodobenzilate (RS IQNB) [36]. The use of receptor-binding radiotracers for external brain imaging differs from their use in vitro. For a radiotracer to be useful *in vivo*, its distribution must be driven by the local receptor concentration rather than by local blood flow or membrane transport properties, so that the images obtained primarily reflect receptor binding. A comprehensive kinetic analysis of RS IQNB in rats was performed by Sawada et al. [37], it showed that the uptake in cerebrum is essentially irreversible during the first 360 minutes after intravenous administration and that the rate of RS IQNB tissue uptake depends on transport across the blood-brain barrier (BBB) and the rate of binding to the receptor. However, after about 24 hours the data showed a sensitivity to receptor concentration. Clinical studies with RS IQNB indicate that it is responsive to changes in receptor concentration at 18–24 hours post-injection [38, 39]. However, it does not demonstrate significant muscarinic subtype selectivity.

Our early work on muscarinic ligands was based on antagonist analogs of quinuclidinyl benzilate (QNB) but, more recently, we have used analogs of 3-alkyl-(1,2,5-thiadiazol-4-yl)-tetrahydro-1-methylpyridine (TZTP). We have investigated both antagonists and agonists, because the former usually have higher affinity and therefore may give more image contrast, while the latter may provide more biologically relevant information by being sensitive to the affinity state of the receptor. The primary goal is the development of tracers with selectivity for the M2 site. In general, M2 subtype specificity cannot be demonstrated by the usual proofs of regional distribution of receptor and the pharmacologic profile. Unlike other receptor systems (e.g., D_2), the distribution of the M2 receptor population is highly uniform, so the radioactivity distribution does not provide compelling evidence. In addition, subtype-selective *in vivo* blocking studies are problematic because of the lack of other high-affinity M2 ligands that cross the BBB. Furthermore, the use of antagonists to block agonist ligands has not been demonstrated. Therefore, we were able to block [^{18}F]-FP-TZTP only with compounds of the same chemical class. Our work on muscarinic ligands described below is based on our hypothesis that subtype selective ligands will be more effective in studying biochemistry *in vivo* than the more nonselective radiotracers used to date.

[^{18}F]-FP-TZTP in vitro and rat studies

We have been working extensively with a muscarinic agonist based on a series first proposed by Sauerberg et al. [40] of Novo Nordisk as potential drugs to treat Alzheimer's disease. These ligands contain a thiadiazolyl moiety attached to various heterocycles, including tetrahydropyridine. Two of these compounds, xanomeline and butylthio-TZTP, demonstrated M1 selectivity and have been labelled with ^{11C} and studied with PET [41]. Another compound, (3-(propylthio)-1,2,5-thiadiazol-4-yl)-tetrahydro-1-methylpyridine (P-TZTP), is M2 selective. In the in vitro NovaScreen assay using brain and heart tissue, it showed a K_i of 23 nM for M1 and 1.5 nM for M2. We expanded upon this work and described the radiosynthesis and preliminary biodistribution of 3-(3-(3-[^{18}F]-fluoropropyl)thio)-1,2,5,thiadiazol-4-yl)-1,2,5,6-tetrahydro-1-methylpyridine ([^{18}F]-FP-TZTP) [42]. [^{18}F]-FP-TZTP showed a K_i of 7.4 nM for M1 and 2.2 nM for M2; it did not bind to M3 receptors or other biogenic amine receptors. *In vivo* studies with [^{18}F]-FP-TZTP in rat brain showed that the early uptake was similar to that obtained with RS-[^{18}F]-FMeQNB, but the net efflux was faster. Autoradiography using no-carrier-added [^{18}F]-FP-TZTP confirmed the uniform distribution of radioactivity characteristic of the M2 pattern of localisation. As shown in Fig. 17.1, at 1 hour after injection, co-injection of P-TZTP at 5, 50 and 500 nM inhibited [^{18}F]-FP-TZTP uptake in a dose dependent manner. The difference in brain regions between each dose level was significant ($p < 0.01$), except for the 5 nM value in medulla ($p < 0.06$). The brain distribution of the agonist [^{18}F]-FP-TZTP was unaffected by co-injection of 5, 50, or 500 nM of the antagonist RR-IQNB. Likewise, the distribution of [^{18}F]-FP-TZTP was unaffected by co-injection of 500 nM of the M2 selective antagonist RS-FMeQNB except in cerebral cortex and hippocampus where the difference was significant ($p < 0.03$). The M2 distribution of [^{18}F]-FP-TZTP was further verified by ^{18}F autoradiography. Binding in the heart was low by 15 min, so earlier time points were studied. At 5 minutes, we observed 55% inhibition of uptake in the heart with co-injection of P-TZTP.

A major concern in PET studies is that the radioligand will metabolize to radioactive metabolites. To quantify receptors accurately, the time course of the parent compound, [^{18}F]-FP-TZTP, in blood must be determined. In addition, it is important to verify that radioactive metabolites do not cross the BBB. [^{18}F]-FP-TZTP metabolizes rapidly *in vivo.* In rats, only 5% of plasma radioactivity was parent compound by 15 min post-injection. One metabolite was almost as lipophilic as the parent compound as measured by Thin-layer Chromatography (TLC), suggesting that it might cross the BBB. However, parent compound was found to represent greater than 95% of extracted radioactivity in rat brain through 30 min and greater than 90% at 45 and 60 min. In summary, our experiments in the rat with [^{18}F]-FP-TZTP, a reversible M2 radioligand with predominantly parent compound in the brain, provided strong support for its use as a PET tracer.

[^{18}F]-FP-TZTP studies in non-human primates

This section summarises the PET studies performed with [^{18}F]-FP-TZTP in monkeys. These experiments were used to develop the methodology and analysis techniques for human studies. They also demonstrated the *in vivo* specific binding of [^{18}F]-FP-TZTP and demonstrated its sensitivity to synaptic acetylcholine levels.

PET studies were performed in isoflurane-anesthetized rhesus monkeys to assess the *in vivo* behaviour of [^{18}F]-FP-TZTP [14]. Control studies (n = 11) were performed first to characterise tracer kinetics and to choose an appropriate mathematical model for the *in vivo* behaviour of the tracer. Application of a model requires measurement of the arterial input function for parent compound. Metabolite correction of the arterial input data was performed using TLC. Parent compound comprised 48–9%, 28–6% and 13–3% of plasma radioactivity at 15, 30 and 90 min, respectively. [^{18}F]-FP-TZTP time-activity curves in brain were well-described by a one-compartment model with three parameters: uptake rate constant K_1, total volume of distribution V and a global brain-to-blood time delay Δt. Models with additional parameters could not be used because reliable parameter estimates could not be obtained. Tracer uptake in the brain was rapid, with K_1 values of 0.4 ml/min/ml to 0.6 ml/min/ml in grey matter. The volume of distribution V represents total tracer binding (.e., free, non-specifically bound and specifically bound) Images of V (Fig. 17.5) were created by pixel-by-pixel fitting using the same approach applied in CBF studies [43]. V values were very similar (22–26 ml/ml) in cortical regions, basal ganglia and thalamus, but they were significantly lower (16 ml/ml, $p < 0.01$) in the cerebellum. Comparing these V values with the receptor distribution reported for rat and monkey, shows an excellent match with the M2 distribution. In rats, the concentration of M2 receptors in cortical structures, basal ganglia and thalamus is highly uniform and is approximately 50% lower in cerebellum [44]. In rhesus monkey, the distribution of M2 receptors is also uniform in cortex, basal ganglia and thalamus [45] (Fig. 17.5). This pattern is unlike that of M1 receptors (the muscarinic subtype to which this tracer shows threefold lower in vitro affinity) for which binding in the basal ganglia is greater than the cortex which in turn is greater than in the thalamus [46. 47].

Blocking Studies

Pre-blocking studies (n = 4) were used to measure non-specific binding. Pre-administering 200 to 400 nM/kg of nonradioactive FP-TZTP produced a dramatic reduction in total binding of ~50% in cerebellum and 60% to 70% in other grey matter regions (Fig. 17.5c). Similar blockade was seen in analogous rat studies [42]. This reduction was highly significant in all regions ($p < 0.001$,

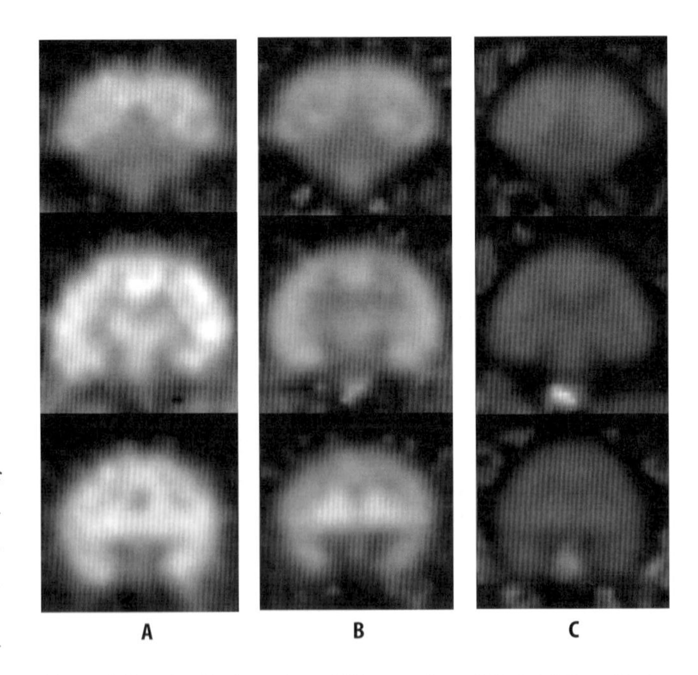

Figure 17.5. Distribution volume (V) images from [^{18}F]-FP-TZTP studies in monkey. Images are coronal slices and all data are scaled to maximum of 27ml/ml. (A) Control study with uniform cortical binding and lower cerebellar binding. (B) Physostigmine (200 mg/kg/hr) started 30 minutes before [^{18}F]-FP-TZTP with significant reduction in binding. **C.** Pre-blocking study with non-radioactive FP-TZTP (400 nM/kg) administered 5 minutes before [^{18}F]-FP-TZTP shows substantial reduction in binding.

one-tailed unpaired *t*-test), and the regional distribution of *V* values after pre-blocking became nearly uniform. Specific binding values were then determined by the difference between control and pre-blocked *V* values. In addition to pre-blocking studies, displacement of [^{18}F]-FP-TZTP with 80 nM/kg of FP-TZTP at 45 minutes post-injection caused a distinct increase in net efflux with decreases of 20%, 36% and 41% in cerebellum, cortex and thalamus, respectively [47] (Fig. 17.5).

Competition Studies with Physostigmine

The sensitivity of [^{18}F]-FP-TZTP binding to changes in brain acetylcholine was assessed by administering physostigmine, an acetylcholinesterase inhibitor, by intravenous infusion beginning 30 min before tracer injection (n = 7). Physostigmine produced a 35% reduction in cortical specific binding ($p < 0.05$), consistent with increased competition from acetylcholine (Fig. 17.5b).

Radiation Dosimetry

In preparation for using [^{18}F]-FP-TZTP in humans, we determined its biodistribution in monkey and obtained radiation dose estimates [48]. There was early activity in liver and kidney, with subsequent appearance of activity in gall bladder and urine, reflecting pathways for tracer excretion. The critical organ was urinary bladder with an absorbed dose of 72.7 µGy/MBq (269 mrad/mCi), followed by pancreas with 46.8 µGy/MBq (173 mrad/mCi) and gall bladder with 39.7 µGy/MBq (147 mrad/mCi). Under NIH radiation safety guidelines, this dosimetry permits a study in humans using up to 370MBq. This dosimetry was found to provide excellent counting statistics for brain imaging (see below).

[^{18}F]-FP-TZTP studies in humans

The studies in monkey indicated that [^{18}F]-FP-TZTP should be useful for the *in vivo* measurement of muscarinic receptors. In collaboration with Drs Connelly, Mentis, Cohen and Sunderland of the National Institute of Mental Health (NIMH), we recently began studies in normal human volunteers, with the eventual goal of studying patients with Alzheimer's disease. PET department investigators collaborated on the design of the protocol and analysed the data from the first human subjects [49], testing and refining the methodology developed in the monkey studies. The goal of this work is to design and implement a reliable, quantitative method for receptor quantification in human subjects. Ideally, the method should minimize the scanning demands on the patient and have a straightforward method for measuring the arterial input function. The initial analysis, based on data from 6 young control subjects, concentrated on determining the appropriate kinetic model for [^{18}F]-FP-TZTP in humans. In each study, CBF was first measured with [^{15}O]-water. Then, a bolus injection of [^{18}F]-FP-TZTP (370MBq) was administered and dynamic scans were acquired in 3D mode on the GE Advance scanner for 120 minutes. The parent fractions in arterial blood samples were analysed by HPLC and TLC and showed excellent agreement (HPLC = 1.02 × TLC – 0.02, r = 0.995). In plasma, parent compound represented 68–8, 41–9 and 14–4% of radioactivity at 20, 40 and 120 minutes post-injection, respectively. The plasma free fraction (f_p) was 5.9–1.2%.

Corrections for subject head motion during the 120 minute acquisition were performed by registering each time frame to the subject's MRI image using the AIR algorithm [50]. As with the monkey data, a global correction for the brain-to-blood delay was applied by analysis of the whole-brain time-activity curve (TAC). A model with one tissue compartment produced an excellent fit for the full 120 minutes of data, so that the additional parameters of a two-compartment model were unidentifiable (see chapter 7 for further discussion of modelling). Functional images of K_1 and V were produced on a pixel-by-pixel basis and had high statistical quality (Fig. 17.6). K_1 values in grey matter regions were high, 0.36 to 0.56 ml/min/ml, and showed excellent correlation with CBF (r = 0.90, K_1 = 0.82 × CBF). V values, representing total tissue binding, were very similar in cortical regions, basal ganglia and thalamus (34–8, 35–9 and 30–7 ml/ml, respectively), but were significantly higher (p < .01) in amygdala (43–10 ml/ml). Unlike the results in the monkey, binding in cerebellum was similar to that in the cerebral cortex (35–9 ml/ml). V values correlated with f_p (r = 0.6 to 0.8 across regions) and normalisation of V by f_p reduced the coefficient of variation of V from 24% to 16%. The methodology and results of our analysis of [^{18}F]-FP-TZTP data in young controls provide the basis for ongoing studies in elderly controls and patients with Alzheimer's disease (Fig. 17.6).

In the first clinical studies, an age-related increase in M2 receptor binding potential was observed using [^{18}F]-FP-TZTP and PET in normal subjects [51]. There

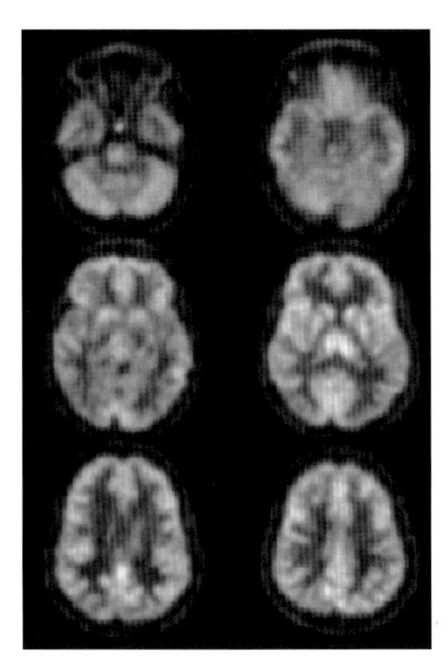

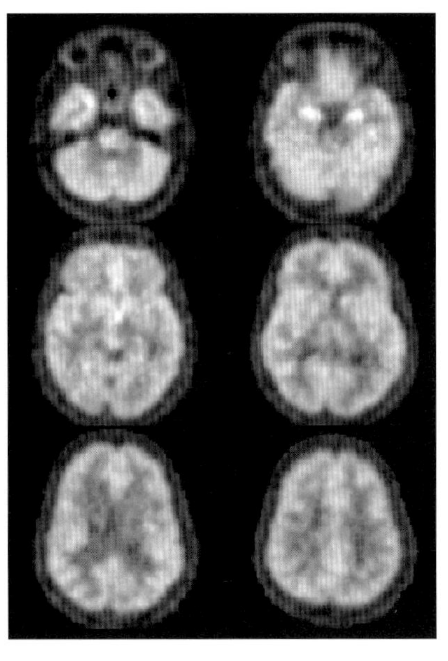

Figure 17.6. Sample K_1 and V images from [^{18}F]-FP-TZTP in a young control subject calculated by pixel-by-pixel fitting of the one tissue compartment model to 120 minutes of data. The K_1 and V images are displayed to a maximum of 0.7 ml/min/ml and 50 ml/ml, respectively. Note the relatively uniform cortical binding in the V image, except for the higher binding in the amygdala.

was a significant increase not only in the average volume of distribution but also variance in the older subjects. This infers that other aspects of the systems biology was coming into play and therefore the data should be modelled by a more complex set of equations. The increase in variance for the volume of distribution was caused by using only age discrimination. This suggested the possibility that the age-related changes might be associated with a specific genotype such as the plasma protein involved in cholesterol transport, Apoe-E4, allele. In the following study, Cohen et al. [52] found that the grey matter distribution volume for [^{18}F]-FP-TZTP was significantly higher in the Apoe-E4 positive normal subjects than in the Apoe-E4 negative normal subjects, whereas there were no differences in global cerebral blood flow. Given that [^{18}F]-FP-TZTP measures the muscarinic system rather than just receptor density because of its agonist properties, changes in receptor binding affinity due to changes in G protein binding and competition with acetylcholine (ACh) at the muscarinic receptor can be monitored. A reasonable hypothesis for the increased volume of distribution in elderly normal subjects with Apoe-E4 positive is a decreased concentration of ACh in the synapse, which would lead to higher binding of [^{18}F]-FP-TZTP. This type of competition was shown to be possible in the studies in monkeys using physostigmine. As a result, the use of [^{18}F]-FP-TZTP can be considered an *in vivo* measurement of muscarinic systems biology, rather than the receptor density alone.

The Length of the Radiolabelling And Validation Process

Validation of the radioligand using the pharmacologic definition of a receptor binding radiotracer is a long process and is not commensurate with the pharmaceutical industries' goal of more NDA approvals per year. Is there a faster method to validate a new radiotracer? We would like to propose the use of gene-manipulated mice as a validation technique and liquid chromatography/mass spectrometry (LC/MS) as a means to show that the experiments in animals were relevant to humans. In this approach, the compound is radiolabelled and then injected in gene-manipulated mice, a procedure that allows validation in a small number of studies. We then use mouse and human freeze-dried hepatocytes to test whether the metabolism pathway is the same in both species. This rapid validation and assurance that the biodistribution in mice will apply to human studies is commensurate with the pharmaceutical industry's goals. The following example, using the

M2 agonist FP-TZTP, shows how quickly a new radiopharmaceutical can be developed.

The Validation of FP-TZTP As A M2 Subtype Selective Radioligand

A subtype-specific muscarinic receptor radioligand would be useful in monitoring changes in receptor density or occupancy as a function of treatment in diseases such as Alzheimer's. In rat, FP-TZTP radiolabelled with the positron-emitting radionuclide ^{18F} ([^{18}F]-FP-TZTP) displayed regional brain distribution of [^{18}F]-FP-TZTP consistent with M2 receptor densities. However, we were interested in developing a rapid validation for the subtype selectivity of [^{18}F]-FP-TZTP using genetically engineered mice that lacked functional M1, M2, M3, or M4 muscarinic receptors [53]. Using ex vivo autoradiography, the regional brain localisation of no-carrier-added [^{18}F]-FP-TZTP in M2 knockout (M2 KO) *vs* wild-type (WT) mice was determined. The relative decrease in [^{18}F]-FP-TZTP uptake in different M2 KO brain regions at 30 min after intravenous injection was similar, ranging from 51.3% to 61.4%, when compared with the distribution of [^{18}F]-FP-TZTP in WT mice (Fig. 17.7). While a significant decrease ($p < 0.01$) of [^{18}F]-FP-TZTP uptake was observed in all brain regions examined for the M2 KO *vs* WT, similar studies with M1 KO, M3 KO and M4 KO *vs* WT did not reveal a significant decrease in grey matter uptake. Residual uptakes most likely represent nonspecific binding since lowering the mass associated with the injected [^{18}F]-FP-TZTP in rat resulted in greater inhibition of brain uptake by P-TZTP, which was co-injected intravenously. The use of knockout mice provides direct evidence for the M2 subtype specificity of [^{18}F]-FP-TZTP and demonstrates the ability of gene-altered mice to give important information on specificity, subtype selectivity and pharmacokinetics in a small number of experiments.

Proof That Mouse and Human Metabolism and Biodistribution Is Similar

To apply the information obtained in the knockout mice to humans, mouse and human metabolism were compared using hepatocytes [54]. To this end, the metabolic profile of FP-TZTP was studied in rat and human hepatocytes using liquid chromatography and mass spectrometry and, when possible, compared with independently synthesised standards. In both human and rat hepatocytes, the major metabolite results from oxidation of the nitrogen in the 1-methyltetrahydropyridine ring. Other metabolites result from sulfur oxidation, demethylation of the tertiary amine and oxidation of the tetrahydropyridine ring. The metabolism of FP-TZTP *in vivo* in rats is similar to that obtained in rat hepatocytes. From our knowledge of the structure of the metabolites, we have developed a two-step extraction sequence that allows the isolation of unmetabolized parent compound. This method allows rapid determination of the parent fraction in plasma and does not require time-consuming chromatographic analysis.

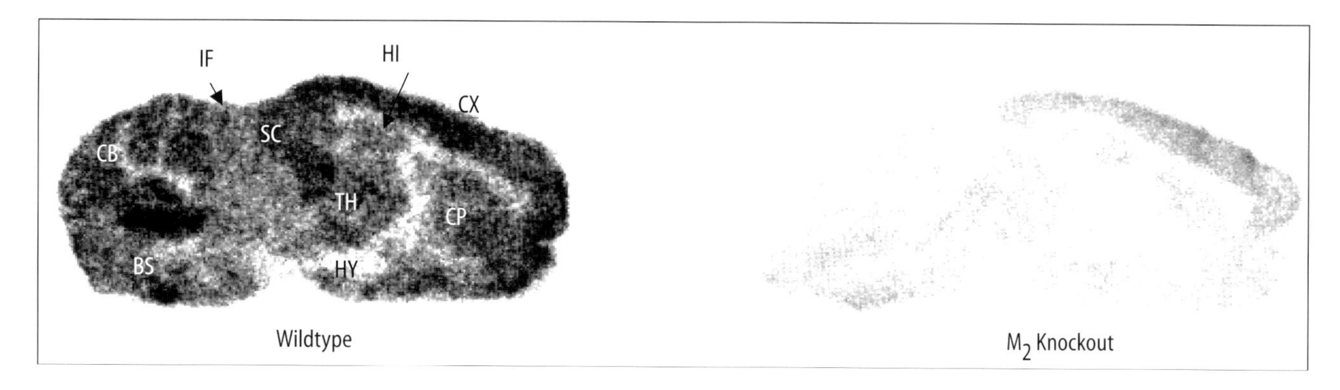

Figure 17.7. Regional brain localisation of [^{18}F]-FP-TZTP in M_2 knockout compared to wild-type mice at 30 minutes post-injection (sagittal slice ~1.8 mm from the midline). In the knockout mice, note the decrease in radioactivity an all grey matter regions with a high percentage of M2 receptors including cerebellum (CB), brain stem (BS) and thalamus (TH).

Conclusion

Drug discovery and drug development are increasingly being accelerated by rapid synthesis of potential drugs and development of high-throughput in vitro tests. Imaging can play a major role in drug development because of its ability to quantify properties of the drug *in vivo*. Radionuclide imaging has been used in all phases of drug discovery and drug development, but pre-clinical studies could be the most effective application of imaging because the pre-clinical drug discovery process may be the slowest step. Piwnica-Worms and Marion have shown that knockout mice can be used to screen a number of potential candidates for clinical efficacy. The value of using these mice was also demonstrated for a series of potential muscarinic agonists using knockout mice with each of the five subtypes knocked out. These studies clearly showed that FP-TZTP was an M2 subtype-specific radioligand with appropriate pharmacokinetics in a minimal number of experiments. Further, the extrapolation from mice to humans was shown to be justified because the metabolic profile was similar, as shown by metabolism experiments with hepatocytes and *in vivo* studies in mice and rats. Thus knockout mice can be used to validate the radiolabelled drug in terms of specificity, saturability and pharmacokinetics, and liquid chromatography and mass spectrometry can predict whether the metabolism data obtained in mice can be extrapolated to humans.

It is clear that PET imaging can be a major force in drug development.

References

1. Hunter Jr, Rosen Dl, Dechristoforo R. How FDA Expedites Evaluation of Drugs for Aids and Other Life-Threatening Illnesses. Wellcome Programs in Pharmacy 1993;January:Ce2–Ce12.
2. Hevesy G, Paneth F. A Manual of Radioactivity, 2nd ed. London: Oxford University Press; 1938.
3. Kamen MD. Isotopic Tracers in Biology. New York: Academic Press; 1957.
4. Lowman J. Techniques of Teaching. San Francisco: Jossey-Bass Publishers; 1995:205–206.
5. Gefvert O, Bergstrom M, Langstrom B, Lundberg T, Lindstrom L, Yates R. Time course of central nervous dopamine-D-2 and 5-Ht2 receptor blockade and plasma drug Concentrations after discontinuation of Quetiapine (Seroquel (R)) in patients with schizophrenia. Psychopharmacology 1998;135(2):119–126.
6. Piwnica-Worms D, Marmion M. Characterization of drug specificity by pharmacokinetic analysis in gene knockout mice. J Clin Pharmacol 1999;Suppl:30s–33s.
7. Wilding IR, Heald DI. Visualization of product performance in the gut: what role in the drug development/regulatory paradigm? J Clin Pharmacol 1999;Suppl:6s–9s.
8. Berridge MS, Heald DI. In vivo characterization of inhaled pharmaceuticals using quantitative positron emission tomography. J Clin Pharmacol 1999;Suppl:25s–29s.
9. Shields AF, Ho PT, Grierson JRr. The role of imaging in the development of oncologic agents. J Clin Pharmacol 1999;Suppl:40s–44s.
10. Salazar DE, Fischman AJ. Central nervous system pharmacokinetics of psychiatric drugs. J Clin Pharmacol 1999;Suppl(39): 10s–12s.
11. Fowler JS, Volkow ND, Ding YS, Wang GjJ Dewey S, Fischman MW, et al. Positron emission tomography studies of dopamine-enhancing drugs. J Clin Pharmacol 1999;Suppl:13s–16s.
12. Offord SJ, Wong DF, Nyberg S. The role of positron emission tomography in the drug development of M100907: a putative antipsychotic with a novel mechanism of action. J Clin Pharmacol 1999;Suppl:17s–24s.
13. Breier A, Su T-P, Saunders R, Carson RE, Kolachana BS, De Bartolomeis A, et al. Schizophrenia is associated with elevated amphetamine-induced synaptic dopamine concentrations: evidence from a novel positron emission tomography method. Proc Natl Acad Sci 1997;94(6):2569–2574.
14. Carson RE, Kiesewetter DO, Jagoda E, Der MG, Herscovitch P, Eckelman WC. Muscarinic cholinergic receptor measurements with [^{18}f]FP-TZTP: control and competition studies. J Cereb Blood Flow Metab 1998;18(10):1130–1142.
15. Campbell DB. The role of radiopharmacological imaging in streamlining the drug development process. Q J Nucl Med 1997;41(2):163–169.
16. Eckelman WC, Waterhouse R, Frank R. Nuclear imaging and biomarkers in drug development using approved radiopharmaceuticals. J Clin Pharmacol 2001;Suppl:4s–6s.
17. Herscovitch P. Can [^{15}o]water be used to evaluate drugs? J Clin Pharmacol 2001;Suppl:11s–20s.
18. Mcculloch J. Mapping functional alterations in the CNS with [14c]Deoxyglucose. In: Handbook of Psychopharmacology. New York: Plenum Press; 1982:321–410.
19. Lammertsma AA. Measurement of tumor response using [^{18}f]-2-Fluoro-2-Deoxy-D-Glucose and positron emission tomography. J Clin Pharmacol 2001;Suppl:104s–106s.
20. Eckelman WC, Tatum JI, Kurdziel KA, Croft BY. Quantitative analysis of tumor biochemisty using PET and SPECT (1). Nucl Med Biol 2000;27(7):633–691.
21. Krohn KA, Mankoff DA, Eary JF. Imaging cellular proliferation as a measure of response to therapy. J Clin Pharmacol 2001; Suppl:96s–103s.
22. Slapak CA, Dahlheimer J, Piwnica-Worms D. Reversal of multidrug resistance with Ly335979: functional analysis of P-glycoprotein-mediated transport activity and its modulation in vivo. J Clin Pharmacol 2001;Suppl:29s–38s.
23. Ballinger JR. 99mtc-tetrofosmin for functional imaging of P-glycoprotein modulation in vivo. J Clin Pharmacol 2001;Suppl:39s–47s.
24. Hendrikse NH, Bart J, De Vries EG, Groen HJ, Van Der Graaf WT, Vaalburg W. P-glycoprotein at the blood-brain barrier and analysis of drug transport with positron emission tomography. J Clin Pharmacol 2001;Suppl:48s–54s.
25. Doudet DJ. Monitoring disease progression in Parkinson's disease. J Clin Pharmacol 2001;Suppl:72s–80s.
26. Kapur S. Neuroimaging and drug development: an algorithm for decision making. J Clin Pharmacol 2001;Suppl:64s–71s.
27. Halldin C. Serotonin 5-Ht1a receptor imaging in the human brain with positron emission tomography: coordination and the standardization and dissemination of methodology. J Clin Pharmacol 2001;Suppl:95s.
28. Fox B. Vaccine design via sentinel node detection. J Clin Pharmacol 2001.
29. Kornblum HI, Cherry SR. The use of micropet for the development of neural repair therapeutics: studies in epilepsy and lesion models. J Clin Pharmacol 2001;Suppl:55s–63s.
30. Quirion R, Aubert I, Labchak PA, Schaum RP, Teolis S, Gauthier S, et al. Muscarinic receptor subypes in human neurodegenerative disorders: focus on Alzheimer's disease. In: Trends In Pharmacologic Science; 1989:80–84.
31. Aubert I, Araujo DM, Cecyre D, Robitaille Y, Gauthier S, Quirion R. Comparative alterations of nicotinic and muscarinic binding sites

in Alzheimer's and Parkinson's diseases. J Neurochem 1992(58): 529–541.
32. Rodriguez-Puertas R, Pascual J, Vilaro T, Pazos A. Autoradiographic distribution of M1, M2, M3 And M4 muscarinic receptor subtypes in Alzheimer's disease. Synapse 1997;26:341–350.
33. Farrow JT, O'Brien RD. Binding of atropine and muscarone to rat brain fractions and its relation to the acetylcholine receptor. Molecular Pharmacol 1973;9(1):33–40.
34. Yamamura HI, Snyder SH. Muscarinic cholinergic binding in rat brain. Proc Natl Acad Sci 1974;71(5):1725–1729.
35. Yamamura HI, Kuhar MJ, Greenberg D, Snyder SH. Muscarinic cholinergic receptor binding: regional distribution in monkey brain. Brain Res 1974(66):541–546.
36. Eckelman WC, Grissom M, Conklin J, Rzeszotarski WJ, Gibson RE, Francis BE, et al. In vivo competition studies with analogues of quinucidinyl benzilate. J Pharm Sci 1984;73:529–533.
37. Sawada Y, Hiraga S, Francis B, Patlak C, Pettigrew K, Ito K, et al. Kinetic analysis of 3-quinuclidinyl 4-I-125 iodobenzilate transport and specific binding to muscarinic acetylcholine receptor in rat-brain in vivo: implications for human studies. J Cereb Blood Flow Metab 1990;10(6):781–807.
38. Weinberger DR, Gibson R, Coppola R, Jones DW, Molchan S, Sunderland T, et al. The distribution of cerebral muscarinic acetylcholine receptors in vivo in patients with dementia: a controlled study with 123iqnb and aingle photon emission computed tomography. Arch Neurol 1991;48(2):169–176.
39. Weinberger DR, Jones D, Reba RC, Mann U, Coppola R, Gibson R, et al. A comparison of FDG PET and IQNB SPECT in normal subjects and in patients with dementia. J Neuropsychiatry Clin Neurosci 1992;4(3):239–248.
40. Sauerberg P, Olesen PH, Nielsen S, Treppendahl S, Sheardown MJ, Honore T, et al. Novel functional M1 selective muscarinic agonists: synthesis and structure-activity relationships of 3-(1,2,5-thiadiazoyl)-1,2,5,6-tetrahydro-1-methylpyridines. J Med Chem 1992(35): 2274–2283.
41. Farde L, Suhara T, Halldin C, Nyback H, Nakashima Y, Swahn CG, et al. PET study of the M1-agonists [11c]xanomeline and [11c]butylthio-TZTP in monkey and man. Dementia 1996; 7(4):187–95.
42. Kiesewetter DO, Lee J, Lang L, Park SG, Paik CH, Eckelman WC. Preparation of 18f-labeled muscarinic agonist with M2 selectivity. J Med Chem 1995;38(1):5–8.
43. Koeppe RA, Holden JE, Ip WR. Performance comparison of parameter estimation techniques for the quantitation of local cerebral blood flow by dynamic positron computed tomography. J Cereb Blood Flow Metab 1985;5:224–234.
44. Li M, Yasuda RP, Wall SJ, Wellstein A, Wolfe BB. Distribution of M2 muscarinic receptors in rat-brain using antisera selective for M2 receptors. Molecular Pharmacol 1991;40(1):28–35.
45. Flynn DD, Mash DC. Distinct kinetic binding properties of N-[3h]-methylscopolamine afford differential labeling and localization of M1, M2, and M3 muscarinic receptor subtypes in primate brain. Synapse 1993;14(4):283–96.
46. Wall SJ, Yasuda RP, Hory F, Flagg S, Martin BM, Ginns EI, et al. Production of antisera selective for M1 muscarinic receptors using fusion proteins: distribution of M1 receptors in rat brain. Molecular Pharmacol 1991;39:643–649.
47. Kiesewetter DO, Carson RE, Jagoda EM, Herscovitch P, Eckelman WC. In vivo muscarinic binding Of 3-(alkylthio)-3-thiadiazolyl tetrahydropyridines. Synapse 1999;31(1):29–40.
48. Herscovitch P, Kiesewetter DO, Carson RE, Eckelman WC. Biodistribution and radiation dose estimates for [F-18]fluoropropyl-TZTP: a new muscarinic cholinergic ligand. J Nucl Med 1998;39:85p.
49. Carson RE, Kiesewetter DO, Connelly K, Mentis MJ, Cohen RM, Herscovitch P, et al. Kinetic analysis of the muscarinic cholinergic ligand [F-18]FP-TZRP in humans. J Nucl Med 1999;40:30p.
50. Woods RP, Mazziotta JC, Cherry SR. MRI-PET registration with automated algorithm. J Comp Assist Tomog 1993;17:536–546.
51. Podruchny TA, Connolly C, Bokde A, Herscovitch P, Eckelman WC, Kiesewetter DO, et al. In vivo muscarinic 2 receptor imaging in cognitively normal young and older volunteers. Synapse 2003;48(1):39–44.
52. Cohen RM, Podruchny TA, Bokde A, Bokde W, Carson RE, Herscovitch P, et al. Higher in vivo muscarinic-2 receptor distribution volumes in aging subjects with an apolipoprotein E-epsilon4 allele (Ms#103pmh). Synapse 2003;48(9):150–156.
53. Eckelman WC. The use of PET and knockout mice in the drug discovery process. Drug Discov Today 2003;8(9):404–10.
54. Ma Y, Kiesewetter D, Lang L, Eckelman WC. Application of Lc-Ms to the analysis of new radiopharmaceuticals. Molecular Imag Biol 2003;5(6):397–403.
55. White R. High-throughput screening in drug metabolism and pharmacokinetic support of drug discovery. Annu Rev Pharmacol Toxicol 2000;40:133–57.

18 PET as a Tool in Multimodality Imaging of Gene Expression and Therapy

Abhijit De and Sanjiv Sam Gambhir

Introduction

Gene therapy holds significant promise in the treatment of many human diseases. Although still in its infancy, it is likely that gene therapy will eventually succeed as a general approach to medical treatment. A key force in the rapid evolution of gene therapy will be the ability to image the location(s), magnitude, and time variation of therapeutic gene expression in animal models as well as in patients. To unravel the complexity and dynamics of molecular and cellular events, it is desirable to image reporter gene expression in individual cells, living animals, and humans with the help of a single construct with multiple reporter genes suitable for various imaging modalities. Positron emission tomography (PET) is likely to play a significant role in multimodality imaging by accelerating animal model development and improving the monitoring of patients in clinical gene therapy trials. In this chapter the fundamentals of gene therapy are reviewed including various viral and non-viral delivery vectors. This is followed by a detailed review of specificity of gene therapy, approaches to cancer gene therapy and safety issues. Details of the Herpes Simplex Virus Type 1 thymidine kinase (HSV1-tk), the Dopamine Type 2 receptor (D2R) and sodium iodide symporter gene as PET reporter genes along with their radiolabeled reporter probes (tracers) are presented. Strategies for coupling a therapeutic gene with a PET reporter gene are also discussed. Finally, examples of recent imaging work in human studies are presented.

Potentials for Imaging in Gene Therapy

The science behind gene therapy relies on introducing genes to cure or retard the progression of a disease. With rapid advances in cellular and molecular biology, gene therapy has great potential to dominate future approaches in treating human diseases. Theoretically, by introducing necessary modifications for the defective part(s) of a gene, one can potentially cure or retard the severity of a disease caused by the effect of a single gene. Today, with the completion of the human genome project [1], we have much of the data available to help identify defective genes that cause various diseases. In parallel, we also have the technical skills to isolate and determine the defective portions of a gene, which may alter gene activity. For example, a tumor-suppressor gene, whose expression normally suppresses tumor growth, that is identified with a defective sequence can potentially be replaced with a functional gene to arrest tumor growth caused by the defective gene. With recent advances in vector design, improvements in transgene (a new or altered gene that is being introduced) and prodrug activation strategies, gene therapy is being applied to a wide variety of diseases and organ systems.

PET imaging can play a critical role in determining the location(s), magnitude, and time variation of therapeutic gene expression. The locations of gene expression are critical for establishing targeting, as well as for making sure that unintended sites do not show significant amounts of gene expression. The ability to monitor the magnitude and time variation can potentially help to titrate gene therapy for optimization of a given approach. A wide variety of imaging techniques including optical imaging, magnetic resonance imaging (MRI), and radionuclide imaging techniques like PET and single photon emission computed tomography (SPECT) can be applied for monitoring gene therapy. Non-invasive and repetitive imaging of molecular events within a single cell to groups of cells within a living subject using different imaging modalities should play a critical role in understanding normal physiology and disease progression. PET is particularly suited because of its very high sensitivity and the ability to easily move from animal models imaged with microPET [2] to patient studies performed with clinical PET [3].

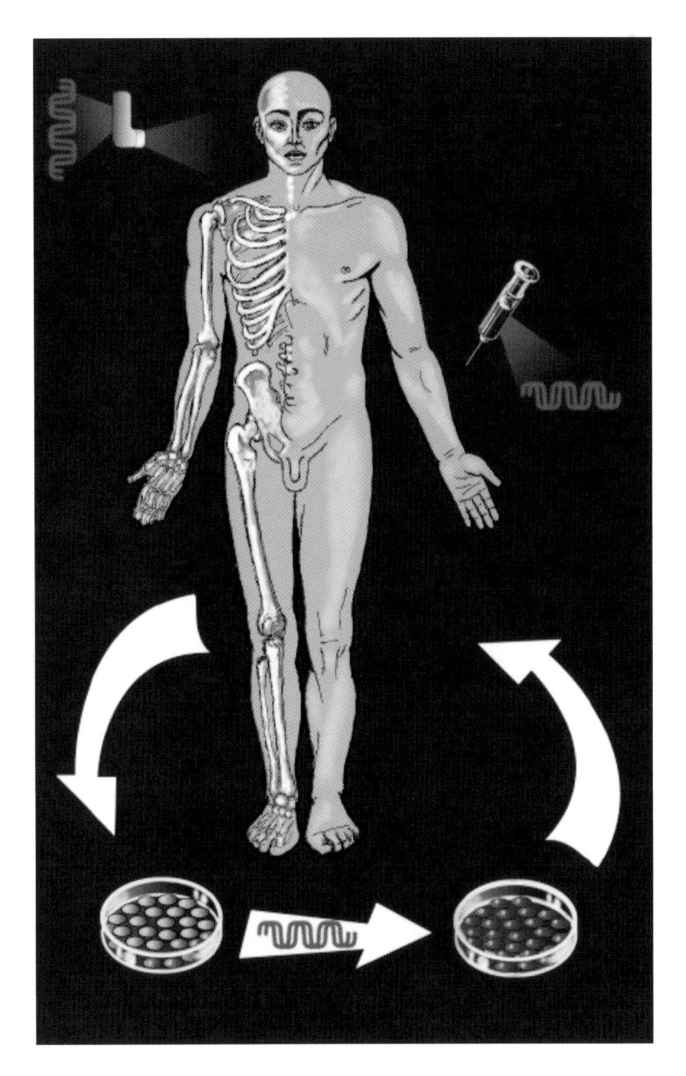

Figure 18.1. Schematic showing different approaches for therapeutic gene delivery to a human subject. With *in vivo* approaches, the gene can be delivered using a viral or nonviral vector such as a liposome. The vector can be introduced into the blood, through inhalation, as well as many other approaches. The vector must be able to target the tissue(s) of interest and/or only be expressed in those tissues. With *ex vivo* approaches, cells are first removed from the subject. These cells are then transfected with the therapeutic gene prior to being re-introduced into the subject. In both the *in vivo* and *ex vivo* approaches it would be very useful to image the location(s), magnitude and time-variation of gene expression.

Technical Issues for Gene Therapy

Gene therapy can be achieved either by *ex vivo* or *in vivo* methods (Fig. 18.1). Technically, *ex vivo* methods are simpler with regard to DNA delivery, but these methods often require minor surgery to harvest cells, which are then replaced after introduction of gene(s) *ex vivo*. With *in vivo* gene therapy protocols, it is desirable to have highly efficient target-specific vectors while minimizing risk. Each disease has its own specific requirements such as target tissues, the amount of gene product required, and similarly each vector has characteristics that may or may not be suitable for a particular application. Furthermore, objective studies comparing the relative merits of different delivery systems are rarely carried out, making comparisons difficult [4–6]. A major goal for the application of gene therapy is to achieve a controlled and effective target-specific expression of genes.

Another important factor in a gene therapy trial is the nature of the disease being treated. For some diseases the pathology affects the function of a particular organ, which must be directly treated, for example, cystic fibrosis in lungs [7] or Parkinson's disease in corpus striatum of brain [8]. Other diseases are systemic (for example, hemophilia or adenosine deaminase deficiency), for which the common sites of therapy are relatively accessible organs like the liver or gut [6, 9]. It is also important whether the disease is caused by a single dominant gene, a single recessive gene or multiple genes. Most genetic disorders are multigenic, with more than one gene being responsible for a disease. Currently, a majority of gene therapy protocols are centered on cancer, which is considered a multigenic disorder. Though many cancers have a genetic predisposition, they all involve acquired mutations, and as they progress the cells become less differentiated and more heterogeneous with respect to the mutations they carry. The range of different cancers encountered and the mutations they carry have led to the adoption of various strategies for gene therapy.

Gene Delivery Approaches

Viral Methods

Viruses are a natural delivery system for delivering their deoxyribonucleic acid (DNA) or ribonucleic acid (RNA) to the host cell. For their own survival they must deliver their DNA or RNA materials, which must be copied successfully within the host cell. Using this delivery system, vector virology has been used to construct several efficient vectors for gene therapy trials. Detailed studies of the viral genome have helped scientists to eliminate disease-causing viral genes before using their natural delivery power for transferring a therapeutic gene of interest. However, the risk factors associated with the integration of the viral genome into patient cells, which can cause unwanted side effects, need always to be critically evaluated before using natural viral particles.

Therapeutic gene transfer vectors have been developed from different groups of viruses, each one with its own set of advantages and disadvantages (Table 18.1). Depending upon the targeted disease and type of treatment needed, one has to decide which vector is suitable for a given application. It is important to keep in mind that development of a viral vector in the laboratory is often done in order to produce a replication-incompetent virus, so that the virus can deliver its genetic contents but cannot replicate. This requires dedicated approaches in which large amounts of virus can be grown through the use of cell lines that carry missing genes or helper viruses that provide the necessary genes needed for viral replication. Without these approaches it would not be possible to replicate the replication-incompetent viruses. Overall gene therapy approaches using

Table 18.1. Comparison chart of major gene delivery vectors.

Vector	Insert size	Transduction efficiency	Major advantages	Major disadvantages
Retrovirus (MMLV)	< 8 Kb	High	Stable expression of dividing cells; low immunogenicity	Poor *in vivo* delivery; unable to transfect nondividing cell; safety concern of insertional mutagenesis
Lentivirus	< 8–9 Kb	High	High and stable expression of both dividing and non-dividing cells	Potential safety concern for generating human disease
Adenovirus	< 7.5 Kb	High	Transfects nearly all cell types, both dividing and nondividing; substantial clinical experience	Strong immune reaction; transient expression
Adeno associated virus	< 4.5 Kb	High	Stable expression; infect wide range of cells; no significant immune response	Capability of carrying small DNA size; Safety concern of insertional mutagenesis; little clinical experience
Herpes simplex virus	< 20 Kb	Low	Capacity to carry large DNA fragment	Neuronal cell specificity; transient expression; Cytotoxic; Safety concern to generate human infections
Liposome	>20 Kb	Low	Larger DNA carrying capacity; low or no immunogenicity; good safety profile; cell specific targeting	Inefficient *in vivo* delivery; transient expression
Naked DNA	< 20 Kb	Low	Simple and cheap; low or no immune reaction; very safe	Very short duration of expression; poor *ex vivo* and *in vivo* delivery
Ballistic DNA	< 20 Kb	Low	Simple; low or no immune reaction; safe	Transient expression; local cellular damage

viral vectors have focused on four major groups of viruses: Retrovirus, Adenovirus, Adeno Associated Virus (AAV) and Herpes Simplex Virus (HSV). Each of these is discussed in the following sections.

Retrovirus

Retroviruses are a class of enveloped viruses, which contain a single-stranded RNA molecule as their genome. After infecting a host cell, the viral genome is reverse transcribed (RNA to DNA) by reverse transcriptase into double-stranded DNA, which integrates into the host genome and is expressed (Fig. 18.2). The retroviral genome consists of little more than the genes essential for viral replication. The prototype and simplest genome is that of the Moloney Murine Leukemia Virus (MoMLV), in contrast to the highly complex genomes of the HTLV (human T-cell leukemia virus) and HIV (human immunodeficiency virus) retro-

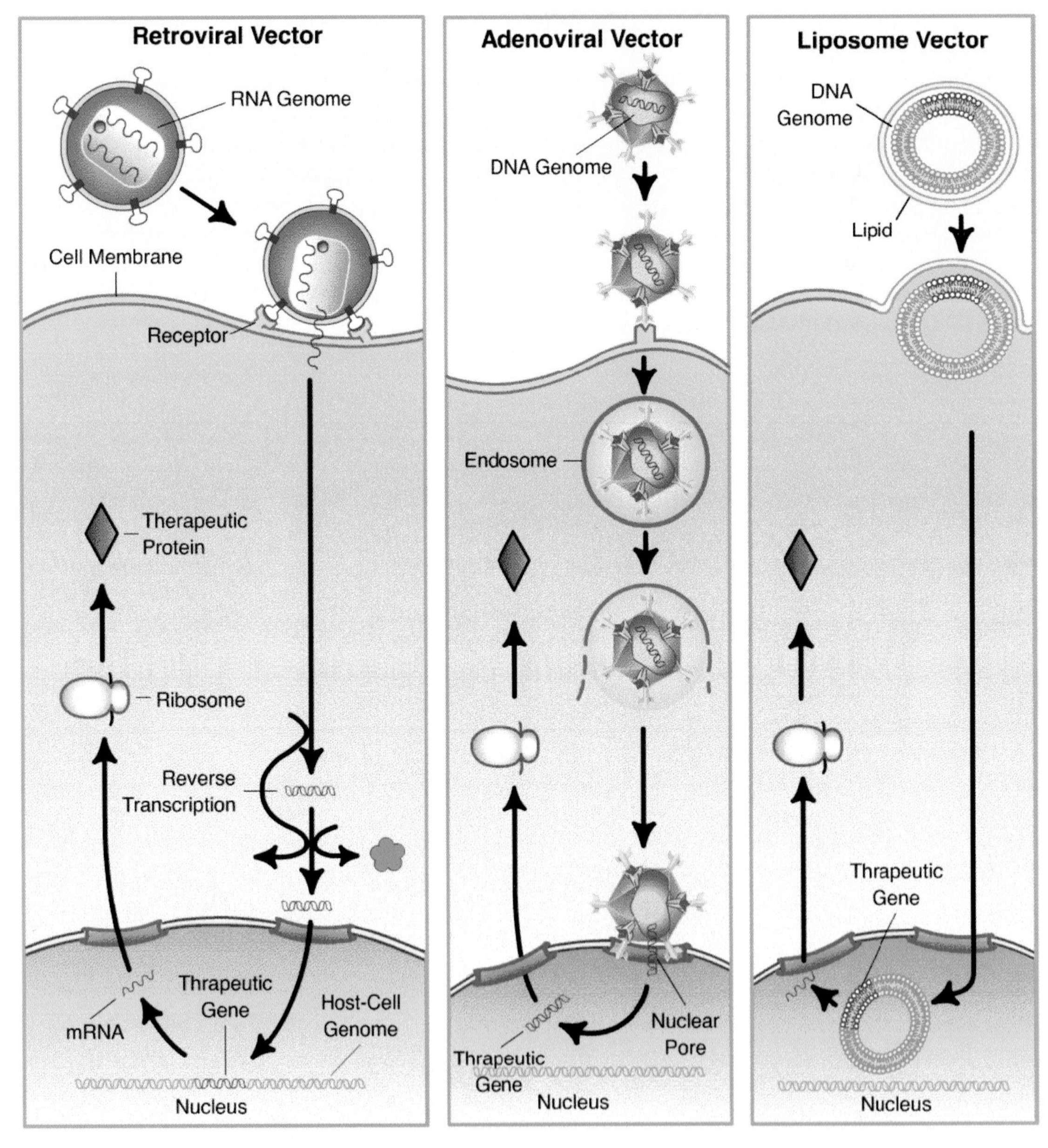

Figure 18.2. Schematic diagrams showing basic differences of three major gene delivery approaches: retrovirus (left), adenovirus (center) and liposome (right). Retrovirus, containing a RNA genome, attaches to a host cell by binding to a specific cell surface receptor and a double-stranded DNA is synthesized from the RNA genome by a process called reverse transcription. This DNA *integrates* into the host genome ensuring sustained expression of the transgene. Adenovirus is comprised of a double-stranded DNA genome. After entering the cell by a receptor mediated endocytosis process, the viral DNA is released within the cell, which then leads to a synthesis of proteins using host cell machinery. The DNA does *not* integrate into the host genome. In liposomal delivery the plasmid DNA containing the therapeutic gene is encased in a lipid bilayer, which upon reaching a host cell, gets fused with the cellular membrane and releases the DNA within the cell. This DNA then synthesizes its protein products along with cellular DNA.

viruses. The basal retroviral genome can be divided into three transcriptional units: *gag, pol* and *env*. The *gag* region encodes genes that comprise the capsid (the virus capsule) proteins; the *pol* region encodes the reverse transcriptase and integrase proteins; and the *env* region encodes the proteins needed for receptor recognition and envelope anchoring. An important feature of the retroviral genome is the long terminal repeat (LTR) regions, which play an important role in initiating viral DNA synthesis, as well as regulating transcription of the viral genes. The recombinant viral vector that was used for the first human clinical gene therapy trial (for severe combined immunodeficiency disease) in 1990 was based on the retrovirus MoMLV [10].

Subsequently, the well-characterized genome and lifecycle of retroviruses has made it possible to utilize them in designing gene delivery vectors for many different purposes [11]. First, retroviruses satisfy one of the major prerequisites for gene therapy applications, which is long-term and relatively stable expression of the transferred gene, due to the fact that their genome is integrated into the host genome. Also, their integration into the host cell does not alter normal cell function. They satisfy the prerequisite of safety for gene delivery since their *cis*-acting elements (i.e., viral genes and the genes responsible for reverse transcriptase, integrase, and packaging signal) can be separated from the coding sequences. Removing these genes from the vector and causing them to be expressed in the packaging cell line allows production of replication-deficient virus particles *ex vivo* and helps to ensure safety. A major disadvantage of using retroviruses as a gene transfer tool is that they only transduce cells that divide shortly after viral infection [12, 13]. Thus, this vector has limited applications to important target organs like the liver, skeletal muscle, hematopoietic stem cells, and neuronal cells. Another potential drawback of retroviruses is the chance of insertional mutagenesis as the virus genome randomly integrates inside the host genome.

Lentivirus

Lentiviruses are a subclass of retroviruses of which HIV are the most recently discovered members. HIV based vectors show promise for future use in clinical studies, but their complex biology and mechanisms of pathogenesis complicate their choice for clinical applications [14]. Like other members of the retroviral family, the HIV genome also contains the *gag, pol* and *env* genes. In addition, six other genes are contained within their genome: the *tat, rev, vpr, vpu, vif,* and *nef* genes, which code for nonstructural regulatory proteins. Lentiviruses have the capability to infect both proliferating and non-proliferating cells, which is considered advantageous since a much larger range of cell types can be targeted for therapeutic applications. Early results using reporter genes have shown promising *in vivo* expression in muscle, liver, and neuronal cells [15–17]. However, exceptions to this ability to transduce non-proliferative tissues have been reported, suggesting that lentivirus vectors cannot infect all non-dividing cells [18]. A major concern in using HIV vectors is the strong possibility of genetic recombination between infectious HIV and the lentiviral vector itself, which would result in the lentiviral vector acting as an infectious HIV particle. One possible method to overcome this risk would be to introduce a suicide gene (see below) into the HIV vector genome. This risk could also be avoided by using equine infectious anemia virus (EIAV) and feline immune deficiency virus (FIV) as vectors, since neither of these is associated with human disease. These lentiviruses do not infect human cells, but the gene determining the host range can be substituted by another gene, allowing the viruses to infect and deliver genes to human tissue.

Adenovirus

Adenoviruses are non-enveloped viruses containing a linear double-stranded DNA genome. There are currently 47 distinct serotypes and as many as 93 different varieties of adenovirus, all of which generally infect the human eye, upper respiratory tract, and gastrointestinal epithelium. The adenoviral genome is about 35 kilobases (kb), and is comprised of four regulatory transcription units, E1, E2, E3, and E4 [19]. For developing an inactivated adenovirus vector, the E1 and E3 genes are deleted from the genome and then supplied in *trans* (gene elements provided in another plasmid/complimentary strand), either by a helper virus or plasmid, when packaging the virus. Graham et al. [20] first developed a cell line, which enabled the production of recombinant adenovirus in a helper-free environment by integrating E1 into the cell genome. Since then, adenoviral vectors have received much attention as gene transfer agents and currently offer a wide variety of gene therapy applications.

The exact mechanism by which an adenovirus targets a host cell is poorly understood. However, internalization occurs via receptor-mediated endocytosis followed by release from an endosome. A primary receptor of the immunoglobulin gene family, CAR (coxsackie adenoviral receptor) has been identified as specific for adenovirus. After endosomal release, the viral capsid undergoes disassembly outside the nucleus (Fig. 18.2). Nuclear entry of the viral DNA is completed upon capsid dissociation, but the viral DNA does not integrate into the host genome; instead it replicates as episomal elements in the nucleus of the host cell. Also, since the viral DNA lacks the ability to integrate, it cannot bring about mutagenic effects caused by random integration into the host genome. As a consequence, their use as a vector is devoid of the risk of generating insertional mutagenesis. Adenoviral vectors have high transduction efficiency, are capable of containing DNA inserts of up to 8 kb, have extremely high viral titers, and infect both replicating and differentiated cells. Disadvantages of adenoviral vectors include: (i) gene expression is transient since the viral DNA does not integrate into the host, (ii) expression of viral proteins that are present in the adenoviral vector leads to an immune response, and (iii) adenoviral vectors are extremely common human pathogens and *in vivo* delivery may be hampered by prior host-immune response to one viral type.

Adeno-Associated Virus

Adeno-associated viruses (AAV) are non-pathogenic satellite viruses of other human viruses and require co-infection with either adenovirus or herpes simplex virus (HSV) for their replication [21, 22]. A unique feature of the DNA molecule contained within the virion is the presence of inverted terminal repeat (ITR) at both ends. These ITRs are important in the site-specific integration of AAV DNA into a specific site in human chromosome 19. The ability of wild-type AAV to selectively integrate into chromosome 19 makes them attractive candidates for the production of a gene therapy vector. For production of virus vectors, cells are first infected with the wild-type adenovirus or HSV and then both the recombinant AAV vector plasmid DNA and the non-rescuable AAV helper plasmids are co-transfected into the cells. The cells produce mature recombinant AAV vectors as well as wild-type adenovirus or HSV. The wild-type adenovirus or HSV is then removed by either density gradient centrifugation or heat inactivation.

These vectors have the potential to produce a therapeutic gene transfer tool with site-specific integration and the ability to infect multiple cell types. Unfortunately, this has not been the case to date for these vectors. Current research focuses on how to regain the missing site-specific integration sequences that are lost when the nonessential genes are removed from the viral genome. These vectors do offer some advantages over other vector systems, including minimization of immune response, stability, and ability to infect a variety of dividing and non-dividing cells. Unfortunately, a major drawback is their inability to incorporate genes larger than 5 kb. Also, these vectors and must be closely screened for adenoviral or HSV contamination [23].

Herpes Simplex Virus

The Herpes simplex virus (HSV) genome consists of one double-stranded DNA molecule, which is 120–200 kb in length. The virus itself is transmitted by direct contact and replicates in the skin or mucosal membranes before infecting cells of the nervous system. It exhibits both lytic and latent function. It is thought that it will be possible to modify this virus for gene therapy, to produce a vector that exhibits only the latent function for long-term gene maintenance and has expression with specific tendency towards cells of the central nervous system.

Key regions of the HSV genome, which must be removed to produce a replication-deficient HSV vector, include the ICP4, ICP0, and ICP27 [24]. The following must be taken into consideration before developing an engineered HSV vector suitable for gene therapy: (i) the vector must be non-cytotoxic to nerve cells, as well to other cell lines; (ii) the vector must be unable to carry out the lytic cycle; (iii) the vector should be able to establish latency; and (iv) there must be persistent and sufficient levels of the desired gene expression once latency has been obtained.

The HSV vectors currently under study have a number of advantages as gene transfer reagents. Because of the large size of the virus, recombinant HSV vectors are capable of containing inserted genes of around 20 kb. They give an extremely high titer, similar to the adenoviruses, and they have been shown to express the transgene for a long period of time in the central nervous system, as long as the lytic cycle does not occur. Major disadvantages for the HSV systems include: (i) expression does appear to be transient with the current vector systems, suggesting either lytic infection or viral protein expression, (ii) transduction efficiency is relatively low such that not many target cells show expression of the desired gene, and (iii) they are far from complete and require much additional engineering to be become efficient vectors for gene therapy.

Amplicon-Based HSV Vectors

Amplicon-based vectors rely upon the natural occurrence of defective interfering (DI) particles. DI particles arise during HSV propagation and are infectious agents, which lack portions of the HSV genome for replication. It was observed that plasmid DNA could be packaged into DI particles if the plasmid DNA contained HSV packaging signals. Such DI particles were referred to as DI vectors. Therefore, amplicon-based HSV or DI vectors rely on two basic components: (i) the amplicon, which contains HSV packaging signals and regions where desired genes can be cloned, and (ii) DI particle production. Following insertion of the desired genes into the amplicon, the amplicon is transfected into any cell line, which allows for efficient HSV propagation. Following amplicon transfection, the cells are infected with helper HSV virus. The DI vector is then generated along with helper virus during the viral propagation process. The system can be manipulated so that the ratio of DI particles to helper virus is high.

Sindbis Virus

The use of virus based expression vectors for gene therapy and vaccine applications is increasing, with a number of diverse virus types and approaches. Alphaviruses, especially Sindbis virus and Semiliki Forest Virus (SFV) are also potential candidates as a gene transfer vector. The viral genome consists of positive single-stranded RNA about 12 Kb that codes for several structural and non-structural proteins. After the virus enters a cell, the nonstructural proteins (NSP1-4) get translated. Then the NS proteins function to produce the sense strand, from which large amounts ($\sim 10^5$) of new genomic and subgenomic RNAs coding for the virus structural proteins get transcribed. Assembly of the structural proteins resulted in the production of progeny viruses. The viruses pass their entire life cycle in the cytoplasm.

Sindbis viral vectors offer obvious advantages as a gene delivery vector in having extremely high transgene expression in a variety of mammalian and insect cell lines. Though this virus has a RNA genome, the viral genome does not integrate, which minimizes the chance of chromosomal recombination to occur in the host genome. Further, relatively small sized genome and rapid production of high-titer virus are advantageous in using this virus as a delivery tool. Typical disadvantages for Sindbis virus is that they have a relatively short-term expression and have a strong cytotoxic effect on host cell. However, both of these characteristics can be utilized in applications such as cancer therapy [25, 26].

Hybrid Vectors

Gene researchers are now studying a class of recombinant virus vectors called "conditional viral vectors" in the hope of designing vectors that selectively deliver transgenes to the site of the disease. This is one of the greatest challenges of gene therapy. The purpose of these vectors is specifically to track tumor cells and kill them by delivering a payload of genes. Using a combination of tissue-specific and tumor-specific elements, researchers have demonstrated that viruses can achieve preferential killing of target tumor cells. Several studies showing the use of adenovirus [27], adeno-associated virus [28, 29] and HSV-1 virus [30, 31] have been reported. In all of these studies, it has been shown that by combining different viral and regulatory elements from different viruses and other genomes, it is possible to obtain regulated transgene expression in a target-specific manner. Greater details of gene therapy applications that involve combining two virus systems (hybrid vector) or combining various specific elements like promoters, enhancers, and insulators within a viral system, can be found elsewhere [32].

Nonviral Methods

The gene therapy field remains divided about optimal delivery methods due to the inherent strengths and weaknesses of different systems. The majority of cancer gene therapy approaches still employs viral-based vectors to express suitable target genes inside cancer cells [33]. However, virus-based approaches cannot overcome certain limitations of expression, such as target specificity, immune response, and safety concerns regarding secondary malignancies and recombination to form replication competent virus. These limitations in fact reinforced efforts to develop other non-infectious *in vivo* gene delivery procedures. As an outcome, several methods have been developed in the past few years, including the use of different plasmid-expression vectors that consist of tissue-specific transcriptional control elements and various other regulatory elements for time-limited vector replication inside tumor cells. Such nonviral vectors, while not as efficient as their viral counterparts for *in vivo* delivery, can avoid triggering an immune response, which is perhaps the biggest obstacle to successful gene therapy [34].

So far, the usefulness of nonviral vectors for gene therapy has been limited by their poor transfection efficiencies. However, their relatively easy production, non-pathogenicity and lack of adventitious contaminants are important advantages (Table 18.1). To develop successful nonviral vectors, the following three cellular barriers need to be overcome: (1) DNA needs to be transported to the cytoplasm, either directly through the plasma membrane or by escape from endosomes; (2) DNA needs to be translocated to the nucleus, and (3) for many applications the DNA has to be integrated into the genome or must be extrachromosomally replicated. In this review, we will discuss only those major nonviral methods that hold promise for application in *in vivo* gene delivery. Other details are reviewed elsewhere [23].

Liposomes

To be useful for clinical application, the methods of gene introduction must be compatible with therapy. This need has led to the development of nonviral liposome-mediated gene transfer techniques (Fig. 18.2). Liposomes were first described in 1965 as a model of cellular membranes, and were quickly applied to the delivery of substances into cells [35]. Methods are reviewed in detail elsewhere [36]. Liposomes entrap DNA by one of two mechanisms, which has resulted in their classification as either cationic liposomes or pH-sensitive liposomes. Cationic liposomes are positively charged and interact with the negatively charged DNA molecules to form a stable complex. Cationic liposomes consist of a positively charged lipid and a colipid. Commonly used colipids include dioleoyl phosphatidylethanolamine (DOPE) or dioleoyl phosphatidylcholine (DOPC). Colipids, also called helper lipids, are in most cases required for stabilization of liposome complex. A variety of positively charged lipid formulations are commercially available, and many other are under active development. One of the most frequently used cationic liposomes for delivery of genes to cells in culture is Lipofectin, which is a mixture of N-[1-(2, 3-dioleyloyx) propyl]-N-N-N-trimethyl ammonia chloride (DOTMA) and DOPE. DNA and Lipofectin interact spontaneously to form complexes that have a 100% loading efficiency, i.e., if enough Lipofectin is available, all DNA molecules form complex with the Lipofectin. It is assumed that the negatively charged DNA molecule interacts with the positively charged groups of DOTMA. The lipid:DNA ratio and overall lipid concentration used in forming these complexes are extremely important for efficient gene transfer and vary with each application. Lipofectin has been used to deliver linear DNA, plasmid DNA, and RNA to a variety of cells in culture [37]. Following intravenous administration of Lipofectin–DNA complexes, both the lung and liver showed marked affinity for uptake of these complexes and transgene expression.

Negatively charged or pH-sensitive liposomes entrap DNA rather than complex with it. Since both the DNA and the lipid are similarly charged, repulsion rather than complex formation occurs. In some cases, these liposomes are destabilized by low pH and hence are called pH-sensitive. To date, cationic liposomes have proved to be much more efficient for gene delivery *in vitro* and *in vivo* than pH-sensitive liposomes. However, pH-sensitive liposomes have the potential to be much more efficient for *in vivo* DNA delivery than their cationic counterparts, and should be able to do so with reduced toxicity and interference from serum proteins.

Liposomes offer several advantages in delivering genes to cells: (1) liposomes can complex with negatively and positively charged molecules; (2) they hold a greater degree of protection of the DNA from degrading processes; (3) liposomes can carry large pieces of DNA, potentially as large as a chromosome; and (4) liposomes potentially can be specifically targeted to any cells or tissues. In addition, liposomes avoid the issues of immunogenicity and replication competent virus contamination. The ability to synthesize chemically a wide variety of liposomes has provided a highly adaptable and flexible system capable of gene delivery, both *in vitro* and *in vivo*. Current limitations regarding *in vivo* application of liposomes revolve around low transfection efficiency and transient gene expression. Also, liposomes display a low degree of cellular toxicity and appear to be inhibited by serum components. As liposome technology is better understood, it should be possible to produce reagents with improved *in vivo* gene delivery into specific tissues.

Naked DNA

A DNA molecule that is not attached or encapsulated by a liposome or virus capsid is referred to as naked DNA. The simplest way of delivering DNA to cells is by direct injection utilizing a syringe. Surprisingly, some tissues exhibit transgene expression following naked plasmid DNA injection, including thymus, skin, cardiac muscle, and skeletal muscle. Of these tissues, long-term transgene expression was observed only in striated muscle [38]. Using this approach, intramascular DNA vaccines are currently being researched as a preventive tool for cancer and infectious diseases. The other way to transfer a segment of DNA inside the cell is by electroporation, a process where the application of an electrical impulse leads to cellular membrane breakdown, giving the negatively charged DNA molecule space to enter the cell. Electroporation has been used for many years in transfecting cells, but has not been used extensively in gene therapy. With the recent development of new devices and procedures, it is now possible to minimize cell damage *in vitro*. For efficient gene transfer, electroporation depends upon the nature of the electrical pulse, distance between the electrodes, ionic strength of the suspension buffer, and nature of the cells. Best results have often been obtained with rapidly proliferating cells. Successful application of this procedure is very difficult, particularly in terms of *in vivo* applications, but it has proved useful to some *ex vivo* applications like local gene therapy applications on the skin.

The greatest advantages of intramuscular plasmid DNA injection are simplicity, low cost, and incorporation of relatively large genes of 2 kb to 19 kb. The application of this technique is still restricted to the skin, thymus, and striated muscle.

Ballistic DNA Injection

Ballistic DNA injection, which is also known as particle bombardment or micro-projectile gene transfer, or the gene-gun, was first developed for gene transfer into plant cells. Since its initial introduction, it has been modified to transfer genes into mammalian cells both *in vitro* and *in vivo* [39]. Plasmid DNA that encodes the gene of interest is used to coat micro-beads, which are then accelerated by motive force to penetrate cell membranes. Specifically, the plasmid DNA is precipitated onto 1–3 micron-sized gold or tungsten particles. These particles are then placed onto a carrier sheet, which is inserted above a discharge chamber. At discharge, the carrier sheet accelerates toward a retaining screen, allowing the particles to continue toward the target surface. The force can be generated by a variety of means; the most common include high-voltage electronic discharge, spark discharge, or helium pressure discharge.

Ballistic DNA injection has been successfully used to transfer genes to a wide variety of cell lines and also to the epidermis, muscle, and liver. In vivo applications have predominantly focused on the liver, skin, muscle or other organs, which can easily be surgically exposed. Ballistic DNA injection also offers the capacity to deliver precise DNA dosages. Unfortunately, genes delivered by this method are expressed transiently and there is considerable cell damage, which occurs at the center of the discharge site. Currently, the most exciting application of intramuscular and ballistic plasmid DNA injection is in DNA-based immunization protocols. DNA-based vaccinations are being developed to prevent infectious diseases and cancer [40]. Cells that express a foreign protein and subsequently present that protein on their surfaces are more likely to produce a cell-mediated immune response. Such a response may be more important in fighting viral infections such as those caused by HIV, HSV, CMV, or RSV. Techniques that inject a foreign antigen usually yield an antibody-mediated response. Both the gene gun and direct intramuscular plasmid DNA injection are being used in protocols to immunize against HIV, hepatitis B and C, HSV, influenza, papilloma, tuberculosis, RSV (Rous Sarcoma virus), CMV, Lyme disease, *Helicobacter pylori*, malaria and Mycoplasma pulmonis.

Specificity of Gene Therapy

A major concern in the application of gene therapy is need to achieve controlled and effective delivery of genes to target cells or their surrounding matrix, and to avoid expression in non-target locations. Ex vivo approaches help to ensure that gene transfer is limited to cells of a particular organ. For example, gene transfer into bone marrow cells provides a means to introduce genes selectively into various blood cell types, including hematopoietic stem cells. Directing gene transfer and/or expression to the appropriate cells would appreciably enhance *in vivo* gene therapy approaches. Tissue-specific gene expression of transferred genes may be accomplished by targeting delivery of the vector primarily to the desired tissues and/or including appropriate regulatory sequences (promoters) in gene transfer vectors. To approach the former problem, research aims to incorporate substrate for cellular receptors into viral envelopes or to achieve cell-specific gene transfer by binding of virus and target cells to particular proteins or fusion proteins [41, 42]. The latter promoter-based approach provides specificity; even if the gene is delivered to non-target tissues, it will not be expressed in those tissues because of the chosen promoter. In general, the major problem is that these promoters suffer from a lack of transcriptional activity, absolute specificity, or both. Some of these regulatory sequences may be responsive to drugs; hence, *in vivo* expression of transferred genes might be regulated by administration of the relevant drug to the host. Many tissue-specific promoter sequences have been characterized and are under evaluation in experimental disease models (reviewed in [43]). However, genes working under a tissue-specific promoter, for example, using prostate-specific antigen promoter (PSA), will also be expressed in the normal tissues from which the tumor originated. Research in these areas within the context of gene therapy strategies is still in its infancy.

To achieve the goal of additional specificity of tumor-selective promoters, a set of genes that shows little or no activity in the cells under non-pathogenic conditions, but are induced or upregulated in certain types of tumors, have been studied. An example of such a gene is alpha-fetoprotein (AFP), which can be activated in hepatoma cells [44, 45] or a gene-encoding carcinoembryonic antigen (CEA), which can be activated in different types of adenocarcinomas [46–48]. Amongst the other promoters that utilize the specific pathological conditions of a disease is hypoxia-inducible factor-1 (HIF-1), which utilizes the typical hypoxic condition of most tumors, and c-erbB-2 promoter, which often is upregulated in altered genetic or signaling pathways in disease states.

Many tissue-specific promoters have relatively weak transcriptional activity and this poses an additional problem. For successful application of gene therapy and for imaging gene therapy, it is necessary to have strong expression of the introduced gene(s). Several strategies for improving promoter strength while maintaining specificity are being investigated. The simplest approach is to eliminate regions of a natural promoter, which do not contribute to its transcriptional strength and specificity, while multimerizing the positive regulatory domain sometimes activating point mutations in the promoter sequence have also been shown to improve the regulatory power of that promoter. However, the occurrence of this second type of promoter is rare. Examples include encoding for a-fetoprotein (AFP) [49] or multiple drug resistance gene 1 (mdr-1) [50]. The most common approach to amplifying promoter activity is by the use of recombinant transcriptional activators.

An ideal example of this approach was reported by Segawa et al. [51], who attached the expression of a prostate-specific PSA promoter to a Gal4-VP16 fusion construct. This recombinant transcription factor then activated the therapeutic gene through a Gal-4 binding site. The VP16 protein acted as a strong transactivator for enhancing gene transcription. This two-step approach allowed the use of an intermediate gene, which could be transcribed in lower amounts and still led to high levels of expression of the gene of interest. Recent work from our group has succeeded in enhancing levels of reporter gene signal (firefly luciferase and HSV1-sr39TK) in prostate cancer cells [52]. Another potential strategy involves enhancing promoter activity by constructing chimeric promoters through combining the transcription regulatory elements from different promoters with specific activity of a particular tissue type. In experimental conditions it has been shown that a chimeric promoter construct produced levels of transgene expression that were much higher than those obtained with a strong constitutive cytomegalovirus (CMV) promoter [43].

Gene Therapy Approaches

Gene therapy protocols for cancer treatment are the most common, but this approach is also being used to treat other diseases, including cardiovascular disorders (Table 18.2). This chapter will focus primarily on

Table 18.2. Distribution of gene transfer protocols by disease or vector type approved by regulatory authorities in North America and Europe[a]

Disease	Number of protocols	Vector	Number of protocols
Cancer	216	Retrovirus	159
Monogenic	49	Adenovirus	58
Infectious	24	Liposome	40
Cardiovascular	8	Naked DNA	16
Rheumatoid arthritis	2	Poxviruses	19
Cubital tunnel syndrome	1	Adeno associated virus	4
Total	300	Herpes simplex virus	1
		Gene gun	1
Ex vivo transfer[b]	156	Electroporation	1
In vivo transfer	144	Naked RNA	1
		Total	300

a Table compiled from References [23, 139]
b Out of 156 *ex vivo* protocols, 32 do not involve any therapeutic intent.

cancer gene therapy; additional material relating to other diseases can be found elsewhere [53–57]. Overall, the protocols for treating cancer can be subcategorized by their mechanisms of action: (1) induction of selective cell sensitivity to a prodrug (for example, ganciclovir) by means of drug-sensitive genes such as herpes simplex viral thymidine kinase, (2) protection of sensitive tissues like bone marrow from otherwise toxic doses of a cancer-treating drug by use of multidrug resistance genes, (3) replacement of lost tumor-suppressor genes or down-regulation of oncogenes, and (4) insertion of a cytokine gene into tumor cells *ex vivo*. Each of these approaches is reviewed briefly.

Suicide Gene Therapy

A suicide gene is one that renders the cells susceptible to a subsequently administered prodrug. Suicide genes include non-mammalian enzymatic genes, which convert nontoxic prodrug into a highly toxic active drug. Systemic application of the nontoxic prodrug results in the production of toxic metabolite only at the tumor site, if suicide gene expression is restricted to the tumor. Monitoring of suicide gene expression in the tumor is a prerequisite for gene therapy using a suicide system for two reasons: to decide if repeated gene transduction of the tumors is necessary, and to find a therapeutic window of maximum gene expression for prodrug administration [58]. Two suicide gene systems have been applied in clinical studies: Herpes Simplex Virus Type 1 Thymidine Kinase (HSV1-*tk*) and Cytosine Deaminase (CD).

Moolten [59] first showed the use of HSV1-*tk* as a suicide gene. The viral thymidine kinase gene has relaxed substrate specificity compared to mammalian thymidine kinase, and is capable of phosphorylating pyrimidine and purine nucleoside derivatives as well as deoxythymidine [60]. These compounds when phosphorylated are trapped intracellularly. The discovery that acyclovir is phosphorylated by HSV1-*tk* led to a potential suicide gene therapy approach. Acyclovir-monophosphate is further phosphorylated by various cellular enzymes with the production of acyclovir-triphosphate. Incorporation of this end product in place of thymidine-triphosphate then leads to DNA chain termination and also interferes with polymerases. Subsequently, ganciclovir and penciclovir have been synthesized as potential substrates for HSV1-*tk* [61]. By introduction of the HSV1-*tk* suicide gene into cancerous cells, followed by prodrug treatment, cellular replication is blocked, leading to a decrease in tumor burden. Multiple clinical trials using the HSV1-*tk* suicide gene are underway. As a delivery vehicle of the HSV1-*tk* gene both retroviral and adenoviral vectors have been used. An advantage of this system is that most of the transduced cells will be killed.

Unlike HSV1-*tk*, cytosine deaminase (CD) is a bacterial gene, which also is expressed in yeast. CD converts 5-fluorocytosine (5-FC) into a highly toxic compound 5-fluorouracil (5-FU). 5-FU exerts its toxic effect by substituting for uracil in cellular RNA and thus interfering with DNA and protein synthesis. It also inhibits thymidilate synthetase through 5-fluorodeoxyuridine monophosphate, resulting in impaired DNA replication [62, 63]. The first report of CD suicide gene therapy was by Nishiyama et al. [64], who showed that treatment of CD-containing rat glioma cells with 5-FC produced significant reduction in tumor growth.

Tumor-Suppressor Gene Therapy

Tumor-suppressor gene therapy involves the introduction of the wild-type (normal) tumor suppressor gene into tumor cells, to overcome the malfunction of the mutated tumor suppressor gene. The logic of how genes function, together with the connection of cell cycle processes to specific gene actions, has led to new concepts of treating tumors by gene therapy. Essentially, cell death may not be the only way to achieve the therapeutic result; changes in growth, behavior, invasiveness or metastatic ability of tumor cells should also be considered as potential approaches. Identification of tumor-suppressor genes that have been lost in specific tumors may allow selective gene reconstitution, and may cause death of tumor cells, but not cells that are normally expressing the gene. Tumor-suppressor genes such as *p53*, retinoblastoma *(Rb)*, and *p16*, which have been implicated in a wide range of cancers, can be targeted for this kind of therapy. Tumor-suppressor gene replacement is a particularly promising method for treating human bladder cancer. Several studies have shown that loss of expression of the *p53* and *Rb* tumor-suppressor genes is common in locally advanced bladder cancer and portends a poor prognosis [65, 66]. Some of the cancer-specific tumor-suppressor genes, such as *APC* (involved in colon cancer) and *WT1* (involved in Wilm's tumor) can also be used. Since *p53* is the most commonly mutated gene in human cancer that causes changes in transcription, cell cycle arrest, DNA repair and apoptosis [67, 68], this gene has drawn a majority of the attention in treating cancers. Transduction of *p53* in cancerous cells can significantly inhibit tumor growth and angiogenesis, and can induce cell death in *p53* mutant cells in different tumor models, including lung and breast cancer [69–71]. Similar studies in lung cancer patients have shown that retroviral mediated transfer of *p53* is nontoxic and suppresses tumor growth [72].

Direct suppression of metastatic growth through *p53* gene therapy has been attempted by systemic delivery of a liposome complex in a breast cancer model [70] and by hepatic artery infusion for colonic liver metastases [73]. However, both the approaches lacked tumor targeting. A novel strategy to address the treatment of *p53* mutant metastases has been the development of an adenovirus that replicates only within *p53* mutant cells, killing through cell lysis [74]. Systemic inoculation of this vector has resulted in significant growth suppression of a primary tumor and deserves further investigation [75].

The field of tumor-suppressor gene therapy is limited by the paucity of identified target genes known to induce or maintain the malignant phenotype. Furthermore, in pre-clinical *in situ* studies, eradication of the treated tumors is rare, due to the difficulty in transduction of sufficient numbers of cells within a cancer to facilitate a cure. A bystander effect, causing the death of more cells than are actually transduced, has been proposed, but is not well understood. This effect may result from cell-to-cell contact, immune-mediated response or other local effects. Mutant or null *p53* status has been associated with increased resistance to radiation therapy and to apoptosis-inducing chemotherapy [76]. Expression of the *p53* protein in tumor cells that lack the gene has been shown to result in increased sensitivity to chemotherapy and radiation [77, 78] and is currently under investigation in clinical trials.

Immunomodulatory Approaches

Immunomodulatory gene therapy is a method to induce cellular immune responses to metastatic lesions. The strategy involves injecting into the skin of a patient a suspension of irradiated tumor cells that have been transduced with a cytokine gene to stimulate a systemic immune response against tumor specific antigens [79]. This system uses a gene-modified vaccine in an attempt to induce specific anti-tumor cytotoxicity. Activated CD-8 T lymphocytes are the primary anti-tumor effector cells of the immune system. Activation of the CD-8 T lymphocyte requires antigen presentation in the context of a major histocompatibility complex-class I receptor and binding of cytokines secreted by CD-4 T-helper lymphocytes [80]. Immunomodulatory gene therapy attempts to augment this system by engineering tumor cells to secrete cytokines, thus bypassing the need for CD-4 T-helper lymphocytes. Studies in animal tumor models have demonstrated that gene-modified vaccines can produce potent, specific, and long-lasting anti-tumor immunity [81]. Several cytokines, including IL-2, IL-4, IL-7, tumor growth factor-beta 2, interferon-gamma, and granulocyte-macrophage colony-stimulating factor (GM-CSF), are currently undergoing phase I and II clinical trials.

In pre-clinical cancer models, vaccination with tumor cells has been demonstrated to generate a cellular immunological response against "challenge" tumors. Animals that became disease-free were resistant to subsequent challenge with highly tumorigenic parental cells. This suggests that long-term immunological memory develops in a subset of animals [82]. However, several problems still need to be addressed to make this strategy successful. Only a few tumor specific antigens act as recognition targets; there has been activity against a relatively low tumor burden in several studies, but the financial and labor costs are high, and efficiencies will need to be improved.

Safety Issues

Gene therapy is a powerful clinical approach, but so far has not produced an unequivocal clinical success [83]. Several requirements must also be fulfilled for the successful use of this technique. These include total understanding of the underlying biology of the disease of interest, the construction of recombinant DNA constructs to correct the abnormality, and finally, a highly efficient delivery system for the transfer of the therapeutic genes into the cells of the patient.

Viral delivery systems have been used in gene therapy applications from the very beginning. In parallel, issues of biological containment are raised by the introduction of new viral vectors. Today, development of viral gene delivery vehicles that are both safe and efficient has become the greatest challenge for the successful use of gene therapy. Genetically modified viruses, influencing the health of human beings, originate from multiple sources, the most important being the use of viral gene therapy and live viral vaccines. There are concerns about the unintended effects of viral vectors (for example, the creation of replication competent viruses, the shedding of viruses from infected individuals, the risk of germ line transmission, etc.). Finally, a continued concern is the lack of standardized assays for detection of potentially replication competent viruses.

Each specific group of viruses has its own limitations and safety issues for use in human gene therapy. Biologically contained viruses for gene therapy have been made by dissection of the viral genomes into crippled and physically separated sub-genomes (for example, helper and vector genomes). The ideal vector genome contains only viral *cis*-acting sequences that are necessary for the vector to infect the target cell and for transfer of the therapeutic gene. The ideal helper genome contains only viral *trans*-acting elements that are necessary for vector packaging, such as sequences coding for the structural proteins and enzymes that are required in generating vector particles. For most systems, the challenge has been to design genomes so as to avoid the creation of replication-competent viruses. This has led to the development of systems with almost no sequence homology between helper and vector genomes, thus minimizing the risk of homologous recombinational events leading to the generation of non-defective viruses. The consequences of possible recombination with HIV-1 in infected individuals, as well as of the introduction of novel sequences into the viral pool infecting humans, should be carefully considered.

Each disease has its own specific requirements, such as the target tissue(s) and the amount of gene-product required. Similarly, each vector has characteristics that may or may not be desirable for a particular application. The major drawback for using retroviral vectors is the risk of adverse event caused by their permanent integration into the genome of the host cell. Among the retroviruses, lentiviruses overcome a lot of limitations, and are starting to receive more attention. The major disadvantage of first- and second-generation lentivirus vectors is their construction on the basis of HIV genomes, and it is still an open question whether or not HIV-derived vectors should be used for treatment outside of AIDS. Hence, a number of non-primate lentiviruses like FIV or EIAV are now being considered as potential gene therapy vectors. Serious inflammatory side effects are the major drawback to using adenoviruses. Recent modifications of the adenovirus backbone have led to the development of helper-dependent vectors completely devoid of viral protein coding regions.

In the field of vaccine technology, naked DNA is presently being tested for its ability to provide immunity against specific antigens. Manufacturers of vaccines are currently looking for easier production systems to provide effective long-lasting immunity. With the aid of DNA recombination techniques, invasive bacteria may be attenuated. The engineering of vaccine bacteria that are invasive to man may create new safety issues. The risk assessments that need to be developed for these types of vaccine productions will most likely also apply to the evaluation of other recombinant DNA procedures.

Imaging Reporter Gene Expression with PET

Imaging therapeutic gene expression involves locating the tissues expressing a therapeutic gene of interest, as well as monitoring the magnitude and time variation of gene expression. Imaging can help to play a critical role in optimizing gene therapy. Two different strategies can be used to image therapeutic gene expression. The first approach involves "direct imaging", where a labeled therapeutic protein substrate is used to image the specific expression of that protein. Direct imaging of the expression of each and every therapeutic gene would require development of a radiolabeled probe for each and every therapeutic protein. However, most therapeutic proteins lack appropriate substrates that could be used as a radiolabeled probe in generating PET or SPECT images. This has led to the novel concept of "indirect imaging", which involves coupling the therapeutic gene to a "reporter gene" and then tracking reporter gene expression by PET or SPECT. As the name suggests, a reporter gene reports about cellular events and has been used for over a decade to monitor gene expression in cell culture. The indirect imaging strategy requires proportional and correlated expression of both the therapeutic and reporter genes. The different ways these genes can be linked are discussed in detail in an upcoming section. In order to develop and validate many of the assays for reporter gene imaging, we and other research groups chose mouse models because of their relatively low cost, short breeding time and successful use in the past by molecular biologists. Small-animal instrumentation such as microPET [2] with a spatial resolution of ~8 mm3 is an ideal tool for imaging mice, and much of our work in developing reporter genes relies on this small-animal PET scanner. This technology and related small-animal imaging technologies are reviewed elsewhere [84].

There are two broad categories of reporter genes that can be targeted using radionuclide imaging approaches,

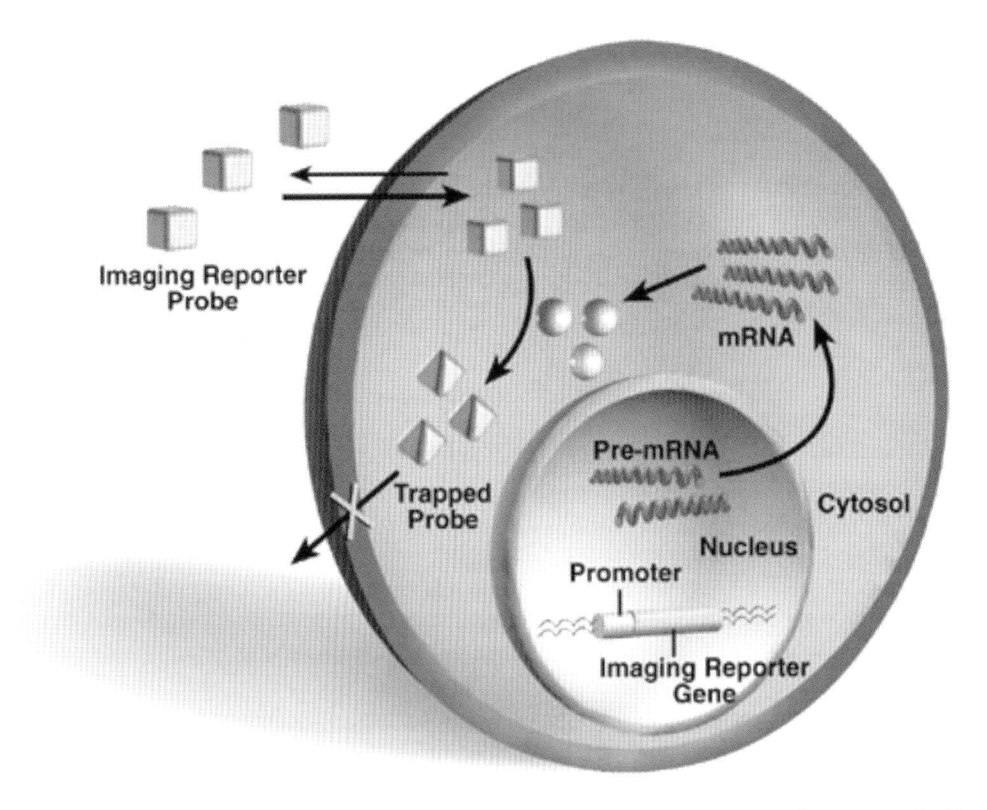

Figure 18.3. Diagram illustrating an approach to image a reporter gene with enzyme mediated trapping of a reporter probe. The reporter gene driven by a promoter of choice must first be introduced into the cells of interest. After transcription and translation, enzymatic protein products (spheres) are available to trap a reporter probe. Once the reporter probes (cubes) enter cells, the enzyme molecules modify and trap them (tetrahedrons) and thus can be used to image reporter gene expression. Cells that do not express the reporter gene will still transport the reporter probe, but it will not be trapped and will eventually efflux from those cells.

one encoding for an intracellular enzyme (Fig. 18.3) and the other for an intracellular and/or cell-surface receptor (Fig. 18.4). The reporter gene can be driven by any promoter of choice, and eventually leads to a protein that can be assayed by a reporter probe (a radiolabeled substrate or ligand). The receptor approach has the potential advantage of not requiring transport of ligand into the cell, but the enzyme approach has the theoretical advantage of signal amplification, because one enzyme can trap many reporter probes. The HSV1-*tk* viral gene has been the most widely exploited enzyme-based reporter gene whose expression is detected by radionuclide product sequestration [85, 86]. Its use as a reporter gene is not directly related to its use as a suicide gene, as described above. The HSV1tk viral gene is an ideal reporter gene because it has relaxed substrate specificity as compared to the mammalian thymidine kinases, which are responsible for phosphorylating thymidine. Therefore, prodrugs can be radiolabeled and systemically administered, and do not accumulate significantly in tissues other than those that express the HSV1-tk reporter gene. One of the reasons HSV1-tk was selected as a reporter gene is because radiolabeled substrates existed or could be synthesized to image its expression [87]. Two major classes of radiolabeled substrates have been explored for HSV1-*tk* reporter gene imaging. These include acycloguanosine derivatives (for example, fluoroganciclovir, fluoropenciclovir) and derivatives of thymidine. Studies initially used fluorine18 derivatives of ganciclovir labeled in the 8-position (FGCV) [88, 89] and penciclovir (FPCV) [90]. Subsequently, it was found that the side-chain labeled ganciclovir 9-[(3-[^{18}F]-fluoro-1-hydroxy-2-propoxy) methyl] guanine (FHPG) [91, 92], and side-chain labeled penciclovir 9-(4-[^{18}F]-fluoro-3-hydroxy-methylbutyl) guanine (FHBG) [93] are better substrates, although they do produce racemic mixtures. The thymidine derivative FIAU (2′-fluoro-2′-deoxy1B-D-arabino-furanosyl-5-iodo-uracil) has been labeled with ^{131I} for SPECT imaging [94] and with 124_I (half-life ~4 days with ~20% positron yield) for PET imaging [95]. It appears that the best substrates for HSV1-*tk* are currently FHBG and FIAU, but which of the two is better will require further investigation [96, 97]. Issues related to transport of these substrates and the K_m and V_{max} of each substrate for various enzymes will need further study. There may be applications in which the shorter half-life and greater

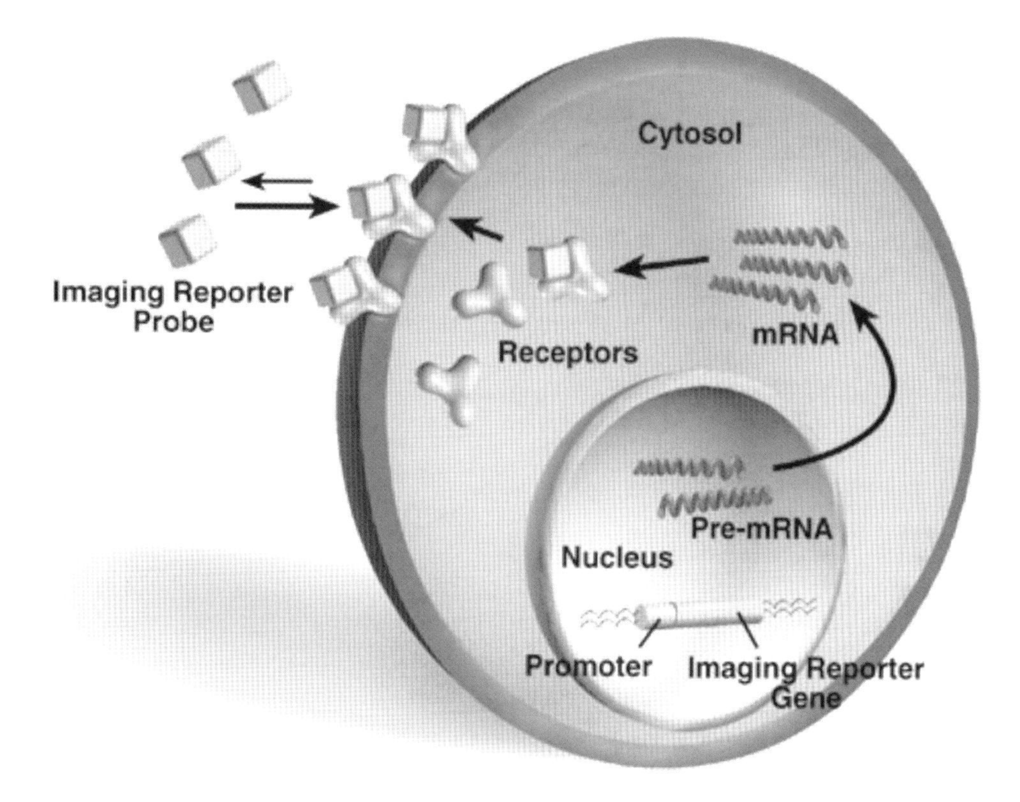

Figure 18.4. Diagram illustrating an approach to image a reporter gene with receptor mediated trapping of a reporter probe. The reporter gene driven by a promoter of choice must first be introduced into the cells of interest. After transcription and translation, receptor protein (Y) is made. The receptor proteins can be either intracellular and/or at the cell surface. These receptor molecules then bind the reporter probes (cubes) and thus can be used to image reporter gene expression. Cells that do not express the receptors will not have any significant retention of the reporter probe (ligand) after some time interval, which allows clearance of ligand.

positron yield of ^{18}F is preferred, and other applications in which the longer half-life of ^{124}I is desirable. Further details of various issues surrounding the substrates for HSV1-*tk* can be found elsewhere [87]. We have also investigated a mutant HSV1-*tk* reporter gene (HSV1-sr39*tk*) that has six point mutations, which leads to a HSV1-sr39*tk* enzyme with better imaging sensitivity when using FGCV and FPCV, as compared to the wild-type HSV1-*tk* [98]. This enzyme was isolated by screening a library generated by site-directed mutagenesis of the HSV1-*tk* gene with acyclovir and ganciclovir [99]. This mutant enzyme also has the distinct advantage of having a lower phosphorylation rate for thymidine, which directly competes with various substrates (for example, penciclovir, FHBG) for HSV1-*tk*/HSV1-sr39*tk*. This is important because levels of thymidine can vary and confound the imaging signal independent of changes in reporter gene expression [100]. Future HSV1-*tk* mutant reporter genes that have been specifically optimized for fluorinated substrates may lead to even better reporter gene/reporter probe pairs.

Significant work has been done to validate the HSV1-*tk* reporter gene assay with various reporter probes. We initially used [^{14}C]-ganciclovir and digital whole-body autoradiography, while using an adenoviral delivery model to image HSV1-*tk* gene expression in the mouse liver [101]. We next used FGCV to demonstrate with microPET that adenoviral-mediated HSV1-*tk* gene expression in the mouse liver could be quantitated (Fig. 18.5). Furthermore, we showed that the trapping of FGCV as measured from microPET images (%ID/g liver) directly correlated with levels of GAPDH normalized HSV1-*tk* message and HSV1-*tk* enzyme activity. We have shown in a mouse tumor xenograft model, in which C6 rat glioma cell lines were stably transfected with the HSV1-*tk* reporter gene *ex vivo*, the ability to image HSV1-*tk* reporter gene expression with FPCV and microPET [90]. We have also recently studied HSV1*tk* gene expression in transgenic mice with FHBG and microPET [102]. These mice have the albumin promoter driving HSV1-*tk* reporter gene expression and therefore expression is confined to hepatocytes. We have also validated tracer kinetic models for FHBG [100, 103], and these should help to improve quantitation of reporter gene expression. The group at Memorial Sloan Kettering Cancer Center has also validated imaging of the HSV1-*tk* reporter gene expression

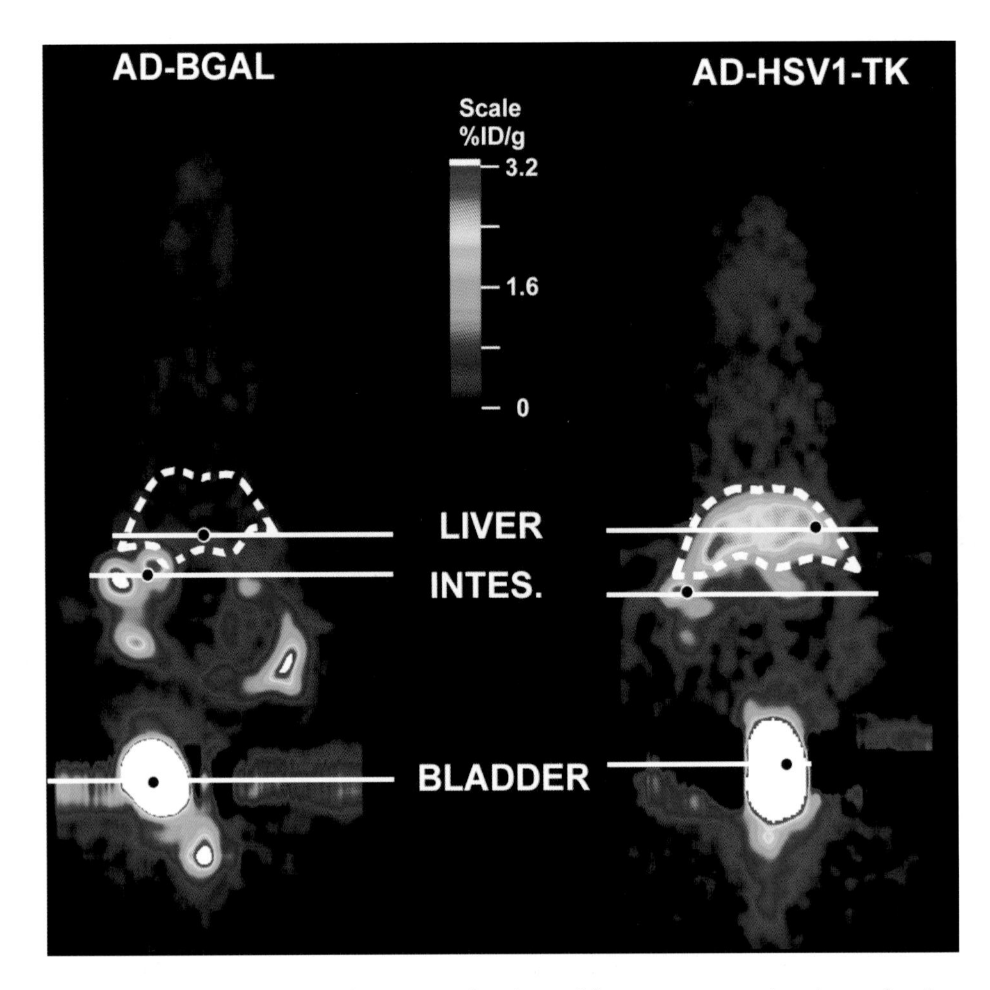

Figure 18.5. MicroPET imaging of two mice with FGCV after using an adenovirus to deliver a reporter gene. In each case, the mice were tail-vein injected with an adenovirus (1.5×10^9 plaque forming units) and 48 hours later imaged in a microPET 1 hour after tail-vein injection of FGCV. The mouse on the left received an adenovirus with a control gene (Ad-βGal) and the mouse on the right with an adenovirus carrying the HSV1-*tk* reporter gene (Ad-HSV1-*tk*). The reporter genes were driven by a cytomegalovirus (CMV) promoter that leads to continuous transcription of the reporter gene. Adenovirus predominantly infects the liver in part due to presence of receptors on hepatocytes. The mouse that received Ad-βGal had minimal hepatic signal whereas the mouse that received Ad-HSV1-*tk* showed significant activity in the liver. Activity in intestine and bladder is due to clearance of FGCV from the hepatobiliary and renal routes. The color scale is in percent injected dose per gram (%ID/g). (Figure reproduced with permission from reference [89]).

in various models with SPECT [104] and PET while utilizing FIAU [31, 104–106].

Two receptor-based reporter gene systems have been described that utilize radionuclide detection technologies. The dopamine D2 receptor (D2R) is expressed primarily in the brain striatum and pituitary gland. The D2R has been used as a reporter gene, both in adenoviral delivery vectors and in stably transfected tumor cell xenografts [107]. The location, magnitude, and duration of D2R reporter gene expression can be monitored by microPET detection of D2R-dependent sequestration of systemically injected 3-(2′-^{18}F-fluoroethyl)-spiperone (FESP), a positron-labeled substrate for the dopamine D2 receptor [108]. Similarly to the HSV1-*tk* reporter gene, the D2R/FESP assay shows excellent correlation between %ID/g liver and levels of GAPDH normalized D2R mRNA and D2R receptor levels [107]. Shown in Fig. 18.6 is an example of imaging both the HSV1-sr39*tk* and D2R reporter genes in the same living animal.

The somatostatin type-2 receptor (SSTr2) is expressed primarily in the pituitary. When used as a reporter gene in adenovirus delivery systems, the location of SSTr2 expression can be monitored by systemic injection of a technetium-99 labeled SSTr2 peptide substrate, with subsequent imaging by planar imaging techniques with a conventional gamma camera [109–111]. One potential problem with receptor-based PET reporter genes is the possibility that endogenous ligands could bind to the receptor and lead to signal transduction. For the D2R reporter gene we have recently reported a mutant reporter D2R80A that uncou-

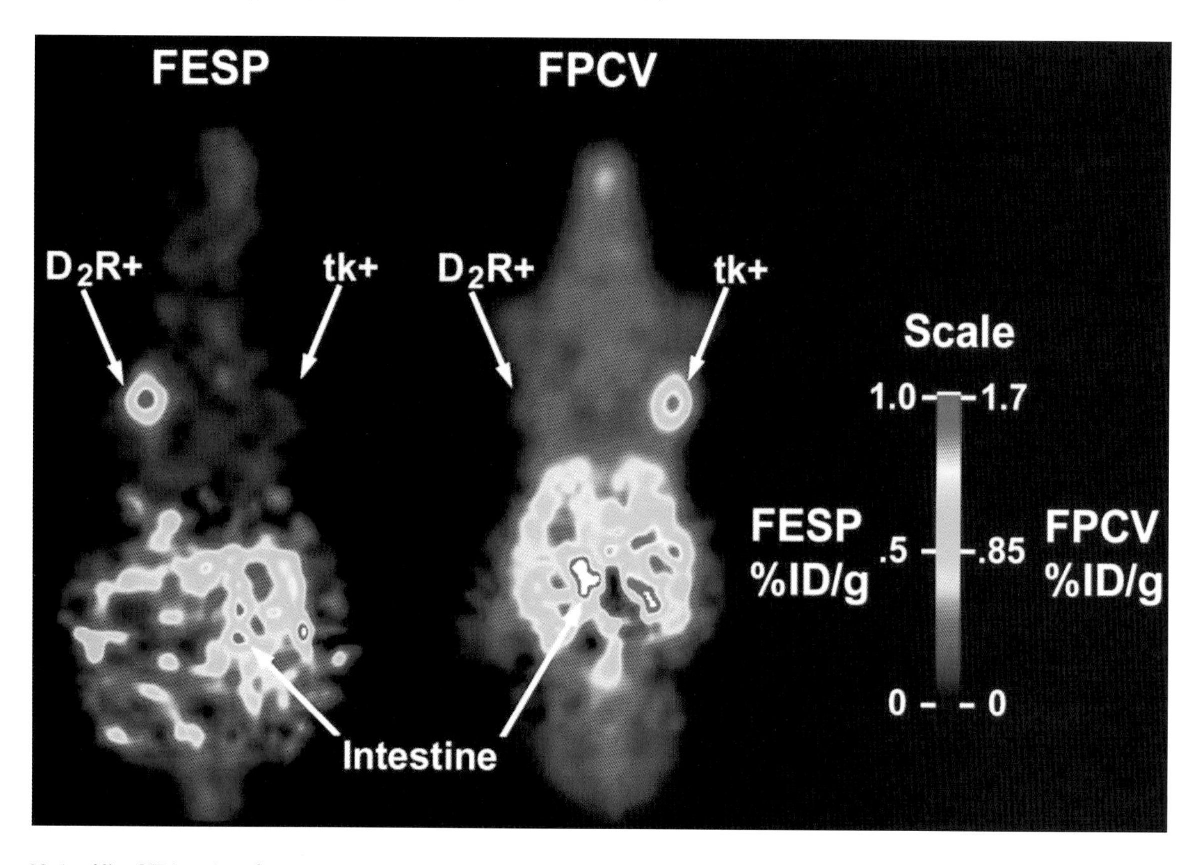

Figure 18.6. MicroPET imaging of a mouse carrying tumors expressing the D2R and HSV1-*tk* reporter genes. The nude mouse was carrying a tumor that expresses D2R on the left shoulder and a tumor that expresses HSV1-*tk* on the right shoulder and was imaged with FESP (day 0) and FPCV (day 1). The tumors were stably transfected with the reporter genes *ex vivo*. FESP accumulates primarily in the D2R positive tumor and FPCV in the HSV1-*tk* positive tumor. Background signal in the intestinal tract is seen due to clearance of both FESP and FPCV from these routes. Images are displayed using the same common global maximum. The color scale is in percent injected dose per gram (%ID/g). (Figure reproduced with permission from reference [87]).

ples signal transduction but maintains affinity for FESP [112]. This should help to minimize perturbations to the system under study.

Driven by the energy of the transmembrane Na ion gradient, iodide is actively transported across the basolateral plasma membrane of thyrocytes together with Na ions by the Sodium/Iodide Symporter (NIS), which has been cloned recently from mouse, rat and human. As down-regulated or mis-targeted NIS expression is commonly found in thyroid carcinomas and frequently correlates with tumor de-differentiation and with loss of radioiodine uptake capacity, NIS represents a key protein for thyroid cancer therapy and diagnosis [113, 114]. Recently, many investigators have shown that Sodium Iodide Symporter (NIS) gene transfer into a variety of cell types confers increased radioiodine uptake up to several hundredfold that of controls [115]. In addition, researchers have shown that NIS gene transfer mediated by viral or nonviral vectors into a variety of cells make them susceptible to being destroyed by the radioisotope ^{131}I in cell culture. Groot-Wassink T et al. [116] also showed that plasmid transfection or adenoviral gene delivery can be monitored *in vitro* on incubation with ^{125}I and intravenous injection of adenovirus carrying human NIS (hNIS) induces gene expression essentially in the liver, adrenal glands, lungs, pancreas, and spleen. Expression of hNIS in tumor xenograft models can also be detected by intratumoral virus injection. Several groups are now exploring imaging possibilities of NIS expression monitored by PET after intravenous injection of ^{124}I, demonstrating the potential of this approach for noninvasive imaging [117–119]. Because NIS activity is the molecular basis of radioiodine therapy, there is great interest in exploring the possibility of NIS gene transfer to facilitate radioiodine therapy for non-thyroidal human cancers. Ectopic expression of this gene in various type of malignant cells has led to radiosensitization and in some cases tumor regression in mouse models, highlighting the therapeutic potential of this approach. Cell culture experiments indicated that NIS infected tumor cells can be selectively killed by the induced accumulation of ^{131}I. The therapeutic efficacy of ^{131}I depends on the radiosensitivity of the tumor tissue, the tumor size and the biologic half-life of ^{131}I, which is dependent on the extent of radioiodine trapping and its retention time in the tumor. Several other reporter genes including the gene encoding for carcinoembryonic (CEA) antigen, as well as reporter genes for magnetic resonance imaging and optical imaging approaches have been developed and are reviewed elsewhere [120].

Indirect Imaging of a Therapeutic Gene via a Reporter Gene

Most therapeutic genes lack appropriate substrates or probes that can be radiolabeled and used to generate images defining the magnitude of therapeutic gene expression. Therefore, it is necessary to develop and validate indirect imaging strategies using a reporter gene and a therapeutic gene in combination (Fig. 18.7). The goal of these approaches is to image quantitatively reporter gene expression, and to quantitate indirectly levels of therapeutic gene expression.

Bicistronic Approach

In many cases, each mRNA is translated into a single protein. The discovery of cap-independent translation and the existence of internal ribosomal entry sites (IRES) in polioviruses and encephalomyocarditis virus [121–123] opened a new gateway to construct bicistronic vectors (vectors containing two genes). IRES sequences are generally 450–800 bp long with complex RNA secondary structures that are required for initiation of translation. Several general translation initiation factors like eIF3, eIF4A, eIF4G and some specific

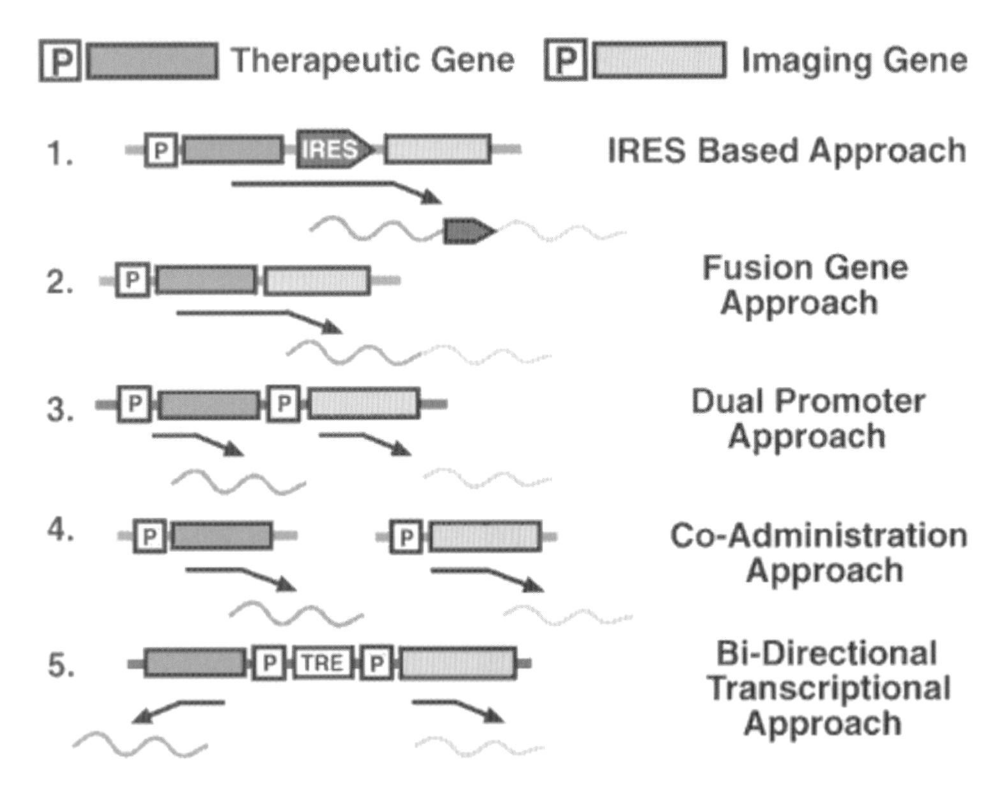

Figure 18.7. Schematic diagram explaining different approaches to link a therapeutic gene and a reporter gene to achieve *indirect* imaging of therapeutic gene expression. In each approach the goal is to image reporter gene expression in order to infer therapeutic gene expression. P is a chosen promoter driving transcription, IRES is an internal ribosomal entry site, and TRE is a tetracycline responsive element. In all cases the gene constructs (Boxes and wide arrow) and the translational protein products (wavy line) are shown. (1) In the bicistronic approach an IRES is used to link the two genes. Transcription occurs under a single promoter to form a single mRNA molecule. Translation of a single mRNA leads to a therapeutic protein and a reporter protein. (2) In a fusion gene construct the two gene sequences are coupled together with a spacer sequence in between them which when transcribed under a single promoter, leads to one mRNA. Translation leads to a single fusion protein capable of therapeutic and reporter protein function. (3) In a dual promoter approach two genes under two identical promoters are used. The resultant protein products form separately from two separate mRNAs. (4) In a co-administration approach the two genes are carried in two identical but distinct vectors. This leads to two mRNAs and two distinct proteins. (5) A bi-directional approach can use a TRE to induce transcription leading to two mRNA's and two proteins and can be titrated through the use of a fusion protein which binds to the TRE in the presence of an antibiotic.

cellular IRES transacting factors (ITAFs) are required to drive translation of the second cistron. We recently reported a bicistronic vector where both D_2R and HSV1-sr39*tk* genes were separated by an EMCV IRES, co-expressed from a common CMV promoter and imaged by microPET in multiple, stably transfected tumors in living mice [124]. Excellent correlation was found between expression of the two reporter genes as measured by %ID/g tumor directly from the corresponding PET images of FHBG and FESP. We have also reported that the gene placed in the second cistron was significantly attenuated. An approach using ~-Gal and HSV1-*tk* has also been reported in mice while imaging with SPECT and [^{131}I]-FIAU [125]. Although the IRES sequence leads to proper translation of the downstream cistron from a bicistronic vector, it can be cell type-specific, and the magnitude of expression of the gene placed distal to the IRES is often attenuated [124, 126, 127]. This can lead to a lower imaging sensitivity, and methods to improve this approach through the use of enhanced IRES sequences [128, 129] are currently under investigation. We have also reported an adenoviral vector carrying the D2R80A-IRES-HSV1sr39*tk* bicistronic vector, and used this vector for microPET imaging with both FESP and FHBG [112]. Again, excellent correlation between D2R80A and HSV1-sr39*tk* gene expression was seen. These data indicate that when one of the two genes in the bicistronic vector is changed to a therapeutic gene, this bicistronic approach is likely to prove useful for indirectly imaging therapeutic gene expression.

Fusion Approach

Fusion gene/protein technology is another approach to introduce simultaneously two or more genes in gene therapy. The strategy for building such constructs lies in joining the coding sequences of the target genes in such a way that they stay in the same reading frame. The genes are usually separated by a small spacer (encoding for a few amino acids) to help preserve function of each individual protein in the fusion [130]. An advantage of this approach is that the expression of the linked genes is absolutely coupled. However, the fusion protein does not always yield functional activity for both of the individual proteins and/or may not localize in an appropriate sub-cellular compartment. This approach therefore cannot be generalized for all therapeutic genes and reporter genes. HSV1-*tk*-GFP [131, 132] and HSV1-*tk* Firefly Luciferase-Neo [133] fusion proteins are good examples of the fusion approach. Recently, we have developed few novel fused reporter genes for multimodality imaging involving a PET reporter, a bioluminescent and a fluorescent optical reporter gene [134]. These fused reporters provides the ability to move between imaging technologies covering fluorescence, bioluminescence, and PET/SPECT imaging needed for almost any reporter gene application. One can utilize these tri-fusion reporter genes to sort cells, to image in small living animals using either fluorescence or bioluminescence, and larger subjects including humans, using PET/SPECT.

Dual Promoter Approach

Two different genes can also be expressed under two separate promoters when incorporated into a single vector (for example, pCMV-D2R-pCMV-HSV1-sr39*tk*) and can potentially be a useful way to couple the expression of two genes. This approach may avoid some of the attenuation and tissue variation problems of the IRES-based approach, which is currently under active investigation [128]. A potential problem of this approach is that the expression of the two genes may become uncoupled if the two identical promoters have different transcriptional activity or if a mutation occurs in one or both promoters that change transcriptional activity.

Two-Vector Administration Approach

The therapeutic and reporter genes can both be imaged by the use of two distinct viral vectors. One vector would carry the reporter gene driven by the promoter of choice, and the other vector would carry the therapeutic gene driven by the same promoter. This approach has recently been tested. It was found that the expression of two PET reporter genes, HSV1-sr39*tk* and D2R, both driven by CMV promoter cloned separately in adenoviral vectors, is well correlated at the macroscopic tissue level when delivered simultaneously [135]. Although some cells are infected by only one of the two viruses, others by both viruses, and some by neither virus, the resolution of PET allows imaging of only a large number of cells. It is important to realize that *trans* effects between promoters on co-administered vectors can potentially affect reporter gene expression.

Bi-Directional Transcriptional Approach

For many gene therapy applications, it would be desirable not only to deliver the therapeutic gene to desired

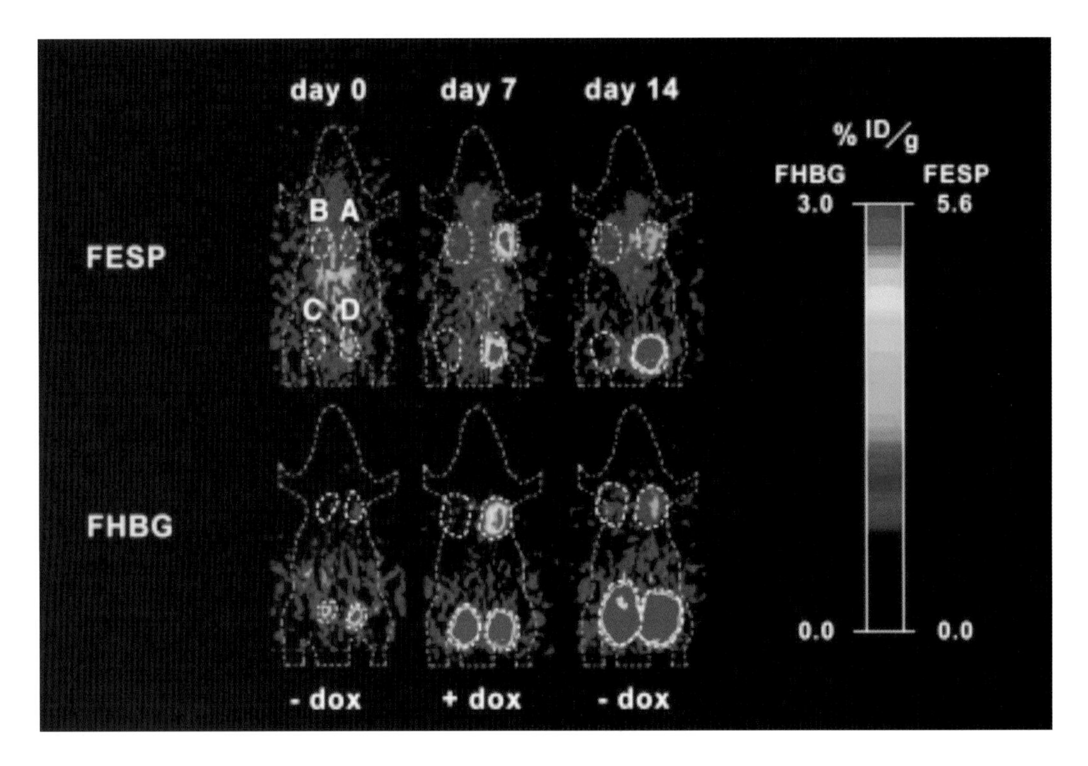

Figure 18.8. Sequential microPET imaging studies of a bi-directional inducible system in a mouse. Out of the four tumors (shown by dotted lines as A, B, C, D) grown in a single mouse, tumor A expresses two genes HSV-1sr39*tk* and D2R under a TRE element (see also Fig. 18.7). Tumor B is a control having none of these genes expressed. Tumor C is expressing only sr39TK gene while tumor D is for D2R gene only. When tumors reached a size of at least 5 mm, the mouse was imaged with FESP and FHBG on sequential days in a microPET scanner (day 0). Doxycycline (500 μg/ml) was added to cage water supply for seven days. Then, the mouse was again scanned with FESP and FHBG on sequential days (day 7). Doxycycline was removed from the cage water supply for the next seven days. The mouse was again scanned with FESP and FHBG on sequential days (day 14). All images are 1–2 mm coronal sections through the 4 tumors. The %ID/g scale for each reporter probe is shown on the right. For tumor A, the top row of images reveal no significant expression of D2R prior to induction of gene expression by doxycycline, high expression following seven days of doxycycline administration, and a return to near baseline low expression following seven days without doxycycline administration. The bottom row of images reveals a nearly identical pattern for induction of HSV1-sr39*tk* gene expression. The tumors appear slightly larger on the FHBG images because of tumor growth during the 24 hours between FESP and FHBG scanning. Control tumor B is a negative control and shows no signal for all images. Tumor C is a positive control for the HSV1-sr39*tk* gene and tumor D is positive control for D2R gene. Correlated signal intensity is observed between the FESP and FHBG images for each tumor for each doxycycline gene expression induction state. (Reproduced with permission from Sun et al. [136]).

target(s), but also to be able to regulate levels of therapeutic gene expression. We have validated an approach in a xenograft tumor model in living mice (Fig. 18.8), in which a simple antibiotic (doxycycline) can be used in combination with a fusion protein and a bi-directional vector to correlatively express two genes of choice [136]. This approach leads to bi-directional transcription, two mRNAs, and two proteins. The unique feature is that the level of transcription can be regulated by the level of doxycycline. This avoids the attenuation and tissue variation problems of the IRES based approach, and may prove to be one of the most robust approaches developed to date. Again, the levels of gene expression for the two genes are highly correlated [136]. This system can be improved further by improving the range of induction. The system is limited by the fact that a fusion protein must also be co-expressed, but future vectors should be able to encode for both the fusion protein and the bi-directional transcriptional system on a single vector.

Human Applications of Imaging Gene Therapy

Studies to image reporter gene expression in human subjects have only recently started. Because HSV1-*tk* is both a reporter and a suicide gene, imaging its expression in humans would be a good initial proof-of principle for the use of PET imaging in gene therapy. We have studied the kinetics, biodistribution, stability, dosimetry and safety of FHBG in healthy human volunteers [137] for eventual use in imaging patients

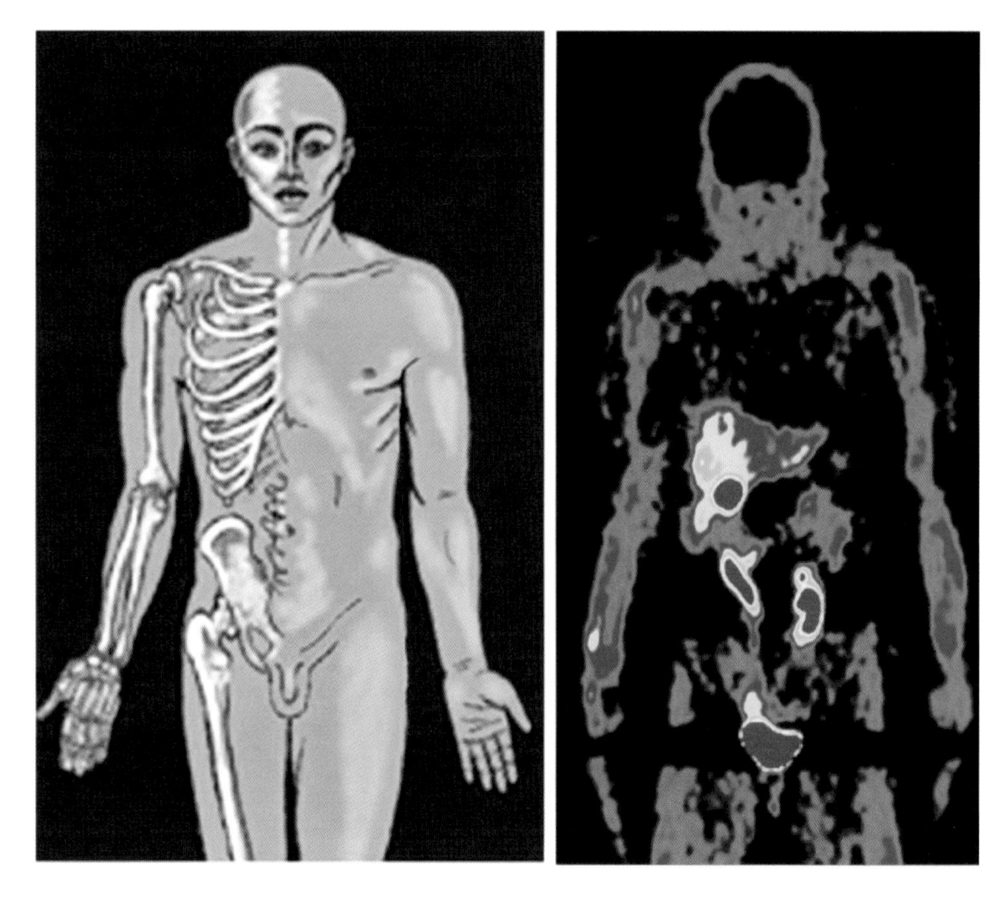

Figure 18.9. Human schematic (left) and corresponding whole body coronal PET image of FHBG biodistribution in a healthy volunteer (right). PET scan was obtained 40 minutes after 167.6 MBq (4.53 μCi) intravenous injection of FHBG. FHBG rapidly clears via the renal and hepatobiliary systems leading to low tissue background signal. Activity is seen in the gall-bladder, kidneys, and bladder. No activity is seen within the brain due to the inability of FHBG to cross the blood-brain-barrier. No significant retention of FHBG in any tissues is seen because the normal volunteer does not express HSV1-*tk*. The relatively low background should allow for specific imaging of HSV1-*tk*/sr39*tk* gene expression if there is enough reporter protein present at a given site.

undergoing HSV1-*tk* suicide gene therapy or therapy where a therapeutic gene is linked to HSV1-*tk* (Fig. 18.9). This study showed that FHBG cleared rapidly via the renal and hepatobiliary systems, and that relatively low signal throughout the whole body was achieved within 60 minutes of injection of tracer. Furthermore, FHBG is stable in blood, and results in sufficiently low radiation exposure to allow multiple FHBG PET scans per year. There are no other safety issues, since FHBG is injected in trace amounts. Using FHBG to image HSV1-*tk* reporter gene expression at sites near the gall bladder, kidneys, or bladder would be difficult due to tracer concentration in these regions. Furthermore, imaging in the brain would be limited by the failure of FHBG to cross the blood–brain barrier.

In another preliminary study, Jacobs et al. [138] has recently shown that [^{124}I]-FIAU PET imaging may be useful in patients with glioblastoma who are undergoing HSV1-*tk* suicide gene therapy. The HSV1-*tk* gene was delivered via intratumoral infusion of cationic liposomes. Five patients undergoing gene therapy with HSV1-*tk* were imaged, and one showed a significant increase in the FIAU accumulation rate post delivery of HSV1-*tk*. In the use of PET for monitoring HSV1-*tk* suicide gene expression, several issues must be kept in mind. First, because the patients will be undergoing treatment with prodrug (for example, ganciclovir) that will compete with the PET reporter probe, timing of the PET scan relative to the plasma concentration of the prodrug will be important. Second, it will be important to demonstrate that tracer accumulation correlates with biopsy-proven reporter gene expression.

Applications in the brain will be particularly challenging because of the blood–brain barrier (BBB). Although tumors may compromise the BBB, it is possible that the tracer can still not reach some portions of tissue that express the reporter gene. Regions near clearance routes of tracer (for example, bladder) will also be challenging for imaging at earlier times. It will be important to image the whole body in order to determine not only expression at intended sites, but also unintended sites of expression that may compromise the safety of a given approach. Correlative anatomic imaging (for example, MRI) and imaging with other PET tracers (for example, FDG) will have to be incorporated into the imaging protocols. Total radiation dosimetry from multiple studies will also have to be considered. It is not clear how accumulation of tracers such as FDG, FHBG or FIAU in tumors may change during treatment, and image interpretation could be confounded by an inflammatory response. Issues regarding accurate quantitation will also have to be considered, especially changes in the tumor size, which may lead to change in appropriate correction for partial volume effects.

Future applications of human gene therapy are likely to involve imaging of the HSV1-*tk* suicide gene directly by various tracers. HSV1-*tk* suicide gene trials are decreasing, but the use of HSV1-*tk* (or a mutant) as a reporter gene is likely to increase. In the next several years, we should see more studies in which HSV1-*tk* is coupled to other genes for indirect imaging of therapeutic gene expression in humans.

This will require modifying the vectors going into existing trials, so that both the therapeutic and reporter genes can be carried. Furthermore, both FHBG and FIAU will likely require investigational new drug (IND) approval from the Food and Drug Administration for trials within the US. These trials will need to be designed for reporter gene imaging, along with tissue biopsy and additional imaging with other tracers or modalities (for example, FDG PET, MRI, etc.) to improve understanding of patient response to treatment. It is likely that other reporter genes (for example, D2R) will also be used in various gene therapy trials. It may be that SPECT will play a role in some trials because of greater availability of some SPECT tracers, even though the technique will be less sensitive and likely less quantitative. It is likely that PET reporter gene imaging will help to optimize various gene therapy approaches and lead to monitoring of human gene therapy for safer utilization.

Future Directions

Clinical gene therapy has been initiated relatively recently and will require many additional refinements prior to being considered a successful therapeutic modality. In the near future the field will have to overcome several problems, such as finding appropriate therapeutic gene(s) and utilizing appropriate delivery vectors for efficient and specific targeting. New methods of molecular engineering, probe design and imaging technology are expected to generate many new imaging approaches. Imaging molecular events with PET should revolutionize current approaches of diagnosing and monitoring diseases treated with gene therapy.

Acknowledgements

We would like to thank the many faculty, students and staff who have contributed to various aspects of developing and validating the imaging technology described in this chapter. Specifically, Drs H Herschman, J Barrio, ME Phelps, SR Cherry, L Wu, A Berk, M Dahlbom, A Chatziioannou, T Toyokuni, N Satyamurthy, M Namavari, A Wu, A Green, X Sun, Q Liang, D MacLaren, S Yaghoubi, K Nguyen, E Bauer, M Iyer, Y Wang, R Goldman, D Pan, J Walsh, N Adonai, Y Yu, A Annala, P Shah and A Borghei. We would like to thank P Ray and M Patel for their help in editing this chapter. We also gratefully acknowledge the funding support provided by the NIH, DoE and CaPCure.

References

1. Jimenez-Sanchez G, Childs B, Valle D. Human disease genes. Nature 2001;409(6822):853–855.
2. Cherry SR et al. MicroPET:A high-resolution PET scanner for imaging small animals. IEEE Transactions on Nuclear Science 1997;44(3):1161–1166.
3. Phelps ME. PET:The merging of biology and imaging into molecular imaging. J Nucl Med 2000;41(4):661–681.
4. Giannoukakis N et al. Infection of intact human islets by a lentiviral vector. Gene Ther 1999;6(9):1545–1551.
5. Hanazono Y et al. In vivo marking of rhesus monkey lymphocytes by adeno-associated viral vectors:direct comparison with retroviral vectors. Blood 1999;94(7):2263–2270.
6. Lozier JN et al. Gut epithelial cells as targets for gene therapy of hemophilia. Hum Gene Ther 1997;8(12):1481–1490.
7. Wheeler CJ et al. A novel cationic lipid greatly enhances plasmid delivery and expression in mouse lung. Proc Natl Acad Sci USA 1996;93:11454–11459.
8. During, MJ et al. Long-term behavioral recovery in Parkinsonian rats by an HSV vector expressing tyrosine hydroxylase. Science 1994;266:1399–1403.
9. Numes FA, Raper SE. Liver-directed gene therapy. Med Clin North Am 1996;80:1201–1213.
10. Blaese RM et al. T lymphocyte-directed gene therapy for ADASCID:initial trial results after 4 years. Science 1995; 270(5235):475–480.

11. Onodera M et al. Development of improved adenosine deaminase retroviral vectors. J Virol 1998;72(3):1769–1774.
12. Miller DG, Adam, MA, MillerAD. Gene transfer by retrovirus vector occurs only in cells that are actively replicating at the time of infection. Mol Cell Biol 1990;10:4139–4142.
13. Roe TY et al. Integration of murine leukemia virus DNA depends on mitosis. EMBO Journal 1993;12(5):2099–2108.
14. Naldini L. Lentiviruses as gene transfer agents for delivery to non-dividing cells. Curr Opin Biotechnol 1998;9(5):457–463.
15. Blmer U et al. Highly efficient and sustained gene transfer in adult neurones with a lentivirus vector. J Virol 1997;71:6641–6649.
16. Kafri T et al. Sustained expression of genes delivered directly into liver and muscle by lentiviral vectors. Nature Genetics 1997; 17(3):314–317.
17. Miyoshi H et al. Stable and efficient gene transfer into the retina using an HIV-based lentiviral vector. Proc Natl Acad Sci USA 1997;94:10319–10323.
18. Emerman M. Learning from lentiviruses [news]. Nature Genetics 2000;24(1):8–9.
19. Verma IM, Somia N. Gene therapy – promises, problems and prospects [news]. Nature 1997;389(6648):239–242.
20. Graham FL et al. Characteristics of a human cell line transformed by DNA from adenovirus type 5. J Gen Virol 1977;36:59–72.
21. Kotin RM et al. Site-specific integration by adeno-associated virus. Proc Natl Acad Sci USA 1990;87:2211–2215.
22. Samulski RJ, Chang L, Shenk T. Helper-free stocks of recombinant adeno-associated viruses:normal integration does not require viral gene expression. J Virol 1989;63:3822–3828.
23. Mountain A. Gene therapy:the first decade. Trends in Biotechnology 2000;18(3):119–128.
24. Wolfe D et al. Design and use of herpes simplex viral vectors for gene therapy. In: Templeton NS, Lasic DD, Editors. Gene therapy: therapeutic mechanisms and strategies. New York:Marcel Dekker 2000;81–108.
25. Tseng JC, Levin B, Hirano T, Yee H, Pampeno C, Meruelo D. In vivo antitumor activity of Sindbis viral vectors. J Natl Cancer Inst 2002;94(23):1790–1802.
26. Tseng JC, Levin B, Hurtado A, Yee H, Perez de Castro I, Jimenez M, Shamamian P, Jin R, Novick RP, Pellicer A, Meruelo D. Systemic tumor targeting and killing by Sindbis viral vectors. Nat Biotechnol 2004;22(1):70–77.
27. Shariat SF et al. Adenovirus-mediated transfer of inducible caspases: a novel "death switch" gene therapeutic approach to prostate cancer. Cancer Res 2001;61(6):2562–2571.
28. Recchia A et al. Site-specific integration mediated by a hybrid adenovirus/adeno-associated virus vector. Proc Natl Acad Sci USA 1999;96(6):2615–2620.
29. Rinaudo D et al. Conditional site-specific integration into human chromosome 19 by using a ligand-dependent chimeric adeno-associated virus/Rep protein. J Virol 2000;74(1):281–294.
30. Constantini LC, Wang S, Fraefel C, Breakefield XO, Isacson O. Gene transfer to the nigrostriatal system by hybrid herpes simplex virus/adeno-associated virus amplicon vectors. Hum Gene Ther 1999;10:2481–2494.
31. Jacobs A et al. Positron emission tomography-based imaging of transgene expression mediated by replication-conditional, oncolytic herpes simplex virus type 1 mutant vectors *in vivo*. Cancer Res 2001;61(7):2983–2995.
32. Lam PY, Breakefield XO. Hybrid vector designs to control the delivery, fate and expression of transgenes. J Gene Med 2000; 2(6):395–408.
33. Wunderbaldinger P, Bogdanov A, Weissleder, R. New approaches for imaging in gene therapy. Eur J Radiol 2000;34(3):156–165.
34. Anderson WF. Human gene therapy. Nat 1998;392(6679(suppl)): 25–30.
35. Felgner PL, Ringold GM. Cationic liposome-mediated transfection. Nature 1989;337:387–388.
36. Roper C. Liposomes as a gene delivery system. Brazilian J Med Biol Res 1999;32:163–169.
37. Loeffler JP, Behr JP. Gene transfer into primary and established mammalian cell lines with lipopolyamine-coated DNA. Methods in Enzymology 1993;217:599–618.
38. Wolff JA et al. Direct gene transfer into mouse muscle *in vivo*. Science 1990;247:1465–1468.
39. Cheng L, Ziegelhoffer PR, Yang NS. In vivo promoter activity and transgene expression in mammalian somatic tissues evaluated by using particle bombardment. Proc Natl Acad Sci USA 1993; 90:4455–4459.
40. Spooner RA, Deonarain MP, Epenetos AA. DNA vaccination for cancer treatment. Gene Ther 1995;2(3):173–180.
41. Matano T et al. Targeted infection of a retrovirus bearing a Cd4-Env Chimera into human cells expressing human immunodeficiency virus type 1. J Gen Virol, 1995;76:3165–3169.
42. Dornburg R, Pomerantz RJ. Gene therapy and HIV-1 infection. In: Templeton NS, Lasic DD, Editors. Gene therapy: Therapeutic mechanisms and strategies. New York: Marcel Dekker 2000; 519–533.
43. Nettelbeck DM, Jérôme V, Müller R. Gene therapy:designer promoters for tumor targeting. Trends in Genetics 2000;16(4):174–181.
44. Huber BE, Richards CA, Krenitsky TA. Retroviral-mediated gene therapy for the treatment of hepatocellular carcinoma:an innovative approach for cancer therapy. Proc Natl Acad Sci USA 1991; 88:8039–8043.
45. Ido A et al. Gene therapy for hepatoma cells using a retrovirus vector carrying herpes simplex virus thymidine kinase gene under the control of human alpha-fetoprotein gene promoter. Cancer Res 1995;55(14):3105–3109.
46. Osaki T et al. Gene therapy for carcinoembryonic antigen-producing human lung cancer cells by cell type-specific expression of herpes simplex virus thymidine kinase gene. Cancer Res 1994; 54(20):5258–5261.
47. Cao G et al. Effective and safe gene therapy for colorectal carcinoma using the cytosine deaminase gene directed by the carcinoembryonic antigen promoter. Gene Ther 1999;6(1):83–90.
48. Cao G et al. Analysis of the human carcinoembryonic antigen promoter core region in colorectal carcinoma-selective cytosine deaminase gene therapy. Cancer Gene Ther 1999;6(6):572–580.
49. Ishikawa H et al. Utilization of variant-type of human alpha-fetoprotein promoter in gene therapy targeting for hepatocellular carcinoma. Gene Ther 1999;(4):465–470.
50. Stein U, Walther W, Shoemaker RH. Vincristine induction of mutant and wild-type human multidrug-resistance promoters is cell-type-specific and dose-dependent. J Cancer Res Clin Oncol 1996;122(5):275–282.
51. Segawa T et al. Prostate-specific amplification of expanded polyglutamine expression:a novel approach for cancer gene therapy. Cancer Res 1998;58(11):2282–2287.
52. Iyer M et al. Two-step transcriptional amplification as a method for imaging reporter gene expression using weak promoters. Proc Natl Acad Sci USA 2001;98(25):14595–14600.
53. Wagner JA, Gardner P. Toward cystic fibrosis gene therapy. Annu Rev Med 1997;48:203–216.
54. Teiger E et al. Gene therapy in heart disease. Biomed Pharmacother 2001;55(3):148–154.
55. Rader DJ. Gene therapy for atherosclerosis. Int J Clin Lab Res 1997;27(1):35–43.
56. Rabinovich GA. Apoptosis as a target for gene therapy in rheumatoid arthritis. Mem Inst Oswaldo Cruz 2000;95(Supp 1): 225–233.
57. Isner JM et al. Assessment of risks associated with cardiovascular gene therapy in human subjects. Circ Res 2001;89(5):389–400.
58. Haberkorn U et al. Monitoring of gene therapy with cytosine deaminase:*in vitro* studies using 3H-5-fluorocytosine. J Nucl Med 1996;37:87–94.
59. Moolten FL. Tumor chemosensitivity conferred by inserted herpes thymidine kinase genes:paradigm for prospective cancer control strategy. Cancer Res 1986;46:5276–5281.
60. De Clercq E. Antivirals for the treatment of herpes virus infections. J Antimicrob Chemothe 1993;32(Suppl. A):1–2132.
61. Alrabiah FA, Sacks SL. New anti-herpes virus agents:their targets and therapeutic potential. Drugs 1996;52(1):17–32.
62. Myers CE. The pharmacology of the fluoropyrimidines. Pharmacol Rev 1981;33:1–15.
63. Scholer HJ. Flucytosine. In: Speller DCE, editor. New York: Wiley 1980;35–106.
64. Nishiyama T, Kawamura Y, Kawamoto KEA. Antineoplastic effects of 5-fluorocytosine in combination with cytosine deaminase capsules. Cancer Res 1985;45:1753–1761.

65. Logothetis CJ et al. Altered expression of retinoblastoma protein and known prognostic variables in locally advanced bladder cancer. J Natl Cancer Inst 1992;84(16):1256–1261.
66. Esrig D et al. Accumulation of nuclear P53 and tumor progression in bladder cancer. N Engl J Med 1994;331(19):1259–1264.
67. Harris CC. Structure and function of the p53 tumor suppressor gene:clues for rational cancer therapeutic strategies. J Natl Cancer Inst 1996;88(20):1442–1455.
68. Hall PA, Lane DP. Tumor suppressors:a developing role for p53? Curr Biol 1997;7(3):R144–R147.
69. Fujiwara T et al. Therapeutic effect of a retroviral wild type p53 expression vector in an orthotopic lung cancer model. J Natl Cancer Inst 1994;86:1458–1462.
70. Lesoon-Wood LA et al. Systemic gene therapy with p53 reduces growth and metastases of a malignant human breast cancer in nude mice. Hum Gene Ther 1995;6(4):395–405.
71. Xu M et al. Parenteral gene therapy with p53 inhibits human breast tumors *in vivo* through a bystander mechanism without evidence of toxicity. Hum Gene Ther 1997;8(2):177–185.
72. Roth JA et al. Retrovirus-mediated wild-type p53 gene transfer to tumors of patients with lung cancer [see comments]. Nat Med 1996;2(9):985–991.
73. Bookstein R et al. p53 gene therapy *in vivo* of herpatocellular and liver metastatic colorectal cancer. Semin Oncol 1996;23(1):66–77.
74. Bischoff JR et al. An adenovirus mutant that replicates selectively in p53-deficient human tumor cells [see comments]. Science 1996;274(5286):373–376.
75. Heise C et al. ONYX-015, an E1B gene-attenuated adenovirus, causes tumor-specific cytolysis and antitumoral efficacy that can be augmented by standard chemotherapeutic agents [see comments]. Nat Med 1997;3(6):639–645.
76. Lowe SW et al. p53 status and the efficacy of cancer therapy *in vivo*. Science 1994;266:807–810.
77. Gjerset RA et al. Characterization of a new human glioblastoma cell line that expresses mutant p53 and lacks activation of the PDGF pathway. In Vitro Cellular and Developmental Biology. Animal 1995;31(3):207–214.
78. Nguyen DM et al. Gene therapy for lung cancer:enhancement of tumor suppression by a combination of sequential systemic cisplatin and adenovirus-mediated p53 gene transfer. Journal of Thoracic and Cardiovascular Surgery 1996;112(5):1372–6; discussion 1376–1377.
79. Hall SJ, Chen SH, Woo SLC. The promise and reality of cancer gene therapy. American J Hum Genet 1997;61(4):785–789.
80. Kupfer A, Singer SJ. The specific interaction of helper T-cells and antigen-presenting B-cells .4. Membrane and cytoskeletal reorganizations in the bound T-cell as a function of antigen dose. J Exp Med 1989;170(5):1697–1713.
81. Dranoff G et al. Vaccination with irradiated tumor cells engineered to secrete murine granulocyte-macrophage colony-stimulating factor stimulates potent, specific and long lasting anti-tumor immunity. Proc Natl Acad Sci USA 1993;90:3539–3543.
82. Connor J et al. Regression of bladder tumors in mice treated with interleukin-2 gene-modified tumor cells. J Exp Med 1993; 177(6):1833.
83. Smith AE. Gene therapy – where are we? Lancet 1999;354 Suppl 1(1):SI1–SI 4.
84. Cherry SR, Gambhir SS. Use of positron emission tomography in animal research. Institute for Laboratory Animal Research Journal 2001;42(3):219–232.
85. Borrelli E et al. Targeting of an inducible toxic phenotype in animal cells. Proc Natl Acad Sci USA 1988;85:7572–7576.
86. Moolten FL, Wells JM. Curability of tumors bearing herpes thymidine kinase genes transfected by retroviral vectors. J Natl Cancer Inst 1990;82:297–300.
87. Gambhir SS et al. Imaging transgene expression with radionuclide imaging technologies. Neoplasia 2000;2(1–2):118–138.
88. Gambhir SS et al. Imaging of adenoviral directed herpes simplex virus Type 1 thymidine kinase gene expression in mice with ganciclovir. J Nucl Med 1998;39(11):2003–2011.
89. Gambhir SS et al. Imaging adenoviral-directed reporter gene expression in living animals with positron emission tomography. Proc Natl Acad Sci USA 1999;96(5):2333–2338.
90. Iyer M et al. 8-[F-18]fluoropenciclovir:An improved reporter probe for imaging HSV1-TK reporter gene expression *in vivo* using PET. J Nucl Med 2001;42(1):96–105.
91. Alauddin MM et al. 9-[(3-[18F]-fluoro-1-hydroxy-2propoxy)-methyl]guanine ([18F]-FHPG):a potential imaging agent of viral infection and gene therapy using PET. Nucl Med Biol 1996; 23(6):787–792.
92. Alauddin MM et al. Evaluation of 9-[(3-18F-fluoro-1-hydroxy-2-propoxy)methyl]guanine ([18F]-FHPG) *in vitro* and *in vivo* as a probe for PET imaging of gene incorporation and expression in tumors. Nuclr Med Biol 1999;26(4):371–376.
93. Alauddin MM, Conti PS. Synthesis and preliminary evaluation of 9-(4-[18F]-fluoro-3-hydroxymethylbutyl)guanine ([18F]FHBG):a new potential imaging agent for viral infection and gene therapy using PET. Nucl Med Biol1998;25(3):175–180.
94. Tjuvajev JG et al. Noninvasive imaging of herpes virus thymidine kinase gene transfer and expression:A potential method for monitoring clinical gene therapy. Cancer Res 1996;56:4087–4095.
95. Tjuvajev JG et al. Imaging herpes virus thymidine kinase gene transfer and expression by positron emission tomography. Cancer Res 1998;58(19):4333–4341.
96. Iyer M et al. Comparison of FPCV, FHBG, and FIAU as reporter probes for imaging Herpes Simplex Virus Type 1 thymidine kinase reporter gene expression. J Nucl Med 2000;41(5 Suppl):80–81.
97. Tjuvajev JG et al. Direct comparison of HSV1-TK PET imaging probes:FIAU, FHPG, FHBG. J Nucl Med 2001;42(5):277.
98. Gambhir SS et al. A mutant herpes simplex virus type 1 thymidine kinase reporter gene shows improved sensitivity for imaging reporter gene expression with positron emission tomography. Proc Natl Acad Sci USA 2000;97(6):2785–2790.
99. Black ME et al. Creation of drug-specific herpes simplex virus type 1 thymidine kinase mutant for gene therapy. Proc Natl Acad Sci USA 1996;93:3525–3529.
100. Green LA et al. Simulation studies of assumptions of a three-compartment FHBG model for imaging reporter gene expression. J Nucl Med 2001;42(5):100.
101. Gambhir SS et al. Imaging of adenoviral-directed herpes simplex virus type 1 thymidine kinase reporter gene expression in mice with radiolabeled ganciclovir. J Nucl Med 1998;39(11):2003–2011.
102. Green LA et al. Indirect monitoring of endogenous gene expression by PET imaging of reporter gene expression in transgenic mice. Molecular Imaging and Biology 2002;4(1):71–81.
103. Green LA et al. Tracer kinetic modeling of FHBG in mice imaged with microPET for quantitation of reporter gene expression. J Nucl Med 2000;41(5 Suppl):58.
104. Tjuvajev JG et al. Imaging the expression of transfected genes *in vivo*. Cancer Res 1995;55:6126–6132.
105. Bennett JJ et al. Positron emission tomography imaging for herpes virus infection:Implications for oncolytic viral treatments of cancer. Nat Med 2001;7(7):861–865.
106. Doubrovin M et al. Imaging transcriptional regulation of p53-dependent genes with positron emission tomography *in vivo*. Proc Natl Acad Sci USA 2001;98(16):9300–9305.
107. MacLaren DC et al. Repetitive, non-invasive imaging of the dopamine D2 receptor as a reporter gene in living animals. Gene Ther 1999;6:785–791.
108. Barrio JB et al. 3-(2′-[^{18}F]fluoroethyl)spiperone:In vivo biochemical and kinetic characterization in rodents, nonhuman primates, and humans. J Cereb Blood Flow Metab 1989;9:830–839.
109. Virgolini I et al. Somatostatin receptor subtype specificity and *in vivo* binding of a novel tumor tracer, 99mTc-P829. Cancer Res 1998;58(9):1850–1859.
110. Rogers BE et al. In vivo localization of [(111)In]-DTPA-D-Phe1-octreotide to human ovarian tumor xenografts induced to express the somatostatin receptor subtype 2 using an adenoviral vector. Clin Cancer Res 1999;5(2):383–393.
111. Rogers BE, Zinn KR, Buchsbaum DJ. Gene transfer strategies for improving radiolabeled peptide imaging and therapy. Q J Nucl Med 2000;44(3):208–223.
112. Liang Q et al. Noninvasive, quantitative imaging in living animals of a mutant dopamine D2 receptor reporter gene in which ligand binding is uncoupled from signal transduction. Gene Ther 2001; 8(19):1490–1498.

113. Chung June-Key. Sodium Iodide Symporter:Its Role in Nuclear Medicine. J Nuc Med. 2002;43(9):1188–1200.
114. Petrich T et al. Establishment of radioactive astatine and iodine uptake in cancer cell lines expressing the human sodium/iodide symporter. European J Nuc Med 2002;29(7):842–854.
115. Shen DHY et al. Sodium iodide symporter in health and disease. Thyroid 2001;11(5):415–425.
116. Groot-Wassink T et al. Adenovirus biodistribution and non-invasive imaging of gene expression *in vivo* by positron emission tomography using human sodium/iodide symporter as reporter gene 2002;13(14):1723–1735.
117. Niu G et al. Multimodality Noninvasive Imaging of Gene Transfer Using the Human Sodium Iodide Symporter. J Nucl Med 2004;45(3):445–449.
118. Shin JH et al. Feasibility of sodium/iodide symporter gene as a new imaging reporter gene:comparison with HSV1-*tk*. Eur J Nucl Med Mol Imaging 2004;31(3):425–432.
119. Groot-Wassink T et al. Quantitative imaging of na/i symporter transgene expression using positron emission tomography in the living animal 2004;9(3):436–442.
120. Ray P et al. Monitoring gene therapy with reporter gene imaging. Sem Nucl Med 2001;31(4):312–320.
121. Sonenberg N, Pelletier J. Poliovirus translation – a paradigm for a novel initiation mechanism. Bioessays 1989;11(5):128–132.
122. Jang SK et al. A segment of the 5′ nontranslated region of encephalomyocarditis virus RNA directs internal entry of ribosomes during *in vitro* translation. J Virol 1988;62(8):2636–2643.
123. Jang SK et al. Initiation of protein synthesis by internal entry of ribosomes into the 5′ nontranslated region of encephalomyocarditis virus RNA *in vivo*. J Virol 1989;63(4):1651–1660.
124. Yu Y et al. Quantification of target gene expression by imaging reporter gene expression in living animals. Nat Med 2000; 6(8):933–937.
125. Tjuvajev JG et al. A general approach to the non-invasive imaging of transgenes using cis-linked herpes simplex virus thymidine kinase. Neoplasia 1999;1(4):315–320.
126. Kamoshita N et al. Genetic analysis of internal ribosomal entry site on hepatitis C virus RNA:implication for involvement of the highly ordered structure and cell type-specific transacting factors. Virology 1997;233(1):9–18.
127. Jackson R, Howell M, Kaminski A. The novel mechanism of initiation of picornavirus RNA translation. Trends Biochem Sci 1990;15(12):477–483.
128. Wang YL et al. New approaches for linking PET & therapeutic reporter gene expression for imaging gene therapy with increased sensitivity. J Nucl Med 2001;42(5):75.
129. Chappell SA, Edelman GM, Mauro VP. A 9-nt segment of a cellular mRNA can function as an internal ribosome entry site (IRES) and when present in linked multiple copies greatly enhances IRES activity. Proc Natl Acad Sci USA 2000;97(4):1536–1541.
130. Wahlfors JJ et al. Evaluation of recombinant alphaviruses as vectors in gene therapy. Gene Ther 2000;7(6):472–480.
131. Loimas S, Wahlfors J, Jänne J. Herpes simplex virus thymidine kinase-green fluorescent protein fusion gene:new tool for gene transfer studies and gene therapy. Biotechniques 1998; 24(4):614–618.
132. Jacobs A et al. Functional coexpression of HSV-1 thymidine kinase and green fluorescent protein:implications for noninvasive imaging of transgene expression. Neoplasia 1999;1(2):154-161.
133. Strathdee CA, McLeod MR, Underhill TM. Dominant positive and negative selection using luciferase, green fluorescent protein and beta-galactosidase reporter gene fusions. Biotechniques 2000; 28(2):210–214.
134. Ray P et al. Imaging tri-fusion multimodality reporter gene expression in living subjects. Cancer Res 2004;64(4):1323–1330.
135. Yaghoubi SS et al. Direct correlation between positron emission tomographic images of two reporter genes delivered by two distinct adenoviral vectors. Gene Ther 2001;8(14): 1072–1080.
136. Sun X et al. Quantitative imaging of gene induction in living animals. Gene Ther 2001;8(20):1572–1579.
137. Yaghoubi SS et al. Human pharmacokinetics and dosiometry studies of [18F]FHBG: A reporter probe for imaging herpes simplex virus type 1 thymidine kinase (hsv1-tk) reporter gene expression. J Nucl Med 2001;42(8):1225–1234.
138. Jacobs A et al. Positron-emission tomography of vector-mediated gene expression in gene therapy for gliomas. Lancet 2001; 358:727–729.
139. Weissleder R, Mahmood U. Molecular imaging. Radiology 2001;219(2):316–333.

Appendix*

Table of Potentially Useful Positron-emitting Radionuclides

Nuclide	Z	Half Life	Production Device	Branching Ratio
^{10}C	6	19.3s	Cyclotron	1.00
^{11}C	6	20.5m	Cyclotron	1
^{13}N	7	9.97m	Cyclotron	1.00
^{14}O	8	70.6s	Cyclotron	1.00
^{15}O	8	122.2s	Cyclotron	1
^{17}F	9	64.5s	Cyclotron	1.00
^{18}F	9	109.8m	Cyclotron	0.97
^{19}Ne	10	17.2s	Cyclotron	> 0.99
^{22}Na	11	2.6y	Reactor	0.91
^{34m}Cl	17	32.2m	Cyclotron	0.54
^{38}K	19	7.7m	Cyclotron	1.00
^{44}Sc	21	3.92h	Generator (from ^{44}Ti)	0.95
^{45}Ti	22	3.09h	Cyclotron	1.00
^{48}V	23	15.97d	Cyclotron	0.50
^{51}Mn	23	46.2m	Cyclotron	1.00
^{52m}Mn	23	21.1m	Generator (from ^{52}Fe)	0.97
^{52}Mn	25	5.59d	Cyclotron	0.29
^{52}Fe	26	8.28h	Cyclotron	0.56
^{55}Co	27	17.5h	Cyclotron	0.76
^{60}Cu	29	23.4m	Cyclotron	0.93
^{61}Cu	29	3.3h	Cyclotron	0.61
^{62}Cu	29	9.74m	Generator (from ^{62}Zn)	0.97
^{64}Cu	29	12.8h	Cyclotron	0.18
^{65}Zn	30	245d	Cyclotron	0.015
^{66}Ga	31	9.45h	Cyclotron	0.57
^{68}Ga	31	68.3m	Generator (from ^{68}Ge)	0.89
^{72}As	33	26.0h	Cyclotron	0.77
^{74}As	33	17.8d	Cyclotron	0.29
^{75}Br	35	1.62h	Cyclotron	0.76
^{76}Br	35	15.9h	Cyclotron	0.54
^{79}Kr	36	34.9h	Cyclotron	0.07
^{81}Rb	37	4.58h	Cyclotron	0.27
^{82}Rb	37	1.27m	Generator (from ^{82}Sr)	0.95
^{82m}Rb	37	6.3h	Cyclotron	0.26
^{83}Sr	38	32.4h	Cyclotron	0.24
^{84}Rb	37	33.0d	Cyclotron	0.22
^{86}Y	39	14.7h	Cyclotron	0.34
^{87}Y	39	80.3h	Cyclotron	0.02
^{89}Zr	40	3.27d	Cyclotron	0.23
^{90}Nb	41	14.6h	Cyclotron	0.53
^{94m}Tc	43	53m	Cyclotron	0.70
^{110}In	49	69m	Generator (from ^{110}Sn)	1.00
^{118}Sb	51	3.5m	Cyclotron	0.78
^{120m}I	53	53m	Cyclotron	1.0
^{122}I	53	3.63m	Generator (from ^{122}Xe)	0.77
^{124}I	53	4.17d	Cyclotron	0.23
^{128}Cs	55	29.9m	Cyclotron	0.45
^{130}Cs	55	29.9m	Cyclotron	0.61
^{204}Tl	81	3.78y	Reactor	0.03

* Reproduced from Valk PE, Bailey DL, Townsend DW, Maisey MN. Positron Emission Tomography: Basic Science and Clinical Practice. Springer-Verlag London Ltd 2003, 869.

Index

B

C

D

K

L

M

N

S

T